Photoprotection, Photoinhibition, Gene Regulation, and Environment

Advances in Photosynthesis and Respiration

VOLUME 21

Series Editor:

GOVINDJEE
University of Illinois, Urbana, Illinois, U.S.A.

Consulting Editors:
Julian EATON-RYE, *Dunedin, New Zealand*
Christine H. FOYER, *Newcastle upon Tyne, U.K.*
David B. KNAFF, *Lubbock, Texas, U.S.A.*
Sabeeha MERCHANT, *Los Angeles, California, U.S.A.*
Anthony L. MOORE, *Brighton, U.K.*
Krishna NIYOGI, *Berkeley, California, U.S.A.*
William PARSON, *Seattle, Washington, U.S.A.*
Agepati RAGHAVENDRA, *Hyderabad, India*
Gernot RENGER, *Berlin, Germany*

The scope of our series, beginning with volume 11, reflects the concept that photosynthesis and respiration are intertwined with respect to both the protein complexes involved and to the entire bioenergetic machinery of all life. *Advances in Photosynthesis and Respiration* is a book series that provides a comprehensive and state-of-the-art account of research in photosynthesis and respiration. Photosynthesis is the process by which higher plants, algae, and certain species of bacteria transform and store solar energy in the form of energy-rich organic molecules. These compounds are in turn used as the energy source for all growth and reproduction in these and almost all other organisms. As such, virtually all life on the planet ultimately depends on photosynthetic energy conversion. Respiration, which occurs in mitochondrial and bacterial membranes, utilizes energy present in organic molecules to fuel a wide range of metabolic reactions critical for cell growth and development. In addition, many photosynthetic organisms engage in energetically wasteful photorespiration that begins in the chloroplast with an oxygenation reaction catalyzed by the same enzyme responsible for capturing carbon dioxide in photosynthesis. This series of books spans topics from physics to agronomy and medicine, from femtosecond processes to season long production, from the photophysics of reaction centers, through the electrochemistry of intermediate electron transfer, to the physiology of whole organisms, and from X-ray crystallography of proteins to the morphology of organelles and intact organisms. The goal of the series is to offer beginning researchers, advanced undergraduate students, graduate students, and even research specialists, a comprehensive, up-to-date picture of the remarkable advances across the full scope of research on photosynthesis, respiration, and related processes.

For other titles published in this series, go to
www.springer.com/series/5599

Photoprotection, Photoinhibition, Gene Regulation, and Environment

Edited by

Barbara Demmig-Adams
*University of Colorado,
Boulder, CO, U.S.A.*

William W. Adams III
*University of Colorado,
Boulder, CO, U.S.A.*

and

Autar K. Mattoo
*Henry A. Wallace Beltsville Agricultural Research Center,
Beltsville, MD, U.S.A.*

Library of Congress Control Number: 2008936125

ISBN 978-1-4020-9281-7 (PB)
ISBN 978-1-4020-3564-7 (HB)
ISBN 978-1-4020-3579-1 (e-book)

Published by Springer,
P.O. Box 17, 3300 AA Dordrecht, The Netherlands.

www.springer.com

Cover Credit:
Cover photograph of Douglas fir on the north-facing slope of Gregory Canyon, Colorado, on a late afternoon during winter by William Adams. Scheme by Barbara Demmig-Adams and William Adams.

Printed on acid-free paper

All Rights Reserved
© 2008 Springer Science+Business Media B.V.
No part of this work may be reproduced, stored in a retrieval system, or transmitted
in any form or by any means, electronic, mechanical, photocopying, microfilming, recording
or otherwise, without written permission from the Publisher, with the exception
of any material supplied specifically for the purpose of being entered
and executed on a computer system, for exclusive use by the purchaser of the work.

From the Series Editor

Advances in Photosynthesis and Respiration, Volume 21

I am delighted to announce the publication, in *Advances in Photosynthesis and Respiration* (AIPH) Series, of **Photoprotection, Photoinhibition, Gene Regulation, and Environment**, a book covering the central role of excess light in how plants monitor, and respond to environmental changes. This volume was edited by three distinguished authorities, all based in the USA, Barbara Demmig-Adams, William W. Adams III, and Autar K. Mattoo. Two earlier AIPH volumes dealt with the topics of Environment and Regulation: *Photosynthesis and the Environment* (1996; edited by Neil R. Baker, from UK); and *Regulation of Photosynthesis* (2001; edited by Eva-Mari Aro and Bertil Andersson, from Finland and Sweden). The current volume follows the 20 volumes listed below.

Published Volumes (1994–2005)

- *Volume 1:* **Molecular Biology of Cyanobacteria** (28 Chapters; 881 pages; 1994; edited by Donald A. Bryant, from USA);
- *Volume 2:* **Anoxygenic Photosynthetic Bacteria** (62 Chapters; 1331 pages; 1995; edited by Robert E. Blankenship, Michael T. Madigan and Carl E. Bauer, from USA);
- *Volume 3:* **Biophysical Techniques in Photosynthesis** (24 Chapters; 411 pages; 1996; edited by the late Jan Amesz and the late Arnold J. Hoff, from The Netherlands);
- *Volume 4:* **Oxygenic Photosynthesis: The Light Reactions** (34 Chapters; 682 pages; 1996; edited by Donald R. Ort and Charles F. Yocum, from USA);
- *Volume 5:* **Photosynthesis and the Environment** (20 Chapters; 491 pages; 1996; edited by Neil R. Baker, from UK);
- *Volume 6:* **Lipids in Photosynthesis: Structure, Function and Genetics** (15 Chapters; 321 pages; 1998; edited by Paul-André Siegenthaler and Norio Murata, from Switzerland and Japan);
- *Volume 7:* **The Molecular Biology of Chloroplasts and Mitochondria in Chlamydomonas** (36 Chapters; 733 pages; 1998; edited by Jean David Rochaix, Michel Goldschmidt-Clermont and Sabeeha Merchant, from Switzerland and USA);
- *Volume 8:* **The Photochemistry of Carotenoids** (20 Chapters; 399 pages; 1999; edited by Harry A. Frank, Andrew J. Young, George Britton and Richard J. Cogdell, from USA and UK);
- *Volume 9:* **Photosynthesis: Physiology and Metabolism** (24 Chapters; 624 pages; 2000; edited by Richard C. Leegood, Thomas D. Sharkey and Susanne von Caemmerer, from UK, USA and Australia);
- *Volume 10:* **Photosynthesis: Photobiochemistry and Photobiophysics** (36 Chapters; 763 pages; 2001; authored by Bacon Ke, from USA);
- *Volume 11:* **Regulation of Photosynthesis** (32 Chapters; 613 pages; 2001; edited by Eva-Mari Aro and Bertil Andersson, from Finland and Sweden);
- *Volume 12:* **Photosynthetic Nitrogen Assimilation and Associated Carbon and Respiratory Metabolism** (16 Chapters; 284 pages; 2002; edited by Christine Foyer and Graham Noctor, from UK and France);
- *Volume 13:* **Light Harvesting Antennas** (17 Chapters; 513 pages; 2003; edited by Beverley Green and William Parson, from Canada and USA);
- *Volume 14:* **Photosynthesis in Algae** (19 Chapters; 479 pages; 2003; edited by Anthony Larkum, Susan Douglas and John Raven, from Australia, Canada and UK);
- *Volume 15:* **Respiration in Archaea and Bacteria: Diversity of Prokaryotic Electron Transport Carriers** (13 Chapters; 326 pages; 2004; edited by Davide Zannoni, from Italy);
- *Volume 16:* **Respiration in Archaea and Bacteria 2: Diversity of Prokaryotic Respiratory Systems** (13 Chapters; 310 pages; 2004; edited by Davide Zannoni, from Italy);
- *Volume 17:* **Plant Mitochondria: From Genome to Function** (14 Chapters; 325 pages; 2004; edited

by David A. Day, A. Harvey Millar and James Whelan, from Australia);
- *Volume 18: Plant Respiration: From Cell to Ecosystem* (13 Chapters; 250 pages; 2005; edited by Hans Lambers, and Miquel Ribas-Carbo, 2005; from Australia and Spain).
- *Volume 19: Chlorophyll a Fluorescence: A Signature of Photosynthesis* (31 Chapters; 817 pages; 2004; edited by George C. Papageorgiou and Govindjee, from Greece and USA); and
- *Volume 20: Discoveries in Photosynthesis* (111 Chapters; 1262 pages; 2005; edited by Govindjee, J. Thomas Beatty, Howard Gest and John F. Allen, from USA, Canada and Sweden (& UK)).

In addition, *Volume 22* (**Photosystem II: The Light-Driven Water:Plastoquinone Oxidoreductase** (34 Chapters, xxvii + 16 color plates + 786 pp., edited by Thomas J. Wydrzynski and Kimiyuki Satoh, from Australia and USA) has already been published in 2005.

Further information on these books and ordering instructions can be found at <http://www.springeronline.com> under the Book Series 'Advances in Photosynthesis and Respiration'. Special discounts are available for members of the International Society of Photosynthesis Research, ISPR (<http://www.photosynthesisresearch.org/>).

Photoprotection, Photoinhibition, Gene Regulation, and Environment

This book was edited by three outstanding authorities in the areas of Photoprotection, Photoinhibition, Gene Regulation, and Environment: Barbara Demmig-Adams and William W. Adams III (both at the University of Colorado, Boulder, Colorado) and Autar K. Mattoo (Henry A. Wallace Beltsville Agricultural Research Center, Beltsville, Maryland).

The topic of the book, as provided by our 3 distinguished editors, is: "*Photoprotection, Photoinhibition, Gene Regulation, and Environment*"; it examines the processes whereby plants monitor environmental conditions and orchestrate their response to change, an ability paramount to the life of all plants. 'Excess light', absorbed by the light-harvesting systems of photosynthetic organisms, is an integrative indicator of the environment, communicating the presence of intense light and any conditions unfavorable for growth and photosynthesis. Key plant responses are photoprotection and photoinhibition. In this volume, the dual role of photoprotective responses in the preservation of leaf integrity and in redox signaling networks modulating stress acclimation, growth, and development is addressed. In addition, the still unresolved impact of photoinhibition on plant survival and productivity is discussed. Specific topics include dissipation of excess energy via thermal and other pathways, scavenging of reactive oxygen by antioxidants, proteins key to photoprotection and photoinhibition, peroxidation of lipids, as well as signaling by reactive oxygen, lipid-derived messengers, and other messengers that modulate gene expression. Approaches include biochemical, physiological, genetic, molecular, and field studies, addressing intense visible and ultraviolet light, winter conditions, nutrient deficiency, drought, and salinity. This book is directed toward advanced undergraduate students, graduate students, and researchers interested in Plant Ecology, Stress Physiology, Plant Biochemistry, Integrative Biology, and Photobiology."

"*Photoprotection, Photoinhibition, Gene Regulation, and Environment*" has 21 authoritative Chapters, and is authored by 57 international authorities from 16 countries. The book begins with three perspectives: Harry Yamamoto (USA) presents a random walk to and through the xanthophyll cycle (*Chapter 1*); Barry Osmond and Britta Förster (Australia) provide an account of Photoinhibition: then and now (*Chapter 2*); Marvin Edelman and Autar Mattoo (Israel and USA) discuss the past and future perspectives of the involvement of the D1 protein in photoinhibition (*Chapter 3*). These perspectives are followed by 18 chapters. In *Chapter 4*, Barbara Demmig-Adams, Volker Ebbert, Ryan Zarter and William Adams (USA) summarize characteristics and species-dependent employment of flexible versus sustained thermal dissipation and photoinhibition. Then, William Adams, C. Ryan Zarter, Kristine Mueh, Véronique Amiard and Barbara Demmig-Adams (USA) discuss details of energy dissipation and photoinhibition as a continuum of protection (*Chapter 5*). In *Chapter 6*, Fermín Morales, Anunciación Abadía and Javier Abadía (Spain) discuss photoinhibition and photoprotection under nutrient deficiencies, drought, and salinity. This is followed by a summary, by Donat-P. Häder (Germany), of photoinhibition and UV responses in the aquatic environment (*Chapter 7*); and a discussion, by Alexander V. Vener (Sweden), of phosphorylation of thylakoid proteins (*Chapter 8*). In *Chapter 9*, Hou-Sung Jung and Krishna K. Niyogi (USA) provide a molecular analysis of photoprotection of photosynthesis. Stefan Jansson (Sweden) discusses the saga of a protein family involved in light harvesting and photoprotection (*Chapter 10*). In *Chapter 11*, a team of 10 authors (Norman Huner, Alexander Ivanov, Prafullachandra Sane, Tessa Pocock, Marianna Król, Andrius Balseris, Dominic Rosso, Leonid Savitch, Vaughan Hurry and Gunnar Öquist (Canada, Sweden and India) discuss the role of reaction center quenching versus antenna quenching in the photoprotection of Photosystem II. Then, Kittisak Yokthongwattana and Anastasios Melis (Thailand and USA) discuss, in *Chapter 12*, the mechanism of a Photosystem II damage and repair cycle involved in photoinhibition (and its recovery) in oxygenic photosynthesis. Subsequently, regulation by environmental conditions of the repair of Photosystem II in cyanobacteria is discussed by Yoshitaka Nishiyama, Suleyman Allakhverdiev and Norio Murata (Japan and Russia) in *Chapter 13*. Tsuyoshi Endo and Kozi Asada (Japan) provide, in *Chapter 14*, an understanding of the role of cyclic electron flow and the so-called water-water cycle in photoprotection, particularly around Photosystem I. This is followed by *Chapter 15* on the integration of signaling in antioxidant defenses by Philip Mullineaux, Stanislaw Karpinski and Gary Creissen (UK and Sweden). *Chapter 16*, by Christine Foyer, Achim Trebst and Graham Noctor (UK, Germany and France), deals with signaling and integration of defense functions of tocopherol, ascorbate, and glutathione. Then, in *Chapter 17*, Sacha Baginsky and Gerhard Link (Switzerland and Germany) discuss redox regulation of chloroplast gene expression. Robert Larkin (USA) provides, in *Chapter 18*, a summary of intracellular signaling and chlorophyll synthesis. Nine authors (Karl-Josef Dietz, Tina Stork, Iris Finkemeier, Petra Lamkemeyer, Wen-Xue Li, Mohamed El-Tayeb, Klaus-Peter Michel, Elfriede Pistorius, and Margarete Baier), from Germany and Egypt, discuss, in *Chapter 19*, the role of peroxiredoxins in oxygenic photosynthesis of cyanobacteria and of higher plants and pose the question of the importance of peroxide detoxification or redox sensing in the process. *Chapter 20*, by Mauro Maccarrone (Italy), reviews lipoxygenases, apoptosis, and the role of antioxidants. The book ends appropriately with *Chapter 21*, by Christiane Reinbothe and Steffen Reinbothe (Germany and France), on the regulation of photosynthetic gene expression by the environment from the seedling de-etiolation stage to leaf senescence.

A Bit of Early History – From there to here

"*It is a noble employment to rescue from oblivion those who deserve to be remembered*" (Pliny the Younger, Letters V).

In 1996, the grand young man of Photosynthesis Jack Myers wrote about the findings in his PhD thesis 65 years ago (Country boy to scientist. Photosynth. Res. 50: 195–208, 1996.) "*Cranking up for my first real experiments was an exciting day. Carefully pipette a cell sample into the Warburg vessel and let it come to temperature in darkness. Then turn on the projection lamp to give a bright light spot already measured at 38 000 foot-candles, almost 4 times as bright as sunlight.... That first experiment was a complete bust. There was only a short burst of the increasing pressure I expected. Thereafter, the pressure change became negative in evidence of oxygen uptake. Something was wrong. So I repeated the procedure with the same result. Only when the intensity was much reduced (1000 foot-candles, by wire screens) did I see the expected high and steady rate of oxygen evolution. Though it took a lot of confirming and polishing experiments, that was an exciting day in the life of a young photosynthetiker. I had made a discovery. I knew something unknown to anyone else in the world. That had been my romantic vision of the fruit of research. And it has not changed in the sixty years since.*" This experiment was published by Jack Myers and George O. Burr (Some effects of high light intensity in Chlorella. Jour. Gen. Physiol. 24: 45–67, 1940)—the discovery of inhibition of photosynthesis by high light, the phenomenon of photoinhibition, but without its name. Only in 1956, did Bessel Kok (On the inhibition of photosynthesis by intense light. Biochim. Biophys. Acta 21: 234–244, 1956) characterize this phenomenon in an elegant manner. Itzhak Ohad and his coworkers (N. Adir, H. Zer, S. Scochat and I. Ohad: Photoinhibition-a historical perspective. Photosynth. Res. 76: 343–370, 2003) have written a current history of photoinhibition. In addition, I mention the personal perspective by Barbara Demmig-Adams (Linking the xanthophyll cycle with thermal energy dissipation. Photosynth. Res. 76: 73–80, 2003). Photographs shown in these two latter papers and in Govindjee and Manfredo Seufferheld (Non-photochemical quenching of chlorophyll a fluorescence: early history and characterization of two xanthophyll cycle mutants of *Chlamydomonas reinhardtii*. Funct. Plant Biol. 29: 1141–1155, 2002) are worth seeing and enjoying. And while much important work on the role of the xanthophyll cycle has been done over many years, it is only now that the nature of the role of zeaxanthin in the de-excitation of chlorophyll is being identified (see N.E. Holt, D. Zigmantas, L. Valkunas, X.-P. Li, K.K. Niyogi and G.R. Fleming: Carotenoid cation formation and the regulation of photosynthetic light harvesting. Science 307: 433–436, 2005).

There have been many books, many chapters in several books, many reviews, and an enormous number of papers in the field of '*Photoinhibition and Photoprotection*'. I do mention, for historical reasons, an edited book, published 18 years ago, in 'Topics of Photosynthesis' (Volume 9, Series Editor James Barber): David Kyle, Barry Osmond and Charles Arntzen (eds) (1987) 'Photoinhibition', Elsevier, Amsterdam (307 pages; Foreword is by Jack Myers; it has 11 Chapters, including chapters by C.B. Osmond, G. Öquist, K. Asada and N. Murata who are also authors in the current book).

Future AIPH Books

The readers of the current series are encouraged to watch for the publication of the forthcoming books (not necessarily arranged in the order of future appearance):

- **Photosystem I: The Light-Driven Plastocyanin: Ferredoxin Oxidoreductase** (Editor: John Golbeck);
- **The Structure and Function of Plastids** (Editors: Robert Wise and J. Kenneth Hoober);
- **Chlorophylls and Bacteriochlorophylls: Biochemistry, Biophysics, Functions and Applications** (Editors: Bernhard Grimm, Robert J. Porra, Wolfhart Rüdiger and Hugo Scheer);
- **Biophysical Techniques in Photosynthesis. II.** (Editors: Thijs J. Aartsma and Jörg Matysik);
- **Photosynthesis: A Comprehensive Treatise; Physiology, Biochemistry, Biophysics, and Molecular Biology, Part 1** (Editors: Julian Eaton-Rye and Baishnab Tripathy); and
- **Photosynthesis: A Comprehensive Treatise; Physiology, Biochemistry, Biophysics, and Molecular Biology, Part 2** (Editors: Baishnab Tripathy and Julian Eaton-Rye)

In addition to these contracted books, we are already in touch with prospective Editors for the following books:

- Anoxygenic Photosynthetic Bacteria. II
- Chloroplast Bioengineering
- Molecular Biology of Cyanobacteria. II.
- Protonation and ATP Synthases
- Genomics and Proteomics
- Sulfur Metabolism in Photosynthetic Organisms

Other books, under discussion, are: Molecular Biology of Stress in Plants; Global Aspects of Photosynthesis and Respiration; and Artificial Photosynthesis. Readers are encouraged to send their suggestions for these and future volumes (topics, names of future editors, and of future authors) to me by E-mail (gov@uiuc.edu) or fax (1-217-244-7246).

In view of the interdisciplinary character of research in photosynthesis and respiration, it is my earnest hope that this series of books will be used in educating students and researchers not only in Plant Sciences, Molecular and Cell Biology, Integrative Biology, Biotechnology, Agricultural Sciences, Microbiology, Biochemistry, and Biophysics, but also in Bioengineering, Chemistry, and Physics.

I take this opportunity to thank Barbara Demmig-Adams, William W. Adams III, and Autar K. Mattoo for their outstanding and painstaking editorial work. I thank all the 57 authors of volume 21: without their authoritative chapters, there would be no such volume. I owe Jacco Flipsen and Noeline Gibson (both of Springer) special thanks for their friendly working relation with us that led to the production of this book. Thanks are also due to Jeff Haas (Director of Information Technology, Life Sciences, University of Illinois at Urbana-Champaign, UIUC), Evan DeLucia (Head, Department of Plant Biology, UIUC) and my dear wife Rajni Govindjee for their constant support.

Govindjee
Series Editor, *Advances in Photosynthesis and Respiration*
University of Illinois at Urbana-Champaign
Department of Plant Biology
Urbana, IL 61801-3707, USA
E-mail: gov@uiuc.edu
URL: http://www.life.uiuc.edu/govindjee

A historical group photograph taken at the wedding of Adam Gilmore and Xiao-Ping Li (29 June, 2003). From left to right: Govindjee; Olle Björkman; Harry Yamamoto; Adam; Xiao-Ping; Krishna Niyogi; and Barry Osmond. Background, Lake Tahoe, California. Photo by Rajni Govindjee.

Govindjee is Professor Emeritus of Biochemistry, Biophysics and Plant Biology at the University of Illinois at Urbana-Champaign (UIUC), Illinois, USA, since 1999. He obtained his B.Sc. (Chemistry and Biology) and M.Sc. (Botany, specializing in Plant Physiology) in 1952 and 1954, respectively, from the University of Allahabad, Allahabad, India. Govindjee was a graduate student, first of Robert Emerson and then of Eugene Rabinowitch, receiving his Ph.D. in Biophysics from the UIUC in 1960. His honors include: Fellow of the American Association for the Advancement of Science (1976); Distinguished Lecturer of the School of Life Sciences, UIUC (1978); President of the American Society of Photobiology (1980–1981); Fulbright Senior Lecturer (1996–1997); and honorary President of the 2004 International Photosynthesis Congress (Montréal, Canada). Govindjee's research has focused on the function of "Photosystem II" (water-plastoquinone oxido-reductase), particularly on its primary photochemistry; the unique role of bicarbonate in electron and proton transport; as well as thermoluminescence, delayed and prompt fluorescence (particularly lifetimes), and their use in understanding electron transport and photoprotection against excess light. His major contribution on the topic of this book has been to use lifetime of chlorophyll a fluorescence measurements, with Adam Gilmore, to demonstrate the 'dimmer-switch' of thermal dissipation (see e.g., Gilmore et al. (1995) Proc. Natl. Acad. Sci. USA 92: 2273–2277; Gilmore et al. (1998) Biochemistry 37: 13582–13593) and monitor, with Oliver Holub and others, zeaxanthin-dependent quenching of chlorophyll a fluorescence using FLIM (Fluorescence Lifetime Imaging Microscopy) in single cells of *Chlamydomonas reinhardtii* (Holub et al. (2000) Photosynthetica 38: 583–601). Some of his reflections on the topic of 'photoprotection' may be found in Govindjee (2002) Plant Cell 14: 1663–1668; and in Govindjee and M. Seufferheld (2002) Functional Plant Biology 29: 1141–1155. Govindjee's scientific interests now include regulation of excitation energy transfer in oscillating light (L. Nedbal et al. (2003) Biochim. Biophys. Acta 1607:5–7). However, his major focus is on the "History of Photosynthesis Research" (see volumes 73 (2002), 76 (2003), and 80 (2004) of *Photosynthesis Research*), and in Photosynthesis Education. In addition to being the Series Editor of *Advances in Photosynthesis and Respiration*, he serves as the "Historical Corner" Editor of "*Photosynthesis Research*". For further information, see his web page at: http://www.life.uiuc.edu/govindjee.

Contents

From the Series Editor	v
Preface	xvii

1 A Random Walk To and Through the Xanthophyll Cycle — 1–10
Harry Y. Yamamoto

Summary	1
I. Introduction	1
II. The Beginnings	1
III. Education	2
IV. The Violaxanthin-Antheraxanthin-Zeaxanthin (VAZ) Pathways Story	3
V. Further Adventures and Advances	4
VI. Many Thanks to Many	9
Notes	9
References	9

2 Photoinhibition: Then and Now — 11–22
Barry Osmond and Britta Förster

Summary	11
I. What Then?	11
II. From Photorespiratory CO_2 Cycling to Photostasis	13
III. Mechanisms of Photoinhibition	14
IV. Photoacclimation: Yin-Yang and the Compromise between Photoinactivation and Photoprotection	14
V. Then There was the Leaf Disc O_2 Electrode, and Now There is Rapid Response Gas Exchange	16
VI. Enlightening the Mechanisms of Photoinhibition in *Chlamydomonas reinhardtii*	17
VII. Quo Vadis?	18
Acknowledgments	19
References	19

3 The D1 Protein: Past and Future Perspectives — 23–38
Marvin Edelman and Autar K. Mattoo

Summary	23
I. The Really Early Days	24
II. Gernot Renger's Shield	24
III. D1 Metabolism is Photoregulated	26
IV. The PEST Sequence	28
V. The Life History of D1	29
VI. The UV-B Story	30
VII. The D1/D2 Heterodimer Takes Center Stage	31
VIII. Phosphorylation–Dephosphorylation	32
IX. Circadian Control	34
X. The Past and Future	35
Acknowledgments	36
References	36

4 Characteristics and Species-Dependent Employment of Flexible Versus Sustained Thermal Dissipation and Photoinhibition   39–48
Barbara Demmig-Adams, Volker Ebbert, C. Ryan Zarter and William W. Adams III

Summary	39
I. Introduction	39
II. Interspecies Differences in the Capacity for Flexible Thermal Dissipation	40
III. Sustained Thermal Dissipation in Photoinhibited Evergreens	40
IV. Two Types of Thermal Energy Dissipation in Evergreens	46
Acknowledgments	46
References	47

5 Energy Dissipation and Photoinhibition: A Continuum of Photoprotection   49–64
William W. Adams III, C. Ryan Zarter, Kristine E. Mueh, Véronique Amiard and Barbara Demmig-Adams

Summary	49
I. Introduction	50
II. Characteristics of Energy Dissipation and Photoinhibition	50
III. Photoprotection and Photoinhibition in Winter	53
IV. Does Photoinhibition Limit the Carbon Available to the Plant?	55
V. An Integrated View of Photoprotection	59
Acknowledgments	61
References	61

6 Photoinhibition and Photoprotection under Nutrient Deficiencies, Drought and Salinity   65–85
Fermín Morales, Anunciación Abadía and Javier Abadía

Summary	65
I. Introduction	66
II. Iron (Fe) Deficiency	66
III. Nitrogen (N) Deficiency	69
IV. Other Nutrient Deficiencies	72
V. Drought	73
VI. Salinity	75
VII. Conclusions and Future Research Directions	77
Acknowledgments	78
References	78

7 Photoinhibition and UV Response in the Aquatic Environment   87–105
Donat-P. Häder

Summary	87
I. Introduction: Life in the Aquatic Environment	87
II. Photoinhibition in the Field	89
III. Effects of Solar UV Radiation	93
IV. Fast Kinetics of Fluorescence Parameters	94
V. Effects on Developmental Stages	95
VI. Pigment Bleaching	95

	VII. Protection Mechanisms against Excessive Radiation Stress	96
	VIII. Conclusions	100
	Acknowledgments	100
	References	100

8 Phosphorylation of Thylakoid Proteins 107–126
Alexander V. Vener

	Summary	107
	I. Introduction	108
	II. Thylakoid Phosphoproteins	109
	III. Reversible Phosphorylation of Photosystem II (PS II) Proteins	113
	IV. PsaD: the First Phosphoprotein in PS I	120
	V. Phosphorylation of Other Thylakoid Proteins	120
	VI. Regulation and Role of Thylakoid Protein Phosphorylation in a Physiological Context	121
	Acknowledgments	123
	References	123

9 Molecular Analysis of Photoprotection of Photosynthesis 127–143
Hou-Sung Jung and Krishna K. Niyogi

	Summary	127
	I. Introduction	128
	II. Avoiding High Light Absorption	128
	III. Coping with Excess Absorbed Light Energy	131
	IV. Gene Expression Responses of Plants to High Light Stress	138
	Acknowledgments	140
	References	140

10 A Protein Family Saga: From Photoprotection to Light-Harvesting (and Back?) 145–153
Stefan Jansson

	Summary	145
	I. Introduction	146
	II. The Light-Harvesting Complexes (LHCs) of Higher Plants	146
	III. The Light-Harvesting Antenna of Lower Plants	147
	IV. The Evolution of LHC Proteins	148
	V. The Evolution of Feedback De-Excitation	150
	VI. Conclusions	152
	Acknowledgments	152
	References	152

11 Photoprotection of Photosystem II: Reaction Center Quenching Versus Antenna Quenching 155–173
Norman P. A. Huner, Alexander G. Ivanov, Prafullachandra V. Sane, Tessa Pocock, Marianna Król, Andrius Balseris, Dominic Rosso, Leonid V. Savitch, Vaughan M. Hurry and Gunnar Öquist

	Summary	155
	I. Introduction	156
	II. Antenna Quenching	157

III.	Reaction Center Quenching	158
IV.	Thermoluminescence	159
V.	Photoprotection through Reaction Center Quenching	160
VI.	Bioenergetics of Reaction Center Quenching	165
VII.	Molecular Mechanisms Regulating Reaction Center Quenching	167
	Acknowledgments	169
	References	169

12 Photoinhibition and Recovery in Oxygenic Photosynthesis: Mechanism of a Photosystem II Damage and Repair Cycle   175–191
Kittisak Yokthongwattana and Anastasios Melis

	Summary	175
I.	Introduction	176
II.	Photosystem II (PS II) Organization	176
III.	PS II Heterogeneity	178
IV.	PS II Damage and Repair Cycle in Chloroplasts	179
V.	DNA Insertional Mutagenesis for the Isolation and Functional Characterization of PS II Repair Aberrant Mutants	185
VI.	Conclusions	186
	Acknowledgments	187
	References	187

13 Regulation by Environmental Conditions of the Repair of Photosystem II in Cyanobacteria   193–203
Yoshitaka Nishiyama, Suleyman I. Allakhverdiev and Norio Murata

	Summary	193
I.	Introduction	194
II.	Effects of Light	194
III.	Effects of Oxidative Stress	197
IV.	Effects of Salt Stress	198
V.	Effects of Low-Temperature Stress	199
VI.	Conclusions and Future Perspectives	200
	Acknowledgments	200
	References	200

14 Photosystem I and Photoprotection: Cyclic Electron Flow and Water-Water Cycle   205–221
Tsuyoshi Endo and Kozi Asada

	Summary	205
I.	Introduction	206
II.	Cyclic Electron Flow around Photosystem I (PS I)	207
III.	The Water-Water Cycle	212
IV.	Comparison between Cyclic Electron Flow and the Water-Water Cycle in Terms of Physiological Role	214
V.	Perspectives	217
	Acknowledgments	217
	References	217

15 Integration of Signaling in Antioxidant Defenses 223–239
Philip M. Mullineaux, Stanislaw Karpinski and Gary P. Creissen

Summary	224
I. Introduction	224
II. Signaling Networks and Cross-Talk	225
III. Reactive Oxygen Species (ROS)-Mediated Signaling	228
IV. Reconfiguration of the Antioxidant Network and the Regulatory Role of Glutathione	231
V. ROS, Antioxidants, and Stress Hormones	232
Acknowledgments	235
References	235

16 Signaling and Integration of Defense Functions of Tocopherol, Ascorbate and Glutathione 241–268
Christine H. Foyer, Achim Trebst and Graham Noctor

Summary	242
I. Introduction	242
II. Singlet Oxygen and Tocopherol Function in PS II	243
III. Ascorbate: A Key Player in Leaf Development and Responses to the Environment	250
IV. Glutathione and the Importance of Cellular Thiol/Disulfide Status	256
V. Conclusions and Perspectives: All for One and One for All?	259
Acknowledgments	260
References	260

17 Redox Regulation of Chloroplast Gene Expression 269–287
Sacha Baginsky and Gerhard Link

Summary	269
I. Introduction	270
II. Posttranscriptional Processes	274
III. Transcription	279
IV. Connections, Outlook and Perspectives	282
Acknowledgments	283
References	283

18 Intracellular Signaling and Chlorophyll Synthesis 289–301
Robert M. Larkin

Summary	289
I. Introduction	289
II. Chlorophyll Biosynthetic Mutant, Inhibitor, and Feeding Studies	290
III. Plastid-to-Nucleus Signaling Mutants Inhibit Mg-Porphyrin Accumulation	293
IV. Mechanism of Mg-Proto/Mg-ProtoMe Signaling	294
V. Plastid and Light Signaling Pathways Appear to Interact	297
VI. Conclusions and Perspectives	298
Acknowledgments	298
References	298

19 The Role of Peroxiredoxins in Oxygenic Photosynthesis of Cyanobacteria and Higher Plants: Peroxide Detoxification or Redox Sensing? 303–319

Karl-Josef Dietz, Tina Stork, Iris Finkemeier, Petra Lamkemeyer, Wen-Xue Li, Mohamed A. El-Tayeb, Klaus-Peter Michel, Elfriede Pistorius, and Margarete Baier

	Summary	303
I.	Oxidative Stress	304
II.	Cyanobacteria as Model Organisms to Study Oxygenic Photosynthesis	304
III.	Peroxiredoxins in Eukaryotes and Their Subcellular Compartmentation	307
IV.	The Reaction Mechanisms of Peroxiredoxins	312
V.	Involvement of Organellar Peroxiredoxins in Stress Response	313
VI.	Peroxiredoxins in Redox Signaling	314
VII.	Conclusions	316
	Acknowledgments	316
	References	316

20 Lipoxygenases, Apoptosis, and the Role of Antioxidants 321–332

Mauro Maccarrone

	Summary	321
I.	Introduction	321
II.	Involvement of Lipoxygenases in Apoptosis	323
III.	Role of Antioxidants in Lipoxygenase-mediated Apoptosis	326
IV.	Concluding Remarks	328
	Acknowledgments	328
	References	328

21 Regulation of Photosynthetic Gene Expression by the Environment: From Seedling De-etiolation to Leaf Senescence 333–365

Christiane Reinbothe and Steffen Reinbothe

	Summary	334
I.	Introduction	334
II.	Control of Photosynthetic Gene Expression during Seedling Etiolation and De-etiolation	335
III.	Photosynthetic Gene Expression during Leaf Senescence	346
IV.	Future Perspectives	354
	Acknowledgments	355
	References	355

Author Index 367

Subject Index 369

Preface

Photosynthesis is integral to plant productivity, both as a source of energy and materials and as a target for feedback regulation by the demand of the whole plant for photosynthate. Photosynthesis is also strongly modulated by the environment; multiple environmental conditions result in the absorption of potentially harmful levels of excess light. Photoprotection and the phenomenon of photoinhibition of photosynthesis are thus key responses of plants to the environment as well. However, it remains unknown whether photoinhibition decreases or increases plant survival, fitness, and productivity. Similarly, interactions among different components of the photoprotective antioxidant network and their roles in cellular signaling and gene expression remain to be fully elucidated. This volume (*Photoprotection, Photoinhibition, Gene Regulation, and Environment*) brings together contributions from widely different areas in the hope of stimulating future research. Several concepts are emphasized in this volume 21 of Advances in Photosynthesis and Respiration. One is that chloroplast defenses against oxidative stress and excess reactive oxygen production are highly integrated with each other and are in communication with the cellular antioxidant network. Furthermore, it is increasingly recognized that antioxidants have a crucial role in redox sensing and signaling, in addition to their role in protecting the integrity of the chloroplast. This is highly significant since cellular redox balance participates in the modulation of a host of key responses in growth and development, such as the regulation of the cell cycle, senescence, and programmed cell death. This volume combines contributions on photoinhibition and photoprotection with those on redox signaling and gene modulation to integrate photoprotective responses into 'the bigger picture'. We bring together the full continuum of processes from protection on one hand to senescence and cell death at the other extreme.

Specific topics covered here include perspectives on the historic development of this research area. Photoprotective thermal dissipation of excess excitation energy is reviewed and its relationship to photoinhibition discussed. What role does photoinhibition play in plant survival, fitness, and productivity? Where does photoinhibition fit into the continuum of plant responses-from photoprotection by a cascade of defense mechanisms to cell death? During seasons with extreme conditions, such as icy winters or scorching, dry summers, the primary reactions of photosynthesis can become disabled. Many studies focus on the inactivation of the photosystem II core protein D1 seen during photoinhibition, and it is currently widely assumed that this inactivation represents damage. However, much of what has been assumed to be damage to chloroplast processes may, in fact, be caused by genetic programs (D. Wagner, D. Przybyla, R. op den Camp, C. Kim, F. Landgraf, K. P. Lee, M. Würsch, C. Laloi, M. Nater, E. Hideg and K. Apel, 2004, The Genetic Basis of Singlet Oxygen–Induced Stress Responses of *Arabidopsis thaliana*. Science 306: 1183–1185). It is noteworthy that the plants exhibiting the greatest degree of photoinhibition of photosynthesis are perennial evergreens adapted to extreme environments. Is photoinhibition a limitation to plant productivity, or might it be a 'talent' of stress-tolerant species? In this volume, widely different views are expressed concerning these questions, making it clear that additional work is needed to resolve them.

A wide range of environmental factors that affect photoprotection and photoinhibition is considered here: intense visible and ultraviolet light, winter conditions in temperate climates, nutrient deficiency, drought, and salinity. A molecular analysis of photoprotection is presented and the role of specific proteins in photoprotection is discussed. In addition to thermal dissipation pathways, alternative pathways for electron flow as well as the scavenging of reactive oxygen species by antioxidant metabolites (such as tocopherol, ascorbate, glutathione, xanthophylls) and enzymatic antioxidant pathways are reviewed. The peroxidation and recycling of lipids is considered together with the signaling functions of these reactions and of lipid-derived messengers. A comprehensive overview of the current understanding of signaling by reactive oxygen species and other signals that modulate gene expression is provided. Experimental approaches range from biochemical, physiological, genetic, and molecular approaches to field studies in a variety of natural environments.

This volume is suitable for advanced students and researchers interested in the general area of redox signaling and antioxidant defenses and their role in responses to the environment as well as the regulation of internal responses of all organisms. In the field of plant biology, the areas of photosynthesis, stress physiology, gene regulation by the environment, and growth and development are particularly relevant.

We express our sincere appreciation to the 57 authors from 15 countries for their outstanding contributions to this book. We are thankful to the Series Editor Govindjee for his invitation to develop this exciting book and for his support. We are grateful to Noeline Gibson and Jacco Flipsen (both of Springer) for their cooperation in producing this book.

Barbara Demmig-Adams
Department of Ecology & Evolutionary Biology
University of Colorado
Boulder, CO 80309-0334
USA
(e-mail: barbara.demmig-adams@colorado.edu)

William W. Adams III
Department of Ecology & Evolutionary Biology
University of Colorado
Boulder, CO 80309-0334
USA
(e-mail: william.adams@colorado.edu)
and
Autar K. Mattoo
USDA-ARS
Sustainable Agricultural Systems Laboratory
Bldg 001, Room 119
The Henry A. Wallace Beltsville Agricultural Research Center
10300 Baltimore Ave
Beltsville, MD 20705-2350, USA
(e-mail: mattooa@ba.ars.usda.gov)

William Adams, Barbara Demmig-Adams, Melanie Adams, and Robert Adams, during a trip to Barbara's native Germany during the summer of 2004. Photograph by Markus Demmig.

Barbara Demmig-Adams and **William W. Adams III** are Professors in the Department of Ecology and Evolutionary Biology at the University of Colorado at Boulder, USA. Barbara, as a native of Germany, received both her undergraduate degree (1979) in biology and chemistry and her graduate degree (1984) in plant physiology (with the late Prof. Hartmut Gimmler) from the Universität Würzburg. She subsequently spent two eventful years (1984–1986) as a postdoctoral fellow in the laboratory of Professor Olle Björkman at the Carnegie Institution of Washington's Department of Plant Biology in Stanford, California. Barbara and Olle Björkman characterized the photoprotective thermal energy dissipation that occurs in the antenna pigments of photosystem II, including the fact that this process can become sustained in evergreens under stress and lower the photon yield of photosystem II for prolonged periods. William attended the University of Kansas, receiving undergraduate degrees in biology (1981) and in atmospheric sciences (1983) as well as an MA degree (1984) in botany (investigating adaptations of epiphytic bromeliads from Mexico). This was followed by 18 months (1984–1985) of research in Reno, Nevada and Death Valley, California as the first half of his PhD work. During this time, William and Barbara began to collaborate personally and professionally across the Sierra Nevada divide that separated Reno and Stanford. Barbara returned to Würzburg in 1986 and provided the first evidence that the xanthophyll zeaxanthin, formed via the xanthophyll cycle under excess light, is involved in thermal energy dissipation. For a more detailed description of the discovery of the role of the xanthophyll cycle in thermal dissipation, see her historical minireview in Photosynthesis Research (2003) 76:73–80. While Barbara was in Germany pursuing the role of zeaxanthin, William completed another 17 months of research in Canberra and at three field sites in Australia, resulting in a PhD (1987) from the Australian National University under Professor C. Barry Osmond's mentorship. This work centered on photoinhibition in CAM plants, and included the first reports of photoinhibition and photoprotection under natural conditions in the field. With the support of a NATO postdoctoral fellowship and a fellowship from the Alexander von Humboldt Foundation, William then spent two years at the Universität Würzburg, where the personal and professional collaboration between him and Barbara Demmig became more firmly established. In the spring of 1988, Barbara completed her habilitation in plant biology (at the Lehrstuhl of Professor Otto L. Lange in Würzburg) and Barbara and William made their union official. They moved to the University of Colorado in 1989, brought two children (Robert, born in 1990, and Melanie, born in 1992) into the world, and continued their collaborative efforts on various aspects of the ecology and physiology of xanthophyll-dependent thermal dissipation. One focus of their work has been the study of unique modifications of photoprotective energy dissipation in evergreen species. They use tropical evergreens as models for the role of sustained thermal dissipation during shade-sun acclimation, and conifers and other evergreens to study its importance in acclimation during Colorado winters. These studies include ecological, comparative, and mechanistic approaches to integrate photoprotective energy dissipation and photoinhibition into whole plant functioning. William and Barbara have also begun to evaluate the influence of foliar carbon export pathways on the acclimation of photosynthesis and photoprotective or photoinhibitory responses. In addition, Barbara has had a long-standing interest in the role of zeaxanthin and other plant protective compounds in human health (see their paper in Science (2002) 298:2149–2153). Their research has been cited frequently by colleagues, leading to their recognition as highly cited researchers in the Plant & Animal Science category by the Institute for Scientific Information (<http://isihighlycited.com/>). Furthermore, William has been honored for his efforts in teaching plant biology by University of Colorado students (Mortar Board Certificate of Recognition for Exceptional Teaching, 2000) and faculty (Boulder Faculty Assembly Excellence in Teaching Award, 2004). For additional information, see their web site: (http://www.colorado.edu/eeb/EEBprojects/Adams_Demmig/).

Autar K. Mattoo

Autar K. Mattoo is a senior scientist with the Henry A. Wallace Beltsville Agricultural Research Center, United States Department of Agriculture (USDA), Agricultural Research Service (ARS), Beltsville, Maryland. He is originally from Kashmir, India and received his M.Sc. in Biochemistry (1965) and Ph.D. in Microbiology (1969) from the Maharaja Sayajirao University of Baroda (India). His Ph.D. advisor was Vinod Modi. He was a postdoctoral fellow with Bruce Keech at the University of Adelaide, a visiting scientist with Bob Vickery at the University of New South Wales in Australia, and later a DAAD Fellow with Marvin Edelman at the Weizmann Institute of Science in Israel. Autar's research activities fall under two diverse programs: (1) Molecular aspects of chloroplast function with particular emphasis on the photosystem II (PS II) reaction center proteins; and (2) Regulation of ethylene biosynthesis and fruit ripening. In collaboration with Marvin Edelman, he identified the rapidly turning over D1 – 32 kDa protein as a diuron-modulated PS II protein, elucidated its complete metabolic life history, demonstrated its reversible, photo-regulated post-translational phosphorylation and palmitoylation, and showed that D1 phosphorylation is regulated by a circadian clock. His current interest is in protein kinases, chloroplast-chromoplast differentiation, and hormonal cross talk in fruit development and ripening. In the field of biotechnology, Autar continues to use molecular genetics, nutritional genomics, and transgenic approaches to produce functional foods. He has incorporated research on interfacing genetically enhanced vegetables in sustainable, alternative agriculture systems. He has published over 200 research articles, twice chaired the Gordon Research Conference on Plant Senescence, and served on the Technical Advisory Committee of the US-Israel BARD. He is a member of the Overseas Standing Advisory Committee for the Department of Biotechnology, Government of India (2004–2007). He has lectured widely in the areas of biochemistry, molecular biology, and physiology. He served as Secretary-Treasurer of the Washington Section of the American Society of Plant Physiologists (1987). He was recognized as the Beltsville Area Scientist of the Year (1998), ARS's Distinguished Senior Scientist of the Year (1998), Secretary of Agriculture's "People Making a Difference" award (1999), and USDA Secretary's Honor Award for Scientific Excellence (1999). For further information, see his web site: (http://www.barc.usda.gov/psi/vl/mattoo.htm).

Chapter 1

A Random Walk To and Through the Xanthophyll Cycle

Harry Y. Yamamoto
Department of Molecular Biosciences and Bioengineering, University of Hawaii, Honolulu, HI 96822, USA

Summary	1
I. Introduction	1
II. The Beginnings	1
III. Education	2
IV. The Violaxanthin-Antheraxanthin-Zeaxanthin (VAZ) Pathways Story	3
V. Further Adventures and Advances	4
VI. Many Thanks to Many	9
Notes	9
References	9

Summary

This is an account of my personal and professional life as a student of the violaxanthin-antheraxanthin-zeaxanthin scheme for the xanthophyll cycle in higher plants. I had no early vision of becoming a scientist, but one circumstance led to another, and what began as a random walk ultimately developed into a life-long study of the biochemistry, physiology, and function of the xanthophyll cycle. The circumstances and people with whom I shared this path are described, with special attention given to the early developments.

I. Introduction

Does anyone accept an autobiographical assignment such as this without some hesitancy? I appreciated the invitation to tell my story, but wondered if I had anything worth contributing. What should I say? Who would care? After some reflection, I thought (or possibly rationalized) that my story, which is best characterized as a random walk to and through the xanthophyll cycle, may give comfort to young people whose vision of what they wish to accomplish in life may not be entirely clear. The circumstances that led to the discovery of light-induced conversions among violaxanthin (V), antheraxanthin (A), and zeaxanthin (Z)–the VAZ pathway for the xanthophyll cycle–may also be of interest. Although the walk was random, with many small and uncertain steps, it almost always carried me forward and ultimately brought me to the "right" path. As a child and even through college, I had no thoughts about becoming a scientist, only an innate desire to seek a better future as my parents had done. With luck and help from many people, I have been privileged to the better life, better than I could have imagined possible as a child. I extend my special thanks to the book's editors for giving me a chance to reflect and open doors to many good memories.

II. The Beginnings

To start, I can thank my father for my good fortune at being born in the U.S. Dad, the second eldest son, had to seek his independent fortune and immigrated to Hawaii while still a teenager. I grew up in the shadows of the famous Moana Hotel on Waikiki Beach. Perhaps some readers who have visited Waikiki remember the large banyan tree in the International Market Place located cater-corner from the hotel. We lived about three hundred yards from that tree; once, I fell out of it while playing Tarzan and broke my arm. Our home was provided by the hotel because Dad, a carpenter, was on call "24-7." We were poor, but I was not aware of it; my parents never complained, and all those around us were also poor. It was a happy and carefree time for me.

I was 8 years old when the Japanese attacked Pearl Harbor. Both my parents were treated as aliens even though Mom was a native-born U.S. citizen. Fortunately, we and many others in Hawaii were not sent to the "relocation camps" in which Americans of Japanese descent were detained on the U.S. mainland. In Hawaii, most of us were spared relocation largely by the actions of John Burns who, as police captain in charge of espionage for the FBI, vouched for the loyalty of Japanese-Americans in Hawaii. Burns was Delegate to Congress when Hawaii became the 50th state and later was elected Governor for three terms. He touched many lives. During the war, I carried a gas mask to school and my club house was the underground shelter Dad had built for our safety. The attack on Pearl Harbor had been led by Admiral Yamamoto; although he was no relation, I avoided problems by assuming the name Harry Chang when around soldiers on "R & R" (rest and recuperation) in Waikiki.

As a child I must have shown an interest in science because one of the best Christmas gifts I recall receiving was a Gilbert Chemistry Set. I can still picture it. It came in a red fold-out wooden case with rows of chemicals in small bottles, a simple balance, a watch glass, and spatulas. It had a manual from which I learned to make, among other things, black powder and "stink bombs." My parents weren't always pleased with the results of my experiments. These types of sets may not be available today and, if they are, their contents are probably more limited given modern concerns about hazardous substances.

III. Education

My friends are the reason I went to college. I took the entrance examination to the University of Hawaii only because they did. A few months later, I enrolled as a freshman and chose medical technology as my major because that is what a friend had selected and, much to my liking, it had a strong science emphasis. The course load was so heavy in zoology, microbiology, and chemistry that it nearly met the major's requirement for each of those fields. However, botany was not required for obvious reasons: medical technology deals with sick people, not sick plants. I didn't know then that I would spend my entire professional life happily working on plants.

The senior year in medical technology consisted of laboratory rounds at hospitals, public health laboratories, and the blood bank. During that year, I took night calls at Kuakini Hospital on alternate nights to earn my tuition for the year. Working as I did, I learned that the field of medical technology, as least at the time, offered limited economic opportunities. This important realization probably came about because by then I had a steady girl friend and was thinking about how to become a good provider. After graduating with a B.S. (1955) and completing a six-month tour of duty as a 2nd lieutenant with the U.S. Army Infantry in Ft. Benning, Georgia and Fort Riley, Kansas as part of my eight-year obligation in the Army Reserves, I embarked on the next leg of my random walk. I enrolled in the M.S. program in the Department of Food Technology at the University of Illinois at Urbana-Champaign. The selection of food technology as a field of study is not as curious as it may seem. I had worked in the Del Monte pineapple cannery for three summers prior to my senior year and was promoted each year to a better paying and more responsible position. I could see that a large food processing company offered many opportunities and thought that an advanced degree in the field would be useful. In changing to food technology, I accepted the possibility of not being able to return to Hawaii since most major food industries, except for pineapple processing, were on the mainland. It was a risk that I was willing to take. As it turned out, the greater risk was the demise of the pineapple canning industry in Hawaii, brought about by foreign competition. With one exception on the island of Maui, the canneries have all since closed.

Attending the University of Illinois was a good decision in several ways. First, I learned that, contrary to what I had assumed, changing fields of study was relatively easy. I discovered it wasn't necessary to complete all the requirements of the previous degree before starting work on a higher degree. Being the only one of three siblings to pursue graduate studies, I hadn't known any better. Next, during my first meeting to discuss my academic program with Reid Milner, Chairman of the Department of Food Technology, he casually asked if I intended to go on for the Ph.D. Me, whose parents had little schooling, who went to college only because his friends were going, and who had decided to pursue the M.S. only as a means for gainful employment? It was an unexpected and welcome expression of confidence in my potential. Thank you Prof. Milner! Finally, while pursuing the M.S., I found that I was more interested in "Why?" than "How?" and preferred fundamentals to applications. The title of my M.S. thesis was "*Kinetic Studies on the Heat Inactivation of Peroxidase in Sweet Corn.*" Peroxidase activity was used, and is

Chapter 1 Random Walk

possibly still being used, as an indicator for adequacy of heat treatment (blanching) of corn prior to freezing. Blanching prevents frozen corn from developing undesirable "off" flavors that result from enzyme activity. I found that there were two types of peroxidases with markedly different heat sensitivities (Q_{10}). The inactivation of both forms followed pseudo first-order kinetics, and the stability of the heat-resistant component made heating to inactivate it almost futile. I think it was this study that awakened by interest in basic science. I have the University of Illinois, and the Department of Food Technology in particular, to thank for giving some direction to my random walk.

The University of California, Davis had a program that seemed ideal for me: a Ph.D. in Comparative Biochemistry within the Department of Food Science and Technology. I was married and had a young son by then. The three of us drove for California, pulling our worldly possessions in a U-Haul trailer. We felt like pioneers traveling cross-country with an infant, but instead of in a covered wagon, we had an aging Ford. In Lincoln, Illinois, just a hundred miles out of Champaign-Urbana, our car broke down and required, so we were told, a complete engine overhaul that strained our resources and delayed our journey by several days. Even with the repairs, we barely made it over Donner Pass at the border between Nevada and California. A few months ago we drove through that region and the incline was hardly noticeable. Was the U-Haul so heavy, the road now less steep, or the rental car that much better? Perhaps yes to all. Thinking back on it now, that trip in the summer of 1958 was a great adventure and a test of endurance. What better preparation could one have for a doctoral program?

At Davis, my research advisor was Clinton "Chi" Chichester, a student of Gordon Mckinney, both of whom were interested in the biosynthesis of carotenoids. Paul Stumpf was my academic advisor. The biosynthetic pathway for carotenoids was not yet clearly established. During my first year I worked on an early step in the pathway and published a note in *Nature* (Yamamoto et al., 1961).

IV. The VAZ Pathways Story

I owe much of what came next to Sputnik, the first satellite, which was placed into orbit by the Soviet Union in October 1957. This achievement shocked the U.S. into giving more support to science and not just to space science. I benefited from this new commitment through a National Science Foundation

Fig. 1. Photograph taken at the National Science Foundation sponsored Carotenoid Symposium in Kyoto, Japan, 1965. Clinton Chichestor is being greeted by Prof. H. Mitsuda. Tom Nakayama is in the background.

predoctoral fellowship that funded the balance of my doctoral program. Besides relieving me of financial worries, the fellowship allowed me more flexibility in selecting a research topic. Also as a result of Sputnik, English translations of Russian articles became available, and a paper by David I. Sapozhnikov was brought to my attention by Tommy Nakayama, a friend and colleague of Chi's (Fig. 1). Sapozhnikov et al. (1957) reported that, in leaves subjected to alternating light and dark treatments, high light induced reciprocal changes in the levels of violaxanthin and lutein. He hypothesized that the reaction was involved in photosynthetic "oxygen transfer," that is, from water to molecular oxygen. I believed the observation merited further study because, unlike most carotenoids that are metabolic end products, this system appeared to be dynamic. Also, if the cycle was indeed involved in photosynthetic oxygen evolution, it would be a very significant discovery. I felt, however, that the reported kinetics made this possibility unlikely. Furthermore, if the reaction was involved in oxygen evolution, it was not an essential pathway since oxygen-evolving organisms such as blue-green algae lack violaxanthin.

Of course, the instrumentation and analytical methods available 45 years ago were crude by today's standards. I used preparative columns packed with powdered sugar to separate the xanthophylls of saponified extracts of leaves. Saponification removed chlorophyll that these columns could not resolve from xanthophylls. To assure complete recovery of xanthophylls after saponification, the xanthophylls were washed into ethyl ether instead of petroleum ether. Safety precautions were not what they are today and I was lucky not to have blown up the lab and myself with it. I would often return to my apartment reeking of ether. The procedures were so slow that I could obtain barely two sets of data points in a day. Despite these limitations, the very first experiment confirmed that high light induced in leaves a decrease in violaxanthin and an apparently corresponding increase in lutein. However, I was still not fully convinced that a symmetrical reactant, violaxanthin, was being converted to an asymmetrical product, lutein. Two mono-de-epoxidase reactions by different enzymes could explain such a conversion but would not be consistent with Sapozhnikov's hypothesis, which implied a single-step double de-epoxidation. Alternatively, the product could be zeaxanthin rather than lutein, leaving open the possibility of a single-step double de-epoxidation. Since the sugar column resolves pigments by normal-phase partitioning, I reasoned that zeaxanthin would, if formed, likely co-migrate with lutein, given that both molecules have similar structures and identical numbers of hydroxyl groups. Rechromatography of the sugar column's lutein band on a magnesium oxide (adsorption) column showed the product to be zeaxanthin and not lutein. The next question presented itself: does the conversion to zeaxanthin occur in one step or two? Antheraxanthin, the expected product of a two-step reaction, was found on the sugar column as a faint band between violaxanthin and lutein. These results established the currently accepted VAZ scheme: the light-induced cyclical conversion, in leaves, of violaxanthin (V) through antheraxanthin (A) to zeaxanthin (Z) (Yamamoto et al., 1962). Today, the pathway for the cycle seems obvious and can be easily demonstrated by HPLC analyses with a column that has mixed partitioning and absorption properties (Gilmore and Yamamoto, 1991). I referred to the pathway as the "violaxanthin cycle" but now "*the* xanthophyll cycle" is more commonly used. I emphasize "the" to acknowledge that other xanthophyll cycles are known, specifically the diadinoxanthin cycle in several algal species (Hager and Stransky, 1970) and the lutein epoxide cycle in mistletoe (Matsubara et al., 2001). It is uncertain whether these other xanthophyll cycles have the same biochemistry and functions as the VAZ cycle.

The VAZ scheme, in my opinion, was strong evidence against Sapozhnikov's hypothesis. Besides the very slow kinetics and incorporation of ^{18}O from O_2 on re-epoxidation of zeaxanthin (Takeguchi and Yamamoto, 1968), stepwise mono-de-epoxidation excluded a single-step removal of the epoxides of violaxanthin as might be expected for a role in oxygen evolution. Prof. Sapozhnikov acknowledged that the product of violaxanthin was zeaxanthin but, as far as I am aware from the literature, he ignored antheraxanthin and continued to suggest a role for the cycle in oxygen evolution (Sapozhnikov, 1973). Regrettably, I did not get to meet Prof. Sapozhnikov to congratulate him for his initial observation that light induces a change in the violaxanthin concentration. Following the VIIIth International Congress on Photosynthesis, which was held in Stockholm, Sweden in 1989, I went on a tour to the Soviet Union. In St. Petersburg I met a few of Prof. Sapozhnikov's former associates and learned that he had passed away in Italy in 1985 on his way to his new adopted home in Canada. Olga Koroleva gave me a photograph of his group taken in 1974; it is a wonderful photograph and I share it as a tribute to him (Fig. 2). In a different vein, I also express my appreciation to C. Stacy French, who was Director of the Carnegie Institution of Washington (Stanford, California), for his encouragement, patience, and graciousness to an aspiring graduate student. I visited his laboratory several times to learn as much as I could about photosynthesis and became friends for life.

V. Further Adventures and Advances

The notion that pursuing higher education would preclude my returning to Hawaii proved wrong. A year before completing the Ph.D., I was offered and accepted a position in the newly formed Department of Food Science and Technology in the College of Tropical Agriculture at the University of Hawaii. By then I had two children, and a secure job was attractive. For the first few years, I pursued research related to agriculture and refrained from working on the xanthophyll cycle, expecting that another student in Chi's lab would take up the work. When it became clear that no one would, I returned to the xanthophyll cycle, focusing on the biochemistry with the long-range objective of gaining insights into function. During my xanthophyll cycle hiatus (1962–65), Achim Hager made significant progress on the cycle's biochemistry. He showed that violaxanthin de-epoxidase (VDE) was localized in the chloroplast lumen and required ascorbate and low pH for activity (Hager, 1966). The cycle's transmembrane organization (Fig. 3) was established when both groups showed that the reverse epoxidation of zeaxanthin to violaxanthin occurred on the stromal side of the thylakoid at near neutral pH in the presence of NADPH and O_2 (Hager, 1975; Siefermann and Yamamoto, 1975).

Working on the xanthophyll cycle in Hawaii was not easy. Funding was limited, and there were no researchers nearby with whom I could interact that were engaged in related work on photosynthesis or carotenoid biosynthesis. Fortunately, grants from the

Chapter 1 Random Walk

Fig. 2. 1974 photograph of David I. Sapozhnikov's group given to me by O. Koroleva when I visited St. Petersburg in 1989. Front row from left: I. Popova, D. I. Sapozhnikov, S. Eidelmann, O. Popova. Back row from left: E. Morkovskaja, M. Gabr, O. Koroleva, T. G. Maslova, and G. Kornjushenko.

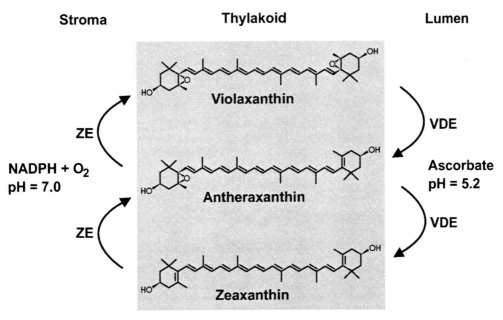

Fig. 3. VAZ transmembrane pathway for the xanthophyll cycle in higher plants.

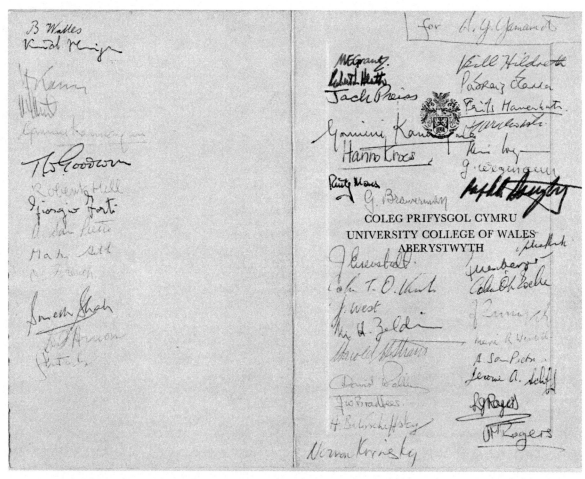

Fig. 4. Autographed banquet menu from the 1965 NATO Advanced Study Institute on the Biochemistry of Chloroplasts in Aberystwyth, Wales. In addition to names already mentioned, signatures by Trevor Goodwin, Giorgio Forti, Martin Gibbs, Norman Krinksy, Harold Strain, Jack Pries and Joseph Bradbeer, among others, are also present. How many signatures can you, the reader, recognize? It was exciting for me to be at this meeting of such notable scientists, most of whom I met for the first time.

National Science Foundation, the U.S.D.A. Competitive Grants Program, and the Department of Energy allowed me to continue research on the VAZ cycle. These grants also enabled me to travel about once a year to a major meeting on photosynthesis. Given my isolation from the mainstream of photosynthesis research, the importance of attending these meetings cannot be overemphasized. The first international meeting I was privileged to attend was the Advance Study Institute on the Biochemistry of Chloroplasts held in Aberystwyth, Wales in 1965, sponsored by the North Atlantic Treaty Organization (NATO). I believe my invitation to attend came from Trevor W. Goodwin. While looking through memorabilia in preparation for this perspective, I found the menu that I had passed around for signatures at the farewell dinner meeting (Fig. 4). I hope readers can make out the names in this marvelous collection of signatures. Among them are Robin Hill, Tony San Pietro, C. Stacy French, Dan Arnon, and many more, with apologies to those I have not mentioned.

Contact with the photosynthesis and plant biochemistry community has been an essential part of my forty-year stroll through the xanthophyll cycle and has created opportunities I might otherwise have missed. For example, in 1968 I spent my first sabbatical with Leo Vernon at the C.F. Kettering Research Laboratory in Yellow Springs, Ohio. There I met Teruo Ogawa, who was completing a postdoctorate with Leo, and with whom I became close personal friends. Teruo introduced me to the "opal glass" spectrophotometric technique perfected by Kazuo Shibata for measurement of light-scattering samples (Shibata, 1973). Upon returning to Hawaii, I applied the technique to chloroplast suspensions and found that violaxanthin de-epoxidation was detectable as a difference spectrum, with a peak at 505 nm, and could also be followed kinetically at 505 *minus* 540 nm (Yamamoto et al., 1972). This sensitive and rapid method for *in situ* measurement of xanthophyll cycle activity in chloroplasts was key for much of the progress we made during the 30 years that followed. Early applications of the spectrophotometirc assay included the discovery of the "availability" phenomenon, the intensity-dependent fractional release of violaxanthin from the total pool (Siefermann and Yamamoto, 1974); inhibition of VDE by dithiothreitol (Yamamoto and Kamite, 1972); and epoxidation of zeaxanthin to violaxanthin (Siefermann and Yamamoto, 1975). The method was also well suited for *in vitro* studies that demonstrated the requirement of lipid for de-epoxidation of pure violaxanthin (Yamamoto et al., 1974) and the substrate stereospecificity of VDE (Yamamoto and Higashi, 1978). The spectrophotometric assay of VDE activity remains useful to this day. It was recently applied to demonstrate that monogalactosyldiacylglycerol (MGDG), the major thylakoid membrane lipid, has a limited capacity to accommodate zeaxanthin and when this capacity is exceeded, stereospecific product feedback inhibition of VDE results (Hieber et al., 2004).

The serendipitous discovery of MGDG as the optimal chloroplast lipid for *in vitro* de-epoxidation of violaxanthin proved important. While I was able to obtain de-epoxidation of violaxanthin bound in washed thylakoid membranes, the same crude VDE preparation had no activity against purified violaxanthin, as had been reported by Hager (1966). Violaxanthin is insoluble in aqueous buffer, and various attempts to suspend or solubilize violaxanthin in a form that yielded activity failed. Isomerization and decomposition of the preparation were excluded as possible reasons. In the course of these tests, I ran out of the violaxanthin preparation that I had been using and, as a matter of convenience, recovered violaxanthin from "fat plates"* that Dorothea Siefermann, then a postdoctoral researcher in my laboratory, happened to be using for analysis of chloroplast pigments. Violaxanthin that was eluted from these plates with acetone and used without further purification gave rapid and nearly complete conversion to zeaxanthin. The reason for this success was traced not to coconut oil from the plates but rather to a lipid component in the unsaponified extract that co-chromatographed with violaxanthin. C. Freeman Allen earlier had separated the lipids in chloroplasts (Allen et al., 1966) and he kindly sent me samples that he still had on hand. All of the lipid samples we received supported de-epoxidation to varying degrees. We subsequently prepared a complete set of the major chloroplast lipids and found that MGDG was the most effective, giving rapid and complete de-epoxidation of violaxanthin in about 5 minutes under optimal conditions (Yamamoto et al., 1974). These results helped define the *in vivo* substrate of VDE: the violaxanthin that is converted to zeaxanthin is free in the membrane lipid phase rather than bound to pigment proteins. Exchanges between protein-bound pigments and free pigments in the lipid phase are implied. Recently, model systems consisting of soybean phosphatidylcholine only (Grotz et al., 1999) or egg phosphatidylchloline combined with MGDG (Latowski et al., 2002) have confirmed that lipid is required for "activation" of pure violaxanthin. However, de-epoxidation in these presumably bilayer systems were relatively slow and incomplete compared to de-epoxidation in the MGDG micelle system. MGDG constitutes a much larger fraction of the total chloroplast lipid: 60% to phosphatidylcholine's 2% or less (Webb and Green, 1991) and thus the micelle system may more closely approximate the *in situ* environment of free violaxanthin. Whatever model system is employed, violaxanthin should be prepared from saponified extracts to avoid artifacts from even trace amounts of contaminating chloroplast lipid.

We used the pH-dependent binding of VDE to the thylakoid membrane and to MGDG to obtain the partial C-terminal sequence (Rockholm and Yamamoto, 1996), which was then used to clone the gene and express the VDE protein (Bugos and Yamamoto, 1996). The complete sequence showed that VDE was a lipocalin enzyme, the first identified in plants** (Bugos et al., 1998; Yamamoto et al., 1999). This finding confirmed conclusions drawn 20 years earlier–before lipocalins were known to exist–that the shape of the VDE active center resembled a deep well (Yamamoto and Higashi, 1978). The cloned VDE carried out the forward VAZ reaction, providing strong evidence that the reaction could be catalyzed by a single enzyme with mono-de-epoxidase function. Evidence that a single gene product accounted for de-epoxidation was shown by Niyogi et al. (1998), in which a deletion mutation in *Arabidopsis* inhibited all de-epoxidase activity. The cysteine rich domain in the N-terminal sequence and highly charged domain in the C-terminal sequence explained, respectively, the DTT inhibition (Yamamoto and Kamite, 1973) and the pH-dependent membrane binding of VDE (Rockholm and Yamamoto, 1996).

My walk through the xanthophyll cycle took several administrative detours from 1980–82, 1982–86, and 1994–96 as Acting Associate Dean of Research, Chair of the Department of Plant Molecular Physiology, and Director of the Hawaii Agricultural Experiment Station, respectively. During the second of these, another chance occurrence caused me to refocus on research. It was popular for a time to hold small, informal bi-national conferences in Hawaii. One such conference, on photoinhibition, was held in Honolulu in 1985. At that time the subject was outside of my field of interest, but I attended on invitation from David Fork, whom I knew from visits to Carnegie during my days as a graduate student. One report by Olle Björkman caught my attention. He showed the kinetics of chlorophyll fluorescence quenching resulting from photoinhibition, which I recognized as being similar to the kinetics of violaxanthin de-epoxidation. After the meeting, I wrote a research proposal, including a request for a pulse-amplitude modulated fluorometer (PAM)*** that I would need to investigate the possible connection between photoinhibition and zeaxanthin formation. The grant proposal was successful but I was "scooped" by publication of a seminal paper by Demmig et al. (1987) that reported the correlation between non-photochemical quenching (NPQ) and zeaxanthin formation. Barbara Demmig had, in fact, noted the possible correlation a few years earlier but had difficulty convincing others of its reality. (For an interesting account of the events surrounding her important discovery, see Demmig-Adams, 2003.) Later, Adam Gilmore showed by a modeling technique that antheraxanthin also contributed to NPQ as effectively as zeaxanthin (Gilmore and Yamamoto, 1993). It is now common practice to express de-epoxidation as the de-epoxidation state (DES), or $(Z + A)/(V + A + Z)$, in conjunction with NPQ. The question of whether the correlation is a direct or indirect effect was recently answered with evidence that zeaxanthin is a direct quencher of excess energy (Ma et al., 2003).

Advances in research often result from the coupling of new analytical instrumentation or methods with the efforts of talented and dedicated individuals. This is certainly the case for contributions my laboratory made regarding the xanthophyll cycle and its relationship to NPQ. In terms of technology, the 505-nm absorbance change associated with de-epoxidation, the MGDG model system, and the HPLC method for resolution of zeaxanthin and lutein made significant differences in our research. The 505-nm change and development of the PAM provided an exceptional opportunity to examine xanthophyll cycle activity and NPQ simultaneously in chloroplasts. This application made it possible to show that although de-epoxidation and NPQ were both induced by light-dependent low pH, the protons for each were localized in different domains of the membrane (Mohanty and Yamamoto, 1996). I have not understood why the relatively simple opal-glass technique for the 505-nm change has not found more use, especially since it can be used simultaneously with NPQ measurement. In contrast, the HPLC method we developed for analysis of plant pigments is in wide use today (Gilmore and Yamamoto, 1991). As with the identification of the 505-nm change and the development of the MGDG model system, we arrived at this method somewhat by circumstance. Thayer and Björkman (1990) had reported an HPLC method that separated lutein and zeaxanthin but the column they used was no longer available. Based on my previous experience in separating lutein and zeaxanthin by sequential partitioning and absorption columns, we looked for a column that had both of these properties. ODS-1 was identified as a possibility because of its light carbon loading and non-endcapping of active silyl groups. The column performed as we hoped.

Although the mechanism of quenching has largely been resolved, numerous questions about the xanthophyll cycle remain. The physiology of the cycle is not well understood. The pool size of violaxanthin and the fraction of the pool that is active in the cycle vary among plant species and growth conditions (Demmig-Adams et al., 1999, this volume). There is a growing body of evidence that the cycle's operations may be related to more than just NPQ (Yokthongwattana and Melis, this volume). If the cycle has multiple functions, how are these functions regulated? Mutant studies suggest that the cycle in not essential for photosynthesis (Jung and Niyogi, this volume) and yet, as far as I am aware, all wild-type plants have the xanthophyll cycle. Why has nature retained this complex, apparently multifunctional system if it is not of some critical advantage? Did the system provide the adaptability to light environments needed for terrestrialization over multiple generations? Is it simply coincidental that the dominant photosynthetic life forms in the ocean and on land have xanthophyll cycles, the diadinoxanthin cycle and the VAZ cycle, respectively? The xanthophyll cycle has been related to photoprotection of plants against sudden and prolonged light stress (Verhoeven et al., 2001) and to improved plant fitness, as indicated by seed production, under fluctuating light intensities of the natural environment (Külheim et al., 2002). Interestingly, only half of the VAZ cycle (to antheraxanthin) appears to be present in a few Rhodophyceae (Aihara and Yamamoto, 1968) and in *Mantoniella* (Goss et al., 1998). Are these species less fit? We recently proposed that zeaxanthin functions as a messenger in a signal-transduction network that operates in the lipid phase of the chloroplast membrane to explain the cycle's multifunctional capabilities (Hieber et al., 2004). As one who has been involved with the xanthophyll cycle for nearly 45 years, I am surprised at how the questions seem never to end.

Fig. 5. Millie and I seated for lunch in Bagan during a recent tour of Myanmar.

VI. Many Thanks to Many

The cycle was the vehicle through which I entered the world of photosynthesis, traveled world wide, and made many good friends. I thank the photosynthesis community, the granting agencies, students, postdoctoral researchers, and colleagues for making my journey such a joy. I have also been blessed with recognition from peers through two awards, the Samuel Cate Prescott Award for Research in 1969 from the Institute of Food Technologists and the Charles Reid Barnes Life Membership Award in 2003 from the American Society of Plant Biologists. I extend special thanks to Govindjee, who offered me encouragement at every stage. Most importantly, I thank my family, especially my wife, Millie, for being understanding and supportive of my "obsession" for these many years. Now that I am retired, we have more time to spend together (Fig. 5). The question that I asked so many years ago as to why higher plants have retained the cycle remains unanswered. To all who will be continuing the walk, I look forward to learning what you find. I hope you enjoy the journey as much as I have!

Notes

*Fat (Egger) plates are Kieselguhr G plates that are dipped in hydrogenated coconut-oil/hexane solution and dried prior to use (Egger, 1962; also see Yamamoto, 1985). The plates, equivalent to a C_{18} endcapped HPLC column, resolve chlorophylls and carotenoids, except for lutein from zeaxanthin, in unsaponified chloroplast extracts. Inasmuch as lutein concentration is not normally affected by light treatments, Egger plates are an inexpensive method for tracking xanthophyll cycle activity.

**Lipoclains are a family of proteins that transport small, hydrophobic molecules such as retinol and porphyrins.

***I met Ulrich Schreiber, the developer of the PAM, at the 1971 International Congress of Photosynthesis meeting in Stresa, Italy. He approached me to discuss my paper on the incorporation of ^{18}O from O_2 into antheraxanthin and violaxanthin (Takeguchi and Yamamoto, 1971). I always appreciated his expression of interest, which was offered long before much was known about fluorescence quenching, let alone its relationship to the xanthophyll cycle.

References

Aihara MS and Yamamoto HY (1968) Occurrence of antheraxanthin in two Rhodophyceae *Acanthophora spicerera* and *Gracilaria lichenoides.* Photochem Photobiol 7: 497–499

Allen CF, Good P, Davis HF, Chisum P and Fowler SD (1966) Methodology for the separation of plant lipids and application to spinach leaf and chloroplast lamellae. J Am Oil Chem Soc 43: 223–231

Bugos RC and Yamamoto HY (1996) Molecular cloning of violaxanthin de-epoxidase from romaine lettuce and expression in *Escherichia coli.* Proc Natl Acad Sci USA 93: 6320–6325

Bugos RC, Hieber AD and Yamamoto HY (1998) Xanthophyll cycle enzymes are members of the lipocalin family, the first identified from plants. J Biol Chem 273: 15321–15324

Demmig B, Winter K, Krüger A and Czygan F-C (1987) Photoinhibition and zeaxanthin formation in intact leaves. A possible role of the xanthophyll cycle in the dissipation of excess light energy. Plant Physiol 84: 218–224

Demmig-Adams B (2003) Linking the xanthophyll cycle with thermal energy dissipation. Photosynth Res 76: 73–80

Demmig-Adams B, Adams WW III, Ebbert V and Logan BA (1999) Ecophysiology of the xanthophyll cycle. In: Frank HA, Young AJ, Britton G and Cogdell RJ (eds) The Photochemistry of Carotenoids, pp 245–268. Kluwer Academic Publishers, Dordrecht

Demmig-Adams B, Ebbert V, Zarter CR and Adams WW III (2005) Characteristics and species-dependent employment of flexible versus sustained thermal dissipation and photoinhibition. In: Demmig-Adams B, Adams WW III and Mattoo AK (eds) Photoprotection, Photoinhibition, Gene Regulation, and Environment, pp 39–48. Springer, Dordrecht

Egger K (1962) Dünnschichtchromatographie der Chloroplastenpigmente. Planta 58: 664–667

Gilmore AM and Yamamoto HY (1991) Resolution of lutein and zeaxanthin using a non-encapped, lightly carbon-loaded C_{18} high-performance liquid chromatographic column. J Chromatogr 543: 137–145

Gilmore AM and Yamamoto HY (1993) Linear models relating xanthophylls and lumen acidity to non-photochemical fluorescence quenching. Evidence that antheraxanthin explains zeaxanthin-independent quenching. Photosynth Res 35: 67–78

Goss R, Böhme K and Wilhelm C (1998) The xanthophyll cycle of *Mantoniella squamata* converts violaxanthin into antheraxanthin but not to zeaxanthin: consequences for the mechanism of enhanced non-photochemical energy dissipation. Planta 205: 613–621

Grotz B, Molnár P, Stransky H and Hager A (1999) Substrate specificity and functional aspects of violaxanthin-de-epoxidase, an enzyme of the xanthophyll cycle. J Plant Physiol 154: 437–446

Hager A (1966) Die Zusammenhänge zwischen lichtinduzierten Xanthophyll-Umwandlungen und Hill-Reaktion. Ber Deutsch Bot Ges 79: 94–107

Hager A (1975) Die reversiblen, lichtabhängigen Xanthophyllumwandlungen im Chloroplasten. Ber Deutsch Bot 88: 27–44

Hager A and Stransky H (1970) Das Carotinoidmuster und die Verbreitung des lichtinduzierten Xanthophyllcyclus in verschiedenen Algenklassen. V. Einzelne vertreter der Cryptophyceae, Euglenophyceae, Bacillariophyceae, Chrysophyseae und Phaeophyceae. Arch Mikrobiol 73: 77–89

Hieber AD, Kawabata O and Yamamoto HY (2004) Significance of the lipid phase in the dynamics and functions of the xanthophyll cycle as revealed by PsbS overexpression in tobacco and in-vitro de-epoxidation in monogalactosyldiaclyglycerol micelles. Plant Cell Physiol 45: 90–102

Jung H-S and Niyogi KK (2005) Molecular analysis of photoprotection of photosynthesis. In: Demmig-Adams B, Adams WW III and Mattoo AK (eds) Photoprotection, Photoinhibition, Gene Regulation, and Environment, pp 127–143. Springer, Dordrecht

Külheim C, Agren J and Jansson S (2002) Rapid regulation of light harvesting and plant fitness in the field. Science 297: 91–93

Latowski D, Kruk J, Burda K, Skrzynecka-Jaskier M, Kostecka-Gugata A and Strzalka D (2002) Kinetics of violaxanthin de-epoxidation by violaxanthin de-epoxidase, a xanthophyll cycle enzyme, is regulated by membrane fluidity in model lipid bilayers. Eur J Biochem 269: 4656–4665

Ma YZ, Holt NE, Li X-P, Niyogi KK and Fleming GR (2003) Evidence for direct carotenoid involvement in the regulation of photosynthetic light harvesting. Proc Natl Acad Sci USA 100: 4377–4382

Matsubara S, Gilmore AM and Osmond CB (2001) Diurnal and acclimatory responses of violaxanthin and lutein epoxide in the Australian mistletoe *Amyema miquelii*. Aust J Plant Physiol 28: 793–800

Mohanty N and Yamamoto HY (1966) Induction of two types of non-photochemical chlorophyll fluorescence quenching in carbon-assimilating intact spinach chloroplasts: the effects of ascorbate, de-epoxidation, and dibucaine. Plant Science 115: 267–275

Niyogi KK, Grossman AR and Björkman O (1998) *Arabidopsis* mutants define a central role for the xanthophyll cycle in the regulation of photosynthetic energy conversion. Plant Cell 10: 1121–1134

Rockholm DD and Yamamoto HY (1996) Purification of a 43-kilodalton lumenal protein from lettuce by lipid-affinity precipitation with monogalactosyldiacylglyceride. Plant Physiol 110: 697–703

Sapozhnikov DI (1973) Investigations of the violaxanthin cycle. Pure Appl Chem 35: 47–61

Sapozhnikov DI, Krasovskaya TA and Maevskaya AN (1957) Change in the interrelationship of the basic carotenoids of the plastids of green leaves under the action of light. Dokl Akad Nauk SSSR 13: 465–467 (or 74–76, English translation)

Shibata K (1973) Dual wavelength scanning of leaves and tissues with opal glass. Biochim Biophys Acta 304: 249–259

Siefermann D and Yamamoto HY (1974) Light-induced de-epoxidation of violaxanthin in lettuce chloroplasts III. Reaction kinetics and effect of light intensity on de-epoxidase activity and substrate availability. Biochim Biophys Acta 357: 144–150

Siefermann D and Yamamoto HY (1975) Properties of NADPH and oxygen-dependent zeaxanthin epoxidation in isolated chloroplasts. Arch Biochem Biophys 171: 70–77

Takeguchi CA and Yamamoto HY (1968) Light-induced $^{18}O_2$ uptake by epoxy xanthophylls in New Zealand spinach leaves (*Teragonia expansa*). Biochim Biophys Acta 153: 459–465

Thayer SS and Björkman O (1990) Leaf xanthophyll content and composition in sun and shade determined by HPLC. Photosynth Res 23: 331–343

Verhoeven AS, Bugos RC and Yamamoto HY (2001) Transgenic tobacco with suppressed zeaxanthin formation is susceptible to stress-induced photoinhibition. Photosyn Res 67: 27–39

Webb MS and Green BR (1991) Biochemical and biophysical properties of thylakoid acyl lipids. Biochim. Biophys Acta 1060: 133–158

Yamamoto H (1985) Xanthophyll cycles. Methods Enzymol 110: 303–312

Yamamoto HY and Higashi RM (1978) Violaxanthin de-epoxidase. Lipid composition and substrate specificity. Arch Biochem Biophys 190: 514–522

Yamamoto HY and Kamite L (1972) The effects of dithiothreitol on violaxanthin de-epoxidation and absorbance changes in the 500-nm region. Biochim Biophys Acta 267: 538–543

Yamamoto H, Yokoyama H, Simpson K, Nakayama TOM and Chichester CO (1961) Incorporation of 5,10,15 ^{14}C-farnesol phyrophosphate into *Phycomyces* carotenoids. Nature 191: 1299–1300

Yamamoto HY, Nakayama TOM and Chichester CO (1962) Studies on the light and dark interconversions of leaf xanthohylls. Arch Biochem Biophys 97: 168–173

Yamamoto HY, Kamite L and Wang YY (1972) An ascorbate-induced absorbance change in chloroplasts from violaxanthin de-epoxidation. Plant Physiol 49: 224–228

Yamamoto HY, Chenchin EE and Yamada DK (1974) Effects of chloroplast lipids on violaxanthin de-epoxidase activity. In: Avron M (ed) Proceedings of the Third International Congress of Photosynthesis, Vol III, pp 1999–2006. Elsevier Scientific Publishing Company, Amsterdam, The Netherlands

Yamamoto HY, Bugos CB and Hieber AD (1999) Biochemistry and molecular biology of the xanthophyll cycle. In: Frank HA, Young AJ, Britton G and Cogdell RJ (eds) The Photochemistry of Carotenoids, pp 293–303. Kluwer Academic Publishers, Dordrecht

Yokthongwattana K and Melis A (2005) Photoinhibition and recovery in oxygenic photosynthesis: mechanism of a photosystem-II damage and repair cycle. In: Demmig-Adams B, Adams WW III and Mattoo AK (eds) Photoprotection, Photoinhibition, Gene Regulation, and Environment, pp 175–191. Springer, Berlin

Chapter 2

Photoinhibition: Then and Now

Barry Osmond* and Britta Förster
*School of Biochemistry and Molecular Biology, The Australian National University,
Canberra ACT 0200, Australia*

Summary	11
I. What Then?	11
II. From Photorespiratory CO_2 Cycling to Photostasis	13
III. Mechanisms of Photoinhibition	14
IV. Photoacclimation: Yin-Yang and the Compromise between Photoinactivation and Photoprotection	14
V. Then There was the Leaf Disc O_2 Electrode, and Now There is Rapid Response Gas Exchange	16
VI. Enlightening the Mechanisms of Photoinhibition in *Chlamydomonas reinhardtii*	17
VII. Quo Vadis?	18
Acknowledgments	19
References	19

Summary

This perspective advocates a holistic view of photoinhibition from the molecule to the biosphere; a view that integrates many biophysical and biochemical processes in antennae and reaction centers of the photosystems that, when acting in concert, allow plants to respond to diverse and dynamic light conditions in many different environments. We take the general view that photoinhibition refers to a reduction in the efficiency of light use in the photosynthetic apparatus (Kok, 1956). Since the 1970s, biochemical, ecophysiological, and genetic studies of photosynthetic functions in strong light, *in vivo* and *in situ*, and their interactions with biotic and abiotic stresses, have significantly advanced our understanding of photoinhibition. We trace some origins of the idea then, that slow dark reactions, such as growth, CO_2 assimilation, photorespiration, and photosynthetic electron transport, ultimately limit light use in photosynthesis, and thus determine whether light is in excess and the magnitude of "excitation pressure" in the photosynthetic apparatus at any moment. This and other ideas are followed through studies of photoacclimation in leaves of plants and algae from diverse terrestrial and marine environments.

We highlight two currently interesting possibilities for the photoprotective dissipation of "excitation pressure" that reduce the efficiency of photosynthesis by changes in structure and function of antenna pigment-protein complexes and in the populations of functional and non-functional PS II centers. We conclude by briefly considering challenges presented now by the discovery of "gain of function", very high light resistant (VHL^R) mutants of *Chlamydomonas*, by the accessory lutein-epoxide cycle, and by technologies for remote sensing of photoinhibition in the field.

I. What Then?

This perspective reviews some hypotheses that have stimulated research in photoinhibition since its renaissance in the 1970s through studies of the slow, dark reactions of photosynthesis, and their implications for the faster, primary light reactions. Then as now, hypotheses were only as good as the experiments they provoked, so these ideas flourished as tools improved for assessing photosynthetic functions in vivo and in situ. Our eclectic pursuit of these ideas would have been impossible, then or now, without generous access to the needed

---

*Author for correspondence, email: barry.osmond@anu.edu.au

infrastructure, made available in a timely fashion by a network of collaborators worldwide.

In the 1970s the Research School of Biological Sciences (RSBS), Institute of Advanced Studies at the Australian National University provided a stimulating setting for creative colleagues and students of great initiative. Barry Osmond drifted into research on photoinhibition unfettered by peer review (that *"well-meaning but narrow-minded nanny of an institution (that) ensures that scientists work according to conventional wisdom and not as curiosity or inspiration moves them"*; Lovelock, 1990). This environment also attracted distinguished visiting scientists and foundations were laid then for many later collaborations and expansion of ideas in other laboratories as one's distractions and responsibilities multiplied. In one of these, a generation later, an exchange program brought Britta Förster from the Freie Universität Berlin to the joint laboratory of John Boynton and Nick Gillham at Duke University. There we were fortunate to find our ideas accepted in an established genetic program where a decade of support from an exceptionally broad-minded, peer-reviewed program of basic energy research in the US Department of Energy allowed us to pursue genetic analyses of key processes in photoinhibition.

Recalling that a picture is worth a thousand words, we would like to bracket this essay with two images created by the processes of photoinhibition. These images carry us from a century old, then comprehensive account of "assimilation inhibition" by strong light (photoinhibition), to visualization now of the questions of photoinactivation and photoprotection in chloroplast grana. Our images were printed by photoinhibiting leaves from a plant (*Cissus*) grown in deep shade, by exposure to full sunlight while covered with 35 mm black and white film (negative and positive; Osmond et al., 1999). Photoinhibition occurred in leaf cells under the more transparent areas of the film, quenching chlorophyll fluorescence so that a positive image could be revealed by a filtered digital camera system, or resolved to the level of granal stacks of chloroplasts in leaves using a confocal microscope.

Text excerpts of Ewart (1896) on a microfiche negative remind us of a century of progress in technologies available for photosynthetic research (Fig. 1). Ewart observed "assimilatory inhibition" (reasonably equated today with photoinhibition) in leaf tissue slices from shade plants after exposure to strong light. Using Engelmann's method, he found that fewer O_2-tactic bacteria congregated adjacent to chlorophyll bodies (chloroplasts) in leaf cells after strong light treatment, indicating reduced O_2 evolution (indicating photoinhi-

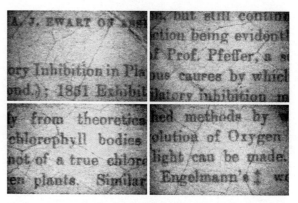

Fig. 1. Photoinhibitory print on a *Cissus* leaf, viewed as chlorophyll fluorescence, showing text excerpts from the introduction of Ewart (1896) that describe the context and methods of his early investigations of what we now refer to as photoinhibition (Osmond et al., 1999, with permission).

bition). Although we have now come a long way technologically, many of the questions addressed by Ewart remain unresolved today. One may ask why so little attention was paid, for so long, to evidence exhaustively assembled by Ewart and many others, indicating exposure to "excess light" impairs the efficiency of photosynthesis?

It may be that most of the chlorophyll on the planet is to be found in the shade and, by focusing on understanding the principles of efficient light utilization in weak light, we took it for granted that plants must have evolved sophisticated means of dealing with exposure to strong light. Negative and positive phototaxis is widespread among motile photosynthetic cyanobacteria and algae, but among immobile higher plants comparatively few have evolved light avoiding leaf or chloroplast movements (Kasahara et al., 2002). Some plants, like *Cotyledon obiculata*, manage external photoprotection with reflective wax and only engage internal, xanthophyll de-epoxidation defenses when brush-wielding, curiosity-driven researchers interfere with the natural order of things (Robinson and Osmond, 1995). Although shade and sun species, and even shade and sun ecotypes, can be distinguished, many plants survive and grow in habitats of glaring sunlight so there seemed to be no obvious problem. As tools emerged for assessing photosynthetic functions of plants in the natural environment, it became clear that photosynthetic efficiency declined in "excess light" and in response to stress. We now know that molecular mechanisms to repair photoinactivation or photodamage, and strategies for "lowering the shades" through photoprotection, have evolved to ensure photoinhibition in strong light is usually reversible. Although

Chapter 2 Photoinhibition: Then and Now

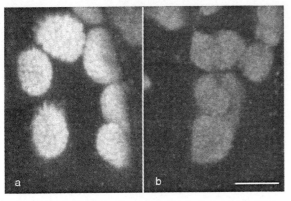

Fig. 2. Photoinhibitory print on a *Cissus* leaf, viewed as chlorophyll fluorescence in a confocal microscope. Cells under the shaded areas of the printed image (a) contain chloroplasts with highly fluorescent PS II centers in grana. Cells under transparent areas of the image were photoinhibited (b), and fluorescence from grana in chloroplasts of these cells is much reduced (Osmond et al., 1999, with permission).

the confocal microscope allowed Owen Schwartz and Brian Gunning to trace chlorophyll fluorescence to the granal stacks of chloroplasts (Fig. 2) in light and dark areas of photoinhibition images (e.g. Fig. 1), we still do not know whether the slow recovery of fluorescence yield from these grana is due to slow repair of PS II reaction centers photoinactivated by strong light, or to sustained photoprotection in the antenna that reduces excitation transfer to the PS II centers, or to both.

II. From Photorespiratory CO$_2$ Cycling to Photostasis

There was a time in the 1960s when some thought that leaves were so complicated that photosynthesis research was best advanced by systematic elimination of artifacts in experiments with cells from algal cultures grown on the window ledge, or with chloroplasts and thylakoids isolated from market spinach. Others were challenged to explore genecological differentiation of sun and shade ecotypes (Björkman and Holmgren, 1963), and Olle Björkman's studies of leaf and thylakoid responses to light, from photoacclimation to photoinhibition, in plants from the rainforest to the desert stimulated a renaissance of research in photoinhibition. During his first sabbatical in RSBS, Olle compared the stability of PS II in thylakoids from sun and shade grown C$_3$ *Atriplex triangularis* (syn. *patula* ssp. *hastata*) following the approach of Jones and Kok (1966). These experiments rekindled interest in the effects of excess light on the efficiency of primary photosynthetic processes, and initiated a long-standing collaboration on photoinhibition.

A comparative study of photorespiratory carbon metabolism in C$_3$ and C$_4$ species of *Atriplex* led to the then improbable speculation that photorespiratory carbon recycling in C$_3$ plants *"provides one mechanism whereby much of the energy input normally associated with net CO$_2$ fixation can be consumed without net carbon gain"*, and that this could mitigate *"a vast excess of excitation energy that would (otherwise) lead to a destruction of the reaction centers of the photosystems"* (Osmond and Björkman, 1972). After some preliminary experiments, and spurred by a report from Cornic (1976), the latent hypothesis was presented as a research topic to Steve Powles, then an aspiring PhD student uninhibited by previous laboratory experience. Steve fast-tracked the hypothesis from first publication (Powles and Osmond, 1978) to Annual Reviews (Powles, 1984). Supporting studies with CO$_2$-fixing chloroplasts of C$_3$ plants (Krause et al., 1978) stimulated further synthesis of the interactions between photorespiration and photoinhibition (Osmond, 1981). Since then, control of CO$_2$ supply in the light has been used in many ways to maximize or mitigate "excitation pressure" and to unravel the dimensions of photoinhibition in diverse organisms, ranging from manipulation of internal CO$_2$ generated by malic acid decarboxylation in CAM plants (Adams and Osmond, 1988), incubation of marine algae in CO$_2$-free seawater (Osmond et al., 1993), and growth of lab cultures of *Chlamydomonas reinhardtii* in air or CO$_2$-enriched atmospheres (Förster et al., 2001a).

The notion that "futile" photorespiratory CO$_2$ cycling, which is an "inevitable consequence" of the evolution of O$_2$-sensitive Rubisco in a CO$_2$-rich but O$_2$-depleted atmosphere (Lorimer and Andrews, 1973), might in part prevent photoinhibition of water-splitting, O$_2$-evolving light reactions of photosynthesis, has become more appealing with time (Osmond and Grace, 1995). Now broadly and more elegantly articulated as the concept of photostasis, *"the predisposition of photosynthetic organisms to maintain a balance between energy input through photochemistry and subsequent energy utilization through metabolism and growth"* (Öquist and Huner, 2003), the idea seems generally accepted that slow dark reactions ultimately limit light use in photosynthesis and determine whether light is in excess, thereby determining the magnitude of "excitation pressure" in the photosynthetic apparatus at any moment.

It is not surprising now that sense and antisense expression of the key photorespiratory enzyme glutamine synthetase in tobacco chloroplasts respectively protects and sensitizes these plants to photoxidative damage in strong light (Kozaki and Takeba, 1996). However, misgivings then as to the extent of photoprotection conferred through photorespiratory carbon cycling during water stress were raised in Powles (1984), and the issue has remained unresolved and controversial. The key question is whether intercellular CO_2 concentrations settle to the compensation point under water stress, and the current view is well articulated by Cornic and Fresneau (2002). Perhaps the most conclusive evidence comes from Olle's experiments with schlerophyllous xerophytes such as *Arbutus menzesii* exposed to extreme water stress, in which 25–35% of absorbed photons seem to be consumed in electron transport associated with photorespiratory CO_2 recycling (zero net CO_2 assimilation) in full midsummer sunlight (Badger et al., 2000). It is now recognised that the rest of the "excitation pressure" is dissipated as heat in processes that were simply beyond our ken in 1972.

III. Mechanisms of Photoinhibition

Plant biologists are no more prone to reductionism than other natural scientists, but the complexity of biological systems and the diversity of organisms and environments tends to encourage prematurely exclusive hypotheses as mechanistic insights deepen. A decade ago it seemed necessary to stand back from the feverish pursuit of two, then largely exclusive, explanations of underlying mechanisms of photoinhibition. One focused on photoinactivation when excess excitation reached the PS II reaction center; the other focused on regulatory processes in the antennae that reduced excitation transfer to PS II, thereby achieving photoprotection.

Molecular insights into the mechanisms of PS II inactivation as an explanation for photoinhibition were greatly stimulated by recognition of the rapid turnover of the D1 polypeptide, the "suicide polypeptide" in algae and thylakoids (Kyle et al., 1984). To the pessimists it seemed that the weak link, the fundamental flaw, at the heart of the photosynthetic apparatus, had been discovered. Charlie Arntzen was quick to recognize the implications for photoinhibition *in vivo* and a bilateral Australia-US workshop held in the East West Center in Hawai'i soon expanded consideration of photoinhibition from reaction center functions to ecosystem processes. It is fair to say that for a few years most researchers then embraced photoinactivation of PS II reaction centers as the underlying feature of photoinhibition *in vivo*. Much subsequent research has been devoted to placing the processes of D1 turnover into context (Aro et al., 1993; Adir et al., 2003) as will be related in the perspective by Edelman and Mattoo (this volume) in Chapter 3.

With 20:20 hindsight, the biggest oversight of the Hawai'i meeting was just around the corner in the laboratory of Harry Yamamoto. But who then could have anticipated that the already well-known xanthophyll pigment interconversions would have such an impact on our thinking about photoinhibition? It is for Harry to relate how the xanthophyll cycles in lettuce leaves and green tomatoes came to assume such a central role in our understanding of "excitation pressure" in photosynthesis (Yamamoto, this volume). The historic demonstration by Demmig et al. (1987) of a correlation between the pool size of zeaxanthin in leaves and fluorescence quenching in strong light in the absence of CO_2 and photorespiration, opened the door to understanding of photoprotection (Demmig-Adams, 2003). Subsequent research in Barbara Demmig-Adams and William Adams' laboratories in Boulder led the way for what is now a near universal mechanism for dissipation of excess light in the photosynthetic apparatus as heat (Adams et al., Demmig-Adams et al., this volume). By and large these insights emerged from in vivo and in situ studies of plant leaves and algae grown under controlled conditions in laboratories or examined by ecophysiologists in the field.

It is still difficult to resolve whether the decline in photosynthetic efficiency after exposure to strong light is due to a sustained reduction in the population of functional PS II centers, or due to reduced transfer of excitation from antennae to an unchanged population of functional centers, or to both. At bottom, if the response was dominated by the former, one would expect that no amount of additional light could restore maximum photosynthetic rate in a photoinhibited system. If it were dominated by the latter, it should be possible to restore maximum photosynthetic rate with increased light (Fig 1.2 in Osmond, 1994). It has proved surprisingly difficult to distinguish these possibilities experimentally. Instead, studies of photoacclimation in the photosynthetic apparatus of plants from different habitats, and in mutants, have brought a more holistic approach to our understanding of photoinhibition.

IV. Photoacclimation: Yin-Yang and the Compromise between Photoinactivation and Photoprotection

The selective pressure of the light environment is well appreciated by ecophysiologists, who recognize that the capacity for acclimation from one light environment

Chapter 2 Photoinhibition: Then and Now

to another varies greatly between and within species. Genecotypic differentiation with respect to shade and sun habitats (Björkman and Holmgren, 1963) remains one of the best (and still under-exploited) contexts for experimental research in photoinhibition, with all the components of photostasis arrayed for assessment. Traditionally, the capacity for acclimation has been assessed in transfer experiments, from low light growth environments to high light treatments (and *vice versa*) that tend to unmask the real significance of components of the photosynthetic apparatus imagined to be of paramount importance to photoinhibition.

Thus the ecotypes of *Solanum dulcamera* ("bittersweet") discovered by Gauhl (1976) were used to uncover the limiting role of nitrogen nutrition in shade-sun acclimation with increased sensitivity to photoinhibition under low nitrogen (Ferrar and Osmond, 1986). Nitrogen dependence was also established for photoacclimation and survival of intertidal *Ulva rotundata* in seawater (Henley et al., 1991) and the possibility emerged that a nitrogen-depleting biotic stress, such as severe virus infection, could also influence photoacclimation (Balachandran and Osmond, 1994) and the fitness of infected plants in shaded and sun habits (Funayama et al., 2001). In general, limiting nitrogen prevents acclimation, i.e. increase in light-saturated photosynthesis, presumably because of the huge nitrogen investment in Rubisco, and renders plants more sensitive to photoinhibition. Indeed, Kato et al. (2003) have now shown that the first order rate constant for PS II photoinactivation is determined by the excess energy not used in photosynthetic electron transport or dissipated as heat, and in turn, that this excess is responsive to nitrogen nutrition.

At about the same time, Jan Anderson initiated a systematic, quantitative biochemical assessment of changes in the components of the photosynthetic apparatus when leaves were transferred from shade to sun (Leong and Anderson, 1983), so that when the organizers of the Hawai'i workshop invited participants with differing perspectives to co-author chapters, another long-standing collaboration was initiated (Anderson and Osmond, 1987). Many studies now confirm the higher capacity of Rubisco, electron transport, and violaxanthin de-epoxidation in nitrogen replete sun plants compared with shade plants, conferring high de-epoxidation state (DES = [antheraxanthin] + [zeaxanthin] / [violaxanthin] + [antheraxanthin] + [zeaxanthin]) and non-photochemical chlorophyll fluorescence quenching (NPQ), a low reduction state in PS II centers, rapid repair of D1 protein-damaged reaction centers, and sustained photoprotection in strong light (given that other electron acceptors and intermediates are not limiting) are required to secure photoacclimation.

Experiments manipulating the excitation pressure on shade and sun grown *Ulva rotundata* using CO_2-depleted seawater, dithiothreitol to inhibit violaxanthin de-epoxidase and the xanthophyll cycle, and chloramphenicol to inhibit chloroplast protein synthesis (Osmond et al., 1993), showed photoprotection and photoinactivation as different interacting processes determining the response to photoinhibition at different time scales. These weekend experiments, with colleagues in the sanctuary of the Duke University Marine Laboratory, catalysed the formulation of the holistic, Yin-Yang interpretation of interactions among many dynamic and interlinked factors that determine the stability of the photosynthetic apparatus during photoacclimation (Osmond, 1994).

Photoacclimation per se has not been pursued to any great depth in unicellular model photosynthetic systems, and interpretations are often complicated by the use of heterotrophic or partially heterotrophic culture conditions. One study with *Chlamydomonas reinhardtii* confirmed the expectation that priority was accorded to restoration of PS II reaction center function on transfer from low to high light treatments. Cell division stopped on transfer to high light, Rubisco large sub-unit synthesis dropped rapidly within minutes and only resumed hours later, and translation of *psbA* mRNA accelerated, facilitating a 10-fold enhancement of D1 protein synthesis within the 12 h required for restoration of photosynthetic efficiency and cell division (Shapira et al., 1997). Other chemostat experiments demonstrated that D1 protein catalysis mutants (herbicide resistant) out-compete D1 protein synthesis mutants (Heifetz et al., 1997), and there is scope for expansion of this approach.

Anticipating that some assessment of the potential for shade-sun acclimation in *Arabidopsis thaliana* also might be useful for the future evaluation of photoinhibition mutants, Russell et al. (1996) undertook a comprehensive set of experiments to test the limits to photoprotection and the extent of photoinactivation in the course of photoacclimation in wildtype. Importantly, these experiments showed that plants grown at 200 μmol photons m^{-2} s^{-1} and transferred to irradiances that saturated photosynthesis (400 μmol photons m^{-2} s^{-1}) experienced little perturbation of photosynthetic parameters during exposures of up to 24 h. Increasing time of exposure to above-saturating irradiances (1,350 and 2,200 μmol photons m^{-2} s^{-1}) elicited a 50% increase in total violaxanthin cycle pigments and increased (but incomplete) de-epoxidation, in spite of which (or because of which?) PS II efficiency, functional PS II centers, and anti-body detectable D1 protein declined markedly. Clearly, in this "lab weed" Yin rose in the face of an inadequate Yang, but initially at least, photoacclimation was not achieved. The Yin-Yang formulation of interactions between these two "cultures" of research seemed to offer promise of a more holistic framework then, but it is for this volume to establish whether fascination with one or other hypothesis, to the exclusion of the other, is now declining.

V. Then There was the Leaf Disc O$_2$ Electrode, and Now There is Rapid Response Gas Exchange

As in most areas of photosynthesis research, progress in understanding photoinhibition in vivo and in situ depended on development of instruments and their creative applications in different contexts. Often these applications and insights emerged in the course of visits to the laboratories of collaborators, ranging from a few days before or after a conference, to months on sabbatical. Thus the network of "safe houses", a concept borrowed from the best spy fiction of the Cold War era, came to feature in the aspirations of colleagues overwhelmed with administrative and other obligations, and with the network came novel "toys". For example, David Walker's first escape to the Canberra laboratories introduced us to the leaf disc O$_2$ electrode system (Delieu and Walker, 1983). Compared with more complex gas exchange systems, this device proved ideal for measuring the efficiency of photosynthetic O$_2$ evolution at CO$_2$ saturation, especially in CAM plants (Adams et al., 1986). When configured for room temperature fluorescence analysis, the device became indispensable for photoinhibition research in vivo, and tales abound as to its impact. David, for example, was dismayed to find that otherwise perfect Sheffield spinach leaves could not be induced to oscillate in the same way as barley, and thus were recalcitrant in revealing regulatory interactions between carbon metabolism, bioenergetics, and photoinhibition in vivo. One visitor of antipodean orientation, impressed by evidence for shade adapted photosynthesis on the underside of leaves (Terashima and Inoue, 1985), presented the leaf disc upside down to the then available light sources, restored oscillatory behaviour, and David's faith in photosynthetic regulation in spinach (Walker and Osmond, 1986).

Olle Björkman had built a simplified apparatus for 77K fluorescence analysis in vivo to assess photoinhibition (Powles and Björkman, 1982) and applied it to evaluate light avoiding leaf movements in *Oxalis* and a water-stressed pasture legume (Ludlow and Björkman, 1984). He also built a better light source for David's leaf disc electrode, deploying a prototype made from a coffee can to measure mangrove photosynthesis on the beach at a North Queensland "safe house" (or so the story goes). When the leaf disc electrode was used in conjunction with 77K fluorescence, the "gold standards" for efficiency of the dark and light reactions of photosynthesis in vivo were established for vascular plants (Björkman and Demmig, 1987). The methods were then used to confirm enhanced sensitivity to photoinhibition under zero CO$_2$ and 2% O$_2$, i.e. in the absence of photorespiration (Demmig and Björkman, 1987) and in CAM plants denied internal CO$_2$ from the decarboxylation of malic acid (Adams and Osmond, 1988).

The research team at Duke University Marine Laboratory made the transition from laborious 77K fluorescence measurements (Henley et al., 1991) to automated O$_2$ electrode and pulse-amplified modulated fluorescence (PAM) analysis of photoinhibition and photoacclimation in *Ulva* (Franklin et al., 1992; Osmond et al., 1993). Plants collected from the channel were grown under light gradients and controlled environments in the ingenious "Beaufort bathtubs" and analyzed in chambers designed to facilitate light response curve measurements of O$_2$ exchange and fluorescence quenching in seawater. More important perhaps, these "sea-trials" with *Ulva* proved to be very valuable for the subsequent analysis of mutants in another algal photosynthetic model organism, *Chlamydomonas reinhardtii*. Little needs to be said here about the impact of PAM instrumentation on research in photoinhibition: it is adequately documented in the Chapters that follow. Devices for pulse modulated analysis of room temperature fluorescence (the PAM fluorometer; Schreiber et al., 1986), changed the way we now go about research in photoinhibition, in the laboratory and in the field.

Fred Chow further adapted David's leaf disc O$_2$ electrode to estimate the single turnover flash-yield of PS II in vivo (Chow et al., 1991) and began a long series of experiments that have yielded remarkable insights into the population biology of PS II centers in vivo. Among other things, Fred's work has shown that PS II reaction centers have a finite functional life in vivo with individual centers being photoinactivated after absorption of 10^6–10^7 photons. Even in weak light, about 10^6 PS

II centers turn over per mm² leaf area per second, from which Anderson et al. (1997) deduced that each PS II center turns over at least once a day. Fred's recent analyses of population dynamics of inactive PS II centers (Chow et al., 2002) lends credence to the proposal of Krause (1988) that such centers serve to protect their neighbors.

Functional analysis of flash-yield in vivo has deepened our understanding of interactions between photoinactivation and photoprotection. For example, Youn-Il Park measured the decline in O_2 flash yield while he "titrated" excitation pressure in vivo against cumulative irradiance (exposure for constant times at different irradiance, or for increasing times at constant irradiance) by changing the gas composition during exposure, thereby altering "sinks" for photosynthetic electron transport in carbon metabolism and/or photoreduction of O_2 (Park et al., 1996). Compared with the slow initial rates of photoinactivation of PS II centers in 1.1% CO_2 and 21% O_2 (saturated CO_2 fixation and O_2 photoreduction), photoinactivation was accelerated by removal of O_2 photoreduction in 1.1% CO_2 alone, was still faster in 60% O_2 alone (with O_2 photoreduction and photorespiration as the only source of CO_2), was further accelerated in 2% O_2 (with the Mehler reaction as the principal remaining sink), and most rapid of all in N_2 alone (in the absence of acceptors for photosynthetic electron transport). These data fit well the earlier ideas of Ulrich Heber and Kozi Asada (reviewed in Asada, 1999) highlighting direct electron flow to O_2 in the Mehler reaction as an additional mechanism for dissipation of excess excitation when carbon assimilation is restricted.

The "safe house" concept took on added reality when, during the declining years of the Soviet Union, the "laboratory family" of Agu Laisk in Estonia went to extraordinary lengths to accommodate visiting scientists and introduce them to the rapid kinetic analysis of photosynthesis in vivo. Experimental and theoretical evaluation of oscillations during shade-sun acclimation in sunflower in Tartu showed that, although the flux control coefficient of Rubisco dominated the rate of sun acclimated leaf photosynthesis, changes in the light absorbing and electron transport components of the photosynthetic apparatus drove the shift in light dependence of regulatory responses (Osmond et al., 1988; Woodrow and Mott, 1988). The potential for molecular biophysical and biochemical evaluation of leaf photosynthesis in vivo using kinetic gas exchange and optical methods with what is now known as the FAST-EST system, seems almost limitless (Laisk and Oja, 1998).

VI. Enlightening the Mechanisms of Photoinhibition in *Chlamydomonas reinhardtii*

Although thallii of the unicellular colonial alga *Ulva rotundata* were convenient vehicles for sorting the Yin-Yang of photoprotection and photoinactivation in the Duke University Marine Laboratory, cultures of *Chlamydomonas reinhardtii* presented a more robust, more acceptable model for further genetic and molecular analysis of these processes in the Boynton-Gillham laboratory on main campus. How the wildtype of this green alga, isolated from soil in the shade of an oak tree, rose to be a major player in photoinhibition research is a tale best told by Elizabeth Harris. However, for over a decade, we collected many and integrated some pieces of the puzzle of photoinhibition and photoacclimation in the course of phenotypic evaluation of *Chlamydomonas* mutants (grown photoautotrophically, it needs to be emphasized; Heifetz et al., 2000). Many physiological responses to excitation pressure, such as cell growth, O_2 evolution and fluorescence parameters, D1 synthesis and turnover, and pigment compositions, were analyzed routinely.

Still intrigued by photoinactivation, Peter Heifetz and others initially constructed "loss of function" mutants affecting different steps in the D1 repair cycle. For example, we learned from mutants inhibited in chloroplast protein synthesis that coarse regulation of D1 protein replacement was more significant for avoiding photoinhibition than fine regulation of D1 function, evident from herbicide resistant mutants with decreased rates of PS II electron transfer (Heifetz et al., 1997). Another herbicide resistant mutant ($A_{251}L$, Lardans et al., 1998) was characterized in detail, and later used for selection of unique "gain of function" mutants that are resistant to very high light (VHL). Peter discovered the first very high light resistant (VHL^R) mutants when they appeared as surviving cells among wild type cells that were dying in very high light (VHL=1,500–2,000 μmol photons m^{-2} s^{-1}). Subsequently, many VHL^R mutants were isolated from wildtype and from the $A_{251}L$ herbicide-resistant mutant with functionally impaired D1 protein, and four of these have been analysed in detail (Förster et al., 1999). The different, single-gene VHL^R mutants show sustained high growth rates at full sun light intensity that kills the parents. Many unexpected insights into photoacclimation have been gained from these mutants, which are still being explored with regard to the physiological and genetic basis of tolerance to high light stress. For example, simple relationships between xanthophyll interconversion and NPQ photoprotection are called into question because mutants achieve maximum growth in VHL irrespective of whether the high DES is or is not expressed as NPQ (Förster et al., 2001). Evidently the 5 x slower PS II electron transfer in mutants from herbicide resistant parents was sufficient to attain high DES (requiring some lumen acidification), but inadequate to sustain the ΔpH necessary for NPQ generation. This perhaps indicates that the role of zeaxanthin as a quencher of reactive oxygen may be more important than its role in excitation dissipation (Niyogi, 1999; Baroli et al., 2003). Confirming earlier interpretations, we found that the mutants survived with high growth rates at VHL in spite of low PS II efficiency (ratio of variable to maximum fluorescence in dark equilibrated samples $F_v/F_m = 0.28 - 0.36$) and slow PS II electron transfer, either because of engagement of photoprotection, and/or in spite of photoinactivation. The basis of the VHL^R phenotype is evidently a combination of traits, including increased zeaxanthin accumulation, maintenance of fast synthesis and degradation rates of the D1 protein, and sustained balanced electron flow into and out of PSI under VHL. It appears to be dominated by enhanced capacity to tolerate reactive oxygen species generated by excess light, methylviologen, rose bengal or additional hydrogen peroxide, and lower levels of ROS in some VHL^R mutants after exposure to very high light (Förster et al., 2005). The VHL^R mutations evidently arise under a selection pressure that favors changes to the regulatory system(s) that coordinate several photoprotective processes, amongst which photoprotection, repair of photoinactivated PSII and enhanced detoxification of reactive oxygen species play seminal roles.

Other surprises had simpler explanations. There was an alarming period when we could not confirm wildtype DES-NPQ observations from the Björkman laboratory. Ironically, at Olle's retirement symposium we were able to report that the much lower DES and NPQ in our cells could be ascribed to the 5% CO_2 growth conditions routinely used in the Boynton-Gillham Lab, compared to the air grown, CO_2 concentrating cells grown in the Björkman laboratory (Förster et al., 2001). Decreased amounts of chlorophyll per biomass in the mutants, without changes in the functional absorption cross section, suggested VHL^R genes may be involved in regulation of stoichiometries of antenna components and photosystems, perhaps as invoked by Allen and Pfannschmidt (2000). Accepting that oxygenic photosynthesis is the most energetic of life processes, and one in which catalytic malfunctions potentially yield a plethora of potentially toxic reactive oxygen species (Asada, 1999), and we were not suprised that the VHL^R mutants also proved more resistant to reactive oxygen species (Förster et al., 2005). The roles of zeaxanthin and ascorbate as anti-oxidants and lipid stabilizers may be as important as their roles in NPQ (Baroli et al., 2003; Müller–Moulé et al., 2004) the roles of zeaxanthin and ascorbate as lipid stabilizers and anti-oxidants (Niyogi, 1999; Müller-Moulé et al., 2003). Based on the above physiological traits, we hypothesize that "regulatory switches" for tuning photochemistry and photoprotective pathways have been altered in these mutants. Once we have identified the VHL^R genes, we will be able to re-evaluate this hypothesis.

VII. Quo Vadis?

From what has been said above, it is already clear that plants confronted with light stress have evolved a variety of photoprotective mechanisms, in the antennae and reaction centers, and although Olle Björkman speaks of the field having matured, he still warms to the revelation of the nuances, and rises enthusiastically to new insights. There is still gold to be found in pigment-associated photoprotection, and perhaps fields are nowhere more promising than when the rapidly reversible violaxanthin-cycle is augmented by the more sluggish lutein-epoxide cycle. Rediscovered in the hemi-parasite dodder (*Cuscuta*; Bungard et al., 1999), in mistletoes (*Amyema*; Matsubara et al., 2001), and in schlerophyllous *Quercus rubra* (Garcia-Plazaola et al., 2003), the significance of the accessory, lutein-epoxide cycle remains conjectural. In deeply shaded leaves of the tropical forest tree legume *Inga sapindoides*, lutein epoxide pools exceed those of violaxanthin 3–5 fold, and although de-epoxidation occurs on exposure to strong light, lutein epoxidation is remarkably slow. Indeed, in simulated sun flecks, maximum NPQ in *I. sapindoides* is initially achieved (within minutes) without pigment interconversion, a perspective (Horton et al., 1996) that tends to have been overlooked in recent treatments. In fact, it seems likely that the one-way conversion of lutein epoxide to lutein serves as a readily available (within hours) source of the latter α-xanthophyll as a first step in photoacclimation, followed by the slower accumulation (over days) of larger β-xanthophyll pools (Matsubara et al., 2005). Where these field discoveries will lead is anyone's guess, but most importantly, the lutein-epoxide cycle reopens questions as to the photoprotective roles of lutein.

Global analysis of PS II fluorescence lifetime distribution in vivo has greatly advanced the population analysis of quenching in pigment protein complexes

(Gilmore 2004), with insights emerging from the phenotypic evaluation of mutants (Li et al., 2002), and from changes in populations of quenching centers during the enhanced thermal dissipation in mistletoe leaves in winter (Matsubara et al., 2002). It has long been proposed that if photoprotection is inadequate, photoinactivated PS II reaction centers may still quench excitation in their own right, protecting their neighbours (Krause, 1988; Chow et al., 2002) in the process of synthesis and reconstruction. It should come as no surprise that the fluorescence lifetime approach has now demonstrated the emergence, in vivo, of distinct populations of short lifetime components that appear sequentially after extensive photoinactivation, at the expense of a slower component from undamaged PS II centers (Matsubara and Chow 2004). These authors believe *"these results provide direct evidence that photoinactivated PS II centers in vivo are able to dissipate excitation energy and avoid further damage to themselves and protect their undamaged neighbors by acting as strong energy sinks"*.

As we scale down to molecular size and rate constants in pursuit of ever more detailed understanding, so we can also scale up to the biosphere and ask whether our insights stand up in the real world. Thus, it has been very exciting, and satisfying, to see the extent to which "remote sensing" devices can assess changes in photosynthetic efficiency and NPQ in canopies. A Laser Induced Fluorescence Transient (LIFT) apparatus, based on the Fast Repetition Rate Fluorometer (Kolber et al., 1998) used so effectively in our *Chlamydomonas* experiments, now can resolve the different NPQ kinetics of *psbS* mutants in *Arabidopsis* to 2 s at a range of 10 m and beyond (Kolber et al., 2005). When suitably housed, it also reports the different diel patterns of photosynthetic electron transport rates in tropical forest canopy. Fluorescence imaging methods have yet to be scaled up to the canopy, but the photosynthetic reflectance index (PRI, Gamon et al., 1997), a "remotely sensible" index of DES, can be observed in canopies with an imaging spectral reflectometer that clearly identifies *psbS* mutants in a lawn of *Arabidopsis* at a range of several meters (U. Rascher, C. Small and C. J. Nichol, unpublished).

To summarize now, perhaps photoinhibition is really best comprehended in terms of the interactions of Yin and Yang. At the molecular level it depends on the ceaseless motion, the interplay of opposites in dark, "forgiving" Yin of inevitable D1-protein degradation, synthesis and replacement in PS II centers in the photosynthetic apparatus following photoinactivation, and the no less restless, "assertive" Yang of photoprotection in the antennae in the light, still largely mysterious as to mechanism, and evidently augmented by photoinactivated reaction centers. At the physiological and ecological levels the dark reactions of CO_2 metabolism, light dependent electron transfer to O_2, and a host of other environmental considerations clearly co-determine the extent of "excitation pressure" and the expression of Yin and Yang in leaves and algae. But what really causes the lights of fluorescence in the grana of chloroplasts in photoinhibited cells in shade leaves of *Cissus* to be switched out after an hour in sunlight (Fig. 2 a,b)? We have a hunch now that quenching in slowly repaired photoinactivated PS II centers might predominate over sustained photoprotection by xanthophyll antenna quenching. After all, Olle Björkman's photoinhibition image on a *Cissus* leaf persisted for 10 days or more in weak light (Osmond et al., 1999), when one might have expected xanthophyll de-epoxidation to have relaxed. But who knows?

Acknowledgments

As we hope this retrospective makes clear, most of the above perspectives derive from a long and much respected association with Olle Björkman, and this paper is dedicated to him on attainment of "three score and ten" years in 2003. Many of the remaining ideas may be attributed to peripatetic admonition from Jan Anderson, for which we remain ever so grateful. The early development of programs (now aborted) for scaling-up these insights for experimental ecosystem and global climate change research at the Biosphere 2 Laboratory was possible through the vision of Dr Michael Crow, Executive Vice Provost, Columbia University, the generosity of Mr Edward P Bass, and the determined effort of many colleagues whose research will be reported elsewhere.

References

Adams WW III and Osmond CB (1988) Internal CO_2 supply during photosynthesis of sun and shade grown CAM plants in relation to photoinhibition. Plant Physiol 86: 117–123

Adams WW III, Nishida K and Osmond CB (1986) Quantum yields of CAM plants measured by photosynthetic O_2 evolution. Plant Physiol 81: 297–300

Adams WW III, Zarter CR, Mueh KE, Amiard V and Demmig-Adams B (2005) Energy dissipation and photoinhibition: A continuum of photoprotection. In: Demmig-Adams B, Adams WW III and Mattoo AK (eds) Photoprotection, Photoinhibition, Gene Regulation, and Environment, pp 49–64. Springer, Dordrecht

Adir N, Zer H, Shochat S and Ohad I (2003) Photoinhibition - a historical perspective. Photosynth Res 76: 343–370

Allen JF and Pfannschmidt T (2000) Balancing the two photosystems: photosynthetic electron transport governs transcription of reaction center genes in chloroplasts. Phil Trans R Soc Lond 355: 1351–1360

Anderson JM and Osmond CB (1987) Sun-shade responses: compromises between acclimation and photoinhibition. In: Kyle DJ, Osmond CB and Arntzen CJ. (eds). Photoinhibition, Topics in Photosynthesis, Vol. 9, pp 1–38. Elsevier, Amsterdam

Anderson JM, Park Y-I and Chow WS (1997) Photoinactivation and photoprotection of photosystem II in nature. Physiol Plant 100: 214–223

Aro E-M, Virgin I and Andersson B (1993) Photoinhibition of photosystem II. Inactivation, protein damage and turnover. Biochim Biophys Acta 1143: 113–134

Asada K (1999) The water-water cycle in chloroplasts: scavenging of active oxygens and dissipation of excess photons. Annu Rev Plant Physiol Plant Mol Biol 50: 601–639

Badger MR, von Caemmerer S, Ruuska S and Nakano H (2000) Electron flow to oxygen in higher plants and algae: rates and control of direct photoreduction (Mehler reaction) and rubisco oxygenase. Phil Trans R Soc Lond B 355: 1433–1446

Balachandran S and Osmond CB (1994) Susceptibility of tobacco leaves to photoinhibition following infection with two strains of tobacco mosaic virus under different light and nitrogen nutrition regimes. Plant Physiol 104: 1051–1057

Baroli I, Do AD, Yamane Y, and Niyogi KK (2003) Zeaxanthin accumulation in the absence of a functional xanthophyll cycle protects *Chlamydomonas reinhardtii* from photooxidative stress. Plant Cell 15, 992–1008.

Björkman O and Demmig B (1987) Photon yield of O_2 evolution and chlorophyll fluorescence characteristics at 77K among vascular plants of diverse origins. Planta 170: 489–504

Björkman O and Holmgren P (1963) Adaptability of the photosynthetic apparatus to light intensity in ecotypes from exposed and shaded habitats. Physiol Plant 16: 889–914

Bungard RA, Ruban AV, Hibberd JM, Press MC, Horton P and Scholes JC (1999) Unusual carotenoid composition and a new type of xanthophyll cycle in plants. Proc Natl Acad Sci USA 96: 1135–1139

Chow WS, Hope AB and Anderson JM (1991) Further studies on quantifying photosystem II in vivo by flash-induced oxygen yield in leaf discs. Aust J Plant Physiol 18: 397–410

Chow WS, Lee H-Y, Park Y-I, Park Y-M, Hong Y-N and Anderson JM (2002) The role of inactive photosystem II- mediated quenching in a last-ditch community defense against high light stress *in vivo*. Phil Trans R Soc Lond B 357: 1441–1450

Cornic G (1976) Effet exercé sur l'activité photosynthetique du *Sinapis alba* L par une inhibition temporaire de la photorespiration se déroulant dans un air sans CO_2. CR Acad Sci D 282: 1955–1958

Cornic G and Fresneau C (2002) Photosynthetic carbon reduction and oxidation cycles are the main electron sinks for photosystem II activity during mild drought. Ann Bot 89: 887–894

Delieu T and Walker DA (1983) Simultaneous measurement of oxygen evolution and chlorophyll fluorescence from leaf pieces. Plant Physiol 73: 534–541

Demmig B and Björkman O (1987) Comparison of the effects of excessive light on chlorophyll fluorescence (77K) and photon yield of O_2 evolution in leaves of higher plants. Planta 171: 171–184

Demmig B, Winter K, Krüger A and Czygan F-C (1987) Photoinhibition and zeaxanthin formation in intact leaves. A possible role of the xanthophyll cycle in the dissipation of excess light energy. Plant Physiol 84: 218–224

Demmig-Adams B (2003) Linking the xanthophyll cycle with thermal energy dissipation. Photosynth Res 76: 73–80

Demmig-Adams B, Ebbert V, Zarter CR and Adams WW III (2005) Characteristics and species-dependent employment of flexible versus sustained thermal dissipation and photoinhibition. In: Demmig-Adams B, Adams WW III and Mattoo AK (eds) Photoprotection, Photoinhibition, Gene Regulation, and Environment, pp 39–48. Springer, Dordrecht

Edelman M and Mattoo AK (2005) The D1 protein: past and future perspectives. In: Demmig-Adams B, Adams WW III and Mattoo AK (eds) Photoprotection, Photoinhibition, Gene Regulation, and Environment, pp 23–38. Springer, Dordrecht

Ewart AJ (1896) On assimilatory inhibition in plants. J Linn Soc 31: 364–461

Ferrar PJ and Osmond CB (1986) Nitrogen supply as a factor influencing photoinhibition and photosynthetic acclimation after transfer of shade grown *Solanum dulcamara* to bright light. Planta 168: 563–570

Förster B, Osmond CB, Boynton JE and Gillham NW (1999) Mutants of *Chlamydomonas reinhardtii* resistant to very high light. J Photochem Photobiol B 48: 127–135

Förster B, Osmond CB and Boynton JE (2001) Very high light resistant mutants of *Chlamydomonas reinhardtii*: responses of photosystem II, nonphotochemical quenching and xanthophyll pigments to light and CO_2. Photosynth Res 67:5–15

Förster B, Osmond CB and Pogson BJ (2005) Improved survival of very high light and oxidative stress is conferred by spontaneous gain-of-function mutations in *Chlamydomonas*. Biochim Biophys Acta (in press)

Franklin LA, Levavasseur G, Osmond CB, Henley WJ and Ramus J (1992) Two components of onset and recovery during photoinhibition of *Ulva rotundata*. Planta 186: 399–408

Funayama S, Terashima I and Yahara T (2001) Effects of virus infection and light environment on the population dynamics of *Eupatorium makinoi* (Asteraceae). Am J Bot 88: 612–622

Gamon JA, Serrano L and Surfas JS (1997) The photochemical reflectance index: an optical indicator of photosynthetic radiation use efficiency across species, functional types and nutrient levels. Oecologia 112: 492–501

Garcia-Plazaola JI, Hernández A, Olano JM and Becerril JM. (2003) The operation of the lutein epoxide cycle correlates with energy dissipation. Functional Plant Biol 30: 319–324

Gauhl E (1976) Photosynthetic response to varying light intensity in ecotypes of *Solanum dulcamara* L. from shaded and exposed habitats. Oecologia 27: 278–286

Gilmore A (2004) Excess light stress: probing excitation dissipation mechanisms through global analysis of time-and wavelength-resolved chlorophyll a fluorescence. In: Chlorophyll a Fluorescence: A Signature of Photosynthesis (Papageorgiou G C and Govindjee, eds), pp. 555–581, Springer, Dordrecht

Heifetz PB, Lers A, Turpin DH, Gillham NW, Boynton JE and Osmond CB (1997) dr and *spr*/sr mutations of *Chlamydomonas reinhardtii* affecting D1 protein function and synthesis define two independent steps leading to chronic photoinhibition and confer differential fitness. Plant Cell Environ 20: 1145–1157

Heifetz PB, Förster B, Osmond CB, Giles LJ and Boynton JE (2000) Effects of acetate on facultative autotrophy in *Chlamydomonas reinhardtii* assessed by photosynthetic measurements and stable isotope analyses. Plant Physiol 122: 1439–1445

Henley WJ, Levavasseur G, Franklin LA, Osmond CB and Ramus J (1991) Photoacclimation and photoinhibition in *Ulva rotundata* as influenced by nitrogen availability. Planta 184: 235–243

Horton P, Ruban AV and Walters RG. (1996) Regulation of light harvesting in green plants. Annu Rev Plant Physiol Plant Mol Biol 47: 655–684

Jones LW and Kok B (1966) Photoinhibition of chloroplast reactions. I. Kinetics and action spectra. Plant Physiol 41: 1037–1043

Kasahara M, kagawa T, Oikawa K, Suetsugu N, Miyao M and Wada M (2002) Chloroplast avoidance movement reduces photodamage in plants. Nature 420: 829–832

Kato MC, Hikosaka K, Hirotsu N, Makino A and Hirose T (2003) The excess light energy that is neither utilized in photosynthesis nor dissipated by photoprotective mechanisms determines the rate of photoinactivation in photosystem II. Plant Cell Physiol 44: 318–325

Kok B (1956) On the inhibition of photosynthesis by intense light. Biochim Biophys Acta 21: 234–244

Kolber ZS, Prasil O and Falkowski PG (1998) Measurements of variable chlorophyll fluorescence using fast repetition rate techniques. I. Defining methodology and experimental protocols. Biochim Biophys Acta 1367: 88–106

Kolber Z, Klimov D, Ananyev G, Rascher U, Berry J and Osmond B. (2005) Measuring photosynthetic parameters at a distance: Laser Induced Fluorescence Transient (LIFT) method for remote measurements of PSII in terrestrial vegetation. Photosynth Res (in press)

Kozaki H and Takeba G (1996) Photorespiration protects plants from photooxidation. Nature 384: 557–560

Krause GH (1988) Photoinhibition of photosynthesis: an evaluation of damaging and protective mechanisms. Physiol Plant 74: 566–574

Krause GH, Kirk M, Heber U and Osmond CB (1978) O_2-dependent inhibition of photosynthetic capacity in intact isolated chloroplasts and isolated cells from spinach leaves illuminated in the absence of CO_2. Planta 142: 229–233

Kyle DJ, Ohad I and Arntzen CJ (1984) Membrane protein damage and repair: selective loss of a quinone-protein function in chloroplast membranes. Proc Natl Acad Sci USA 81: 4070–4074

Laisk A and Oja V (1999) Dynamics of leaf photosynthesis: Rapid response measurements and their interpretation. CSIRO Publishing, Collingwood, Australia

Lardans A, Förster B, Prásil O, Falkowski PG, Sobolev V, Edelman M, Osmond CB, Gillham NW and Boynton JE (1998) Biophysical, biochemical and physiological characterization of *Chlamydomonas reinhardtii* mutants with amino acid substitutions at the Ala$_{251}$ residue in the D1 protein having varying levels of photosynthetic competence. J Biol Chem 272: 11082–11091

Lee H-Y, Hong Y-N and Chow WS (2001) Photoinactivation of photosystem II complexes and photoprotection by non-functional neighbours in *CaPSIcum annuum* L. leaves. Planta 212: 332–342

Li X-P, Gilmore AM and Niyogi K (2002) Molecular and global time-resolved analysis of a *psbS* gene dosage effect on pH- and xanthophyll cycle-dependent nonphotochemical quenching in photosystem II. J Biol Chem 277: 33590–33597

Lorimer GH and Andrews TJ (1973) Plant photorespiration – an inevitable consequence of the existence of atmospheric oxygen. Nature 243: 359

Lovelock J (1990) The Ages of Gaia. Bantam, New York, pp. xiii

Ludlow MM and Björkman O (1984) Paraheliotropic leaf movement in Siratro as a protectective mechanism against drought-induced damage to primary photosynthetic reactions by excessive light and heat. Planta 61: 505–518

Matsubara S and Chow WS (2004) Populations of photoinactivated photosystem II reaction centers characterized by chlorophyll a fluorescence lifetime in vivo. Proc Natl Acad Sci USA 101, 18234–18239

Matsubara S, Gilmore AM and Osmond CB (2001) Diurnal and acclimatory responses of violaxanthin and lutein epoxide in the Australian mistltoe *Amyema miquelii*. Aust J Plant Physiol 28: 793–800

Matsubara S, Gilmore AM, Ball MC, Anderson JM and Osmond CB (2002) Sustained down regulation of photosystem II in mistletoes during winter depression of photosynthesis. Functional Plant Biol 29: 1157–1169

Matsubara S, Naumann M, Martin R, Rascher U, Nichol C, Morosinotto T, Bassi R and Osmond B. (2005) Slowly reversible de-epoxidation of lutein-epoxide in deep shade leaves of a tropical tree legume may "lock-in" lutein-based photoprotection during acclimation to strong light. J Exp Bot 56, 461–468

Müller-Moulé P, Havaux M and Niyogi KK (2003) Zeaxanthin deficiency enhances the high light sensitivity of an ascorbate-deficient mutant of *Arabidopsis*. Plant Physiol 133: 748–760

Nichol CJ, Huemmrich KF, Black TA, Jarvis PJ, Walthall CL, Grace J and Hall FG (2000) Remote sensing of photosynthesis-light-use efficiency of Boreal Forest. Agric For Meterol 101: 131–141

Niyogi KK (1999) Photoprotection revisited: genetic and molecular approaches. Annu Rev Plant Physiol Plant Mol Biol, 50: 333–359

Osmond CB (1981) Photorespiration and photoinhibition, some implications for the energetics of photosynthesis. Biochim Biophys Acta 639: 77–98

Osmond CB (1994) What is photoinhibition? Some insights from comparisons of shade and sun plants. In: Baker NR and Bowyer JR (eds) Photoinhibition: Molecular Mechanisms to the Field pp 1–24. Bios Scientific Publications, Oxford

Osmond CB and Björkman O (1972) Simultaneous measurements of O_2 effects on net photosynthesis and glycolate metabolism in C_3 and C_4 species of *Atriplex*. Carnegie Inst Wash Yearbook 71: 141–148

Osmond CB and Chow WS (1988) Ecology of photosynthesis in the sun and shade: summary and prognostications. Aust J Plant Physiol 15: 1–9

Osmond CB and Grace SC (1995) Photoinhibition and photorespiration: the quintessential inefficiencies of the light and dark reactions of terrestrial oxygenic photosynthesis. J Expt Bot 46: 1351–1362

Osmond CB, Oja V and Laisk A (1988) Regulation of carboxylation and photosynthetic oscillations during sun-shade acclimation in *Helianthus annuus* measured with a rapid-response gas exchange system. Aust J Plant Physiol 15: 239–251

Osmond CB, Ramus J, Levavasseur G, Franklin LA and Henley WJ (1993) Fluorescence quenching during photosynthesis and photoinhibition of *Ulva rotundata*. Planta 190: 97–106

Osmond B, Schwartz O and Gunning B (1999) Photoinhibitory printing on leaves, visualised by chlorophyll fluorescence imaging and confocal microscopy, is due to diminished fluorescence from grana. Aust J Plant Physiol 26: 717–724

Öquist G, Huner NPA (2003) Photosynthesis of overwintering evergreen plants. Annu Rev Plant Biol 54: 329–355

Park Y-I, Chow WS. Osmond CB and Anderson JM (1996) Electron transport to oxygen mitigates against the photoinactivation of photosystem II in vivo by both enhanced utilization, and increased non-radiative dissipation, of excess photons. Photosynth Res 50: 23–32

Powles SB (1984) Photoinhibition of photosynthesis. Annu Rev Plant Physiol 35: 15–44

Powles SB and Björkman O (1982) Leaf movement in the shade species *Oxalis oregana* II. Role in the protection against injury by intense light. Carnegie Inst Wash Yearbook 81: 63–66

Powles SB and Osmond CB (1978) Inhibition of the capacity and efficiency of photosynthesis in bean leaflets illuminated in absence of CO_2 at low O_2 concentrations – a protective role for photorespiration. Aust J Plant Physiol 5: 619–629

Robinson SA and Osmond CB (1994) Internal gradients of chlorophyll and carotenoid pigments in relation to photoprotection in thick leaves of plants with Crassulacean acid metabolism. Aust J Plant Physiol 21: 497–506

Russell AW, Critchley C, Robinson SA, Franklin LA, Seaton GGR, Chow WS, Anderson JM and Osmond CB (1996) Photosystem II regulation and dynamics of the chloroplast D1 protein in Arabidopsis leaves during photosynthesis and photoinhibition. Plant Physiol 107: 943–952

Schreiber U, Schliwa U and Bilger W (1986) Continuous recording of photochemical and nonphotochemical chlorophyll fluorescence quenching with a new type of modulation fluorometer. Photosynth Res 10: 51–62

Shapira M, Lers A, Heifetz PB, Osmond CB, Gillham NW and Boynton JE (1997) Differential regulation of chloroplast gene expression in *Chlamydomonas reinhardtii* during photoacclimation. Light stress transiently suppresses synthesis of Rubisco LSU protein while enhancing synthesis of the PS II D1 protein. Plant Mol Biol 33: 1001–1011

Terashima I and Inoue Y (1985) Vertical gradient in photosynthetic properties of spinach chloroplasts dependent on intraleaf light environment. Plant Cell Physiol 26: 781–785

Walker DA and Osmond CB (1986) Measurement of photosynthesis i *nvivo* with a leaf disc electrode: correlations between light dependence of steady state photosynthetic O_2 evolution and chlorophyll a fluorescence transients. Proc R Soc Lond B 227: 267–280

Woodrow IE and Mott KA (1998) Quantitative assessment of the degree to which ribulosebisphosphate carboxylase/oxygenase determines the steady–state rate of photosynthesis during sun-shade acclimation in *Helianthus annuus* L. Aust J Plant Physiol 15: 253—262

Yamamoto HY (2005) A random walk to and through the xanthophyll cycle. In: Demmig-Adams B, Adams WW III and Mattoo AK (eds) Photoprotection, Photoinhibition, Gene Regulation, and Environment, pp 1–10. Springer, Dordrecht

Chapter 3

The D1 Protein: Past and Future Perspectives[†]

Marvin Edelman*[1] and Autar K. Mattoo[2]
[1]*Department of Plant Sciences, The Weizmann Institute of Science, Rehovot 76100, Israel;* [2]*Henry A. Wallace Beltsville Agricultural Research Center, USDA/ARS, Beltsville, MD 20705–2350, USA*

Summary		23
I.	The Really Early Days	24
II.	Gernot Renger's Shield	24
III.	D1 Metabolism is Photoregulated	26
IV.	The PEST Sequence	28
V.	The Life History of D1	29
VI.	The UV-B Story	30
VII.	The D1/D2 Heterodimer Takes Center Stage	31
VIII.	Phosphorylation–Dephosphorylation	32
IX.	Circadian Control	34
X.	The Past and Future	35
Acknowledgments		36
References		36

Summary

The chloroplast-coded D1 protein of Photosystem II (PS II) is the major membrane protein synthesized within the plastid. It is involved in light-dependent electron transport, is a major target for photosynthesis herbicides and is universal to oxygenic phototrophs. The defining feature of D1 is its rapid turnover in spite of its being a structural component of the PS II reaction center core. Processing of nascent D1 precursor (33.5–34 kDa) occurs on unstacked stromal lamellae. The mature protein (32 kDa) then migrates to the grana where an initial scission occurs producing a 23 kDa N-terminal degradation fragment. Post-translational and reversible palmitoylation and phosphorylation accompany the protein along its life cycle. Both anabolism and catabolism of D1 are photoregulated, with synthesis coupled to phosphorylation but degradation coupled to PS II electron transport. Dephosphorylation of D1, in turn, is regulated by PS I excitation. Thus, the phosphorylation state of the protein is sensitive to the relative energy distribution between the two photosystems. Beyond redox regulation of D1 phosphorylation, an internal, circadian clock exerts overriding control. Two photosensitizers are involved in D1 degradation: chlorophyll pigments in the visible and far-red regions of the spectrum, and plastosemiquinone in the UV-B region. D1 degradation in visible light is a process only marginally overlapping with photoinhibition and overwhelmingly associated with fluences limiting for photosynthesis. Mixing physiological levels of visible and UV-B radiances leads to synergistic effects such that above a critical threshold of UV-B, the D1 as well as its sister protein, D2, both are targeted for accelerated degradation. These and other D1 protein studies, mainly carried out with intact *Spirodela* plants during the past 25 years in the authors' laboratories, are presented in a historical perspective.

---

*Author for correspondence, email: marvin.edelman@weizmann.ac.il
[†]This account does not intend to survey the field but rather to present a retrospective view of our D1 protein studies over the years. We gratefully acknowledge all of our collaborators whose names appear in the Chapter. In addition, we wish to acknowledge those who contributed to other aspects of D1 research in our laboratories: Adi Avni, Alessandra Cona, Bharat Chattoo, Martine Devic, Yoram Eyal, Robert Fluhr, Hillel Fromm, Maria Teresa Giardi, Richard B. Hallick, Dina Heller, Karl Jakob, Michael Koblizek, Vinod Kumar, Chiara Leonardi, Alexander Raskind, Judy St. John, and William Wergin.

B. Demmig-Adams, William W. Adams III and A.K. Mattoo (eds), Photoprotection, Photoinhibition, Gene Regulation, and Environment, 23–38.
© Springer Science+Business Media B.V. 2008

I. The Really Early Days

The D1 protein had its research debut in the mid 1970s. Several groups studying protein synthesis in isolated chloroplasts reported a membrane-associated polypeptide with an apparent molecular mass of 32 kDa on polyacrylamide gels. Eagelsham & Ellis (1974), working with pea chloroplasts at the University of Warwick, termed this protein "Peak D". Soon afterwards, it was found that chloroplast RNA could stimulate translation of an approximately 32 kDa protein in a heterologous in vitro system (Wheeler and Hartley, 1975). While this was unfolding in England, Arie Rosner, working in Jonathan Gressel's laboratory at the Weizmann Institute of Science in Israel, together with Daphna Sagher working in ours, reported a 0.5×10^6 Da chloroplast RNA fraction forming the bulk of the discrete pulse-labeled RNA molecules produced in vivo after transfer of steady-state dark grown *Spirodela* plants to light (Rosner et al., 1975). Actually, the biological system worked on in the Edelman laboratory at that time was *Euglena*, brought over from Jerome Schiff's Laboratory at Brandeis University. We mention this because shortly thereafter, Edelman left for a Sabbatical year in the United States. No sooner did he depart, than Avi Reisfeld, the other student in his group, impressed by the results obtained with *Spirodela*, quickly switched to the latter system and tied the 0.5×10^6 Da mRNA to the precursor of the 32 kDa protein in time to submit a paper to the first international plant molecular biology meeting, organized by Laurence Bogorad and Jacques Weil, in Strasburg, France in the Summer of 1976. To this day we are grateful to Reisfeld for his Israeli 'hutzpa' of switching systems when the boss was away. *Euglena* and *Spirodela* share a rare trait for photosynthetic organisms; both can grow indefinitely in the dark when supplied with an organic carbon source, thus enabling the study of chloroplast development from a state of true heterotrophy. Although *Euglena* is a zooflagellate of uncertain pedigree, *Spirodela* is a true blooded, if eccentric, monocot. In a series of articles, Reisfeld and others in the Edelman laboratory showed that the 32k Da polypeptide in *Spirodela* is the major membrane protein synthesized within the chloroplast (Edelman and Reisfeld, 1978), is derived from a rapidly synthesized 33.5 kDa precursor polypeptide (co-discovered in maize [Grebanier et al., 1978)]), and lacks lysine residues (Edelman and Reisfeld, 1980; Reisfeld et al., 1982). Soon afterwards, it was shown to occur universally and to be structurally similar in various photosynthetic organisms (Hoffman-Falk et al., 1982).

In one of the quirks of the field, the absence of lysine from the 32 kDa protein in *Spirodela* played an important role in the early days. In the summer meetings of 1982, Paul Whitfeld's group reported the very first sequence of the *psbA* gene (coding for the 32 kDa protein) in spinach and in tobacco, the model system of the time (Zurawski et al., 1982). When, at the end of their talk, they stated that the gene lacked codons for lysine, Charles Arntzen, who was chairing the session, jumped up and said "Edelman, you were right, no lysine." For several years following, the lack of lysine remained the criterion by which a chloroplast membrane protein was, or was not, deemed to be the 32 kDa protein. Indeed, the 32 kDa protein in most species investigated lacks lysine. However, in maize and some of the grains, lysine replaces arginine at codon 238.

II. Gernot Renger's Shield

Autar Mattoo arrived at the Weizmann Institute as a DAAD Scholar in 1979 to study chloroplast molecular biology, then an emerging discipline within plant molecular biology (Edelman et al., 1982 Preface). In the summer of 1980, Mattoo and a M.Sc. student, Hedda Hoffman-Falk, had succeeded in obtaining SDS-PAGE and electron transport results tying the 32 kDa protein to photosystem II (PS II) electron transport and herbicide sensitivity (Mattoo et al., 1981). Edelman was on annual army reserve duty when he received a field call from Mattoo. The message was terse: "The 32 kDa protein is the 'proteinaceous shield' of Renger!" The use of mild trypsin digestion to probe the structure-function relationship of surface exposed thylakoid membrane proteins (Regitz and Ohad, 1975) led Gernot Renger in 1976 to infer the existence of a "proteinaceous shield" covering the primary electron acceptor of PS II and acting as a regulator of electron flow between PS II and PS I (Renger, 1976). Mattoo found that mild trypsin treatment of *Spirodela* thylakoid membranes led to partial digestion of the 32 kDa protein. Under these conditions, photoreduction of ferricyanide becomes insensitive to diuron (DCMU, 3-(3,4-dichlorophenyl)-1,1-dimethylurea), the well-known inhibitor of PS II electron transport. When the thylakoids were preincubated with diuron, however, expression of insensitivity was prevented. The clincher came with the SDS-PAGE results, which brought on the phone call. Preincubation with diuron caused some conformational change in the 32 kDa protein that modified its trypsin digestibility and produced a different banding pattern

Chapter 3 The D1 Protein: Past and Future Perspectives

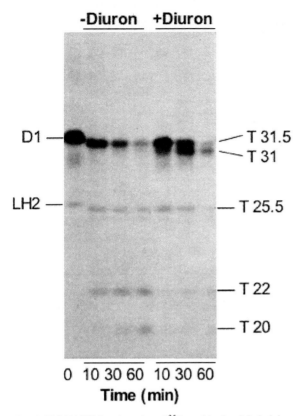

Fig. 1. SDS-PAGE fractionation of ^{35}S methionine-labeled thylakoid proteins after trypsinization in the presence of diuron. The major labeled proteins in the non-digested control lane (0 time) are the D1 (32 kDa) and LH2 (26 kDa) proteins. In the presence of trypsin, for the times indicated, both proteins are attacked and various trypsinized (T) fragments are produced. Note the different digestion pattern for D1 when incubation is in the presence of diuron, an inhibitor of PS II electron transport. Diuron induces a conformational change in D1 retarding its trypsin digestibility, fulfilling a requirement of Rengers "proteinaceous shield". (Adapted from Mattoo et al., 1981; with designations updated as in Marder et al., 1984)

upon gel electrophoresis (Fig. 1). Significantly, light affected the susceptibility of the 32 kDa protein to digestion by trypsin. Moreover, together with Uri Pick, we showed that in thylakoids selectively depleted of the 32 kDa protein [a system developed in *Spirodela* by Steve Weinbaum (Weinbaum et al., 1979)] electron transport was deficient on the reducing side of PS II, but not the oxidizing side or in PS I activities (Mattoo et al., 1981). Taken together, these were the very characteristics postulated by Renger for his "proteinaceous shield"!

We wrote up the results and ran a draft through Mordhay Avron, whose laboratory was two floors above ours in the Life Sciences Building at the Weizmann Institute. The discussions with Avron, whose knowledge of photosynthetic electron transport was encyclopedic, and his critical reading of the manuscript, were immensely helpful. Edelman knew Martin Gibbs, a member of the National Academy of Sciences, USA, from his graduate studies at Brandeis University. We sent the manuscript to Gibbs and asked him to communicate it for us to PNAS. Over the years, Martin Gibbs communicated four of our manuscripts to PNAS on 32 kDa protein turnover. We were both at the Beltsville Agricultural Research Center when the third manuscript was ready for submission. Edelman stopped in at Brandeis University to deliver a seminar on the 32 kDa work and to ask Gibbs if he would mind communicating a third manuscript for us! In his wry sense of humor, Gibbs' comment was something like "It's your name that goes on the article not mine. So if you write something foolish it's you who will suffer not me. All I'm doing is acting as a conduit to pass your page charges on to PNAS so us National Academy members can continue getting the journal for free."

The 1981 PNAS article made quite a splash. Klaus Pfister, Katherine Steinback, Gary Gardner, and Charles Arntzen had just published a paper in the same journal describing the photoaffinity labeling of an herbicide receptor protein in chloroplast membranes (Pfister et al., 1981). Atrazine (2-chloro-4-ethylamino-6-isopropylamino-s-triazine) inhibits photosynthetic electron transport virtually at the same site affected by diuron (Sobolev and Edelman, 1995). A radiolabeled atrazine analog was used to identify the herbicide receptor protein from thylakoid membranes isolated from a triazine-susceptible and triazine-resistant biotype of the weed *Amaranthus*. Analysis of the thylakoid polypeptides demonstrated specific association of the herbicide with the 32 kDa protein in the sensitive but not resistant biotype. The studies from *Amaranthus* and *Spirodela* were a one-two punch that put the 32 kDa protein at center stage. A rapidly metabolized chloroplast membrane protein involved in light-dependent electron transport and acting as a target for photosynthesis herbicides was bound to be a spotlight grabber.

When Autar Mattoo left the laboratory in November 1980 to take a position at the USDA's Beltsville Agricultural Research Center (BARC) in Maryland, it was clear to us that we would continue collaborating closely. Indeed, since then, our respective laboratories have acted as two parts of a single unit to pursue PS II protein turnover studies. Easy communication was not a trivial matter in the '80's. Faxes, "Bitnet", Sabbaticals, summer visits, and the long-term support of BARD (US-Israel Binational Agricultural Research and

Development fund) kept the collaboration viable and robust.

III. D1 Metabolism is Photoregulated

The 32 kDa protein was featured in News & Views in Nature in 1984 (Bennett, 1984). Again, it was our group and Arntzen's that caused a stir with another set of papers in PNAS. From measurements of electron transport capacity in isolated chloroplast membranes of *Chlamydomonas*, David Kyle, Itzhak Ohad, and Charles Arntzen suggested that quinone anions, which may interact with molecular oxygen to produce an oxygen radical, selectively damage the apoprotein of the secondary acceptor of PS II (i.e., the 32 kDa protein), thus rendering it inactive and thereby blocking photosynthetic electron flow under conditions of high (photoinhibitory) photon flux densities (Kyle et al., 1984). Working with *Spirodela*, we showed that the rates of both synthesis and degradation of the 32 kDa protein are controlled by light intensity in vivo. Light-driven synthesis, but not degradation, was dependent on ATP. Furthermore, degradation was blocked by herbicides inhibiting PS II electron transport, such as diuron and atrazine (Fig. 2). Thus, both anabolism and catabolism of the 32 kDa protein were shown to be photoregulated, with degradation coupled to electron transport rather than phosphorylation (Fig. 3).

In Mattoo et al. (1984), the state of the art was summarized as follows: "Results of fluorescence experiments, carried out with intact *Spirodela* plants can be interpreted as showing that partial depletion of the 32 kDa protein from the photosynthetic membranes

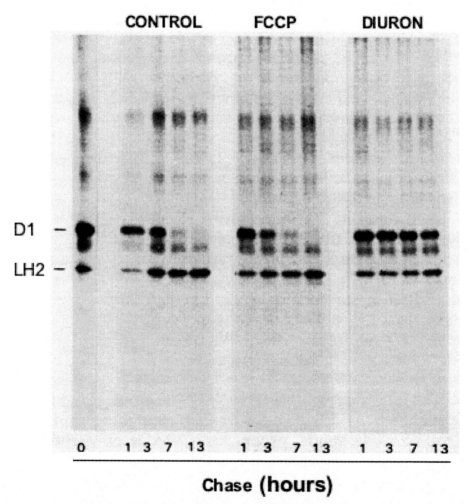

Fig. 2. Degradation of the D1 (32 kDa) protein in the presence of an inhibitor of ATP formation (FCCP) or PS II electron transport (Diuron). *Spirodela* plants were pulse labeled with ^{35}S methionine for 4 h and chased as indicated. Uncoupling of phosphorylation from electron transport does not effect D1 degradation while inhibition of oxygen evolution does. (From Mattoo et al., 1984a)

Chapter 3 The D1 Protein: Past and Future Perspectives

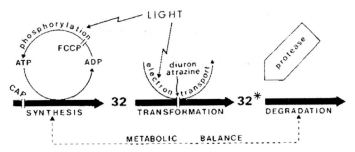

Fig. 3. Regulation of D1 (32 kDa) protein metabolism by light. All events occur on the thylakoid membranes. Synthesis is driven by ATP, which is derived from photophosphorylation. Inhibition occurs either upon uncoupling of phosphorylation (with FCCP) or interference with chloroplast ribosomal function (CAP [D-*threo*-chloramphenicol], Weinbaum et al., 1979). A conformational change (Transformation) occurs during light driven electron transport through the photosystems (Mattoo et al., 1981). The conformational rearrangements may involve redox events or substrate binding (Satoh et al., 1983). As a consequence of these molecular rearrangements, the transformed protein (32*) is prone to degradation. Herbicides such as diuron and atrazine inhibit both electron flow through PSII (Trebst, 1979; Pfister et al., 1979; Mattoo et al., 1981) and the susceptibility of D1 to degradation. Degradation of D1 is catalyzed by a membrane-bound protease. Metabolic balance for D1 is normally maintained over a wide range of light intensities. In the dark, both synthesis and degradation are minimal. (From Mattoo et al., 1984a)

directly affects the photochemistry at the reducing side of PS II. This would support theories regarding this protein as a functional component of the secondary acceptor of PS II. On the other hand, maintenance of the low, steady-state level of the 32 kDa protein in the light, in spite of its massive synthesis, represents a programming of gene expression designed for regulation. Expression of such regulatory genes is characterized by high energy costs, which purchase an ability to respond quickly, dramatically, and temporarily to perturbations in the steady state. Such features are characteristic of the 32 kDa protein. Thus, further experimentation will be required to decipher the exact role of this protein in photosynthetic electron transport."

The role was deciphered soon enough. In pioneering work, the molecular structure of the photosynthetic reaction center from *Rhodopseudomonas viridis*, a purple bacterium, was resolved to a resolution of 3 Å, earning Hartmut Michel, Johann Deisenhofer, and Robert Huber a Nobel Prize in Chemistry. The resolved structure showed a central part consisting of two subunits, L and M, each spanning the membrane five times and sharing a special pair of chlorophyll molecules (Deisenhofer et al., 1985). At the end of their paper, the authors noted that sequence homologies exist between L and M and the D1 and D2 proteins of PS II. Particularly, the histidines that ligand the special pair chlorophylls and the non-heme iron are conserved. Deisenhofer et al. (1985) went on to propose that the D1 and D2 proteins form the core of the PS II reaction center. And this generated a name change! Back in 1977, Nam-Hai Chua and Nicholas Gillham had published a numbering scheme for the functionally unidentified chloroplast membrane proteins of *Chlamydomonas rheinhardtii* (Chua and Gillham, 1977). When they came to two radiolabeled proteins producing fuzzy bands on SDS-PAGE, they named them D1 and D2 (D for "diffuse"). With the realization that the 32 kDa protein and the D2 protein shared significant sequence identity (Rochaix et al., 1984), and that they were, in fact, sister components of the PS II reaction center core [the proposal of Deisenhofer et al. (1985) was proved correct by Nanba & Satoh (1987)], "D1" slowly replaced the more cumbersome "32 kDa" designation.

Resolution of the photosynthetic reaction center also advanced the understanding of D1 turnover in an indirect way. Reports of rapid turnover of the L subunit were not forthcoming. At the same time, it was noticed that the decoded sequence of the *psb*A gene (i.e., the D1 protein) putatively possessed two significantly enlarged loop regions *vis a vis* the L subunit of the photosynthetic reaction center (Mattoo et al., 1989). Could these two bits of information be related? The structure of the *psb*A gene product was by this point under intensive investigation. Hirschberg and McIntosh (1983) at the DOE Laboratory at Michigan State University, East Lansing had shown that atrazine resistance in a field biotype of *Amaranthus* was astonishingly the result of a single nucleotide change in the *psb*A gene, resulting in a serine to glycine change in codon 264 of the D1 sequence! In Rehovot, Jonathan Marder, in a tour de force, radiolabeled *Spirodela* separately with each of 14 high specific activity ^{3}H-amino acids and followed partial enzymatic degradation of the D1 protein. Together with Pierre Goloubinoff, he established a cleavage map for trypsin and additional proteolytic enzymes, pinpointing the locations of the cuts via the decoded sequence of tobacco D1, previously established by Zurawski et al. (1982). This unequivocally showed that the 33.5 kDa precursor of the D1 protein underwent processing at its carboxy terminus and established the amino–carboxy polarity of the molecule (Marder et al., 1984).

IV. The PEST Sequence

In 1985, Bruce Greenberg joined Edelman's laboratory as an NIH postdoctoral fellow. He tackled the issue of the initial scission of the D1 protein in the degradation pathway using the *Spirodela* system to search for the breakdown products in vivo. Building on the proteolytic techniques developed by Marder, Greenberg demonstrated that an in vivo precursor–product relationship existed between the D1 protein and a 23.5 kDa degradation product through kinetic pulse-chase experiments. This primary cleavage site was mapped to the large hydrophilic loop between helices *D* and *E* of the D1 protein (Fig 4). We had a clean result, but the significance of the scission region was not apparent to us, so we held back. Everything fell in place a few months later with the publication by Martin Rechsteiner's laboratory that alpha-helix destabilizing oligopeptide regions abundant in proline, glutamate, serine, and threonine, and bordered by positively charged residues (i.e., "PEST" regions), were primary determinants for rapid degradation of proteins (Rogers et al., 1986). Greenberg moved quickly, phoning Rechsteiner for the computer program to locate PEST sequences. When he analyzed all the known proteins of the fully sequenced tobacco chloroplast genome (Shinozaki et al., 1986) and found a positive score only for a single region, which further turned out to be in the large loop between helices *D* and *E* of the D1 protein, we knew we were ready to publish.

Greenberg's paper (Greenberg et al., 1987) provoked experiments by several groups who deleted the PEST region from one of the *psb*A copies in the cyanobacterium *Synechocystis* 6803 with little effect on the degradation rate of the D1 protein under the conditions used (Nixon et al., 1995; Mulo et al., 1997). However, it is worth emphasizing that PEST-like regions are facilitators in initiating protein breakdown (Rogers et al., 1986), not pathway switches. From their deletion experiments, Mulo et al. (1997) narrowed attention to the sequence NIV$_{247-249}$ as a possible determinant for rapid proteolysis. This sequence is at the carboxy end of the region proposed by Greenberg et al. (1987) as producing the 23.5 kDa fragment, and down stream of the PEST-like region. In addition, N$_{247}$ was already

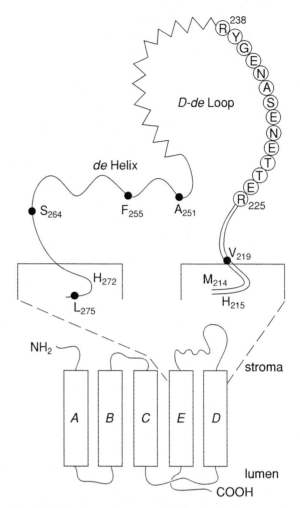

Fig. 4. Molecular environment of the primary cleavage site in D1 yielding the 23.5 kDa polypeptide. The D1 protein is shown spanning the membrane 5 times in analogy to the data of Deisenhofer et al. (1985), and as modeled by Trebst (1986) and Barber and Marder (1986). The *D-de* loop and *de* helix between transmembrane helices *D* and *E* is presented in the enlargement. The proposed degradation site is shown as a saw-toothed region. The site is adjacent to the carboxy end of the PEST-like region (shown as circled amino acid residues). The photoaffinity binding domain of azidoatrazine (residue 214 to 225; Wolber et al., 1986) borders the amino acid end of the PEST-like region. The known sites of genetic mutation linked to herbicide tolerance at the time (Hirschberg and McIntosh, 1983; Erickson et al., 1985; Johanningmeier et al., 1987) are indicated as black circles. Also shown are histidine residues 215 and 272 that bind iron and quinone acceptor in the analogous L protein of the *Rhodopseudomonas viridis* reaction center (Deisenhofer et al., 1985). (Adapted from Greenberg et al., 1987)

implicated as being involved in QB binding (Kless et al., 1994). Thus, the region of the D1 protein bounded by helices *D* and *E*, and dubbed by Achim Trebst "the 86 residues of life" (personal communication), contained not only the special pair chlorophylls and quinone and herbicide binding sites, but also a site, specific to oxygenic photosynthetic organisms, heralding rapid breakdown of the protein.

V. The Life History of D1

The field was now ripe for linking D1 protein molecular architecture with the kinetics of its rapid turnover. While reading a paper by Sefton et al. (1982) on acylation of viral transforming proteins during a visiting scientist stint at NIH, Mattoo hatched the idea that D1 may be acylated and thus might move within the membrane. Feeding ^{3}H-palmitic acid to *Spirodela*, he and his technician Cathy Conlon confirmed post-translational pamitoylation of D1. Around the same time, a paper appeared showing that some proteins were redistributed between stromal and granal membranes in a high fluorescing mutant of maize versus wild type (Leto et al., 1985). Mattoo drew a connection between these two findings and used the method developed for maize to separate stromal and granal membranes. Edelman's Sabbatical came about then and the team was together again for an extended period, this time at Beltsville. We quickly showed that membrane attachment of the D1 precursor and processing to the mature D1 form occurred in the unstacked stromal lamellae. Once processed, the D1 protein migrated rapidly, within the thylakoids, to the topologically distinct granal lamellae where it remained and where the initial scission to a 23.5 kDa fragment occurred (Mattoo and Edelman, 1987). The metabolic life history of the D1 protein was presented in detail (Fig 5). Posttranslational palmitoylation of the processed D1 protein in a membrane-protected domain on the granal lamellae was also documented, the first posttranslational acylation event shown for a plant protein (Mattoo and Edelman, 1987).

During this study, Frank Callahan joined the Mattoo laboratory as a postdoctoral fellow. Callahan had an ability to coax an extra mile out of a polyacrylamide gel. He used this talent to show that the D1 protein in *Spirodela* is distributed between stromal and granal lamellae in the steady state as well (Callahan et al., 1989). Later, Maria Luisa Ghirardi took charge in Mattoo's laboratory and teamed up with Sudha Mahajan and Sudhir Sopory to reveal intramembranal cycling of the D1/D2 heterodimer (Ghirardi et al., 1990), nailing this concept with the isolation and characterization of a discrete PS II reaction center particle from stromal lamellae (Ghirardi et al., 1993). The metabolic and physical dynamics of D1 within the photosynthetic membrane, and the structural/functional regions of the protein that are likely to be involved in initiation of rapid degradation, were reviewed by us in The Cell (Mattoo et al., 1989).

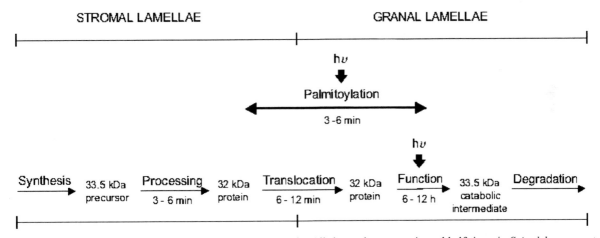

Fig. 5. Membrane-associated events in the life of the D1 protein. All time values are estimated half-times in *Spirodela* grown at 30 μmol photons·m^{-2}·s^{-1} PAR. Light-dependent synthesis (Mattoo et al., 1984a; Fromm et al., 1985) of nascent 33.5 kDa precursor takes place on 70S thylakoid-bound ribosomes (Ellis, 1977; Minami and Watanabe, 1984; Herrin and Michaelis, 1985), most likely on unstacked stromal lamellae (Yamamoto et al., 1981). Upon completion of translation, the 33.5 kDa precursor is associated exclusively with the stromal lamellae. There, carboxy-terminal processing (Marder et al., 1984) occurs with a half-time of 3–6 min (Reisfeld et al., 1982). After processing, the 32 kDa form is translocated to the stacked granal lamellae with a half-time of 6–12 min. At this location, D1 is functionally active and gains the ability to bind herbicides (Wettern, 1986). The 32 kDa D1 protein remains in the granal lamellae with a half-time of 6–12 h. Light-dependent degradation of the protein occurs there with production of a 23.5 kDa membrane bound catabolic intermediate (Greenberg et al., 1987). At some point between translocation and formation of the catabolic intermediate, reversible, light-induced palmitoylation of the protein occurs. (From Mattoo and Edelman, 1987)

VI. The UV-B Story

There was one more D1 protein facet uncovered during the 1980s by our groups, the most amazing of them all—the UV story. Victor Gaba, on a Royal Society postdoctoral fellowship, arrived at the Weizmann Institute in the mid 1980s with a background in photobiology. Gaba teemed up with Greenberg to uncover the photosensitizers that mediate D1 protein degradation. Early on, we had realized that D1 protein degradation is radiance driven, the protein being stable in the dark (Mattoo et al., 1984a). However, little was known of how the cleavage site became activated by light. In an initial study, Gaba concluded that phytochrome was not a photoreceptor for D1 degradation in the visible or far-red regions of the spectrum (Gaba et al., 1987). On the other hand, we and others were speculating that the action of a semiquinone anion radical, normally formed in the QB pocket during photosynthesis, might play a role in promoting degradation of the protein. We noted that quinone, semiquinone, and quinol all have characteristic UV spectra. Also, from classical studies (Jones and Kok, 1966) it was known that UV irradiation could inhibit PS II electron flow. We thus decided to give UV a try. Greenberg and Gaba (Greenberg et al., 1989) quantified the rate of D1 protein degradation over a broad spectral range from 250 to 730 nm (Fig 6). Unexpectedly, the quantum yield for degradation proved highest in the UV-B (280-320 nm) region! Several lines of evidence clearly demonstrated two distinct photosensitizers for D1 protein degradation: the bulk photosynthetic pigments (primarily chlorophyll) in the visible and far-red regions, and plastosemiquinone in the UV. A final, important point made by Greenberg was that a significant portion (>30%) of D1 protein degradation in sunlight could be attributed to UV-B irradiance.

UV-B driven degradation of the D1 protein took an additional turn with the arrival of Marcel Jansen at Rehovot. Jansen, with a background in PS II inhibitors from his MSc degree with Jack van Rensen in Wageningen, compared the mechanisms of UV-B driven and visible-light driven degradation of D1. He showed that all inhibitors of electron flow that replace bound QB from its niche on D1 inhibit UV-B-driven D1 degradation, but only some inhibit visible-light-driven degradation. Thus, UV-B-driven D1 protein degradation, but not that driven by visible light, requires plastoquinone

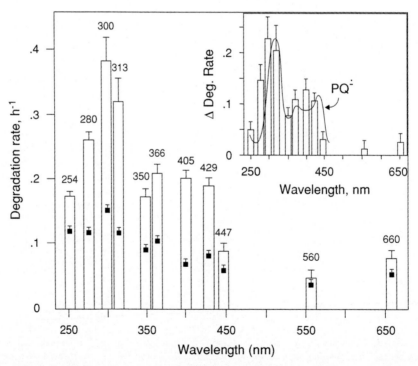

Fig. 6. Enhanced degradation of D1 protein in the UV in antennae-chlorophyll-deficient plants (white histogram bars). Degradation rates in control plants are indicated by black squares within the histogram bars. The INSET shows the difference spectrum for degradation of D1. It was obtained by subtracting the D1 degradation rate of continuously-illuminated control plants from those of intermittently-illuminated, antennae-chlorophyll-deficient plants. The curved line is the absorbance spectrum of the plastosemiquinone anion radical (Amesz, 1977), which is liganded to the D1 protein at the QB niche. The D1 degradation spectrum in the UV resembles the absorption spectrum of plastosemiquinone. (From Greenberg et al., 1989)

in the QB niche to proceed. By using a series of nitrophenol inhibitors varying in side chain bulkiness, and analyzing effects on D1 mutants of *Chlamydomonas*, Jansen showed that D1 degradation in visible light occurs as long as specific regions at the end of helix D and in the *D-de* loop of the protein are not engaged. These regions were proposed to regulate rapid degradation of the D1 protein through substrate-(i.e., QB)-mediated stabilization (Jansen at al., 1993). Achim Trebst was intimately involved in this study. It took place partly during the 1993 Iraq war. Jansen caught the last plane out of Israel before the Iraqi scuds started falling on Tel-Aviv and Riyad (what a combination!). The day before, Edelman had phoned Trebst, at Ruhr University of Bochum, who agreed on the spot to accept Jansen for as long as necessary in his laboratory. Two months later, when the Mid East returned to its normal simmering self, Jansen was back at his Rehovot bench with the *Chlamydomonas* mutant part of the study that was carried out in Bochum.

VII. The D1/D2 Heterodimer Takes Center Stage

Next came two seminal findings by Jansen: UV-B drives not only D1 but also D2 protein degradation (Jansen et al., 1996a); moreover, mixing visible and UV-B radiances leads to synergistic effects (Jansen et al., 1996b). The physiological implications were considerable. The PS II reaction center has at its core a heterodimer made up of two proteins, D1 and D2 (Trebst, 1986; Nanba and Satoh, 1987; Marder et al., 1987). While D1 rapidly degrades under PAR (photosynthetically active radiation), the D2 protein is relatively stable. What Jansen showed is that when *Spirodela* plants are exposed to UV-B radiation, D2 degradation accelerates markedly and half-life times approach those of the D1 protein (Jansen et al., 1996a). Moreover, in the presence of an environmentally relevant background of PAR, low fluxes of UV-B (but not UV-A) radiation synergistically stimulated degradation of the D2 protein within functional reaction centers (Jansen et al., 1996b). Thus, above a critical threshold, UV-B specifically targets the D1/D2 heterodimer for accelerated degradation (Fig. 7). The acceleration effect (as opposed to the effects of PAR or UV-B alone) was tightly coupled to the redox status of PS II (Babu et al., 1999).

It is not known if D1/D2 heterodimer degradation is significant in the field under current UV-B levels. Sunlight contains approximately 7.5 μmol photons m^{-2} s^{-1} of UV-B irradiation at a photon flux density of 1000 μmol m^{-2} s^{-1} of PAR. However, it is difficult to extrapolate from the UV-B threshold value for heterodimer degradation obtained with laboratory-grown plants (<1.0 μmol photons m^{-2} s^{-1}; Fig. 7) to conditions outdoors. In nature, selective absorption by UV-absorbing pigments in the epidermis and cuticle can severely (90–99%) diminish transmission of UV-B irradiance to the chloroplasts in the mesophyll. Moreover, some plants respond to increased UV-B irradiance by elevating their UV-B screening capacity (Teramura and Sullivan, 1994), thereby protecting PS II activity and stabilizing the D1 protein (Wilson and Greenberg, 1993). These factors, along with screening from antenna chlorophylls (Greenberg et al., 1989), may normally lower UV-B irradiance reaching PS II reaction centers to below the threshold for accelerated D1/D2 heterodimer degradation. However, as UV-B increases in the environment, this threshold may be more frequently breached, especially in UV-B sensitive species. Indeed, we showed that under PAR, two cultivars of soybeans, one UV-B sensitive (cv. CNS) and the other (cv. Williams) UV-B tolerant, have similar kinetics of degradation both for D1 and D2. However, when UV-B is mixed with PAR, degradation of the D1/D2 heterodimer was significantly enhanced in cv. CNS compared to cv. Williams (Booij et al., 1995). Later, Isabelle Booij-James used the power of *Arabidopsis* mutants to reveal the involvement of secondary, phenolic metabolites as protectants of the UV-B radiation-mediated D1 degradation (Booij-James et al., 2000). These studies showed that the contributions of sinapate esters and flavonoids to the total screening capacity of the leaf varied with the genetic background.

In an attempt to capture the big picture, Marcel Jansen set out to determine the quantitative relationship between rates of D1 and D2 degradation and photon fluences in *Spirodela* plants over the entire light intensity range from darkness to full sunlight (~1600 μmol photons m^{-2} s^{-1}). This was a Herculean task, involving over 800 points collected from pulse chase experiments. He found that kinetics for D2 catabolism essentially mirror those for D1, except that the actual half-life times of the D2 protein are about 3 times longer than those of D1 (Jansen et al., 1999). Thus, the catabolic ratio, D2/D1, is fluence independent, supporting the proposal (Jansen et al., 1996a) that catabolism of the two proteins is coupled. Analyzing D1 degradation in detail, he showed that it increases with photon flux in a distinctly multiphasic manner. Four phases were uncovered over the fluence range from 0 to 1600 μmol photons m^{-2} s^{-1}. The physiological processes associated with each phase remain to be determined. However, we note that a fluence as low as 5 μmol photons m^{-2} s^{-1} elicited a reaction constituting >25% of

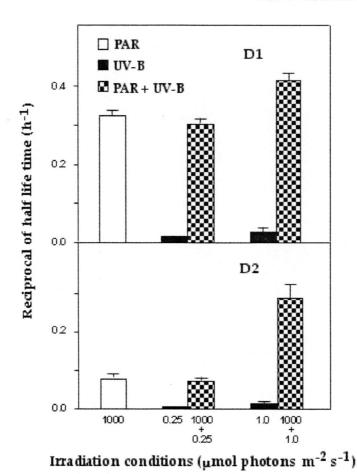

Fig. 7. Synergistic rates of D2 and D1 protein degradation in physiological mixtures of PAR and UV-B irradiation. *Spirodela* plants labeled with ^{35}S methionine were chased under radiances of PAR, UV-B, or a mixture of both. Protein degradation was determined from SDS-PAGE fluorograms. Bars represent averaged data from several experiments. (From Jansen et al., 1996b)

the total degradation rate response, while >90% of the rate potential was attained at intensities below saturation for photosynthesis (~750 μmol photons m^{-2} s^{-1}). In *Spirodela*, oxygen evolution is significantly impeded after exposure to photon fluences >600 μmol m^{-2} s^{-1}. At these photoinhibitory fluences, however, less than 25% of the D1 catabolic potential remains unaccounted for (Fig. 8). Thus, in intact plants, D1 degradation is a process only marginally overlapping with photoinhibition and overwhelmingly associated with fluences limiting for photosynthesis. This point has since been corroborated in *Chlamydomonas* (Keren et al., 1997).

VIII. Phosphorylation–Dephosphorylation

While the 1990s saw our UV-B and D1/D2 heterodimer studies maturing and taking physiological/ecological direction, this period was also consumed by ambiguity concerning the role played by phosphorylation in D1 function. Frank Callahan was following the spatial separation of the different forms of D1 in thylakoids when in early 1990 he discovered a new form of D1, which we designated 32* (32 star). This form transiently accumulated in the grana regions of thylakoids. He showed that its appearance was light dependent and inhibited by diuron (Callahan et al., 1990). A prolonged electrophoresis run of a polyacrylamide gel facilitated resolution of the three known forms of D1: the 33.5 kDa precursor, the modified 32*, and the 32 kDa-D1 protein (Fig. 9).

Tedd Elich joined the Mattoo laboratory at that time and methodically showed that the transiently appearing 32* was actually the phosphorylated form of D1 (Elich et al., 1992). Previously, Bennett and co-workers (Michel et al., 1988) showed that several PS II proteins undergo phosphorylation at the N-terminus. However, because they used a nickel column which irreversibly bound the isolated phosphorylated proteins, they could

Chapter 3 The D1 Protein: Past and Future Perspectives

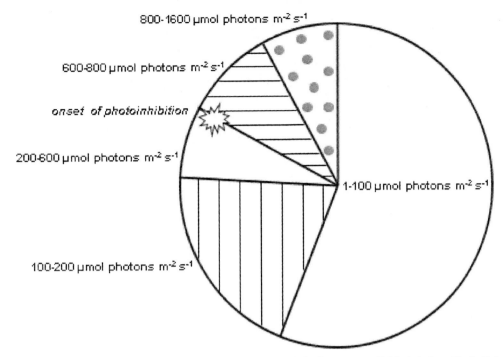

Fig. 8. The relative contribution of fluence ranges to D1 degradation. *Spirodela* plants were radiolabeled, chased in the light at various intensities, and rates of protein degradation determined following SDS-PAGE. The total pie represents the rate of D1 degradation at 1600 μmol photons m^{-2} s^{-1} (0.344 h^{-1}). Segments indicate which percentage of this rate is contributed by a given range of photon fluencies. Onset of photoinhibition in *Spirodela* occurs at 600 μmol photons m^{-2} s^{-1}. Thus, D1 degradation in vivo is mainly a low intensity event. (Adapted from Jansen et al., 1999)

not have anticipated that phosphorylation was, in fact, only transiently associated with the grana localized form of D1. The important role that phosphorylation plays in the life cycle of the D1 protein led to a flurry of activity, with several investigators confirming these observations with studies on other plants (Aro et al., 1992; Rintamäki et al., 1995).

Elich soon came up with a whopper of a discovery. Dephosphorylation of 32* in *Spirodela* was significantly stimulated by far-red (720 nm) light (Elich et al., 1993)! In the chloroplast, this wavelength is absorbed almost exclusively by PS I. The involvement of PS II-driven linear electron transport was ruled out by adding diuron. In the presence of this inhibitor, cyclic PS I electron transport (Jansen et al., 1992) and dephosphorylation of the D1 protein (Elich et al., 1993) continued unabated in *Spirodela* plants under far-red light. Elich, always the careful worker, further probed the involvement of PS I by introducing DBMIB (2,5-dibromo-3-methyl-6-isopropylbenzoquinone), a plastoquinone antagonist and in vivo inhibitor of PS I-cyclic electron transport. DBMIB inhibited light-stimulated dephosphorylation of 32* in a concentration-dependent manner. It also inhibited light-stimulated dephosphorylation of other PS II core proteins, D2, and CP43, but not LH2, the light harvesting chlorophyll a/b binding protein associated with the PS II antennae. This was taken as evidence for multiple phosphatases involved in thylakoid protein dephosphorylation. Out of these studies (Elich et al., 1993) came what at first glance is an amazing proposal: dephosphorylation of PS II core proteins is regulated in vivo by PS I excitation! However, upon reflection, one realizes that PS I-regulated dephosphorylation would provide the perfect complement to PS II-regulated phosphorylation, and would render the phosphorylation state of PS II core proteins exquisitely sensitive to the relative energy distribution between the two photosystems.

Soon, results in different laboratories made it likely that multiple light-regulated protein kinases are also present and active in the photosynthetic organelle. Elich took advantage of the fact that LH2 phosphorylation in vivo is effectively inhibited by propylgallate (Sopory et al., 1990) and demonstrated that reversible protein phosphorylation of PS II reaction center proteins is more strictly light dependent than that of the light harvesting chlorophyll a/b binding protein (Elich et al., 1997). We were tempted to suggest that such a cross talk in the chloroplast may regulate chromatic adaptation of plants.

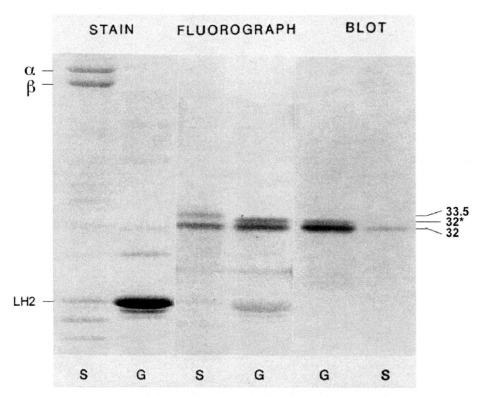

Fig. 9. Stromal (S) and granal (G) lamellae were isolated from ^{35}S-methionine labeled *Spirodela* plants and resolved by SDS-PAGE. The Coomassie-stained gel shows the overall protein profiles of the fractions used. The highly resolved positions of the 33.5, 32*, and 32 kDa D1 bands are shown in the fluorograph (S lane- 3min pulse, no chase; G lane- 3min pulse, 2 h chase). Duplicated lanes of the same gel were electroblotted onto nitrocellulose and probed with polyclonal antibody to the D1 protein. (From Callahan et al., 1990)

IX. Circadian Control

While searching for a D1-specifc protein kinase, Mark Swegle and Isabelle Booij-James, at the Beltsville laboratory, added yet another dimension to the D1 phosphorylation saga. Beyond the known redox regulation of D1 phosphorylation, they discovered an overriding control by an internal, circadian clock (Booij-James et al., 2002). In greenhouse grown plants under natural diurnal cycles, the peak of D1 phosphorylation was found to occur at about 10 am, 4 hours after daybreak but 2 hours before maximal light intensity (Fig. 10)! Moreover, the up and down cycling of D1 phosphorylation could occur at light intensities well below those saturating for photosynthesis or initiation of photoinhibition. Once induced, the D1 phosphorylation cycle maintained its oscillations under free running conditions in continuous light, one of several signs for circadian control that were found (Booij-James et al., 2002).

A number of questions surfaced as we were revising the manuscript on circadian entrainment of D1 phosphorylation for Plant Physiology and considering the reviewers' comments. Over long hours of deliberation, we synthesized the discussion that was presented in Booij-James et al. (2002): Is the phosphorylation state of PS II reaction core proteins a consequence or a determinant of the relative energy distribution between the two photosystems in oxygenic photosynthesis? Do plants use the phosphorylation index of D1 as a sensor to anticipate the onset of higher light intensities (Mattoo and Edelman, 1985)? We suggested that circadian regulation of a nuclear D1-kinase gene product could provide a cross-talk mechanism by which a biological clock uses a nuclear-encoded protein to regulate PS II core function and that light-mediated reversible phosphorylation is a means for the chloroplast to anticipate environmental change. We pointed out that the manner in which D1 metabolic regulation is achieved likely differs among different photosynthetic organisms. In higher plants, where D1 is reversibly phosphorylated, circadian regulation of metabolism may be at the phosphorylation level. In cyanobacteria, redox-regulated phosphorylation of D1 does not appear to occur; but these oxygenic bacterial phototrophs

Chapter 3 The D1 Protein: Past and Future Perspectives

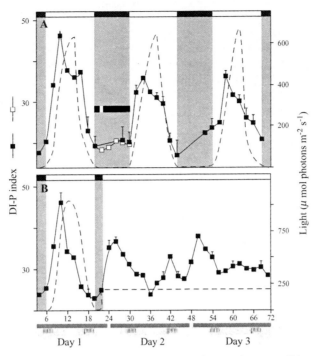

Fig. 10. Rhythmic behavior of the phosphorylated D1 level in *Spirodela* under greenhouse conditions. **A.** The light intensity (broken lines) and the D1-P index ([D1−P]/[D1] + [D1−P]) are shown at indicated times over 3 day/night cycles (filled rectangles). At the end of the light period and 2 h into darkness, a set of plants was exposed for 5 min to 300 μmol photons m^{-2} s^{-1} fluorescent light and thereafter returned to darkness (empty rectangles). **B.** D1-P index maintains oscillations in free running conditions in continuous light. Plants were grown in the greenhouse under natural light/dark cycles for a week in medium lacking sucrose. Then, at the end of the light cycle and 2 h into darkness, a set of plants was left in the greenhouse until the end of the experiment while another was brought into the laboratory and incubated in continuous light at 200 μmol photons m^{-2} s^{-1} until the end of the experiment. Error bars indicate standard errors. (From Booij-James et al., 2002)

often possess multiple copies of the *psbA* gene coding for D1 (Golden, 1995; Chen et al., 1999), with different D1 iso-forms adapted in vivo to varying photon irradiation, one dominant at lower and another at higher light intensities (Bustos et al., 1990; Clarke et al., 1993; Kulkarni and Golden, 1994). Several studies showed that light-induced transcription of cyanobacterial *psbA* occurs within the larger framework of circadian control (Liu et al., 1995; Chen et al., 1999). We thus put forward a generalized hypothesis that reversible phosphorylation of D1 (and maybe, other PS II proteins) in higher plants evolutionarily replaced multiple DNA copies in cyanobacteria as a more energy efficient substrate for circadian clock regulation of PS II core metabolism.

X. The Past and Future

Thus, over three decades, considerable structural and functional information was amassed concerning the life cycle of the D1 protein and the region of the molecule where primary light-dependent cleavage occurred. Two major steps were uncovered regarding radiance-driven D1 degradation. First, depending on incident radiation, one or more photoreceptors, characterized by specific action spectra, are activated. Then, the energized photoreceptor, directly or indirectly, activates a cleavage site resulting in the appearance of a specific breakdown product. While these findings provided some detail concerning mechanism, they skirted the issue of dual functionality of the D1 protein: a central scaffolding protein of the PS II reaction center core that inexplicably turns over more rapidly than other core and PS II proteins. We spent many coffee breaks discussing this. Back in 1981, John Ellis (Ellis, 1981) put forward the idea that D1 turnover was a protective mechanism. In 1985 we wrote (Mattoo & Edelman, 1985): "The coordination in the rates of synthesis and breakdown of this regulatory protein may be one of the mechanisms by which plants adapt to changing light conditions. Light adaptations are known to be associated with specific alterations at several levels in the chloroplast, resulting in plants with shade-type or sun-type plastids. Herbicides that inhibit electron flow through PS II also induce shade-type chloroplasts in plants grown in high light, while in certain plants, high light can affect photosynthetic efficiency and cause losses. Indeed, when *Spirodela* is cultivated on a sublethal dose of atrazine, alterations seen in the ultrastructure and lipid composition of thylakoids resemble those seen in shade-type chloroplasts and triazine-resistant weeds. Under these conditions, the degradation of the D1 protein is retarded compared to control plants (Mattoo et al., 1984b). It would appear that the lipid environment of the D1 protein affects conformation, orientation and function of the protein in PS II. Of particular interest is a plausible but entirely speculative suggestion that a signal whose amplitude varies with the rate of D1 protein turnover acts as the plant's light meter, setting in motion a process that helps the plant to adjust to changing light conditions. This adjustment could involve rapid reorganization of the lipid and protein components of the chloroplast for optimal photosynthetic efficiency by maintaining a functional interaction of the light harvesting chlorophyll a/b protein, the regulatory components (e.g., the D1 protein) and lipids of the PS II complex."

Looking back now at 25 years of our joint D1 protein research, we are at one and the same time awed and pleased, but yet unfulfilled. Awed at the hundreds of publications researching various aspects of D1 metabolism, regulation, structure, and function that have ensued partially due to our early findings. Pleased that D1 and D2 research still rules the roost (witness the tens of abstracts in the Program Book of the 2004 symposium on Photosynthesis and Post-Genomic Era dealing with the D1/D2 heterodimer (Carpentier & Allakhverdiev, 2004). But yet unfulfilled, as we suspect that a fundamental regulatory function associated with rapid D1 degradation still awaits resolution. In this regard, we are encouraged by current (as yet unpublished) studies in Avigdor Scherz's group which point to conformational flexibility in the apposition of the D and E helices (involving reversible hydrogen bond and other weak associations at their crossing point) as regulating protein-gated electron transfer (personal communication). This dynamic breathing of the D1/D2 heterodimer not far from the QB binding niche and D1 degradation site may yet hook up with our speculation above concerning a light meter function for rapid D1 degradation.

And that's what its all about; being unsatisfied with the present state of knowledge and planning for the next experiment.

Acknowledgments

ME acknowledges the support of the Avron-Minerva Center for Photosynthesis at Rehovot. AKM thanks James D. Anderson for support and incorporating research on the D1 protein at Beltsville.

References

Andronis C, Kruse O, Deak Z, Vass I, Diner BA and Nixon PJ (1998) Mutation of residue threonine-2 of the D2 polypeptide and its effect on PS II function in *Chlamydomonas reinhardtii*. Plant Physiol 117: 515–524

Aro E-M, Kettunen R and Tyystjärvi E (1992) ATP and light regulate D1 protein modification and degradation: Role of D1* in photoinhibition. FEBS Lett 297: 29–33

Babu TS, Jansen MAK, Greenberg BM, Gaba V, Malkin S, Mattoo AK and Edelman M (1999) Amplified degradation of PS II D1 and D2 proteins under a mixture of photosynthetically active radiation and UV-B radiation: dependence on redox status of PS II. Photochem Photobiol 69: 553–559

Bennett J (1984) Photosynthesis: Control of protein turnover by photosynthetic electron transport. Nature 310: 547–548

Booij IS, Swegle M, Dube S, Edelman M and Mattoo AK (1995) Photodegradation of D1-D2 PS II reaction center heterodimer. In: Mathis P (ed) Phostosynthesis: From Light to Biosphere, pp 487–490. Kluwer, Dordrecht

Booij-James IS, Dube SK, Jansen MAK, Edelman M and Mattoo AK (2000) Ultraviolet-B radiation impacts light-mediated turnover of the PS II reaction center heterodimer in Arabidopsis mutants altered in phenolic metabolism. Plant Physiol 124: 1275–1283

Booij-James IS, Swegle M, Edelman M and Mattoo AK (2002) Phosphorylation of the D1 PS II reaction center protein is controlled by an endogenous circadian rhythm. Plant Physiol 130: 2069–2075

Bustos SA, Schaefer MR and Golden SS (1990) Different and rapid responses of four cyanobacterial psbA transcripts to changes in light intensity. J Bacteriol 172: 1998–2004

Callahan FE, Wergin WP, Nelson N, Edelman M and Mattoo AK (1989) Distribution of thylakoid proteins between stromal and granal lamellae in Spirodela. Dual location of PS II components. Plant Physiol 91: 629–635

Callahan FE, Ghirardi ML, Sopory SK, Mehta AM, Edelman M and Mattoo AK (1990) A novel metabolic form of the 32kDa-D1 protein in the grana-localized reaction center of PS II. J Biol Chem 265: 15357–15360

Carpentier R and Allakhverdiev S (2004) Photosynthesis & Post Genomic Era: From Biophysics to Molecular Biology, a Path in the Research of PS II. Trois-Rivieres, Abstracts: pp 1–205

Chen Y-B, Dominic B, Zani S, Mellon MT and Zehr JP (1999) Expression of photosynthesis genes in relation to nitrogen fixation in the diazo trophic filamentous nonheterocystous cyanobacterium Trichdesmium sp. IMS 101. Plant Mol Biol 41: 89–104

Chua N-H and Gilham NW (1977) The sites of synthesis of the principal thylakoid membrane polypeptides in *Chlamydomonas reinhardtii*. J Cell Biol 74: 441–452

Clarke AK, Soitamo A, Gustafsson P and Öquist G (1993) Rapid interchange between two distinct forms of cyanobacterial PS II

reaction center protein D1 in response to photoinhibition. Proc Natl Acad Sci USA 90: 9973–9977
Deisenhofer J, Epp O, Miki K, Huber R and Michel H (1985) Structure of the protein subunits in the photosynthetic reaction center of *Rhodopseudomonas viridis* at 3A resolution. Nature 318: 618–624
Eaglesham ARJ and Ellis RJ (1974) Protein synthesis in chloroplasts. II. Light-driven synthesis of membrane proteins by isolated pea chloroplasts. Biochim Biophys Acta 335: 396–407
Edelman M and Reisfeld A (1978) Characterization, translation and control of the 32,000 Dalton chloroplast membrane protein in Spirodela. In: Akoyunoglou G and Argyroudi-Akoyunoglou JH (eds) Chloroplast Development, pp 641–652. Elsevier/North Holland, New York
Edelman M and Reisfeld A (1980) Synthesis processing and functional probing of P-32000, the major membrane protein translated within the chloroplast. In: Leaver C (ed) Genome Organization and Expression in Plants, pp 353–362. Plenum, New York
Edelman M, Hallick RB and Chua N-H (eds) (1982) Methods in Chloroplast Molecular Biology, Preface: pp v. Elsevier, Amsterdam
Elich TD, Edelman, M and Mattoo AK (1992) Identification, characterization and resolution of the in vivo phosphorylated form of the D1 PS II reaction center protein. J Biol Chem 267: 3523–3529
Elich TD, Edelman M and Mattoo AK (1993) Dephosphorylation of PS II core proteins is light regulated in vivo. EMBO J 12: 4857–4862
Elich TD, Edelman M and Mattoo AK (1997) Evidence for light-dependent and light-independent protein dephosphorylation in chloroplasts. FEBS Lett 411: 236–238
Ellis RJ (1981) Chloroplast proteins: synthesis, transport and assembly. Annu Rev Plant Physiol 32: 111–137
Gaba V, Marder JB, Greenberg BM, Mattoo AK and Edelman M (1987) Degradation of the 32 kDa herbicide binding protein in far red light. Plant Physiol 84: 348–352
Garcia-Fernandez JM, Hess WR, Houmard J and Partensky F (1998) Expression of the psbA gene in the marine oxyphotobacteria *Prochlorococcus* Spp. Arch Biochem Biophys 359: 17–23
Ghirardi ML, Callahan FE, Sopory SK, Elich TD, Edelman M and Mattoo AK (1990) Cycling of the PS II reaction center core between grana and stromal lamellae. In: Baltschefssky M (ed) Current Research in Photosynthesis, 2: 733–738, Kluwer, the Netherlands
Ghirardi ML, Mahajan S, Sopory SK, Edelman M and Mattoo AK (1993) PS II reaction center particle from Spirodela stroma lamellae. J Biol Chem 268: 5357–5360
Golden SS (1995) Light responsive gene expression in cyanobacteria. J Bacteriol 177: 1651–1654
Grebanier AE, Coen DM, Rich A and Bogorad L (1978) Membrane proteins synthesized but not processed by isolated maize chloroplasts. J Cell Biol 78: 734–746
Greenberg BM, Gaba V, Mattoo AK and Edelman M (1987) Identification of a primary in vivo degradation product of the rapidly-turning-over 32 kDa protein of PS II. EMBO J 6: 2865–2869
Greenberg BM, Gaba V, Canaani O, Malkin S, Mattoo AK and Edelman M (1989) Separate photosynthesizers mediate degradation of the 32-kDa PS II reaction center protein in the visible and UV spectral regions. Proc Natl Acad Sci USA 86: 6617–6620

Hirschberg J, McIntosh L (1983) Molecular basis of herbicide resistance in *Amaranthus hybridus*. Science 222: 1346–1349
Hoffman-Falk H, Mattoo AK, Marder JB, Edelman M and Ellis RJ (1982) General occurrence and structural similarity of the rapidly synthesized, 32,000-dalton protein of the chloroplast membrane. J Biol Chem 257: 4583–4587
Hwang S, Kawazoe R and Herrin DL (1996) Transcription of tufA and other chloroplast-encoded genes is controlled by a circadian clock in Chlamydomonas. Proc Natl Acad Sci USA 93: 996–1000
Jansen MAK, Driesenaar A, Kless H, Malkin S, Mattoo AK and Edelman M (1992) PS II inhibitor binding, QB-mediated electron flow and rapid degradation are separable properties of the D1 reaction center protein. In: Argyroudi-Akoyunoglou JH (ed) Regulation of Chloroplast Biogenesis, pp 303–311. Plenum, New York
Jansen MAK, Depka B, Trebst A and Edelman M (1993) Engagement of specific sites in the plastoquinone niche regulates degradation of the D1 protein in PS II. J Biol Chem 268: 21246–21252
Jansen MAK, Gaba V, Greenberg BM, Mattoo AK and Edelman M (1996a) Low threshold levels of ultraviolet-B in a background of photosynthetically active radiation trigger rapid degradation of the D2 protein of PS II. Plant J 9: 693–699
Jansen MAK, Greenberg BM, Edelman M, Mattoo AK and Gaba V (1996b) Accelerated degradation of the D2 protein of PS II under ultraviolet radiation. Photochem Photobiol 63: 814–817
Jansen MAK, Mattoo AK and Edelman M (1999) Photodynamics of D1-D2 protein catabolism in the chloroplast. Eur J Biochem 260: 527–532
Jones LW and Kok B (1966) Photoinhibition of chloroplast reactions: kinetics and action spectra. Plant Physiol 41: 1037–1043
Kawazoe R, Hwang S and Herrin DL (2000) Requirement for cytoplasmic protein synthesis during circadian peaks of transcription of chloroplast-encoded genes in Chlamydomonas. Plant Mol Biol 44: 699–709
Keren N, Berg A, van Kan PJM, Levanon H and Ohad I (1997) Mechanism of photosystem II photoinactivation and D1 protein degradation at low light: The role of back electron flow. Proc Natl Acad Sci USA 94: 1579–1584
Kless H, Oren-Shamir M, Malkin S, McIntosh L and Edelman M (1994) The D-E region of the D1 protein is involved in multiple quinone and herbicide interactions in PS II. Biochemistry 33: 10501–10507
Kulkarni RD and Golden SS (1994) Adaptation to high light intensity in *Synechococcus* sp. strain PCC 7942: regulation of three psbA genes and two forms of D1 protein. J Bacteriol 176: 959–965
Kyle DJ, Ohad I and Arntzen CJ (1984) Membrane protein damage and repair: Selective loss of a quinone-protein function in chloroplast membranes. Proc Natl Acad Sci USA 81: 4070–4074
Leto KJ, Bell E and McIntosh L (1985) Nuclear mutation leads to an accelerated turnover of chloroplast-encoded 48 kDa and 34.5 kDa polypeptides in thylakoids lacking PS II. EMBO J 4: 1645–1653
Liu Y, Golden SS, Kondo T, Ishiura M and Johnson CH (1995) Bacterial luciferase as a reporter of circadian gene expression in cyanobacteria. J Bacteriol 177: 2080–2086

Marder JB, Goloubinoff P and Edelman M (1984) Molecular architecture of the rapidly metabolized 32-kilodalton protein of photosystem II: Indications for COOH-terminal processing of a chloroplast membrane polypeptide. J Biol Chem 259: 3900–3908

Marder JB, Chapman DJ, Telfer A, Nixon PJ and Barber J (1987) Identification of psbA and psbD gene products, D1 and D2, as reaction centre proteins of PS II. Plant Mol Biol 9: 325–333

Mattoo AK, Edelman M (1985) Photoregulation and metabolism of a thylakoidal herbicide-receptor protein. In: St John JB, Berlin E and Jackson PC (eds) Frontiers of Membrane Research in Agriculture, pp 23–34. Rowman & Allanheld, Totowa

Mattoo AK and Edelman M (1987) Intramembrane translocation and posttranslational palmitoylation of the chloroplast 32-kDa herbicide-binding protein. Proc Natl Acad Sci USA 84: 1497–1501

Mattoo AK, Pick U, Hoffman-Falk H and Edelman M (1981) The rapidly metabolized 32,000-dalton polypeptide of the chloroplast is the "proteinaceous shield" regulating PS II electron transport and mediating diuron herbicide sensitivity. Proc Natl Acad Sci USA 78: 1572–1576

Mattoo AK, Hoffman-Falk H, Marder JB and Edelman M (1984a) Regulation of protein metabolism: Coupling of photosynthetic electron transport to in-vivo degradation of the rapidly metabolized 32 kDa protein of the chloroplast membrane. Proc Natl Acad Sci USA 81: 1380–1384

Mattoo AK, St John JB and Wergin WP (1984b) Adaptive reorganization of protein and lipid components in chloroplast membranes as associated with herbicide binding. J Cellular Biochem 24: 163–175

Mattoo AK, Marder JB and Edelman M (1989) Dynamics of the PS II reaction center. Cell 56: 241–246

Michel H, Hunt DF, Shabanowitz J and Bennett J (1988) Tandem mass spectrometry reveals that three photosystem II proteins of spinach chloroplasts contain N-acetyl-O-phosphothreonine at their NH2 termini. J Biol Chem 263: 1123–1130

Mulo P, Tystjärvi T, Tystjärvi E, Govindjee, Maenpaa P and Aro E-M (1997) Mutagenesis of the D-E loop of photosystem II reaction centre protein D1. Function and assembly of PS II. Plant Mol Biol 33: 1059–1071

Nanba O and Satoh K (1987) Isolation of a photosystem II reaction center consisting of D-1 and D-2 polypeptides and cytochrome b-559. Proc Natl Acad Sci USA 84: 109–112

Nixon PJ, Komenda J, Barber J, Deak Z, Vass I and Diner BA (1995) Deletion of the PEST-like region of PS II modifies the QB-binding pocket but does not prevent rapid turnover of D1. J Biol Chem 270: 14919–14927

Pfister K, Steinback KE, Gardner G and Arntzen CJ (1981) Photoaffinity labeling of an herbicide receptor protein in chloroplast membranes. Proc Natl Acad Sci USA 78: 981–985

Regitz G and Ohad I (1975) Changes in the protein organization in developing thylakoids of Chlamydomonas reinhardtii Y-1 as shown by sensitivity to trypsin. In: Avron M (ed) Proc 3rd Intl Congr Photosynthesis, 3: 1615–1625. Elsevier, Amsterdam

Reisfeld A, Mattoo AK and Edelman M (1982) Processing of a chloroplast-translated membrane protein in vivo. Analysis of the rapidly synthesized 32000-dalton shield protein and its precursor in Spirodela oligorrhiza. Eur J Biochem 124: 125–129

Renger G (1976) Studies on the structural and functional organization of system II of photosynthesis. The use of trypsin as a structurally selective inhibitor at the outer surface of the thylakoid membrane. Biochim Biophys Acta 440: 287–300

Rintamäki E, Kettunen R, Tyystjärvi E and Aro, E-M (1995) Light dependent phosphorylation of D1 reaction centre protein of PS II: hypothesis for the functional role in vivo. Physiol Plant 93: 191–195

Rochaix JD, Dron M, Rahire M and Malone P (1984) Sequence homology between the 32 kDa and the D2 chloroplast membrane polypeptide of Chlamydomonas reinhardii. Plant Mol Biol 3: 363–370

Rogers S, Wells R and Rechsteiner M (1986) Amino acid sequences common to rapidly degraded proteins: The PEST hypothesis. Science 234: 364–368

Rosner A, Jakob KM, Gressel J and Sagher D (1975) The early synthesis and possible function of a 0.5 x 106 Mr RNA after transfer of dark-grown Spirodela plants to light. Biochem Biophys Res Commun 67: 383–391

Sefton BM, Trowbridge IS, Cooper JA and Scolnik EM (1982) The transforming protein of Rous sarcoma virus, Harvey sarcoma virus and Abelson virus contain tightly bound lipid. Cell 31: 465–474

Shinozaki K, Ohme M, Tanaka M, Wakasugi T, Hayashida N, Matsubayashi T, Zaita N, Chunwongse J, Obokata J, Yamaguchi-Shinozaki K, Ohto C, Torazawa K, Meng BY, Sugita M, Deno H, Kamogashira T, Yamada K, Kusuda J, Takaiwa F, Kato A, Tohdoh N, Shimada H and Sugiura M (1986) The complete nucleotide sequence of the tobacco chloroplast genome: Its gene organization and expression. EMBO J 5: 2043–2049

Sobolev V and Edelman M (1995) Modeling the quinone-B site of the PS II reaction center using notions of complementarity and contact surface between atoms. Proteins 21: 214–225

Sopory SK, Greenberg BM, Mehta RA, Edelman M and Mattoo AK (1990) Free radical scavengers inhibit light-dependent degradation of the 32 kDa PS II reaction center protein. Z Naturforsch 45c: 412–417

Teramura AH and Sullivan JH (1994) Effects of UV-B radiation on photosynthesis and growth of terrestrial plants. Photosynth Res 39: 463–473

Trebst A (1986) The topology of plastoquinone and herbicide binding peptides of PS II in the thylakoid membrane. Z Naturforsch 41c: 240–245

Weinbaum SA, Gressel J, Resifeld A and Edelman M (1979) Characterization of the 32000 Dalton chloroplast membrane protein. III. Probing its biological function in Spirodela. Plant Physiol 64: 828–832

Wheeler AM and Hartley MR (1975) Major mRNA species from spinach chloroplasts do not contain poly(A). Nature 257: 66–67

Wilson MI and Greenberg BM (1993) Protection of the D1 PS II reaction center protein from degradation in ultraviolet radiation following adaptation of Brassica napus L. to growth in ultraviolet-B. Photochem Photobiol 57: 556–563

Zurawski G, Bohnert HJ, Whitfeld PR and Bottomley W (1982) Nucleotide sequence of the gene for the Mr 32000 thylakoid protein from Spinacia oleracea and Nicotiana debneyi predicts a totally conserved primary translation product of Mr 38950. Proc Natl Acad Sci USA 79: 7699–7703

Chapter 4

Characteristics and Species-Dependent Employment of Flexible Versus Sustained Thermal Dissipation and Photoinhibition

Barbara Demmig-Adams*, Volker Ebbert, C. Ryan Zarter and William W. Adams III
*Department of Ecology & Evolutionary Biology, University of Colorado, Boulder,
CO 80309-0334, USA*

Summary	39
I. Introduction	39
II. Interspecies Differences in the Capacity for Flexible Thermal Dissipation	40
III. Sustained Thermal Dissipation in Photoinhibited Evergreens	40
A. Photoinhibition, Xanthophyll Cycle Arrest, and Sustained Dissipation	40
B. PsbS and its Relatives	42
C. Sustained Protein Phosphorylation	44
IV. Two Types of Thermal Energy Dissipation in Evergreens	45
Acknowledgments	46
References	47

Summary

Photoprotective energy dissipation is a process in which xanthophylls (particularly zeaxanthin and antheraxanthin; Z + A) facilitate the dissipation of excess absorbed light energy as heat. This process can occur in different versions that meet the demands of different environments. Flexible thermal dissipation (qE type) responds to intra-thylakoid pH, is controlled by the PsbS protein, and relaxes rapidly upon darkening at warm temperatures. This flexible, PsbS/ΔpH-dependent dissipation is the predominant form of energy dissipation under environmental conditions favorable for growth, irrespective of plant species. In comparison with short-lived species, perennial evergreens not only have slower growth and lower photosynthetic capacities but also possess higher capacities for flexible thermal dissipation and show greater increases in PsbS level and Z + A in full sun versus moderate growth light intensity. Furthermore, a sustained form of thermal dissipation (qI type) is observed predominantly in photoinhibited evergreens under environmental conditions unfavorable for growth, irrespective of the environmental factor(s) involved. Sustained thermal dissipation is associated with Z + A retention, but is not ΔpH-dependent and does not relax rapidly in darkness even at warm temperatures. The role of PsbS in sustained dissipation versus other factors is discussed. Moreover, a correlation between sustained thermal dissipation in Z + A-retaining photoinhibited leaves and sustained phosphorylation of the photosystem II (PS II) core's D1 protein is shown. In overwintering Douglas fir, this is associated with an upregulation of an inhibitor (TLP40) of PS II core protein phosphatase.

I. Introduction

When plants absorb more light than they can utilize in photochemistry, they dissipate the excess as thermal energy in order to limit the formation of reactive oxygen species that can ultimately cause cell death (Niyogi, 2000; Demmig-Adams and Adams, 2002). In this chapter, two different forms of photoprotective thermal energy dissipation (assessed from non-photochemical chlorophyll fluorescence quenching or

---

*Author for correspondence, email: barbara.demmig-adams@colorado.edu

NPQ) are compared and contrasted. These two forms differ in their flexibility (e.g. speed of recovery), and are often referred to as flexible, rapidly reversible dissipation (qE type of NPQ or "feedback de-excitation") versus sustained dissipation (qI type of NPQ), respectively. Pronounced differences among plant species in the employment of these two forms of thermal dissipation are documented here. It is concluded that evergreen species employ both forms of thermal dissipation to a greater extent than species with shorter-lived leaves. In addition, sustained thermal dissipation is shown to be closely associated with photoinhibition of photosynthesis that is, furthermore, demonstrated to be considerably more common in evergreen species than in short-lived species.

II. Interspecies Differences in the Capacity for Flexible Thermal Dissipation

Long-term growth in different light environments, but under otherwise optimal conditions, reveals different strategies in an annual crop versus a perennial tropical evergreen (Fig. 1). Only the highly shade-tolerant evergreen species is able to survive and grow in deep shade. The crop species grows best at full sunlight and has a higher photosynthetic capacity in full sun compared to an intermediate light level (Fig. 1A). In contrast, the evergreen does not increase its photosynthetic capacity between moderate and full sunlight, and thus utilizes less of the extra energy of full sunlight for increased growth and photosynthesis than the crop species (Fig. 1B). On the other hand, the crop species does not increase its capacity for thermal energy dissipation between moderate and full sunlight, whereas the evergreen shows a strong increase in this capacity. The allocation of full sunlight to photosynthesis versus thermal dissipation is thus very different in the fast-growing crop versus the slow-growing evergreen species. Consistent with the increase in thermal dissipation capacity at full sunlight in the evergreen, the levels of the PsbS protein as well as the maximal level of zeaxanthin and antheraxanthin (Z + A) are also considerably increased in full versus moderate sunlight in the evergreen (Fig. 1D,F) but not in the crop species (Fig. 1C,E).

The PsbS protein and the de-epoxidized, excess-light-induced forms (Z + A) of the xanthophyll cycle are both required for ΔpH-dependent, rapidly reversible NPQ (Niyogi, 2000; Jung and Niyogi, this volume). The thermal dissipation process employed in either plant species under these favorable conditions is this flexible, reversible form of NPQ (qE type). This is supported by the fact that the majority of NPQ under these conditions can be abolished with uncouplers as well as by the absence of sustained increases in NPQ or decreases in predawn PS II efficiency (from the ratio of variable to maximal Chl fluorescence, F_v/F_m; Fig. 1 I,K). Under high light in the absence of additional environmental stresses, the PsbS/ΔpH-dependent form of thermal dissipation thus predominates irrespective of a species' inherent growth rate. It can also be concluded that the capacity for PsbS/ΔpH-dependent NPQ is higher in evergreens with lower growth and maximal photosynthesis rates than in species with short-lived leaves. Maximal NPQ capacity in sun-grown *Monstera deliciosa* was as high as that of PsbS-overexpressing *Arabidopsis* strains (Li et al., 2002a). The authors of the latter study brought up the question of potential costs of a greater constitutive expression of PsbS that may not be compatible with maximal light harvesting, growth, and/or reproduction. In short-lived species, high growth rates are necessary for rapid completion of their life cycle. Since perennial evergreens do not possess, and would not appear to require, the high growth rates of annuals, their higher level of PsbS in full sun would seem to offer only benefits (in the form of maximal photoprotection).

III. Sustained Thermal Dissipation in Photoinhibited Evergreens

A. Photoinhibition, Xanthophyll Cycle Arrest, and Sustained Dissipation

Excess light can lead to a lasting inactivation of PS II that is termed "photoinhibition" of photosynthesis and of PS II (see e.g. Powles, 1984; Demmig-Adams and Adams, 2003) under certain conditions. These include (a) a sudden transfer of shade-grown leaves to high light levels and (b) sun exposure in the presence of additional environmental stresses such as e.g. cold temperature during seasonal transition to winter in temperate climates. However, not all plant species respond in the same fashion. During summer-winter transition, overwintering annuals and biennials typically continue growing and maintain a high photosynthetic capacity. On the other hand, overwintering evergreens frequently cease growth and downregulate their photosynthetic capacity (Adams et al., 2002, 2004, this volume; Öquist and Huner, 2003; Huner et al., this volume). In such an inactivated state, the continuously light-absorbing leaves of evergreens must be protected by formidable photoprotection to prevent excess excitation energy from leading to a massive production of reactive oxygen species.

Chapter 4 Two Types of Photoprotective Thermal Dissipation

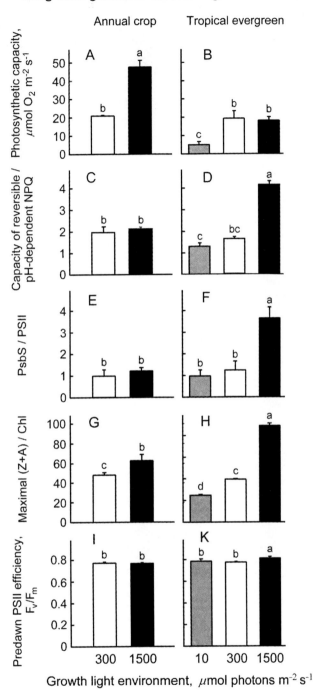

Fig. 1. Effect of growth light environment on photosynthetic capacity (light- and CO_2-saturated rate of photosynthetic oxygen evolution; A,B), NPQ capacity (C,D), PsbS level (E,F), maximal zeaxanthin and antheraxanthin (Z + A) level (G,H), and maximal PS II efficiency F_v/F_m at predawn (I,K) in the annual crop spinach versus the tropical evergreen *Monstera deliciosa*. Growth light environments were either continuous 10 μmol photons m^{-2} s^{-1} during the photoperiod or in a glasshouse where the peak PFD levels were 300 or 1500 μmol photons m^{-2} s^{-1} for several hours during midday. The parameters shown were measured as described in Ebbert et al. (2005; for PsbS), Demmig-Adams (1998; for NPQ capacity), and Adams et al. (2002; for all others). NPQ capacity and maximal Z + A levels were determined under conditions allowing only low electron transport rates (2%O_2, 98% N_2). Data shown are means ($\pm$SD, n = 3). Significant differences are indicated by different letters; $p < 0.001$ for photosynthetic capacity; $p < 0.05$ for NPQ; $p < 0.001$ for PsbS/PS II (ratio of PsbS level to the level of the oxygen-evolving complex); $p < 0.01$ for both Z + A (zeaxanthin and antheraxanthin) and F_v/F_m.

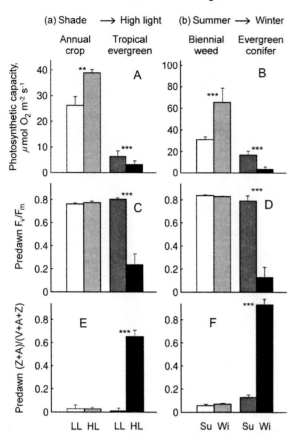

Fig. 2. Effects of photoinhibition caused by either a sudden transfer from low (LL) to high (HL) growth light intensity (A,C,E) or by summer (Su) to winter (Wi) transition (B,D,F), respectively, on photosynthetic capacity (light- and CO_2-saturated rate of oxygen evolution), maximal PS II efficiency, F_v/F_m, at predawn, and predawn zeaxanthin and antheraxanthin, $(Z + A)/(V + A + Z)$, levels in two pairs of species with short-lived versus evergreen leaves. Data on *Malva neglecta* and Douglas fir (B,D,F) are from Adams et al. (2002). Data on spinach (A,C), pumpkin (E), and *Monstera deliciosa* (A,C,E) were collected before and after five days upon transfer. LL was 10 and 100 μmol photons m^{-2} s^{-1} for *M. deliciosa* and crops, respectively, and HL was 1200 μmol photons m^{-2} s^{-1} for 10–12 hours per day each. V = violaxanthin. Data shown are means ($\pm$SD, n = 3). Significant differences are indicated by stars; ** for $p < 0.01$; *** for $p < 0.001$.

Upon sudden transfer of shade-grown plants to high light, mature leaves of annual crops increase their photosynthetic capacity and exhibit no (lasting) decreases in predawn PS II efficiency F_v/F_m (Fig. 2A,C). In contrast, mature leaves of a tropical evergreen respond to a sudden increase in growth light intensity with photoinhibition, as characterized by a further decrease in their already low photosynthetic capacity as well as a sustained depression of PS II efficiency at predawn (Fig. 2A,C). Very similar patterns are observed between summer and winter in another comparison of an evergreen versus a species with shorter-lived leaves (Fig. 2B,D). Once again, the (weed) species with the shorter-lived leaves shows an increased photosynthetic capacity, exhibits no sustained depressions of PS II efficiency at predawn, and thus shows no signs of photoinhibition in winter (Fig. 1B,D). In contrast, the overwintering evergreen exhibits strong decreases in both photosynthetic capacity and PS II efficiency at predawn and is thus strongly photoinhibited in the winter (Fig. 2B,D). Leaves of photoinhibited evergreens also show sustained overnight retention of high levels of Z + A that is absent in crops and weeds (Fig. 2E,F; Adams et al., this volume). The xanthophyll cycle is thus arrested in its high light state (as Z + A) in photoinhibited evergreens. This overnight retention of Z + A is associated with an overnight retention of high levels of NPQ in photoinhibited shade leaves of evergreens transferred to high light (Demmig-Adams et al., 1998; Ebbert et al., 2001). This is likely to also be the case in overwintering evergreens where NPQ levels, however, cannot be compared between summer and winter (Adams et al., this volume). Sustained NPQ in photoinhibited evergreens is not ΔpH-dependent, as was demonstrated in an overwintering evergreen (Verhoeven et al. 1998) and as is also the case for shade-grown *Monstera deliciosa* transferred to high light.

It is useful to clearly separate this ΔpH-independent form of NPQ that is sustained at warm temperatures, from the maintenance of ΔpH-dependent NPQ (in both annuals and evergreens) at low temperatures. In the latter, the trans-thylakoid pH gradient is maintained in darkness (Gilmore and Björkman, 1995) and overnight in the field (Verhoeven et al., 1998; Adams et al., 2002) at low temperatures.

B. PsbS and its Relatives

From the discussion of sustained NPQ in photoinhibited evergreens in the previous section, the question arises as to whether or not ΔpH-independent NPQ is facilitated by PsbS. PsbS facilitates ΔpH-dependent NPQ (Li et al., 2000, 2002b; Jung and Niyogi, this volume) and increases plant fitness under moderate, intermittent levels of excess light (Külheim et al., 2002) as shown in the model for short-lived species, *Arabidopsis*. PsbS is protonated during the development of excess light, and is thus thought to act as the pH sensor that triggers ΔpH-dependent, flexible NPQ (Li et al., 2002b; Jung and Niyogi, this volume). But does

PsbS also facilitate the ΔpH-independent, sustained NPQ under severely and continuously excessive light as seen in shade leaves of evergreens transferred to high light, or in overwintering evergreens? In photoinhibited shade leaves of the tropical evergreen *M. deliciosa* transferred to high light and exhibiting high levels of ΔpH-independent, sustained NPQ, the level of PsbS declined to about half of the already low levels present in the shade (cf. Fig. 1F). However, the shade leaves transferred to high light require an even greater level of photoprotective dissipation than that utilized by sun leaves (cf. Fig. 1). While the levels of PsbS decrease in photoinhibited *M. deliciosa* leaves, the levels of another protein are upregulated. This protein has an apparent molecular weight of 15-17 kDa, cross-reacts with an anti-HLIP (high-light-inducible protein) generated against *Arabidopsis* HLIP, and continues to increase for several days after sudden transfer to high light.

Similar features are observed in overwintering evergreens. The evergreen ground-cover bearberry (*Arctostaphylos uva-ursi*) occurs over a wide range of altitudes and light environments e.g. in the Front Range of the Rocky Mountains (Fig. 3). Bearberry becomes photoinhibited only at the highest altitude (of near 3000 m), as evidenced by the pronounced decreases in photosynthetic capacity and predawn PS II efficiency F_v/F_m (Fig. 3A,B) in this environment. The levels of PsbS, however, increase in winter only in the shaded, lower altitude bearberry population that does not become photoinhibited (Fig. 3C). Photoinhibited populations of bearberry at the highest altitude show an increase in another protein, or set of proteins, of about 15–17 kDa recognized by an antibody against pea ELIP (for early-light-inducible protein; Fig. 3D) as well as the same antibody (against *Arabidopsis* HLIP) that recognizes the light-stress-induced protein in *M. deliciosa*. Thus, the greatest increase in PsbS in winter occurs in non-photoinhibited leaves with high levels of reversible NPQ whereas the greatest increase in the ELIP/HLIP-reactive protein(s) takes place in photoinhibited leaves with high levels of sustained NPQ.

The proteins upregulated in photoinhibited leaves are likely to be members of the light-stress-responsive subgroup of light-harvesting proteins that includes PsbS (four-helix protein), ELIPs (three-helix proteins), HLIPs (one-helix proteins), as well as two-helix proteins (Jansson et al., 2000; Heddad and Adamska, 2000; He et al., 2001; Jansson, this volume). A transient upregulation of ELIPs was observed during transition from early to late spring in a Scots pine boreal forest (Ensminger et al., 2004) as well as during intermittent cooler temperatures in field-grown pea (Norén

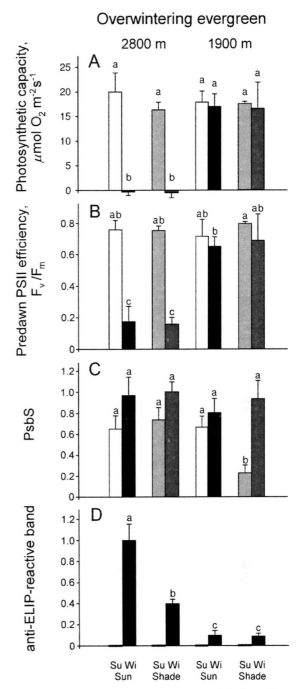

Fig. 3. Effect of season, altitude, and growth light environment on photosynthetic capacity (A), maximal PS II efficiency, F_v/F_m, at predawn (B), and levels of PsbS (C) as well as a 15–17 kDa, anti ELIP-reactive protein in bearberry (*Arctostaphylos uva-ursi* or kinnikinnick). Protein levels are given as relative optical densities. Analyses were as described in Ebbert et al. (2005). Data shown are means (±SD, n = 3). Significant differences are indicated by different letters; $p < 0.001$ for (A), $p < 0.05$ for (B), (C), and (D).

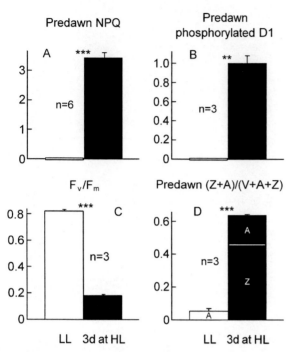

Fig. 4. Effects of a sudden transfer from low (LL) to high (HL) growth light intensity on (A) predawn levels of D1 protein phosphorylation (from relative optical densities of anti-phosphothreonine = anti-P-Thr), (B) zeaxanthin and antheraxanthin, $(Z + A)/(V + A + Z)$, and (C) thermal dissipation (NPQ) in *Monstera deliciosa* (data from Ebbert et al., 2001). LL was 8 μmol photons m^{-2} s^{-1} and HL was 700 μmol photons m^{-2} s^{-1} for 12 hours per day. NPQ = non-photochemical quenching; V = violaxanthin. Data shown are means ($\pm$SD, n = 3 or 6). Significant differences are indicated by stars; ** for $p < 0.01$; *** for $p < 0.001$.

et al., 2003). Overexpression of ELIPs in *Arabidopsis* increases photoprotective capacity but does not affect the levels of reversible NPQ of the qE type (Hutin et al., 2003). This, however, does not exclude the possibility that members of this subgroup other than PsbS may play a role in ΔpH-independent, sustained NPQ in photoinhibited evergreens. Do evergreens possibly possess additional genes of the above-mentioned subgroup of light stress-inducible proteins that are not present in short-lived species?

C. Sustained Protein Phosphorylation

As discussed above, sustained decreases in predawn PS II efficiency in photoinhibited leaves of evergreens are associated with a retention of high levels of Z + A and sustained high levels of thermal dissipation. These changes are, furthermore, accompanied by sustained phosphorylation of the D1 protein of the PS II core in both *M. deliciosa* transferred to a higher PFD (Fig. 4) and overwintering Douglas fir (Fig. 5). Evidence for sustained D1 phosphorylation is provided (Fig. 5) by the combination of (i) decreased levels of unphosphorylated D1 (detected by a specific antibody against unphosphorylated D1; Booij-James et al., 2002) and (ii) increased levels of phosphorylated D1 (detected by an anti-phosphothreonine antibody). A close correlation exists between sustained D1 phosphorylation and Z + A retention, sustained NPQ maintenance, and decreases in PS II efficiency (Fig. 6; Ebbert et al., 2005). The sustained high level of D1 phosphorylation in overwintering Douglas fir is, furthermore, associated with an increase in the level of the TLP40 protein (Fig. 5)

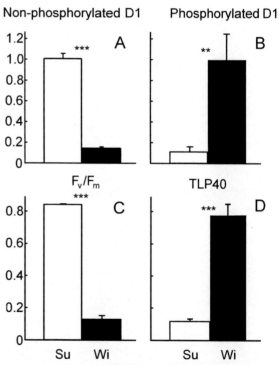

Fig. 5. Effects of summer (Su) to winter (Wi) transition on predawn levels of (A) of non-phosphorylated D1 (as anti-SP1), (B) phosphorylated D1 (as anti-P-Thr = anti-phosphothreonine), (C) maximal PS II efficiency, F_v/F_m, and (D) the inhibitor of PS II protein phosphatase, TLP40 (data from Ebbert et al., 2005). Protein levels are given as relative optical densities. Data shown are means ($\pm$SD, n = 3). Significant differences are indicated by stars; ** for $p < 0.01$; *** for $p < 0.001$.

Chapter 4 Two Types of Photoprotective Thermal Dissipation

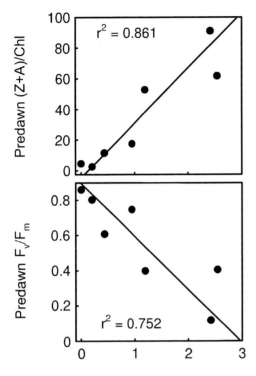

Fig. 6. Correlation between D1 phosphorylation state (from anti-phosphothreonine as relative optical densities) and predawn levels of zeaxanthin and antheraxanthin (Z + A) per Chl or predawn PS II efficiency F_v/F_m (data from Ebbert et al., 2005).

that is thought to act as an inhibitor of a PS II core protein phosphatase (Fulgosi et al., 1998; Vener et al., 1999; Rokka et al., 2000).

In summary, evergreen species have a greater propensity for photoinhibition than species with shorter-lived leaves. Photoinhibition is manifest as sustained decreases in predawn PS II efficiency and in the maximal rates of linear photosynthetic electron flow (photosynthetic capacity). This inactivation is accompanied by maintenance of high levels of thermal dissipation with a sustained retention of the xanthophylls Z + A required for dissipation. Under favorable summer conditions, high levels of Z + A and NPQ are observed exclusively at maximal light intensities. The continuous maintenance, and perhaps maximization of thermal dissipation in overwintering leaves of evergreens should provide protection against the formation of reactive oxygen species, such as singlet oxygen formed by transfer of excess excitation energy from chlorophyll to oxygen. In addition, the D1 protein of PS II cores is continuously maintained in the phosphorylated state that, under favorable conditions, is observed only at high light intensities (Rintamäki et al., 1997; Ebbert et al., 2001). PS II core and light-collecting antennae are thus maintained in their high light conformation throughout the day/night cycle in photoinhibited leaves. The increased levels of TLP40 in overwintering Douglas fir suggest that the sustained, elevated D1 phosphorylation represents an active acclimation to winter conditions.

The role of D1 phosphorylation, and PS II core protein phosphorylation, is not well understood (see Booij-James et al., 2002). Suggestions for its function include a possible role in regulating D1 turnover (Baena-Gonzalez et al., 1999) as well as a possible role in signal transduction. The results presented here suggest a possible function of sustained high D1/PS II core protein phosphorylation in photoprotection in overwintering leaves. One may speculate that sustained PS II protein phosphorylation could be involved in converting PS II cores to dissipating centers in which thermal dissipation may be facilitated by zeaxanthin. Some zeaxanthin had previously been localized to PS II cores of an evergreen at low temperature (Verhoeven et al., 1999). It has been suggested that over the course of evolution, reversible phosphorylation of D1 has replaced the multiple D1 genes present in cyanobacteria (Booij-James et al., 2002). In cyanobacteria, transition to a different D1 form at lower temperature was suggested to favor thermal energy dissipation within the PS II core (Sane et al., 2002). A conversion of PS II cores to photochemically inactive, dissipating centers (Matsubara et al., 2002) may counteract not only

Two types of photoprotective thermal dissipation

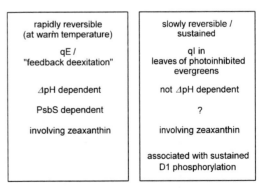

Fig. 7. Summary of the characteristics of the two different forms of thermal dissipation, i.e. flexible versus sustained dissipation.

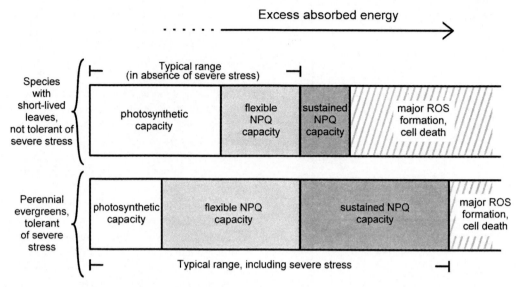

Fig. 8. Schematic depiction of the different capacities of species with short-lived leaves and perennial evergreens for photosynthesis as well as flexible (qE) and sustained (qI) thermal energy dissipation (NPQ). Despite their lower photosynthetic capacity for photosynthesis, evergreens' greater capacity for qE and, particularly qI, is speculated to afford a greater protection from massive reactive oxygen (ROS) production and cell death (including programmed cell death) under severe stress and thus severely excessive light.

singlet oxygen formation but also charge separation and transfer of electrons to oxygen, leading to superoxide formation in the chloroplast. Sustained D1 phosphorylation, F_v/F_m depression, and Z + A retention are observed irrespective of the type of stress that caused photoinhibition and are thus a characteristic feature of photoinhibited leaves of evergreens in general. Being more prominent in evergreens with their high stress resistance, photoinhibition may thus be one of the many "talents" of evergreens.

IV. Two Types of Thermal Energy Dissipation in Evergreens

In summary (Fig. 7), there are two distinct forms of thermal energy dissipation that differ in their properties, such as pH-dependence, level of flexibility (rapidly versus slowly reversible), and association with PS II core modification (D1 phosphorylation). Evergreens employ greater maximal levels of pH-dependent NPQ (qE type) and also have a much greater propensity for pH-independent NPQ (qI type) than shorter-lived species (Fig. 8). In addition, leaves of evergreen species exhibit very similar features of photoinhibition irrespective of what stress induced the photoinhibition. These features include pH-independent NPQ, xanthophyll cycle arrest (as Z + A), and PS II core modification by sustained phosphorylation. Future research should examine whether PS II core rearrangement (via sustained phosphorylation) and/or light-stress-inducible proteins related to PsbS are involved in sustained thermal dissipation or other photoprotective processes. These factors may contribute to the high stress resistance of evergreens and their ability to maintain green, yet completely photochemically inactive leaves for a whole season without suffering any apparent adverse effects.

Acknowledgments

We are indebted to Drs. Autar K. Mattoo (anti-SP1), Anna Sokolenko (anti-TLP40), Iwona Adamska (anti-ELIP), Christiane Funk (anti-PsbS used in spinach), Krishna K. Niyogi (anti-PsbS used in *M. deliciosa*), and Stefan Jansson (anti-HLIP) for making antibodies available to us. Technical assistance of Lisa Schaffer and David Mellman is also appreciated. This work has been supported by grants from the United States Department of Agriculture (Award No. 00-35100-9564), the National Science Foundation (Award No. IBN-9974620), and the Andrew W. Mellon Foundation (Award No. 20200747).

References

Adams WW III, Demmig-Adams B, Rosenstiel TN, Brightwell AK and Ebbert V (2002) Photosynthesis and photoprotection in overwintering plants. Plant Biology 4: 545–557

Adams WW III, Zarter CR, Ebbert V and Demmig-Adams B (2004) Photoprotective strategies of overwintering evergreens. BioScience 54: 41–49

Adams WW III, Zarter CR, Mueh KE, Amiard V and Demmig-Adams B (2005) Energy dissipation and photoinhibition: A continuum of photoprotection. In: Demmig-Adams B, Adams WW III and Mattoo AK (eds) Photoprotection, Photoinhibition, Gene Regulation, and Environment, pp 49–64. Springer, Dordrecht

Baena-Gonzalez E, Barbato R and Aro E-M (1999) Role of phosphorylation in the repair cycle and oligomeric structure of photosystem II. Planta 208: 196–204

Booij-James IS, Swegle WM, Edelman M and Mattoo AK (2002) Phosphorylation of the D1 photosystem II reaction center protein is controlled by an endogenous circadian rhythm. Plant Physiol 130: 2069–2075

Demmig-Adams B (1998) Survey of thermal energy dissipation and pigment composition in sun and shade leaves. Plant Cell Physiol 39: 474–482

Demmig-Adams B and Adams WW III (2002) Antioxidants in photosynthesis and human nutrition. Science 298: 2149–2153

Demmig-Adams B and Adams WW III (2003) Photoinhibition. In: Thomas B, Murphy D and B Murray B (eds) Encyclopedia of Applied Plant Science, pp 707–714. Academic Press, London

Demmig-Adams B, Moeller DL, Logan BA and Adams WW III (1998) Positive correlation between levels of retained zeaxanthin + antheraxanthin and degree of photoinhibition in shade leaves of *Schefflera arboricola* (Hayata) Merrill. Planta 205: 367–374

Ebbert V, Demmig-Adams B, Adams WW III, Mueh KE and Staehelin LA (2001) Correlation between persistent forms of zeaxanthin-dependent energy dissipation and thylakoid protein phosphorylation. Photosynth Res 67: 63–78

Ebbert V, Adams WW III, Mattoo AK, Sokolenko A and Demmig-Adams B (2005) Upregulation of a PSII core protein phosphatase inhibitor and sustained D1 phosphorylation in zeaxanthin-retaining, photoinhibited needles of overwintering Douglas fir. Plant Cell Environ, in press

Ensminger I, Sceshnikov D, Campbell DA, Funk C, Jansson S, Lloyd J, Shibistova O and Öquist G (2004) Intermittent low temperatures constrain spring recovery of photosynthesis in boreal Scots pine forests. Global Change Biol 10: 995–1008

Fulgosi H, Vener AV, Altschmied L, Herrmann R and Andersson B (1998) A novel multi-functional chloroplast protein: identification of a 40 kDa immunophilin-like protein located in the thylakoid lumen. EMBO J 17: 1577–1587

Gilmore AM and Björkman O (1995) Temperature-sensitive coupling and uncoupling of ATPase-mediated, nonradiative energy dissipation – similarities between chloroplasts and leaves. Planta 197: 646–654

He Q, Dolganov N, Björkman O and Grossman AR (2001) The high light-inducible polypeptides in *Synechocystis* PCC6803. J Biol Chem 276: 306–314

Heddad M and Adamska I (2000) Light stress-regulated two-helix proteins in *Arabidopsis thaliana* related to the chlorophyll *a/b*-binding family. Proc Natl Acad Sci USA 97: 3741–3746

Huner NPA, Ivanov AG, Sane PV, Pocock T, Król M, Balserus A, Rosso D, Savitch LV, Hurry VM and Öquist G (2005) Photoprotection of photosystem II: reaction center quenching versus antenna quenching. In: Demmig-Adams B, Adams WW III and Mattoo AK (eds) Photoprotection, Photoinhibition, Gene Regulation, and Environment, pp 155–173. Springer, Dordrecht

Hutin C, Nussaume L, Moise N, Moya I, Kloppstech K and Havaux M (2003) Early light-induced proteins protect *Arabidopsis* from photooxidative stress. Proc Natl Acad Sci USA 100: 4921–4926

Jansson J (2005) A Protein Family Saga: From Photoprotection to Light-harvesting (and Back?) In: Demmig-Adams B, Adams WW III and Mattoo AK (eds) Photoprotection, Photoinhibition, Gene Regulation, and Environment, pp 145–153. Springer, Dordrecht

Jansson S, Andersson J, Kim SJ and Jackowski G (2000) An *Arabidopsis thaliana* protein homologous to cyanobacterial high-light-inducible proteins. Plant Mol Biol 42: 345–351

Jung H-S and Niyogi KK (2005) Molecular analysis of photoprotection and photosynthesis. In: Demmig-Adams B, Adams WW III and Mattoo AK (eds) Photoprotection, this volume. Springer, Berlin

Külheim C, Agren J and Jansson S (2002) Rapid regulation of light harvesting and plant fitness in the field. Science 297: 91–93

Li X-P, Björkman O, Shih C, Grossman AR, Rosenquist M, Jansson S and Niyogi KK (2000) A pigment-binding protein essential for regulation of photosynthetic light harvesting. Nature 403: 391–395

Li X-P, Müller-Moulé P, Gilmore AM and Niyogi KK (2002a) PsbS-dependent enhancement of feedback de-excitation protects photosystem II from photoinhibition. Proc Natl Acad Sci USA 99: 15222–15227

Li X-P, Phippard A, Pasari J and Niyogi KK (2002b) Structure-function analysis of photosystem II subunit S (PsbS) in vivo. Funct Plant Biol 29: 1131–1139

Matsubara S, Gilmore AM, Ball MC, Anderson JM and Osmond CB (2002) Sustained downregulation of photosystem II in mistletoes during winter depression of photosynthesis. Funct Plant Biol 29: 1157–1169

Niyogi KK (2000) Safety valves for photosynthesis. Curr Opin Plant Biol 3: 455–560

Norén H, Svensson P, Stegmark R, Funk C, Adamska I and Andersson B (2003) Expression of the early light-induced protein but not the PsbS protein is influenced by low temperature and depends on the developmental stage of the plant in field-grown pea cultivars. Plant Cell Environ 26: 245–253

Öquist G and Huner NPA (2003) Photosynthesis of overwintering evergreen plants. Ann Rev Plant Biol 54: 329–355

Powles SB (1984) Photoinhibition of photosynthesis induced by visible light. Ann Rev Plant Physiol 35: 15–44

Rintamäki E, Salonen M, Suoranta U-M, Carlberg I, Andersson B and Aro E-M (1997) Phosphorylation of light-harvesting complex II and Photosystem II core proteins shows different irradiance-dependent regulation in vivo – Application of phosphothreonine antibodies to analysis of thylakoid phosphoproteins. J Biol Chem 272: 30476–30482

Rokka A, Aro E-M, Herrmann RG, Andersson B and Vener AV (2000) Dephosphorylation of photosystem II reaction center proteins in plant photosynthetic membranes as an immediate response to abrupt elevation of temperature. Plant Physiol 123: 1525–1535

Sane PV, Ivanov AG, Sveshnikov D, Huner NPA and Öquist G (2002) A transient exchange of the photosystem II reaction center protein D1:1 with D1:2 during low temperature stress of Synechococcus sp. PCC 7942 in the light lowers the redox potential of Q_B. J Biol Chem 277: 32739–32745

Vener AV, Rokka A, Fulgosi H, Andersson B and Herrmann RG (1999) A cyclophilin-regulated PP2A-like protein phosphatase in thylakoid membranes of plant chloroplasts. Biochemistry 38: 14955–14965

Verhoeven AS, Adams WW III and Demmig-Adams B (1998) Two forms of sustained xanthophyll cycle-dependent energy dissipation in overwintering *Euonymus kiautschovicus*. Plant Cell Environ 21: 893–903

Verhoeven AS, Adams WW III, Demmig-Adams B, Croce R and Bassi R (1999) Xanthophyll cycle pigment localization and dynamics during exposure to low temperatures and light stress in *Vinca major*. Plant Physiol 120: 727–737

Chapter 5

Energy Dissipation and Photoinhibition: A Continuum of Photoprotection

William W. Adams III*, C. Ryan Zarter, Kristine E. Mueh, Véronique Amiard
and Barbara Demmig-Adams
*Department of Ecology & Evolutionary Biology, University of Colorado, Boulder, CO
80309-0334, USA*

Summary	49
I. Introduction	50
II. Characteristics of Energy Dissipation and Photoinhibition	50
A. Flexible Energy Dissipation	50
B. Photoinhibition	52
III. Photoprotection and Photoinhibition in Winter	53
IV. Does Photoinhibition Limit the Carbon Available to the Plant?	55
V. An Integrated View of Photoprotection	59
Acknowledgments	61
References	61

Summary

The photosynthetic apparatus is exquisitely adapted to capture light energy and convert it into reduced carbon compounds while also protecting against the potential deleterious effects of excessive excitation energy. The latter is achieved through fine regulation of thermal energy dissipation over multiple time scales and in response to many different environmental stresses. Over short time scales in the absence of additional stress, control is exerted through pH regulation of the enzymatic conversion of violaxanthin to zeaxanthin (and its return to violaxanthin) and engagement of zeaxanthin in thermal energy dissipation. Under more extreme exposure to excess light (transfer of shade leaves to high light or the imposition of additional stresses in the presence of high light), greater levels of zeaxanthin are retained and may also be maintained in a dissipative configuration even in darkness. Engagement of zeaxanthin in thermal energy dissipation lowers the maximal efficiency of photosystem II (PS II) as the excess excitation energy is diverted away from the reaction centers and harmlessly released as heat. Thus, maximal PS II efficiency exhibits decreases and increases with varying degrees of light absorption. Under prolonged and/or pronounced exposure to excess light, maximal PS II efficiency can furthermore exhibit nocturnally sustained decreases as the potential for photoprotective zeaxanthin-dependent energy dissipation is maintained. Zeaxanthin-dependent energy dissipation that is sustained at moderate temperatures is also typically accompanied by downregulation of photosynthesis, including photosynthetic electron transport. Decreases in photosynthetic electron transport presumably lower the likelihood of electrons reducing molecular oxygen to superoxide, and sustained zeaxanthin-dependent energy dissipation mitigates the formation of singlet excited oxygen. Thus, while sustained decreases in maximal PS II efficiency and photosynthetic capacity are key characteristics of photoinhibition, they are also the features that provide powerful photoprotection against the formation of toxic reactive oxygen species.

---

*Author for correspondence, email: william.adams@colorado.edu

I. Introduction

The photosynthetic portions of all plants exposed to sunlight regularly face the potential problem of excess excitation energy. Although a decrease in photosynthesis was recognized as one response to excess light more than a century ago (see Osmond and Förster, this volume) and subsequently termed photoinhibition fifty years ago (Kok, 1956), the more ubiquitous occurrence of photoprotective energy dissipation as a common response to even moderately excessive light has only been recognized for the past decade-and-a-half. In this Chapter, we summarize information about how the two phenomena are inextricably linked, and represent photoprotective responses along a continuum of adjustments in response to excess light.

II. Characteristics of Energy Dissipation and Photoinhibition

A. Flexible Energy Dissipation

Under physiologically normal conditions, dark adaptation (e.g. a night of darkness) returns the photosynthetic apparatus to its most oxidized and relaxed state in leaves. If the interval of darkness is relatively short, progression into the oxidized state can be facilitated with far-red radiation (Schreiber et al., 1984; Adams and Demmig-Adams, 2004). In such a state, the efficiency of excitation energy transfer within PS II light-harvesting antennae and of photochemical charge separation within PS II reaction centers is maximal. This state is reflected in an elevated ratio of variable to maximal chlorophyll fluorescence F_v/F_m (typically 0.78–0.87 in C3 and CAM plants, but lower in C4 plants) that is emitted primarily from photosystem II (PS II), and a photon yield of photosynthesis that is also maximal (0.106 O_2 evolved per absorbed photon, but lower in C4 plants) (Kitajima and Butler, 1975; Björkman and Demmig, 1987; Adams et al., 1990; Adams and Demmig-Adams, 2004). This situation is illustrated by the light response curve depicted in Fig. 1A, where PS II efficiency F_v/F_m is maximal at 0 μmol photons m^{-2} s^{-1}, and the slope of the light response curve of photosynthesis in the light-limited region (= photon yield) is steep (Björkman and Demmig, 1987).

Increasing light levels lead to a number of responses within the leaf (Fig. 1A). Photosynthesis increases proportionally to the increases in PFD until its rate begins to saturate. As saturation is approached, the concentration of protons in the thylakoid lumen increases, activating the enzyme violaxanthin de-epoxidase that converts violaxanthin (V, not shown) into antheraxanthin (A) and zeaxanthin (Z) (Yamamoto, 1979, this volume; Hager, 1980; Demmig-Adams et al., 1989a). There is also a concurrent protonation of specific sites on the PsbS protein, resulting in a conformational change that presumably facilitates the engagement of zeaxanthin (and antheraxanthin) in photoprotective thermal energy dissipation (Li et al., 2000, 2002; Ma et al., 2003; Jung and Niyogi, this volume). The latter can be assessed through changes in nonphotochemical quenching (NPQ) of chlorophyll fluorescence calculated as $F_m/F_m'-1$ (Bilger and Björkman, 1990). A strong linear correlation has been demonstrated between the foliar content of Z + A and the total level of NPQ during active engagement (Bilger and Björkman, 1991, 1994; Demmig-Adams and Adams, 1994a,b, 1996). Thus, as a proportionally greater fraction of the absorbed light cannot be utilized in photosynthesis at higher light levels, there is a compensatory increase in the level of Z + A, which is then engaged in dissipation of the excess excitation energy as heat. There is furthermore a concomitant decrease in maximal PS II efficiency as the level of energy dissipation increases (Adams et al., 1989, 1995, 1999; Björkman and Demmig-Adams, 1994; Demmig-Adams et al., 1995, 1996a; Demmig-Adams and Adams, 1996; Demmig-Adams et al., this volume), as predicted from the analysis by Kitajima and Butler (1975), reflecting a diversion of the excess excitation energy away from PS II reaction centers.

For a high light-grown leaf, such increases in thermal energy dissipation and decreases in maximal PS II efficiency are flexible; they are rapidly reversible upon transition to non-excessive light or darkness as the deprotonation of PsbS presumably leads to a rapid disengagement of Z + A from their thermal dissipating function and a return to high PS II efficiency in limiting light. Hence NPQ is designated as NPQ$_{flex}$ in Fig. 1A. Upon such transitions, however, Z + A are converted much more slowly to V by zeaxanthin epoxidase, and under these conditions the linear correlation between the amount of Z + A and the level of thermal energy dissipation no longer exists. On the other hand, retention of Z + A under such conditions permits a more rapid engagement of energy dissipation upon a subsequent exposure to excessive light (Demmig-Adams et al., 1989b; Barker et al., 2002), since it only requires the rapid protonation of PsbS without the (slower) enzymatic conversion of V to Z + A. This rapid modulation of thermal energy dissipation is particularly physiologically relevant e.g. under conditions of intermittent cloud cover or in the understory of a forest (Fig. 2A–D; Adams et al., 1999). In fact, reproductive fitness

Chapter 5 Photoprotection and Photoinhibition

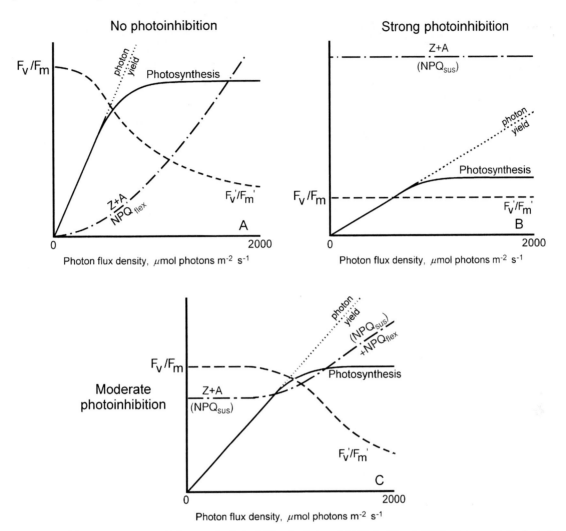

Fig. 1. Idealized light response curves of photosynthesis (solid lines), photon yield (dotted lines), PS II efficiency (F_v/F_m in darkness, F_v'/F_m' in light; dashed lines), and levels of zeaxanthin + antheraxanthin (Z + A) and of energy dissipation activity (depicted as nonphotochemical quenching of chlorophyll fluorescence, or NPQ) (mixed dashed line) in a system in which there is no photoinhibition (A), a system with strong photoinhibition (B), and one in which there is moderate photoinhibition (C). For the situation with no photoinhibition, NPQ is rapidly and completely reversible and is denoted as NPQ_{flex}. For moderate and strong photoinhibition, NPQ can be sustained and show no reversibility and is denoted as NPQ_{sus}. NPQ_{sus} is bracketed by parentheses because this value cannot, under most circumstances, be calculated correctly (see text).

was shown to be lower in the absence of such rapidly modulated thermal energy dissipation in PsbS-deficient *Arabidopsis* mutants (Külheim et al., 2002). The term "dynamic photoinhibition" (Osmond, 1994; Osmond and Grace, 1995; Osmond and Förster, this volume) has been adopted by some to describe the transient and rapidly reversible decreases in maximal PS II efficiency that result from this photoprotective energy dissipation process, even though rates of photosynthetic electron transport remain maximal (e.g. Adams et al., 1999) and photosynthesis is not inhibited.

In sun-exposed leaves under otherwise favorable conditions, the xanthophyll cycle conversion state [(Z + A)/(V + A + Z)], level of thermal energy dissipation activity (NPQ), and maximal PS II efficiency change in a very predictable manner over the course of the day, paralleling increases and decreases in PFD (Figs. 2E–L). However, the magnitude of these changes differs among species depending on the proportion of the absorbed excitation energy that is used for photosynthesis. For instance, the rapidly growing annual mesophyte sunflower has a high light- and CO_2-saturated rate of photosynthetic oxygen evolution of typically 50 to 60 μmol O_2 m^{-2} s^{-1}, whereas the evergreen shrub *Euonymus kiautschovicus* utilizes only half as much of the midday light with a photosynthetic

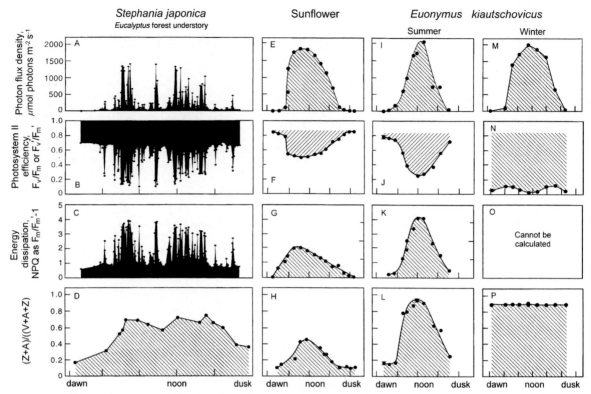

Fig. 2. Changes in photon flux density, PS II efficiency (F_v/F_m predawn and F_v'/F_m' during daylight illumination), level of photoprotective energy dissipation calculated as NPQ, and the level of zeaxanthin + antheraxanthin (depicted as a fraction of the total xanthophyll cycle carotenoids) in *Stephania japonica* growing in the understory of a forest in Australia, in sunflower growing in full sunlight, and in full-sunlight exposed *Euonymus kiautschovicus* in summer and in winter. Data are redrawn from Adams et al. (1999), Demmig-Adams et al. (1996b), and Verhoeven et al. (1998).

capacity that ranges between 20 and 30 μmol O_2 m^{-2} s^{-1} (Adams et al., 1992, 2002; Demmig-Adams et al., 1992). As a consequence of these different levels of photosynthetic utilization of absorbed light energy, *Euonymus kiautschovicus* converts a greater proportion of the xanthophyll cycle to Z + A (Fig. 2L vs. 2H) and employs higher levels of photoprotective thermal energy dissipation (Fig. 2K vs. 2G), resulting in greater midday depressions in maximal PS II efficiency (Fig. 2J vs. 2F) compared to sunflower. Or, as some prefer to view it, the more stress-tolerant evergreen species experiences a greater level of "dynamic photoinhibition".

B. Photoinhibition

Characteristics of photoinhibition include several phenomena that have been the subject of intense study for the past several decades (see Powles, 1984; Kyle et al., 1987; Barber and Andersson, 1992; Aro et al., 1993; Baker and Bowyer, 1994; Long et al., 1994; Adams et al., 1995, 2002; Barber, 1995; Osmond and Grace, 1995; Osmond et al., 1997; Melis, 1999; Telfer et al., 1999; Marshall et al., 2000; Demmig-Adams and Adams, 2003; Adir et al., 2003). Under physiologically relevant conditions, such phenomena typically arise in response to conditions of prolonged or pronounced excess light absorption; either exposure of low light-grown plants or leaves to high light, or exposure to light in the presence of one or more additional stresses. Several of these characteristics are depicted in Fig. 1B in comparison to the non-photoinhibited state shown in Fig. 1A.

The most frequently assessed parameter in this regard is the maximal efficiency of PS II, F_v/F_m. This arises not because F_v/F_m is necessarily the most relevant parameter for understanding photoinhibition, but because it is easily and rapidly determined with little perturbation to the system. In fact, decreases in F_v/F_m can arise from several different changes in the photosynthetic apparatus (Kitajima and Butler, 1975; Björkman, 1987; Adams and Demmig-Adams, 2004). Two common changes leading to decreased levels of F_v/F_m are increases in photoprotective zeaxanthin-dependent thermal dissipation and decreases in the competence of PS II reaction centers to carry out photochemical charge separation. In intact leaves under physiologically relevant conditions, there tends to be a strong component of the former.

Strong photoinhibition is characterized by a PS II efficiency (F_v/F_m) as low as 0.1 and 0.2 in a dark-adapted, fully relaxed state, which corresponds to 12 to 24% of the F_v/F_m observed in non-photoinhibited leaves. From this low efficiency state, there is typically no or little further decrease in PS II efficiency F_v'/F_m' upon exposure to light (Fig. 1B). Another measure of the efficiency of energy conversion in photosynthesis, the photon yield (or the slope of the linear portion of the light response curve of photosynthesis), is also typically lower in photoinhibited leaves (Figs. 1B vs. 1A) (i.e. there is a strong correlation between F_v/F_m and the photon yield of photosynthesis; Björkman and Demmig, 1987; Demmig and Björkman, 1987; Adams et al., 1990). Following the light response curve of photosynthesis up to light saturation reveals that the capacity for photosynthesis is also typically lower in photoinhibited leaves compared to non-photoinhibited leaves (Figs. 1B vs. 1A). This feature is not what is expected from a simple increase in thermal dissipation. Instead, an inactivation of PS II photochemistry may contribute to this photoinhibition and likely involves inactivation and/or disassembly of PS II cores (especially the D1 protein; see Edelman and Mattoo, Häder, Huner et al., Nishiyama et al., Yokthongwattana and Melis, this volume). In addition, strongly photoinhibited leaves (Fig. 1B) have most of the pool of the xanthophyll cycle carotenoids retained as zeaxanthin, and high levels of sustained thermal energy dissipation (although calculation of NPQ in leaves that are experiencing photoinhibition over longer time periods [days to months] can be problematic in the absence of a valid control value for F_m; see Adams and Demmig-Adams 1995, 2004; Adams et al., 1995a, b). Thus strongly photoinhibited leaves are apparently in a photochemically inactive, albeit highly photoprotected state, diverting the majority of absorbed excitation energy into zeaxanthin-dependent thermal energy dissipation.

Moderate levels of photoinhibition typically involve characteristics that are intermediate between those for non-photoinhibited leaves and strongly photoinhibited leaves (Fig. 1C). Dark-adapted PS II efficiency F_v/F_m may range between 0.4 and 0.7, and PS II efficiency in the light F_v'/F_m' shows further decreases as the light becomes more excessive. Both photon yield and photosynthetic capacity may be lower than those of non-photoinhibited leaves. Furthermore, intermediate levels of Z + A are typically retained in darkness and remain engaged in a state primed for thermal energy dissipation. Thus, there is a certain level of sustained NPQ in darkness (which may be impossible to quantify; see above), and additional increases in NPQ (characterized by decreases in F_v'/F_m', but impossible to quantify as NPQ since a true F_m control cannot be obtained) occur in response to increasing levels of excess light. At midday or during exposure to light, moderately photoinhibited leaves therefore apparently rely on a combination of sustained and rapidly reversible zeaxanthin-dependent thermal energy dissipation for photoprotection.

III. Photoprotection and Photoinhibition in Winter

In the midst of winter, leaves of many evergreen species can be found in various states of photosynthetic down-regulation (for reviews, see Adams et al., 2001a, 2002, 2004; Öquist and Huner, 2003). Such species typically cease growth during the autumn and may or may not exhibit decreases in the capacity for photosynthesis depending on species, light environment, and the severity of the conditions to which the plants are exposed. Most do, however, exhibit nocturnally sustained depressions in PS II efficiency that are associated with the retention of Z + A (e.g. Figs. 2N, 2P). In fact, the close association between Z + A level and PS II efficiency that is similar in leaves transiently exposed to high light under otherwise favorable conditions (experiencing "dynamic photoinhibition") and in photoinhibited leaves/needles in the winter (compare Figs. 3A and 3B), suggests that the latter are actually in the same highly protected state.

A portion of winter-induced, nocturnally sustained decreases in PS II efficiency in evergreen species is rapidly reversible upon warming (see Verhoeven et al., 1998). This portion is presumably due to maintenance of thylakoid lumen acidification at low temperature that keeps zeaxanthin in an engaged state (see Demmig-Adams et al., 1996b; Gilmore, 1997). The remaining portion of the decrease in PS II efficiency reverses only slowly over days at warmer temperatures (Adams et al., 1995; Verhoeven et al., 1996) and does not seem to involve a pH gradient across the thylakoid membrane (Verhoeven et al., 1998; Gilmore and Ball, 2000). Herbaceous annual and biennial species that maintain leaves during the winter exhibit only small nocturnally sustained decreases in PS II efficiency (see Fig. 3B, spinach), and such minor depressions reverse rapidly upon warming (Adams et al., 1995b; Verhoeven et al., 1999). In contrast to evergreen species that exhibit no change or a decrease in photosynthetic capacity in the winter compared to the summer, the herbaceous species

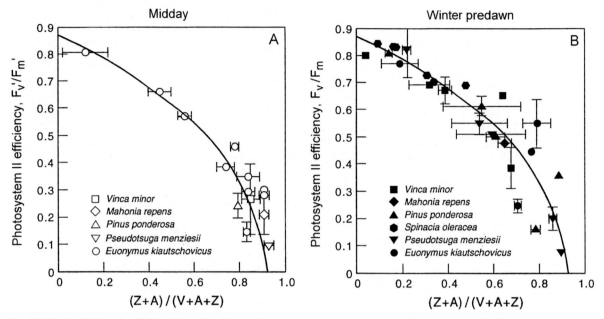

Fig. 3. Relationship between the level of zeaxanthin + antheraxanthin (depicted as a fraction of the total xanthophyll cycle carotenoids) and photosystem II efficiency as either F_v'/F_m' during exposure to light (A) or as F_v/F_m that is sustained during winter in the predawn darkness (B). The midday values for *Euonymus kiautschovicus* were determined from 10 different leaves of varying exposure to light during mild conditions in the summer (open circles), whereas all other values were determined during winter. The line fitted through the data is identical for each panel. Error bars represent standard deviations. Data redrawn from Adams et al. (1995a).

that retain their leaves typically exhibit an upregulation of photosynthetic capacity under winter conditions (see reviews of the literature in Adams et al., 2001a, 2002, 2004; Öquist and Huner, 2003; Huner et al., this volume).

In evergreen species that downregulate photosynthesis during the winter, there is a strong correlation between the capacity for photosynthesis and predawn PS II efficiency within a species both during the winter and during the period of transition from winter through spring (closed circles in Fig. 4). On the other hand, there is no correlation between the two parameters under favorable conditions during the summer. Different needles or leaves can have very different capacities for photosynthesis while all have a high PS II efficiency (open circles in Fig. 4). This is true among sun-exposed needles and leaves, between sun (high capacity) and shade (low capacity) needles/leaves of the same species, and among different species in which the capacity for photosynthesis varies greatly with differences in sink activity (utilization of the carbohydrates produced through photosynthesis for growth, storage, and respiration) but where PS II efficiency is equally high.

The strong correlation between photosynthetic capacity and PS II efficiency in evergreen species under winter conditions and during the transition from winter to spring suggests a tight link between the capacity for utilizing light energy for photosynthetic electron transport and the capacity for dissipating excess excitation energy thermally. Under favorable conditions, this involves pH modulation of (1) the activity of the enzymes responsible for converting violaxanthin to zeaxanthin and zeaxanthin to violaxanthin (see Yamamoto, this volume) and (2) PsbS, the protein facilitating chlorophyll de-excitation by zeaxanthin and antheraxanthin in energy dissipation (Li et al., 2000, 2002; see also Jung and Niyogi, this volume). Such regulation is depicted schematically in Fig. 5A, with violaxanthin present as the predominant carotenoid of the xanthophyll cycle and the PsbS "valve" closed during the night. During the winter, on the other hand, evergreen species may enter a downregulated state in which photosynthetic electron transport is greatly diminished, zeaxanthin is retained in large amounts all of the time, and Z is continuously engaged in a state that can facilitate thermal energy dissipation whenever light energy is absorbed by the light-harvesting chlorophyll (Fig. 5B). We have recently suggested that the downregulation of photosynthetic electron transport coupled with sustained engagement of zeaxanthin in energy dissipation may provide an effective means of preventing the formation of the reactive oxygen species superoxide and singlet oxygen

Chapter 5 Photoprotection and Photoinhibition

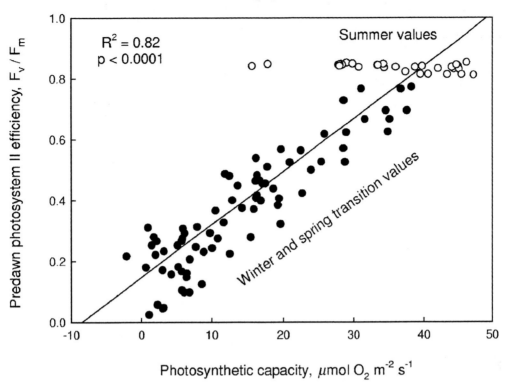

Fig. 4. Relationship between the light- and CO_2-saturated capacity for photosynthetic oxygen evolution (determined at 25°C) and photosystem II efficiency determined predawn in needles of lodgepole pine growing at approximately 3000 m in the Rocky Mountains of Colorado. Values were determined in the summers of 2001 through 2003 (open circles) or during the winter through spring transition of 2003 (closed circles). The photosynthetic capacities during the summer ranged from 15.6 to 47.2 μmol O_2 m^{-2} s^{-1}, whereas photosystem II efficiency varied little (mean summer value ± SD of F_v/F_m = 0.836 ± 0.013, n = 30). Unpublished data of CR Zarter, WW Adams, and B Demmig-Adams.

(Demmig-Adams and Adams, 2003; Adams et al., 2004). Furthermore, several studies have suggested that proteins related to PsbS, such as early light-inducible proteins (ELIPs) or high light-inducible proteins (HLIPs), may play a role in xanthophyll-dependent photoprotection under more severe conditions (Norén et al., 2003; Ensminger et al., 2004; Demmig-Adams et al., this volume).

IV. Does Photoinhibition Limit the Carbon Available to the Plant?

There have been many claims that photoinhibition is likely to lead to decreases in carbon gain and reduced plant productivity, due either to photodamage or photoinactivation of the D1 protein and the attendant decrease in electron transport capacity or even to sustained photoprotective energy dissipation that continues to siphon off absorbed energy upon a return to non-excessive light conditions (e.g. Ball et al., 1991; Long et al., 1994; Melis, 1999; Werner et al., 2001; Zhu et al., 2004). This view has persisted for many years due primarily to the perspective that "damage" to D1 is "suffered" during photoinhibition (see Adir et al., 2003 for an historical review; Häder, Huner et al., Nishiyama et al., Yokthongwattana and Melis, this volume) and that this "lesion" of the photosynthetic apparatus, or "impairment" of photosynthesis, must limit the supply of carbohydrates to the rest of the plant. Thus, there is considerable support for the view that photoinhibition is something that should be protected against (e.g. Endo and Asada, this volume), when in reality photoinhibition may be a means by which plants sustain photoprotection.

While there is little doubt that D1 can be inactivated by reactive oxygen species (ROS) under strongly excessive light, this may reflect a photoprotective

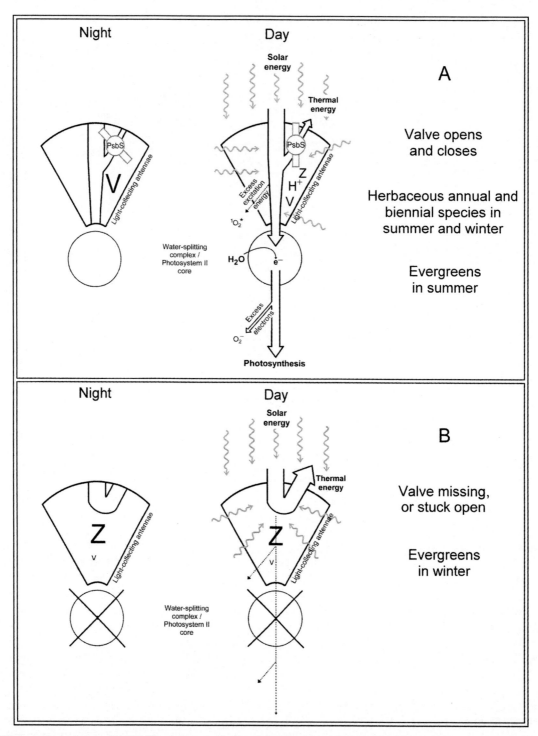

Fig. 5. Schematic depiction of (A) flexible photoprotection involving diurnal conversion of violaxanthin (V) to zeaxanthin (Z) and its engagement in energy dissipation through the protonation of the PsbS protein, thus minimizing formation of singlet excited oxygen, and (B) sustained photoprotection involving the nocturnal retention of zeaxanthin in a configuration engaged for energy dissipation and the downregulation of electron transport to minimize the formation of superoxide.

Chapter 5 Photoprotection and Photoinhibition

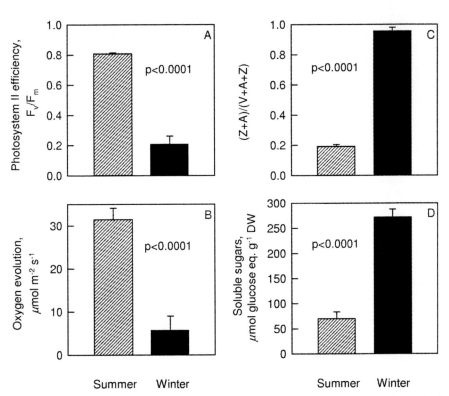

Fig. 6. Summer and winter levels of (A) predawn photosystem II efficiency, (B) light- and CO_2-saturated capacity for photosynthetic oxygen evolution (determined at 25°C), (C) predawn conversion state of the xanthophyll cycle $(Z+A)/(V+A+Z)$, and (D) total soluble sugars in leaves of *Vinca minor* growing in full sunlight. Standard deviations are depicted, and all parameters were statistically different at the $p < 0.001$ level as determined from the Student's t-test. Data from Adams et al. (2001b, 2002).

downregulation of its photochemical activity via oxidative modification rather than damage (Sopory et al., 1990). In the general field of oxidative stress physiology, the classic view of protein damage by ROS and subsequent protein repair has been replaced by the realization that the proteins most susceptible to oxidation by ROS are those in the service of cellular regulation and/or signal transduction (Weindruch and Sohal, 1997; Maher and Schubert, 2000).

In addition, a number of studies purporting to show that photoinhibition involves photodamage to the D1 protein have relied on the utilization of inhibitors of chloroplast-encoded protein synthesis (see, e.g., the reviews by Melis, 1999; Nishiyama et al., this volume). However, these inhibitors can have effects beyond the simple inhibition of D1 synthesis. It is quite reasonable to assume that the photosynthetic apparatus is induced to undergo adjustments in the presence of such inhibitors that would not otherwise occur. For instance, chloramphenicol can inhibit photosynthesis directly (Okada et al., 1991), and lincomycin and streptomycin have both been found to influence calcium channels and to alter transmembrane ion gradients (Fiekers et al., 1979; Prior et al., 1990). Furthermore, some chloroplast-encoded protein synthesis inhibitors can influence the operation of the xanthophyll cycle and the engagement of zeaxanthin in energy dissipation. Lincomycin inhibits the recovery of PS II efficiency from winter photoinhibition and from high light photoinhibitory treatment, but this appears to be due to an inhibition of the disengagement of zeaxanthin from sustained photoprotective energy dissipation (Verhoeven et al., 1998; Bachmann et al., 2004) rather than to an inhibition of D1 synthesis (Bachmann et al., 2004).

The initial characterization of the impact of light on the D1 protein did not invoke the view of "damage" to the protein at all. Instead, it was simply recognized that this protein is turned over rapidly (Mattoo et al., 1984; Edelman and Mattoo, this volume). One of the hallmarks of proteins that serve as control points in regulation is that they turn over rapidly, thus permitting rapid adjustment of their levels. There is no question that, upon exposure to excess light levels for prolonged periods, the D1 protein becomes inactivated, its levels can decrease, and the capacity for photosynthesis can decrease in turn. However, is this a response that has negative consequences for the plant, and if it could be prevented, would plant productivity actually be higher? Or is it an appropriate response of the plant to a situation where, due to a lack of opportunity for growth imposed by unfavorable environmental conditions, the demand for carbohydrates is either very low or a sufficient supply of carbohydrates to meet the maintenance and growth demands of the plant can continue to be generated while permitting photosynthesis to be downregulated in order to prevent excessive damage due to the generation of toxic reactive oxygen species (Fig. 5)?

Mesophytic species (annual and biennial crops and herbaceous species in general) exhibit little propensity for photoinhibition under high light or during winter conditions, due to high rates of utilization of the absorbed light for photosynthesis and continued growth (Adams et al., 2001a, 2002, 2004; Öquist and Huner, 2003; Huner et al., this volume). On the other hand, evergreen species, which typically cease growth in the autumn, readily experience photoinhibition in high light during the winter (Adams et al., 2001a, 2002, 2004; Öquist and Huner, 2003; Demmig-Adams et al., this volume). Furthermore, exposure of evergreen species to high levels of CO_2 throughout the winter, increasing the source to sink ratio, also result in a greater downregulation of photosynthesis (Hymus et al., 1999; Roden et al., 1999). Winter-induced photoinhibition always involves sustained decreases in PS II efficiency, and often decreases in photosynthetic capacity as well. However, overwintering leaves and needles of evergreen species also contain high levels of soluble carbohydrates (as cryoprotectants). For the example shown in Fig. 6, both photosynthetic efficiency and photosynthetic capacity were downregulated to an extreme degree in the winter, and yet the level of soluble carbohydrates was four times greater in the winter compared to the summer. Does the photoinhibition experienced by this plant limit the availability of carbohydrates? Or does the demand for carbohydrates diminish to such an extent (cessation of growth under the short, cold days of winter) that the plant can supply all of the carbohydrates that are necessary for maintenance activities and cryoprotection with a much lower rate of photosynthesis?

One might argue that, in response to low temperatures, carbohydrates are diverted and maintained in the tissues for cryoprotection and that this diversion, coupled with the lower rates of photosynthesis, does limit the supply of carbohydrates that might otherwise be available to these plants to continue growing in the freezing and subfreezing conditions of winter. However, those plants experiencing the greatest levels of photoinhibition during winter were found to exhibit the greatest rates of growth during the subsequent spring (Blennow et al., 1998; Roden et al., 1999). Furthermore, accumulation of carbohydrates also occurs in leaves of plants under photoinhibitory conditions that do not involve a particular requirement for cryoprotection or osmotic adjustment. A good correlation between the level of photoinhibition (decreases in PS II efficiency) and increased starch accumulation was found in two species of *Eucalyptus* in response to excess light under conditions of water stress and/or high temperatures (Roden and Ball, 1996). Starch also accumulates in leaves subjected to the classic photoinhibitory transfer of shade plants to high light, as illustrated in the following example.

The leaves of *Monstera deliciosa* (a neotropical evergreen hemi-epiphyte) from plants grown under low light (10 μmol photons m^{-2} s^{-1}) and then transferred to high light (700 μmol photons m^{-2} s^{-1} for 10 h per day) experienced photoinhibition as determined from a 50% decrease in light-saturated electron transport during the first day (not shown) and nocturnally-sustained depressions in PS II efficiency to below 0.6 (Fig. 7A). Nonetheless, leaf carbohydrate content increased fourfold over the next five days (Fig. 7B), the majority of which accumulated as starch in the chloroplasts. It does not seem unreasonable to conclude that, faced with the sudden wealth of available light following transfer from low to high light, PS II and photosynthesis can be downregulated (or experience photoinhibition) and still provide a greater income of carbohydrates than was possible during the low light growth conditions. The persistent photoinhibition in such transferred shade leaves may be related to a carbon export capacity that cannot be increased to the level of that typically found in a high light acclimated leaf. For instance, vein density of *Monstera deliciosa* leaves was significantly different between those grown in low light (2.8 ± 0.3) versus those grown in a sunlight-exposed glasshouse (4.9 ± 0.5), and vein density cannot be increased in fully expanded leaves (not shown). Upon transfer from low to high light, growth rates will increase in response to the appropriate signals in the elevated light environment, new leaves will be produced with higher rates of photosynthesis, and carbon supply is unlikely to be a limiting factor.

Chapter 5 Photoprotection and Photoinhibition

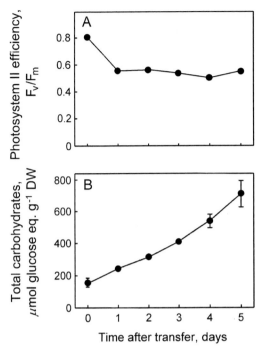

Fig. 7. Predawn determinations of photosystem II efficiency (A) and total non-structural carbohydrates (B) in low-light grown (10 μmol photons m^{-2} s^{-1}) leaves of *Monstera deliciosa* determined prior to (0 day) and daily upon transfer to high light (700 μmol photons m^{-2} s^{-1} for 10 h per day). Starch represented the greatest fraction of carbohydrates. Error bars represent standard deviations. Unpublished data of B Demmig-Adams, BA Logan, TR Rosenstiel, V Ebbert and WW Adams.

V. An Integrated View of Photoprotection

Zeaxanthin-dependent thermal energy dissipation thus spans many scales, providing photoprotection under the most benign conditions (e.g. understory of a rainforest; Logan et al., 1997) to the most severe that plants experience. This photoprotective process is modulated through several means to provide the level of thermal dissipation required 1) for flexible engagement and disengagement whenever the level of excess absorbed excitation energy varies rapidly (e.g. Figs. 1A and 2A–D) or 2) when photoprotection must be engaged in a sustained manner under long-term photoinhibitory conditions (e.g. Figs. 1B and 2M–P). The latter appears to occur readily in evergreen species during winter stress (e.g. Adams et al., 2001a, 2002, 2004; Öquist and Huner, 2003; Demmig-Adams et al., this volume) and upon exposure of shade-acclimated plants to high light (e.g. Demmig-Adams et al., 1998, this volume). For both of these scenarios, the capacity for detoxification of reactive oxygen species and other radicals is likely to be limited due to either low levels of antioxidants (shade-acclimated leaves) or an inhibition of the activity of enzymatic antioxidants by the low temperatures. On the other hand, under conditions of limiting nutrients (Verhoeven et al., 1997; Logan et al., 1999; Morales et al., this volume), and low water availability and/or high temperatures (Barker et al., 2002), photosynthesis can be downregulated, and zeaxanthin retained nocturnally, but without being maintained in an engaged state primed for thermal energy dissipation. Instead, the retained zeaxanthin remains poised for engagement (presumably upon protonation of the PsbS protein) and can thus respond more rapidly than if violaxanthin had to first be enzymatically converted to zeaxanthin, yet the system maintains complete flexibility in terms of engagement and disengagement. These three possible scenarios are depicted schematically in the context of whole plant source sink relationships in Figure 8.

Downregulation (or repression) of photosynthesis is a well-characterized response to conditions in which the supply of carbohydrates by source leaves exceeds the export and utilization of those sugars (Krapp and Stitt, 1995; Koch, 1996; Paul and Foyer, 2001). No one has ever suggested that rubisco or any of the other enzymes involved in the fixation and reduction of CO_2 to sugars are damaged when their levels decrease under sink-limiting conditions. In addition, some components of photosynthetic electron transport and ATP synthesis are downregulated in response to sugar repression or sink-limiting conditions (Krapp and Stitt, 1995; Dijkwel et al., 1996). Furthermore, levels of the D1 protein decrease dramatically under low light when spinach leaves are fed glucose (Kilb et al., 1996). It seems only logical that, in a situation where carbohydrates are in abundance and the biochemistry of photosynthesis is downregulated, primary photochemistry and photosynthetic electron transport should also be downregulated to reduce the likelihood of electrons being passed on to oxygen to form toxic superoxide (Fig. 5). This downregulation should be most easily achieved through the D1 protein that is turned over more rapidly than any other protein in the thylakoid membranes (Mattoo et al., 1984; Edelman and Mattoo, this volume). Is this one of the functions of light-mediated D1 turnover?

Some have been puzzled by the fact that a transgenically altered tobacco line with reduced levels of the cytochrome $b_6 f$ complex (and thus impaired

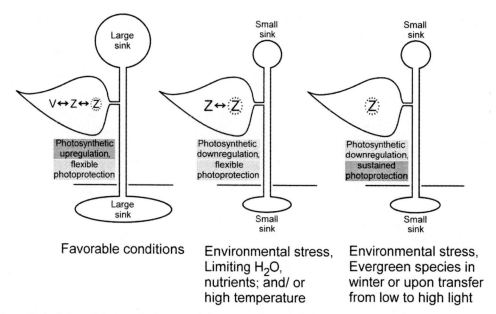

Fig. 8. Schematic depiction of photosynthetic upregulation versus downregulation and the engagement of flexible versus sustained zeaxanthin-dependent energy dissipation to effect photoprotection under different environmental conditions. Z surrounded by lines represents its engagement in energy dissipation. Under favorable conditions, growth and utilization of the products of photosynthesis is maximal (large sinks for carbohydrates), photosynthesis is upregulated or maintained at a high level for maximal utilization of the absorbed energy, and violaxanthin is converted to zeaxanthin and zeaxanthin and is engaged in energy dissipation as needed to siphon off excess absorbed excitation energy that is released harmlessly as heat. Under the less favorable conditions of reduced water availability, limiting nutrients, or high temperatures, growth is reduced (smaller sinks for carbohydrates), photosynthesis is often downregulated, and zeaxanthin may be retained nocturnally and engaged as required in energy dissipation only when absorbed excitation energy exceeds that which can be utilized by photosynthesis. In evergreen species that experience winter conditions or that are acclimated to low light and then suddenly exposed to high light, growth and utilization of carbohydrates is low (small sinks), photosynthesis is downregulated, and large amounts of zeaxanthin are retained in a state engaged for energy dissipation.

electron transport) experienced less photoinhibition (based upon decreases in PS II efficiency) compared to the wild type (Hurry et al., 1996; Yokthongwattana and Melis, this volume). If sustained decreases in PS II efficiency are interpreted to reflect sustained engagement of zeaxanthin in photoprotective energy dissipation, then such findings are entirely predictable. Any impairment in electron transport should be expected to limit the conversion of violaxanthin to zeaxanthin and the latter's engagement in energy dissipation due to an inability to acidify the thylakoid lumen to the same extent as the wild type tobacco.

What about photoprotection and photoinhibition in cyanobacteria and algae that do not export carbon to distant sinks? All algae employ xanthophylls in photoprotective energy dissipation, e.g. either zeaxanthin as part of the xanthophyll cycle in green, brown, and some red algae (Demmig-Adams et al., 1990; Uhrmacher et al., 1995; Gevaert et al., 2003; Ursi et al., 2003), diatoxanthin in diatoms and dinoflagellates as part of the diadinoxanthin-diatoxanthin cycle (Evens et al., 2001; Lavaud et al., 2004), or zeaxanthin that accumulates constitutively under high light in cyanobacteria and some red algae (Demmig-Adams et al., 1990; Cunningham et al., 1989). It has been suggested that zeaxanthin protects photodamaged PS II centers of algae exposed to photoinhibitory conditions (Jin et al., 2003). Is it not also possible that PS II centers are inactivated and/or disassembled under high light in response to signals exchanged among algae/bacteria when their densities are high and thus resources for growth and division are potentially limited? It is well known that bacteria decrease their rate of growth in response to signals from neighbors in close proximity, and it has now been established that both cyanobacteria and green algae produce signaling compounds that bacteria respond to in such quorum sensing (Braun and Bachofen, 2004; Teplitski et al., 2004). Whenever light is in excess, there are two possible responses at either end of a spectrum of potential adjustments: upregulate photosynthesis to utilize the additional light energy, or downregulate photosynthesis to minimize the possibility of forming reactive oxygen species (Figs. 1, 5, and 8; Demmig-Adams and Adams, 2003; Adams et al., 2004).

Photoprotection of photosynthesis over different time scales and in response to many different environmental conditions thus involves finely tuned adjustments in the capacity to utilize the absorbed excitation energy through photosynthetic electron transport and modulation of zeaxanthin-dependent energy dissipation. The adjustments vary from the highly flexible regulation of the xanthophyll cycle enzymes and PsbS protein protonation to the retention of zeaxanthin and its sustained configuration in a dissipative state under more severe stress. Although each of these has been categorized as being distinct from one another based upon the kinetics of engagement (qE or energy dependent quenching for that which is flexible versus qI or inhibitory quenching for that which is sustained), they truly represent extremes of a continuum of zeaxanthin-dependent photoprotection that is critical to the maintenance of the photosynthetic apparatus under conditions of excess light.

Acknowledgments

Our work has been supported by awards from the Andrew W. Mellon Foundation (Award No. 20200747), the National Science Foundation (Award No. IBN-0235351), and the Cooperative State Research, Education, and Extension Service, U.S. Department of Agriculture, under Agreement No. 00-35100-9564.

References

Adams WW III and Demmig-Adams B (1992) Operation of the xanthophyll cycle in higher plants in response to diurnal changes in incident sunlight. Planta 186: 390–398

Adams WW III and Demmig-Adams B (1995) The xanthophyll cycle and sustained thermal energy dissipation activity in *Vinca minor* and *Euonymus kiautschovicus* in winter. Plant Cell Environ 18: 117–127

Adams WW III and Demmig-Adams B (2004) Chlorophyll fluorescence as a tool to monitor plant response to the environment. In: Papageorgiou GC and Govindjee (eds) Chlorophyll a Fluorescence: A Probe of Photosynthesis, pp 583–604. Springer, Dordrecht

Adams WW III, Díaz M and Winter K (1989) Diurnal changes in photochemical efficiency, the reduction state of Q, radiationless energy dissipation, and non-photochemical fluorescence quenching in cacti exposed to natural sunlight in northern Venezuela. Oecologia 80: 553–561

Adams WW III, Demmig-Adams B, Winter K and Schreiber U (1990) The ratio of variable to maximum chlorophyll fluorescence from photosystem II, measured in leaves at ambient temperature and at 77K, as an indicator of the photon yield of photosynthesis. Planta 180: 166–174

Adams WW III, Demmig-Adams B, Verhoeven AS and Barker DH (1995a) 'Photoinhibition' during winter stress: Involvement of sustained xanthophyll cycle-dependent energy dissipation. Aust J Plant Physiol 22: 261–276

Adams WW III, Hoehn A and Demmig-Adams B (1995b) Chilling temperatures and the xanthophyll cycle. A comparison of warm-grown and overwintering spinach. Aust J Plant Physiol 22: 75–85

Adams WW III, Demmig-Adams B, Logan BA, Barker DH and Osmond CB (1999) Rapid changes in xanthophyll cycle-dependent energy dissipation and photosystem II efficiency in two vines, *Stephania japonica* and *Smilax australis*, growing in the understory of an open *Eucalyptus* forest. Plant Cell Environ 22: 125–136

Adams WW III, Demmig-Adams B, Rosenstiel TN and Ebbert V (2001a) Dependence of photosynthesis and energy dissipation activity upon growth form and light environment during the winter. Photosynth Res 67: 51–62

Adams WW III, Demmig-Adams B, Rosenstiel TN, Ebbert V, Brightwell AK, Barker DH, Zarter CR (2001b) Photosynthesis, xanthophylls, and D1 phosphorylation under winter stress. In: PS2001 Proceedings: 12th International Congress on Photosynthesis. CSIRO Publishing: Melbourne, Australia. Available at http://www.publish.csiro.au/ps2001

Adams WW III, Demmig-Adams B, Rosenstiel TN, Brightwell AK and Ebbert V (2002) Photosynthesis and photoprotection in overwintering plants. Plant Biol 4: 545–557

Adams WW III, Zarter CR, Ebbert V and Demmig-Adams B (2004) Photoprotective strategies of overwintering evergreens. BioScience 54: 41–49

Adir N, Zer H, Shochat S and Ohad I (2003) Photoinhibition – a historical perspective. Photosynth Res 76: 343–370

Aro E-M, Virgin I and Andersson B (1993) Photoinhibition of photosystem 2 – inactivation, protein damage and turnover. Biochim Biophys Acta 1143: 113–134

Bachmann KM, Ebbert V, Adams WW III, Verhoeven AS, Logan BA and Demmig-Adams B (2004) Effects of lincomycin on PSII efficiency, non-photochemical quenching, D1 protein and xanthophyll cycle during photoinhibition and recovery. Func Plant Biol 31: 803–813

Baker NR and Bowyer JR (eds) (1994) Photoinhibition of Photosynthesis from Molecular Mechanisms to the Field. Bios Scientific Publishers, Oxford

Ball MC, Hodges VS and Laughlin GP (1991) Cold-induced photoinhibition limits regeneration of snow gum at tree-line. Funct Ecol 5: 663–668

Barber J (1995) Molecular basis of the vulnerability of photosystem II to damage by light. Aust J Plant Physiol 22: 201–208

Barber J and Andersson B (1992) Too much of a good thing – light can be bad for photosynthesis. Trends Biochem Sci 17: 61–66

Barker DH, Adams WW III, Demmig-Adams B, Logan BA, Verhoeven AS and Smith SD (2002) Nocturnally retained zeaxanthin does not remain engaged in a state primed for energy dissipation during the summer in two Yucca species growing in the Mojave Desert. Plant Cell Environ 25: 95–103

Bilger W and Björkman O (1990) Role of the xanthophyll cycle in photoprotection elucidated by measurements of light-induced absorbance changes, fluorescence and photosynthesis in leaves of *Hedera canariensis*. Photosynth Res 25: 173–185

Bilger W and Björkman O (1991) Temperature dependence of violaxanthin de-epoxidation and non-photochemical fluorescence quenching in intact leaves of *Gossypium hirsutum* L. and *Malva parviflora* L. Planta 184: 226–234

Bilger W and Björkman O (1994) Relationships among violaxanthin deepoxidation, thylakoid membrane conformation, and nonphotochemical chlorophyll fluorescence quenching in leaves of cotton (*Gossypium hirsutum* L.). Planta 193: 238–246

Björkman O (1987) Low-temperature chlorophyll fluorescence in leaves and its relationship to photon yield of photosynthesis in photoinhibition. In: Kyle DJ, Osmond CB and Arntzen CJ (eds) Photoinhibition, pp 123–144. Elsevier, Amsterdam

Björkman O and Demmig B (1987) Photon yield of O_2 evolution and chlorophyll fluorescence characteristics at 77K among vascular plants of diverse origin. Planta 170: 489–504

Blennow K, Lang ARG, Dunne P and Ball MC (1998) Cold-induced photoinhibition and growth of seedling snow gum (*Eucalyptus pauciflora*) under differing temperature and radiation regimes in fragmented forests. Plant Cell Environ 21: 407–416

Braun E and Bachofen R (2004) Homoserine lactones and microcystin in cyanobacterial assemblages in Swiss lakes. Hydrobiol 522: 271–280

Cunningham FX Jr, Dennenberg RJ, Mustardy L, Jursinic PA and Gantt E (1989) Stoichiometry of photosystem I, photosystem II, and phycobilisomes in the red alga *Porphyridium cruentum* as a function of growth irradiance. Plant Physiol 91: 1179–1187

Demmig B and Björkman O (1987) Comparison of the effect of excessive light on chlorophyll fluorescence (77 K) and photon yield of O_2 evolution in leaves of higher plants. Planta 171: 171–184

Demmig-Adams B and Adams WW III (1992) Carotenoid composition in sun and shade leaves of plants with different life forms. Plant Cell Environ 15: 411–419

Demmig-Adams B and Adams WW III (1994a) Light stress and photoprotection related to the xanthophyll cycle. In: Foyer CH and Mullineaux PM (eds) Causes of Photooxidative Stress and Amelioration of Defense Systems in Plants, pp 105–126. CRC Press, Boca Raton

Demmig-Adams B and Adams WW III (1994b) Capacity for energy dissipation in the pigment bed in leaves with different xanthophyll cycle pools. Aust J Plant Physiol 21: 575–588

Demmig-Adams B and Adams WW III (1996) Xanthophyll cycle and light stress in nature: uniform response to excess direct sunlight among higher plant species. Planta 198: 460–470

Demmig-Adams B and Adams WW III (2003) Photoinhibition. In: Thomas B, Murphy D and Murray B (eds) Encyclopedia of Applied Plant Science, pp 707–714. Academic Press, London

Demmig-Adams B, Winter K, Krüger A and Czygan F-C (1989a) Light response of CO_2 assimilation, dissipation of excess excitation energy, and zeaxanthin content of sun and shade leaves. Plant Physiol 90: 881–886

Demmig-Adams B, Winter K, Krüger A and Czygan F-C (1989b) Zeaxanthin and the induction and relaxation kinetics of the dissipation of excess excitation energy in leaves in 2% O_2, 0% CO_2. Plant Physiol 90: 887–893

Demmig-Adams B, Adams WW III, Czygan F-C, Schreiber U and Lange OL (1990) Differences in the capacity for radiationless energy dissipation in the photochemical apparatus of green and blue-green algal lichens associated with differences in carotenoid composition. Planta 180: 582–589

Demmig-Adams B, Adams WW III, Logan BA and Verhoeven AS (1995) Xanthophyll cycle-dependent energy dissipation and flexible PS II efficiency in plants acclimated to light stress. Aust J Plant Physiol 22: 249–260

Demmig-Adams B, Adams WW III, Barker DH, Logan BA, Verhoeven AS and Bowling DR (1996a) Using chlorophyll fluorescence to assess the allocation of absorbed light to thermal dissipation or excess excitation. Physiol Plant 98: 253–264

Demmig-Adams B, Gilmore AM, Adams WW III (1996b) In vivo functions of carotenoids in higher plants. FASEB J 10: 403–412

Demmig-Adams B, Moeller DL, Logan BA and Adams WW III (1998) Positive correlation between levels of retained zeaxanthin + antheraxanthin and degree of photoinhibition in shade leaves of *Schefflera arboricola* (Hayata) Merrill. Planta 205: 367–374

Demmig-Adams B, Ebbert V, Zarter CR and Adams WW III (2005) Characteristics and species-dependent employment of flexible versus sustained thermal dissipation and photoinhibition. In: Demmig-Adams B, Adams WW III and Mattoo AK (eds) Photoprotection, Photoinhibition, Gene Regulation, and Environment, pp 39–48. Springer, Dordrecht

Dijkwel PP, Kock PAM, Bezemer R, Weisbeek PJ and Smeekens SCM (1996) Sucrose represses the developmentally controlled transient activation of the plastocyanin gene in *Arabidopsis thaliana* seedlings. Plant Physiol 110: 455–463

Edelman M and Mattoo AK (2005) D1 protein retrospective: the sizzling '80s. In: Demmig-Adams B, Adams WW III and Mattoo AK (eds) Photoprotection, Photoinhibition, Gene Regulation, and Environment, pp 23–38. Springer, Dordrecht

Endo T and Asada K (2005) Photosystem I and photoprotection: cyclic electron flow and water-water cycle. In: Demmig-Adams B, Adams WW III and Mattoo AK (eds) Photoprotection, Photoinhibition, Gene Regulation, and Environment, pp 205–221. Springer, Dordrecht

Ensminger I, Sveshnikov D, Campbell DA, Funk C, Jansson S, Lloyd J, Shibistova O and Öquist G (2004) Intermittent low temperatures constrain spring recovery of photosynthesis in boreal Scots pine forests. Glob Change Biol 10: 995–1008

Evens TJ, Kirkpatrick GJ, Millie DF, Chapman DF and Schofield OME (2001) Photophysiological response of the red-tide dinoflagellate *Gymnodinium breve* (Dinophyceae) under natural sunlight. J Plankton Res 23: 1177–1193

Fiekers JF, Marshal IB and Parsons RL (1979) Clindamycin and lincomycin alter miniature endplate current decay. Nature 281: 680–682

Gevaert F, Creach A, Davoult D, Migne A, Levavasseur G, Arzel P, Holl AC and Lemoine Y (2003) *Laminaria saccharina* photosynthesis measure in situ: photoinhibition and xanthophyll cycle during a tidal cycle. Mar Ecol Prog Ser 247: 43–50

Gilmore AM (1997) Mechanistic aspects of xanthophyll cycle-dependent photoprotection in higher plant chloroplasts and leaves. Physiol Plant 99: 197–209

Gilmore AM and Ball MC (2000) Protection and storage of chlorophyll in overwintering evergreens. Proc Natl Acad Sci USA 97: 11098–11101

Häder D-P (2005) Photoinhibition and UV response in the aquatic environment. In: Demmig-Adams B, Adams WW III and Mattoo AK (eds) Photoprotection, Photoinhibition, Gene Regulation, and Environment, pp 87–105. Springer, Dordrecht

Hager A (1980) The reversible, light-induced conversions of xanthophylls in the chloroplast. In: Czygan F-C (ed) Pigments in Plants, pp 57–79. Fischer, Stuttgart

Huner NPA, Ivanov AG, Sane PV, Pocock T, Król M, Balserus A, Rosso D, Savitch LV, Hurry VM and Öquist G (2005) Photoprotection of photosystem II: reaction center quenching versus antenna quenching. In: Demmig-Adams B, Adams WW III and Mattoo AK (eds) Photoprotection, Photoinhibition, Gene Regulation, and Environment, pp 155–173. Springer, Dordrecht

Hurry VM, Anderson JM, Badger MR and Price GD (1996) Reduced levels of cytochrome b-f in transgenic tobacco increases the excitation pressure on photosystem II without increasing sensitivity to photoinhibition in vivo. Photosynth Res 50: 159–169

Hymus GJ, Ellsworth DS, Baker NR and Long SP (1999) Does free-air carbon dioxide enrichment affect photochemical energy use by evergreen trees in different seasons? A chlorophyll fluorescence study of mature loblolly pine. Plant Physiol 120: 1183–1191

Jin ES, Yokthongwattana K, Polle JEW and Melis A (2003) Role of the reversible xanthophyll cycle in the photosystem II damage and repair cycle in *Dunaliella salina*. Plant Physiol 132: 352–364

Jung H-S and Niyogi KK (2005) Molecular analysis of photoprotection and photosynthesis. In: Demmig-Adams B, Adams WW III and Mattoo AK (eds) Photoprotection, Photoinhibition, Gene Regulation, and Environment, pp 127–143. Springer, Dordrecht

Kilb B, Wietoska H and Godde D (1996) Changes in the expression of photosynthetic genes precede the loss of photosynthetic activities and chlorophyll when glucose is supplied to mature spinach leaves. Plant Sci 115: 225–235

Kitajima M and Butler WL (1975) Quenching of chlorophyll fluorescence and primary photochemistry in chloroplasts by dibromothymoquinone. Biochim Biophys Acta 376: 105–115

Koch KE (1996) Carbohydrate-modulated gene expression in plants. Annu Rev Plant Physiol Plant Mol Biol 47: 509–540

Kok B (1956) On the inhibition of photosynthesis by intense light. Biochim Biophys Acta 21: 234–244

Krapp A and Stitt M (1995) An evaluation of direct and indirect mechanisms for the 'sink-regulation' of photosynthesis in spinach: Changes in gas exchange, carbohydrates, metabolites, enzyme activities and steady-state transcript levels after cold-girdling source leaves. Planta 195: 313–323

Külheim C, Agren J and Jansson S (2002) Rapid regulation of light harvesting and plant fitness in the field. Science 297: 91–93

Kyle DJ, Osmond CB and Arntzen CJ (eds) (1987) Photoinhibition. Elsevier, Amsterdam

Lavaud J, Rousseau B and Etienne AL (2004) General features of photoprotection by energy dissipation in planktonic diatoms (Bacillariophyceae). J Phycol 40: 130–137

Li X-P, Björkman O, Shih C, Grossman AR, Rosenquist M, Jansson S and Niyogi KK (2000) A pigment-binding protein essential for regulation of photosynthetic light harvesting. Nature 403: 391–395

Li X-P, Phippard A, Pasari J and Niyogi KK (2002) Structure-function analysis of photosystem II subunit S (PsbS) in vivo. Func Plant Biol 29: 1131–1139

Logan BA, Barker DH, Adams WW III and Demmig-Adams B (1997) The response of xanthophyll cycle-dependent energy dissipation in *Alocasia brisbanensis* to sunflecks in a subtropical rainforest. Aust J Plant Physiol 24: 27–33

Logan BA, Demmig-Adams B, Rosenstiel TN and Adams WW III (1999) Effect of nitrogen limitation on foliar antioxidants in relationship to other metabolic characteristics. Planta 209: 213–220

Long SP, Humphries S and Falkowski PG (1994) Photoinhibition of photosynthesis in nature. Annu Rev Plant Physiol Plant Mol Biol 45: 633–662

Ma Y-Z, Holt NE, Li X-P, Niyogi KK and Fleming GR (2003) Evidence for direct carotenoid involvement in the regulation of photosynthetic light harvesting. Proc Natl Acad Sci USA 100: 4377–4382

Maher P and Schubert D (2000) Signaling by reactive oxygen species in the nervous system. Cell Mol Life Sci 57: 1287–1305

Marshall HL, Geider RJ and Flynn KJ (2000) A mechanistic model of photoinhibition. New Phytol 145: 347–359

Mattoo AK, Hoffman-Falk H, Marder JB and Edelman M (1984) Regulation of protein metabolism: coupling of photosynthetic electron transport to *in vivo* degradation of the rapidly metabolized 32-kDa protein of the chloroplast membrane. Proc Natl Acad Sci USA 81:1380–1384

Melis A (1999) Photosystem II damage and repair cycle in chloroplasts: what modulates the rate of photodamage in vivo? Trends Plant Sci 4: 130–135

Morales F, Abadía A and Abadía J (2005) Photoinhibition and photoprotection under nutrient deficiencies, drought, and salinity. In: Demmig-Adams B, Adams WW III and Mattoo AK (eds) Photoprotection, Photoinhibition, Gene Regulation, and Environment, pp 65–85. Springer, Dordrecht

Nishiyama Y, Allakhverdiev SI and Murata N (2005) Regulation by environmental conditions of the repair of photosystem II in cyanobacteria. In: Demmig-Adams B, Adams WW III and Mattoo AK (eds) Photoprotection, Photoinhibition, Gene Regulation, and Environment, pp 193–203. Springer, Dordrecht

Norén H, Svensson P, Stegmark R, Funk C, Adamska I and Andersson B (2003) Expression of early light-induced protein but not the PsbS protein is influenced by low temperature and depends on the development state of the plant in field-grown pea cultivars. Plant Cell Environ 26: 245–253

Okada K, Satoh K and Katoh S (1991) Chloramphenicol is an inhibitor of photosynthesis. FEBS Lett 295: 155–158

Osmond CB (1994) What is photoinhibition? Some insights from comparisons of shade and sun plants. In: Baker NR and Bowyer JR (eds) Photoinhibition of Photosynthesis from Molecular Mechanisms to the Field, pp 1–24. Bios Scientific Publishers, Oxford

Osmond CB and Förster B (2005) Photoinhibition: then and now. In: Demmig-Adams B, Adams WW III and Mattoo AK (eds) Photoprotection, Photoinhibition, Gene Regulation, and Environment, pp 11–22. Springer, Dordrecht

Osmond CB and Grace SC (1995) Perspectives on photoinhibition and photorespiration in the field. Quintessential inefficiencies of the light and dark reactions of photosynthesis. J Exp Bot 46: 1351–1362

Osmond CB, Badger M, Maxwell K, Björkman O and Leegood R (1997) Too many photons: Photorespiration, photoinhibition and photooxidation. Trends Plant Sci 2: 119–121

Paul MJ and Foyer CH (2001) Sink regulation of photosynthesis. J Exp Bot 52: 1383–1400

Powles SB (1984) Photoinhibition of photosynthesis induced by visible light. Annu Rev Plant Physiol 35: 15–44

Prior C, Fiekers JF, Henderson F, Dempster J, Marshall IG and Parsons RL (1990) End-plate ion channel block produced by lincosamide antibiotics and their chemical analogs. J Pharm Exp Ther 255: 1170–1176

Roden JS and Ball MC (1996) The effect of elevated [CO_2] on growth and photosynthesis of two Eucalyptus species exposed to high temperatures and water deficits. Plant Physiol 111: 909–919

Roden JS, Egerton JJ and Ball MC (1999) Effect of elevated (CO_2) on photosynthesis and growth of snow gum (*Eucalyptus pauciflora*) seedlings during winter and spring. Aust J Plant Physiol 26: 37–46

Schreiber U, Bilger W and Neubauer C (1994) Chlorophyll fluorescence as a nonintrusive indicator for rapid assessment of in vivo photosynthesis. In: Schulze E-D and Caldwell MM (eds) Ecophysiology of Photosynthesis, pp 49–70. Springer, Berlin

Sopory SK, Greenberg BM, Mehta RA, Edelman M and Mattoo AK (1990) Free radical scavengers inhibit light-dependent degradation of the 32 kDa PS II reaction center protein. Z Naturforsch 45c: 412–417

Telfer A, Oldham TC, Phillips D and Barber J (1999) Singlet oxygen formation detected by near-infrared emission from isolated photosystem II reaction centres: direct correlation between P680 triplet decay and luminescence rise kinetics and its consequences for photoinhibition. J Photochem Photobiol B Biol 48: 89–96

Teplitski M, Chen HC, Rajamani S, Gao MS, Merighi M, Sayre RT, Robinson JB, Rolfe BG and Bauer WD (2004) *Chamydomonas reinhardtii* secretes compounds that mimic bacterial signals and interfere with quorum sensing regulation in bacteria. Plant Physiol 134: 137–146

Uhrmacher S, Hanelt D and Nultsch W (1995) Zeaxanthin content and the degree of photoinhibition are linearly correlated in the brown alga *Dictyota dichotoma*. Mar Biol 123: 159–165

Ursi S, Pedersen M, Plastino E and Snoeijs P (2003) Intraspecific variation of photosynthesis, respiration and photoprotective carotenoids in *Gracilaria birdiae* (Gracilariales: Rhodophyta). Mar Biol 142: 997–1007

Verhoeven AS, Adams WW III and Demmig-Adams B (1996) Close relationship between the state of the xanthophyll pigments and photosystem II efficiency during recovery from winter stress. Physiol Plant 96: 567–576

Verhoeven AS, Demmig-Adams B and Adams WW III (1997) Enhanced employment of the xanthophyll cycle and thermal energy dissipation in spinach exposed to high light and N stress. Plant Physiol. 113: 817–824

Verhoeven AS, Adams WW III and Demmig-Adams B (1998) Two forms of sustained xanthophyll cycle-dependent energy dissipation in overwintering *Euonymus kiautschovicus*. Plant Cell Environ 21: 893–903

Verhoeven AS, Adams WW III and Demmig-Adams B (1999) The xanthophyll cycle and acclimation of *Pinus ponderosa* and *Malva neglecta* to winter stress. Oecologia 118: 277–287

Weindruch R and Sohal RS (1997) Caloric intake and aging. New Engl J Med 337: 986–994

Werner C, Ryel RJ, Correia O and Beyschlag W (2001) Effects of photoinhibition on whole-plant carbon gain assessed with a photosynthesis model. Plant Cell Environ 24: 27–40

Yamamoto HY (1979) Biochemistry of the violaxanthin cycle in higher plants. Pure Appl Chem 51: 639–648

Yamamoto HY (2005) A random walk to and through the xanthophyll cycle. In: Demmig-Adams B, Adams WW III and Mattoo AK (eds) Photoprotection, Photoinhibition, Gene Regulation, and Environment, pp 1–10. Springer, Dordrecht

Yokthongwattana K and Melis A (2005) Photoinhibition and recovery in oxygenic photosynthesis: Mechanism of a photosystem-II damage and repair cycle. In: Demmig-Adams B, Adams WW III and Mattoo AK (eds) Photoprotection, Photoinhibition, Gene Regulation, and Environment, pp 175–191. Springer, Dordrecht

Zhu XG, Ort DR, Whitmarsh J and Long SP (2004) The slow reversibility of photosystem II thermal energy dissipation on transfer from high to low light may cause large losses in carbon gain by crop canopies: a theoretical analysis. J Exp Bot 55: 1167–1175

Chapter 6

Photoinhibition and Photoprotection under Nutrient Deficiencies, Drought and Salinity

Fermín Morales*, Anunciación Abadía and Javier Abadía
Department of Plant Nutrition, Aula Dei Experimental Station, Consejo Superior de Investigaciones Científicas (CSIC), Apdo. 202, E-50080 Zaragoza, Spain

Summary		65
I.	Introduction	66
II.	Iron (Fe) Deficiency	66
	A. Effects of Fe Deficiency on Photosynthesis	66
	B. Fe Deficiency and Photoinhibition	67
	C. Fe Deficiency and Photoprotection	67
III.	Nitrogen (N) Deficiency	69
	A. Effects of N Deficiency on Photosynthesis	70
	B. Nitrogen Deficiency and Photoinhibition	70
	C. Nitrogen Deficiency and Photoprotection	71
IV.	Other Nutrient Deficiencies	72
V.	Drought	73
	A. Effects of Drought on Photosynthesis	73
	B. Drought and Photoinhibition	73
	C. Drought and Photoprotection	73
VI.	Salinity	75
	A. Effects of Salinity on Photosynthesis	75
	B. Salinity and Photoinhibition	75
	C. Salinity and Photoprotection	75
VII.	Conclusions and Future Research Directions	77
Acknowledgments		78
References		78

Summary

Some of the more frequent abiotic stresses in plants are limited availability of nutrients and water, as well as salinity. All these situations occur both in natural habitats and in crops. Stressed plants often experience decreases in photosynthetic rates, whereas they still harvest sunlight. Environmental stresses such as those may decrease the efficiency with which solar energy is harvested and used by plants in photosynthetic reactions. This feature is what the scientific community has often called photoinhibition. Some researchers tacitly assume that photoinhibition may result from photodamage, whereas others believe that it is more the integration of a series of regulatory and protective adjustments. The aim of this review is to summarize the current knowledge concerning photoinhibition- and photoprotection-related processes under nutrient deficiencies, drought, and salinity stress, and to discuss the role that photoinhibition could play under such environmental stresses.

---

*Author for correspondence, email: fmorales@eead.csic.es

I. Introduction

The terminology concerning photoinhibition in plants is still a matter of debate and cause of misunderstandings (Adams et al., Demmig-Adams et al., this volume). For instance, in environmental and agronomic forums it is still very common to translate and understand any reference to photoinhibition as a damage-type occurrence, with little or no consideration to the possible photoprotective sides of those processes.

The aim of the present paper is to summarize the current knowledge on photoinhibition and photoprotection occurring under the most common nutrition-related abiotic stresses in plants. This includes deficiencies of Fe, N, and other elements, and also drought and elevated salinity. We have considered as photoinhibition-related those situations where one or several of the following observations have been made: i) a decrease in quantum yield of photosynthesis, ii) decreases in F_v/F_m ratios after dark adaptation, iii) changes in D1 protein amount or turnover rate, and iv) a permanent "lock-in" of the xanthophyll cycle pigments in the de-epoxidized state. We have considered as photoprotection-related observations those reporting non-permanent changes in xanthophyll cycle-mediated thermal dissipation, elicitation of antioxidative systems, and decreases in leaf Chl concentrations. In this latter case, however, it should be taken into account that the role of smaller antenna size in photoprotection seems to be small (Baroli et al., 2003).

II. Iron (Fe) Deficiency

Iron is part of many plant components, and therefore is required for plant growth. Iron is abundant in the earth's crust, but under oxygenic conditions and at the pH values prevailing in many environments, the existing Fe(III) equilibrium concentrations are far lower than those required for plant growth (Marschner, 1995). As a consequence, Fe deficiency is a common abiotic stress that affects many photosynthetic organisms (Terry and J. Abadía, 1986; Geider and La Roche, 1994; Straus, 1994). Species affected range from sea phytoplankton (Behrenfeld et al., 1996) to high value crops in arid and semiarid environments (Mortvedt, 1991). Iron deficiency is a potential problem in Calcisol soils, that cover approximately 800 million ha worldwide (FAO, 1988). Since the most visual effect of Fe deficiency in plants is the yellowing of young leaves, Fe deficiency is usually named Fe chlorosis (from the Greek word "chloros", yellow-green). Iron chlorosis is one of the yield limiting factors for some crops. A good example is fruit tree crops in Mediterranean environments, since growers not using Fe fertilization face major fruit yield and quality losses (Álvarez-Fernández et al., 2003) and also marked reductions in orchard longevity (Sanz et al., 1992).

A. Effects of Fe Deficiency on Photosynthesis

Approximately 80% of the plant Fe is located in the chloroplast, where it is a constituent of a number of photosynthetic machinery components, including cytochromes, Fe-S centers, and others (Terry and J. Abadía, 1986). When Fe is in low supply, the amount of photosynthetic membranes per chloroplast decreases. This is accompanied by decreases in all membrane components, including electron carriers of the photosynthetic electron transport chain (Terry and J. Abadía, 1986 and references therein) and the thylakoid pigments Chls and carotenoids (Morales et al., 1990, 1994; J. Abadía and A. Abadía, 1993). Because of these changes, thylakoids from Fe-deficient plants show characteristics of a "diluted" photosynthetic membrane (Terry and J. Abadía, 1986). Iron deficiency also decreases RuBP carboxylation capacity, both through reduced Rubisco enzyme activation (Taylor and Terry, 1986) and down-regulation of gene expression (Winder and Nishio, 1995). The Fe deficiency-mediated decreases in light harvesting, electron transport, and carbon fixation capacities seem to be well coordinated (Winder and Nishio, 1995).

As a consequence of all these changes, Fe deficient leaves have low photosynthetic rates and can dissipate in this way a limited amount of energy. The decreases in leaf pigment concentrations occurring with Fe deficiency may provide some protection through decreases in light harvesting capacity (see below). Pigment

---

Abbreviations: A – antheraxanthin; C_a – external CO_2 concentration; Chl – chlorophyll; C_i – substomatal CO_2 concentration; D – fraction of light absorbed by PS II that is dissipated thermally in the antenna; ETR – electron transport rate; NPQ – non-photochemical quenching; P – fraction of light absorbed by PS II that is used in photochemistry; Pc – fraction of P that Rubisco uses for RuBP carboxylation; Po – fraction of P that Rubisco uses for RuBP oxygenation; PPFD – photosynthetic photon flux density; PS I – Photosystem I; PS II – photosystem II; Rubisco – ribulose-1,5-*bis*phosphate carboxylase oxygenase; RuBP – ribulose *bis*phosphate; V – violaxanthin; X – fraction of light absorbed by PS II that is neither used in photochemistry nor dissipated thermally in the antenna; Z – zeaxanthin.

decreases, however, may also involve some dangers, since these pigments have a general protective function, acting as "sunglasses" for the abaxial leaf cell layers (Nishio, 2000). In fact, the number of blue and red photons absorbed per Chl increases with Fe deficiency, because decreases in absorptance at these wavelengths are less marked than those found for photosynthetic pigments (Morales et al., 1991; J. Abadía et al., 1999b). Therefore, Fe-deficient photosynthetic organisms are prone to be exposed to an excessive photosynthetic photon flux density (PPFD).

B. Fe Deficiency and Photoinhibition

Some reports indicate that severe Fe deficiency could induce PS II photoinhibition in photosynthetic organisms. For instance, the abundance of the reaction center polypeptide D1 was reduced in an Fe-limited diatom (Geider et al., 1993) and in *Dunaliella tertiolecta* (Vassiliev et al., 1995). Also, in isolated thylakoids of severely Fe-deficient peach, grapevine, and tomato, decreases in the abundance of D1 and other PS II polypeptides have been found (Nedunchezhian et al., 1997; Bertamini et al., 2002; Ferraro et al., 2003). The turnover of D1 was also increased in Fe-deficient maize (Jiang et al., 2001). These studies, however, were carried out with SDS-PAGE of isolated thylakoids, which may not be fully representative of the whole leaf cell population (Quílez et al., 1992). Recent studies on Cu excess, which causes a Cu-induced Fe deficiency, indicate that Fe-deficient, low-Chl plants could be photoinhibited faster than the controls in the presence of lincomycin (Patsikka et al., 2002). In isolated thylakoids from the same plants, however, susceptibility to photoinhibition was not affected (Patsikka et al., 2002). This suggests that Fe deficiency increases the susceptibility of PS II towards photoinhibition mainly through the decrease in leaf Chl. This is likely related to the increases in the amount of photons absorbed per Chl, which is markedly higher in Fe-deficient leaves.

Moderately Fe-deficient leaves have no signs of significant photoinhibitory damage, as judged by both Chl fluorescence and gas exchange parameters. Evidence for the maintenance of a good maximum PS II energy conversion efficiency comes from measurements of the F_v/F_m ratio after dark adaptation, both in controlled environments (Morales et al., 2001) and in the field (Morales et al., 2000a). Also, high quantum yields of CO_2 fixation (Terry, 1980) and O_2 evolution occur in moderately Fe-deficient leaves (Morales et al., 1991). Even when Fe deficiency occurs in fruit trees concomitantly with other stresses such as mild water stress, high temperatures, low air relative humidity and high PPFDs, no sustained decreases in F_v/F_m were found (Morales et al., 2000b). Some data in the literature had previously suggested that Fe deficiency could cause general decreases in maximum potential PS II efficiency. For instance, markedly low F_v/F_m ratios were reported in Fe-deficient cyanobacteria (Guikema, 1985), sugar beet (Morales et al., 1991), and eukaryotic marine algae (Greene et al., 1992; Falkowski et al., 1995). These values, however, were likely due to inaccurately high F_o values, caused by the reduction of the plastoquinone pool occurring in Fe-deficient organisms during dark adaptation, which is only prevented by FR pre-illumination (Belkhodja et al., 1998). No detailed study has been carried out so far on the effects of different levels of Fe deficiency on specific photoinhibition-related sites as a consequence of excessive PPFD levels.

In Fe-deficient leaves of pear trees grown at high PPFDs in the field, a significant part of the zeaxanthin pool is maintained ("locked in") overnight (Morales et al., 1994, 2000a). Similar data have been reported in *Yucca* species experiencing high temperature and drought stress (Barker et al., 2002). This might suggest that Fe-deficient plants may experience processes that could be considered as photoinhibitory. In these cases, however, dark-adapted F_v/F_m ratios still remained high (Morales et al., 2000a).

C. Fe Deficiency and Photoprotection

Iron-deficient leaves are first protected to some extent against excess PPFD by decreases in light absorptance, associated with the decreases in the concentrations of the photosynthetic pigments Chls and carotenoids (Terry, 1980; Morales et al., 1991; Masoni et al., 1996; J. Abadía et al., 1999). Also, Fe-deficient leaves show increases in light reflectance and transmittance (Masoni et al., 1996; J. Abadía et al., 1999). Iron deficiency decreases absorptance from control values of 80% (of the incident PPFD) to approximately 20, 40, and 60% in Fe-deficient leaves of sugar beet, pear, and peach, respectively (Morales et al., 1991; J. Abadía et al., 1999). Therefore, 40–80% of the incident PPFD is simply not absorbed by the low-Chl leaves. The origin of leaf reflectance increases with Fe deficiency is not known, but it may be related to changes both in cuticle composition and in leaf optical properties.

Evidence for a different mechanism of photoprotection has arisen from studies with very severe Fe-deficient leaves. These leaves show sustained decreases in dark-adapted F_v/F_m ratios, which cannot be relieved by FR pre-illumination. This finding has been recently attributed to the presence of a constant PS II emission, with a lifetime of approximately 3.3 ns, which would account for approximately 15% of the total fluorescence (Morales et al., 2001). When F_v/F_m values were corrected to eliminate such emission, the measured F_v/F_m values of 0.66 become very similar to those of the controls, in the range 0.73–0.83. Only in some leaves F_v/F_m are still lower than the controls, possibly due to the presence of some closed PS II reaction centers in dark-adapted Fe-deficient leaves (Belkhodja et al., 1998). It has been hypothesized that the constant PS II fluorescence emission in severely Fe-deficient sugar beet leaves would come from internal PS II antenna complexes fully disconnected from the PS II reaction center (Morales et al., 2001). The possible relationship of this mechanism with the altered emission characteristics of a "cold hard band" (CHB) reported in overwintering evergreens and with Gilmore's model of disconnected PS II cores (Gilmore et al., 2003) is under investigation.

Cyanobacteria grown under Fe deficiency have a supercomplex consisting of a trimeric PS I reaction center encircled by a ring of 18 copies of the Chl *a*-binding, light harvesting complex protein CP43', encoded by the *isiA* gene (Bibby et al., 2001a; Boekema et al., 2001). The CP43' ring increases PS I antenna size by at least 70% (Bibby et al., 2001b) and PS I absorption cross-section by approximately 2-fold (Andrizhiyevskaya et al., 2002), serving as an efficient antenna for PS I (Melkozernov et al., 2003). In addition to its functions in Chl *a* storage and energy absorption, CP43' has been implicated in protection against excessive light, through non-radiative energy dissipation (Park et al., 1999; Sandstrom et al., 2001, 2002). Also, *isiA* has been identified as a salt-regulated gene in cyanobacteria (Vinnemeier et al., 1998; Vinnemeier and Hagemann, 1999). The presence of CP43' seems to be exclusive to cyanobacteria that, unlike green algae and higher plants, do not contain Chl *b*. However, it has recently been reported that Chl *b* is able to bind to the CP43' protein (Duncan et al., 2003), suggesting that *isiA* could exist in organisms other than cyanobacteria. However, the *isiA* gene products have not been detected in Fe-deficient marine algae (Marquardt et al., 1999) and there is no evidence of their presence in higher plants. In *Chlamydomonas*, the PS I reaction centers seem to be protected from excess PPFD by a disconnection of the LHCI antenna (Moseley et al., 2002).

A new modulated Chl fluorescence approach to assess the relative importance of different photoprotective mechanisms was proposed by Demmig-Adams et al. (1996a). This technique can be used to allocate the energy absorbed by PS II in three different fractions; i.e. a fraction that is dissipated thermally by the antenna (D), a fraction used in photochemistry (P), and a third fraction not used in photochemistry nor dissipated by the PS II antenna (X). The fraction of the absorbed PPFD used in photochemistry would include electrons used in RuBP carboxylation and oxygenation, Mehler reaction, chloro-respiration, and any other electron-consuming processes. The size of the X fraction (which could be slightly overestimated by this method, see Schreiber et al., 1994 and Verhoeven et al., 1997) would probably determine the rate of PS II photo-inactivation (Kato et al., 2002). The mechanisms underlying X are so far not understood, although they may include potentially deleterious mechanisms, such as the formation of triplet Chl excited state (Demmig-Adams et al., 1996b), that may lead to singlet oxygen formation (Demmig-Adams et al., 1996a; Foyer et al., this volume) and in turn to PS II damage by photo-oxidative and degradation processes.

Using this approach it can be concluded that in Fe-deficient plants a large part of the energy absorbed by PS II is dissipated thermally by the PS II antenna. This occurs both at low PPFDs and in plants grown at high PPFDs in the field (Fig. 1). With severe Fe deficiency, D may reach up to 75–80% of the light absorbed by PS II at midday (Fig. 1). Similar D values may also occur even early in the morning, if a mild water stress is present (Morales et al., 2000b). In Fe-sufficient controls, D only accounts for approximately 25 and 54–57% under low- and high-PPFD, respectively (Fig. 1). The Fe deficiency-mediated relative increases in D have been related to the increases in the Z + A to Chl molar ratio and to the extent of de-epoxidation of V into A + Z, which occur in daily cycles in Fe-deficient plants (Morales et al., 1990, 1994, 1998, 2000a). At low PPFDs, photochemistry (P) accounts for 66, 34, and 21% of the light absorbed by PS II in control, moderate, and severely Fe-deficient leaves, respectively (Fig. 1). Under full sunlight field conditions, P decreases from 29–38% in control leaves to 16–19 and 10% under moderate and severe Fe deficiency, respectively (Fig. 1). Both in Fe-deficient and control leaves, a significant amount of light (X; accounting for 4–14% of the PPFD absorbed by PS II) is predicted to be neither dissipated thermally nor used in photochemistry. The xanthophyll cycle-related thermal dissipation found in leaves of Fe-deficient sugar beet grown at low PPFDs is not "locked-in" overnight (Morales et al., 1990, 1998), indicating that it constitutes a photoprotective mechanism. In pear grown in the field at high PPFDs, xanthophyll cycle-related thermal dissipation

Chapter 6 Nutrients, Drought, Salinity

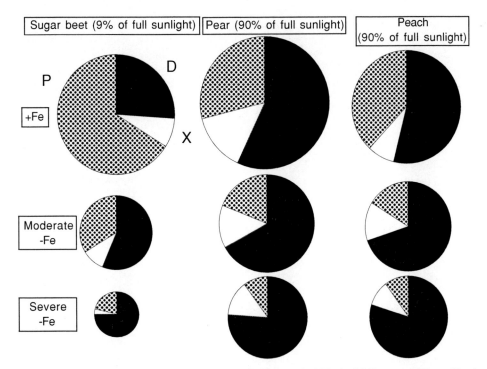

Fig. 1. Fractions of the total light absorbed by PS II that are thermally dissipated within the PS II antenna (D), used in photochemistry (P), and not used in photochemistry nor thermally dissipated within the PS II antenna (X) in Fe-sufficient (+Fe), moderate, and severe Fe-deficient sugar beet (grown in growth chamber at 9% full sunlight), pear, and peach (grown in field conditions; data are for 90% of full sunlight) leaves. Total pie areas are proportional to the leaf absorptance. The allocation of energy to D, P, and X follows the methodology proposed by Demmig-Adams et al. (1996a).

is not maintained overnight in moderately Fe-deficient leaves, but appears to be "locked in" in severely deficient leaves (Morales et al., 1994; 2000a).

In spite of the excess PPFD experienced by Fe-deficient leaves, the extent of oxidative damage seems very limited. For instance, Fe-deficient pea leaves did not show accumulation of oxidatively-damaged lipids and proteins (Iturbe-Ormaetxe et al., 1995). Of course, the concentrations of some anti-oxidative elements containing Fe, including Fe-SOD (Kurepa et al., 1997) and ascorbate peroxidase (Tobías, 1999; Ranieri et al., 2001) are reduced in Fe-deficient plants. The decrease in some of these elements is compensated by increases in alternative enzymes, such as Mn- and Cu/Zn-SODs, that may take on the role of Fe-SOD (Tobías, 1999). Iron-deficient leaves could be protected against reactive oxygen species by relatively high concentrations of antioxidant molecules, such as ascorbate and glutathione (Iturbe-Ormaetxe et al., 1995; Tobías, 1999). The possible photoprotective role of zeaxanthin as an antioxidant (Havaux and Niyogi, 1999; Baroli et al., 2003; Müller-Moulé et al., 2003) or as a reactive oxygen species signaling modulating substance (Demmig-Adams and Adams, 2002) has not been investigated in Fe-deficient leaves. The ratio of protective enzymes and antioxidants to Chl could become very large in Fe-deficient plants (Tobías, 1999). Also, the ratios carotenoids/Chl increase markedly, and these pigments can directly de-excite triplet Chl (Foyer et al., 1994). All these data suggest that Fe-deficient leaves are well protected against oxidative damage, and contribute to explain the observation that chlorotic leaves of Fe-deficient trees can remain stable in the field for a few months. Most of these leaves appear otherwise healthy, and only extremely deficient leaves show necrotic spots.

III. Nitrogen (N) Deficiency

Nitrogen is a constituent of many plant cell components, such as amino acids, nucleoside bases, chlorophylls, and others (Marschner, 1995). Therefore, plant growth requires a continuous supply of N. Only some species can benefit from symbiotic fixation of atmospheric N, and most of them depend on the availability of N in the soil, which is not sufficient in many environments, and therefore limits plant growth.

This occurs both in plants in nature and also in most crop plants, where fertilization with N-containing compounds is carried out worldwide.

A. Effects of N Deficiency on Photosynthesis

The restricted development of N-deficient plants has usually been ascribed to lower rates of leaf expansion, rather than to declines in photosynthesis per unit of leaf area (Sage and Pearcy, 1987a). Nitrogen shortage, however, does result in a marked decrease in plant photosynthesis. This is to be expected, because more than half of the total leaf N is allocated to the photosynthetic apparatus (Makino and Osmond, 1991). Photosynthetic capacity and total amount of leaf N per unit leaf area are often correlated (Field and Mooney, 1986; Sage and Pearcy, 1987b; Walcroft et al., 1997). Thylakoid membrane properties, however, are not modified by N deficiency in spinach, whereas the amount of thylakoids and thylakoid protein per chloroplast are decreased (Evans and Terashima, 1987; Terashima and Evans, 1988). As it occurs in other stresses, changes in different components are coordinated, and both Chl and Rubisco concentrations per unit leaf area decrease with N deficiency. The effects of N deficiency on Rubisco, however, are often larger than those on Chl (Ferrar and Osmond, 1986; Evans and Terashima, 1987; Seemann et al., 1987), and part of the decrease in photosynthetic capacity occurring with N deficiency can therefore be ascribed to the diminished amounts of Calvin cycle enzymes (Terashima and Evans, 1988; Sugiharto et al., 1990). Also, there is nowadays clear evidence that N deficiency induces sink limitation within the whole plant, due to decreased growth (Paul and Driscoll, 1997; Logan et al., 1999; Paul and Foyer, 2001). This leads, in turn, to feedback down-regulation of photosynthesis.

Since N-deficient plants have decreases in photosynthetic rates, when they are exposed to high PPFDs, such as those found in the Mediterranean area in summer at midday (2200 μmol photons m^{-2} s^{-1}), they are prone to experience photoinhibition. The effects of N deficiency are generally more marked at high than at low PPFDs (Seemann et al., 1987; Terashima and Evans, 1988).

B. Nitrogen Deficiency and Photoinhibition

Plants growing under N limitation exhibit decreases not only in photosynthetic capacity, but also in leaf Chl concentrations and in the quantum yield of photosynthesis (Terashima and Evans, 1988; Khamis et al., 1990; Verhoeven et al., 1997; Lu and Zhang, 2000; Chen et al., 2001; Demmig-Adams and Adams, 2003). These features are found in sun-exposed, fast-growing leaves but not in shaded ones (Terashima and Evans, 1988; J. He et al., 2000), suggesting that photodamage may occur. Alternatively, this observation may reflect a differential N deficiency-mediated feedback down-regulation of photosynthesis in sun and shade leaves.

The susceptibility to photoinhibition is larger in plants grown with low N than in those grown with high N (Balachandran and Osmond, 1994; Skillman and Osmond, 1998; Bungard et al., 2000; Grassi et al., 2001). Also, when transferred from low to high PPFDs, plants grown under N limitation are more susceptible to photoinhibition, and furthermore the recovery is slower than in N-sufficient controls (Ferrar and Osmond, 1986; Khamis et al., 1990; Henley et al., 1991). Seedlings fertilized with N also show faster recovery rates from photoinhibition than the unfertilized ones (Close et al., 2003). Photoinhibition was also enhanced in N-deficient, field-grown rice, when incident PPFD was increased by altering the leaf angle (Chen et al., 2003). In N-limited phytoplankton, a decreased midday PS II efficiency was found to be associated with photoinduced damage of D1 (Bergmann et al., 2002). In the diatom *Phaeodactylum tricornutum* and in the cyanobacterium *Prochlorococcus marinus*, N deficiency decreased the abundance of D1 (Geider et al., 1993; Steglich et al., 2001). These D1 changes do not necessarily indicate damage of PS II reaction centers. Levels of D1 can also decrease as a regulatory adjustment, when there are more products of photosynthesis than those required (i.e., a feedback down-regulation of D1 levels). Even at low PPFD, D1 levels decrease in response to increased levels of sugars (Kilb et al., 1996). A decrease in D1 would decrease the likelihood of superoxide production (Iglesias et al., 2002; Adams et al., 2004).

No sustained photoinhibition occurs under N deficiency in spinach, as judged by the high F_v/F_m ratios after dark adaptation (Verhoeven et al., 1997). Also, Bungard et al. (1997) and Lu and Zhang (2000) reported values ranging from 0.83–0.80 to 0.77 in control and N-deficient *Clematis vitalba* and maize plants, respectively, the lowest values corresponding to plants growing at the highest PPFDs. These slightly low F_v/F_m ratios at predawn are likely related to the measurable overnight retention ("lock-in") of Z + A and sustained NPQ occurring in N-deficient leaves (Verhoeven et al., 1997). Nitrogen availability, however, did not change D1 turnover rates in leaves of *Chenopodium album*

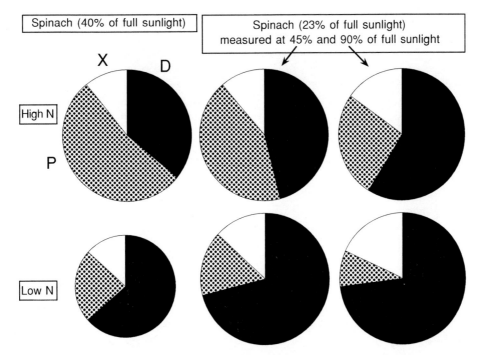

Fig. 2. Fractions of the total light absorbed by PS II that are thermally dissipated within the PS II antenna (D), used in photochemistry (P), and not used in photochemistry nor thermally dissipated within the PS II antenna (X) in leaves of spinach grown in growth chambers with high or low N and at 40% of full sunlight (measurements at actual PPFD, 900 μmol photons m^{-2} s^{-1}) and 23% of full sunlight (measurements at 1000 or 2000 μmol photons m^{-2} s^{-1}). Pie areas are proportional to the leaf absorptance (considering that Chl decreases affect light absorption only in the low-N plants grown at 40% of full sunlight; see Morales et al., 1991 and Verhoeven et al., 1997). The allocation of energy to D, P, and X follows the methodology proposed by Demmig-Adams et al. (1996a). Data were re-calculated from Verhoeven et al. (1997).

(Kato et al., 2002), and the amounts of D1 were similar in N-replete and N-starved *Anacystis nidulans* (Biswal et al., 1994). It seems therefore that the potential occurrence of photoinhibition in response to N deficiency, with net loss of D1 protein, will depend much on species and on the previous acclimation of the organism to high PPFDs, which may confer a better capacity for D1 renewal (Long et al., 1994).

C. Nitrogen Deficiency and Photoprotection

In N-deficient spinach some photoprotection occurs through a decrease in light harvesting capacity, associated with the decrease in leaf Chl (Fig. 2; Bungard et al., 1997; Verhoeven et al., 1997; Demmig-Adams and Adams, 2003). Light absorptance, however, is less affected than both leaf Chl concentrations and the capacity of energy utilization in photosynthesis. This results in increases in the number of photons absorbed by each Chl molecule, and also in an excess of absorbed PPFD over the amount that can be used in photosynthesis (Verhoeven et al., 1997).

A rapidly reversible, xanthophyll cycle-dependent, thermal energy dissipation within the Chl pigment bed occurs in N-deficient leaves (Verhoeven et al., 1997; Chen et al., 2003; Demmig-Adams and Adams, 2003). In N-deficient spinach leaves, thermal dissipation may account for 64–73% of the light absorbed by the PS II antenna, whereas in controls it is in the range 36-59% (Fig. 2; Verhoeven et al., 1997). In apple leaves with low N, thermal dissipation accounts for 35 and 60% of the absorbed light at limiting and saturating light, respectively, whereas those with high N dissipate 25 and 40%, respectively (data re-calculated from Chen et al., 2001). Similar increases in thermal energy dissipation with N deficiency have been found in maize growing at low PPFDs (Lu and Zhang, 2000).

The fraction of light absorbed in PS II and used in photochemistry (P) decreases with N deficiency (Fig. 2; Verhoeven et al., 1997). In spinach grown at 40% of full sunlight, P accounted for 53 and 23% of the absorbed light in control and N-deficient leaves, respectively. When grown at lower PPFDs (23% of full sunlight) and measured at higher PPFDs (90% of full sunlight) differences increase, and P accounted for 26 and 9% in control and N-deficient leaves, respectively. The partitioning of electron flow between CO_2 assimilation and photorespiration was not affected by N deficiency in apple leaves (Chen et al., 2001). Also, the C_i/C_a ratios either increase or did not change under N limitation (Sage and Pearcy, 1987b; see also Le Roux et al., 2001). The absorbed light not thermally dissipated and not used in photochemistry (X) in spinach and maize would be approximately 10–15 and 13-19% in high- and low-N leaves, respectively (re-calculated from Verhoeven et al., 1997 and Lu and Zhang, 2000).

Antioxidative plant systems are activated in N-deficient leaves. Both increases in Cu-ZnSOD activity and decreases of ascorbate peroxidase and glutathione reductase activities have been reported in response to an enhanced formation rate of superoxide radical in N-deficient coffee plants (Ramalho et al., 1998). The fatty acid composition of the chloroplastic membranes of these plants was also modified by N-deficiency, which may have made them less susceptible to peroxidation and better protected against photo-oxidative stress (Ramalho et al., 1998). Strong, long-term N-limitation in spinach caused decreases in the concentrations of Chl, carotenoid pools, ascorbate peroxidase, and glutathione reductase when data were expressed on a leaf area or dry weight basis (Logan et al., 1999). When the same data were expressed on a Chl basis, however, leaf concentrations of all these components were similar in N-deficient and N-replete plants (Logan et al., 1999). Furthermore, on a total protein basis the antioxidant enzyme activities were higher in N-limited plants (Logan et al., 1999). Nitrogen-deficient, low-Chl leaves also have increases in the carotenoid to Chl molar ratios (Khamis et al., 1990; Verhoeven et al., 1997), which may help to prevent singlet oxygen formation within the photosynthetic apparatus.

IV. Other Nutrient Deficiencies

Decreases in photosynthetic rates with K and phosphorus deficiencies have been reported for several species (Egilla and Davies, 1995; Bednarz and Oosterhuis, 1999; Basile et al., 2003). Since the light-dependent uptake of K into the guard cells is a critical step for stomatal opening (Schroeder, 2003), it would be likely that stomatal limitations may arise under K deficiency. However, there is no experimental evidence to date showing stomatal limitations in plants affected by K deficiency. Inorganic P is also required for the biosynthesis of ATP, which is involved in stomatal physiology (Agbariah and Roth-Bejerano, 1990). Phosphorus deficiency leads to significant decreases in leaf stomatal conductance (Rao and Terry, 1989a) and also to biochemical changes associated with decreases in photosynthetic rates (Rao and Terry, 1989b, 1990). Therefore, phosphorus deficiency limits photosynthesis through both stomatal and non-stomatal processes.

In almond, the development of K deficiency symptoms depends on leaf position, since leaves from the upper part of the canopy develop symptoms, whereas those of the lower part of the canopy do not (Basile et al., 2003). This might suggest that the part of the canopy more exposed to sunlight could be affected by photoinhibitory damage, although a feedback down-regulation of photosynthesis similar to that occurring in N-deficient leaves cannot be excluded. Under phosphorus deficiency, Chl leaf concentrations usually increase, whereas photosynthesis saturates at lower PPFD values than in the phosphorus-sufficient controls (Rao and Terry, 1989a). Both factors would result in an excess of energy excitation when phosphorus-deficient leaves are exposed to relatively high PPFD, leading in turn to a potential photoinhibition.

There are not enough data to conclude how K- or phosphorus-deficient plants protect themselves from the possible energy excesses, although dark respiration and photorespiration are unlikely to contribute to the protection of PS II. In both cases, dark respiration rates (Terry and Ulrich, 1973b) and photorespiration (Terry and Ulrich, 1973a, 1973b; Peoples and Koch, 1979) have been shown to decrease with respect to the controls. Therefore, further research is required to elucidate the role that other electron-consuming processes may play in photoprotection in these deficiencies. So far no studies have tackled the functioning of the xanthophyll cycle and its associated energy dissipation in plants affected by K or phosphorus deficiency.

A combined Mg and S deficiency leads initially to decreases in leaf Chl, without loss of PS II activity but with increases in D1 turnover (Godde and Dannehl, 1994; Dannehl et al., 1995). After 4–8 weeks of deficiency, PS II efficiency (estimated with the F_v/F_m Chl fluorescence ratio) was significantly reduced and a net loss of D1 was observed (Godde and Hefer, 1994; Dannehl et al., 1995). When plants were transferred back to a nutrient medium containing Mg and S, D1 turnover increased, resynthesis of PS II centers occurred, and F_v/F_m ratios increased (Dannehl et al., 1996). These observations do not necessarily mean that a combined Mg and S deficiency induces photoinhibitory damage, since the deficiencies in these essential elements might also lead to decreased growth, resulting in feedback down-regulation of photosynthesis, which may include a possible down-regulation of D1 levels (Kilb et al., 1996).

V. Drought

Drought limits the production of agricultural soils in large areas of the world. Irrigation is possibly the oldest agricultural management technique aimed to increase crop yield. Worldwide, approximately 1 billion ha of cultivated land are irrigated (Toenniessen, 1984). Water supply is one of the major economic constraints in a large part of the developing countries, and drought has had a dramatic impact on many human communities in the last several decades. Developed countries in America and Europe suffer from water shortages in many of their cultivated areas. Drought is also of extreme importance in natural habitats, and climatic change trends could make it even more important in the future.

A. Effects of Drought on Photosynthesis

Drought markedly decreases net photosynthetic rates in plants (Dubey, 1997; Medrano et al., 1998; Flexas and Medrano, 2002, and references therein). Decreases in photosynthesis with drought are initially due to stomatal closure, which decreases or prevents water loss but reduces CO_2 availability for the chloroplasts. When drought progresses, both stomatal closure and decreases in PS II photochemistry could contribute to the decreases in photosynthesis (Flexas and Medrano, 2002). Photosynthesis saturates at a lower PPFD in plants grown under drought than in well-watered controls (Lawlor, 1995). Therefore, when high PPFDs occur concomitantly with water stress, plants could be exposed to an excess of excitation energy (Demmig-Adams and Adams, 1992, 1996), increasing the risk of photodamage (Powles, 1984). Also, high PPFDs in the presence of water stress could lead to an accumulation of sugars and a down-regulation of photosynthesis (see above).

B. Drought and Photoinhibition

Some reports indicate that in plants suffering water stress for long periods, excitation pressure on PS II may lead to photoinhibitory processes. For instance, the amount of D1 was shown to decrease in water-stressed wheat (J. He et al., 1995) and spruce seedlings (Eastman et al., 1997). Leaves from drought-stressed plants had an enhanced degradation of the light-harvesting antenna CP43 and the reaction center D1 protein (Giardi et al., 1996, 1997). Water stress may also increase the susceptibility of PS II to photoinactivation (Lu and Zhang, 1998).

There is no evidence, however, for major sustained photodamage in plants suffering water stress, as judged by the lack of effects of drought on the maximum potential PS II efficiency, estimated from the dark-adapted F_v/F_m Chl fluorescence ratio (Epron and Dreyer, 1993; Faria et al., 1998; Flexas and Medrano, 2002). Photoinhibition seems to occur only when drought is very severe and stomata are almost completely closed. The decrease in the predawn F_v/F_m ratios in this case is still quite limited, since non-irrigated plants have ratios of 0.72–0.78 (Flexas et al., 1998). Organisms such as the moss *Rhizomnium punctatum* are also well protected against water stress, since no D1 degradation occurred during extensive desiccation (Bartoskova et al., 1999).

C. Drought and Photoprotection

In response to water stress, some species such as barley, coffee, grapevine, and others maintain high leaf Chl concentrations, and therefore high light-harvesting capacities, while having a decreased capacity of utilization of solar energy for photosynthesis (Da Matta et al., 1997; Flexas et al., 2002; Flexas and Medrano, 2002, and references therein; Bukhow and Carpentier, 2004). Conversely, other species could prevent the absorption of an excess of light in the presence of water stress by decreasing leaf Chl concentrations (Kyparissis et al., 2000; Munné-Bosch and Alegre, 2000b), thereby diminishing the capacity for light harvesting. Recent evidences, however, suggest that changes in light harvesting capacity play only a small role in photoprotection (Baroli et al., 2003).

The relative contribution of different processes to total PS II energy dissipation under different drought conditions has been reviewed recently (Flexas and Medrano, 2002). These authors have found a general response pattern of C_3 plants to water stress, relatively independent of species, environmental conditions during imposition of drought (temperature, PPFD, etc.), rate of drought imposition (from days to weeks or months) and also of any specific acclimation to drought (Flexas and Medrano, 2002). It has been proposed that such a general pattern could arise from the low availability of CO_2 in the chloroplasts, which would trigger most of the drought-mediated changes in dissipation pathways (Flexas and Medrano, 2002). In the case of water stress, some studies have given simultaneous data sets of Chl fluorescence and gas exchange, permitting a discrimination among the major components of P (the fraction used in photochemistry), photosynthesis, photorespiration, and Mehler reaction, assuming that the proportion of P allocated to other minor components is negligible (Valentini et al., 1995; Medrano et al., 2002). This type of approach cannot be done yet with Fe- and N-deficiencies, where no such data sets are available.

From the study of Flexas and Medrano (2002), approximately 54–72% of the absorbed light is dissipated thermally (D) under well-watered conditions and saturating light, whereas photosynthesis and photorespiration would utilize 12–35% and 6–9% of the PPFD absorbed by PS II, respectively (Table 1; data from Flexas and Medrano, 2002). Under mild drought there would be no change in the extent of D, whereas the contribution of photorespiration would increase somewhat at the expense of photosynthesis (Table 1). This increase in photorespiration with mild water stress is probably the most common situation in crops and in nature. This occurs, for instance, in plants growing in Mediterranean habitats. Every midday in summer, high vapor-pressure deficits lead to temporary, mild leaf water deficits, which occur irrespective of the degree of soil water availability.

When moderate to severe drought occurs, however, D increases to reach values as high as 70–92% of the absorbed PPFD, and the contribution of photosynthesis and photorespiration decreases (Table 1), possibly due to a decreased RuBP availability and/or decreased Rubisco activity (Lawlor, 1995). Most of this thermal dissipation is associated with the functioning of the xanthophyll cycle (de-epoxidation of V into A + Z), whereas dissipation through photoinactivated photosystem units is only a minor proportion of it (Björkman and Demmig-Adams, 1994; Niyogi, 1999; Flexas and Medrano, 2002). Some species may retain some A + Z (15–20% of the total V + A + Z carotenoids) overnight in summer (Björkman and Demmig-Adams, 1994; Faria et al., 1998; Chaves et al., 2002; Medrano et al., 2002), but the extent of this retention does not seem to be changed by water stress (Chaves et al., 2002). Other species, however, retain large amounts of A + Z overnight in summer under drought (Demmig et al., 1988; Barker et al., 2002) and this retention matches large increases in the rate constant for energy dissipation (Demmig et al., 1988). Björkman (1987) also reported drought-mediated increases in energy dissipation.

The contribution of the Mehler reaction to energy dissipation appears to be very low both under irrigation and drought (Table 1; data from Flexas and Medrano, 2002). Values as high as 9% of total absorbed PPFD found previously in severely stressed wheat (Biehler and Fock, 1996) could have been overestimated due to methodological reasons (Flexas and Medrano, 2002). The contributions of other pathways proposed to dissipate excess light under drought, such as PS II and PS I cyclic electron transport (Canaani and Havaux, 1990; Katona et al., 1992; Fork and Herbert, 1993), are considered to be minor. Although it is known that re-assimilation of CO_2 evolved from photorespiration may increase as drought progresses (Haupt-Herting et al., 2001), estimated rates of dissipation are similar whether or not re-assimilation is considered (Flexas and Medrano, 2002).

Water stress may induce oxidative stress, and in turn trigger anti-oxidative defenses in plants (Smirnoff, 1993; Loggini et al., 1999). For instance, oxidative stress occurred in rice plants subjected to water deficit. These plants showed lipid peroxidation, Chl bleaching and losses of small-molecule antioxidants, such as ascorbate, glutathione, α-tocopherol, and carotenoids (Boo and Jung, 1999). Drought also enhanced lipid peroxidation in *Salvia officinalis* (Munné-Bosch et al., 2001). Protective mechanisms against oxidative stress, such as an increased SOD activity and an activated ascorbate-glutathione cycle (Foyer et al., Mullineaux et al., this volume), have been found in different plant species (Castillo, 1996; Boo and Jung, 1999; Alonso et al., 2001). In species such as *Fagus sylvatica*, *Rosmarinus officinalis*, *Melissa officinalis*, and *Cistus* the α-tocopherol pool increased in response to

Table 1. Fractions of the total light absorbed by PS II that are thermally dissipated within the PS II antenna (D), used in photochemistry, either in RuBP carboxylation (P_C) or oxygenation (P_O), in Mehler reaction or other minor electron-consuming processes in well watered (control) plants and in plants with mild, moderate, or severe drought at saturating light (see Flexas and Medrano, 2002 for further details).

	D	P_C	P_O	Mehler	Other
Control	54–72	12–35	6–9	<1–3	0–3
Mild drought	55–64	5–30	12	<1	0–3
Moderate drought	60–75	10–18	10	3	0–2
Severe drought	70–92	2–10	3–13	1–9	0–6

drought (García-Plazaola and Becerril, 2000; Munné-Bosch and Alegre, 2000a, 2000b; Munné-Bosch et al., 2003). Also, drought increased carotenoid as well as total glutathione and α-tocopherol concentrations in wheat (Herbinger et al., 2002). The antioxidant roles of carotenoids and tocopherols may help to avoid irreversible damage to the photosynthetic apparatus under severe drought. The chloroplast small heat-shock protein has also been shown to play a role in protecting PS II against oxidative stress and photoinhibition during water stress (Downs et al., 1999).

VI. Salinity

Salinity affects many plant processes (Hasegawa et al., 2000) and reduces the productivity of irrigated agriculture worldwide. Salinity-affected areas are estimated to occupy approximately 1 billion ha (Long and Baker, 1986). Salinity has been a traditional problem in many agricultural areas, and irrigation with salt-containing water annually adds millions of tons of salt to cultivated soils (Kingsbury et al., 1984). Modern agriculture management practices often worsen the extent of salinity by remobilizing salts from deep soil layers, previously undisturbed by rainfall. In saline areas or areas irrigated with saline waters, most crop plant species exhibit marked reductions in yield.

A. Effects of Salinity on Photosynthesis

Salinity reduces plant photosynthesis (Long and Baker, 1986). The diminished productivity observed for many plant species subjected to salinity is often associated with a reduced photosynthetic capacity, due both to stomatal limitations and non-stomatal processes (Long and Baker, 1986; Brugnoli and Björkman, 1992; Belkhodja et al., 1999; Delfine et al., 1999; Loreto et al., 2003). Recent evidence suggests, however, that salt stress in olive trees needs to be further studied in terms of diffusional and non-diffusional limitations (Centritto et al., 2003). These recently published results indicate that salinity does not modify the biochemical capacity of olive leaves to assimilate CO_2, and that estimates of photosynthetic rates obtained with salt-stressed leaves would be artifactually low unless stomatal limitation is removed by forcing stomata to open completely (Centritto et al., 2003).

It should also be kept in mind that decreases in crop photosynthesis could be indirect consequences of the impaired physiology of the plants growing under salt stress, since salts taken up by plants may not directly control growth by affecting turgor, photosynthesis, or enzyme activities. Instead, the build-up of salt may hasten death of old leaves, in turn affecting growth by decreasing the supply of assimilates or hormones to the growing regions (Munns, 1993). The possibility that the decreases in photosynthesis under salinity were due to inhibited growth and feedback down-regulation deserves further investigation.

In barley, salinity decreases both photosynthetic capacity and the PPFD at which photosynthesis saturates (Rawson, 1986; Sharma and Hall, 1991; Belkhodja et al., 1999). Chlorophyll concentration, however, is not decreased by salinity in barley, and therefore light absorption by leaves is unaffected (A. Abadía et al., 1999). Therefore, leaves of plants grown under saline conditions are exposed to an excess of excitation energy and consequently to a potential photoinhibition (Mishra et al., 1991; Sharma and Hall, 1991; Belkhodja et al., 1994).

B. Salinity and Photoinhibition

Salinity does not affect the F_v/F_m ratios, measured after dark adaptation, in plants grown under high PPFDs in the field, indicating that salinity does not induce sustained photodamage. This has been shown to occur in halophytes such as *Artimisia anethifolia* (Lu et al., 2003) and also in the crop plant species barley (Belkhodja et al., 1999). Similar conclusions have been obtained at lower PPFDs in species such as cowpea, wheat, barley, spinach, cotton, and bean (Larcher et al., 1990; Brugnoli and Lauteri, 1991; Mishra et al., 1991; Brugnoli and Björkman, 1992; Morales et al., 1992; Belkhodja et al., 1994; Delfine et al., 1999).

However, preliminary data suggest that some perennial species could be more susceptible to salinity than annual plants. For instance, salt-stressed leaves from olive trees could have F_v/F_m ratios as low as 0.5 (Loreto et al., 2003). The underlying causes of these decreases in F_v/F_m ratios have not been explored yet.

C. Salinity and Photoprotection

We have used fluorescence and photosynthesis data published by Brugnoli and Björkman (1992) to assess the partitioning of energy dissipation in cotton grown under salinity, following the approach of Demmig-Adams et al. (1996a). When cotton was grown in a controlled environment at 50% of full sunlight, salinity did not cause changes in the amount of light dissipated thermally by the PS II antenna (D) (Fig. 3). Under low, moderate, and severe salinity, D would account for 37–41%

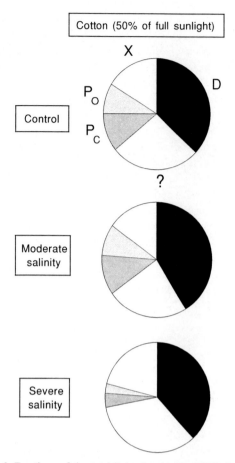

Fig. 3. Fractions of the total light absorbed by PS II that are thermally dissipated within the PS II antenna (D), used in photochemistry, either in RuBP carboxylation (P_C) or oxygenation (P_O), not used in photochemistry nor thermally dissipated within the PS II antenna (X), and used in an unidentified electron-consuming processes (?) in control, moderate, and severe salt-affected cotton leaves. Plants were grown in a growth chamber at approximately 50% full sunlight. Pie areas are proportional to the leaf absorptance. The allocation of energy to D, P, and X follows the methodology proposed by Demmig-Adams et al. (1996a). The separation of P into P_C and P_O are made following Brugnoli and Björkman (1992). The fraction devoted to "?" was calculated as the excess of ETR not dedicated to P_C and P_O, assuming that $P_C + P_O$ consume approximately 4–4.5 mol e- per mol CO_2 (von Caemmerer and Farquhar, 1981). Data were re-calculated from Brugnoli and Björkman (1992).

of the PPFD absorbed by PS II (Fig. 3). These unaffected D values are in good agreement with the lack of effect of salinity on the de-epoxidation state of the xanthophyll cycle pigments in the same plants (Brugnoli and Björkman, 1992). The fraction used in photochemistry, P, did not decrease much with salinity, although a fraction due to unidentified processes included in this fraction would increase with severe salinity at the expense of those dedicated to RuBP carboxylation (P_C)

and oxygenation (P_O) (Fig. 3). Under control, moderate, and severe salinity, unidentified processes would use approximately 27, 24, and 34% of the PPFD absorbed by PS II, respectively, whereas photosynthesis would account for 11, 11, and 4% and photorespiration for 9, 9, and 3% of the PPFD absorbed by PS II (Fig. 3). The existence of unidentified processes consuming one third of the absorbed light may be related to the unusually high electron transport/net photosynthesis ratios in these cotton leaves, which were approximately 15, 18, and 31 under no salinity, moderate, and severe salinity, respectively. The fraction not thermally dissipated nor used in photochemistry (X) increased from 15 to 21% of the PS II absorbed light with salinity (Fig. 3).

Increases of the electron transport/net photosynthesis ratios with salinity have been reported in other plant species, supporting the existence of dissipation processes other than CO_2 fixation. At this stage it is not possible to differentiate between respiration, photorespiration, Mehler reaction, PS II cyclic electron transport, or any other electron-consuming processes. The electron transport/net photosynthesis ratios also increased with salinity in barley, from control values of 6.0 to approximately 12.0 (data recalculated from Belkhodja et al., 1999). Also, salt stress increased this ratio in spinach from control values of 7–8 to approximately 11 (re-calculated from Delfine et al., 1999). Dark respiration does not seem to be an alternative electron-consuming process under salinity. For instance, dark respiration rates were not affected by salinity in several cultivars of olive trees (Centritto et al., 2003; Loreto et al., 2003). One possible way of dissipating energy is through ATP consumption in salt pumping, which has been suggested as an alternative electron-consuming process in plants affected by salinity (Yeo, 1983).

Oxidative stress symptoms have been found in plants grown in saline conditions. Lipid peroxidation has been found in several plant species submitted to salinity (Mishra et al., 1993; Gueta-Dahan et al., 1997; Gómez et al., 1999; Sairam and Srivastava, 2002). Oxidative damage to proteins has been also reported (Hernández et al., 2000). In addition, H_2O_2 concentrations have been reported to increase in plants under salinity (Gómez et al., 1999; Lee et al., 2001; Sairam and Srivastava, 2002). Anti-oxidative mechanisms are triggered by salinity, possibly in response to oxidative stress. For instance, Fe-, Mn-, and Cu/Zn-SOD activities have been found to increase with salt stress in different plant species (Gómez et al., 1999; Lee et al., 2001; Almansa et al., 2002; Broetto et al., 2002; Luna et al., 2002; Sairam and Srivastava, 2002). The ascorbate-glutathione cycle was also activated in salt-stressed plants (Gómez et al., 1999; Lee et al., 2001; Luna et al., 2002; Sairam and Srivastava, 2002). In *Nicotiana plumbaginifolia*, salinity stimulated catalase activity (Savouré et al., 1999), whereas in rice salt stress enhanced ascorbate and guaiacol peroxidase activities while catalase activity decreased (Lee et al., 2001).

VII. Conclusions and Future Research Directions

When plants are under stress, their photosynthetic rates decrease for a variety of reasons. At the same time, they often cannot avoid continuing to gather sunlight. In some abiotic stresses, such as Fe and N deficiency, the amount of Chl decreases, in turn decreasing leaf absorptance. In other common abiotic stresses, such as P deficiency or salinity stress, Chl concentration generally does not change. As a consequence of the imbalance between energy harvesting and the capacity for energy utilization, at high PPFDs plants experience what has been called in a wide sense "photoinhibitory" processes. With the current data available, we can say that although some signs of photodamage are often found, photoprotection mechanisms are far more important. Of course, photoprotective mechanisms in plants are part of a continuum, and the relative importance of each mechanism would depend on the type of abiotic stress and also of the plant species in question.

Plants under stress generally dissipate a large part of the light absorbed by PS II as heat, a process mediated by the xanthophyll cycle pigments within the PS II antenna pigment bed. Cases where very high levels of thermal dissipation occur are frequently, but not always, those involving decreases in leaf Chl concentrations. This has been confirmed for N and Fe deficiencies and also for water stress. Thermal dissipation also usually becomes increasingly important when the intensity of the stress increases. In the case of salinity, however, thermal dissipation does not appear to increase, although this should be confirmed in further experiments. To have an overall assessment of the general importance of thermal dissipation in abiotic stress cases, it is necessary to obtain additional information on the extent of thermal dissipation in other nutrient deficiencies, which are so far unexplored in this respect.

Part of the energy is dissipated by using O_2 as an electron sink in processes that produce active oxygen species. Some symptoms of oxidative damage are frequent in stressed plants, including lipid and protein peroxidation and the presence of detectable amounts of H_2O_2. However, the combined action of energy dissipation processes, antioxidants, and antioxidant enzymatic systems in water stressed, nutrient-limited, and salinity-affected plants seems to maintain a large part of PS II undamaged, and in most cases there is no evidence for major sustained photodamage under such environmental conditions. In general, the concentrations of most oxidative defense components tend to increase with the stress when they are expressed on a basis that considers the capacity to produce reactive oxygen species (i.e. on a chlorophyll basis). In any case, oxidative stress-related processes, which are traditionally regarded as damage, could also be re-interpreted as part of the redox-regulated signal transduction pathways (Demmig-Adams and Adams, 2003). Future research should be focused to investigate these systems in plants under limited nutrient or water availability.

An often-overlooked fact in studies of the effects of abiotic stress on photosynthetic parameters is the existence of major gradients within the leaf (Sun et al., 1996b; Nishio, 2000). Superficial cell layers are exposed to higher effective PPFDs than inner ones, and the composition of cells in the gradient is also different (Sun et al., 1996a). This heterogeneity within the leaf can cause at least two major problems that should be considered. First, some techniques often used to study effects of abiotic stress, such as gas exchange or pigment composition, offer integrative results of all cell layers within the leaf. Chlorophyll fluorescence studies, however, give an inherently larger weight to the effects of stress on the superficial cell layers. Therefore, combining and understanding data obtained using these very different techniques is not always straightforward, and further work is needed in this area.

The combined use of Chl fluorescence and gas exchange techniques (Valentini et al., 1995; Medrano et al., 2002) has been successful for the separation of the electron transport rate into several fractions, including that dedicated to photosynthesis, dark respiration, and photorespiration, in drought-affected plants (Flexas and Medrano, 2002). However, using this approach with Fe- or N-deficient plants has not been possible so far. This is because the ratio between electron transport and the sum of net photosynthetic rates, dark respiration, and photorespiration decreases with Fe and N deficiency (our unpublished observations and data recalculated from the literature) below the theoretical value of 4.5–5 (von Caemmerer and Farquhar, 1981). A possible explanation for this is that the electron transport rate (estimated with Chl fluorescence) and the photosynthesis rate (measured with gas exchange techniques) could originate from different cell populations. The uppermost cell layers, contributing more to the electron transport rate estimation, may be more affected than deeper cell layers, which contribute more to the estimation of gas exchange rates. In the case of water stress, the ratio between electron transport and net photosynthesis never decreases below 5 (Flexas et al., 1999). A second problem arising from leaf heterogeneity is related to thylakoid isolation. Indeed, some of the evidence for photodamage comes from studies carried out with isolated thylakoids. Isolated thylakoids, however, may or may not be representative of the leaves used for thylakoid isolation (Quílez et al., 1992). Because of these reasons, methodologies should be established to assess the extent of photodamage from whole leaf materials instead of isolated thylakoids.

Excessive light interception is avoided in some species by having a canopy structure providing decreases in the exposure of leaf surfaces to sunlight (J. He et al., 1996; Valladares and Pugnaire, 1999; Werner et al., 2002) or by the presence of epidermal trichomes that may decrease incoming light (Morales et al., 2002). For instance, it has been reported that in *Heliconia* a change in leaf orientation from 10 to 80° from the horizontal decreases the PPFD incident on the leaf surface by 41% and results in F_v/F_m ratio increases from 0.63 to 0.72 (He et al., 1996). Some species may also change the light orientation during the day. For instance, both wild soybean grown under low water availability (Kao and Tsai, 1998) and *Arbutus unedo* (Gratani and Ghia, 2002) exhibit at midday more vertical leaf angles. The importance of mechanisms such as leaf movements and leaf folding, exhibited by some species to cope with sustained photoinhibition and PS II damage under water stress, should be also a matter of future research.

Down-regulation of photosynthesis at midday, associated with an enhanced xanthophyll cycle-related thermal energy dissipation process, is advantageous for the survival of plants under abiotic stress conditions, but it may compete with light use efficiency. Does thermal energy dissipation decrease plant productivity? More research is needed to answer this question. On one hand, it has been estimated that midday photosynthesis depression may decrease crop productivity by at least 30–50% (Xu and Shen, 1997). This contention arises from studies in which midday mist irrigation increased photosynthesis, plant biomass, and crop yield in several species (Xu and Shen, 1997, and references therein). However, the possibility that sink limitation may be the origin of the midday photosynthesis depression has not been excluded, and therefore, unless one can demonstrate that the products of photosynthesis (sugars) are limiting under these conditions, it is not possible from these data to conclude that thermal energy dissipation can cause decreases in crop productivity. On the other hand, however, in the first hours after Fe-resupply to Fe-deficient sugar beet plants, xanthophyll cycle pigments shift rapidly towards violaxanthin and photosynthetic rates increase concomitantly, without significant changes in leaf Chl concentration (Larbi et al., 2004). These data show that the conversion of V into A + Z, together with its associated energy dissipation, does compete with PS II photochemistry, decreasing photosynthetic rates and potentially crop productivity. Data from Fe-deficient sugar beet plants show that growth is not much affected (Terry, 1979) and that their leaves have sugar concentrations largely reduced (Arulanantham et al., 1990), which makes the existence of sink limitations in these leaves very unlikely. In any case, more data are needed to ascertain the real impact of photoinhibitory processes in crop productivity, considering both the deleterious impacts and the protective, beneficial ones.

Acknowledgments

This work was supported by Spanish grants AGL2000-1720 and AGL2003-01999 to A.A. and BOS2001-2343 to J.A. The authors gratefully acknowledge Dr. Jaume Flexas for valuable comments and a critical reading of the manuscript.

References

Abadía A, Belkhodja R, Morales F and Abadía J (1999) Effects of salinity on the photosynthetic pigment composition of barley (*Hordeum vulgare* L.) grown under a triple-line-source sprinkler system in the field. J Plant Physiol 154: 392–400

Abadía J and Abadía A (1993) Iron and plant pigments. In: Barton LL, Hemming BC (eds) Iron Chelation in Plants and Soil Microorganisms, pp 327–344. Academic Press, San Diego, California

Abadía J, Morales F and Abadía A (1999) Photosystem II efficiency in low chlorophyll, iron-deficient leaves. Plant Soil 215: 183–192

Adams WW III, Zarter CR, Ebbert V and Demmig-Adams B (2004) Photoprotective strategies of overwintering evergreens. BioScience 54: 41–49

Adams WW III, Zarter CR, Mueh KE, Amiard V and Demmig-Adams B (2005) Energy dissipation and photoinhibition: a continuum of photoprotection. In: Demmig-Adams B, Adams WW III and Mattoo AK (eds) Photoprotection, Photoinhibition, Gene Regulation, and Environment, pp 49–64. Springer, Dordrecht

Agbariah K-T and Roth-Bejerano N (1990) The effect of blue light on energy levels in epidermal strips. Physiol Plant 78: 100–104

Almansa MS, Hernández JA, Jiménez A, Botella MA and Sevilla F (2002) Effect of salt stress on the superoxide dismutase activity in leaves of *Citrus limonum* in different rootstock-scion combinations. Biol Plant 45: 545–549

Alonso R, Elvira S, Castillo FJ and Gimeno BS (2001) Interactive effects of ozone and drought stress on pigments and activities of antioxidative enzymes in *Pinus halepensis*. Plant Cell Environ 24: 905–916

Álvarez-Fernández A, Paniagua P, Abadía J and Abadía A (2003) Effects of Fe deficiency chlorosis on yield and fruit quality in peach (*Prunus persica* L. Batsch). J Agric Food Chem 51: 5738–5744

Andrizhiyevskaya EG, Schwabe TME, Germano M, D'Haene S, Kruip J, van Grondelle R and Dekker JP (2002) Spectroscopic properties of PSI-IsiA supercomplexes from the cyanobacterium *Synechococcus* PCC 7942. Biochim Biophys Acta 1556: 265–272

Arulanantham AR, Rao M and Terry N (1990) Limiting factors in photosynthesis. VI. Regeneration of ribulose 1,5-*bis*phosphate limits photosynthesis at low photochemical capacity. Plant Physiol 93: 1466–1475

Balachandran S and Osmond CB (1994) Susceptibility of tobacco leaves to photoinhibition following infection with two strains of tobacco mosaic virus under different light and nitrogen nutrition regimes. Plant Physiol 104: 1051–1057

Barker DH, Adams WW III, Demmig-Adams B, Logan BA, Verhoeven AS and Smith SD (2002) Nocturnally retained zeaxanthin does not remain engaged in a state primed for energy dissipation during the summer in two *Yucca* species growing in the Mojave Desert. Plant Cell Environ 25: 95–103

Baroli I, Do AD, Yamane T and Niyogi KK (2003) Zeaxanthin accumulation in the absence of a functional xanthophyll cycle protects *Chlamydomonas reinhardtii* from photooxidative stress. Plant Cell 15: 992–1008

Bartoskova H, Komenda J and Naus J (1999) Functional changes of photosystem II in the moss *Rhizomnium punctatum* (Hedw.) induced by different rates of dark desiccation. J Plant Physiol 154: 597–604

Basile B, Reidel EJ, Weinbaum SA and DeJong TM (2003) Leaf potassium concentration, CO_2 exchange and light interception in almond trees (*Prunus dulcis* (Mill) D.A. Webb). Sci Hort 98: 185–194

Bednarz CW and Oosterhuis DM (1999) Physiological changes associated with potassium deficiency in cotton. J Plant Nutr 22: 303–313

Behrenfeld MJ, Bale AJ, Kolber ZS, Aiken J and Falkowski PG (1996) Confirmation of iron limitation of phytoplankton photosynthesis in the equatorial Pacific Ocean. Nature 383: 508–511

Belkhodja R, Morales F, Abadía A, Gómez-Aparisi J and Abadía J (1994) Chlorophyll fluorescence as a possible tool for salinity tolerance screening in barley (*Hordeum vulgare* L.). Plant Physiol 104: 667–673

Belkhodja R, Morales F, Quílez R, López-Millán AF, Abadía A and Abadía J (1998) Iron deficiency causes changes in chlorophyll fluorescence due to the reduction in the dark of the Photosystem II acceptor side. Photosynth Res 56: 265–276

Belkhodja R, Morales F, Abadía A, Medrano H and Abadía J (1999) Effects of salinity on chlorophyll fluorescence and photosynthesis of barley (*Hordeum vulgare* L.) grown under a triple-line-source sprinkler system in the field. Photosynthetica 36: 375–387

Bergmann T, Richardson TL, Paerl HW, Pinckney JL and Schofield O (2002) Synergy of light and nutrients on the photosynthetic efficiency of phytoplankton populations from the Neuse River Estuary, North Carolina. J Plankton Res 24: 923–933

Bertamini M, Muthuchelian K and Nedunchezhian N (2002) Iron deficiency induced changes on the donor side of PS II in field grown grapevine (*Vitis vinifera* L. cv. Pinot noir) leaves. Plant Sci 162: 599–605

Bibby TS, Nield J and Barber J (2001a) Iron deficiency induces the formation of an antenna ring around trimeric photosystem I in cyanobacteria. Nature 412: 743–745

Bibby TS, Nield J and Barber J (2001b) Three-dimensional model and characterization of the iron stress-induced CP43'-photosystem I supercomplex isolated from the cyanobacterium *Synechocystis* PCC 6803. J Biol Chem 276: 43246–43252

Biehler K and Fock H (1996) Evidence for the contribution of the Mehler-peroxidase reaction in dissipating excess electrons in drought-stressed wheat. Plant Physiol 112: 265–272

Biswal B, Smith AJ and Rogers LJ (1994) Changes in carotenoids but not in D1 protein in response to nitrogen depletion and recovery in a cyanobacterium. FEMS Microbiol Lett 116: 341–347

Björkman O (1987) High-irradiance stress in higher plants and interaction with other stress factors. In: Biggins J (ed) Progress in Photosynthesis Research, Vol 4, pp 11–18. Martinus Nijhoff, Dordrecht

Björkman O and Demmig-Adams B (1994) Regulation of photosynthetic light energy capture, conversion, and dissipation in leaves of higher plants. In: Schulze E-D, Caldwell MM (eds) Ecophysiology of Photosynthesis, pp 17–47. Springer-Verlag, Berlin

Boekema EJ, Hifney A, Yakushevska AE, Piotrowski M, Keegstra W, Berry S, Michel KP, Pistorius EK and Kruip J (2001) A giant chlorophyll-protein complex induced by iron deficiency in cyanobacteria. Nature 412: 745–748

Boo Y and Jung J (1999) Water deficit-induced oxidative stress and antioxidative defenses in rice plants. J Plant Physiol 155: 255–261

Broetto F, Lüttge U and Ratajczak R (2002) Influence of light intensity and salt-treatment on mode of photosynthesis and enzymes of the antioxidative response system of *Mesembryanthemum crystallinum*. Funct Plant Biol 29: 13–23

Brugnoli E and Björkman O (1992) Growth of cotton under continuous salinity stress: influence on allocation pattern, stomatal and non-stomatal components of photosynthesis and dissipation of excess light energy. Planta 187: 335–347

Brugnoli E and Lauteri M (1991) Effects of salinity on stomatal conductance, photosynthetic capacity, and carbon isotope discrimination of salt tolerant (*Gossypium hirsutum* L.) and salt sensitive (*Phaseolus vulgaris* L.) C_3 non-halophytes. Plant Physiol 95: 628–635

Bukhow NG and Carpentier R (2004) Effects of water stress on the photosynthetic efficiency of plants. In: Papageorgiou GC and Govindjee (eds) Chlorophyll a Fluorescence: A Signature of Photosynthesis, in press. Springer, Berlin

Bungard RA, McNeil D and Morton JD (1997) Effects of nitrogen on the photosynthetic apparatus of *Clematis vitalba* grown at several irradiances. Aust J Plant Physiol 24: 205–214

Bungard RA, Press MC and Scholes JD (2000) The influence of nitrogen on rain forest dipterocarp seedlings exposed to a large increase in irradiance. Plant Cell Environ 23: 1183–1194

Canaani O and Havaux M (1990) Evidence for a biological role in photosynthesis for cytocrome b-559, a component of Photosystem II reaction center. Proc Natl Acad Sci USA 87: 9295–9299

Castillo FJ (1996) Antioxidative protection in the inducible Cam plant *Sedum album* L. following the imposition of severe water stress and recovery. Oecologia 107: 469–477

Centritto M, Loreto F and Chartzoulakis K (2003) The use of low [CO_2] to estimate diffusional and non-diffusional limitations of photosynthetic capacity of salt-stressed olive saplings. Plant Cell Environ 26: 585–594

Chaves MM, Pereira JS, Maroco J, Rodrigues ML, Ricardo CPP, Osório ML, Carvalho I, Faria T and Pinheiro C (2002) How plants cope with water stress in the field. Photosynthesis and growth. Ann Bot 89: 907–916

Chen L, Fuchigami LH and Breen PJ (2001) The relationship between Photosystem II efficiency and quantum yield for CO_2 assimilation is not affected by nitrogen content in apple leaves. J Exp Bot 52: 1865–1872

Chen YZ, Murchie EH, Hubbart S, Horton P and Peng SB (2003) Effects of season-dependent irradiance levels and nitrogen-deficiency on photosynthesis and photoinhibition in field-grown rice (*Oryza sativa*). Physiol Plant 117: 343–351

Close DC, Beadle CL and Hovenden MJ (2003) Interactive effects of nitrogen and irradiance on sustained xanthophyll cycle engagement in *Eucalyptus nitens* leaves during winter. Oecologia 134: 32–36

Da Matta FM, Maestri M and Barros RS (1997) Photosynthetic performance of two coffee species under drought. Photosynthetica 34: 257–264

Dannehl H, Herbik A and Godde D (1995) Stress-induced degradation of the photosynthetic apparatus is accompanied by changes in thylakoid protein turnover and phosphorylation. Physiol Plant 93: 179–186

Dannehl H, Wietoska H, Heckmann H and Godde D (1996) Changes in D1-protein turnover and recovery of photosystem II activity precede accumulation of chlorophyll in plants after release from mineral stress. Planta 199: 34–42

Delfine S, Alvino A, Villani MC and Loreto F (1999) Restrictions to CO_2 conductance and photosynthesis in spinach leaves recovering from salt stress. Plant Physiol 119: 1101–1106

Demmig B, Winter K, Krüger A and Czygan F-C (1988) Zeaxanthin and the heat dissipation of excess light energy in *Nerium oleander* exposed to a combination of high light and water stress. Plant Physiol 87: 17–24

Demmig-Adams B and Adams WW III (1992) Photoprotection and other responses of plants to high light stress. Annu Rev Plant Physiol Plant Mol Biol 43: 599–626

Demmig-Adams B and Adams WW III (1996) Xanthophyll cycle and light stress in nature: uniform response to excess direct sunlight among higher plant species. Planta 198: 460–470

Demmig-Adams B and Adams WW III (2002) Antioxidants in photosynthesis and human nutrition. Science 298: 2149–2153

Demmig-Adams B and Adams WW III (2003) Photoinhibition. In: Thomas B, Murphy D and Murray B (eds) Encyclopedia of Applied Plant Science, pp 707–714. Academic Press, New York

Demmig-Adams B, Adams WW III, Barker DH, Logan BA, Bowling DR and Verhoeven AS (1996a) Using chlorophyll fluorescence to assess the fraction of absorbed light allocated to thermal dissipation of excess excitation. Physiol Plant 98: 253–264

Demmig-Adams B, Gilmore AM and Adams WW III (1996b) *In vivo* functions of carotenoids in higher plants. FASEB J 10: 403–412

Demmig-Adams B, Ebbert V, Zarter CR and Adams WW III (2005) Characteristics and species-dependent employment of flexible versus sustained thermal dissipation and photoinhibition. In: Demmig-Adams B, Adams WW III and Mattoo AK (eds) Photoprotection, Photoinhibition, Gene Regulation, and Environment, pp 39–48. Springer, Dordrecht

Downs CA, Ryan SL and Heckathorn SA (1999) The chloroplast small heat-shock protein: Evidence for a general role in protecting photosystem II against oxidative stress and photoinhibition. J Plant Physiol 155: 488–496

Dubey RS (1997) Photosynthesis in plants under stressful conditions. In: Pessarakli M (ed) Handbook of Photosynthesis, pp 859–875. Marcel Dekker Inc, New York

Duncan J, Bibby T, Tanaka A and Barber J (2003) Exploring the ability of chlorophyll b to bind to the CP43' protein induced under iron deprivation in a mutant of *Synechocystis* PCC 6803 containing the cao gene. FEBS Lett 541: 171–175

Eastman PAK, Rashid A and Camm EL (1997) Changes of the photosystem 2 activity and thylakoid proteins in spruce seedlings during water stress. Photosynthetica 34: 201–210

Egilla JN and Davies FT Jr (1995) Response of *Hibiscus rosa-sinensis* L. to varying levels of potassium fertilization: growth, gas exchange and mineral concentration. J Plant Nutr 18: 1765–1783

Epron D and Dreyer E (1993) Photosynthesis of oak leaves under water stress: maintenance of high photochemical efficiency of photosystem II and occurrence of non-uniform CO_2 assimilation. Tree Physiol 13: 107–117

Evans JR and Terashima I (1987) Effects of nitrogen nutrition on electron transport components and photosynthesis in spinach. Aust J Plant Physiol 14: 59–68

Falkowski PG, Behrenfeld M and Kolber Z (1995) Variations in photochemical energy conversion efficiency in oceanic phytoplankton: scaling from reaction center to the global ocean. In: Mathis P (ed) Photosynthesis: from Light to Biosphere, Vol V, pp 755–759. Kluwer Academic Publishers, Dordrecht

FAO (1988) FAO-UNESCO Soil map of the world, revised legend. FAO, Rome

Faria T, Silvério D, Breia E, Cabral R, Abadía A, Abadía J, Pereira JS and Chaves MM (1998) Differences in the response of carbon assimilation to summer stress (water deficits, high light and temperature) in four Mediterranean tree species. Physiol Plant 102: 419–428

Ferrar PJ and Osmond CB (1986) Nitrogen supply as a factor influencing photoinhibition and photosynthetic acclimation after transfer of shade-grown *Solanum dulcamara* to bright light. Planta 168: 563–570

Ferraro F, Castagna A, Soldatini GF and Ranieri A (2003) Tomato (*Licopersicon esculentum* M.) T3238*FER* and T3238*fer* genotypes. Influence of different iron concentrations on thylakoid pigment and protein composition. Plant Sci 164: 783–792

Field C and Mooney H (1986) The photosynthesis-nitrogen relationship in wild plants. In: Givnish GT (ed) On the Economy of Plant Form and Function, pp 25–55. Cambridge University Press, London

Flexas J and Medrano H (2002) Energy dissipation in C_3 plants under drought. Funct Plant Biol 29: 1209–1215

Flexas J, Escalona JM and Medrano H (1998) Down-regulation of photosynthesis by drought under field conditions in grapevine leaves. Aust J Plant Physiol 25: 893–900

Flexas J, Escalona JM and Medrano H (1999) Water stress induces different levels of photosynthesis and electron transport rate regulation in grapevines. Plant Cell Environ 22: 39–48

Flexas J, Bota J, Escalona JM, Sampól B and Medrano H (2002) Effects of drought on photosynthesis in grapevines under field conditions: an evaluation of stomatal and mesophyll limitations. Funct Plant Biol 29: 461–471

Fork DC and Herbert SK (1993) Electron transport and photophosphorylation by Photosystem I *in vivo* in plants and cyanobacteria. Photosynth Res 36: 149–168

Foyer CH, Lelandais M and Kunert KJ (1994) Photooxidative stress in plants. Physiol Plant 92: 696–717

Foyer CH, Trebst A and Noctor G (2005) Signaling and integration of defense functions of tocopherol, ascorbate, and glutathione. In: Demmig-Adams B, Adams WW III and Mattoo AK (eds) Photoprotection, Photoinhibition, Gene Regulation, and Environment, pp 241–268. Springer, Dordrecht

García-Plazaola JI and Becerril JM (2000) Effects of drought on photoprotective mechanisms in European beech (*Fagus sylvatica* L.) seedlings from different provenances. Trees 14: 485–490

Geider RJ and La Roche J (1994) The role of iron in phytoplankton photosynthesis, and the potential for iron-limitation of primary productivity in the sea. Photosynth Res 39: 275–301

Geider RJ, Laroche J, Greene RM and Olaizola M (1993) Response of the photosynthetic apparatus of *Phaeodactylum tricornutum* (Bacillariophyceae) to nitrate, phosphate, or iron starvation. J Phycol 29: 755–766

Giardi MT, Cona A, Geiken B, Kucera T, Masojidek J and Mattoo AK (1996) Long-term drought stress induces structural and functional reorganization of photosystem II. Planta 199: 118–125

Giardi MT, Masojidek J and Godde D (1997) Effects of abiotic stresses on the turnover of the D-1 reaction centre II protein. Physiol Plant 101: 635–642

Gilmore AM, Matsubara S, Ball MC, Barker DH and Itoh S (2003) Excitation energy flow at 77 K in the photosynthetic apparatus of overwintering evergreens. Plant Cell Environ 26: 1021–1034

Godde D and Dannehl H (1994) Stress-induced chlorosis and increase in D1-protein turnover precede photoinhibition in spinach suffering under magnesium/sulphur deficiency. Planta 195: 291–300

Godde D and Hefer M (1994) Photoinhibition and light-dependent turnover of the D1 reaction-centre polypeptide of photosystem-II are enhanced by mineral-stress conditions. Planta 193: 290–299

Gómez JM, Hernández JA, Jiménez A, del Río LA and Sevilla F (1999) Differential response of antioxidative enzymes of chloroplasts and mitochondria to long-term NaCl stress of pea plants. Free Radical Res 31 (Suppl S): S11–S18

Grassi G, Colom MR and Minotta G (2001) Effects of nutrient supply on photosynthetic acclimation and photoinhibition of one-year-old foliage of *Picea abies*. Physiol Plant 111: 245–254

Gratani L and Ghia E (2002) Adaptive strategy at the leaf level of *Arbutus unedo* L. to cope with Mediterranean climate. Flora 197: 275–284

Greene RM, Geider RJ, Kolber Z and Falkowski PG (1992) Iron-induced changes in light harvesting and photochemical energy conversion processes in eukaryotic marine algae. Plant Physiol 100: 565–575

Gueta-Dahan Y, Yaniv Z, Zilinskas BA, Benhayyim G (1997) Salt and oxidative stress: similar and specific responses and their relation to salt tolerance in Citrus. Planta 203: 460-469

Guikema JA (1985) Fluorescence induction characteristics of *Anacystis nidulans* during recovery from iron deficiency. J Plant Nutr 8: 891–908

Hasegawa PM, Bressan RA, Zhu, JK and Bohnert HG (2000) Plant cellular and molecular responses to high salinity. Annu Rev Plant Physiol Plant Mol Biol 51: 463-499

Haupt-Herting S, Klug K and Fock H (2001) A new approach to measure gross CO_2 fluxes in leaves. Gross CO_2 assimilation, photorespiration, and mitochondrial respiration in the light in tomato under drought stress. Plant Physiol 126: 388–396

Havaux M and Niyogi KK (1999) The violaxanthin cycle protects plants from photooxidative damage by more than one mechanism. Proc Natl Acad Sci USA 96: 8762–8767

He J, Chee CW and Goh CJ (1996) "Photoinhibition" of *Heliconia* under natural tropical conditions: the importance of leaf orientation for light interception and leaf temperature. Plant Cell Environ 19: 1238–1248

He J, Tan LP and Goh CJ (2000) Alleviation of photoinhibition in *Heliconia* grown under tropical natural conditions after release from nutrient stress. J Plant Nutr 23: 181–196

He JX, Wang J and Liang HG (1995) Effects of water stress on photochemical function and protein metabolism of photosystem II in wheat leaves. Physiol Plant 93: 771–777

Henley WJ, Levavasseur G, Franklin LA, Osmond CB and Ramus J (1991) Photoacclimation and photoinhibition in *Ulva rotundata* as influenced by nitrogen availability. Planta 184: 235–243

Herbinger K, Tausz M, Wonisch A, Soja G, Sorger A and Grill D (2002) Complex interactive effects of drought and ozone stress on the antioxidant defence systems of two wheat cultivars. Plant Physiol Biochem 40: 691–696

Hernández JA, Jiménez A, Mullineaux P and Sevilla F (2000) Tolerance of pea (*Pisum sativum* L.) to long term salt stress is associated with induction of antioxidant defences. Plant Cell Environ 23: 853–862

Iglesias DJ, Lliso I, Tadeo FR and Talón M (2002) Regulation of photosynthesis through source:sink imbalance in citrus is mediated by carbohydrate content in leaves. Physiol Plant 116: 563–572

Iturbe-Ormaetxe I, Morán JF, Arrese-Igor C, Gogorcena Y, Klucas RV and Becana M (1995) Activated oxygen and antioxidant defences in iron-deficient pea plants. Plant Cell Environ 18: 421–429

Jiang CD, Gao HY and Zou Q (2001) Enhanced thermal energy dissipation depending on xanthophyll cycle and D1 protein turnover in iron-deficient maize leaves under high irradiance. Photosynthetica 39: 269–274

Kao W-Y and Tsai T-T (1998) Tropic leaf movements, photosynthetic gas exchange, leaf $\delta^{13}C$ and chlorophyll a fluorescence of three soybean species in response to water availability. Plant Cell Environ 21: 1055–1062

Kato MC, Hikosaka K and Hirose T (2002) Photoinactivation and recovery of photosystem II in *Chenopodium album* leaves grown at different levels of irradiance and nitrogen availability. Funct Plant Biol 29: 787–795

Katona E, Neimais S, Schönknechst G and Heber U (1992) Photosystem I-dependent cyclic electron transport is important in controlling Photosystem II activity in leaves under water stress. Photosynth Res 34: 449–469

Khamis S, Lamaze T, Lemoine Y and Foyer CH (1990) Adaptation of the photosynthetic apparatus in maize leaves as a result of nitrogen limitation. Relationships between electron transport and carbon assimilation. Plant Physiol 94: 1436–1443

Kilb B, Wietowska K and Godde D (1996) Changes in the expression of photosynthetic genes precede the loss of photosynthetic activities and chlorophyll when glucose is supplied to mature spinach leaves. Plant Sci 114: 225–235

Kingsbury RW, Epstein E and Pearcy RW (1984) Physiological responses to salinity in selected lines of wheat. Plant Physiol 74: 417–423

Kurepa J, Bueno P, Kampfenkel K, Van Montagu M and Vandenbulcke M (1997) Effects of iron deficiency on iron superoxide dismutase expression in *Nicotiana tabacum*. Plant Physiol Biochem 35: 467–474

Kyparissis A, Drilias P and Manetas Y (2000) Seasonal fluctuations in photoprotective (xanthophyll cycle) and photoselective (chlorophylls) capacity in eight Mediterranean plant species belonging to two different growth forms. Aust J Plant Physiol 27: 265–272

Larbi A, Abadía A, Morales F and Abadía J (2004) Fe resupply to Fe-deficient sugar beet plants leads to rapid changes in the violaxanthin cycle and other photosynthetic characteristics without significant *de novo* chlorophyll synthesis. Photosynth Res 79: 59–69

Larcher W, Wagner J and Thammathaworn A (1990) Effects of superimposed temperature stress on *in vivo* chlorophyll fluorescence of *Vigna unguiculata* under saline stress. J Plant Physiol 136: 92–102

Lawlor DW (1995) The effects of water deficit on photosynthesis. In: Smirnoff M (ed) Environment and Plant Metabolism. Flexibility and Acclimation, pp 129–160. BIOS Scientific, Oxford

Le Roux X, Walcroft AS, Daudet FA, Sinoquet H, Chaves MM, Rodrigues A and Osorio L (2001) Photosynthetic light acclimation in peach leaves: importance of changes in mass:area ratio, nitrogen concentration, and leaf nitrogen partitioning. Tree Physiol 21: 377–386

Lee DH, Kim YS and Lee CB (2001) The inductive responses of the antioxidant enzymes by salt stress in the rice (*Oryza sativa* L.). J Plant Physiol 158: 737–745

Logan BA, Demmig-Adams B, Rosenstiel TN and Adams WW III (1999) Effect of nitrogen limitation on foliar antioxidants in relationship to other metabolic characteristics. Planta 209: 213–220

Loggini B, Scartazza A, Brugnoli E and Navari-Izzo F (1999) Antioxidative defense system, pigment composition, and photosynthetic efficiency in two wheat cultivars subjected to drought. Plant Physiol 119: 1091–1099

Long SP and Baker NR (1986) Saline terrestrial environments. In: Baker NR and Long SP (eds) Photosynthesis in Contrasting Environments, pp 63–102. Elsevier, New York

Long SP, Humphries S and Falkowski PG (1994) Photoinhibition of photosynthesis in nature. Annu Rev Plant Physiol Plant Mol Biol 45: 633–662

Loreto F, Centritto M and Chartzoulakis K (2003) Photosynthetic limitations in olive cultivars with different sensitivity to salt stress. Plant Cell Environ 26: 595–601

Lu CM and Zhang JH (1998) Effects of water stress on photosynthesis, chlorophyll fluorescence and photoinhibition in wheat plants. Aust J Plant Physiol 25: 883–892

Lu CM and Zhang JH (2000) Photosystem II photochemistry and its sensitivity to heat stress in maize plants as affected by nitrogen deficiency. J Plant Physiol 157: 124–130

Lu CM, Jiang G, Wang B and Kuang T (2003) Photosystem II photochemistry and photosynthetic pigment composition in salt-adapted halophyte *Artimisia anethifolia* grown under outdoor conditions. J Plant Physiol 160: 403–408

Luna C, de Luca M and Taleisnik E (2002) Physiological causes for decreased productivity under high salinity in Boma, a tetraploid *Chloris gayana* cultivar. II. Oxidative stress. Aust J Agr Res 53: 663–669

Makino A and Osmond CB (1991) Effects of nitrogen nutrition on nitrogen partitioning between chloroplast and mitochondria in pea and wheat. Plant Physiol 96: 355–362

Marquardt J, Schultze A, Rosenkranz V and Wehrmeyer W (1999) Ultrastructure and photosynthetic apparatus of *Rhodella violacea* (Porphyridiales, Rhodophyta) grown under iron-deficient conditions. Phycol 38: 418–427

Marschner H (1995) Mineral nutrition of higher plants. Academic Press, London

Masoni A, Ercoli L and Mariotti M (1996) Spectral properties of leaves deficient in iron, sulfur, magnesium and manganese. Agron J 88: 937–943

Medrano H, Bota J, Abadía A, Sampól B, Escalona JM and Flexas J (2002) Effects of drought on light-energy dissipation mechanisms in high-light-acclimated, field-grown grapevines. Funct Plant Biol 29: 1197–1207

Medrano H, Parry MAJ, Socías X and Lawlor DW (1998) Long term water stress inactivates Rubisco in subterranean clover. Ann Appl Biol 131: 491–501

Melkozernov AN, Bibby TS, Lin S, Barber J and Blankenship RE (2003) Time-resolved absorption and emission show that the CP43' antenna ring of iron-stressed *Synechocystis* sp PCC6803 is efficiently coupled to the photosystem I reaction center core. Biochemistry 42: 3893–3903

Mishra SK, Subrahmanyam D and Singhal GS (1991) Interrelationship between salt and light stress on primary processes of photosynthesis. J Plant Physiol 138: 92–96

Mishra SK, Dogra JVV and Singhal GS (1993) Influence of high salt on protein contents and lipid peroxidation and their interaction with high photon flux density on isolated chloroplasts of mustard. J Plant Biochem Biotech 2: 39–42

Morales F, Abadía A and Abadía J (1990) Characterization of the xanthophyll cycle and other photosynthetic pigment changes induced by iron deficiency in sugar beet (*Beta vulgaris* L.). Plant Physiol 94: 607–613

Morales F, Abadía A and Abadía J (1991) Chlorophyll fluorescence and photon yield of oxygen evolution in iron-deficient sugar beet (*Beta vulgaris* L.) leaves. Plant Physiol 97: 886–893

Morales F, Abadía A, Gómez-Aparisi J and Abadía J (1992) Effects of combined NaCl and CaCl₂ salinity on photosynthetic parameters of barley grown in nutrient solution. Physiol Plant 86: 419–426

Morales F, Belkhodja R, Abadía A and Abadía J (1994) Iron deficiency-induced changes in the photosynthetic pigment composition of field-grown pear (*Pyrus communis* L.) leaves. Plant Cell Environ 17: 1153–1160

Morales F, Abadía A and Abadía J (1998) Photosynthesis, quenching of chlorophyll fluorescence and thermal energy dissipation in iron-deficient sugar beet leaves. Aust J Plant Physiol 25: 403–412

Morales F, Belkhodja R, Abadía A and Abadía J (2000a) Photosystem II efficiency and mechanisms of energy dissipation in iron-deficient, field-grown pear trees (*Pyrus communis* L.). Photosynth Res 63: 9–21

Morales F, Belkhodja R, Abadía A and Abadía J (2000b) Energy dissipation in the leaves of Fe-deficient pear trees grown in the field. J Plant Nutr 23: 1709–1716

Morales F, Moise N, Quílez R, Abadía A, Abadía J and Moya I (2001) Iron deficiency interrupts energy transfer from a disconnected part of the antenna to the rest of Photosystem II. Photosynth Res 70: 207–220

Morales F, Abadía A, Abadía J, Montserrat G and Gil-Pelegrín E (2002) Trichomes and photosynthetic pigment composition changes: responses of *Quercus ilex* subsp. *ballota* (Desf.) Samp. and *Quercus coccifera* L. to Mediterranean stress conditions. Trees 16: 504–510

Mortvedt JJ (1991) Correcting iron deficiencies in annual and perennial plants: Present technologies and future prospects. Plant Soil 130: 273–279

Moseley JL, Allinger T, Herzog S, Hoerth P, Wehinger E, Merchant S and Hippler M (2002) Adaptation to Fe-deficiency requires remodeling of the photosynthetic apparatus. EMBO J 21: 6709–6720

Müller-Moulé P, Havaux M and Niyogi KK (2003) Zeaxanthin deficiency enhances the high light sensitivity of an ascorbate-deficient mutant of *Arabidopsis*. Plant Physiol 133: 748–760

Mullineaux PM, Karpinski S and Creissen GP (2005) Integration of signaling in antioxidant defenses. In: Demmig-Adams B, Adams WW III and Mattoo AK (eds) Photoprotection, Photoinhibition, Gene Regulation, and Environment, this volume. Springer, Berlin

Munné-Bosch S and Alegre L (2000a) Changes in carotenoids, tocopherols and diterpenes during drought and recovery, and the biological significance of chlorophyll loss in *Rosmarinus officinalis* plants. Planta 210: 925–931

Munné-Bosch S and Alegre L (2000b) The significance of β-carotene, α-tocopherol and the xanthophyll cycle in droughted *Melissa officinalis* plants. Aust J Plant Physiol 27: 139–146

Munné-Bosch S, Jubany-Mari T and Alegre L (2001) Drought-induced senescence is characterized by a loss of antioxidant defences in chloroplasts. Plant Cell Environ 24: 1319–1327

Munné-Bosch S, Jubany-Mari T and Alegre L (2003) Enhanced photo- and antioxidative protection, and hydrogen peroxide accumulation in drought-stressed *Cistus clusii* and *Cistus albidus* plants. Tree Physiol 23: 1–12

Munns R (1993) Physiological processes limiting plant growth in saline soils: some dogmas and hypotheses. Plant Cell Environ 16: 15–24

Nedunchezhian N, Morales F, Abadía A and Abadía J (1997) Decline in photosynthetic electron transport activity and changes in thylakoid protein pattern in field grown iron deficient peach (*Prunus persica* L.). Plant Sci 129: 29–38

Nishio JN (2000) Why are higher plants green? Evolution of the higher plant photosynthetic pigment complement. Plant Cell Environ 23: 539–448

Niyogi KK (1999) Photoprotection revisited: genetic and molecular approaches. Annu Rev Plant Physiol Plant Mol Biol 50: 333–359

Park YI, Sandstrom S, Gustafsson P and Öquist G (1999) Expression of the isiA gene is essential for the survival of the cyanobacterium *Synechococcus* sp. PCC 7942 by protecting photosystem II from excess light under iron limitation. Mol Microbiol 32: 123–129

Patsikka E, Kairavuo M, Sersen F, Aro E-M and Tyystjärvi E (2002) Excess copper predisposes photosystem II to photoinhibition *in vivo* by outcompeting iron and causing decrease in leaf chlorophyll. Plant Physiol 129: 1359–1367

Paul MJ and Driscoll SP (1997) Sugar repression of photosynthesis: the role of carbohydrates in signalling nitrogen deficiency through source:sink imbalance. Plant, Cell Environ 20: 110–116

Paul MJ and Foyer CH (2001) Sink regulation of photosynthesis. J Exp Bot 52: 1383–1400

Peoples TR and Koch DW (1979) Role of potassium in carbon dioxide assimilation in *Medicago sativa* L. Plant Physiol 63: 878–881

Powles SB (1984) Photoinhibition of photosynthesis induced by visible light. Annu Rev Plant Physiol 35: 15–44

Quílez R, Abadía A and Abadía J (1992) Characteristics of thylakoids and photosystem II membrane preparations from iron deficient and iron sufficient sugar beet (*Beta vulgaris* L.). J Plant Nutr 15: 1809–1819

Ramalho JC, Campos PS, Teixeira M and Nunes MA (1998) Nitrogen dependent changes in antioxidant system and in fatty acid composition of chloroplast membranes from *Coffea arabica* L. plants submitted to high irradiance. Plant Sci 135: 115–124

Ranieri A, Castagna A, Baldan B and Soldatini GF (2001) Iron deficiency differently affects peroxidase isoforms in sunflower. J Exp Bot 52: 25–35

Rao IM and Terry N (1989a) Leaf phosphate status, photosynthesis and carbon partitioning in sugar beet. I. Changes in growth, gas exchange and Calvin cycle enzymes. Plant Physiol 90: 814–819

Rao IM and Terry N (1989b) Leaf phosphate status, photosynthesis and carbon partitioning in sugar beet. II. Diurnal changes in sugar phosphates, adenylates and nicotinamide nucleotides. Plant Physiol 90: 820–826

Rao IM and Terry N (1990) Leaf phosphate status, photosynthesis and carbon partitioning in sugar beet. III. Diurnal changes in carbon partitioning and carbon export. Plant Physiol 92: 29–36

Rawson HM (1986) Gas exchange and growth in wheat and barley grown in salt. Aust J Plant Physiol 13: 475–489

Sage RF and Pearcy RW (1987a) The nitrogen use efficiency of C₃ and C₄ plants. I. Leaf nitrogen, growth and biomass partitioning in *Chenopodium album* L. and *Amaranthus retroflexus* L. Plant Physiol 84: 954–958

Sage RF and Pearcy RW (1987b) The nitrogen use efficiency of C₃ and C₄ plants. II. Leaf nitrogen effects on the gas exchange characteristics of *Chenopodium album* L. and *Amaranthus retroflexus* L. Plant Physiol 84: 959–963

Sairam RK and Srivastava GC (2002) Changes in antioxidant activity in sub-cellular fractions of tolerant and susceptible wheat genotypes in response to long term salt stress. Plant Sci 162: 897–904

Sandstrom S, Park YI, Öquist G and Gustafsson P (2001) CP43', the isiA gene product, functions as an excitation energy dissipator in the cyanobacterium *Synechococcus* sp PCC 7942. Photochem Photobiol 74: 431–437

Sandstrom S, Ivanov AG, Park YI, Öquist G and Gustafsson P (2002) Iron stress responses in the cyanobacterium *Synechococcus* sp PCC7942. Physiol Plant 116: 255–263

Sanz M, Cavero J and Abadía J (1992) Iron chlorosis in the Ebro river basin, Spain. J Plant Nutr 15: 1971–1981

Savouré A, Thorin D, Davey M, Hua X-J, Mauro S, Van Montagu M, Inzé D and Verbruggen N (1999) NaCl and CuSO₄ treatments trigger distinct oxidative defence mechanisms in *Nicotiana plumbaginifolia* L. Plant Cell Environ 22: 387–396

Schreiber U, Bilger W and Neubauer C (1994) Chlorophyll fluorescence as a nonintrusive indicator for rapid assessment of *in vivo* photosynthesis. In: Schulze E-D and Caldwell MM (eds) Ecophysiology of Photosynthesis, pp 49–70. Springer-Verlag, Berlin

Schroeder JI (2003) Knockout of the guard cell K⁺ out channel and stomatal movements. Proc Natl Acad Sci USA 100: 4976–4977

Seemann JR, Sharkey TD, Wang JL and Osmond CB (1987) Environmental effects on photosynthesis, nitrogen-use efficiency, and metabolite pools in leaves of sun and shade plants. Plant Physiol 84: 796–802

Sharma PK and Hall DO (1991) Interaction of salt stress and photoinhibition on photosynthesis in barley and sorghum. J Plant Physiol 138: 614–619

Skillman JB and Osmond CB (1998) Influence of nitrogen supply and growth irradiance on photoinhibition and recovery in *Heuchera americana* (Saxifragaceae). Physiol Plant 103: 567–573

Smirnoff N (1993) The role of active oxygen in the response of plants to water deficit and desiccation. New Phytol 125: 27–58

Steglich C, Behrenfeld M, Koblizek M, Claustre H, Penno S, Prasil O, Partensky F and Hess WR (2001) Nitrogen deprivation strongly affects Photosystem II but not phycoerythrin level in the divinyl-chlorophyll b-containing cyanobacterium *Prochlorococcus marinus*. Biochim Biophys Acta 19: 341–349

Straus NA (1994) Iron deprivation: Physiology and gene regulation. In: Bryant DA (ed) The Molecular Biology of Cyanobacteria, pp 731–750. Kluwer Academic Publishers, Dordrecht

Sugiharto B, Miyata K, Nakamoto H, Sasakawa H and Sugiyama T (1990) Regulation of expression of carbon-assimilating enzymes by nitrogen in maize leaf. Plant Physiol 92: 963–969

Sun J, Nishio JN and Vogelmann TC (1996a) 35S-Methionine incorporates differentially into polypeptides across leaves of spinach (*Spinacia oleracea*). Plant Cell Physiol 37: 996–1006

Sun J, Nishio JN and Vogelmann TC (1996b) High-light effects on CO_2 fixation gradients across leaves. Plant Cell Environ 19: 1261–1271

Taylor SE and Terry N (1986) Variation in photosynthetic electron transport capacity and its effect on the light modulation of ribulose bisphosphate carboxylase. Photosynth Res 8: 249–256

Terashima I and Evans JR (1988) Effects of light and nitrogen nutrition on the organization of the photosynthetic apparatus in spinach. Plant Cell Physiol 29: 143–155

Terry N (1979) The use of mineral nutrient stress in the study of limiting factors in photosynthesis. In: Marcelle R, Clijsters H and Van Poucke M (eds) Photosynthesis and Plant Development, pp 151–160. Dr. Junk W Publishers, The Hague

Terry N (1980) Limiting factors in photosynthesis. I. Use of iron stress to control photochemical capacity *in vivo*. Plant Physiol 65: 114–120

Terry N and Abadía J (1986) Function of iron in chloroplasts. J Plant Nutr 9: 609–646

Terry N and Ulrich A (1973a) Effects of potassium deficiency on the photosynthesis and respiration of leaves of sugar beet. Plant Physiol 51: 783–786

Terry N and Ulrich A (1973b) Effects of phosphorus deficiency on the photosynthesis and respiration of leaves of sugar beet. Plant Physiol 51: 43–47

Tobías D (1999) Efectos de la deficiencia de hierro sobre los sistemas antioxidantes en hojas de remolacha y peral. PhD Thesis, University of Zaragoza, Spain

Toenniessen GH (1984) Review of the world food situation and the role of salt-tolerant plants. In: Staples RC and Toenniessen GH (eds) Salinity Tolerance of Plants, pp 399–413. Wiley-Interscience, New York

Valentini R, Epron D, De Angelis P, Matteucci G and Dreyer E (1995) *In situ* estimation of net CO_2 assimilation, photosynthetic electron flow and photorespiration in Turkey oak (*Quercus cerris* L.) leaves: diurnal cycles under different levels of water supply. Plant Cell Environ 18: 631–640

Valladares F and Pugnaire FI (1999) Tradeoffs between irradiance capture and avoidance in semi-arid environments assessed with a crown architecture model. Ann Bot 83: 459–469

Vassiliev IR, Kolber Z, Wyman KD, Mauzerall D, Shukla VK and Falkowski PG (1995) Effects of iron limitation on photosystem II composition and light utilization in *Dunaliella tertiolecta*. Plant Physiol 109: 963–972

Verhoeven AS, Demmig-Adams B and Adams III WW (1997) Enhanced employment of the xanthophyll cycle and thermal energy dissipation in spinach exposed to high light and N stress. Plant Physiol 113: 817–824

Vinnemeier J and Hagemann M (1999) Identification of salt-regulated genes in the genome of the cyanobacterium *Synechocystis* sp strain PCC 6803 by subtractive RNA hybridization. Arch Microbiol 172: 377–386

Vinnemeier J, Kunert A and Hagemann M (1998) Transcriptional analysis of the isiAB operon in salt-stressed cells of the cyanobacterium *Synechocystis* sp. PCC 6803. FEMS Microbiol Lett 169: 323–330

von Caemmerer S and Farquhar GD (1981) Some relationships between the biochemistry of photosynthesis and the gas exchange of leaves. Planta 153: 376–387

Walcroft AS, Whitehead D, Silvester WB and Kelliher FM (1997) The response of photosynthetic model parameters to temperature and nitrogen concentration in *Pinus radiata* D. Don. Plant Cell Environ 20: 1338–1348

Werner C, Correia O and Beyschlag W (2002) Characteristic patterns of chronic and dynamic photoinhibition of different functional groups in a Mediterranean ecosystem. Funct Plant Biol 29: 999–1011

Winder TL and Nishio J (1995) Early iron deficiency stress response in leaves of sugar beet. Plant Physiol 108: 1487–1494

Xu D-Q and Shen Y-K (1997) Midday depression of photosynthesis. In: Pessarakli M (ed) Handbook of Photosynthesis, pp 451–459. Marcel Dekker Inc, New York

Yeo AR (1983) Salinity resistance: physiologies and prices. Physiol Plant 58: 214–222

Chapter 7

Photoinhibition and UV Response in the Aquatic Environment

Donat-P. Häder*
Institut für Botanik und Pharmazeutische Biologie, Friedrich-Alexander-Universität, Staudtstr. 5, D-91058 Erlangen, Germany

Summary	87
I. Introduction: Life in the Aquatic Environment	87
II. Photoinhibition in the Field	89
III. Effects of Solar UV Radiation	93
IV. Fast Kinetics of Fluorescence Parameters	94
V. Effects on Developmental Stages	95
VI. Pigment Bleaching	95
VII. Protection Mechanisms against Excessive Radiation Stress	96
A. UV Protection by Absorbing Substances	97
B. Repair Mechanisms	97
VIII. Conclusions	100
Acknowledgments	100
References	100

Summary

This chapter summarizes the effects of excessive solar radiation on aquatic primary producers with an emphasis on macroalgae. The introductory paragraphs deal with the aquatic environment and the specific implications for sessile algae and their vertical distribution on the coast. Macroalgae are exposed to dramatically changing irradiances and complicated light patterns governed by the diel solar cycle, the tidal rhythm, and changing cloud cover. The following sections concentrate on the phenomenon of photoinhibition with specific reference to in situ measurements with as little disturbance of the specimens on site as possible. Despite its low percentage contribution in solar radiation, short wavelength ultraviolet is a major component in photoinhibition of algae in their natural habitat. Fast kinetics of fluorescence parameters demonstrates the rapid adaptation of the organisms to their changing photic environment. Early developmental stages are more prone to inhibitory effects of excessive solar radiation. Pigment bleaching and resynthesis are important consequences of solar exposure. Macroalgae have developed several strategies for protection against excessive light stress. UV-absorbing substances, which they share with cyanobacteria and phytoplankton, limit the amount of UV photons reaching the photosynthetic apparatus and the nucleus. They include carotenoids, mycosporine-like amino acids, as well as several chemically not yet identified substances. In addition, the fast turnover of the D1 protein in photosystem II allows rapid recovery from photoinhibition.

I. Introduction: Life in the Aquatic Environment

Macroalgae and seagrasses are major biomass producers in marine ecosystems inhabiting coastal regions and continental shelves (Häder et al., 2003a). They are ecologically important for providing the basis of the intricate food web in coastal habitats and giving shelter to adult and larval stages of fish, crustaceans, mollusks, and other animals. In addition, they are economically

---

*Author for correspondence, email: dphaeder@biologie.uni-erlangen.de

important and are being exploited for food production, as fertilizers, and as raw materials for gelling substances such as agar and carragheenan (Jensen, 1995; Lüning, 1990).

Incident solar radiation at the surface and the depth of penetration into the water column are the decisive factors controlling photodamage to aquatic photosynthetic organisms. Aquatic ecosystems differ substantially in their transparency and thus the depth of solar penetration (Laurion et al., 2000). Especially in eutrophic freshwater systems and coastal areas of the oceans, absorbing and scattering substances limit the transparency of the water, while solar radiation penetrates to greater depths in clear oceanic waters (Conde et al., 2000a; Kuhn et al., 1999). In addition, there is a pronounced seasonal variability in the transparency (Dring et al., 2001a; Kuwahara et al., 2000). Inorganic particulate substances, dissolved and particulate organic carbon (DOC and POC), humic substances, and suspended organisms are the main absorbers of (especially short wavelength) solar radiation (Arts et al., 2000).

Phytoplankton and other pelagic organisms are free to move in the water column by active swimming or actively changing their buoyancy. E.g., several Dinophyceae have been found to undergo diel active vertical migrations of up to 30 m (Tyler et al., 1981; Yentsch et al., 1964). Diatoms and floating cyanobacteria, which do not have the capability of propelling themselves, resort to flotation by producing gas vacuoles (Walsby et al., 1992) or lipid droplets (Gosink et al., 1993). In open oceanic waters, vertical migration is superimposed by the action of waves and wind (Häder et al., 2003a). Therefore, the impact of solar radiation is modified by the depth and rate of the mixing layer (Huot et al., 2000).

In contrast to phytoplankton organisms, most marine macroalgae are confined to the coastal areas of the continental shelves. Also, while phytoplankton are motile in the water column (Häder, 1995), most macroalgae are sessile and therefore restricted to their growth site (Lüning, 1990). Macroalgae show a distinct pattern of vertical distribution in their habitat that is mainly controlled by light penetration (Dring et al., 1996; Hanelt et al., 1997; Larkum et al., 1993). The photoprotective

---

Abbreviations: DOC – dissolved organic carbon; MAA – mycosporine-like amino acid; PAM – pulse amplitude modulated (fluorescence); PAR – photosynthetically active radiation, 400 – 700 nm; POC – particulate organic carbon; PS II – Photosystem II; UV – ultraviolet radiation; UV-A – ultraviolet radiation in the wavelength range 315 – 400 nm; UV-B – ultraviolet radiation in the wavelength range 280 – 315 nm

capabilities define the upper growth limit of a species. Some of these plants populate the supralittoral (coast above high water mark), where they are exposed only to the spray from the surf. Others inhabit the eulittoral (intertidal zone), which is characterized by the regular tidal change (Häder, 1997). Subtidal macroalgae are never exposed to air since they thrive below the tidal zone. The range in solar radiation can be substantial, from over 1000 W m^{-2} (total solar radiation) at the surface to less than 0.01% of this value, which penetrates to the understory of e.g. a kelp habitat (Markager et al., 1994).

The phenomenon of photoinhibition protects the photosynthetic apparatus by affecting photosystem II in such a way that excess absorbed energy is rendered harmless by thermal dissipation (Krause and Weis, 1991; Adams et al., this volume). This effect can be measured using pulse amplitude modulated (PAM) fluorescence. The ratio F_v/F_m (see below) is defined as the photosynthetic quantum yield. In a dark-adapted organism this is interpreted as the "optimal quantum yield" and in a light-exposed organism the "effective photosynthetic yield". Note that this definition is in contrast to others in which the quantum yield describes the amount of oxygen released per photon or the amount of CO_2 fixed per absorbed photon. A decrease in the quantum yield is interpreted as a sign of photoinhibition (Hanelt, 1995b; see Osmond and Förster, this volume). After exposure of the plants to excessive light, several researchers found a strong degradation of the D1 protein in the reaction center of photosystem II (Critchley and Russell, 1994; Hanelt, 1998; see Edelman and Mattoo, this volume). This process is defined as chronic photoinhibition. However, D1 degradation was not determined in all experiments with aquatic photosynthetic organisms. Since the D1 levels can also be adjusted by acclimation to the prevailing light conditions (e.g. sun vs. shade), the determination of capacity is the only clear criterion for the assessment of this phenomenon.

Many algae (but also higher plants) show a typical midday depression in their photosynthetic yield between noon and the early afternoon hours. This diurnal pattern is attributed mainly to dynamic photoinhibition (Hanelt et al., 1994a, 1994b; Hanelt, 1998; Henley et al., 1992). This process is believed to be due to the xanthophyll cycle (see Yamamoto, this volume), involved in the regulation of photosynthetic quantum yield by reversibly increasing thermal dissipation of excess energy (Demmig-Adams and Adams, 1992). In marine algae this pattern is complicated by the tidal rhythm: the highest irradiance stress occurs when low tides coincide with high solar angles (Häder et al.,

2000b; Hanelt et al., 1994b; Hanelt, 1998). Depending on the tidal change, eulittoral macroalgae can be submerged under several meters of turbid water during high tide or be fully exposed to solar radiation when the water retreats. As a consequence, macroalgae are subjected to a complicated pattern of irradiances controlled by the daily cycle, the tidal rhythm, which is phase shifted from day to day, and the changing cloud cover. It will be interesting in the future to determine if this pattern is endogenously regulated, e.g. by an entrainment to the tidal rhythm. Judging from the drastically changing exposure conditions in their habitat, algae need to adapt to widely changing irradiation levels. Most accomplish positive net photosynthesis at low and intermediate irradiances like shade plants in terrestrial habitats. This is indicated by the fact that the PS II photochemical efficiency (F_v/F_m) decreases even at moderate irradiances and non-photochemical quenching is observed (Porst et al., 1997). During excessive irradiation, the algae need effective preventive mechanisms to protect themselves from photooxidative damage. In macroalgae, the regulatory mechanisms designed to ameliorate light stress are similar to those in higher plants and include thermal dissipation of excess excitation energy, antioxidant systems, chloroplast movements, adjustment of the antenna size, and the fast repair of photooxidative damage (Niyogi et al., 1998). These putative mechanisms are utilized by algae to different degrees. For instance, in the unicellular green alga, *Chlamydomonas reinhardtii* antenna size has little impact on the absorption of light (Baroli et al., 2003). Fast induction and relaxation kinetics of the fluorescence parameters impressively demonstrate how fast the photosynthetic apparatus can adapt to the ambient light conditions (Häder et al., 2000a, 2001a).

II. Photoinhibition in the Field

Under high light stress as well as other stresses, including elevated temperatures, desiccation, and excessive salinity, macroalgae undergo a similar photoinhibition as described for higher plants defined as a decrease in the PS II photochemical efficiency (see Nishiyama et al., this volume). The underlying mechanisms are generally believed to be identical to those identified for terrestrial photosynthetic organisms (Neidhardt et al., 1998; Jung and Niyogi, this volume). Photoinhibition can be quantified by oxygen exchange (Häder et al., 1994; Hanelt et al., 1995) or by measurements of chlorophyll fluorescence (Schreiber et al., 1986). While some researchers failed to see a correlation between oxygen production and PAM fluorescence parameters (Hanelt and Nultsch, 1995), others found a close relationship (Björkman and Demmig, 1986; Adams et al., 1987, 1990a; Demmig-Adams et al., 1989; Häder et al., 1996a–c, 1997a; Herrmann et al., 1995).

Pulse amplitude modulated (PAM) fluorescence also allows one to determine photochemical and non-photochemical quenching (Schreiber et al., 1986; Büchel et al., 1993). For aquatic ecosystems, an underwater PAM instrument was developed for in situ measurement of the quantum yield, which proved very useful to study the ecophysiology of macroalgae in the field during diving (diving PAM, Walz, Effeltrich, Germany) (Häder et al., 2001c).

Macroalgae inhabiting the upper intertidal zone, such as several brown (*Cystoseira*, *Padina*, *Fucus*) and green (*Ulva*, *Enteromorpha*) algae, show a maximum of oxygen production at or close to the surface, whereas algae adapted to lower irradiances usually thrive best when exposed deeper in the water column (the green algae *Cladophora*, *Caulerpa*, most red algae) (Häder et al., 1997a; Herrmann et al., 1995). Oxygen production in this context means net O_2 release and reflects the difference of photosynthetic production minus respiratory uptake. When exposed to excessive solar radiation at the surface, the photosynthetic oxygen production decreases within minutes or hours in most species (Fig. 1). Exclusion of UV-B partially reduced the effects, indicating its significant role in photoinhibition (see below). In contrast to photosynthesis, respiration is inhibited much less during photoinhibition. As a consequence of the inhibited photosynthetic oxygen production and less impaired respiration, increasing exposure to solar radiation results in a shift of the compensation point to higher irradiances, e.g., in the green alga *Ulva laetevirens* (Herrmann et al., 1995). The compensation point defines the irradiance at which photosynthetic oxygen production and respiratory oxygen consumption balance each other. The lowest light compensation points for photosynthesis have been reported for Arctic and Antarctic algae (Gómez et al., 1995; Gómez and Wiencke, 1996; Wiencke, 1996).

Green macroalgae freshly harvested from their growth site early in the morning or after dark adaptation show high optimal PS II photochemical efficiency (measured as F_v/F_m, using PAM fluorescence) comparable to that of higher plants (Häder et al., 1999; 2000b). Rhodophytes (red algae) and Phaeophytes (brown algae) usually have slightly lower PS II photochemical efficiency (Bischof et al., 1999; Häder et al., 1997b, 2001d). Marine, crust-forming cyanobacteria are usually characterized by even lower values. When exposed

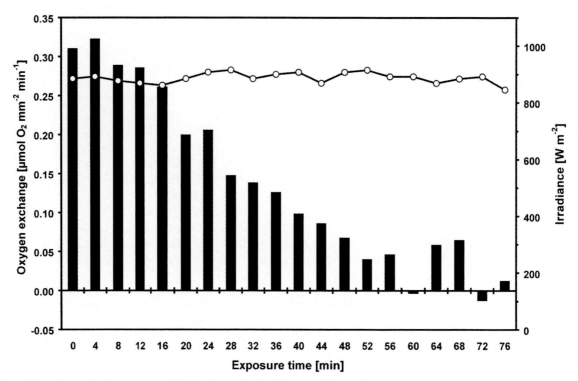

Fig. 1. Net photosynthetic oxygen production (bars) in *Cladophora prolifera* affected by solar radiation at the surface and PAR irradiance (open circles and solid line). In order to keep the CO_2 concentration constant, 10 mM $NaHCO_3$ was added. Temperature was 23°C (Häder et al., 1997a).

to high solar radiation, e.g. during noon at the surface, the PS II photochemical efficiency decreases rapidly, especially when the thalli are mounted in a fixed position so that they face the sun with the same side up. This may be an artificial stress situation since in their natural habitat the organisms are free to sway in the water so that different sides of their thallus are exposed. When this situation is mimicked by allowing the algae to float in a larger volume of water near the surface, such as in a rock pool, photoinhibition sets on after longer exposure times and usually is not as pronounced as in the fixed-position experiment (Fig. 2). When, after exposure, the algae are subsequently transferred to dim light in the shade, they recover from their photoinhibition in a matter of hours. The degree of photoinhibition and their recovery time depend on the species, its natural position in the water column, duration of exposure, and irradiance during exposure. Deep water species are more prone to photoinhibition than surface-growing algae and require longer periods of dim light or darkness to recover (Herrmann et al., 1995; Häder et al., 1996c; Bischof et al., 1998; Hanelt, 1998; Altamirano et al., 2000). One extreme example is the red alga *Peyssonnelia* that is found in the Mediterranean at greater depths and often in caves or in the shade of overhanging rocks (Häder et al., 1998c). When exposed at the surface, net oxygen release ceased within 15 min. Optimal photosynthesis was found when the thalli where exposed at 5 m depth, but even at 4 m depth there was a considerable decrease in the PS II photochemical efficiency after 2 h of exposure.

Closely related species can show considerable differences in photoinhibition and recovery. The green Mediterranean *Cladophora prolifera* and *C. pellucida* are found in the same habitat at moderate depth. When exposed to full solar radiation, *C. prolifera* showed a higher degree of photoinhibition than *C. pellucida*, but the latter took longer times for recovery than the first (Häder et al., 1997a). Thalli of the same species can grow at different depths in the water column. The thalli often develop morphological and physiological differences, when growing at different depths, which are believed to be due to the ambient light conditions they experience in their habitat. Thalli of the common crustaceous red alga *Corallina elongata* (Corallinaceae) are characterized by a sun morphotype when growing at the surface, while algae growing at 5 m depth and in caves display a typical shade morphotype (Algarra and Niell, 1990). The shade type has a higher assimilative surface,

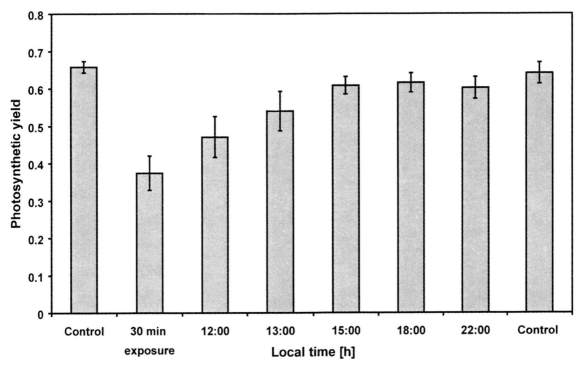

Fig. 2. Effective photosynthetic quantum yield (F_v/F_m) of *Corallina officinalis* (harvested from the mid intertidal) measured after 30 min dark adaptation, 30 min exposure to unfiltered solar radiation during local noon free floating in a large volume of water, and after increasing recovery times in the shade calculated as $(F'_m - F_t)/F'_m$. The control sample was treated as the other specimens except for solar exposure. For each data point n was equal to 8 (the lines on top of the bars indicate one SD) (after Häder et al., 2003b).

higher chlorophyll *a* and phycoerythrin contents, but a lower carotenoid content than the sun type. The latter also shows a higher canopy density. These phenotypes are not different genetic strains since both morphotypes are easily interconvertable upon transplantation (Algarra and Niell, 1990). Thalli of the same species harvested from different depths show different degrees of photoinhibition when exposed to surface solar radiation and differences in the time required for recovery; photoinhibition is less extensive in surface-adapted algae (Fig. 3) (Häder et al., 1996d, 1996c). In the siphonal green alga *Halimeda*, transplantation experiments confirmed that the differences in morphology and physiology are not due to genetic differentiation of strains, but rather to phenotypically induced adaptations at different depths (de Beer et al., 2001).

One can still argue that the experimental procedures described above are artificial and cause stress to the specimens when taking algal samples from their growth site to the surface where they are exposed to excessive radiation. In order to reveal the real behavior of macroalgae in their natural habitat, it is desirable to use non-invasive techniques on site without removing the thalli from their habitat or changing their growth conditions. PAM fluorescence can now be performed by a diver on specimens under water. For this purpose the selected plants need to be marked, e.g. by a buoy, in order to perform time-series experiments on the same specimens (Häder et al., 2001d). Figure 4 shows the PS II photochemical efficiency measured by this technique in the common brown alga *Dictyota dichotoma* at 2 m depth off the Atlantic island of Gran Canaria. In the morning an initial yield of >0.8 was determined which dropped during the day to 0.65 and increased again later in the afternoon, but full recovery of optimal quantum yield did not occur until about 6 p.m. (local time). When measuring the PS II photochemical efficiency at 1-h intervals from dawn to dusk in the Mediterranean siphonal Chlorophyte *Codium bursa*, a pronounced photoinhibition (PS II photochemical efficiency from 0.58 to 0.2) was found at local noon and the early afternoon hours. The alga did not fully recover until evening (Häder et al., 1999). While noon photoinhibition is quite common in rock pool algae and specimens growing near the surface, some reduction in the PS II photochemical efficiency can be detected even at several meters depth (Fig. 5) (Häder et al., 1996d; Porst et al., 1997).

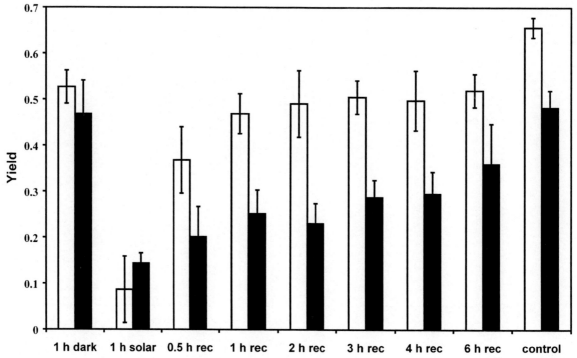

Fig. 3. Photosynthetic quantum yield (F_v/F_m) in *Corallina elongata*, harvested from the surface (open bars) and from a depth of 5 m (black bars), measured after dark adaptation, after 1 h exposure to solar radiation in a rock pool and during recovery in the shade. The controls were treated as the other specimens except for the solar exposure. Each data point represents the mean of eight measurements and S.E. (Häder et al., 1997b).

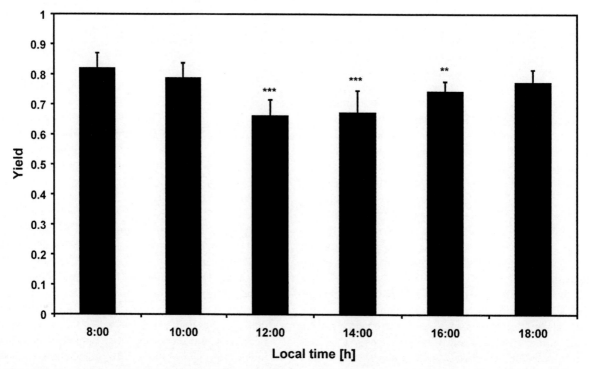

Fig. 4. Photosynthetic quantum yield (F_v/F_m) of *Dictyota dichotoma* at 2 m depth off the Atlantic island of Gran Canaria measured with a diving PAM fluorometer on site. Each data point represents the mean of eight measurements and S.E. Values are statistically different from the initial value (Student's *t*-test) with $P < 0.001$ (***) or $P < 0.01$ (**) (Häder et al., 2001d).

Chapter 7 Photoinhibition in Aquatic Ecosystems

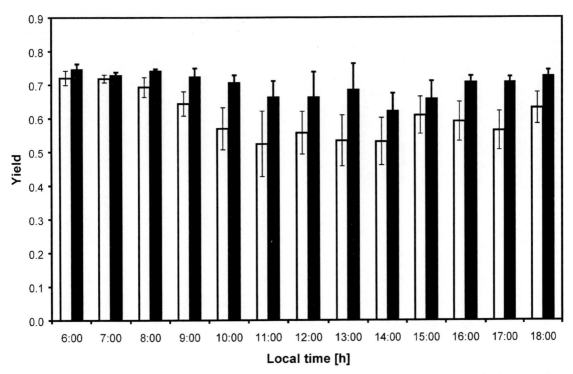

Fig. 5. Photosynthetic quantum yield of *Dictyota dichotoma* growing at the surface (open bars) and at 6 m depth measured on site from dawn to dusk at 1-h intervals. D.-P. Häder, 2006 unpublished

III. Effects of Solar UV Radiation

While most of the photoinhibitory effects in macroalgae and phytoplankton are due to the white light component (PAR, photosynthetic active radiation, 400–700 nm), a considerable fraction is due to short-wavelength solar UV radiation (Smith et al., 1980; Neale et al., 1993; Benet et al., 1994; Dring et al., 2001b; van de Poll et al., 2001). This is an interesting phenomenon since UV-A (315–400 nm, C.I.E. definition) contributes only about 5 % of the total solar radiation and UV-B (280–315 nm) less than 0.2 % (Webb, 1998).

The depletion in stratospheric ozone resulting from anthropogenically released atmospheric pollutants such as chlorinated fluorocarbons has caused an increase in UV-B radiation reaching the Earth's surface (Crutzen, 1992; Kerr et al., 1993; Lubin et al., 1995). Biologically effective UV-B radiation can penetrate deep into the water column (Smith et al., 1979; Smith et al., 1992) and may affect the aquatic ecosystems (Häder et al., 2003a). Both UV-A and UV-B cause depression of the photosynthetic rate in the brown alga *Laminaria digitata* (Foster et al., 1996). UV-B penetration into the water column depends on the optical properties. 10 % transmission of solar radiation at 310 nm varies from about 20 m in the clearest oceanic water to a few centimeters in brown humic water. Benthic algae may experience deleterious UV radiation to depths below 20 m (Booth et al., 1997).

The effect of short wavelength UV radiation can clearly be demonstrated with UV exclusion studies. (Gunasekera et al., 1997). Experiments were conducted with algal thalli submerged in a volume of sea water in an open tray floating on top of a larger reservoir of water for thermal stabilization. The trays were covered with filter foils which removed UV-B or total UV, respectively. The control was covered with foil with a cut-off at 290 nm in order to warrant comparable microclimatic conditions and evaporation in all parallel samples. The experimental procedure was similar to the one outlined above: the specimens were dark-adapted, then exposed to a brief period of solar radiation, and subsequently recovery was studied in dim light. Figure 6 shows the PS II photochemical efficiency in the common Atlantic Chlorophyte *Enteromorpha* (Häder et al., 2001a). Exposing the thalli to full solar radiation resulted in a statistically significant higher inhibition than exposure to radiation deprived of UV-B or total UV. This effect is even noticed during recovery, indicating that the effects of solar UV radiation are taking longer to be reversed than those induced by visible radiation. Similar

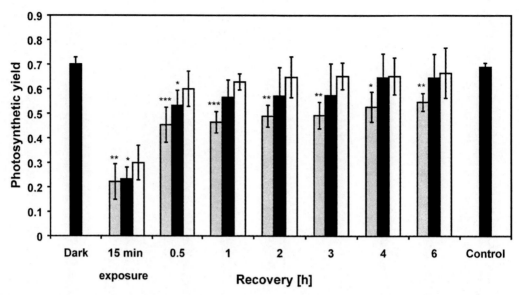

Fig. 6. Effective photosynthetic quantum yield of *Enteromorpha linza* measured after 30 min dark adaptation, 15 min exposure to solar radiation, and after increasing recovery times in the shade. Gray bars, specimens exposed to unfiltered solar radiation. Black bars, specimens exposed to UV-A + PAR. White bars, specimens exposed to PAR only. For each data point n was equal to 8 (the lines on top of the bars indicate one SD). The values for unfiltered solar radiation and under the 320 nm cut-off filter treatments are statistically significantly different (Student's *t*-test) from the PAR-only values (395 nm cut-off) in each set with $P < 0.001$ (***), $P < 0.01$ (**), or $P < 0.1$ (*), respectively, as indicated by the Student's *t*-test. Modified from Häder et al., (2001a)

effects were also found in other green, red, and brown algae (Franklin et al., 1997, 1998; Häder et al., 1997a, 1998b, 2001c, 2001b, 2001d). It could be shown that the primary damage to the D1 protein is qualitatively different when excessive UV or visible radiation is used (Etienne et al., 1996; Mate et al., 1998). A more detailed study using a series of Schott cut-off filters confirmed the notion that the effect of solar radiation is more pronounced with shorter wavelengths (Häder et al., 1998c).

Susceptibility to UV radiation is highly variable among marine macroalgae which results in a specific depth distribution of species (van de Poll et al., 2001). Subtidal species are generally more sensitive to UV radiation than eulittoral species. This can be easily demonstrated by transplantation experiments of algae from deep to shallow waters (Karsten et al., 2001). Intertidal species readily tolerate or acclimate to UV. (Altamirano et al., 2000). In some Arctic fjords, algal primary productivity is strongly affected by the availability of solar radiation (Hanelt et al., 2001). Even though the solar UV-B radiation at the surface never exceeded 0.27 W m^{-2}, it deeply penetrated into the water column and impaired growth to a depth of 5–6 m. During summer this inhibition was less pronounced due to high influx of turbid fresh water from melting snow and glacier ice into the fjord water, decreasing the transparency in the top water layer.

IV. Fast Kinetics of Fluorescence Parameters

Induction curves of fluorescence parameters reveal the velocity with which the specimens can adapt to their changing environment. Induction curves with quenching analysis measured at 10 ms sampling rate showed first a fast and then a gradual drop of the current fluorescence (F_t) and maximal fluorescence (F_m) in the common Rhodophyte *Corallina officinalis* when the thalli were exposed to a bright halogen light source (Fig. 7) (Häder et al., 2003b). Non-photochemical quenching followed this pattern antagonistically and reached high values within less than 100 s. During this time the PS II photochemical efficiency (Y) hardly increased above zero and photochemical quenching (qP) increased only slightly. Both parameters showed a significant oscillation. This is even more pronounced in the subsequently measured relaxation kinetics where Y and qP followed a parallel oscillation that dampened out after some time. The PS II photochemical efficiency and the photochemical quenching stepped up within a few seconds while the non-photochemical quenching gradually decreased

Chapter 7 Photoinhibition in Aquatic Ecosystems

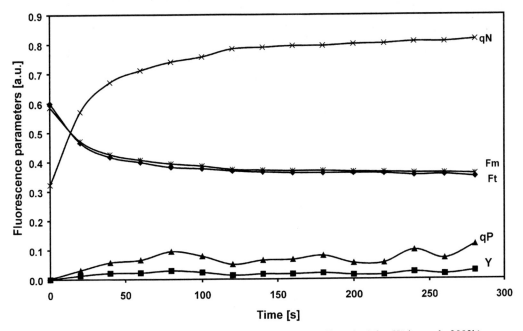

Fig. 7. Induction curve with quenching analysis in *Corallina officinalis* (after Häder et al., 2003b).

with time. It is interesting to note that adaptation of the fluorescence parameters occurred in such a short time.

Fast induction kinetics at 1000 μs per data point in *Corallina* show the involvement of several components of the PS II redox system (Fig. 8) following dark adaptation for 4 h. After determination of F_o, the actinic halogen light was switched on for 2 s (at time 0.5 s). At least two distinct rise components can be distinguished which lead to a transient fluorescent maximum with a subsequent decrease, indicating the involvement of several redox components. The biphasic fluorescence rise during light exposure can be attributed to a reduction of Q_A and Q_B followed by a subsequent, slower reduction of the plastoquinone pool. The gradual decrease to lower fluorescence values indicates that it takes more than a second until the electron transport chain operates smoothly. Exposure to unfiltered solar radiation with increasing durations (15, 30, 45 or 60 min) before determination of the fast induction kinetics strongly decreased the fluorescence signal in this species. Similar results were obtained for other Rhodophytes (Häder et al., 2003b) and Phaeophytes (Häder et al., 2001b).

V. Effects on Developmental Stages

Excessive solar radiation inhibits growth in adult stages of many species of macroalgae both in short- and long-term exposure (Franklin et al., 1998; Grobe et al., 1998; Häder, 1998; Dring et al., 2001b). Young developmental stages of macroalgae (zoospores, gametes, zygotes, and young germlings) are even more susceptible to UV radiation stress (Wiencke et al., 2000; Coelho et al., 2001): Mortality of Phaeophyte zoospores from southern Spain was induced by full solar UV radiation, with short wavelength UV-B being more effective than UV-A. The effects include loss of viability, cellular disintegration, and inhibition of motility and phototaxis. It is interesting to note that zoospores of the shallow water species *Chordaria flagelliformis* require higher UV doses than the mid-sublittoral species *Laminaria saccharina* to suffer mortality (Wiencke et al., 2000). In addition to photosynthesis, polarity induction, mitosis, and cytokinesis during the development of brown algae zygotes are affected.

The Chlorophyte *Chara* is regarded as a link to higher plants, which makes it interesting for research on UV-related effects. Under elevated UV-B radiation, vegetative reproduction was stimulated while generative reproduction was suppressed (de Bakker et al., 2001a,b).

VI. Pigment Bleaching

The pigment content in macroalgae depends on the irradiance levels in solar radiation, and, as is the case in higher plants, it differs between sun and shade type

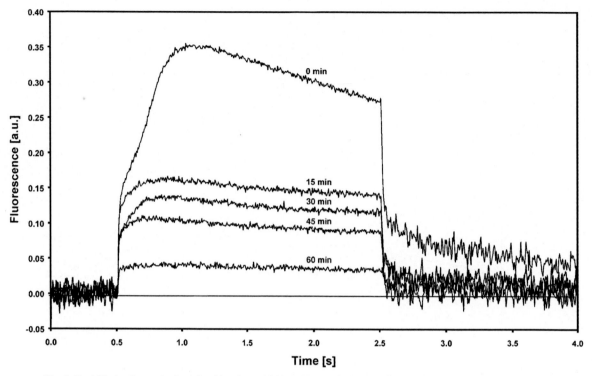

Fig. 8. Rapid induction and relaxation kinetics at 1000 μs/data point in *Corallina officinalis* (Häder et al., 2003b).

algae (Herrmann et al., 1996). Furthermore, it is known to be governed by a diurnal rhythm (Häder et al., 1997a; Talarico, 1996). These daily patterns indicate a rapid turnover of the pigments in the natural environment of algae (Grumbach et al., 1978; Algarra and Niell, 1990).

Absorption spectra of methanolic extracts of two strains of the Rhodophyte *Corallina officinalis* showed that specimens collected from the lower intertidal had a considerably lower concentration of all photosynthetic pigments than mid intertidal samples. When exposed to unfiltered solar radiation over the day, both morphotypes had a significantly lower absorption than controls kept in the shade (Häder et al., 2003b). There was no selective decrease of any of the photosynthetic pigments.

During a 2-d exposure experiment, selected species of green, red, and brown algae all showed a pronounced variation in their pigmentation. The green alga *Ulva rigida,* adapted to excessive solar radiation since it grows in shallow rock pools high in the intertidal range, showed a significant increase in the chlorophyll concentration in the extract during the first day and a decrease during the subsequent night (Häder et al., 2002). However, on the second day the experienced cumulative radiation dose prevented any further increase in pigmentation. In contrast, two other species, the Rhodophyte *Porphyra columbina* and the Phaeophyte *Dictyota dichotoma,* which grow lower in the intertidal range, showed a drastic decrease in the absorption at 665 nm which was even more pronounced on the second day.

VII. Protection Mechanisms against Excessive Radiation Stress

The decrease in PS II efficiency and photosynthetic capacity can be regarded as an active adaptation to the changing irradiance conditions (Hanelt et al., 1994a). It is not only found in algae from the tropics and mid latitudes (Wood, 1989; Dring et al., 1996; Franklin et al., 1996a) but also in Arctic and Antarctic macroalgae (Hanelt et al., 1994a, 1995a; Gomez et al., 1995). While most of the photoinhibition is due to PAR, a statistically significant percentage of the effect is caused by UV-B (and less so by UV-A) in the top few meters of the water column. This high photoinhibition by solar UV has been found in a number of marine algae (Häder et al., 1997a, 1998a, b) and phytoplankton species in the North Atlantic and Mediterranean Sea (Jimenez et al., 1996; Figueroa et al., 1997a, b). The regulatory mechanisms designed to ameliorate light stress include thermal dissipation of excess excitation energy, antioxidant systems, chloroplast movements, adjustment of the antenna size, and the fast repair of photooxidative damage (Niyogi et al., 1998). The fast induction and relaxation kinetics clearly demonstrate how fast the photosynthetic apparatus can adapt to the ambient light conditions.

A. UV Protection by Absorbing Substances

One efficient protective mechanism against UV radiation (UV-A and UV-B) is the production of one or several UV-absorbing substances such as carotenoids or UV-absorbing mycosporine-like amino acids (MAAs) (Karsten et al., 1999). Most MAA-producing species belong to the Rhodophytes, some are found among the Phaeophytes, while only a few Chlorophytes produce MAAs (Gröniger et al., 2000). Three different strategies of protection by UV-screening pigments have been found: sublittoral algae, not likely exposed to higher doses of solar UV, do not synthesize UV-absorbing pigments at all (Karsten et al., 1998b); another group, mainly consisting of supralittoral and high eulittoral algae, which find themselves exposed to high irradiances of solar UV, produces large amounts of MAAs, but cannot be induced by natural or artificial radiation (e.g. the red alga *Porphyra,* Gröniger et al., 1999). In the third group, MAA production can be induced by solar radiation or through experimental manipulation (Gröniger and Häder, 2002).

MAAs are water-soluble compounds characterized by a cyclohexenone or cyclohexenimine chromophore conjugated with the nitrogen substituent of an amino acid or its imino alcohol (Fig. 9). They show absorption maxima ranging from 310 to 360 nm, are transparent to visible light, and have an average molecular weight of around 300 (Cockell et al., 1999; Shick et al., 2002). Approximately 20 MAAs have been found in diverse organisms; they were first identified in fungi where they have a role in UV-induced sporulation (Favre-Bonvin et al., 1976). Related substances, such as asterine-330, porphyra-334, mycosporine-glycine, palythine, palythene, palythinol, shinorine etc., have since been found and characterized in a wide variety of organisms from cyanobacteria (Garcia-Pichel et al., 1993; Sinha et al., 1998), algae (Karsten et al., 1998a; Gröniger et al., 2000), phytoplankton (Carreto et al., 1990; Klisch et al., 2000), lichens (Karentz et al., 1991), marine invertebrates (Shick et al., 1991; Gleason et al., 1993; Adams et al., 1996), and vertebrates including fish (Karentz et al., 1991). The ubiquitous occurrence of MAAs across a large geographical and taxonomic range suggests their early phylogenetic appearance and their importance as natural UV-screening/absorbing compounds (Gröniger et al., 2000). MAAs are very stable compounds and are not easily modified by heat, UV radiation, or extreme pH (Conde et al., 2000b; Gröniger et al., 2000).

A polychromatic action spectrum for the induction of MAAs in the Chlorophyte *Prasiola stipitata* shows a clear maximum at 300 nm (Gröniger et al., 2002). In the red alga *Chondrus crispus*, blue and UV-A radiation induce the synthesis of MAAs; however, the induction by UV-B was not investigated (Franklin et al., 2001). While most algae use MAAs, a few produce different types of UV-absorbing compounds (Perez-Rodriguez et al., 1998).

MAAs have been found in Rhodophytes, Phaeophytes, and Chlorophytes from tropical, temperate, and polar regions (Häder et al., 2003a). Since these compounds are chemically very stable, they accumulate in the sediments of lakes and have been used as a permanent record for past ultraviolet radiation environments (Leavitt et al., 1997). In tropical macroalgae, higher levels of carotenoids and UV-absorbing compounds were detected in surface-adapted specimens as compared to those from understory locations in turf-forming rhodophytes (Beach et al., 1996a, b). Other UV-absorbing compounds were found to be induced by UV-A (Yamazawa et al., 1999). A database on photoprotective compounds in cyanobacteria, phytoplankton, and macroalgae is available at www.biologie.uni-erlangen.de/botanik1/index.html (Gröniger et al., 2000).

B. Repair Mechanisms

Chronic photoinhibition is thought to be due to structural change and subsequent proteolysis of the D1 protein in the reaction center of photosystem II (Hui et al., 2000; Krieger-Liszkay et al., 2000). Visible light and solar UV seem to cause different lesions in the protein (Etienne et al., 1996). To be sure, also regulatory D1 protein pool size adjustment needs to be considered. During recovery this protein is synthesized de novo (Etienne et al., 1996; Mate et al., 1998) which is facilitated by a fast turnover (Polle et al., 1999; Komenda, 2000). In order to prove this hypothesis for photoinhibited macroalgae, inhibitors of chloroplast protein synthesis, streptomycin or chloramphenicol, were applied during inhibition and recovery. Streptomycin inhibits protein biosynthesis by impairing the initiation of translation and causing misreading of mRNA in prokaryotes and plastids, while chloramphenicol inhibits the peptidyl transferase activity of the 50S ribosomal subunits

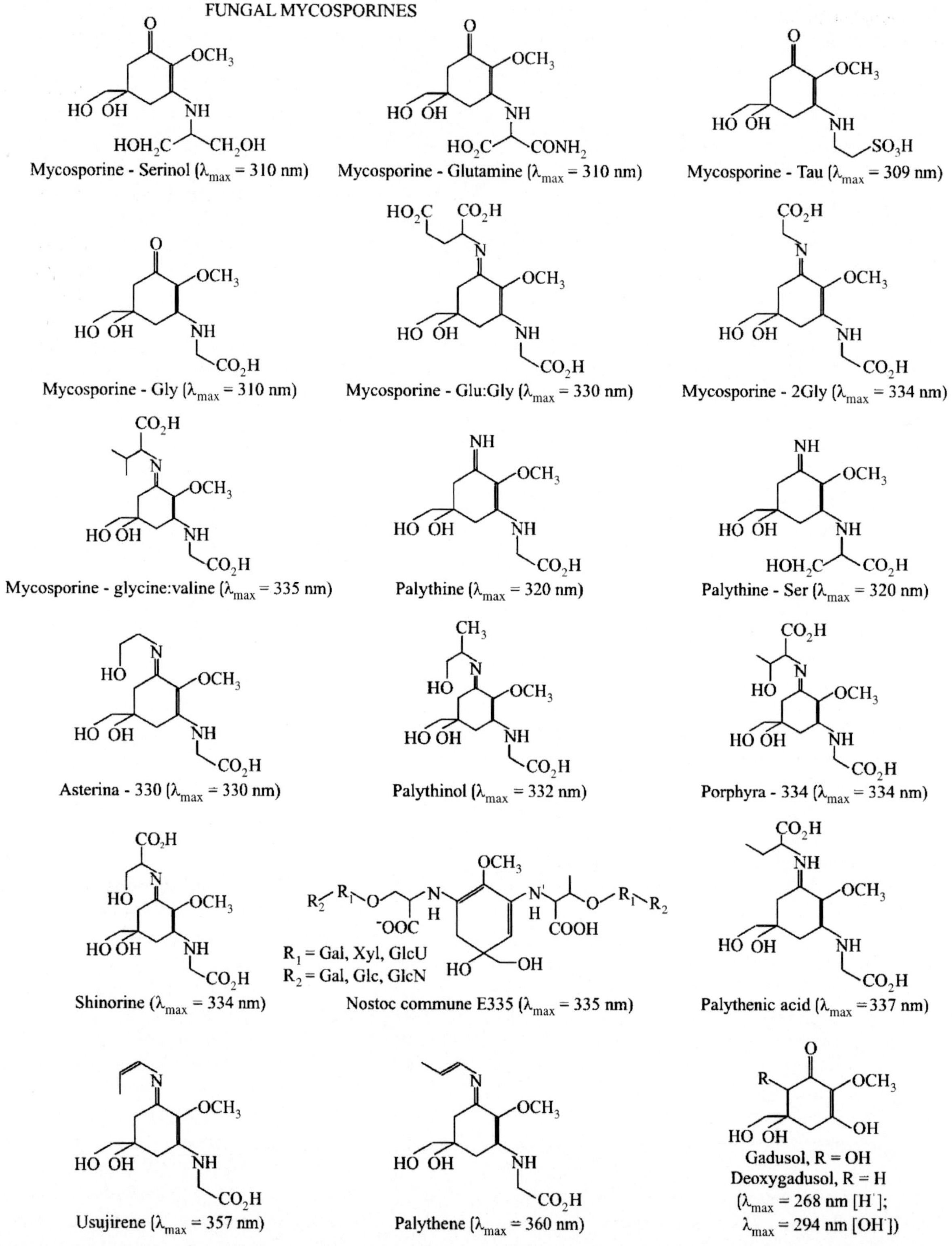

FUNGAL MYCOSPORINES

Mycosporine - Serinol (λ_{max} = 310 nm)

Mycosporine - Glutamine (λ_{max} = 310 nm)

Mycosporine - Tau (λ_{max} = 309 nm)

Mycosporine - Gly (λ_{max} = 310 nm)

Mycosporine - Glu:Gly (λ_{max} = 330 nm)

Mycosporine - 2Gly (λ_{max} = 334 nm)

Mycosporine - glycine:valine (λ_{max} = 335 nm)

Palythine (λ_{max} = 320 nm)

Palythine - Ser (λ_{max} = 320 nm)

Asterina - 330 (λ_{max} = 330 nm)

Palythinol (λ_{max} = 332 nm)

Porphyra - 334 (λ_{max} = 334 nm)

Shinorine (λ_{max} = 334 nm)

Nostoc commune E335 (λ_{max} = 335 nm)
R_1 = Gal, Xyl, GlcU
R_2 = Gal, Glc, GlcN

Palythenic acid (λ_{max} = 337 nm)

Usujirene (λ_{max} = 357 nm)

Palythene (λ_{max} = 360 nm)

Gadusol, R = OH
Deoxygadusol, R = H
(λ_{max} = 268 nm [H$^+$];
λ_{max} = 294 nm [OH$^-$])

Fig. 9. Structures of several mycosporine-like amino acids (MAAs) found in cyanobacteria, phytoplankton, and macroalgae.

Chapter 7   Photoinhibition in Aquatic Ecosystems

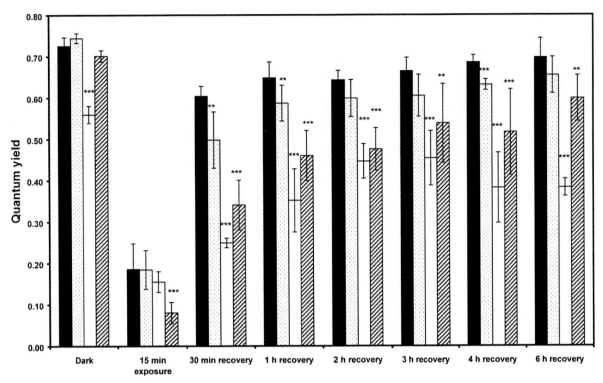

Fig. 10. Photosynthetic quantum yield in *Ulva rigida* after a dark period, after 15 min of exposure to solar radiation during local noon, and during recovery in dim light in the control (black bars), with the addition of 500 μg/ml streptomycin (dotted bars), 1 mM DTT (open bars), or 2 mg/ml chloramphenicol (striped bars). For each data point eight measurements were averaged and the standard deviation calculated. The values for the inhibited samples are statistically significantly different from the controls in each set with p < 0.001 (***) or < 0.01 (**), respectively, as indicated by the Student's *t*-test. Modified from Häder et al., (2002)

in prokaryotes and plastids and also impairs chloroplast protein synthesis.

Both agents retarded the repair process indicating an inhibition of the D1 protein de novo synthesis during recovery after the damage the specimens experienced during excessive light exposure (Häder et al., 2002). After dark adaptation over night, *Ulva* thalli had an optimal PS II photochemical efficiency of over 0.7. After exposure to unfiltered solar radiation the effective PS II photochemical efficiency decreased to below 0.2 (Fig. 10). While streptomycin had no statistically significant effect, addition of chloramphenicol further decreased the PS II photochemical efficiency to below 0.1. Because of the short exposure time, the PS II photochemical efficiency quickly recovered to almost its initial value within 30 min under dim light conditions. During recovery, the inhibitory effects of both chloroplast protein synthesis blockers are evident and statistically significant. The effect of chloramphenicol was even stronger than that of streptomycin and retarded recovery to the initial value for more than 6 h in dim light. Similar effects were found for the brown alga *Dictyota dichotoma* and the red alga *Porphyra columbina* (Häder et al., 2002).

Both in higher plants and in a number of macroalgae, the onset of nonphotochemical quenching is related to the xanthophyll cycle, which is believed to quench excess excitation energy (Demmig-Adams and Adams, 1992; Häder and Figueroa, 1997; Niyogi, 2000). The protective role of carotenoids against excessive excitation energy is based on quenching and prevention of singlet oxygen formation (Noctor et al., 1991). Zeaxanthin formation in the xanthophyll cycle during non-photochemical quenching is well established (Demmig-Adams et al., 1999). Zeaxanthin and change in quantum yield were found to be closely correlated in brown algae (Uhrmacher et al., 1995). Algae with a large xanthophyll pool are characterized by an efficient de-epoxidation of violaxanthin; they have been shown to tolerate higher irradiances and longer periods of emersion (Harker et al., 1999). The protective mechanism of the xanthophyll cycle has been investigated mostly in phytoplankton (Schubert et al., 1994) and in some macroalgae, e.g., the green alga *Ulva lactuca* (Grevby, 1996) and the brown algae *Dictyota dichotoma* (Uhrmacher et al., 1995) and *Lobophora variegata* (Franklin et al., 1996b). Rhodophytes were thought not to use the xanthophyll cycle (Häder et al., 1997a). However, recently, the unicellular red alga *Rhodella violacea* was found to exhibit photoprotective energy dissipation (Ritz et al., 1999) and the presence of photoprotective carotenoids was demonstrated in *Gracilaria birdiae* (Ursi et al., 2003).

The functioning of the xanthophyll cycle can be demonstrated by inhibiting the violaxanthin deepoxidase, by e.g. dithiothreitol (DTT) (Bilger et al., 1989; Adams et al., 1990b; Demmig-Adams and Adams, 1990; Demmig-Adams et al., 1990a, b; Bilger and Björkman, 1990; Öquist et al., 1992; Osmond et al., 1993). As expected, the drug strongly augmented photoinhibition but also retarded recovery in green and brown algae (Fig. 10.). However, it also had effects in the red alga *Porphyra columbina*, indicating that this drug has significant side effects. Furthermore, DTT treatment significantly affected the optimal PS II photochemical efficiency in *Porphyra* and *Ulva* even before light exposure, again indicating strong side effects independent of the desired target. DTT is routinely used to separate polypeptide subunits linked by disulfide bonds. The results indicate that this inhibitor should be used with caution when trying to affect the xanthophyll cycle in macroalgae. The use of the protein synthesis inhibitors streptomycin and chloramphenicol should also be viewed with caution. Their presence can induce responses not found in their absence. In addition, some protein synthesis inhibitors have been shown to influence the xanthophyll cycle and photoprotective energy dissipation.

VIII. Conclusions

Aquatic primary producers such as cyanobacteria, phytoplankton, and macroalgae experience even more rapid and excessive changes in solar radiation than higher plants due to the combined effects of wave action, tides, changing cloud cover, and diurnal changes. In order to thrive under low light conditions, most of these organisms operate as shade organisms, with some specialists surviving at extremely low irradiances, e.g. at 200 m water depth. As a consequence, algae in the tidal zone have to compensate the high light conditions when exposed to full solar radiation during low tide. The adaptation mechanisms to high light, and especially high-energy UV radiation, include dynamic photoinhibition based on the involvement of the xanthophyll cycle and chronic photoinhibition in which the electron flux through the reaction center of photosystem II is reduced by reversible D1 protein degradation. Fast induction kinetics document the rapid adaptation of the photosynthetic apparatus to the fast and drastically changing light conditions that the organisms experience in the water column. Another response to diurnal changes and faster alterations in light conditions are chloroplast movements, changes in pigmentation, and pigment bleaching. Adaptive protection mechanisms against high UV include the synthesis of UV-screening pigments such as MAAs and efficient repair processes of damage to the DNA, the photosynthetic apparatus, and other vital structures in the cell.

Acknowledgments

The work discussed here was financially supported by the European Union (DG XII, Environmental Program, ENV4-CT97-0580). The author thanks M. Schuster for excellent technical assistance.

References

Adams NL and Shick JM (1996) Mycosporine-like amino acids provide protection against ultraviolet radiation in eggs of the green sea urchin *Strongylocentrotus droebachiensis*. Photochem Photobiol 64: 149–158

Adams WW III, Smith SD and Osmond CB (1987) Photoinhibition of the CAM succulent *Opuntia basilaris* growing in Death Valley: evidence from 77K fluorescence and quantum yield. Oecologia 71: 221–228

Adams WW III, Demmig-Adams B, Winter K and Schreiber U (1990a) The ratio of variable to maximum chlorophyll fluorescence from photosystem II, measured in leaves at ambient temperature and at 77K, as an indicator of the photon yield of photosynthesis. Planta 180: 166–174

Adams WW III, Demmig-Adams B and Winter K (1990b) Relative contributions of zeaxanthin-related and zeaxanthin-unrelated types of 'high-energy-state' quenching of chlorophyll fluorescence in spinach leaves exposed to various environmental conditions. Plant Physiol 92: 302–309

Adams WW III, Zarter CR, Mueh KE, Amiard V and Demmig-Adams B (2005) Energy dissipation and photoinhibition: A continuum of photoprotection. In: Demmig-Adams B, Adams WW III and Mattoo AK (eds) Photoprotection, Photoinhibition, Gene Regulation, and Environment, pp 49–64. Springer, Dordrecht

Algarra P and Niell P (1990) Short-term pigment response of *Corallina elongata* Ellis et Solander to light intensity. Aquat Bot 36: 127–138

Altamirano M, Flores-Moya A, Conde F and Figueroa FL (2000) Growth seasonality, photosynthetic pigments, and carbon and nitrogen content in relation to environmental factors: a field study of *Ulva olivascens* (Ulvales, Chlorophyta). Phycologia 39: 50–58

Arts MT, Robarts RD, Kasai F, Waiser MJ, Tumber VP, Plante AJ, Rai H and de Lange HJ (2000) The attenuation of ultraviolet radiation in high dissolved organic carbon waters of wetlands and lakes on the northern Great Plains. Limnol Oceanogr 45: 292–299

Baroli I, Do AD, Yamane T, Niyogi KK (2003) Zeaxanthin accumulation in the absence of a functional xanthophyll cycle protects *Chlamydomonas reinhardtii* from photooxidative stress. Plant Cell 15: 992–1008

Beach KS and Smith CM (1996a) Ecophysiology of tropical rhodophytes. I. Microscale acclimation in pigmentation. J Phycol 32: 701–710

Beach KS and Smith CM (1996b) Ecophysiology of tropical rhodophytes. II. Microscale acclimation in photosynthesis. J Phycol 32: 710–718

Benet H, Bruss U, Duval J-C and Kloareg B (1994) Photosynthesis and photoinhibition in protoplasts of the marine brown alga *Laminaria saccharina*. J Exp Bot 45: 211–220

Bilger W and Björkman O (1990) Role of the xanthophyll cycle in photoprotection elucidated by measurements of light-induced absorbance changes, fluorescence and photosynthesis in leaves of *Hedera canariensis*. Photosynth Res 25: 173–185

Bilger W, Björkman O and Thayer SS (1989) Light-induced spectral absorbance changes in relation to photosynthesis and the epoxidation state of the xanthophyll cycle components in cotton leaves. Plant Physiol 91: 542–551

Bischof K, Hanelt D and Wiencke C (1998) UV-radiation can affect depth-zonation of Antarctic macroalgae. Mar Biol 131: 597–605

Bischof K, Hanelt D and Wiencke C (1999) Acclimation of maximal quantum yield of photosynthesis in the brown alga *Alaria esculenta* under high light and UV radiation. Plant Biol 1: 435–444

Björkman O and Demmig B (1986) Photon yield of O_2 evolution and chlorophyll fluorescence characteristics at 77K among vascular plants of diverse origin. Planta 170: 489–504

Booth CR and Morrow JH (1997) The penetration of UV into natural waters. Photochem Photobiol 65: 254–257

Büchel C and Wilhelm C (1993) *In vivo* analysis of slow chlorophyll fluorescence induction kinetics in algae: progress, problems and perspectives. Photochem Photobiol 58: 137–148

Carreto JI, Carignan MO, Daleo G and De Marco SG (1990) Occurrence of mycosporine-like amino acids in the red tide dinoflagellate *Alexandrium excavatum*: UV-protective compounds? J Plankton Res 12: 909–921

Cockell CS and Knowland J (1999) Ultraviolet radiation screening compounds. Biol Rev 74: 311–345

Coelho S, Rijstenbil JW, Sousa-Pinto I and Brown MT (2001) Cellular responses to elevated light levels in *Fucus spiralis* embryos during the first days after fertilization. Plant Cell Environ 24: 801–810

Conde D, Aubriot L and Sommaruga R (2000a) Changes in UV penetration associated with marine intrusions and freshwater discharge in a shallow coastal lagoon of the Southern Atlantic Ocean. Mar Ecol Prog Ser 207: 19–31

Conde FR, Churio MS and Previtali CM (2000b) The photoprotector mechanism of mycosporine-like amino acids. Excited-state properties and photostability of porphyra-334 in aqueous solution. J Photochem Photobiol B: Biol 56: 139–144

Crutzen PJ (1992) Ultraviolet on the increase. Nature 356: 104–105

de Bakker N, Rozema J and Aerts R (2001a) UV effects on a charophycean algae, *Chara aspera*. Plant Ecol 154: 205–212

de Bakker NVJ, van Beem AP, van de Staaij JWM, Rozema J and Aerts R (2001b) Effects of UV-B radiation on a charophycean alga, *Chara aspera*. In: Rozema J, Manetas Y and Björn LO (eds) Plant Ecology Special Issue: Responses of Plants to UV-B Radiation, pp 237–246. Kluwer Academic Publishers, Dordrecht

de Beer D and Larkum AWD (2001) Photosynthesis and calcification in the calcifying algae *Halimeda discoidea* studied with microsensors. Plant Cell Environ 24: 1209–1217

Demmig-Adams B and Adams WW III (1990) The carotenoid zeaxanthin and 'high-energy-state quenching' of chlorophyll fluorescence. Photosynth Res 25: 187–197

Demmig-Adams B, Adams WW III, Winter K, Meyer A, Schreiber U, Pereira JS, Krüger A, Czygan F-C and Lange OL (1989) Photochemical efficiency of photosystem II, photon yield of O_2 evolution, photosynthetic capacity, and carotenoid composition during the midday depression of net CO_2 uptake in *Arbutus unedo* growing in Portugal. Planta 177: 377–387

Demmig-Adams B, Adams WW III, Heber U, Neimanis S, Winter K, Krüger A, Czygan F-C, Bilger W and Björkman O (1990a) Inhibition of zeaxanthin formation and of rapid changes in radiationless energy dissipation by dithiothreitol in spinach leaves and chloroplasts. Plant Physiol 92: 293–301

Demmig-Adams B, Adams WW III, Czygan F-C, Schreiber U and Lange OL (1990b) Differences in the capacity for radiationless energy dissipation in the photochemical apparatus of green and blue-green algal lichens associated with differences in carotenoid composition. Planta 180: 582–589

Demmig-Adams B, Adams III WW, Ebbert V and Logan BA (1999) Ecophysiology of the xanthophyll cycle. In: Frank HA, Young AJ, Britton G and Cogdell RJ (eds) The Photochemistry of Carotenoids, Advances in Photosynthesis, Vol 8, pp 245–269. Kluwer Academic Publishers, Dordrecht

Dring MJ, Wagner A, Boeskop J and Lüning K (1996) Sensitivity of intertidal and subtidal red algae to UV-A and UV-B radiation, as monitored by chlorophyll fluorescence measurements: influence of collection, depth and season and length of irradiation. Eur J Phycol 31: 293–302

Dring MJ, Wagner A, Franklin LA, Kuhlenkamp R and Lüning K (2001a) Seasonal and diurnal variations in ultraviolet-B and ultraviolet-A irradiances at and below the sea surface at Helgoland (North Sea) over a 6-year period. Helgol Mar Res 55: 3–11

Dring MJ, Wagner A and Lüning K (2001b) Contribution of the UV component of natural sunlight to photoinhibition of photosynthesis in six species of subtidal brown and red seaweeds. Plant Cell Environ 24: 1153–1164

Edelman M and Mattoo AK (2005) D1 protein retrospective: the sizzling '80s. In: Demmig-Adams B, Adams WW III and Mattoo AK (eds) Photoprotection, Photoinhibition, Gene Regulation, and Environment, pp 23–38. Springer, Dordrecht

Etienne A-L, Kirilovsky D and Vass I (1996) Report on the ESF workshop on visible and UV light stress. Plant Sci 115: 5–7

Favre-Bonvin J, Arpin N and Brevard C (1976) Structure de la mycosporine (P 310). Can J Chem 54: 1105–1113

Figueroa FL, Blanco JM, Jimenez-Gomez F and Rodriguez J (1997a) Effects of ultraviolet radiation on carbon fixation in Antarctic nanophytoflagellates. Photochem Photobiol 66: 185–189

Figueroa FL, Mercado J, Jimenez C, Salles S, Aguilera J, Sanchez-Saavedra MP, Lebert M, Häder D-P, Montero O and Lubian L (1997b) Relationship between bio-optical characteristics and photoinhibition of phytoplankton. Aquatic Bot 59: 237–251

Foster R and Lüning K (1996) Photosynthetic response of *Laminaria digitata* to ultraviolet A and B radiation. In: Figueroa FL, Jiménez C and Pérez-Lloréns JL (eds) Underwater Light and Algal Photobiology, 60 (Suppl. 1), pp 65–71. Scientia Marina, Barcelona, Spain

Franklin LA and Forster RM (1997) The changing irradiance environment: consequences for marine macrophyte physiology, productivity and ecology. Eur J Phycol 32: 207–232

Franklin L, Schäfer C and Osmond CB (1996a) Degradation of the DI protein in *Ulva* during photoinhibition in the lab and in the field. Abstract of the I European Phycological Congress: 12

Franklin LA, Seaton GGR, Lovelock CE and Larkum AWD (1996b) Photoinhibition of photosynthesis on a coral reef. Plant Cell Environ 19: 825–836

Franklin LA, Forster RM and Lüning K (1998) UVB radiation and macroalgae: Present effects and future directions. In: Nolan CV and Häder D-P (eds) Role of Solar UV-B Radiation on Ecosystems, Ecosystem Research Report No. 30, pp 134–146. European Communities, Belgium

Franklin LA, Kräbs G and Kuhlenkamp R (2001) Blue light and UV-A radiation control the synthesis of mycosporine-like amino acids in *Chondrus crispus* (Florideophyceae). J Phycol 37: 257–270

Garcia-Pichel F and Castenholz RW (1993) Occurrence of UV-absorbing, mycosporine-like compounds among cyanobacterial isolates and an estimate of their screening capacity. Appl Environm Microbiol 59: 163–169

Gleason DF and Wellington GM (1993) Ultraviolet radiation and coral bleaching. Nature 365: 836–838

Gomez I, Thomas DN and Wiencke C (1995) Longitudinal profiles of growth, photosynthesis and light independent carbon fixation in the Antarctic brown alga *Ascoseira mirabilis*. Bot Marina 38: 157–164

Gosink JJ, Irgens RL and Staley JT (1993) Vertical distribution of bacteria in Arctic sea ice. FEMS Microbiol Ecol 102: 85–90

Grevby C (1996) Organisation of the light harvesting complex in fucoxanthin containing algae (PhD thesis). Göteborg University Sweden

Grobe CW and Murphy TM (1998) Solar ultraviolet-B radiation effects on growth and pigment composition of the intertidal alga *Ulva expansa* (Setch.) S. & G. (Chlorophyta). J Exp Mar Biol Ecol 225: 39–51

Gröniger A and Häder D-P (2000) Stability of mycosporine-like amino acids. Recent Res Devel Photochem Photobiol 4: 247–252

Gröniger A and Häder D-P (2002) Induction of the synthesis of an UV-absorbing substance in the green alga *Prasiola stipitata*. J Photochem Photobiol B: Biol 66: 54–59

Gröniger A, Hallier C and Häder D-P (1999) Influence of UV radiation and visible light on *Porphyra umbilicalis*: Photoinhibition and MAA concentration. J Appl Phycol 11, 437–445

Gröniger A, Sinha RP, Klisch M and Häder D-P (2000) Photoprotective compounds in cyanobacteria, phytoplankton and macroalgae - a database. J Photochem Photobiol B: Biol 58: 115–122

Grumbach KH, Lichtenthaler HK and Erismann KH (1978) Incorporation of $^{14}CO_2$ in photosynthetic pigments of *Chlorella pyrenoidosa*. Planta 140: 37–43

Gunasekera TS, Paul ND and Ayres PG (1997) The effects of ultraviolet-B (UV-B: 290-320 nm) radiation on blister blight disease of tea (*Camellia sinensis*). Plant Pathol 46: 179–185

Häder D-P (1995) Influence of ultraviolet radiation on phytoplankton ecosystems. In: Wiessner W, Schnepf E and Starr RC (eds) Algae, Environment and Human Affairs, pp 41–55. Biopress Limited, Bristol, England

Häder D-P (1997) Penetration and effects of solar UV-B on phytoplankton and macroalgae. Plant Ecol 128: 4–13

Häder D-P (1998) Are seaweeds affected by solar UV? In: Hönigsmann H, Knobler RM, Trautinger F and Jori G (eds) Landmarks in Photobiology, Proceedings of the 12th Congress on Photobiology (ICP '96), pp 31–33. OEMF spa, Milano, Italy

Häder D-P and Figueroa FL (1997) Photoecophysiology of marine macroalgae. Photochem Photobiol 66: 1–14

Häder D-P and Schäfer J (1994) Photosynthetic oxygen production in macroalgae and phytoplankton under solar irradiation. J Plant Phys 144: 293–299

Häder D-P, Herrmann H and Santas R (1996a) Effects of solar radiation and solar radiation deprived of UV-B and total UV on photosynthetic oxygen production and pulse amplitude modulated fluorescence in the brown alga *Padina pavonia*. FEMS Microbiol Ecol 19: 53–61

Häder D-P, Herrmann H, Schäfer J and Santas R (1996b) Photosynthetic fluorescence induction and oxygen production in corallinacean algae measured on site. Bot Acta 109: 285–291

Häder D-P, Lebert M, Mercado J, Aguilera J, Salles S, Flores-Moya A, Jimenez C and Figueroa FL (1996c) Photosynthetic oxygen production and PAM fluorescence in the brown alga *Padina pavonica* (Linnaeus) Lamouroux measured in the field under solar radiation. Mar Biol 127: 61–66

Häder D-P, Porst M, Herrmann H, Schäfer J and Santas R (1996d) Photoinhibition in the Mediterranean green alga *Halimeda tuna* Ellis et Sol measured in situ. Photochem Photobiol 64: 428–434

Häder D-P, Herrmann H, Schäfer J and Santas R (1997a) Photosynthetic fluorescence induction and oxygen production in two Mediterranean *Cladophora* species measured on site. Aquatic Bot 56: 253–264

Häder D-P, Lebert M, Flores-Moya A, Jimenez C, Mercado J, Salles S, Aguilera J and Figueroa FL (1997b) Effects of solar radiation on the photosynthetic activity of the red alga *Corallina elongata* Ellis et Soland. J Photochem Photobiol B: Biol 37: 196–202

Häder D-P, Kumar HD, Smith RC and Worrest RC (1998a) Effects on aquatic ecosystems. J Photochem Photobiol B: Biol 46: 53–68

Häder D-P, Lebert M, Figueroa FL, Jiménez C, Viñegla B and Perez-Rodriguez E (1998b) Photoinhibition in Mediterranean macroalgae by solar radiation measured on site by PAM fluorescence. Aquatic Bot 61: 225–236

Häder D-P, Porst M and Santas R (1998c) Photoinhibition by solar radiation in the Mediterranean alga *Peyssonnelia squamata* measured on site. Plant Ecol 139: 167–175

Häder D-P, Lebert M, Jimenez C, Salles S, Aguilera J, Flores-Moya A, Mercado J, Vinegla B and Figueroa FL (1999) Pulse amplitude modulated fluorescence in the green macrophytes, *Codium adherens*, *Enteromorpha muscoides*, *Ulva gigantea* and *Ulva rigida*, from the Atlantic coast of Southern Spain. Environ Exp Bot 41: 247–255

Häder D-P, Lebert M and Helbling EW (2000a) Photosynthetic performance of the chlorophyte *Ulva rigida* measured in Patagonia on site. Recent Res Devel Photochem Photobiol 4: 259–269

Häder D-P, Porst M and Lebert M (2000b) On site photosynthetic performance of Atlantic green algae. J Photochem Photobiol B: Biol 57: 159–168

Häder D-P, Lebert M and Helbling EW (2001a) Effects of solar radiation on the Patagonian macroalga *Enteromorpha linza* (L.) J. Agardh - Chlorophyceae. J Photochem Photobiol B: Biol 62: 43–54

Häder D-P, Lebert M and Helbling EW (2001b) Photosynthetic performance of marine macroalgae measured in Patagonia on site. Trends Photochem Photobiol 8: 145–152

Häder D-P, Porst M and Lebert M (2001c) Photoinhibition in common Atlantic macroalgae measured on site in Gran Canaria. Helgol Mar Res 55: 67–76

Häder D-P, Porst M and Lebert M (2001d) Photosynthetic performance of the Atlantic brown macroalgae, *Cystoseira abies-marina*, *Dictyota dichotoma* and *Sargassum vulgare*, measured in Gran Canaria on site. Environ Exp Bot 45: 21–32

Häder D-P, Lebert M, Sinha RP, Barbieri ES and Helbling EW (2002) Role of protective and repair mechanisms in the inhibition of photosynthesis in marine macroalgae. Photochem Photobiol Sci 1: 809–814

Häder D-P, Kumar HD, Smith RC and Worrest RC (2003a) Aquatic ecosystems: effects of solar ultraviolet radiation and interactions with other climatic change factors. Photochem Photobiol Sci 2: 39–50

Häder D-P, Lebert M and Helbling EW (2003b) Effects of solar radiation on the Patagonian Rhodophyte, *Corallina officinalis* (L.). Photosynth Res 78: 119–132

Hanelt D (1998) Capability of dynamic photoinhibition in Arctic macroalgae is related to their depth distribution. Mar Biol 131: 361–369

Hanelt D and Nultsch W (1995) Field studies of photoinhibition show non-correlations between oxygen and fluorescence measurements in the Arctic red alga *Palmaria palmata*. J Plant Physiol 145: 31–38

Hanelt D, Jaramillo JM, Nultsch W, Senger S and Westermeier R (1994a) Photoinhibition as a regulative mechanism of photosynthesis in marine algae of Antarctica. Serie Cient Inst Antarct 44: 67–77

Hanelt D, Li J and Nultsch W (1994b) Tidal dependence of photoinhibition of photosynthesis in marine macrophytes of the South China Sea. Bot Acta 107: 66–72

Hanelt D, Uhrmacher S and Nultsch W (1995) The effect of photoinhibition on photosynthetic oxygen production in the brown alga *Dictyota dichotoma*. Bot Acta 108: 99–105

Hanelt D, Melchersmann B, Wiencke C and Nultsch W (1997) Effects of high light stress on photosynthesis of polar macroalgae in relation to depth distribution. Mar Ecol Progr Ser 149: 255–266

Hanelt D, Tüg H, Bischof K, Groß C, Lippert H, Sawall T and Wiencke C (2001) Light regime in an Arctic fjord: a study related to stratospheric ozone depletion as a basis for determination of UV effects on algal growth. Mar Biol 138: 649–658

Harker M, Berkaloff C, Lemoine Y, Britton G, Young AJ, Duval J-C and Rmiki N-E (1999) Effects of high light and desiccation on the operation of the xanthophyll cycle in two marine brown algae. Eur J Phycol 34: 35–42

Henley WJ, Lindley ST, Levavasseur G, Osmond C and Ramus J (1992) Photosynthetic response of *Ulva rotundata* to light and temperature during emersion on an intertidal sand flat. Oecologia 89: 516–523

Herrmann H, Ghetti F, Scheuerlein R and Häder D-P (1995) Photosynthetic oxygen and fluorescence measurements in *Ulva laetevirens* affected by solar irradiation. J Plant Physiol 145: 221–227

Herrmann H, Häder D-P, Köfferlein M, Seidlitz HK and Ghetti F (1996) Effects of UV radiation on photosynthesis of phytoplankton exposed to solar simulator light. J Photochem Photobiol B: Biol 34: 21–28

Hui Y, Jie W and Carpentier R (2000) Degradation of the photosystem I complex during photoinhibition. Photochem Photobiol 72: 508–512

Huot Y, Jeffrey WH, Davis RF and Cullen JJ (2000) Damage to DNA in bacterioplankton: a model of damage by ultraviolet radiation and its repair as influenced by vertical mixing. Photochem Photobiol 72: 62–74

Jensen A (1995) Production of alginate. In: Wiessner W, Schnepf E and Starr RC (eds) Algae, Environment and Human Affairs, pp 79–92. Biopress Ltd., Bristol, England

Jimenez C, Figueroa FL, Aguilera J, Lebert M and Häder D-P (1996) Phototaxis and gravitaxis in *Dunaliella bardawil*: Influence of UV radiation. Acta Protozool 35: 287–295

Jung H-S and Niyogi KK (2005) Molecular analysis of photoprotection and photosynthesis. In: Demmig-Adams B, Adams WW III and Mattoo AK (eds) Photoprotection, Photoinhibition, Gene Regulation, and Environment, pp 127–143. Springer, Dordrecht

Karentz D, McEuen FS, Land MC and Dunlap WC (1991) A survey of mycosporine-like amino acid compounds in Antarctic marine organisms: potential protection from ultraviolet exposure. Mar Biol 108: 157–166

Karsten U and Wiencke C (1999) Factors controlling the formation of UV-absorbing mycosporine-like amino acids in the marine red alga *Palmaria palmata* from Spitsbergen (Norway). J Plant Physiol 155: 407–415

Karsten U, Franklin LA, Lüning K and Wiencke C (1998a) Natural ultraviolet radiation and photosynthetically active radiation induce formation of mycosporine-like amino acids in the marine macroalga *Chondrus crispus* (Rhodophyta). Planta 205: 257–262

Karsten U, Sawall T and Wiencke C (1998b) A survey of the distribution of UV-absorbing substances in tropical macroalgae. Phycol Res 46: 271–279

Karsten U, Bischof K and Wiencke C (2001) Photosynthetic performance of Arctic macroalgae after transplantation from deep to shallow waters. Oecologia 127: 11–20

Kerr JB and McElroy CT (1993) Evidence for large upward trends of ultraviolet-B radiation linked to ozone depletion. Science 262: 1032–1034

Klisch M and Häder D-P (2000) Mycosporine-like amino acids in the marine dinoflagellate *Gyrodinium dorsum*: induction by ultraviolet irradiation. J Photochem Photobiol B: Biol 55: 178–182

Komenda J (2000) Role of two forms of the DI protein in the recovery from photoinhibition of photosystem II in the cyanobacterium *Synechococcus* PCC 7942. Biochim Biophys Acta 1457: 243–252

Krieger-Liszkay A, Kienzler K and Johnson GN (2000) Inhibition of electron transport at the cytochrome $b_6 f$ complex protects photosystem II from photoinhibition. FEBS Lett 486: 191–194

Kuhn P, Browman H, McArthur B and St-Pierre J-F (1999) Penetration of ultraviolet radiation in the waters of the estuary and Gulf of St. Lawrence. Limnol Oceanogr 44: 710–716

Kuwahara VS, Ogawa H, Toda T, Kikuchi T and Taguchi S (2000) Variability of bio-optical factors influencing the seasonal attenuation of ultraviolet radiation in temperate coastal waters of Japan. Photochem Photobiol 72: 193–199

Larkum AWD and Wood WF (1993) The effect of UV-B radiation on photosynthesis and respiration of phytoplankton, benthic macroalgae and seagrasses. Photosynth Res 36: 17–23

Laurion I, Ventura M, Catalan J, Psenner R and Sommaruga R (2000) Attenuation of ultraviolet radiation in mountain lakes: factors controlling the among- and within-lake variability. Limnol Oceanogr 45: 1274–1288

Leavitt PR, Vinebrooke RD, Donald DB, Smol JP and Schindler DW (1997) Past ultraviolet radiation environments in lakes derived from fossil pigments. Nature 388: 457–459

Lubin D and Jensen EH (1995) Effects of clouds and stratospheric ozone depletion on ultaviolet radiation trends. Nature 377: 710–713

Lüning K (ed) (1990) Seaweeds. Their environment, biogeography and ecophysiology. New York, Wiley

Markager S and Sand-Jensen K (1994) The physiology and ecology of light-growth relationship in macroalgae. In: Round FE and Chapman DJ (eds) Progress in Phycological Research, 10, pp 209–298. Biopress Ltd., Bristol

Mate Z, Sass L, Spetea C, Nagy F and Vass I (1998) De Novo synthesis of the D1 and D2 reaction center subunits protects against UV-B induced damage of photosystem II in *Synechocystis* sp. PCC 6803. In: Nolan CV and Häder D-P (eds) Role of Solar UV-B Radiation on Ecosystems, Ecosystem Research Report No. 30, pp 159–166. European Communities, Belgium

Neale PJ, Cullen JJ, Lesser MP and Melis A (1993) Physiological bases for detecting and predicting photoinhibition of aquatic photosynthesis by PAR and UV radiation. In: Yamamoto HY and Smith CM (eds) Photosynthetic Responses to the Environment, pp 61–77. American Society of Plant Physiologists Rockville Maryland

Neidhardt J, Benemann JR, Zhang L and Melis A (1998) Photosystem-II repair and chloroplast recovery from irradiance stress: relationship between chronic photoinhibition, light-harvesting chlorophyll antenna size and photosynthetic productivity in *Dunalialla salina* (green algae). Photosynth Res 56: 175–184

Nishiyama Y, Allakhverdiev SI and Murata N (2005) Regulation by environmental conditions of the repair of photosystem II in cyanobacteria. In: Demmig-Adams B, Adams WW III and Mattoo AK (eds) Photoprotection, Photoinhibition, Gene Regulation, and Environment, pp 193–203. Springer, Dordrecht

Niyogi KK (2000) Safety valves for photosynthesis. Curr Opin Plant Biol 3: 455–460

Niyogi KK, Grossman AR and Björkman O (1998) *Arabidopsis* mutants define a central role for the xanthophyll cycle in the regulation of photosynthetic energy conversion. Plant Cell 10: 1121–1134

Noctor G, Rees D, Young A and Horton P (1991) The relationship between zeaxanthin, energy-dependent quenching of chlorophyll fluorescence, and trans-thylakoid pH gradient in isolated chloroplasts. Biochim Biophys Acta 1057: 320–330

Öquist G, Anderson JM, McCaffery S and Chow WS (1992) Mechanistic differences in photoinhibition of sun and shade plants. Planta 188: 422–431

Osmond CB and Förster B (2005) Photoinhibition: then and now. In: Demmig-Adams B, Adams WW III and Mattoo AK (eds) Photoprotection, Photoinhibition, Gene Regulation, and Environment, pp 11–22. Springer, Dordrecht

Osmond CB, Ramus J, Levavasseur G, Franklin LA and Henley WJ (1993) Fluorescence quenching during photosynthesis and photoinhibition of *Ulva rotundata* Blid. Planta 190: 97–106

Perez-Rodriguez E, Gomez I, Karsten U and Figueroa FL (1998) Effects of UV radiation on photosynthesis and excretion of UV-absorbing compounds of *Dasycladus vermicularis* (Dasycladales, Chlorophyta) from southern Spain. Phycologia 37: 379–387

Polle JEW and Melis A (1999) Recovery of the photosynthetic apparatus from photoinhibition during dark incubation of the green alga *Dunaliella salina*. Aust J Plant Physiol 26: 679–686

Porst M, Herrmann H, Schäfer J, Santas R and Häder D-P (1997) Photoinhibition in the mediterranean green alga *Acetabularia mediterranea* measured in the field under solar irradiation. J Plant Physiol 151: 25–32

Ritz M, Neverov KV and Etienne AL (1999) Delta pH-dependent fluorescence quenching and its photoprotective role in the unicellular red alga *Rhodella violacea*. Photosynthetica 37: 267–280

Schreiber U, Endo T, Mi H and Asada K (1986) Quenching analysis of chlorophyll fluorescence by saturation pulse method: particular aspects relating to eukariotic algae and cyanobacteria. Plant Cell Physiol 36: 873–882

Schubert H, Kroon BMA and Matthijs HCP (1994) In vivo manipulation of the xanthophyll cycle and the role of zeaxanthin in the protection against photodamage in the green alga *Chlorella pyrenoidosa*. J Biol Chem 269: 7267–7272

Shick JM, Lesser MP and Stochaj WR (1991) Ultraviolet radiation and photooxidative stress in zooxanthelate anthozoa: the sea anemone *Phyllodiscus semoni* and the octocoral *Clavularia* sp. Symbio 10: 145–173

Shick JM, Dunlap WC, Pearse JS and Pearse VB (2002) Mycosporine-like amino acid content in four species of sea anemones in the genus *Anthopleura* reflects phylogenetic but not environmental or symbiotic relationships. Biol Bull 203: 315–330

Sinha RP, Klisch M, Gröniger A and Häder D-P (1998) Ultraviolet-absorbing/screening substances in cyanobacteria, phytoplankton and macroalgae. J Photochem Photobiol B: Biol 47: 83–94

Smith RC and Baker KS (1979) Penetration of UV-B and biologically effective dose-rates in natural waters. Photochem Photobiol 29: 311–323

Smith RC, Baker KS, Holm-Hansen O and Olson R (1980) Photoinhibition of photosynthesis in natural waters. Photochem Photobiol 31: 585–592

Smith RC, Prezelin BB, Baker KS, Bidigare RR, Boucher NP, Coley T, Karentz D, MacIntyre S, Matlick HA, Menzies D, Ondrusek M, Wan Z and Waters KJ (1992) Ozone depletion:

ultraviolet radiation and phytoplankton biology in Antarctic waters. Science 255: 952–959

Talarico L (1996) Phycobiliproteins and phycobilisomes in red algae: adaptive responses to light. In: Figueroa FL, Jiménez C and Pérez-Lloréns JL (eds) Underwater Light and Algal Photobiology. Scientia Marina 60 (Suppl. 1), pp 205–222. Barcelona, Spain

Tyler MA and Seliger HH (1981) Selection for a red tide organism: physiological responses to the physical environment. Limnol Oceanogr 26: 310–324

Uhrmacher S, Hanelt D and Nultsch W (1995) Zeaxanthin content and the degree of photoinhibition are linearly correlated in the brown alga *Dictyota dichotoma*. Mar Biol 123: 159–165

Ursi S, Pedersen M, Plastino E and Snoeijs P (2003) Intraspecific variation of photosynthesis, respiration and photoprotective carotenoids in *Gracilaria birdiae* (Gracilariales: Rhodophyta). Mar Biol 142: 997–1007

van de Poll WH, Eggert A, Buma AGJ and Breeman AM (2001) Effects of UV-B-induced DNA damage and photoinhibition on growth of temperate marine red macrophytes: habitat-related differences in UV-B tolerance. J Phycol 37: 30–37

Walsby AE, Kinsman R and George KI (1992) The measurement of gas volume and buoyant density in planktonic bacteria. J Microbiol Meth 15: 293–309

Webb AR (1998) UV-B radiation measurements: Data and methods for ecosystem research. In: Nolan CV and Häder D-P (eds) Role of Solar UV-B Radiation on Ecosystems, Ecosystem Research Report No. 30: pp 7–16. European Communities, Belgium

Wiencke C, Gómez I, Pakker H, Flores-Moya A, Altamirano M, Hanelt D, Bischof K and Figueroa FL (2000) Impact of UV radiation on viability, photosynthetic characteristics and DNA of brown algal zoospores: implications for depth zonation. Mar Ecol Progr Ser 197: 217–229

Wood WF (1989) Photoadaptive responses of the tropical red alga *Eucheuma striatum* Schmitz (Gigartinales) to ultra-violet radiation. Aquat Bot 33: 41–51

Yamamoto HY (2005) A random walk to and through the xanthophyll cycle. In: Demmig-Adams B, Adams WW III and Mattoo AK (eds) Photoprotection, Photoinhibition, Gene Regulation, and Environment, pp 1–10. Springer, Dordrecht

Yamazawa A, Takeyama H, Takeda D and Matsunaga T (1999) UV-A-induced expression of GroEL in the UV-A-resistant marine cyanobacterium *Oscillatoria* sp. NKBG 091600. Microbiol 145: 949–954

Yentsch CS, Backus RH and Wing A (1964) Factors affecting the vertical distribution of bioluminescence in the euphotic zone. Limnol Oceanogr 9: 519–524

Chapter 8

Phosphorylation of Thylakoid Proteins

Alexander V. Vener*
Division of Cell Biology, Linköping University, SE-581 85, Linköping, Sweden

Summary	107
I. Introduction	108
II. Thylakoid Phosphoproteins	109
A. Detection Techniques	109
B. Identified Thylakoid Phosphoproteins	110
C. Characteristics and Classification of Thylakoid Phosphoproteins	111
D. Assessing Changes in the Stoichiometry of Protein Phosphorylation	112
III. Reversible Phosphorylation of Photosystem II (PS II) Proteins	113
A. What is the Role of PS II Phosphorylation?	113
B. Reversible Phosphorylation and Turnover of D1 Protein	114
C. Reversible Phosphorylation of D2, CP43, and PsbH Proteins During PS II Turnover	115
D. Phosphorylation of PS II During Adaptive Responses	116
E. PS II-Specific Protein Phosphatase and TLP40	117
F. Trans-Membrane Signaling by the Phosphatase and PS II Biogenesis	119
G. Reversible Phosphorylation of LHCII Polypeptides and State Transitions	119
H. Differential Phosphorylation of CP29	120
IV. PsaD: the First Phosphoprotein in PS I	120
V. Phosphorylation of Other Thylakoid Proteins	120
A. TMP14: a Previously Unknown Thylakoid Phosphoprotein	120
B. TSP9 and Light-Induced Cell Signaling	121
VI. Regulation and Role of Thylakoid Protein Phosphorylation in a Physiological Context	121
Acknowledgments	123
References	123

Summary

Application of novel techniques for the characterization of in vivo protein phosphorylation has revealed sixteen distinct phosphorylation sites in ten integral and two peripheral proteins in photosynthetic thylakoid membranes. In addition to phosphorylation of the photosystem II (PS II) proteins D1, D2, CP43, and PsbH, and the light-harvesting antenna polypeptides LHCII and CP29, phosphorylation has been found in photosystem I (PS I) protein PsaD and in two recently identified proteins TSP9 and TMP14. The accumulated knowledge favors an involvement of reversible phosphorylation in adaptive stress responses and cellular signaling, but not in direct regulation of photosynthetic activities like electron transfer or oxygen evolution. Enhancement of PS II protein phosphorylation by abiotic stress maintains the integrity of PS II before it migrates to the stroma regions of the thylakoids where dephosphorylation and subsequent protein turnover take place. Specific dephosphorylation of the D1, D2, and CP43 polypeptides is performed by a heat shock-inducible protein phosphatase intrinsic to the thylakoid membrane. The phosphatase activity is regulated by the lumenal peptidyl-prolyl isomerase TLP40. This regulation may coordinate the protein folding activity of TLP40 in the lumen with the protein dephosphorylation at the opposite side of the thylakoid membrane. Reversible phosphorylation of LHCII in vivo is under complex redox and metabolic control and is probably involved in regulation of the size of the PS II antennae. Cold- and high light-induced phosphorylation

---

*Author for correspondence, email: aleve@ibk.liu.se

of CP29 may facilitate photoprotective energy dissipation by changing PS II-LHCII interactions under stress conditions. Phosphorylation of PsaD protein could be involved in regulation of PS I stability and ferredoxin reduction by PS I. The light-induced phosphorylation of TSP9, followed by its release from thylakoids, is implicated in plant cell signaling. The exact physiological roles of the protein phosphorylation events in thylakoids should be revealed by studies with appropriate mutants of plants and algae.

I. Introduction

Light- and redox-induced protein phosphorylation in chloroplast membranes was discovered by Bennett in 1977 (Bennett, 1977). The prevailing hypothesis during much of the last two decades has been that reversible phosphorylation of LHCII is involved in state transitions, i.e. in balancing the distribution of absorbed light energy between the two photosystems, PS II and PS I (Bennett et al., ; 1980; Allen et al., 1981; Allen, 1992, 2002, 2003; Allen and Forsberg, 2001). This hypothesis has further evolved through studies of the redox sensing that connects electron transfer and protein kinase activity in photosynthetic membranes (Allen, 1992; Vener et al., 1998; Aro and Ohad, 2003). Studies of the molecular aspects of redox-dependent thylakoid protein phosphorylation have revealed it to be an extremely complex process. A multiple factor-dependent regulation of LHCII phosphorylation has been demonstrated. In addition to the requirement of plastoquinone reduction for activation of LHCII kinase (Allen et al., 1981), the Qo site of the cytochrome bf complex operates as the redox sensor for induction of the kinase activity (Vener et al., 1995; Vener et al., 1997; Zito et al., 1999). Light-induced changes in LHCII also affect its phosphorylation (Zer et al., 1999; Zer et al., 2003) as does the thiol redox state and the ferredoxin-thioredoxin system of chloroplasts (Carlberg et al., 1999; Rintamäki et al., 2000). The latter mechanism for control of LHCII phosphorylation was uncovered largely due to the finding of an initially surprising irradiance-dependence for the amount of phospho-LHCII in vivo (Rintamäki et al., 1997). In plant leaves, LHCII was found phosphorylated only at light intensities lower than those during normal plant growth (Rintamäki et al., 1997; Rintamäki and Aro, 2001). These findings and the measurements of the excitation energy transfer between the two photosystems in plant leaves have seriously questioned the role of LHCII phosphorylation in state transitions (Elich et al., 1997; Haldrup et al., 2001; Rintamäki and Aro, 2001). Accordingly, the physiological function of LHCII phosphorylation remains an open question.

More than 1100 genes encode for protein kinases in the genome of *Arabidopsis thaliana* (The *Arabidopsis* Genome Initiative, 2000). At present there are five candidate *Arabidopsis* genes for membrane protein kinases that could phosphorylate thylakoid proteins: a family of three TAK kinases (Snyders and Kohorn, 1999, 2001) and two kinases homologous to Stt7 kinase from *Chlamydomonas reinhardtii* (Depege et al., 2003). The mechanism for the redox regulation of these kinases is elusive as is the identity and number of other possible thylakoid protein kinases. With respect to the genes for protein phosphatases operating in thylakoid membranes, the situation is even less clear. No gene or protein sequence information has yet been published concerning the enzymes involved in dephosphorylation of thylakoid phosphoproteins. The additional challenge in elucidating the redox-dependent system for thylakoid protein phosphorylation lies in the fact that it requires the integrity of the membrane and the electron transfer chain for operation. Nevertheless, there has been steady progress in the decoding of the molecular mechanisms for redox regulation of thylakoid protein phosphorylation, which has been periodically reviewed (Vener et al., 1998; Ohad et al., 2001; Rintamäki and Aro, 2001; Aro and Ohad, 2003; Zer and Ohad, 2003).

Three recent groundbreaking developments provided unprecedented possibilities for revealing the functions of protein phosphorylation in the regulation of photosynthesis. Firstly, the sequencing of plant genomes allowed the full-power application of proteomic approaches to study protein modifications in these species. Secondly, plant lines with knockouts of individual genes became commercially available. Thirdly, new analytical techniques permitted the detection of protein modifications in vivo in variable

---

Abbreviations: D1 – photosystem II reaction center protein; D2 – photosystem II reaction center sister protein; of 29 kDa – minor chlorophyll *a/b*-binding protein of photosystem II; of 43 kDa – chlorophyll *a* binding protein of photosystem II; LHCII – light harvesting chlorophyll *a/b*-binding proteins of photosystem II; PsbH – 9 kDa *psbH* gene product; PS I – photosystem I; PS II – photosystem II; PPIase – peptidyl-prolyl cis-trans isomerase; PP2A – protein phosphatase 2A; TLP20 – thylakoid lumen PPIase of 20 kDa; TLP40 – thylakoid lumen PPIase of 40 kDa; TMP14 – thylakoid membrane phosphoprotein of 14 kDa; TSP9 – thylakoid soluble phosphoprotein of 9 kDa

environmental conditions. The latter development has already brought new insights in the field of thylakoid protein phosphorylation, showing that this process could be crucially involved in the response of the photosynthetic machinery to stress. In this chapter, I review phosphorylation of thylakoid proteins in relation to different physiological conditions in plants. Special attention is paid to the in vivo phosphorylation sites found in the individual thylakoid proteins, which form the basis for the studies of environmentally dependent changes in these distinct modifications.

II. Thylakoid Phosphoproteins

A. Detection Techniques

Five different approaches have been used for the detection of phosphorylation of thylakoid membrane proteins: 1) radioactive labeling; 2) detection of the shift in the electrophoretic mobility of individual proteins; 3) immunological analysis with phosphoamino acid antibodies; 4) measurement of the phosphorylation-induced increase in the mass of intact proteins by mass spectrometry; 5) identification and sequencing of phosphorylated peptides obtained after proteolytic degradation of proteins. The experimental protocols for determination of phosphoproteins in higher plant thylakoids by some of these methods have recently been published (Aro et al., 2004). It is important to keep in mind that the ultimate evidence for phosphorylation requires identification of the phosphorylated amino acid in the sequence of the corresponding protein. In this respect, the first four techniques listed above are limited by their inability to determine the residue(s) phosphorylated and should therefore be complemented by protein sequencing revealing the phosphorylation sites. All five approaches provided valuable information about the status of protein phosphorylation in thylakoids from different species. However, each of these methods has its own disadvantages that should be taken into account when evaluating the accumulated literature.

Detection of phosphoproteins labeled with radioactive isotopes ^{32}P or ^{33}P is the most sensitive and common technique for studies on protein phosphorylation. Radioactive labeling of proteins also provides a dependable avenue to localize the labeled phosphoamino acids (Michel and Bennett, 1987). Radio-labeled phosphate has been used in studies of thylakoid protein phosphorylation in organello (Bennett, 1977) as well as in vivo (Elich et al., 1992; Elich et al., 1993; Elich et al., 1997; Fleischmann et al., 1999; Fleischmann and Rochaix, 1999; Depege et al., 2003). Radio-labeled ATP has been widely used for phosphorylation of proteins in isolated thylakoids (Bhalla and Bennett, 1987; Cheng et al., 1994; Vener et al., 1995; Snyders and Kohorn, 1999). The limitations of radioactive labeling consist of uneven uptake of the label in different plant tissues, the large pools of endogenous phosphate, and the presence of pre-existing phosphorylation in the proteins.

The phosphorylated D1, D2, and CP43 proteins of PS II have been found to have a slightly slower electrophoretic mobility than the corresponding nonphosphorylated proteins (Callahan et al., 1990; de Vitry et al., 1991; Elich et al., 1992; Rintamäki et al., 1997). Use of this electrophoretic property in combination with specific antibodies against each individual protein has allowed studies of changes in protein phosphorylation status under different conditions (Elich et al., 1992; Rintamäki et al., 1997). This approach has mostly been used for studies of D1 protein phosphorylation. Successful application of this technique is limited to well-characterized proteins and requires the use of specific antibodies that cross-react exclusively with either phospho- or dephospho-forms of the protein.

Phosphothreonine antibodies have also been used in studies of thylakoid protein phosphorylation (Rintamäki et al., 1997; Rintamäki and Aro, 2001). This method has an advantage over labeling experiments in allowing the detection of endogenous levels of thylakoid protein phosphorylation in vivo under particular environmental conditions. It is also suitable for studies on the regulation of thylakoid protein phosphorylation using intact chloroplasts, isolated thylakoid membranes, or membrane subfractions. However, serious attention should be paid to the fact that the immunoreactivity with different commercial antibodies differs between various phosphoproteins, and the linearity of the immunoreactivity should be monitored in each case (Rintamäki et al., 1997). Moreover, the use of phosphothreonine antibodies is rather limited to the detection of only four or five major thylakoid phosphoproteins (Aro et al., 2004).

Recent developments in proteomics and mass spectrometry have allowed the detection of phosphoproteins by measuring the phosphorylation-induced change in the mass of intact proteins. Analysis of the chloroplast grana proteome by liquid chromatography mass spectrometry (LCMS) has confirmed the phosphorylation of full-length D1, D2, CP43, PsbH, and two LHCII polypeptides (Gomez et al., 2002). Three different phosphorylated forms of the recently characterized thylakoid-associated chloroplast phosphoprotein

TSP9 have also been detected by matrix-assisted laser desorption/ionization (MALDI) mass spectrometry (Carlberg et al., 2003). Measurement of the intact protein mass provides strong indication for protein modification, but requires complementary mapping of the phosphorylation site(s) in the protein.

Sequencing of phosphopeptides obtained by proteolysis of phosphorylated thylakoid proteins has provided most of the presently available information on the exact phosphorylation sites (Michel and Bennett, 1987; Michel et al., 1988, 1991; Vener et al., 2001; Carlberg et al., 2003; Hansson and Vener, 2003). Conventional chemical sequencing of phosphopeptides corresponding to the N-termini of thylakoid proteins has been rather limited (Michel and Bennett, 1987) because many of these peptides are N-terminally blocked (Michel et al., 1988, 1991; Vener et al., 2001; Hansson and Vener, 2003; Turkina et al., 2004). Application of mass spectrometry has been most efficient for mapping of the phosphorylation sites in thylakoid proteins. Mass spectrometry analysis can now be combined with plant genomic sequence information. The sequencing of phosphopeptides selected from a complex peptide mixture identifies both the site of the phosphorylation and the parent phosphoprotein. This has led to direct analyses of the phosphopeptides from thylakoid membranes without prior isolation of individual proteins or protein complexes (Vener et al., 2001; Hansson and Vener, 2003; Turkina et al., 2004). Identification of nine in vivo protein phosphorylation sites in the thylakoids of *Arabidopsis thaliana* has been achieved by "shaving" of the surface-exposed domains of thylakoid proteins with trypsin, following enrichment of the phosphopeptides by immobilized metal affinity chromatography and their subsequent sequencing using mass spectrometry (Vener et al., 2001; Hansson and Vener, 2003).

B. Identified Thylakoid Phosphoproteins

A different number of phosphoproteins has been detected in thylakoids from different plants (Bennett, 1977; Cheng et al., 1994; Vener et al., 1995; Rintamäki et al., 1997; Gomez et al., 2002; Hansson and Vener, 2003) and algal species (de Vitry et al., 1991; Fleischmann et al., 1999; Depege et al., 2003). Decisive proof for phosphorylation of any protein should include the determination of the phosphoamino acid in the protein sequence. This task has been accomplished for the major thylakoid phosphoproteins. A summary of the known phosphorylation sites in thylakoids is presented in Table 1. These principal biochemical data form the basis for addressing the questions on the physiological significance of the distinct phosphorylation events by means of molecular biology and reversed genetics. Accordingly, I make a classification of these phosphorylated proteins and then analyze the published information on each of these phosphoproteins in the following parts of this chapter.

In addition to the proteins with the mapped phosphorylation sites (Table 1), another set of proteins has been advocated to undergo phosphorylation in the photosynthetic membranes. The subunit V of cytochrome bf complex in *Chlamydomonas reinhardtii* has been reported to be reversibly phosphorylated upon state transitions (Hamel et al., 2000). It was radio-labeled in the alga incubated with radioactive phosphate and identified as a 15.2-kDa polypeptide encoded by the nuclear gene *PETO* (Hamel et al., 2000; Finazzi et al., 2001). Phosphorylation of subunit V was proposed to be involved in signal transduction during redox-controlled short and long term adaptation of the photosynthetic apparatus in eukaryotes (Hamel et al., 2000; Finazzi et al., 2001).

The thylakoid protein kinases (TAKs), represented by three family members in *A. thaliana,* have been shown phosphorylated by immunoblotting with anti-phosphothreonine and anti-phosphoserine antisera (Snyders and Kohorn, 1999). The thylakoid kinase Stt7 in *C. reinhardtii* has also been suggested to undergo in vivo phosphorylation (Depege et al., 2003). This proposal was based on indirect evidence showing a shift in the electrophoretic mobility of the protein after treatment of thylakoids with a phosphatase (Depege et al., 2003). TAKs and Stt7 have been implied in phosphorylation of LHCII and state transitions (Snyders and Kohorn, 1999, 2001; Depege et al., 2003). Phosphorylation of other thylakoid proteins, including LHCII, was reduced in antisense *TAK1 Arabidopsis* mutants (Snyders and Kohorn, 2001). The activity of all of these protein kinases and their regulation was suggested to be part of a possible cascade of redox-controlled thylakoid protein phosphorylation (Snyders and Kohorn, 1999, 2001; Depege et al., 2003). Thus, elucidation of the phosphorylation sites in these enzymes and of the distinct protein kinases involved in modification of each of them may provide a key to understanding the complex regulatory network in thylakoid and chloroplast signal transduction. The low abundance of these enzymes is a major challenge in characterization of their posttranslational modifications.

Application of LCMS for measurement of masses of intact integral membrane proteins from pea thylakoids gave a strong indication for phosphorylation of the

Chapter 8 Phosphorylation of Thylakoid Proteins

Table 1. **Phosphorylation sites in thylakoid proteins.** A single letter amino acid code is used in the sequences with the low case t and s designating phosphorylated threonine and serine, correspondingly. *Ac-* designates the N-terminal acetylation of the peptides.

Protein	Species	Phosphopeptide sequence	Reference
D1	Spinach	*Ac*-tAILGRR	(Michel et al., 1988)
	Arabidopsis	*Ac*-tAILER	(Vener et al., 2001)
D2	Spinach	*Ac*-tIAVGK	(Michel et al., 1988)
	Arabidopsis	*Ac*-tIALGK	(Vener et al., 2001)
CP43	Spinach	*Ac*-tLFNGTLTLAGR	(Michel et al., 1988)
	Arabidopsis	*Ac*-tLFNGTLALAGR	(Vener et al., 2001)
PsbH	Spinach	AtGTVESSSR	(Michel and Bennett, 1987)
	Arabidopsis	AtQTVEDSSR	(Vener et al., 2001)
	ArabidopsisArabidopsis	AtQtVEDSSR	(Vener et al., 2001)
LHCII	SpinachSpinach	*Ac*-RKtAGKPKT	(Michel et al., 1991)
	Spinach	*Ac*-RKtAGKPKN	(Michel et al., 1991)
	Spinach	*Ac*-RKsAGKPKN	(Michel et al., 1991)
	Spinach	*Ac*-RRtVKSAPQ	(Michel et al., 1991)
	Arabidopsis	*Ac*-RKtVAKPK	(Vener et al., 2001)
CP29	Maize	AGGIItRFESSE	(Testi et al., 1996)
	Arabidopsis	*Ac*-RFGFGtK	(Hansson and Vener, 2003)
	C. reinhardtii	*Ac*-VFKFPtPPGTQK	(Turkina et al., 2004)
PsaD	Arabidopsis	EKtDSSAAAAAAPATK	(Hansson and Vener, 2003)
TMP14	Arabidopsis	ATtEVGEAPATTTEAETTE	(Hansson and Vener, 2003)
TSP9	Spinach	GGtTSGK	(Carlberg et al., 2003)
	Spinach	KGtVSIPSK	(Carlberg et al., 2003)
	Spinach	SSGStSGK	(Carlberg et al., 2003)

PsbT protein (Gomez et al., 2002). In this study, the mature PsbT protein of PS II was assigned a mass of 4,032 Da. The putative phosphorylated form of the protein was found to have a mass increased by 80 Da, corresponding to incorporation of a phosphoryl group in the protein (Gomez et al., 2002). Detection of this previously unidentified thylakoid phosphoprotein shows the power of intact protein analysis by LCMS. The site of phosphorylation in PsbT, as well as the role of its phosphorylation in the function of PS II remains to be determined.

Tyrosine phosphorylation of a set of thylakoid proteins including LHCII has been reported (Tullberg et al., 1998). This suggestion was based mainly on the results of immunoblotting analysis with an anti-phosphotyrosine antibody. However, the recent use of antibodies against phosphotyrosine did not reveal any change in the immunoreactivity of thylakoid proteins under changing environmental conditions or in response to redox state of chloroplasts (Rintamäki and Aro, 2001). Tyrosine phosphorylation of thylakoid proteins has not been confirmed by phosphoamino acid analysis or phosphopeptide sequencing. Thus, the specificity of cross reaction of anti-phosphotyrosine antibodies with thylakoid proteins has been questioned (Rintamäki and Aro, 2001). At present, there are no convincing data on tyrosine phosphorylation of thylakoid proteins.

C. Characteristics and Classification of Thylakoid Phosphoproteins

All of the known thylakoid proteins shown to be phosphorylated undergo this modification at threonine residues (Table 1). Only one LHCII polypeptide from spinach (Table 1) has been found phosphorylated at a serine residue (Michel et al., 1991). Selective threonine phosphorylation of proteins is rather unusual for the majority of eukaryotic serine/threonine kinases and may be considered a unique feature of the thylakoid protein phosphorylation system.

Most of the thylakoid membrane proteins phosphorylated at the N-terminal threonine residues are also amino-acetylated (Table 1). Formation of N-acetyl-O-phosphothreonine is a specific trait of D1, D2, and CP43 that classifies them as a distinct group of thylakoid phosphoproteins. Importantly, these features have been suggested as determinants of the substrate specificity of the PP2A-like thylakoid membrane protein phosphatase (Vener et al., 1999, 2001) and the rapid dephosphorylation of these PS II core proteins at elevated temperatures (Rokka et al., 2000). In contrast, the PsbH protein of PS II can be phosphorylated at threonine 2 and threonine 4 (Table 1) and also differs from D1, D2, and CP43 in the environmentally dependent changes in phosphorylation, as discussed below.

The light-harvesting proteins of PS II are the most abundant phosphoproteins in thylakoid membranes and include different subunits of LHCII and the minor chlorophyll-binding protein CP29. Most of the characterized LHCII polypeptides are phosphorylated at the N-terminal threonine 3 (Table 1), and the presence of basic amino acid residues on both sides of the phosphorylation sites seems important (Table 1). This requirement was shown by experiments with synthetic peptides used as the substrates for redox-dependent thylakoid kinases (Michel et al., 1991). In the same study, it was concluded that acetylation of the amino termini of LHCII-like peptides was not required for their phosphorylation. CP29 is phosphorylated at position 6 both in *Arabidopsis* (Hansson and Vener, 2003) and the green alga *C. reinhardtii* (Turkina et al., 2004). The amino acid sequences around these phosphorylation sites are similar to those for phosphorylated LHCII polypeptides (Table 1). A second phosphorylation site, corresponding to position 83, was reported in CP29 from maize (Table 1). The phosphorylated sequence has been considered unique among thylakoid proteins since it met the phosphorylation site requirements for casein kinases (Testi et al., 1996).

Three recently identified phosphoproteins, TSP9, TMP14, and PsaD (the first phosphoprotein found in PSI), do not have sequences around their phosphorylation sites that are similar to those of either PS II-core proteins or LHCII polypeptides (Table 1). TMP14 is an intrinsic membrane protein (Hansson and Vener, 2003). In contrast, TSP9 and PsaD are peripheral proteins associated with the stromal side of thylakoid membranes. TSP9 differs from the other thylakoid proteins in that it can be phosphorylated at three different threonine residues that are situated in the middle portion, and not at the N-terminus, of the protein (Carlberg et al., 2003).

D. Assessing Changes in the Stoichiometry of Protein Phosphorylation

The first determination of the extent of phosphorylation was done for Lhcb1 and Lhcb2 polypeptides of LHCII by quantitative SDS-PAGE and scintillation counting after phosphorylation of isolated spinach thylakoids with radioactive ATP (Islam, 1987). Successful measurements of the phosphorylation level for the PS II protein D1 has been also achieved by densitometric quantification of the immunoblots and autoradiograms after separation of phosphorylated and non-phosphorylated forms of the protein by SDS-PAGE (Callahan et al., 1990; Elich et al., 1992; Rintamäki et al., 1996a). The stoichiometry of in vivo D1 phosphorylation under different conditions was assessed by immunoblotting with anti-phosphothreonine antibody as well (Rintamäki et al., 1997). All of these experiments were based on SDS-PAGE separation of thylakoid proteins prior to the analyses. An alternative approach is based on LCMS analyses of the peptides released by trypsin from the surface of thylakoid membranes (Vener et al., 2001). In this case, the stoichiometry of in vivo phosphorylation for individual proteins is determined by measuring the ratio of the phosphorylated to non-phosphorylated peptide originating from the same protein present in thylakoid membranes isolated in the presence of phosphatase inhibitors. This technique allowed monitoring of phosphorylation changes in D1, D2, CP43, and PsbH proteins, but was not quantitative for LHCII phosphorylation because of alternative cleavage of these polypeptides by trypsin (Vener et al., 2001). Assessing the extent of phosphorylation by all of the methods listed above confirmed the dynamic nature of thylakoid protein phosphorylation and demonstrated that, normally, none of the thylakoid proteins is completely phosphorylated (Elich et al., 1992; Rintamäki et al., 1996a, 1997; Vener et al., 2001; Booij-James et al., 2002).

The maximal phosphorylation level of Lhcb1 and Lhcb2 polypeptides of LHCII in vitro corresponds to 22–25% of the total amount of these proteins (Islam, 1987). Importantly, in vivo phosphorylation of LHCII has been found to occur only at light intensities lower than those used for normal plant growth, which has led to the question of the physiological role of LHCII phosphorylation (Rintamäki et al., 1997; Haldrup et al., 2001; Rintamäki and Aro, 2001). The highest extent of D1 protein phosphorylation, corresponding to 80–90% of protein content, has been detected only under reducing conditions in vitro (Elich et al., 1992) or at high light intensities in vivo (Rintamäki et al., 1997). The extent of in vivo phosphorylation for D1, D2, CP43, and PsbH proteins in *A. thaliana* under standard growth conditions corresponded to about 30–50%, being higher in the leaves harvested during the daytime (Vener et al., 2001). The only rapid and significant changes during light-dark transition were found for the phosphorylation of threonine 4 in PsbH (Table 1). The level of in vivo phosphorylation of PS II core proteins was found to be highly susceptible to elevated temperatures, indicating an involvement of this reversible protein modification in response of plants to heat stress (Rokka et al., 2000; Vener et al., 2001). In this respect, the measurements of in vivo protein phosphorylation levels has started to shift the concept for thylakoid protein phosphorylation in photosynthesis from the direct regulation of the electron flow to the involvement in adaptive responses under stress conditions (Giardi et al., 1996, 1997; Rokka et al., 2000; Vener et al., 2001).

III. Reversible Phosphorylation of Photosystem II (PS II) Proteins

A. What is the Role of PS II Phosphorylation?

Phosphorylation of PS II core proteins was first suggested to regulate the electron transfer activity in this photosystem. Decrease of the maximum capacity of PS II electron transfer upon phosphorylation of chloroplast thylakoids has been reported (Horton and Lee, 1984). In contrast, a study on thylakoid membranes with differentially phosphorylated PS II or LHCII proteins led other authors to conclude that phosphorylated PS II polypeptides continued to support high rates of electron transport (Harrison and Allen, 1991). The binding properties of the secondary quinone acceptor site of PS II, Q(B), were found to be unaffected by phosphorylation (Harrison and Allen, 1991). However, reduced binding of photosynthetic herbicides at this site due to the phosphorylation of PS II core proteins was reported in another study (Giardi et al., 1992). It was proposed that phosphorylation of PS II polypeptides modifies the Q(B) pocket and regulates electron transfer by changing the quinone binding affinity of PS II (Giardi et al., 1992, 1995). In a more recent study focused on the relationship between PS II phosphorylation and electron transport activity, individual electron transport reactions in PS II were measured in PS II membranes with different levels of protein phosphorylation (Mamedov et al., 2002). The extent of D1 protein phosphorylation in the analyzed PS II membranes varied from 10% to 58%. Despite some minor changes in the properties of the differentially phosphorylated PS II, the major conclusion of this study was that there was no direct link between the phosphorylation of PS II core polypeptides and electron transfer activity or oxygen evolution by PS II.

PS II undergoes photoinhibition at light intensities beyond those saturating for photosynthesis (Greer et al., 1986; Cleland et al., 1990; Aro et al., 1993). It was reported that phosphorylation of thylakoid membrane proteins partially protects PS II against photoinhibition (Horton and Lee, 1985). The study of differentially phosphorylated thylakoids also suggested that phosphorylation of PS II polypeptides is required for PS II function at high light intensities (Harrison and Allen, 1991). On the other hand, the phosphorylation of D1, D2, CP43, and PsbH proteins was correlated with the decline of PS II activity during high irradiance treatments (Giardi et al., 1994). Phosphorylation of PS II proteins was also proposed as an early stage of photoinhibition involved in the disassembly of the photosystem prior to degradation of D1 protein (Giardi, 1993). However, in later studies the rate of photoinactivation of PS II electron transport and oxygen evolution was not found to be affected by PS II protein phosphorylation in thylakoids from different plants (Koivuniemi et al., 1995; Rintamaki et al., 1996b). A consensus in the interpretation of the contradictory data in this field could probably be found if the dynamics of photoinhibition and its relation to the repair of the damaged PS II are taken into account. At present, it looks plausible that PS II core protein phosphorylation does not directly influence PS II susceptibility to photoinactivation and damage but is crucial for the repair cycle of the damaged PS II.

Continuous functioning of PS II requires coordinated processes of lateral migration of the inactivated photosynthetic units from grana to stroma regions of the thylakoid membrane and their replacement (Ghirardi et al., 1990, 1993; Andersson and Aro, 1997; van Wijk et al., 1997; Baena-Gonzalez et al., 1999). The inactivation of PS II is directly proportional to light intensity (Tyystjarvi and Aro, 1996), but the photoinhibition of PS II appears only at high light intensities when the rate of inactivation exceeds that of PS II replacement (Andersson and Aro, 1997, 2001; Baena-Gonzalez et al., 1999). The light-induced inactivation of PS II is a consequence of the inactivation of the D1 reaction center protein, which has to be degraded and substituted with a newly synthesized copy (Ohad et al., 1984; Mattoo and Edelman, 1987; Andersson and Aro, 2001). This turnover of D1 is a key event in the turnover of PS II. Importantly, dephosphorylation of the photoinactivated D1 has been revealed as a prerequisite for its degradation (Koivuniemi et al., 1995; Rintamäki et al., 1996a; Andersson and Aro, 1997). The dephosphorylation of D1 as a control step for proteolysis of the damaged protein is so far the most feasible regulatory role suggested for the reversible phosphorylation of the PS II core proteins. The dephosphorylation of D2, CP43, and PsbH proteins has also been found to be involved in the partial disassembly of PS II monomers that migrate to the stroma regions of thylakoids (Baena-Gonzalez et al., 1999). Accordingly, the present paradigm considers the physiological role for reversible phosphorylation of PS II polypeptides in

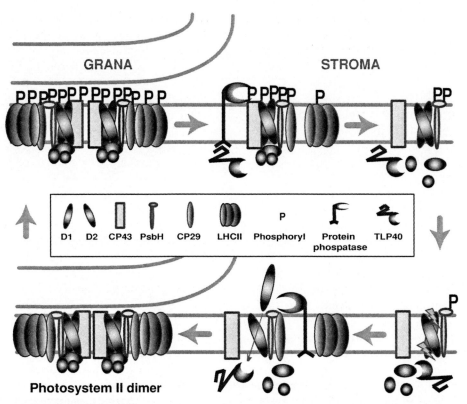

Fig. 1. Reversible protein phosphorylation during repair cycle of PS II in thylakoids of higher plants. The cycle includes light-induced phosphorylation of PS II proteins in the grana regions of thylakoids, migration of phosphorylated PS II monomers with the damaged D1 protein to the stroma membrane domains, sequential dephosphorylation of PS II polypeptides that leads to partial disassembly of PS II core complexes, proteolysis of the photodamaged D1, co-translational insertion of new D1 polypeptide in the remaining PS II complex, final assembly of the repaired PS II and its migration in the grana. The scheme shows only the PS II proteins that undergo reversible phosphorylation during the cycle. The membrane protein phosphatase responsible for dephosphorylation of PS II core polypeptides and luminal PPIase TLP40 regulating the phosphatase and probably assisting in folding of PS II polypeptides are also shown in the scheme.

the context of the PS II turnover cycle (Andersson and Aro, 2001; Rintamäki and Aro, 2001; Aro and Ohad, 2003).

B. Reversible Phosphorylation and Turnover of D1 Protein

The D1 protein is a central functional subunit of PS II, with a light-induced turnover rate higher than that of any other thylakoid polypeptide (Mattoo et al., 1981). D1 and PS II assembly as a whole undergoes a cyclic process, which is schematically outlined in Fig. 1. The "repair" cycle includes migration of PS II units containing the photoinactivated D1 from grana to stroma regions of thylakoids, partial disassembly of PS II complexes, degradation of D1 by specific proteases, co-translational insertion of the newly synthesized D1 copy in the remaining PS II complex (Andersson and Aro, 1997, 2001; Zhang et al., 1999; Rintamäki and Aro, 2001), and final assembly into functional PS II that migrates back to the stacked grana regions of thylakoids (Mattoo and Edelman, 1987). It has been demonstrated that degradation of the D1 protein necessitates this repair cycle (Aro et al., 1992; Ebbert and Godde, 1994, 1996; Rintamäki et al., 1995). The study of D1 turnover in isolated chloroplasts revealed that only the dephosphorylated protein was degraded, while D1 phosphorylated by light-activated kinase was not a subject for proteolysis (Ebbert and Godde, 1996). The proteolytic stability of phosphorylated D1 was also increased under conditions of reduced phosphatase activity in vitro (Koivuniemi et al., 1995). In leaves, the degradation of damaged D1 is prevented by sodium fluoride (Rintamäki et al., 1996a), an inhibitor of thylakoid protein phosphatases (Bennett, 1980). Accordingly, it has been suggested that dephosphorylation is a prerequisite for degradation of the damaged D1 (Rintamäki et al., 1996a). After dephosphorylation, D1 is degraded (Fig. 1). Two different proteases, DegP2 (Haussuhl et al., 2001) and FtsH (Lindahl et al., 2000), have been implicated in the D1 degradation process. Thus, dephosphorylation of D1 allows for its degradation, necessitating the insertion of a new D1 copy into the temporarily dysfunctional PS II unit, thereby completing the 'repair' cycle (Andersson and Aro, 1997).

So far, only a single PS II-specific thylakoid membrane protein phosphatase has been identified (Vener et al., 1999). This PP2A-like protein phosphatase catalyzes the rapid and complete dephosphorylation of D1 (Vener et al., 1999; Rokka et al., 2000) and is presumably localized in the non-appressed membranes (Fig. 1) where it is anticipated to interact with a cyclophilin, TLP40 (Fulgosi et al., 1998; Vener et al., 1999; Rokka et al., 2000), as described below.

Reversible phosphorylation of the D1 protein has been found only in seed plants but not in mosses, liverworts, ferns, algae, or cyanobacteria (Rintamaki et al., 1996b; Pursiheimo et al., 1998; Rintamäki and Aro, 2001). Comparison of D1 turnover rates in higher plant and moss thylakoids under conditions promoting protein phosphorylation and PS II photoinhibition revealed a faster degradation of the moss D1 protein (Rintamaki et al., 1996). The specific phosphorylation of the D1 protein, causing retardation of its proteolysis in higher plants, has been suggested to have evolved in order to adapt the PS II repair cycle to the highly organized structure of the higher plant thylakoids with stacked grana and unstacked stroma membrane domains (Rintamäki and Aro, 2001; Aro and Ohad, 2003). It is worth noting that other hypothesis (Booij-James et al., 2002) proposed that the reversible phosphorylation of D1 in higher plants evolutionary replaced multiple D1 DNA copies in cyanobacteria for regulation of PS II core metabolism. Phosphorylation of D1 in the grana regions (Callahan et al., 1990) may serve to maintain the integrity of inactivated PS II during migration to the stroma membrane regions for dephosphorylation, degradation, and substitution of the inactivated protein with the newly synthesized polypeptide. Reversible phosphorylation of the other PS II core proteins proceeds along with that for D1 in the same cycle (Fig. 1) and likely serves a complementary role in the PS II repair process.

C. Reversible Phosphorylation of D2, CP43, and PsbH Proteins During PS II Turnover

As is the case for the D1 protein, phosphorylation of the other polypeptides of the PS II core, namely D2, CP43, and PsbH, does not appear to be involved in a direct regulation of electron transport and oxygen evolution by this photosystem. Particularly, this was demonstrated by modification of the phosphorylation sites in D2 (Andronis et al., 1998; Fleischmann and Rochaix, 1999) and PsbH (O'Connor et al., 1998) of *C. reinhardtii*. On the other hand, reversible protein phosphorylation of these proteins in thylakoids of higher plants was found to be closely related to the lateral migration of PS II between the different structural domains of the membrane system during the repair cycle. The phosphorylation of all PS II core polypeptides occurs in the grana regions of thylakoids (Ebbert and Godde, 1996; Baena-Gonzalez et al., 1999; Rintamäki and Aro, 2001). The conversion of isolated dimers of PS II to monomers was found to occur in their non-phosphorylated but not the phosphorylated forms, when studied in vitro (Kruse et al., 1997). However, equivalent phosphorylation was found in both dimers and monomers of PS II in vivo (Baena-Gonzalez et al., 1999), as illustrated in Fig. 1. The phosphorylated monomers migrated to the stroma regions of thylakoids where a stepwise dephosphorylation of CP43, D2, and D1 proteins has been demonstrated (Baena-Gonzalez et al., 1999). First, CP43 was dephosphorylated and then detached from the PS II core (Fig. 1). Second, D2 and D1 were dephosphorylated. Thus, it was suggested that phosphorylation of the PS II core proteins ensures the integrity of the monomers until repair can proceed, while dephosphorylation of CP43 and D2 proteins opens the complex for dephosphorylation of D1, its proteolysis, and the attachment of ribosomes inserting a new polypeptide (Baena-Gonzalez et al., 1999).

The reversible phosphorylation of PsbH has not been followed in the study of oligomeric PS II structures isolated from grana and stroma regions of thylakoids (Baena-Gonzalez et al., 1999). Nevertheless, an earlier study reported that dephosphorylation of both PsbH and CP43 proteins in thylakoids by exogenous alkaline phosphatase resulted in an extreme sensitivity of PS II to strong illumination (Giardi et al., 1994). It is reasonable to assume that the PsbH protein plays a role in the process of PS II disassembly and assembly (Fig. 1). The requirement of PsbH for assembly and stability of PS II was demonstrated in *C. reinhardtii* (Summer et al., 1997). A functional PsbH was also found to be necessary for rapid degradation of photoinactivated D1 and insertion of newly synthesized D1 molecules into thylakoid membrane of cyanobacteria (Bergantino et al., 2003). While the N-terminal extension containing the phosphorylation sites is typical only for eukaryotic PsbH proteins (Vener et al., 2001;

Gomez et al., 2002; Hansson and Vener, 2003), the characteristics of this phosphoprotein suggest its involvement in the assembly and disassembly of PS II (Giardi, 1993). Phosphorylation of PsbH differs from that of other PS II core proteins in two aspects. Firstly, PsbH has two phosphorylation sites (Table 1) and threonine 4 of the protein undergoes rapid reversible phosphorylation in response to light/dark transitions. This process is considerably faster than reversible phosphorylation of D1, D2, CP43 as well as of the threonine 2 in PsbH (Vener et al., 2001). Secondly, the phosphorylation sites in PsbH show principal sequence differences from those in D1, D2, and CP43 (Table 1). This difference may explain the slower dephosphorylation rates for phosphorylated PsbH compared to other PS II core phosphoproteins by the specific heat-shock-induced protein phosphatase, as discussed below.

D. Phosphorylation of PS II During Adaptive Responses

Phosphorylation of PS II polypeptides inhibits D1 protein turnover and increases PS II stability upon high light stress (Giardi, 1993; Ebbert and Godde, 1996; Baena-Gonzalez et al., 1999). Additionally this phosphorylation is likely involved in the maintenance and regulation of PS II turnover under different stress conditions. Thus, subjecting spinach to combined magnesium and sulfur deficiency was found to be accompanied by changes in D1 protein phosphorylation and turnover (Dannehl et al., 1995). In the first stages of the deficiency, the turnover of D1 was increased and D1 phosphorylation maintained in the dark. This led to a higher stability of active PS II supported by the efficient turnover of D1 protein (Dannehl et al., 1995). Prolonged stress for a few weeks, however, led to the degradation of the photosynthetic apparatus and chlorosis. Increase in the protein phosphorylation of PS II has been also found in plants in response to water deficient conditions (Giardi et al., 1996).

The phosphorylation of D1 and D2 proteins has been related to the process of photoprotective energy dissipation in plants under different environmental conditions. Energy dissipation is important for plant survival when leaves absorb more light energy than can be utilized for photosynthesis. The dissipation of the excess energy involves the xanthophyll cycle and the accumulation of deepoxidized pigments antheraxanthin and zeaxanthin (Niyogi et al., 1998; Muller et al., 2001). Chilling treatment of rice was found to decrease the photochemical efficiency of PS II in parallel with an increase in the level of zeaxanthin. Phosphatase inhibitors increased the rate of zeaxanthin accumulation under these conditions, as well as during dark-incubation of leaves at the normal temperature after chilling (Xu et al., 1999). A correlation was also found between dark-sustained phosphorylation of D1 and D2 proteins and dark-sustained zeaxanthin retention and maintenance of PS II in a state primed for energy dissipation in plants subjected to high light stress (Ebbert et al., 2001). Nocturnal retention of zeaxanthin and antheraxanthin, and their sustained engagement in a state primed for energy dissipation, have also been observed in the leaves/needles of sun-exposed evergreen species during winter (Adams et al., 2001, 2002). Phosphorylation of D1 and D2 polypeptides has been found retained along with retention of zeaxanthin and antheraxanthin and PS II remained primed for energy dissipation during nights with subfreezing temperatures, while this was rapidly reversed upon exposure to increased, non-freezing temperatures (Adams et al., 2001). In contrast to the wintertime, no nocturnal retention of zeaxanthin and antheraxanthin was found prior to sunrise on warm summer mornings (Barker et al., 2002). It is plausible that sustained phosphorylation of PS II polypeptides favors a structure of PS II and its interaction with light harvesting antennae that facilitate better nonphotochemical energy dissipation under stressful environmental conditions, particularly high light and low temperature.

The CP29 antenna protein of PS II has been found phosphorylated after exposure of *Zea mays* plants to high light in the cold (Bergantino et al., 1995). Phosphorylation of this minor light-harvesting polypeptide following chilling treatment in the light has been associated with the resistance of maize plants to cold stress (Bergantino et al., 1995). Induction of CP29 phosphorylation by cold has also been found in barley (Bergantino et al., 1998), as well as in winter rye upon high light and cold treatment (Pursiheimo et al., 2001). Recent work on the structure of the PS II supercomplexes has localized CP29 between PS II dimers and LHCII trimers associated with the photosystem (Yakushevska et al., 2003). This localization also supports an important role of CP29 in the stabilization of oligomeric PS II structure (Yakushevska et al., 2003). Thus, cold- and high light-induced phosphorylation of this protein may stabilize PS II in a way similar to that of PS II core protein phosphorylation following high light treatment. Moreover, localization of CP29 between the PS II core and the major LHCII antennae may also contribute to more efficient photoprotective energy dissipation upon phosphorylation of CP29 under stressful conditions.

Contrary to the sustained phosphorylation of PS II proteins in chilling and cold stress conditions,

Chapter 8   Phosphorylation of Thylakoid Proteins

extremely fast dephosphorylation of these polypeptides has been observed during short heat shock treatments of both plant leaves and isolated thylakoids. Studies on protein dephosphorylation in isolated thylakoids from spinach and *A. thaliana* revealed specific acceleration of dephosphorylation for PS II core proteins at elevated temperatures (Rokka et al., 2000; Vener et al., 2001). Raising the temperature from 22°C to 42°C resulted in a more than ten-fold increase in the dephosphorylation rates of D1 and D2 and CP43 proteins in spinach thylakoids. In contrast, the dephosphorylation rates for PsbH and LHCII polypeptides were accelerated only 2- to 3-fold (Rokka et al., 2000). The use of a phosphothreonine antibody to measure in vivo phosphorylation levels in spinach leaves revealed a more than 20-fold acceleration in D1, D2, and CP43 dephosphorylation induced by abrupt elevation of temperature, but no increase in LHCII dephosphorylation (Rokka et al., 2000). A specific dephosphorylation of D1, D2, and CP43 has also been observed by mass spectrometric techniques after a short heat shock treatment of *A. thaliana* leaves (Vener et al., 2001), although this dephosphorylation was less pronounced than in spinach. It is important to emphasize that heat-induced dephosphorylation of PS II core proteins occurs on a time scale of minutes (Rokka et al., 2000; Vener et al., 2001), while continuously high steady-state phosphorylation of these polypeptides under chilling lasts over days (Adams et al., 2001). Thus, dynamics of PS II dephosphorylation could be very flexible depending on the environmental conditions. The molecular mechanism for the adaptive response of PS II to an abrupt elevation of temperature consists of the rapid dephosphorylation of CP43, D1, and D2, which primes the photosystem to a fast turnover in response to heat shock. The fast decrease in PS II core protein phosphorylation at high temperatures is accomplished by the heat-shock-induced PS II-specific protein phosphatase regulated by the cyclophilin TLP40.

E. PS II-Specific Protein Phosphatase and TLP40

A number of different protein phosphatases have been implied in dephosphorylation of phosphoproteins in thylakoid membranes (Bennett, 1980; Sun et al., 1989; Carlberg and Andersson, 1996; Hast and Follmann, 1996; Elich et al., 1997; Hammer et al., 1997; Vener et al., 1999), although none of them were characterized at the molecular/gene level. Most soluble protein phosphatases isolated from chloroplasts were efficient in dephosphorylation of LHCII polypeptides (Sun et al., 1989; Hast and Follmann, 1996; Hammer et al., 1997). A PS II-specific membrane protein phosphatase from spinach thylakoids was purified over a thousand-fold, using detergent-engaged FPLC and thylakoid phosphopeptides for the enzyme assay (Vener et al., 1999). The purified enzyme exhibited characteristics typical of eukaryotic Ser/Thr phosphatase of the PP2A family in that it was inhibited by okadaic acid and tautomycin, irreversibly bound to microcystin-agarose, and recognized by a polyclonal antibody raised against a recombinant catalytic subunit of human PP2A. Interestingly, okadaic acid has not inhibited protein phosphatase activity in the intact thylakoid membranes, while the anti-PP2A antibody inhibited protein dephosphorylation (Vener et al., 1999). When the isolated enzyme was added to the phosphorylated thylakoid membranes, increased rates of dephosphorylation were observed for D1, D2, and CP43 proteins (Vener et al., 1999). A common trait for these three phosphoproteins is the presence of N-terminal acetylated and phosphorylated threonine residues (Table 1), which was proposed as the reason for the substrate specificity of the protein phosphatase towards these PS II core proteins (Rokka et al., 2000; Vener et al., 2001). The other distinct characteristic of this phosphatase was its association with and regulation by TLP40, a cyclophilin-like peptidyl-prolyl isomerase (PPIase) located in the thylakoid lumen (Fulgosi et al., 1998; Vener et al., 1999).

TLP40 was discovered due to its copurification with the protein phosphatase from the thylakoid membrane (Fulgosi et al., 1998). The presence of PPIases in plant chloroplasts was established more then a decade ago (Breiman et al., 1992; Mattoo, 1998). However, TLP40 was the first complex, multi-domain cyclophilin-like PPIase found in chloroplasts as well as in plant species (reviewed in (Vener, 2001; He et al., 2004). The structure of TLP40 includes a cyclophilin-like C-terminal segment of 20 kDa, a predicted N-terminal leucine zipper, and potential phosphatase-binding sites flanking the leucine zipper (Fulgosi et al., 1998; Vener et al., 1999; Vener, 2001). The isolated protein possesses peptidyl-prolyl cis-trans isomerase protein folding activity (Fulgosi et al., 1998). TLP40 is localized in the thylakoid lumen, interacts with the inner surface of the thylakoid membrane, and regulates the activity of the PS II-specific protein phosphatase (Fig. 1). As judged from immunoblotting analyses, TLP40 appears to be confined predominantly to the unstacked thylakoid regions, the site of protein integration into the photosynthetic membrane (Fulgosi et al., 1998; Vener et al., 1999). It was proposed that TLP40 has a dual role in protein folding catalysis and in trans-membrane regulation of the PS II-specific protein phosphatase (Fulgosi et al., 1998; Vener et al., 1998, 1999; Vener, 2001). The recent finding that the major PPIase protein-folding activity in the soluble lumen is associated with the cyclophilin TLP20 (Edvardsson et al., 2003; Romano et al., 2004) but not with TLP40 supports the suggestion of a specialized regulatory function for TLP40.

It has been found that binding of Cyclosporin A, an inhibitor of PPIases, to TLP40 activated thylakoid phosphatase, while PPIase substrates, prolyl-containing oligopeptides inhibited protein dephosphorylation (Fulgosi et al., 1998; Vener et al., 1999). These experiments required thylakoids be ruptured first to expose the lumenal membrane surface where TLP40 is located (Vener et al., 1999). Thus, TLP40 may act as a regulatory subunit of the PP2A-like membrane phosphatase, modulating activity of the latter at the outer thylakoid surface. This regulation likely operates via reversible binding of TLP40 to the inner membrane surface (Fig. 2). Indeed, significant heat shock-induced activation of the phosphatase coincided with a temperature-induced release of TLP40 from the membrane into the thylakoid lumen (Rokka et al., 2000). Moreover, induction of the phosphatase activity by TLP40 release from the membranes was confirmed

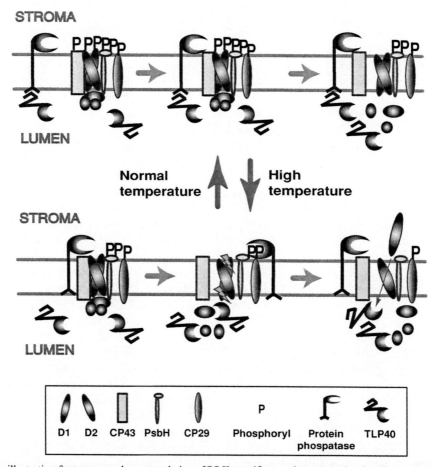

Fig. 2. Schematic illustration for trans-membrane regulation of PS II-specific membrane protein phosphatase by cyclophilin TLP40. TLP40 is present in the lumen in soluble and membrane-bound forms. The membrane-bound TLP40 interacts with the lumen-exposed region of the membrane phosphatase, which suppress the phosphatase activity at the opposite stroma-exposed surface of the membrane. At high temperatures (35°C and 42°C, for spinach), TLP40 is released from the membrane into the lumen, the phosphatase becomes highly active and dephosphorylates CP43, D2, and D1 polypeptides of PS II in a time scale of several minutes. The dephosphorylation leads to acceleration of PS II repair cycle. The release of TLP40 from the membrane also increases its PPIase protein folding activity in the lumen, which can support functional conformation of oxygen evolving complex polypeptides, as well as folding of newly inserted D1 protein at elevated temperature. Thus, coordination of the phosphatase and PPIase activities at the opposite sides of the thylakoid membrane may orchestrate PS II biogenesis.

by phosphatase assays using intact thylakoids, solubilized membranes, and the isolated protein phosphatase (Rokka et al., 2000).

F. Trans-Membrane Signaling by the Phosphatase and PS II Biogenesis

Signaling from TLP40 to the protein phosphatase has been proposed to be involved in the coordination of PS II dephosphorylation with protein folding (Rokka et al., 2000; Vener, 2001). Both processes are required for protein turnover of PS II reaction centers. A model outlined in Fig. 2 summarizes the experimental data on the possible trans-membrane signaling by the PS II-specific thylakoid protein phosphatase. The protein phosphatase is suggested to have a single trans-membrane span and a short extension into the thylakoid lumen (Vener et al., 1999; Vener, 2001). The active sites of the phosphatase and TLP40 PPIase are situated on opposite sides of the thylakoid membrane (Fig. 2). Accordingly, the dephosphorylation of D1, D2, and CP43 proteins by the phosphatase proceeds at the stroma-exposed surface of the membrane. TLP40 is localized in the thylakoid lumen and distributed between the membrane and the soluble fractions (Fulgosi et al., 1998; Vener et al., 1999). Thus, the PPIase protein folding activity of TLP40 may be restricted to the proteins of the lumen and the lumen-exposed domains of integral thylakoid membrane proteins. The dissociation of TLP40 from the membrane surface upon abrupt elevation of temperature activates the PS II-specific protein phosphatase (Fig. 2). The phosphatase may rapidly dephosphorylate CP43, D2, and D1 proteins and allow for disassembly of the PS II monomers migrated to the non-appressed regions of thylakoids. The dephosphorylation may trigger degradation of D1, its substitution by a newly synthesized polypeptide, and the following steps of biogenesis and final assembly of functional PS II. In parallel with these events, the release of TLP40 from the membrane may increase protein-folding activity in the lumen that may ensure the sustained active conformation of the oxygen evolving complex polypeptides at high temperatures, as well as accelerate folding of the proteins newly inserted into the luminal space. The reversible interaction of TLP40 and the membrane phosphatase provides a potential molecular mechanism for trans-membrane signaling and synchronization of the degradation, synthesis, assembly, and folding of PS II polypeptides at both sides of the thylakoid membrane. A strong coordination of the numerous steps of PS II biogenesis and assembly is obviously required for maintenance of the functional photosynthetic machinery under stressful environmental conditions.

G. Reversible Phosphorylation of LHCII Polypeptides and State Transitions

The major polypeptides of the PS II light harvesting antenna, LHCII, undergo dynamic light- and redox-dependent reversible phosphorylation both in vitro and in vivo (Allen, 1992; Vener et al., 1995; Rintamäki et al., 1997; Fleischmann et al., 1999; Haldrup et al., 2001). For two decades, the major role for LHCII phosphorylation was ascribed to balancing of absorbed light energy distribution between the two photosystems, also called state transitions (Allen et al., 1981; Allen, 1992; Allen and Mullineaux, 2004; Allen and Forsberg, 2001). According to this concept, phosphorylation of the mobile pool of LHCII polypeptides leads to their detachment from PS II and migration to PS I (Allen, 1992). However, later studies have revealed that phosphorylation of LHCII in vivo occurs only at rather low light intensities and decreases dramatically at higher irradiances corresponding to normal plant growth (Rintamäki et al., 1997; Haldrup et al., 2001; Rintamäki and Aro, 2001). Thus, it has been proposed that LHCII phosphorylation could regulate the balance of excitation energy distribution just under moderate light intensities (Aro and Ohad, 2003). Moreover, the discovery of a deficiency in state transitions in *Arabidopsis* plants lacking the PSI-H or PSI-L subunits but exhibiting functional LHCII phosphorylation questioned a regulation of state transitions by LHCII phosphorylation (Lunde et al., 2000; Haldrup et al., 2001). When PSI-H was absent, LHCII was not able to attach to PS I and state transitions did not occur even though LHCII was highly phosphorylated (Lunde et al., 2000). In agreement with this, two recent studies have demonstrated that a significant fraction of LHCII interacting with PS I was not phosphorylated (Snyders and Kohorn, 2001; Zhang and Scheller, 2004). The dephosphorylation of LHCII associated with PS I may also result from the action of the LHCII-specific protein phosphatase (Elich et al., 1997). Despite the findings listed above, involvement of LHCII phosphorylation in state transitions cannot be completely ruled out (Allen and Forsberg, 2001). Nevertheless, the original paradigm of a mobile LHCII bound to PS II or PS I in dephosphorylated or phosphorylated form, respectively, certainly has to be modified.

Transient phosphorylation of the mobile LHCII polypeptides may be required to overcome a potential barrier for their detachment from PS II. The fate of the released LHCII in the membrane may then be unrelated to its phosphorylation state. It is probable that significant light-induced structural changes in LHCII polypeptides (Zer et al., 1999, 2003; Garab et al., 2002) could be more important than phosphorylation for the determination of the interacting partners of free LHCII. In this respect, the light-dependent dynamics of the protein complexes in the photosynthetic membrane are not compatible with the oversimplified model featuring only two states, with LHCII attached to PS II and phospho-LHCII attached to PS I. Phosphorylation of LHCII polypeptides may regulate just some individual steps in the reversible and flexible interactions of these antenna proteins with both photosystems.

A regulatory role for reversible phosphorylation has also been proposed in the process of LHCII degradation when the antenna size of PS II is reduced upon acclimation of plants from low to high light intensities. Phosphorylated LHCII polypeptides were found to be poor substrates for proteolytic degradation during this process in comparison with the unphosphorylated LHCII (Yang et al., 1998). In this respect, phosphorylation may delay degradation of LHCII in a way similar to that demonstrated for D1 protein. However, an involvement of LHCII phosphorylation in the regulation of the size of light-harvesting antennae during acclimation still remains to be proven.

H. Differential Phosphorylation of CP29

The phosphorylated CP29 isolated from cold-treated maize were N-terminally blocked against N-terminal chemical sequencing (Bergantino et al., 1995), and was probably acetylated as was shown later for CP29 (LHCb4.2) from *A. thaliana* (Hansson and Vener, 2003) and for CP29 from the green alga *C. reinhardtii* (Turkina et al., 2004). Nevertheless, the phosphorylation site in maize CP29 has been localized to threonine residue number 83 (Table 1) by mapping of proteolytic fragments of the protein (Testi et al., 1996). The site of phosphorylation in CP29 (LHCb4.2) from *Arabidopsis thaliana* and in CP29 from *C. reinhardtii* (Turkina et al., 2004) has been localized to a threonine residue at position 6 of the mature proteins (Table 1). Interestingly, CP29 from *C. reinhardtii* revealed a unique characteristic. In contrast to all known nuclear-encoded thylakoid proteins, the transit peptide in the mature algal CP29 was not removed but processed by methionine excision, N-terminal acetylation, and phosphorylation on threonine 6 (Turkina et al., 2004). The N-termini of the mature CP29 from *Chlamydomonas* and *Arabidopsis* have a significant sequence similarity around their phosphorylation sites. The importance of this phosphorylation was proposed as the reason for the unique retention of the transit peptide in the mature algal CP29 (Turkina et al., 2004). The difference between phosphorylation of CP29 in *Arabidopsis* and in green algae versus phosphorylation of CP29 in maize lies in the physiological conditions inducing these modifications. Phosphorylation of *Arabidopsis* and algal proteins was found under standard growth conditions, while phosphorylation of maize CP29 was induced only by high light in the cold and on the other threonine residue. Notably, we have recently detected multiple and differential phosphorylation of CP29 depending on the environmental conditions (M.V. Turkina and A.V. Vener, unpublished data). Thus, phosphorylation of CP29 protein that is localized between PS II dimers and LHCII trimers in this photosystem (Yakushevska et al., 2003) may likely be involved in a number of responses to the changing environment.

IV. PsaD: the First Phosphoprotein in PS I

In a recent study on in vivo thylakoid protein phosphorylation in *A. thaliana* plants under normal growth light conditions, PsaD was identified as the first phosphoprotein in PS I (Hansson and Vener, 2003). The site of PsaD phosphorylation in vivo was mapped to the first threonine at the N-terminus of the mature protein (Table 1). The PsaD protein is essential for a functional PS I in plants and is required for the proper assembly and stability of this photosystem (Haldrup et al., 2003). PsaD is a hydrophilic protein that has no stable three-dimensional structure in solution (Antonkine et al., 2003), but forms a well-defined three-dimensional structure when bound to PS I (Fromme et al., 2001; Antonkine et al., 2003). Thus, the significant structural changes and flexibility of PsaD, together with its control position at the electron donor site of PS I, may rely on regulatory mechanisms operating via protein phosphorylation. The finding of PsaD phosphorylation opens a new direction in the investigation of possible PS I regulation by protein phosphorylation.

V. Phosphorylation of Other Thylakoid Proteins

A. TMP14: a Previously Unknown Thylakoid Phosphoprotein

During mass spectrometric characterization of phosphorylated peptides released from thylakoid membranes of *A. thaliana* by trypsin (Hansson and Vener, 2003), a phosphopeptide from a previously uncharacterized protein, TMP14, was identified (Table 1). This protein is annotated as "expressed protein" in the *Arabidopsis* database and as a potential membrane protein with two trans-membrane regions in the database for *Arabidopsis* membrane proteins. It was named TMP14 for thylakoid membrane phosphoprotein of 14 kDa (Hansson and Vener, 2003). TMP14 is encoded by the nuclear gene At2g46820 in *Arabidopsis* and has homologous proteins encoded in the genomes of other plants and cyanobacteria. All of these proteins contain two potential trans-membrane helices and well-defined signaling peptides with a high predicted probability for chloroplast targeting. Experimental results confirmed localization of TMP14 in the thylakoid membranes of chloroplasts. Topology prediction for the plant proteins places the N-terminus of TMP14 on the stromal side of the thylakoid membrane, which is in agreement with phosphorylation of TMP14 at the membrane surface exposed to chloroplast stroma. The phosphorylation site in TMP14 was confined to one of two N-terminal threonine residues in the sequenced peptide. These residues correspond to the positions 65 and 66 in the sequence of the precursor protein. The discovery of phosphorylated TMP14 in the photosynthetic membrane raises questions about the function of this protein, its binding partners, association with the photosynthetic protein complexes, and the role of its in vivo phosphorylation.

B. TSP9 and Light-Induced Cell Signaling

The nature of a phosphoprotein with a relative electrophoretic mobility of 12 kDa (Bhalla and Bennett, 1987) has remained elusive during two decades of studies on redox-dependent protein phosphorylation in plant thylakoid membranes. This protein has recently been characterized as a novel plant specific protein and called TSP9 for thylakoid soluble phosphoprotein of 9 kDa (Carlberg et al., 2003). Genes encoding homologous "unknown" proteins were found on chromosome 3 of *Arabidopsis* and rice as well as in ESTs from more than 20 different plant species but not in any other organisms. TSP9 is a very basic protein. Mass spectrometric analyses revealed the existence of non-, mono-, di-, and tri-phosphorylated forms of TSP9 and phosphorylation of three distinct threonine residues in the central part of the protein (Table 1). The nature of this modification was transient with steady increase of the protein phosphorylation level upon illumination (Bhalla and Bennett, 1987; Carlberg et al., 2003). Light-induced phosphorylation of the protein was associated with partial release of phosphorylated TSP9 from the thylakoid membrane, which is in contrast to all other known thylakoid phosphoproteins tightly bound to the membrane (Carlberg et al., 2003).

The phosphorylation-dependent transient association of TSP9 with the photosynthetic membrane suggests a possible role of this protein in chloroplast signaling (Fig. 3). Plants acclimate to changes in environmental light quality and intensity by an adjustment of photosystem stoichiometry and size of the light-harvesting antennae. Regulation of expression of both chloroplast-encoded and nuclear-encoded photosynthetic genes in photosynthetic organisms by chloroplast redox signals, involving the redox state of plastoquinone, has been demonstrated (Escoubas et al., 1995; Pfannschmidt et al., 1999, 2001). Since plastoquinone also controls the activation of membrane protein kinases (Vener et al., 1998), repression of nuclear-encoded *cab* genes for light-harvesting polypeptides may thus be coupled to the redox status of plastoquinone via a thylakoid protein kinase that phosphorylates a protein dissociating from thylakoids (Escoubas et al., 1995). TSP9 is a potential candidate for such a plant cell signaling component (Fig. 3).

VI. Regulation and Role of Thylakoid Protein Phosphorylation in a Physiological Context

The recent application of analytical techniques for detection of thylakoid protein phosphorylation in vivo has uncovered a complex regulation of this process. The cooperative regulation of LHCII phosphorylation by plastoquinone reduction and the ferredoxin-thioredoxin system has been demonstrated (Rintamäki et al., 2000; Martinsuo et al., 2003). Feeding of pea leaves with glucose induced LHCII phosphorylation in darkness, which demonstrated that sugar metabolism or signaling exerted a control over phosphorylation of LHCII polypeptides (Hou et al., 2002). Phosphorylation of CP29 in winter rye has been shown to be induced by high light specifically at low temperatures (Pursiheimo et al., 2001). Phosphorylation of the D1 protein in the higher plant *Spirodela oligorrhiza* grown under natural diurnal cycles of solar irradiation has been shown to undergo circadian oscillation (Booij-James et al., 2002). These oscillations were out of phase with the period of maximum light intensity. However, light resetted the phase in the circadian rhythm of D1 phosphorylation

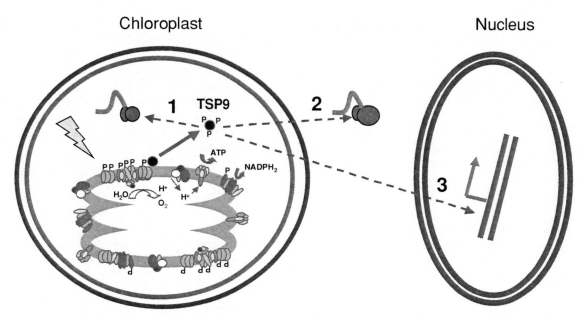

Fig. 3. Potential model for light-induced cell signaling by phosphorylation of TSP9 and its release from the thylakoid membrane. Upon illumination of thylakoids, TSP9 is phosphorylated at three distinct threonine residues and then released from the membrane. TSP9 is a basic protein that can potentially regulate transcription or translation of photosynthetic proteins. The model suggests three signaling options for the released TSP9: 1) regulation of protein expression in the chloroplast; 2) regulation of protein synthesis in the cytosol; and 3) regulation of gene expression in the nucleus.

(Booij-James et al., 2002). Specific fast dephosphorylation of D1, D2, and CP43 polypeptides of PS II in response to abrupt elevation of temperature has been demonstrated (Rokka et al., 2000; Vener et al., 2001). These findings clearly call for further studies of thylakoid protein phosphorylation in vivo to understand the physiological implications of this phenomenon. The differential changes in protein phosphorylation under variable environmental conditions are also indicative of a multifunctional involvement of this posttranslational modification in regulation and adaptive responses of the photosynthetic apparatus.

The complexity of thylakoid protein phosphorylation in a physiological context implies additional hurdles to revealing the enzymes and other molecular factors involved in reversible phosphorylation of thylakoid proteins. A multiple control of LHCII phosphorylation at the redox level alone has already revealed involvement of (i) plastoquinone reduction (Allen et al., 1981; Allen, 1992), (ii) plastoquinol binding at the Qo site of the cytochrome bf complex (Vener et al., 1995, 1997; Zito et al., 1999; Finazzi et al., 2001), (iii) plastoquinol oxidation at the Qo site and rapid reoccupation of the site with a new plastoquinol molecule (Hou et al., 2003), and (iv) the thiol redox state (Rintamäki et al., 1997, 2000; Carlberg et al., 1999; Martinsuo et al., 2003). At present, it is not clear how these events may influence any of the three TAK kinases (Snyders and Kohorn, 1999, 2001) or the two Stt7-like kinases (Depege et al., 2003) shown to be important for phosphorylation of LHCII polypeptides. The most feasible strategy to understand the role of each individual protein kinase is to use kinase-gene knockout plants for in vivo characterization of thylakoid protein phosphorylation to reveal the substrates for each kinase. The regulation of the individual kinase(s) involved in phosphorylation of distinct thylakoid proteins could then be studied in a more focused way. Examination of the in vivo protein phosphorylation patterns in mutant plants lacking individual regulatory components, like TLP40, TSP9, subunits of cytochrome *bf* complex, and thioredoxin-like proteins, should reveal the regulatory network for the reversible phosphorylation of photosynthetic proteins and signaling cascades controlling expression of these proteins.

The present level of knowledge on thylakoid protein phosphorylation favors different regulatory, adaptive, and signaling functions for reversible phosphorylation of individual proteins upon changing physiological conditions including stress. The recent progress in creating publicly available plant knockout lines and development of the analytical techniques for characterization of in vivo protein phosphorylation will likely lead to rapid progress in the understanding of

Chapter 8 Phosphorylation of Thylakoid Proteins

the multiple physiological regulatory functions of reversible and environmentally-modulated phosphorylation of thylakoid proteins.

Acknowledgments

I thank Inger Carlberg for helpful discussions and critical reading of the manuscript. The author's work was supported by grants from The Swedish Research Council, The Swedish Research Council for Environment, Agriculture and Spatial Planning (Formas) and Nordiskt Kontaktorgan för Jordbruksforskning (NKJ).

References

Adams WW III, Demmig-Adams B, Rosenstiel TN and Ebbert V (2001) Dependence of photosynthesis and energy dissipation activity upon growth form and light environment during the winter. Photosynth Res 67: 51–62

Adams WW III, Demmig-Adams B, Rosenstiel TN, Brightwell AK and Ebbert V (2002) Photosynthesis and photoprotection in overwintering plants. Plant Biol 4: 545–557

Allen JF (1992) Protein phosphorylation in regulation of photosynthesis. Biochim Biophys Acta 1098: 275–335

Allen JF (2002) Plastoquinone redox control of chloroplast thylakoid protein phosphorylation and distribution of exitation energy between photosystems: discovery, background, implications. Photosynth Res 73: 139–148

Allen JF (2003) Botany. State transitions–a question of balance. Science 299: 1530–1532

Allen JF and Forsberg J (2001) Molecular recognition in thylakoid structure and function. Trends Plant Sci 6: 317–326

Allen JF and Mullineaux CW (2004) Probing the mechanism of state transitions in oxygenic photosynthesis by chlorophyll fluorescence spectroscopy, kinetics and imaging. In: Papageorgiou GC and Govindjee (eds) Chlorophyll a Fluorescence: A Signature of Photosynthesis, pp 447–461

Allen JF, Bennett J, Steinback KE and Arntzen CJ (1981) Chloroplast protein phosphorylation couples plastoquinone redox state to distribution of excitation energy between photosystems. Nature 291: 25–29

Andersson B and Aro E-M (1997) Proteolytic activities and proteases of plant chloroplasts. Physiol Plant 100: 780–793

Andersson B and Aro E-M (2001) Photodamage and D1 protein turnover in photosystem II. In: Aro E-M and Andersson B (eds) Regulation of Photosynthesis, pp 377–393. Kluwer Acad. Publ, Dordrecht

Andronis C, Kruse O, Deak Z, Vass I, Diner BA and Nixon PJ (1998) Mutation of residue threonine-2 of the D2 polypeptide and its effect on photosystem II function in *Chlamydomonas reinhardtii*. Plant Physiol 117: 515–524

Antonkine ML, Jordan P, Fromme P, Krauss N, Golbeck JH and Stehlik D (2003) Assembly of protein subunits within the stromal ridge of photosystem I. Structural changes between unbound and sequentially PS I-bound polypeptides and correlated changes of the magnetic properties of the terminal iron sulfur clusters. J Mol Biol 327: 671–697

Aro E-M and Ohad I (2003) Redox regulation of thylakoid protein phosphorylation. Antioxid Redox Signal 5: 55–67

Aro E-M, Kettunen R and Tyystjarvi E (1992) ATP and light regulate D1 protein modification and degradation. Role of D1* in photoinhibition. FEBS Lett 297: 29–33

Aro E-M, Virgin I and Andersson B (1993) Photoinhibition of photosystem II. Inactivation, protein damage and turnover. Biochim Biophys Acta 1143: 113–134

Aro E-M, Rokka A and Vener AV (2004) Determination of phosphoproteins in higher plant thylakoids. In: Carpentier R (ed) Methods in Molecular Biology, Vol 274, pp 271–285. Humana Press Inc., Totowa, NJ

Baena-Gonzalez E, Barbato R and Aro E-M (1999) Role of phosphorylation in repair cycle and oligomeric structure of photosystem two. Planta 208: 196–204

Barker DH, Adams WW III, Demmig-Adams B, Logan BA, Verhoeven AS and Smith SD (2002) Nocturnally retained zeaxanthin does not remain engaged in a state primed for energy dissipation during the summer in two Yucca species growing in the Mojave Desert. Plant Cell Environ 25: 95–103

Bennett J (1977) Phosphorylation of chloroplast membrane polypeptides. Nature 269: 344–346

Bennett J (1980) Chloroplast phosphoproteins. Evidence for a thylakoid-bound phosphoprotein phosphatase. Eur J Biochem 104: 85–89

Bennett J, Steinback KE and Arntzen CJ (1980) Chloroplast phosphoproteins: regulation of excitation energy transfer by phosphorylation of thylakoid membrane polypeptides. Proc Natl Acad Sci USA 77: 5253–5257

Bergantino E, Dainese P, Cerovic Z, Sechi S and Bassi R (1995) A post-translational modification of the photosystem II subunit CP29 protects maize from cold stress. J Biol Chem 270: 8474–8481

Bergantino E, Sandona D, Cugini D and Bassi R (1998) The photosystem II subunit CP29 can be phosphorylated in both C3 and C4 plants as suggested by sequence analysis. Plant Mol Biol 36: 11–22

Bergantino E, Brunetta A, Touloupakis E, Segalla A, Szabo I and Giacometti GM (2003) Role of the PSII-H subunit in photoprotection: novel aspects of D1 turnover in Synechocystis 6803. J Biol Chem 278: 41820–41829

Booij-James IS, Swegle WM, Edelman M and Mattoo AK (2002) Phosphorylation of the D1 photosystem II reaction center protein is controlled by an endogenous circadian rhythm. Plant Physiol 130: 2069–2075

Bhalla P and Bennett J (1987) Chloroplast phosphoproteins: phosphorylation of a 12-kDa stromal protein by the redox-controlled kinase of thylakoid membranes. Arch Biochem Biophys 252: 97–104

Breiman A, Fawcett TW, Ghirardi ML and Mattoo AK (1992) Plant organelles contain distinct peptidylprolyl cis,trans-isomerases. J Biol Chem 267: 21293–21296

Callahan FE, Ghirardi ML, Sopory SK, Mehta AM, Edelman M and Mattoo AK (1990) A novel metabolic form of the 32 kDa-D1 protein in the grana-localized reaction center of photosystem II. J Biol Chem 265: 15357–15360

Carlberg I and Andersson B (1996) Phosphatase activities in spinach thylakoid membranes - effectors, regulation and location. Photosynth Res 47: 145–156

Carlberg I, Rintamäki E, Aro E-M and Andersson B (1999) Thylakoid protein phosphorylation and the thiol redox state. Biochemistry 38: 3197–3204

Carlberg I, Hansson M, Kieselbach T, Schroder WP, Andersson B and Vener AV (2003) A novel plant protein undergoing light-induced phosphorylation and release from the photosynthetic thylakoid membranes. Proc Natl Acad Sci USA 100: 757–762

Cheng L, Spangfort MD and Allen JF (1994) Substrate specificity and kinetics of thylakoid phosphoprotein phosphatase reactions. Biochim Biophys Acta 1188: 151–157

Cleland RE, Ramage RT and Critchley C (1990) Photoinhibition causes loss of photochemical activity without degradation of D1 protein. Aust J Plant Physiol 17: 641–651

Dannehl H, Herbik A and Godde D (1995) Stress-induced degradation of the photosynthetic apparatus is accompanied by changes in thylakoid protein-turnover and phosphorylation. Physiol Plant 93: 179–186

de Vitry C, Diner BA and Popo JL (1991) Photosystem II particles from *Chlamydomonas reinhardtii*. Purification, molecular weight, small subunit composition, and protein phosphorylation. J Biol Chem 266: 16614–16621

Depege N, Bellafiore S and Rochaix JD (2003) Role of chloroplast protein kinase Stt7 in LHCII phosphorylation and state transition in Chlamydomonas. Science 299: 1572–1575

Ebbert V and Godde D (1994) Regulation of thylakoid protein phosphorylation in intact chloroplasts by the activity of kinases and phosphatases. Biochim Biophys Acta 1187: 335–346

Ebbert V and Godde D (1996) Phosphorylation of PS II polypeptides inhibits D1 protein-degradation and increases PS II stability. Photosynth Res 50: 257–269

Ebbert V, Demmig-Adams B, Adams WW III, Mueh KE and Staehelin LA (2001) Correlation between persistent forms of zeaxanthin-dependent energy dissipation and thylakoid protein phosphorylation. Photosynth Res 67: 63–78

Edvardsson A, Eshaghi S, Vener AV and Andersson B (2003) The major peptidyl-prolyl isomerase activity in thylakoid lumen of plant chloroplasts belongs to a novel cyclophilin TLP20. FEBS Lett 542: 137–141

Elich TD, Edelman M and Mattoo AK (1992) Identification, characterization, and resolution of the in vivo phosphorylated form of the D1 photosystem II reaction center protein. J Biol Chem 267: 3523–3529

Elich TD, Edelman M and Mattoo AK (1993) Dephosphorylation of photosystem II core proteins is light-regulated in vivo. EMBO J 12: 4857–4862

Elich TD, Edelman M and Mattoo AK (1997) Evidence for light-dependent and light-independent protein dephosphorylation in chloroplasts. FEBS Lett 411: 236–238

Escoubas JM, Lomas M, LaRoche J and Falkowski PG (1995) Light intensity regulation of cab gene transcription is signaled by the redox state of the plastoquinone pool. Proc Natl Acad Sci USA 92: 10237–10241

Finazzi G, Zito F, Barbagallo RP and Wollman FA (2001) Contrasted effects of inhibitors of cytochrome b6f complex on state transitions in *Chlamydomonas reinhardtii*: the role of Qo site occupancy in LHCII kinase activation. J Biol Chem 276: 9770–9774

Fleischmann MM and Rochaix JD (1999) Characterization of mutants with alterations of the phosphorylation site in the D2 photosystem II polypeptide of *Chlamydomonas reinhardtii*. Plant Physiol 119: 1557–1566

Fleischmann MM, Ravanel S, Delosme R, Olive J, Zito F, Wollman FA and Rochaix JD (1999) Isolation and characterization of photoautotrophic mutants of *Chlamydomonas reinhardtii* deficient in state transition. J Biol Chem 274: 30987–30994

Fromme P, Jordan P and Krauss N (2001) Structure of photosystem I. Biochim Biophys Acta 1507: 5–31

Fulgosi H, Vener AV, Altschmied L, Herrmann RG and Andersson B (1998) A novel multi-functional chloroplast protein: identification of a 40 kDa immunophilin-like protein located in the thylakoid lumen. EMBO J 17: 1577–1587

Garab G, Cseh Z, Kovacs L, Rajagopal S, Varkonyi Z, Wentworth M, Mustardy L, Der A, Ruban AV, Papp E, Holzenburg A and Horton P (2002) Light-induced trimer to monomer transition in the main light-harvesting antenna complex of plants: Thermo-optic mechanism. Biochemistry 41: 15121–15129

Ghirardi ML, Callahan FE, Sopory SK, Elich TD, Edelman M and Mattoo AK (1990) Cycling of the photosystem II reaction center core between grana and stroma lamellae. In: Baltschefssky M (ed) Current Research in Photosynthesis, Vol 2, pp 733–738. Kluwer Acad. Publ., Dordrecht

Ghirardi ML, Mahajan S, Sopory SK, Edelman M and Mattoo AK (1993) Photosystem II reaction center particle from Spirodela stroma lamellae. J Biol Chem 268: 5357–5360

Giardi MT (1993) Phosphorylation and disassembly of the photosystem II core as an early stage of photoinhibition. Planta 190: 107–113

Giardi MT, Rigoni F and Barbato R (1992) Photosystem II core phosphorylation heterogeneity, differential herbicide binding, and regulation of electron-transfer in photosystem II preparations from spinach. Plant Physiol 100: 1948–1954

Giardi MT, Komenda J and Masojidek J (1994) Involvement of protein phosphorylation in the sensitivity of photosystem II to strong illumination. Physiol Plant 92: 181–187

Giardi MT, Cona A and Geiken B (1995) Photosystem II core phosphorylation heterogeneity and the regulation of electron transfer in higher plants - a review. Bioelectrochemistry and Bioenergetics 38: 67–75

Giardi MT, Cona A, Geiken B, Kucera T, Masojidek J and Mattoo AK (1996) Long-term drought stress induces structural and functional reorganization of Photosystem II. Planta 199: 118–125

Giardi MT, Masojidek J and Godde D (1997) Effects of abiotic stresses on the turnover of the D-1 reaction centre II protein. Physiol Plant 101: 635–642

Gomez SM, Nishio JN, Faull KF and Whitelegge JP (2002) The chloroplast grana proteome defined by intact mass measurements from liquid chromatography mass spectrometry. Mol Cell Proteomics 1: 46–59

Greer DH, Berry JA and Björkman O (1986) Photoinhibition of photosynthesis in intact bean-leaves - role of light and temperature, and requirement for chloroplast-protein synthesis during recovery. Planta 168: 253–260

Haldrup A, Jensen PE, Lunde C and Scheller HV (2001) Balance of power: a view of the mechanism of photosynthetic state transitions. Trends Plant Sci 6: 301–305

Haldrup A, Lunde C and Scheller HV (2003) *Arabidopsis thaliana* plants lacking the PSI-D subunit of photosystem I suffer severe photoinhibition, have unstable photosystem I complexes, and altered redox homeostasis in the chloroplast stroma. J Biol Chem 278: 33276–33283

Hamel P, Olive J, Pierre Y, Wollman FA and de Vitry C (2000) A new subunit of cytochrome b6f complex undergoes reversible phosphorylation upon state transition. J Biol Chem 275: 17072–17079

Hammer MF, Markwell J and Sarath G (1997) Purification of a protein phosphatase from chloroplast stroma capable of dephosphorylating the light-harvesting complex-II. Plant Physiol 113: 227–233

Hansson M and Vener AV (2003) Identification of three previously unknown in vivo protein phosphorylation sites in thylakoid membranes of *Arabidopsis thaliana*. Mol Cell Proteomics 2: 550–559

Harrison MA and Allen JF (1991) Light-dependent phosphorylation of photosystem II polypeptides maintains electron transport at high light intensity - separation from effects of phosphorylation of LhcII. Biochim Biophys Acta 1058: 289–296

Hast T and Follmann H (1996) Identification of two thylakoid-associated phosphatases with protein phosphatase activity in chloroplasts of the soybean (*Glycine max*). J Photochem Photobiol B 36: 313–319

Haussuhl K, Andersson B and Adamska I (2001) A chloroplast DegP2 protease performs the primary cleavage of the photodamaged D1 protein in plant photosystem II. EMBO J 20: 713–722

He Z, Li L and Luan S (2004) Immunophilins and parvulins. Superfamily of peptidyl prolyl isomerases in Arabidopsis. Plant Physiol 134: 1248–1267

Horton P and Lee P (1984) Phosphorylation of chloroplast thylakoids decreases the maximum capacity of photosystem II electron transfer. Biochim Biophys Acta 767: 563–567

Horton P and Lee P (1985) Phosphorylation of chloroplast membrane proteins partially protects against photoinhibition. Planta 165: 37–42

Hou CX, Pursiheimo S, Rintamäki E and Aro E-M (2002) Environmental and metabolic control of LHCII protein phosphorylation: revealing the mechanisms for dual regulation of the LHCII kinase. Plant Cell Environ 25: 1515–1525

Hou CX, Rintamäki E and Aro E-M (2003) Ascorbate-mediated LHCII protein phosphorylation–LHCII kinase regulation in light and in darkness. Biochemistry 42: 5828–5836

Islam K (1987) The rate and extent of phosphorylation of the 2 light-harvesting chlorophyll a/b binding-protein complex (Lhc-II) polypeptides in isolated spinach thylakoids. Biochim Biophys Acta 893: 333–341

Koivuniemi A, Aro E-M and Andersson B (1995) Degradation of the D1- and D2-proteins of photosystem II in higher plants is regulated by reversible phosphorylation. Biochemistry 34: 16022–16029

Kruse O, Zheleva D and Barber J (1997) Stabilization of photosystem two dimers by phosphorylation: implication for the regulation of the turnover of D1 protein. FEBS Lett 408: 276–280

Lindahl M, Spetea C, Hundal T, Oppenheim AB, Adam Z and Andersson B (2000) The thylakoid FtsH protease plays a role in the light-induced turnover of the photosystem II D1 protein. Plant Cell 12: 419–431

Lunde C, Jensen PE, Haldrup A, Knoetzel J and Scheller HV (2000) The PSI-H subunit of photosystem I is essential for state transitions in plant photosynthesis. Nature 408: 613–615.

Mamedov F, Rintamäki E, Aro E-M, Andersson B and Styring S (2002) Influence of protein phosphorylation on the electron-transport properties of Photosystem II. Photosynth Res 74: 61–72

Martinsuo P, Pursiheimo S, Aro E-M and Rintamäki E (2003) Dithiol oxidant and disulfide reductant dynamically regulate the phosphorylation of light-harvesting complex II proteins in thylakoid membranes. Plant Physiol 133: 37–46

Mattoo AK (1998) Peptidylprolyl cis-trans-isomerases from plant organelles. Methods Enzymol 290: 84–100

Mattoo AK and Edelman M (1987) Intramembrane translocation and posttranslational palmitoylation of the chloroplast 32-kDa herbicide-binding protein. Proc Natl Acad Sci USA 84: 1497–1501

Mattoo AK, Pick U, Hoffman-Falk H and Edelman M (1981) The rapidly metabolized 32,000-dalton polypeptide of the chloroplast is the "proteinaceous shield" regulating photosystem II electron transport and mediating diuron herbicide sensitivity. Proc Natl Acad Sci USA 78: 1572–1576

Michel HP and Bennett J (1987) Identification of the phosphorylation site of an 8.3 kDa protein from photosystem II of spinach. FEBS Lett 212: 103–108

Michel HP, Hunt DF, Shabanowitz J and Bennett J (1988) Tandem mass spectrometry reveals that three photosystem II proteins of spinach chloroplasts contain N-acetyl-O-phosphothreonine at their NH2 termini. J Biol Chem 263: 1123–1130

Michel HP, Griffin PR, Shabanowitz J, Hunt DF and Bennett J (1991) Tandem mass spectrometry identifies sites of three post-translational modifications of spinach light-harvesting chlorophyll protein II. Proteolytic cleavage, acetylation, and phosphorylation. J Biol Chem 266: 17584–17591

Muller P, Li XP and Niyogi KK (2001) Non-photochemical quenching. A response to excess light energy. Plant Physiol 125: 1558–1566

Niyogi KK, Grossman AR and Björkman O (1998) Arabidopsis mutants define a central role for the xanthophyll cycle in the regulation of photosynthetic energy conversion. Plant Cell 10: 1121–1134

O'Connor HE, Ruffle SV, Cain AJ, Deak Z, Vass I, Nugent JH and Purton S (1998) The 9-kDa phosphoprotein of photosystem II. Generation and characterisation of Chlamydomonas mutants lacking PSII-H and a site- directed mutant lacking the phosphorylation site. Biochim Biophys Acta 1364: 63–72

Ohad I, Kyle DJ and Arntzen CJ (1984) Membrane protein damage and repair: removal and replacement of inactivated 32-kilodalton polypeptides in chloroplast membranes. J Cell Biol 99: 481–485

Ohad I, Vink M, Zer H, Herrmann RG and Andersson B (2001) Novel aspects on the regulation of thylakoid protein phosphorylation. In: Aro E-M and Andersson B (eds) Regulation of Photosynthesis, pp 419–432. Kluwer Acad. Publ, Dordrecht

Pfannschmidt T, Nilsson A and Allen JF (1999) Photosynthetic control of chloroplast gene expression. Nature 397: 625–628.

Pfannschmidt T, Schutze K, Brost M and Oelmuller R (2001) A novel mechanism of nuclear photosynthesis gene regulation by redox signals from the chloroplast during photosystem stoichiometry adjustment. J Biol Chem 276: 36125–36130

Pursiheimo S, Rintamäki E, Baena-Gonzalez E and Aro E-M (1998) Thylakoid protein phosphorylation in evolutionarily divergent species with oxygenic photosynthesis. FEBS Lett 423: 178–182

Pursiheimo S, Mulo P, Rintamäki E and Aro E-M (2001) Coregulation of light-harvesting complex II phosphorylation and lhcb mRNA accumulation in winter rye. Plant J 26: 317–327

Rintamäki E and Aro E-M (2001) Phosphorylation of photosystem II proteins. In: Aro E-M and Andersson B (eds) Regulation of Photosynthesis, pp 395–418. Kluwer Acad. Publ, Dordrecht

Rintamäki E, Salo R, Lehtonen E and Aro E-M (1995) Regulation of D1 protein-degradation during phoroinhibition of photosystem-II. In vivo phosphorylation of the D1 protein in various plant groups. Planta 195: 379–386

Rintamäki E, Kettunen R and Aro E-M (1996a) Differential D1 dephosphorylation in functional and photodamaged photosystem II centers. Dephosphorylation is a prerequisite for degradation of damaged D1. J Biol Chem 271: 14870–14875

Rintamäki E, Salo R, Koivuniemi A and Aro E-M (1996b) Protein phosphorylation and magnesium status regulate the degradation of the D1 reaction centre protein of Photosystem II. Plant Science 115: 175–182

Rintamäki E, Salonen M, Suoranta UM, Carlberg I, Andersson B and Aro E-M (1997) Phosphorylation of light-harvesting complex II and photosystem II core proteins shows different irradiance-dependent regulation in vivo. Application of phosphothreonine antibodies to analysis of thylakoid phosphoproteins. J Biol Chem 272: 30476–30482

Rintamäki E, Martinsuo P, Pursiheimo S and Aro E-M (2000) Cooperative regulation of light-harvesting complex II phosphorylation via the plastoquinol and ferredoxin-thioredoxin system in chloroplasts. Proc Natl Acad Sci USA 97: 11644–11649

Rokka A, Aro E-M, Herrmann RG, Andersson B and Vener AV (2000) Dephosphorylation of photosystem II reaction center proteins in plant photosynthetic membranes as an immediate response to abrupt elevation of temperature. Plant Physiol 123: 1525–1536

Romano PG, Edvardsson A, Ruban AV, Andersson B, Vener AV, Gray JE and Horton P (2004) Arabidopsis AtCYP20-2 is a light-regulated cyclophilin-type peptidyl-prolyl cis-trans isomerase associated with the photosynthetic membranes. Plant Physiol 134: 1244–1247

Snyders S and Kohorn BD (1999) TAKs, thylakoid membrane protein kinases associated with energy transduction. J Biol Chem 274: 9137–9140

Snyders S and Kohorn BD (2001) Disruption of thylakoid-associated kinase 1 leads to alteration of light harvesting in Arabidopsis. J Biol Chem 276: 32169–32176

Summer EJ, Schmid VH, Bruns BU and Schmidt GW (1997) Requirement for the H phosphoprotein in photosystem II of *Chlamydomonas reinhardtii*. Plant Physiol 113: 1359–1368

Sun G, Bailey D, Jones MW and Markwell J (1989) Chloroplast thylakoid protein phosphatase is a membrane surface-associated activity. Plant Physiol 89: 238–243

Testi MG, Croce R, Polverino-De Laureto P and Bassi R (1996) A CK2 site is reversibly phosphorylated in the photosystem II subunit CP29. FEBS Lett 399: 245–250

The Arabidopsis Genome Initiative (2000) Analysis of the genome sequence of the flowering plant *Arabidopsis thaliana*. Nature 408: 796–815

Tullberg A, Hakansson G and Race HL (1998) A protein tyrosine kinase of chloroplast thylakoid membranes phosphorylates light harvesting complex II proteins. Biochem Biophys Res Commun 250: 617–622

Turkina MV, Villarejo A and Vener AV (2004) The transit peptide of CP29 thylakoid protein in *Chlamydomonas reinhardtii* is not removed but undergoes acetylation and phosphorylation. FEBS Lett 564: 104–108

Tyystjärvi E and Aro E-M (1996) The rate constant of photoinhibition, measured in lincomycin-treated leaves, is directly proportional to light intensity. Proc Natl Acad Sci USA 93: 2213–2218

van Wijk KJ, Roobol-Boza M, Kettunen R, Andersson B and Aro E-M (1997) Synthesis and assembly of the D1 protein into photosystem II: processing of the C-terminus and identification of the initial assembly partners and complexes during photosystem II repair. Biochemistry 36: 6178–6186

Vener AV (2001) Peptidyl-prolyl isomerases and regulation of photosynthetic functions. In: Aro E-M and Andersson B (eds) Regulation of Photosynthesis, pp 177–193. Kluwer Academic Publishers, Dordrecht

Vener AV, Van Kan PJ, Gal A, Andersson B and Ohad I (1995) Activation/deactivation cycle of redox-controlled thylakoid protein phosphorylation. Role of plastoquinol bound to the reduced cytochrome bf complex. J Biol Chem 270: 25225–25232

Vener AV, Van Kan PJM, Rich PR, Ohad I and Andersson B (1997) Plastoquinol at the quinol oxidation site of reduced cytochrome bf mediates signal transduction between light and protein phosphorylation: Thylakoid protein kinase deactivation by a single-turnover flash. Proc Natl Acad Sci USA 94: 1585–1590

Vener AV, Ohad I and Andersson B (1998) Protein phosphorylation and redox sensing in chloroplast thylakoids. Cur Opin Plant Biol 1: 217–223

Vener AV, Rokka A, Fulgosi H, Andersson B and Herrmann RG (1999) A cyclophilin-regulated PP2A-like protein phosphatase in thylakoid membranes of plant chloroplasts. Biochemistry 38: 14955–14965

Vener AV, Harms A, Sussman MR and Vierstra RD (2001) Mass spectrometric resolution of reversible protein phosphorylation in photosynthetic membranes of *Arabidopsis thaliana*. J Biol Chem 276: 6959–6966

Xu CC, Jeon YA, Hwang HJ and Lee CH (1999) Suppression of zeaxanthin epoxidation by chloroplast phosphatase inhibitors in rice leaves. Plant Sci 146: 27–34

Yakushevska AE, Keegstra W, Boekema EJ, Dekker JP, Andersson J, Jansson S, Ruban AV and Horton P (2003) The structure of photosystem II in Arabidopsis: Localization of the CP26 and CP29 antenna complexes. Biochemistry 42: 608–613

Yang DH, Webster J, Adam Z, Lindahl M and Andersson B (1998) Induction of acclimative proteolysis of the light-harvesting chlorophyll a/b protein of photosystem II in response to elevated light intensities. Plant Physiol 118: 827–834

Zer H and Ohad I (2003) Light, redox state, thylakoid-protein phosphorylation and signaling gene expression. Trends Biochem Sci 28: 467–470

Zer H, Vink M, Keren N, Dilly-Hartwig HG, Paulsen H, Herrmann RG, Andersson B and Ohad I (1999) Regulation of thylakoid protein phosphorylation at the substrate level: reversible light-induced conformational changes expose the phosphorylation site of the light-harvesting complex II. Proc Natl Acad Sci USA 96: 8277–8282

Zer H, Vink M, Shochat S, Herrmann RG, Andersson B and Ohad I (2003) Light affects the accessibility of the thylakoid light harvesting complex II (LHCII) phosphorylation site to the membrane protein kinase(s). Biochemistry 42: 728–738

Zhang L, Paakkarinen V, van Wijk KJ and Aro E-M (1999) Co-translational assembly of the D1 protein into photosystem II. J Biol Chem 274: 16062–16067

Zhang S and Scheller HV (2004) Light-harvesting complex II binds to several small subunits of photosystem I. J Biol Chem 279: 3180–3187

Zito F, Finazzi G, Delosme R, Nitschke W, Picot D and Wollman FA (1999) The Qo site of cytochrome b6f complexes controls the activation of the LHCII kinase. EMBO J 18: 2961–2969

Chapter 9

Molecular Analysis of Photoprotection of Photosynthesis

Hou-Sung Jung and Krishna K. Niyogi*
*Department of Plant and Microbial Biology,
University of California, Berkeley, CA 94720-3102, USA*

Summary	127
I. Introduction	128
II. Avoiding High Light Absorption	128
A. Chloroplast Avoidance Movement	128
B. Adjustment of Antenna Size	130
III. Coping with Excess Absorbed Light Energy	131
A. Feedback De-Excitation	131
1. Xanthophyll Cycle	133
2. PsbS Protein	134
3. Light-Harvesting Complex (LHC)	134
B. Electron Transport	135
1. Linear Electron Transport	135
2. Cyclic Electron Transport	136
C. Antioxidants	136
1. Carotenoids	136
2. Ascorbate	137
3. Tocopherols	137
D. Repair of Damaged D1	138
IV. Gene Expression Responses of Plants to High Light Stress	138
A. Early Light-Induced Proteins, Stress-Enhanced Proteins, and One-Helix Proteins	138
B. Other Proteins Induced by High Light	139
Acknowledgments	140
References	140

Summary

Plants have diverse defense mechanisms against high light stress. Plants can reduce absorption of light energy through chloroplast avoidance and antenna size reduction. However, the capacity of the avoidance and the antenna size reduction for protection is limited, so that plants often absorb more energy than they can use. Therefore, plants need mechanisms to deal with this excess absorbed light energy, such as harmless thermal dissipation by feedback de-excitation. The transthylakoid pH gradient, xanthophyll cycle, PsbS, and other light-harvesting complex proteins are required for this thermal dissipation. In addition, alternative electron transport allows electrons to pass to acceptors other than CO_2, thereby relieving overreduction of electron transport components in high light conditions. To detoxify reactive oxygen species that are inevitably produced during high light stress, plants have antioxidants including carotenoids, ascorbate, and tocopherols. In spite of these photoprotective mechanisms, photodamage may still occur, and efficient repair of damaged systems could be a photoprotective mechanism. In this chapter, recently published molecular genetics studies on each step of photoprotection have been reviewed. Genes

---

*Author for correspondence, email: niyogi@nature.berkeley.edu

required for each defense mechanism that have been identified thus far are introduced, and cloned genes that can possibly be related to photoprotection are discussed.

I. Introduction

In nature, photosynthesis is indispensable for sustaining much of the life on earth. Photosynthesis begins with light energy absorption and its conversion into chemical energy. During this reaction, ATP and NADPH are produced, and oxygen is also generated from H_2O as a byproduct. This converted chemical energy is then used to assimilate CO_2 into carbohydrates.

By definition, photosynthesis requires light energy. The light energy is collected mainly by chlorophylls in light-harvesting complexes (LHCs). After absorbing the light energy, a chlorophyll becomes excited to its singlet excited state and then transfers the absorbed energy in one of several ways including photochemistry indicated by photochemical quenching of chlorophyll fluorescence. The energy is delivered to reaction centers of photosystems, where it drives the initial charge separation reactions of photosynthesis (photochemistry). Besides photochemistry, fluorescence emission, de-excitation by thermal dissipation, and decay through triplet state are the other ways by which excited chlorophylls return to ground state.

The light energy, however, is not always a good thing, because too much light may cause damage in plants. When plants receive more light than they can utilize, the lifetime of singlet excited chlorophyll extends, and the chance of returning to ground state through triplet state chlorophyll is increased. This pathway can dissipate excess energy (Foyer and Harbinson, 1999; Niyogi, 2000); however, the generated triplet chlorophyll can transfer its energy to oxygen so that singlet oxygen is produced. Singlet oxygen is a harmful type of reactive oxygen species (ROS) that can cause degradation of membrane and protein structure of photosystems (Barber and Andersson, 1992; Melis, 1999).

Plants have various levels of photoprotective mechanisms (Fig. 1) (Barber and Andersson, 1992; Long

Abbreviations: ELIP – early light-induced protein; ETR – electron transport rate; FNR – ferredoxin-NADP⁺ reductase; FQR – ferredoxin-plastoquinone oxidoreductase; HPT – homogentisate phytyltransferase; HSP – heat shock protein; LHC – light-harvesting complex; Mg-ProtoIX – Mg-protoporphyrin IX; NDH – NADPH/NADH dehydrogenase; NPQ – nonphotochemical quenching; OHP – one-helix protein; PQ – plastoquinone; PS II – photosystem II; ROS – reactive oxygen species; SEP – stress-enhanced protein; Φ_{PSII} – PS II efficiency (quantum yield of Photosystem II)

et al., 1994; Niyogi, 1999). First of all, plants can protect themselves from excess light by avoiding absorption of the high light. Second, plants can reduce the amount of absorbed energy by thermal dissipation. Third, electrons can be transported through alternative pathways to relieve excitation pressure (Asada, 1999; Ort, 2001). Plants also have antioxidants to detoxify ROS (Bartley and Scolnik, 1995; Smirnoff, 2000). Finally, plants can efficiently repair damaged photosystems. Each mechanism plays a role depending on how strong the incident light is. In most cases, these mechanisms are well coordinated to protect plants from not only just steady high light, but also sudden light intensity changes, such as strong sunshine between clouds.

To understand the photoprotective mechanisms in detail, molecular analysis has been used to identify and characterize the function of various genes in photoprotection. Most of these genes have been identified in plant model systems including Synechocystis, *Chlamydomonas*, and especially *Arabidopsis*, using forward and reverse genetics (Golan et al., 2004). Forward genetics begins with a mutant phenotype and then proceeds to identification of the affected gene; reverse genetics starts with a gene sequence followed by generation and characterization of knockout mutants.

In this chapter, we present a review of the molecular analysis of photoprotection including descriptions of mutants and cloned genes involved in photoprotection. We have tried to focus on recently identified factors for photoprotection as well as to update some results concerning previously known factors.

II. Avoiding High Light Absorption

Avoidance of high light absorption can be a photoprotective mechanism because it can simply reduce the input of light energy (Fig. 1). Although plants are sessile, they are able to adjust the amount of energy absorption through chloroplast movement and antenna size reduction in chloroplasts.

A. Chloroplast Avoidance Movement

In limiting light, plant chloroplasts move to the periclinal walls that are perpendicular to the incident light

Chapter 9 Molecular Genetics of Photoprotection

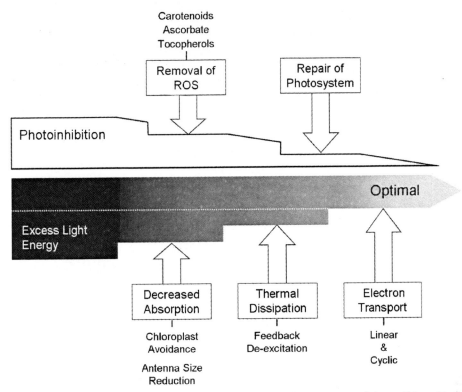

Fig. 1. Photoprotective mechanisms in plants. To decrease light energy absorption in an excess light condition, chloroplasts can move away from the light (chloroplast avoidance), and antenna size can be reduced (antenna size reduction). To get rid of excess absorbed light energy, feedback de-excitation can dissipate excess energy as heat. In addition, efficient electron transport in both linear and cyclic electron transport pathways can relieve excitation pressure generated by high light. However, reactive oxygen species (ROS) are still generated, and ROS can be detoxified by antioxidant systems including carotenoids, ascorbate, and tocopherols. Inevitable damage to photosystems still occurs, and efficient regeneration can be a photoprotective mechanism.

direction to maximize photosynthesis, whereas, when the light intensity is too high, chloroplasts move to anticlinal walls that are parallel to the incident light direction (Fig. 2) (Kagawa and Wada, 2002). Chloroplast movements toward light and away from light are termed accumulation response and avoidance response or movement, respectively (Wada et al., 2003). The chloroplast avoidance movement results in reduction in light energy absorption; therefore, a plant can be protected from high light stress. It has been known that the signal for chloroplast avoidance movement is generated mainly by blue light (Kagawa and Wada, 1999; Kagawa and Wada, 2000) and, recently, the blue light receptor has been identified.

Two research groups have almost simultaneously identified the blue light receptor required for chloroplast avoidance movement in *Arabidopsis* using forward (Kagawa et al., 2001) and reverse genetics approaches (Jarillo et al., 2001). For the forward genetics, mutants were identified using a strip assay in which a section of leaf was illuminated with high light, while the rest of the leaf was masked (Kagawa et al., 2001). In the wild type, the illuminated area turned pale green following the high light treatment, while the masked area stayed green; however, the illuminated area of *cav1* (*chloroplast avoidance movements 1*), affected in chloroplast avoidance, stayed as green as the masked area. Using map-based cloning, it has been found that mutation in the *PHOT2* (originally, *NPL1*) gene is responsible for the absence of chloroplast avoidance movement in the *cav1* mutant.

The second group (Jarillo et al., 2001) used reverse genetics, taking advantage of high sequence similarity (58% identity and 67% similarity) between the PHOT2 and the PHOT1 (originally, NPH1) protein, the blue light photoreceptor for phototropic bending (Huala et al., 1997). The knockout mutant of *PHOT2* was identified from T-DNA insertion lines in *Arabidopsis* with PCR-based screening. Following identification of the *phot2* mutant, they tested the chloroplast avoidance movement by measuring the amount of light transmission through leaves and confirmed that PHOT2

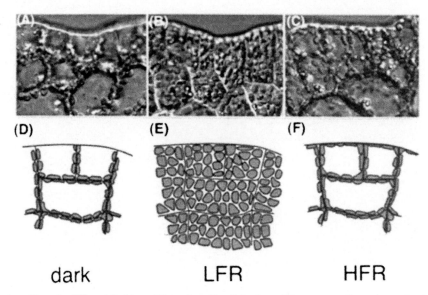

Fig. 2. Chloroplast positions in different light conditions in cells of the fern *Adiantum capillus-veneris*. In the dark (A and D), chloroplasts locate along anticlinal cell walls. Under the low fluence rate (LFR) white light (B and E), chloroplasts move and locate along periclinal cell walls that are perpendicular to the direction of light in order to maximize photosynthesis; however, under the high fluence rate (HFR) white light (C and F), chloroplasts move to anticlinal cell walls that are parallel to light direction to avoid light absorption. These images are from Kagawa and Wada (2002).

is required for the chloroplast avoidance movement mechanism.

To tell the differences between the roles of PHOT1 and PHOT2 in chloroplast movement, the *phot1 phot2* double mutant was generated. It was determined that for chloroplast movement towards a light source, both PHOT1 and PHOT2 are required, whereas for chloroplast movement away from a light source, only PHOT2 is responsible (Sakai et al., 2001).

To determine how much the absorption reduction by chloroplast avoidance movement contributes to photoprotection, sensitivity to high light stress was tested in the *phot2* mutant (Kasahara et al., 2002). The extent of light stress can be monitored by the appearance of bleached leaves in a high light condition and also by measuring the maximum quantum yield of photosystem II (PS II), F_v/F_m (ratio of variable to maximum fluorescence). The dark-adapted value of F_v/F_m reflects photosynthetic performance, and values that are lower than the optimal value of 0.83 indicate a decrease in PS II efficiency (Björkman and Demmig, 1987; Adams et al., 1990; K. Maxwell and Johnson, 2000). The *phot2* mutant clearly had bleached leaves after 22 hours of high light treatment, while leaves of wild type and the *phot1* mutant did not exhibit visible symptoms of high light stress (Kasahara et al., 2002). The F_v/F_m decreased in both the wild type and the *phot2* mutant; however, in *phot2*, the reduction of F_v/F_m was more drastic, and it took a much longer time to recover back to the normal level in a low light condition following high light treatment. Taken together, these results demonstrate that there is a correlation between lack of chloroplast avoidance movement and photodamage under high light stresses.

B. Adjustment of Antenna Size

The antenna consisting of LHC proteins absorbs light energy using pigments and then transfers the absorbed energy to reaction centers for photosynthesis. The size of the antenna can be adjusted as a photosynthetic acclimation response to regulate light energy absorption (Anderson, 1986; Bailey et al., 2001). In high light, the antenna size can be reduced to absorb less light energy thereby protecting plants from high light stress (Park et al., 1997; Baroli and Melis, 1998). The LHC proteins are encoded in the nucleus and translated in the cytoplasm (Wollman et al., 1999). Thus the high light signal should be perceived and then transduced to effect changes in the LHC protein level in chloroplasts.

The *Arabidopsis ape* (*acclimation of photosynthesis to the environment*) mutants are deficient in photosynthetic acclimation (Walters et al., 2003). The *ape* mutant screening took advantage of differences in efficiency of PS II photochemistry (Φ_{PSII}) between high light- and low light-acclimated plants. The *ape1* mutant showed the wild-type level of Φ_{PSII} in low light, but following transfer to high light, Φ_{PSII} of *ape1* did not increase as much as that of the wild type. The *ape1* mutant is also affected in the antenna size reduction by high light, because increases in the chlorophyll *a/b* ratio in *ape1* following transfer to high light were significantly smaller than in the wild type (Walters et al., 2003). Because the *ape1* mutant was screened from a T-DNA insertion population, the interrupted gene in *ape1* was cloned by determining the flanking DNA sequences of the T-DNA using a thermal asymmetric interlaced PCR technique. The *APE1* (At5g38660) gene is conserved in other photosynthetic organisms, but it encodes a protein of unknown function. Thus it would be interesting to determine its physiological role in the antenna size reduction by high light.

The expression of LHC genes in the nucleus is regulated by signals generated in chloroplasts. The signals include chlorophyll biosynthetic intermediates and the redox state of electron transport components (Surpin et al., 2002; Pfannschmidt et al., 2003). In Dunaliella, it has been reported that LHCII gene expression is regulated by the redox state of the plastoquinone (PQ) pool (Escoubas et al., 1995) and by the redox state of intersystem electron transport (D.P. Maxwell et al., 1995) (Fig. 3A). In *Arabidopsis*, it has also been shown that photosynthetic electron transport is involved in the transcriptional regulation of LHCII gene expression and that this signal can override a sugar-related signal for gene expression (Oswald et al., 2001).

The evidence that the chlorophyll biosynthetic intermediates regulate the LHC gene expression has come from studies on the *Arabidopsis gun* (*genomes uncoupled*) mutants (Susek et al., 1993; Mochizuki et al., 2001). Compared to the wild type, the *gun* mutants showed higher *LHC* (originally *CAB*) gene expression in a norflurazon-mediated photobleaching condition in which *LHC* gene expression was suppressed in the wild type (Susek et al., 1993). A molecular genetics approach has identified that four genes, *GUN2~GUN5*, encode enzymes in the tetrapyrrole biosynthetic pathway; for example, *GUN5* encodes the H subunit of Mg-chelatase, an enzyme in the chlorophyll biosynthetic branch of the pathway (Mochizuki et al., 2001; Larkin et al., 2003). Interestingly, the amount of Mg-protoporphyrin IX (Mg-ProtoIX), the product of Mg-chelatase, was several-fold higher in norflurazon-treated plants than in untreated plants; therefore, the correlation between accumulation of Mg-ProtoIX and reduction of gene expression was suggested (Strand et al., 2003). This suggestion has been confirmed by the observations that accumulation of Mg-ProtoIX by Dipyridyl treatment abolished the *gun5* mutant phenotype and feeding Mg-ProtoIX to the wild-type protoplasts repressed the *LHC* gene expression. Taken together, the mutations in the *gun2~gun5* mutants prevent the accumulation of Mg-ProtoIX that is required for repression of the LHC gene expression in the absence of proper chloroplast development (Fig. 3A) (Strand et al., 2003). Identification of factors involved in transduction of these signals from chloroplasts to the nucleus would be interesting. Additionally, it needs to be determined whether Mg-ProtoIX is also accumulated under high light conditions.

III. Coping with Excess Absorbed Light Energy

Chloroplast movements and antenna size adjustment can protect plants from high light stress by decreasing the amount of light absorption. However, chloroplasts moved to anticlinal cell walls are still exposed to strong incident light, and the antenna size reduction is not fast enough to respond to light intensity fluctuations such as sudden light appearance between clouds. Therefore, plants often receive light energy that exceeds their photosynthetic capacity, and this excess absorbed energy can cause photodamage. As a photoprotection against the excess absorbed energy, plants have several defense mechanisms (Fig. 1). Feedback de-excitation dissipates excess absorbed light energy as heat, thereby shortening the lifetime of excited singlet chlorophyll and reducing the generation of singlet oxygen that causes photodamage. Alternative electron transport also contributes to photoprotection by relieving overreduction of the electron transport pathway. Even though the feedback de-excitation and the alternative electron transport can reduce the excess energy, ROS are still produced so that plants need antioxidants to protect the photosynthetic apparatus. Despite these multiple defense mechanisms, plants inevitably experience photodamage. Therefore the efficient turnover of damaged D1 protein and repair could be a critical photoprotective mechanism.

A. Feedback De-Excitation

Feedback de-excitation is a photoprotective mechanism that dissipates excess absorbed light energy as heat, thereby protecting plants from high light stress. The amount of feedback de-excitation is determined by measuring nonphotochemical quenching of chlorophyll fluorescence (NPQ). The amount of chlorophyll

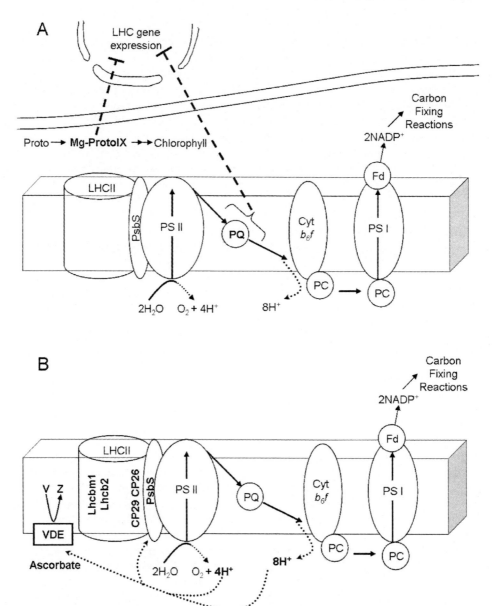

Fig. 3. Photoprotection factors involved in antenna size reduction (A), thermal dissipation (B), electron transport (C), and removal of reactive oxygen species (ROS) (D). (**A**) Mg-protoporphyrin IX (Mg-ProtoIX) and the redox states of plastoquinone (PQ) and intersystem electron transport are plastid signals involved in regulation of light-harvesting complex (LHC) gene expression. (**B**) For thermal dissipation, feedback de-excitation dissipates excess absorbed light energy as heat, and it requires zeaxanthin (Z) converted from violaxanthin (V) by violaxanthin de-epoxidase (VDE), and PsbS (NPQ4). In higher plants, LHC components such as Lhcb2, CP26, and CP29 are indirectly involved in NPQ; however, in *Chlamydomonas*, one of the major light-harvesting proteins, Lhcbm1, is required for NPQ. (*Continued on next page*)

fluorescence that is quenched non-photochemically can be separated from photochemical quenching by applying a pulse of saturating light (Schreiber et al., 1986). NPQ has multiple components including feedback de-excitation, state-transition quenching, and photoinhibitory quenching (Müller et al., 2001).

Among them, feedback de-excitation is responsible for thermal dissipation and shows the fastest relaxation following disappearance of the high light condition. This kind of regulation must occur rapidly because, in a moderate light condition, constitutive feedback de-excitation can diminish the amount of photosynthesis

Chapter 9 Molecular Genetics of Photoprotection

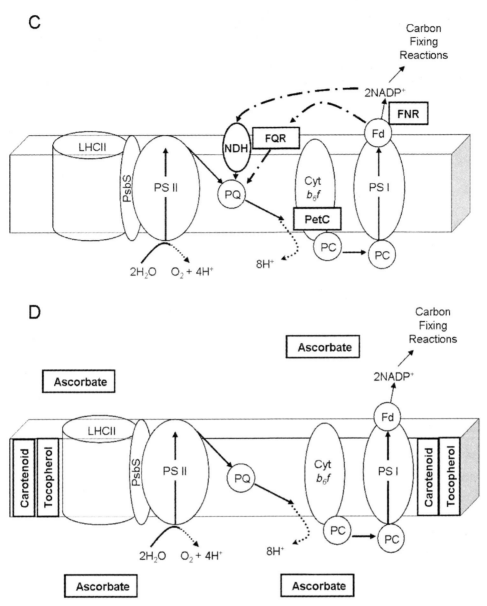

Fig. 3. (Continued from previous page) (**C**) PetC (PGR1) and Ferredoxin-NADP⁺ reductase (FNR) play photoprotective roles in linear electron transport pathway, while NADPH/NADH dehydrogenase (NDH) and ferredoxin-plastoquinone oxidoreductase (FQR) mediate cyclic electron transport thereby relieving excitation pressure especially in stress conditions. (**D**) Carotenoids and tocopherol present in the thylakoid membrane and ascorbate in stroma and thylakoid lumen remove ROS to protect photosystems.

through competition with photochemical quenching. Mutants defective in the feedback de-excitation have been screened by using video imaging systems that digitize chlorophyll fluorescence data into NPQ images (Niyogi et al., 1997, 1998; Shikanai et al., 1999). Video imaging has made it possible to screen many mutant lines simultaneously instead of measuring NPQ one by one. This approach has provided genetic evidence that the xanthophyll cycle, PsbS protein, and LHC are critical factors for feedback de-excitation (Fig. 3B). In this chapter, hereafter, NPQ is used as a synonym for feedback de-excitation.

1. Xanthophyll Cycle

The xanthophyll cycle pigments consist of zeaxanthin, antheraxanthin, and violaxanthin. Under moderate light conditions, violaxanthin is the most abundant pigment; however, under high light conditions, violaxanthin de-epoxidase converts violaxanthin to zeaxanthin through antheraxanthin as an intermediate pigment (Fig. 3B) (Yamamoto et al., 1962). Previous biochemical studies using a violaxanthin de-epoxidase inhibitor indicated that the xanthophyll cycle is required for NPQ (Bilger and Björkman, 1990; Demmig-Adams et al., 1990). Molecular genetics studies with low NPQ mutants have confirmed that the xanthophyll cycle is involved in NPQ (Niyogi et al., 1997, 1998).

The *npq1* mutant was screened from both *Chlamydomonas* and *Arabidopsis* mutants as a low NPQ mutant by using the video imaging system (Niyogi et al., 1997, 1998). Pigment analysis showed that the *npq1* mutant is unable to convert violaxanthin to zeaxanthin under high light in both *Chlamydomonas* and *Arabidopsis*. However, interestingly, there are differences in the degree of dependency of NPQ induction on the xanthophyll cycle between the algal and plant system. In *Chlamydomonas*, the *npq1* mutant still has a rapidly induced NPQ component, while in *Arabidopsis*, the *npq1* mutant is severely deficient in the induction of NPQ. In *Arabidopsis*, map-based cloning allowed the identification of the affected gene in the *npq1* mutant (Niyogi et al., 1998). The affected gene in *npq1* is, as predicted, violaxanthin de-epoxidase, the enzyme mediating the conversion of violaxanthin to zeaxanthin (Fig. 3B). The results were confirmed by complementation of the low NPQ phenotype to a wild-type level of NPQ by introduction of the violaxanthin de-epoxidase gene into the *npq1* mutant. These studies in *Chlamydomonas* and *Arabidopsis* have provided concrete molecular genetic evidence that the xanthophyll cycle has an important function in NPQ. Recently it has been reported that zeaxanthin plays a role as a direct quencher of excess energy based on femtosecond transient absorption experiments in which excitation of zeaxanthin was detected following excitation of chlorophyll (Ma et al., 2003). Among the screened low NPQ mutants, some of them still exhibited normal xanthophyll pigment conversion by high light (Niyogi et al., 1997; Li et al., 2000). These reports reflected that there were more factors than the xanthophyll cycle involved in NPQ.

2. PsbS Protein

The *Arabidopsis npq4* mutant has a normal xanthophyll cycle, but lacks most of NPQ (Li et al., 2000). Despite the lack of NPQ, the *npq4* mutant retains a functional photosynthetic apparatus determined by measuring photosynthetic parameters including oxygen evolution. To clone the affected gene, a map-based cloning approach was used. Following determination of linkage group with molecular and visible markers, PCR of candidate genes and DNA gel blotting showed the absence of the *PsbS* gene (At1g44575) in the *npq4* mutant (Fig. 3B) (Li et al., 2000). The PsbS protein is a member of the LHC protein family. However, while most LHC proteins consist of three trans-membrane domains, PsbS protein has four trans-membrane domains (Kim et al., 1994), and instead of light harvesting, PsbS is involved in light energy dissipation.

To understand the function of PsbS in NPQ, molecular biological and biochemical studies have been conducted. In the molecular biological approach, carboxyl-containing glutamates, candidate amino acid residues for protonation in PsbS, were changed into neutral amino acids by site-directed mutagenesis and then introduced into the *npq4* mutant (Li et al., 2002). In the *npq4* mutant background, the transgenic plants expressing a mutant form of PsbS lacking two specific glutamates did not show NPQ, whereas plants transformed with a wild-type *PsbS* gene copy could reconstitute a high level of NPQ. This study has provided evidence that the protonation of the PsbS protein is essential for NPQ (Fig. 3B). In addition, zeaxanthin binding to isolated PsbS has been reported (Aspinall-O'Dea et al., 2002). It has thus been suggested that low pH in the thylakoid lumen caused by high light would lead to protonation of glutamate residues in PsbS, and in turn this protonation may induce conformational changes. These conformational changes may allow binding of zeaxanthin that is converted from violaxanthin by violaxanthin de-epoxidase.

3. Light-Harvesting Complex (LHC)

Forward genetic approaches have successfully identified genes (e.g. *NPQ1* and *NPQ4*) necessary for NPQ. To determine the involvement of LHC in NPQ, reverse genetics approaches have been utilized. In *Arabidopsis*, antisense constructs against LHCII genes were introduced into plants, and then the photosynthetic phenotypes of each knockout line were characterized.

A relatively minor effect on NPQ was observed in Lhcb2, CP26, and CP29 antisense plants (Fig. 3B) (Andersson et al., 2001; Andersson et al., 2003a). Lhcb2 is a major component of the PS II peripheral antenna, while CP26 and CP29 are minor components of the PS II antenna (Wollman et al., 1999). Several lines of evidence had suggested that CP26 and CP29 may play a role in NPQ (Jahns and Krause, 1994; Jahns

Chapter 9 Molecular Genetics of Photoprotection

and Schweig, 1995; Walters et al., 1996; Bassi et al., 1997; Ruban et al., 1998). In each antisense line, the expression of the target gene was suppressed. In addition, the absence of Lhcb1 and CP24 was also observed in Lhcb2 and CP29 antisense lines, respectively. The lower NPQ in these lines indicated that LHCII might play a role in feedback de-excitation. However, F_v/F_m was decreased in these antisense lines, and the protein level of PsbS, an essential factor for the NPQ, was also decreased in CP29 and Lhcb2 antisense lines (Andersson et al., 2001; Andersson et al., 2003a). These results have indicated that in *Arabidopsis*, individual LHCII components are not directly involved in NPQ; instead LHCII organization in *Arabidopsis* might affect NPQ.

In *Chlamydomonas*, however, forward genetics has identified a role of LHCII in NPQ (Elrad et al., 2002). In this study, mutants were generated by random insertion of a selectable marker, and then the *npq5* mutant was isolated as a low NPQ mutant by using the video imaging system. Detailed mutant phenotype characterization showed that the *npq5* mutant was almost completely defective in NPQ, although *npq5* exhibited a wild-type level of violaxanthin de-epoxidation. In addition, *npq5* had a smaller antenna and lower chlorophyll *b* content. As the *npq5* mutant was generated by insertional mutagenesis, the interrupted gene in *npq5* was determined by isolating flanking DNA of the insertion site. The interrupted gene in *npq5* encodes a component of the major LHCII so that it is designated as *Lhcbm1* (Fig. 3B). The low NPQ phenotype of *npq5* was complemented by the genomic DNA region encompassing *Lhcbm1*. Taken together, LHCII appears to play a major role in NPQ in *Chlamydomonas*, whereas in *Arabidopsis*, LHCII may be indirectly involved in NPQ.

B. Electron Transport

It has been suggested that electron transport from H_2O oxidation to various acceptors ($NADP^+$, O_2, etc.) plays a role in photoprotection (Asada, 1999; Niyogi, 1999; Ort and Baker, 2002). Although all of these electron transport reactions are driven by light energy absorbed by LHC, the pathways that the electron takes are quite diverse. Electrons can be transported to many kinds of acceptors that are involved in either assimilatory or nonassimilatory pathways. Electrons can also return to the electron transport pathway through PS I without reducing any electron acceptors. In this section, we will introduce the role of electron transport in photoprotection.

1. Linear Electron Transport

In the linear electron transport pathway, electrons are transported through PS II, Cyt $b_6 f$, and PS I. Following PS I, the electrons reduce $NADP^+$ to NADPH, and this reducing power can be used for assimilatory CO_2 fixation. However, under stress conditions such as low CO_2, O_2 can be incorporated into RuBP instead of CO_2 in the Rubisco-catalyzed reaction, or O_2 can also directly accept electrons from PS I. The former pathway is called photorespiration, while the latter has been defined with various terms including the water-water cycle (Asada, 1999; Ort and Baker, 2002). Both of them could play a role in photoprotection by allowing electron transport to continue, thereby relieving excitation pressure that is generated by excess absorbed light energy (Asada, 1999; Niyogi, 1999).

Molecular genetics studies have provided evidence that linear electron transport is apparently involved in photoprotection via generation of the transthylakoid pH gradient. In *Arabidopsis*, the *pgr1* (*proton gradient regulation 1*) mutant showed lower NPQ and was very sensitive to high light (Munekage et al., 2001). Map-based cloning showed that the mutation is in *PetC* (*At4g03280*), encoding the cytochrome $b_6 f$ Rieske iron-sulfur subunit (Fig. 3C). This result is consistent with reduced electron transport rate and higher reduction state of Q_A in *pgr1*. The pH gradient across the thylakoid membrane, generated by electron transport, is a prerequisite for activation of violaxanthin de-epoxidase and protonation of PsbS that are required for feedback de-excitation (Fig. 3B). In *pgr1*, the pH gradient was lower than in the wild type, and the de-epoxidation state was also lower.

The involvement of electron transport in photoprotection has also been supported by experiments with tobacco ferredoxin-$NADP^+$ reductase (FNR) knockout lines (Palatnik et al., 2003). The gene knockout lines were generated by introducing antisense DNA of FNR into tobacco plants. FNR catalyzes the final step of the linear electron transport pathway from ferredoxin to $NADP^+$ (Fig. 3C). FNR-deficient plants suffered severe photooxidative damage. Especially grown in photoautotrophic conditions, FNR knockout lines showed bleached phenotypes, while, in the same condition, wild-type plants were still intact. It has been suggested that this photodamage results from accumulation of singlet oxygen following overreduction of the electron transport pathway (Palatnik et al., 2003). However, the phenotype of linear electron transport mutant lines could be attributed to a deficiency of the transthylakoid pH gradient that is required for zeaxanthin accumulation under high light and for ATP synthesis.

2. Cyclic Electron Transport

PS I cyclic electron transport is believed to provide an alternative pathway for electron transport under stress conditions such as a low CO_2, but still allowing proton pumping for both NPQ induction and ATP formation. It has been known that the cyclic electron transport may be conducted through more than one pathway (Bendall and Manasse, 1995).

Reverse genetics with the plastid transformation technique have identified a cyclic electron transport pathway that includes the NADPH/NADH dehydrogenase (NDH) complex (Fig. 3C) (Burrows et al., 1998; Shikanai et al., 1998). In tobacco chloroplast, *NDH* genes were disrupted by homologous recombination. These *NDH* gene disruptions specifically affected cyclic electron transport as was determined by the absence of a transient increase of chlorophyll fluorescence following light to dark transition. This was also confirmed by measuring differences in the redox kinetics of P700 between wild type and *ndh* knockout tobacco leaves. Although, under optimal growth conditions, these mutant plants did not show any apparent phenotype, the *NDH*-deficient mutant was more susceptible to repeated application of strong light (Endo et al., 1999). Therefore, it has also been suggested that the cyclic electron transport pathway including the NDH complex is involved in photoprotection (Endo et al., 1999).

Another cyclic electron transport pathway that is sensitive to antimycin A, an inhibitor of cyclic electron transport flow, was observed in the *NDHB*-deficient mutant background (Joët et al., 2001). This antimycin A-sensitive pathway was predicted to include ferredoxin-plastoquinone oxidoreductase (FQR). The evidence for the presence of cyclic electron transport including FQR has come from a forward genetics approach with the *Arabidopsis* mutant *pgr5* (*proton gradient regulation 5*) (Munekage et al., 2002).

The *pgr5* mutant is affected in NPQ and this is probably caused by a reduction of electron transport rate (ETR). This phenomenon is very similar to *pgr1* in which cytochrome b_6f complex is affected (Munekage et al., 2001). However, detailed characterization of the mutant phenotype has determined that in *pgr5*, cyclic electron transport between ferredoxin and PQ is affected (Fig. 3C) (Munekage et al., 2002). In addition, *pgr5* lacks induction of NPQ under limited carbon fixation conditions such as in CO_2-free air. Thus it has been proposed that the PGR5-dependent cyclic electron transport pathway allows pH gradient generation for induction of NPQ especially when carbon fixation is limited. Without this cyclic electron pathway, plants suffer much more photodamage, especially to PS I because plants cannot prevent overreduction of the acceptor side of PS I.

However, the exact role of PGR5 in cyclic electron transport remains to be defined. PGR5 was detected in a thylakoid membrane fraction in immunoblot analysis, and it did not contain a metal binding domain that might be involved in electron transport (Munekage et al., 2002). Thus, instead of being an electron transporter, PGR5 may be a component of an electron transport protein complex or a docking protein for a soluble electron transport protein to thylakoid membrane. It would be interesting to identify PGR5-interacting proteins to fully understand the PGR5-dependent cyclic electron transport mechanism. As *ndh* and *pgr5* single mutants did not show any apparent phenotype in a normal light condition, complete knockout of both cyclic electron transport pathways would allow an assessment of the contribution of both cyclic electron transport pathways for photoprotection and the identification of possible additional cyclic pathways.

C. Antioxidants

Even though excess absorbed light energy can be dissipated as heat by feedback de-excitation and excitation pressure can be relieved by alternative electron transport, triplet chlorophyll *a* and ROS are still generated in chloroplasts, especially when much more energy is absorbed than both protection systems can process (Niyogi, 1999). Triplet chlorophyll should be quenched to ground state before interacting with oxygen, and singlet oxygen produced by energy transfer from triplet chlorophyll should be detoxified. Otherwise, the singlet oxygen can damage protein and lipid structure, especially the PS II reaction center. These harmful molecules can be detoxified by many kinds of antioxidants including carotenoids, tocopherols, and ascorbate (Fig. 1).

1. Carotenoids

Carotenoids are lipid-soluble pigments present in chloroplasts (Fig. 3D), and they have several major roles. In addition to serving in feedback de-excitation, carotenoids play roles as accessory pigments for photosynthesis, as structural components of photosynthetic complexes, and as antioxidants to protect plants from photooxidative damage (Bartley and Scolnik, 1995). It has already been discussed that xanthophyll cycle pigments, formed from β-carotene by adding hydroxyl groups followed by epoxidation, are involved in NPQ. Besides NPQ, xanthophylls can play a role of an antioxidant to quench singlet oxygen. After quenching, xanthophylls decay to ground state without any harmful effects (Frank and Cogdell, 1993; Baroli and Niyogi, 2000). To characterize the function of the xanthophylls as antioxidants, however, specific experimental conditions are required to distinguish their effects as antioxidants from their contribution to NPQ.

The antioxidant role of xanthophylls was studied with *Arabidopsis* transgenic plants that contain higher amounts of xanthophyll cycle pigments than the wild type (Davison et al., 2002). Transgenic plants were generated by introducing a gene encoding β-carotene hydroxylase, a key enzyme in the carotenoid biosynthesis pathway, under the control of cauliflower mosaic virus 35S promoter, a strong constitutive promoter. The β-carotene hydroxylase catalyzes the conversion of β-carotene to zeaxanthin, followed by conversion of zeaxanthin to violaxanthin resulting in transgenic plants containing more than twice as much violaxanthin as the wild type. The contents of other carotenoids, such as β-carotene and lutein, were similar to the wild-type level. When these plants were subjected to stress, such as higher light intensity and increased growth temperature, the transgenic plants containing more violaxanthin showed elevated stress tolerance (Davison et al., 2002). Under high light conditions, the amount of zeaxanthin was twice as large in the transgenic plants as in the wild type, although the de-epoxidation states were similar, and NPQ was also not different. Thus, it has been suggested that the increased tolerance may not come from NPQ, but from a protective role of zeaxanthin (Davison et al., 2002).

A specific antioxidant role of zeaxanthin has also been deduced from suppressor screening studies of the *npq1 lor1* double mutant in *Chlamydomonas* (Baroli et al., 2003). The *npq1 lor1* double mutant lacked both the xanthophyll cycle and lutein and showed a bleached phenotype in a high light condition. Interestingly, suppressors of the *npq1 lor1* double mutant showed normal growth in HL. Although NPQ was restored in the suppressors, the extent of NPQ wan not increased that much. Instead, the suppressors accumulated zeaxanthin, and this accumulation was caused by mutations in the zeaxanthin epoxidase. Taken together, this study has shown that accumulated zeaxanthin as an antioxidant can protect *Chlamydomonas* from high light stress.

2. Ascorbate

Ascorbate has multiple functions in photoprotection (Fig. 3D) (Smirnoff, 2000; Conklin, 2001). Ascorbate can remove ROS directly and act as a cofactor of ascorbate peroxidases in the elimination of H_2O_2. In addition, ascorbate is a cofactor of violaxanthin de-epoxidase in the xanthophyll cycle (Fig. 3B) (Müller-Moulé et al., 2002), and it is also involved in the regeneration of oxidized tocopherol.

Molecular analysis of ascorbate's role in photoprotection has been possible with the isolation of ascorbate-deficient mutants. The ascorbate-deficient *Arabidopsis* mutant *vtc1* (*vitamin c1*) was originally isolated as an ozone-sensitive mutant (Conklin et al., 1996). With the development of an ascorbate detection method using the electron transfer dye, nitroblue tetrazolium, three more ascorbate-deficient mutants (*vtc2*, *vtc3*, and *vtc4*) were isolated (Conklin et al., 2000). Under oxidative stress conditions, these mutants were bleached faster than wild type indicating an important role of ascorbate in photoprotection (Smirnoff, 2000; Müller-Moulé et al., 2003).

3. Tocopherols

Tocopherols are lipophilic compounds that can protect membranes from oxidative stress (Fig. 3D) (Fryer, 1992; Niyogi, 1999; Munné-Bosch and Alegre, 2002). Mutants affected in tocopherol biosynthesis were selected by an HPLC-based screening procedure. The selected mutants accumulated tocopherol precursors at the expense of tocopherols. Map-based cloning has identified MPBQ/MSBQ methyltransferase (At3g63410) and tocopherol cyclase (At4g32770) in *Arabidopsis* (Porfirova et al., 2002; Cheng et al., 2003; Sattler et al., 2003). The gene encoding homogentisate phytyltransferase (HPT), catalyzing the first committed step in tocopherol biosynthesis, has also been identified in *Synechocystis* sp. PCC6803 and *Arabidopsis* (Collakova and DellaPenna, 2001; Savidge et al., 2002).

However, the tocopherol biosynthetic mutants did not show any visible phenotype under optimal light conditions (Porfirova et al., 2002; Cheng et al., 2003). In addition, high light treatment did not cause any significant differences between wild type and the mutants; therefore, it was suggested that other antioxidants might compensate for the absence of tocopherol (Collakova and DellaPenna, 2001; Porfirova et al., 2002). To test this statement, it would be interesting to generate double/triple deficient mutants of antioxidants and assess their degree of sensitivity to high light. For example, a *vte1 vtc1* double mutant deficient in both tocopherol and ascorbate would be useful to test their antioxidant roles. Also *Arabidopsis* lines that accumulate more tocopherols than the wild type have recently been generated by overexpressing HPT and γ-tocopherol methyltransferase (Collakova and DellaPenna, 2003). These lines would be useful to determine the function of tocopherols in photoprotection.

D. Repair of Damaged D1

In spite of these photoprotective mechanisms, D1 of PS II is still easily damaged in high light (Melis, 1999; Noguchi, 2002). In some cases such as a continued stress condition, D1 inactivation can be a photoprotection mechanism as the inactivation reduces the formation of high-energy electrons and superoxide (Adams et. al, 2004). When such stress conditions are over, the damaged D1 should be removed and replaced by a newly synthesized D1 protein to continue photosynthesis; therefore, an efficient repair cycle is critical as a photoprotective mechanism (Fig. 1). Biochemical studies have indicated that the FtsH protease, an ATP-dependent metalloprotease, may be responsible for the degradation of damaged D1 (Lindahl et al., 2000; Silva et al., 2003). The FtsH protease is one of the major protease families found in higher plants, and nine FtsH isomers have been found in *Arabidopsis* (Lindahl et al., 1996; Adam et al., 2001). Recent molecular genetics studies have provided evidence for the role of FtsH protease in photoprotection.

Map-based cloning and T-DNA tagged gene cloning have already found that FtsH protease genes are affected in the *var* (*yellow variegated*) mutants (Chen et al., 2000; Takechi et al., 2000; Sakamoto et al., 2002). The *var* mutants were treated with high light in order to determine the role of FtsH protease in photoprotection. Under stress conditions, the FtsH protease-deficient mutants were much more susceptible than the wild type, and recovery following high light stress was also slower compared to that of the wild type (Bailey et al., 2002; Sakamoto et al., 2002). Based on these findings, it has been suggested that FtsH protease is involved in photoprotection by efficiently removing damaged D1 proteins and thereby allowing faster reconstitution of PS II.

IV. Gene Expression Responses of Plants to High Light Stress

A. Early Light-Induced Proteins, Stress-Enhanced Proteins, and One-Helix Proteins

If gene expression is up-regulated by high light treatment instead of repression like LHC protein, it has been proposed that these up-regulated genes are possibly involved in photoprotection (Montané and Kloppstech, 2000). Transcription of early light-induced protein (ELIP) genes is induced by high light in mature leaves of pea, and the induction level is proportional to light intensity (Adamska et al., 1992). The *Arabidopsis* genome contains two ELIPs (ELIP1: At3g22840 and ELIP2:At4g14690), and both of them are induced by high light (Heddad and Adamska, 2000). Recent studies with chlorophyll fluorescence lifetime measurements in protein-targeting mutants, which are unable to accumulate ELIPs in high light, have shown that ELIPs may play a photoprotective role (Hutin et al., 2003). As a role of ELIP, it was suggested that, under high light conditions, ELIPs might bind to free chlorophyll that is released from degraded photosynthetic systems. Otherwise, the free chlorophyll that has a longer lifetime than bound chlorophyll might generate singlet oxygen and thereby cause damage to photosystems (Montané and Kloppstech, 2000; Hutin et al., 2003).

Sequence database searches using conserved domains of ELIPs has identified stress-enhanced proteins (SEPs) and one-helix proteins (OHPs) in *Arabidopsis* (Heddad and Adamska, 2000; Jansson et al., 2000; Andersson et al., 2003b). The expression of SEP and OHP genes are also increased by high light. However, SEPs and OHPs differ from ELIPs in the number of transmembrane domains and expression level in low light conditions. ELIPs are predicted to contain three transmembrane domains, while SEPs and OHPs are predicted to contain two and one transmembrane domains, respectively (Heddad and Adamska, 2000; Andersson et al., 2003b). The expression of SEP and OHP but not ELIP genes is detected in low light, and expression levels are significantly increased by high light treatment (Heddad and Adamska, 2000; Jansson et al., 2000; Andersson et al., 2003b). However, the physiological functions of SEPs and OHPs need to be investigated. For example, the phenotype of knockout mutants of SEPs and OHPs would be interesting to determine whether SEPs and OHPs are involved in photoprotection. In addition, whether SEPs and OHPs bind pigments remains an open question.

B. Other Proteins Induced by High Light

Approaches using mRNA differential display and microarray were applied to understand responses of plants to the high light treatment. In the differential display, cDNA fragments were amplified by using an oligo(dT) or an arbitrary primer set, and then genes that were up-regulated and down-regulated by high light were identified based on changes in amplified band intensity on agarose gels (Dunaeva and Adamska, 2001). Using the differential display, five genes (*Lsr1~Lsr5*) were shown to increase after 2 hours of high light treatment. Among them, *Lsr1~Lsr4* were previously characterized genes. For example, *Lsr4* encodes the metallothionein class 1a (MT1a) protein. Interestingly, their encoded proteins are either predicted or determined to be localized in the cytoplasm. *Lsr5*, encoding a protein with high similarity to β-1,3-galactosyl transferase previously reported only in vertebrates, is predicted to be present in the Golgi body. As these gene products are located outside of chloroplasts, their physiological roles under the high light condition would be interesting to determine (Dunaeva and Adamska, 2001).

Microarray technology allows analysis of gene expression profiles of thousands of genes simultaneously in response to environmental stimuli. The responses of plants to high light stress have also been studied using microarray technology (Hihara et al., 2001; Rossel et al., 2002; Kimura et al., 2003).

Synechocystis cells grown in low light (20 μmol photons m^{-2} sec^{-1}) were transferred to high light (300 μmol photons m^{-2} sec^{-1}) and RNA samples were collected after 15 min, 1 hours, 6 hours, and 15 hours of the high light treatment (Hihara et al., 2001). To analyze expression profiles of these RNA samples, microarray slides containing 3079 PCR fragments that cover almost all open reading frames in the genome of *Synechocystis* sp. PCC6803 were used. *Arabidopsis* plants grown under moderate light (100 μmol photons m^{-2} sec^{-1}) for 24 days were treated in high light (1000 μmol photons m^{-2} sec^{-1}) and also in filtered high light (1000 μmol photons m^{-2} sec^{-1}) that contains half the amount of the infrared spectrum (Rossel et al., 2002). For these *Arabidopsis* samples, microarray slides of 6000 clones plus an additional 220 clones encoding genes involved in pigment biosynthesis, antioxidant biosynthesis, and photosystem from expressed sequence tags or reverse transcriptase PCR were used. Using another *Arabidopsis* microarray slides of 7000 full-length cDNAs, transcription profiles of *Arabidopsis* seedlings grown for 10 days in low light (ca 30 μmol photons m^{-2} sec^{-1}) and treated in high light (ca 800 μmol photons m^{-2} sec^{-1}) for 3 hours were investigated (Kimura et al., 2003).

In these studies, down-regulated expression of genes related to photosynthetic antennae was observed. In Synechocystis, the *APC* genes and the *CPC* genes encoding allophycocyanin and phycocyanin, respectively, were down-regulated by high light (Hihara et al., 2001). In addition, some genes related to the biosynthesis of photosynthetic pigments were also down-regulated. In *Arabidopsis*, most of the photosysnthesis-related genes including LHC genes were repressed (Rossel et al., 2002; Kimura et al., 2003). Additionally, many proteins with unknown function were also repressed. Unexpectedly, the gene expression of violaxanthin de-epoxidase was also repressed (Rossel et al., 2002).

In contrast, many other genes were up-regulated by high light. In Synechocystis, expression of genes homologous to heat shock genes in other organisms were significantly induced by high light (Hihara et al., 2001). Genes encoding scavenging enzymes for ROS, such as glutathione peroxidase, and two FtsH homologs were also up-regulated by high light. In *Arabidopsis*, genes involved in ROS detoxification such as ascorbate peroxidase 1 and glutathione-*S*-transferase 6 were up-regulated (Rossel et al., 2002; Kimura et al., 2003). In addition, chalcone synthase gene in the anthocyanin pathway and the β-carotene hydroxylase II gene in xanthophyll biosynthesis were up-regulated (Rossel et al., 2002). The accumulation of anthocyanin and lignin by high light stress was determined (Kimura et al., 2003). Interestingly, gene expression of many heat shock proteins (HSPs) such as HSP70-3, HSP90, and HSP81-2 was increased by high light (Rossel et al., 2002; Kimura et al., 2003) and even by the filtered high light (Rossel et al., 2002). It has been concluded that the induction was not caused by elevated temperature, but by oxidative stress in the high light condition (Rossel et al., 2002). Moreover, gene expression of many unknown proteins was also up-regulated by high light. It would be interesting to determine the functions of these up-regulated heat shock proteins and unknown proteins in photoprotection. Gene knockout lines that interrupt the expression of the HSPs and the unknown proteins could be used to test high light sensitivity.

Acknowledgments

This work was supported by the Director, Office of Science, Office of Basic Energy Sciences, Chemical Science Division, of the U.S. Department of Energy under Contract No. DE-AC03-765F00098, by the U.S. Department of Agriculture National Research Initiative (grant no. 98-35306-6600), and by the National Institutes of Health (GM58799).

References

Adam Z, Adamska I, Nakabayashi K, Ostersetzer O, Haussuhl K, Manuell A, Zheng B, Vallon O, Rodermel SR, Shinozaki K and Clarke AK (2001) Chloroplast and mitochondrial proteases in Arabidopsis. A proposed nomenclature. Plant Physiol 125: 1912–1918

Adams WW III, Demmig-Adams B, Winter K and Schreiber U (1990) The ratio of variable to maximum chlorophyll fluorescence from photosystem II, measured in leaves at ambient temperature and at 77 K, as an indicator of the photon yield of photosynthesis. Planta 180: 166–174

Adams WW III, Zarter CR, Ebbert V and Demmig-Adams B (2004) Photoprotective strategies of overwintering evergreens. BioScience 54: 41–49

Adamska I, Ohad I and Kloppstech K (1992) Synthesis of the early light-inducible protein is controlled by blue light and related to light stress. Proc Natl Acad Sci USA 89: 2610–2613

Anderson JM (1986) Photoregulation of the composition, function, and structure of thylakoid membranes. Annu Rev Plant Physiol 37: 93–136

Andersson J, Walters RG, Horton P and Jansson S (2001) Antisense inhibition of the photosynthetic antenna proteins CP29 and CP26: implications for the mechanism of protective energy dissipation. Plant Cell 13: 1193–1204

Andersson J, Wentworth M, Walters RG, Howard CA, Ruban AV, Horton P and Jansson S (2003a) Absence of the Lhcb1 and Lhcb2 proteins of the light-harvesting complex of photosystem II - effects on photosynthesis, grana stacking and fitness. Plant J 35: 350–361

Andersson U, Heddad M and Adamska I (2003b) Light stress-induced one-helix protein of the chlorophyll a/b-binding family associated with photosystem I. Plant Physiol 132: 811–820

Asada K (1999) The water-water cycle in chloroplasts: scavenging of active oxygens and dissipation of excess photons. Annu Rev Plant Physiol Plant Mol Biol 50: 601–639

Aspinall-O'Dea M, Wentworth M, Pascal A, Robert B, Ruban A and Horton P (2002) In vitro reconstitution of the activated zeaxanthin state associated with energy dissipation in plants. Proc Natl Acad Sci USA 99: 16331–16335

Bailey S, Walters RG, Jansson S and Horton P (2001) Acclimation of Arabidopsis thaliana to the light environment: the existence of separate low light and high light responses. Planta 213: 794–801

Bailey S, Thompson E, Nixon PJ, Horton P, Mullineaux CW, Robinson C and Mann NH (2002) A critical role for the Var2 FtsH homologue of Arabidopsis thaliana in the photosystem II repair cycle in vivo. J Biol Chem 277: 2006–2011

Barber J and Andersson B (1992) Too much of a good thing: light can be bad for photosynthesis. Trends Biochem Sci 17: 61–66

Baroli I and Melis A (1998) Photoinhibitory damage is modulated by the rate of photosynthesis and by the photosystem II light-harvesting chlorophyll antenna size. Planta 205: 288–296

Baroli I and Niyogi KK (2000) Molecular genetics and xanthophyll-dependent photoprotection in green algae and plants. Phil Trans R Soc Lond B 355: 1385–1394

Baroli I, Do AD, Yamane T and Niyogi, KK (2003) Zeaxanthin accumulation in the absence of a functional xanthophyll cycle protects Chlamydomonas reinhardtii from photooxidative stress. Plant Cell 15: 992–1008

Bartley GE and Scolnik PA (1995) Plant carotenoids: pigments for photoprotection, visual attraction, and human health. Plant Cell 7: 1027–1038

Bassi R, Sandonà D and Croce R (1997) Novel aspects of chlorophyll a/b-binding proteins. Physiol Plant 100: 769–779

Bendall DS and Manasse RS (1995) Cyclic photophosphorylation and electron transport. Biochim Biophys Acta 1229: 23–38

Bilger W and Björkman O (1990) Role of the xanthophyll cycle in photoprotection elucidated by measurements of light-induced absorbance changes, fluorescence and photosynthesis in leaves of Hedera canariensis. Photosynth Res 25: 173–185

Björkman O and Demmig B (1987) Photon yield of O_2 evolution and chlorophyll fluorescence characteristics at 77 K among vascular plants of diverse origins. Planta 170: 489–504

Burrows PA, Sazanov LA, Svab Z, Maliga P and Nixon PJ (1998) Identification of a functional respiratory complex in chloroplasts through analysis of tobacco mutants containing disrupted plastid ndh genes. EMBO J 17: 868–876

Chen M, Choi YD, Voytas DF and Rodermel S (2000) Mutations in the Arabidopsis VAR2 locus cause leaf variegation due to the loss of a chloroplast FtsH protease. Plant J 22: 303–313

Cheng Z, Sattler S, Maeda H, Sakuragi Y, Bryant DA and DellaPenna D (2003) Highly divergent methyltransferases catalyze a conserved reaction in tocopherol and plastoquinone synthesis in cyanobacteria and photosynthetic eukaryotes. Plant Cell 15: 2343–2356

Collakova E and DellaPenna D (2001) Isolation and functional analysis of homogentisate phytyltransferase from Synechocystis sp. PCC 6803 and Arabidopsis. Plant Physiol 127: 1113–1124

Collakova E and DellaPenna D (2003) Homogentisate phytyltransferase activity is limiting for tocopherol biosynthesis in Arabidopsis. Plant Physiol 131: 632–642

Conklin PL (2001) Recent advances in the role and biosynthesis of ascorbic acid in plants. Plant Cell Environ 24: 383–394

Conklin PL, Williams EH and Last RL (1996) Environmental stress sensitivity of an ascorbic acid-deficient Arabidopsis mutant. Proc Natl Acad Sci USA 93: 9970–9974

Conklin PL, Saracco SA, Norris SR and Last RL (2000) Identification of ascorbic acid-deficient Arabidopsis thaliana mutants. Genetics 154: 847–856

Davison PA, Hunter CN and Horton P (2002) Overexpression of β-carotene hydroxylase enhances stress tolerance in Arabidopsis. Nature 418: 203–206

Demmig-Adams B, Adams WW, Heber U, Neimanis S, Winter K, Krüger A, Czygan F-C, Bilger W and Björkman O (1990) Inhibition of zeaxanthin formation and of rapid changes in radiationless energy dissipation by dithiothreitol in spinach leaves and chloroplasts. Plant Physiol 92: 293–301

Dunaeva M and Adamska I (2001) Identification of genes expressed in response to light stress in leaves of *Arabidopsis thaliana* using RNA differential display. Eur J Biochem 268: 5521–5529

Elrad D, Niyogi KK and Grossman AR (2002) A major light-harvesting polypeptide of photosystem II functions in thermal dissipation. Plant Cell 14: 1801–1816

Endo T, Shikanai T, Takabayashi A, Asada K and Sato F (1999) The role of chloroplastic NAD(P)H dehydrogenase in photoprotection. FEBS Lett 457: 5–8

Escoubas J-M, Lomas M, LaRoche J and Falkowski PG (1995) Light Intensity regulation of *cab* gene transcription is signaled by the redox state of the plastoquinone pool. Proc Natl Acad Sci USA 92: 10237–10241

Foyer CH and Harbinson J (1999) Relationships between antioxidant metabolism and carotenoids in the regulation of photosynthesis. In: Frank HA, Young AJ, Britton G and Cogdell RJ (eds) The Photochemistry of Carotenoids, pp 305–325. Kluwer Academic Publishers, Dordrecht

Frank HA and Cogdell RJ (1993) The photochemistry and function of carotenoids in photosynthesis. In: Young A and Britton G (eds) Carotenoids in Photosynthesis, pp 252–326. Champman & Hall, London

Fryer MJ (1992) The antioxidant effects of thylakoid vitamin E (α-tocoherol). Plant Cell Environ 15: 381–392

Golan T, Li X-P, Müller-Moulé P and Niyogi KK (2004) Using mutants to understand light stress acclimation in plants. In: Papageorgiou C and Govindjee (eds) Chlorophyll a Fluorescence: A Signature of Photosynthesis, pp 525–554. Kluwer Academic Publishers, Dordrecht, Dordrecht

Heddad M and Adamska I (2000) Light stress-regulated two-helix proteins in *Arabidopsis thaliana* related to the chlorophyll *a/b*-binding gene family. Proc Natl Acad Sci USA 97: 3741–3746

Hihara Y, Kamei A, Kanehisa M, Kaplan A and Ikeuchi M (2001) DNA microarray analysis of cyanobacterial gene expression during acclimation to high light. Plant Cell 13: 793–806

Huala E, Oeller PW, Liscum E, Han I-S, Larsen E and Briggs WR (1997) Arabidopsis NPH1: a protein kinase with a putative redox-sensing domain. Science 278: 2120–2123

Hutin C, Nussaume L, Moise N, Moya I, Kloppstech K and Havaux M (2003) Early light-induced proteins protect Arabidopsis from photooxidative stress. Proc Natl Acad Sci USA 100: 4921–4926

Jahns P and Krause GH (1994) Xanthophyll cycle and energy-dependent fluorescence quenching in leaves from pea plants grown under intermittent light. Planta 192: 176–182

Jahns P and Schweig S (1995) Energy-dependent fluorescence quenching in thylakoids from intermittent light grown pea plants: evidence for an interaction of zeaxanthin and the chlorophyll *a/b* binding protein CP26. Plant Physiol Biochem 33: 683–687

Jansson S, Anderson J, Kim SJ and Jackowski G (2000) An *Arabidopsis thaliana* protein homologous to cyanobacterial high-light-inducible proteins. Plant Mol Biol 42: 345–351

Jarillo JA, Gabrys H, Capel J, Alonso JM, Ecker JR and Cashmore AR (2001) Phototropin-related NPL1 controls chloroplast relocation induced by blue light. Nature 410: 952–954

Joët T, Cournac L, Horvath EM, Medgyesy P and Peltier G (2001) Increased sensitivity of photosynthesis to antimycin A induced by inactivation of the chloroplast ndhB gene. Evidence for a participation of the NADH-dehydrogenase complex to cyclic electron flow around photosystem I. Plant Physiol 125: 1919–1929

Kagawa T and Wada M (1999) Chloroplast-avoidance response induced by high-fluence blue light in prothallial cells of the fern *Adiantum capillus-veneris* as analyzed by microbeam irradiation. Plant Physiol 119: 917–923

Kagawa T and Wada M (2000) Blue light-induced chloroplast relocation in *Arabidopsis thaliana* as analyzed by microbeam irradiation. Plant Cell Physiol 41: 84–93

Kagawa T and Wada M (2002) Blue light-induced chloroplast relocation. Plant Cell Physiol 43: 367–371

Kagawa T, Sakai T, Suetsugu N, Oikawa K, Ishiguro S, Kato T, Tabata S, Okada K and Wada M (2001) Arabidopsis NPL1: a phototropin homolog controlling the chloroplast high-light avoidance response. Science 291: 2138–2141

Kasahara M, Kagawa T, Oikawa K, Suetsugu N, Miyao M and Wada M (2002) Chloroplast avoidance movement reduces photodamage in plants. Nature 420: 829–832

Kim S, Pichersky E and Yocum CF (1994) Topological studies of spinach 22 kDa protein of photosystem II. Biochim Biophys Acta 1188: 339–348

Kimura M, Yamamoto YY, Seki M, Sakurai T, Sato M, Abe T, Yoshida S, Manabe K, Shinozaki K and Matsui M (2003) Identification of Arabidopsis genes regulated by high light-stress using cDNA microarray. Photochem Photobiol 77: 226–233

Larkin RM, Alonso JM, Ecker JR and Chory J (2003) GUN4, a regulator of chlorophyll synthesis and intracellular signaling. Science 299: 902–906

Li X-P, Björkman O, Shih C, Grossman AR, Rosenquist M, Jansson S and Niyogi KK (2000) A pigment-binding protein essential for regulation of photosynthetic light harvesting. Nature 403: 391–395

Li X-P, Phippard A, Pasari J and Niyogi KK (2002) Structure–function analysis of photosystem II subunit S (PsbS) in vivo. Func Plant Biol 29: 1131–1139

Lindahl M, Tabak S, Cseke L, Pichersky E, Andersson B and Adam Z (1996) Identification, characterization, and molecular cloning of a homologue of the bacterial FtsH protease in chloroplasts of higher plants. J Biol Chem 271: 29329–29334

Lindahl M, Spetea C, Hundal T, Oppenheim AB, Adam Z and Andersson B (2000) The thylakoid FtsH protease plays a role in the light-induced turnover of the photosystem II D1 protein. Plant Cell 12: 419–432

Long SP, Humphries S and Falkowski PG (1994) Photoinhibition of photosynthesis in nature. Annu Rev Plant Physiol Plant Mol Biol 45: 633–662

Ma Y-Z, Holt NE, Li X-P, Niyogi KK and Fleming GR (2003) Evidence for direct carotenoid involvement in the regulation of photosynthetic light harvesting. Proc Natl Acad Sci USA 100: 4377–4382

Maxwell DP, Laudenbach DE and Huner N (1995) Redox regulation of light-harvesting complex II and cab mRNA abundance in *Dunaliella salina*. Plant Physiol 109: 787–795

Maxwell K and Johnson GN (2000) Chlorophyll fluorescence-a practical guide. J Exp Bot 51: 659–668

Melis A (1999) Photosystem-II damage and repair cycle in chloroplasts: what modulates the rate of photodamage in vivo? Trends Plant Sci 4: 130–135

Mochizuki N, Brusslan JA, Larkin R, Nagatani A and Chory J (2001) Arabidopsis *genomes uncoupled 5* (*GUN5*) mutant reveals the involvement of Mg-chelatase H subunit in plastid-to-nucleus signal transduction. Proc Natl Acad Sci USA 98: 2053–2058

Montané M-H and Kloppstech K (2000) The family of light-harvesting-related proteins (LHCs, ELIPs, HLIPs): was the harvesting of light their primary function? Gene 258: 1–8

Müller P, Li X-P and Niyogi KK (2001) Non-photochemical quenching. A response to excess light energy. Plant Physiol 125: 1558–1566

Müller-Moulé P, Conklin PL and Niyogi KK (2002) Ascorbate deficiency can limit violaxanthin de-epoxidase activity in vivo. Plant Physiol 128: 970–977

Müller-Moulé P, Havaux M and Niyogi KK (2003) Zeaxanthin deficiency enhances the high light sensitivity of an ascorbate-deficient mutant of Arabidopsis. Plant Physiol 133: 748–760

Munekage Y, Takeda S, Endo T, Jahns P, Hashimoto T and Shikanai T (2001) Cytochrome $b_6 f$ mutation specifically affects thermal dissipation of absorbed light energy in Arabidopsis. Plant J 28: 351–359

Munekage Y, Hojo M, Meurer J, Endo T, Tasaka M and Shikanai T (2002) PGR5 is involved in cyclic electron flow around photosystem I and is essential for photoprotection in Arabidopsis. Cell 110: 361–371

Munné-Bosch S and Alegre L (2002) The function of tocopherols and tocotrienols in plants. Crit Rev Plant Sci 21: 31–57

Niyogi KK (1999) Photoprotection revisited: genetic and molecular approaches. Annu Rev Plant Physiol Plant Mol Biol 50: 333–359

Niyogi KK (2000) Safety valves for photosynthesis. Curr Op Plant Biol 3: 455–460

Niyogi KK, Björkman O and Grossman AR (1997) Chlamydomonas xanthophyll cycle mutants identified by video imaging of chlorophyll fluorescence quenching. Plant Cell 9: 1369–1380

Niyogi KK, Grossman AR and Björkman O (1998) Arabidopsis mutants define a central role for the xanthophyll cycle in the regulation of photosynthetic energy conversion. Plant Cell 10: 1121–1134

Noguchi T (2002) Dual role of triplet localization on the accessory chlorophyll in the photosystem II reaction center: photoprotection and photodamage of the D1 protein. Plant Cell Physiol. 43: 1112–1116

Ort DR (2001) When there is too much light. Plant Physiol 125: 29–32

Ort DR and Baker NR (2002) A photoprotective role for O_2 as an alternative electron sink in photosynthesis? Curr Op Plant Biol 5: 193–198

Oswald O, Martin T, Dominy PJ and Graham IA (2001) Plastid redox state and sugars: interactive regulators of nuclear-encoded photosynthetic gene expression. Proc Natl Acad Sci USA 98: 2047–2052

Palatnik JF, Tognetti VB, Poli HO, Rodríguez RE, Blanco N, Gattuso M, Hajirezaei M-R, Sonnewald U, Valle EM and Carrillo N (2003) Transgenic tobacco plants expressing antisense ferredoxin-NADP(H) reductase transcripts display increased susceptibility to photo-oxidative damage. Plant J 35: 332–341

Park Y, Chow WS and Anderson JM (1997) Antenna size dependency of photoinactivation of photosystem II in light-acclimated pea leaves. Plant Physiol 115: 151–157

Pfannschmidt T, Schütze K, Fey V, Sherameti I and Oelmüller R (2003) Chloroplast redox control of nuclear gene expression-a new class of plastid signals in interorganellar communication. Antioxid Redox Signal 5: 95–101

Porfirova S, Bergmüller E, Tropf S, Lemke R and Dörmann P (2002) Isolation of an Arabidopsis mutant lacking vitamin E and identification of a cyclase essential for all tocopherol biosynthesis. Proc Natl Acad Sci USA 99: 12495–12500

Rossel JB, Wilson IW and Pogson BJ (2002) Global changes in gene expression in response to high light in Arabidopsis. Plant Physiol 130: 1109–1120

Ruban AV, Pesaresi P, Wacker U, Irrgang KD, Bassi R and Horton P (1998) The relationship between the binding of dicyclohexylcarbodiimide and quenching of chlorophyll fluorescence in the light-harvesting proteins of photosystem II. Biochemistry 37: 11586–11591

Sakai T, Kagawa T, Kasahara M, Swartz TE, Christie JM, Briggs WR, Wada M and Okada K (2001) Arabidopsis nph1 and npl1: blue light receptors that mediate both phototropism and chloroplast relocation. Proc Natl Acad Sci USA 98: 6969–6974

Sakamoto W, Tamura T, Hanba-Tomita Y, Sodmergen and Murata M (2002) The VAR1 locus of Arabidopsis encodes a chloroplastic FtsH and is responsible for leaf variegation in the mutant alleles. Genes Cells 7: 769–780

Sattler SE, Cahoon EB, Coughlan SJ and DellaPenna D (2003) Characterization of tocopherol cyclases from higher plants and cyanobacteria. Evolutionary implications for tocopherol synthesis and function. Plant Physiol 132: 2184–2195

Savidge B, Weiss JD, Wong Y-H H, Lassner MW, Mitsky TA, Shewmaker CK, Post-Beittenmiller D and Valentin HE (2002) Isolation and characterization of homogentisate phytyltransferase genes from *Synechocystis* sp. PCC 6803 and Arabidopsis. Plant Physiol 129: 321–332

Schreiber U, Schliwa U and Bilger W (1986) Continuous recording of photochemical and non-photochemical chlorophyll fluorescence quenching with a new type of modulation fluorometer. Photosynth Res 10: 51–62

Shikanai T, Endo T, Hashimoto T, Yamada Y, Asada K and Yokota A (1998) Directed disruption of the tobacco ndhB gene impairs cyclic electron flow around photosystem I. Proc Natl Acad Sci USA 95: 9705–9709

Shikanai T, Munekage Y, Shimizu K, Endo T and Hashimoto T (1999) Identification and characterization of Arabidopsis mutants with reduced quenching of chlorophyll fluorescence. Plant Cell Physiol 40: 1134–1142

Silva P, Thompson E, Bailey S, Kruse O, Mullineaux CW, Robinson C, Mann NH and Nixon PJ (2003) FtsH is involved in the early stages of repair of photosystem II in *Synechocystis* sp. PCC 6803. Plant Cell 15: 2152–2164

Smirnoff N (2000) Ascorbate biosynthesis and function in photoprotection. Phil Trans R Soc Lond B 355: 1455–1464

Strand Å, Asami T, Alonso J, Ecker JR and Chory J (2003) Chloroplast to nucleus communication triggered by accumulation of Mg-protoporphyrinIX. Nature 421: 79–83

Surpin M, Larkin RM and Chory J (2002) Signal transduction between the chloroplast and the nucleus. Plant Cell 14: S327–338

Susek RE, Ausubel FM and Chory J (1993) Signal transduction mutants of Arabidopsis uncouple nuclear *CAB* and *RBCS* gene expression from chloroplast development. Cell 74: 787–799

Takechi K, Sodmergen, Murata M, Motoyoshi F and Sakamoto W (2000) The *YELLOW VARIEGATED* (*VAR2*) locus encodes a homologue of FtsH, an ATP-dependent protease in Arabidopsis. Plant Cell Physiol 41: 1334–1346

Wada M, Kagawa T and Sato Y (2003) Chloroplast Movement. Annu Rev Plant Biol 54: 455–468

Walters RG, Ruban AV and Horton P (1996) Identification of proton-active residues in a higher plant light-harvesting complex. Proc Natl Acad Sci USA 93: 14204–14209

Walters RG, Shephard F, Rogers JJM, Rolfe SA and Horton P (2003) Identification of mutants of Arabidopsis defective in acclimation of photosynthesis to the light environment. Plant Physiol 131: 472–481

Wollman F-A, Minai L and Nechushtai R (1999) The biogenesis and assembly of photosynthetic proteins in thylakoid membranes. Biochim Biophys Acta 1411: 21–85

Yamamoto HY, Nakayama TO and Chichester CO (1962) Studies on the light and dark interconversions of leaf xanthophylls. Arch Biochem Biophys 97: 168–173

Chapter 10

A Protein Family Saga: From Photoprotection to Light-Harvesting (and Back?)

Stefan Jansson*
Umeå Plant Science Centre,
Department of Plant Physiology, Umeå University, SE-901 87 Umeå, Sweden

Summary	145
I. Introduction	146
II. The Light-Harvesting Complexes (LHCs) of Higher Plants	146
A. Ten, Twelve, or Fourteen LHC Proteins?	146
B. Secondary and Tertiary Structure	146
C. The Cousins: One, Two, and Four-Helix Proteins	147
III. The Light-Harvesting Antenna of Lower Plants	147
A. Cyanobacteria	147
B. Green Algae	147
C. Liverworts, Mosses, and Ferns	148
D. Conifers	148
IV. The Evolution of LHC Proteins	148
A. One, Two, Four, Three	148
B. From Photoprotection to Light Harvesting?	150
V. The Evolution of Feedback De-Excitation	150
A. Cyanobacteria	151
B. Xanthophyll Conversions are Old	151
C. What is the "Quenching Protein" in *Chlamydomonas*?	152
VI. Conclusions	152
Acknowledgments	152
References	152

Summary

Photoprotection seems to be an intrinsic property of light-harvesting systems, and an interesting question to address is whether the light-harvesting or the photoprotection function was the "original" function, and which function evolved subsequently. It appears that the cyanobacterial one-helix proteins, the presumed ancestors to the LHC proteins, were not designed as antenna proteins but were involved in photoprotection and/or pigment metabolism. Some intermediate steps (two- and four-helix proteins) also seem to have photoprotective functions. The antenna function appeared later in evolution, and many different LHC proteins with somewhat diversified functions arose. To some extent, this happened before the lineages leading to *Chlamydomonas* and higher plants separated, but further diversification also took place following the split, and some of the proteins may have evolved in a direction away from optimizing light harvesting. When the evolution of feedback de-excitation is put into this evolutionary scheme, it is likely that xanthophyll conversions, that evolved previously to optimize photoprotection, were starting to be used as indicators of light stress and regulators of antenna function.

*Author for correspondence, email: stefan.jansson@plantphys.umu.se

I. Introduction

Novel genes do not fall from heaven, but evolve from pre-existing genes. This means that when a new biological function arises through evolution, this occurs through mutations that make a protein with one enzymatic or structural function gain a new function. A classical example of this is the evolution of crystallins, the structural proteins of eyes, where different enzymes have become major lens components in different vertebrate species (Wistow and Piatigorsky, 1987), probably as a consequence of a need for different ocular optics. The eye is a structure that has evolved several times during animal evolution, since the ability to interpret light signals from the surroundings and to change behaviour is, of course, expected to confer a huge evolutionary advantage to an animal. Higher plants use the information in light as signals to change developmental patterns but, and perhaps more importantly, the energy of light to drive the photosynthetic reactions. Photosynthetic reaction centers seem to have evolved only once; PS I, PS II, and the corresponding reaction centers of photosynthetic prokaryotes share all structural elements and are likely to be homologous structures. In contrast, the proteins of the light-harvesting systems that are present in the different taxa share no sequence, and very little structural, similarity and it is most likely that the light-harvesting systems of higher plants, cyanobacteria, purple bacteria, and green sulphur bacteria evolved independently from each other (Green, 2001). In this process, different proteins have been recruited to fulfill the function of coordinating the photosynthetic pigment molecules into ordered arrays that enable efficient transfer of excitation energy into the reaction centers where charge separation takes place. In the following, the evolution of the higher plant light-harvesting antenna structure into its present form, where light harvesting and light dissipation are intimately coupled processes, will be discussed.

II. The Light-Harvesting Complexes (LHCs) of Higher Plants

A. Ten, Twelve, or Fourteen LHC Proteins?

The proteins of the higher plant photosynthetic antenna, the light-harvesting chlorophyll a/b-binding (LHC) proteins, make up a protein family of ten principal members, plus a couple of related proteins (Jansson, 1994). The gene products of the Lhca1-4 genes associate with PS I and the Lhcb1-6 genes primarily with PS II, although the Lhcb1 and Lhcb2 proteins that, at least in low and intermediate light conditions, make up the bulk of the antenna, distribute between PS I and PS II to balance the flow-through of electrons in the photosystems. Several names have been used to designate the gene products (see Jansson, 1994 for a compilation). In the following, the names relating to gene names (e.g. Lhca1) will primarily be used, although e.g. Lhcb4 is commonly designated as CP29 in the photosynthesis literature, an informative name reflecting the mobility of the protein during electrophoresis. In terms of evolution, the focus of this paper, the Lhc acronyms are more consistent. Several of the LHC proteins are encoded by multiple genes; in *Arabidopsis thaliana* there are e.g. 5 and 3 genes encoding Lhcb1 and Lhcb2, respectively (Leutweiler et al., 1986; McGrath et al., 1992; Legen et al., 2001; Andersson et al., 2003). Although the individual Lhcb1 proteins have slightly different amino acid sequences, these differences are not conserved among plant species and thus they probably do not represent differences in functions, which would likely be conserved. This distinction is, however, not straightforward to make based on sequence differences alone. For instance, an *Arabidopsis* gene encoding a protein originally named Lhca2 (Zhang et al., 1992) encodes a distinct protein (Lhca6), and the same is probably true for the "third *Lhcb4* gene", *Lhcb4.3* (Jansson, 1999). Recently, a few additional related genes have been identified in *Arabidopsis*. Under normal conditions, these have a much lower expression level than the "normal" proteins, and these will be more thoroughly discussed later in the chapter.

B. Secondary and Tertiary Structure

Although LHC II (in this case a mixture of Lhcb1 and Lhcb2) is the only LHC protein whose tertiary structure has been experimentally determined (Kühlbrandt et al. 1994), all LHC proteins are believed to fold in an identical way. They consist of three membrane-spanning helices (MSHs), of which the first and the third are homologous to each other, and form a dimeric "core" of the protein. The LHC proteins bind pigment in various stoichiometries; chlorophyll a, chlorophyll b, xanthophylls, and the pigments of the xanthophyll cycle associate with most, if not all, of the polypeptides. The pigments are bound by non-covalent bonds and, although there is some conservation of the

---

Abbreviations: ELIP – Early light inducible proteins; HLIP – High light inducible proteins; LHC – Light-harvesting Complex; MSH – Membrane spanning helices; SCP – Small cab-like proteins

pigment-binding amino acid residues, conservation is not strict. Mutational analyses have been employed to describe the properties of the different pigment-binding sites (e.g. Morosinotto et al., 2002).

C. The Cousins: One, Two, and Four-Helix Proteins

Higher plants also contain proteins related to the LHC proteins. The first to be described were the Early Light Inducible Proteins (ELIPs) that accumulate e.g. during early thylakoid development and during light stress (Adamska et al., 1992). The ELIPs also have three MSHs and are likely to fold in a similar way as the LHC proteins. The related PsbS protein, however, has four MSH, and the "extra" helix is found at the N-terminus and is homologous to helix 2. More recently, two related one-helix and two two-helix proteins have also been described in *Arabidopsis* (Jansson, 1999; Jansson et al., 2000; Heddad and Adamska, 2000; Andersson et al., 2003b). The single MSH of the one-helix proteins, and the first of the helices in the two-helix proteins, share similarity to the first and the third MSHs of LHCs, ELIPs, and PsbS. With the exception of PsbS (Li et al., 2000), the functions of the other proteins are not well defined. It is striking that they all exhibit regulatory responses opposite to those of the ten normal LHC proteins; under high light conditions when the expression of the LHC proteins is repressed, these other proteins are upregulated or at least are not repressed. This indicates a function in high-light protection in a broad sense, but whether they provide direct protection, are pigment carriers, or have other, novel functions is not yet known. From an evolutionary perspective, it is likely that they all bind pigments (chlorophylls and carotenoids), although undisputable evidence for this is lacking.

III. The Light-Harvesting Antenna of Lower Plants

In order to understand how the higher plant light-harvesting antenna has evolved, one could study the corresponding structures of the ancestors and relatives of higher plants. The genome of *Synechocystis* sp. strain PCC6803 (Kaneko et al., 1996) was the first to be sequenced, and has recently been followed by several others (Kaneko et al., 2001; Nakamura et al., 2002); for an updated list see the homepage of Masahiko Ikeuchi (http://bio.c.u-tokyo.ac.jp/~ikeuchi/c_genomeE.html). What are the characteristics of the proteins homologous and analogous to the higher plant light-harvesting antenna in these organisms?

A. Cyanobacteria

The main light-harvesting structure of cyanobacteria, the phycobilisome, is a soluble antenna system that coordinates linear tetrapyrroles and shares no homology with the higher plant LHC proteins. However, one group of cyanobacterial proteins is the possible ancestors of the LHC proteins, the High-Light-Inducible Proteins (HLIPs) (Dolganov et al., 1995), also named Small Cab-like Proteins (SCPs) (Funk and Vermaas, 1999). Since their discovery, their primary function has been assumed to be something other than light harvesting; for example, they tend to accumulate under high light conditions when light harvesting does not need to be optimized. Not surprisingly, Rhodophytes (red algae) also contain related proteins; e.g. the unicellular Red Alga *Cyanidioschyuzon merolae* has a plastid-encoded one-helix protein that shares homology with HLIPs/SCPs (Ohta et al., 2003).

Disruption of the genes encoding HLIPs/SCPs leads to a photosensitive phenotype (He et al., 2003). There has, however, been a considerable debate about their precise role in photoprotection that could either be direct through some mechanism for light dissipation (Havaux et al., 2003) or indirect, since data have also indicated a role in the regulation of chlorophyll biosynthesis (Xu et al., 2002). For both of these suggested functions, pigments are likely to bind to the proteins, but direct proof of pigment binding properties of HLIPs/SCPs is still absent. In any case, the HLIPs/SCPs do not seem to be light-harvesting proteins, but are rather involved in photoprotection.

B. Green Algae

The green alga *Chlamydomonas reinhardtii* has, not surprisingly, a light-harvesting apparatus that is similar to that of higher plants. Genomic efforts have recently allowed for identification of 28 genes coding for proteins related to LHCs (Elrad and Grossman, 2004). There is not a clear-cut correlation between the orthologues of *Chlamydomonas* and higher plant LHC proteins. It seems as if Lhcb4 and Lhcb5 diverged prior to the separation of higher plant and green algal lineages, but that the separation into the four (or six) Lhca proteins in the higher plant lineage, as well as into the nine Lhca proteins in the *Chlamydomonas* lineage, took place after this separation (Elrad and Grossman, 2004).

The same holds true for the other Lhcb genes; the nine *Chlamydomonas* genes do not correspond to the nine *Arabidopsis* genes coding for Lhcb1, Lhcb2, Lhcb3, and Lhcb6. In fact, the encoded proteins of these genes are so similar that it is quite likely that the functions will be very similar. In addition, *Chlamydomonas* contains two genes coding for a protein (Ll818) with no clear homologue in *Arabidopsis* (Savard et al., 1996), as well as genes coding for ELIP-like proteins and a PsbS homologue (Elrad and Grossman, 2004). Interestingly, no genes coding for proteins similar to the HLIPs of cyanobacteria or the one- or two-helix proteins of higher plants have been identified yet in the *Chlamydomonas* genome (Elrad and Grossman, 2004). It is possible that they have so far escaped detection, but it is also possible that they were lost in the lineage leading to *Chlamydomonas*, but retained in the lineage leading to higher plants.

Thus, it seems as if the last common ancestor of *Chlamydomonas* and higher plants had *Lhcb4, Lhcb5*, and at least one gene coding for LHCI polypeptides and one for the major LHCII polypeptide. It is quite possible that other subtypes have existed, but these must, in that case, have been lost during evolution.

C. Liverworts, Mosses, and Ferns

We only have fragmentary information about the LHC protein content from these taxa, but some bits and pieces of information could be put together. In the liverwort *Marchantia polymorpha*, the distinction between the Lhcb1 and Lhcb2 has not yet occurred, but Lhcb3 and Lhcb6 seem to have separated out (Groke I, PhD thesis, University of Bremen). The Lhca proteins of higher plants also seem to have orthologues in *Marchantia*. The scarce data available from mosses and ferns are consistent with these assumptions.

D. Conifers

Lhc genes have been sequenced from numerous species of gymnosperms (mainly conifers) but only in the case of Scots pine (*Pinus sylvestris*) to some depth (Jansson et al., 1992). The conifers probably separated from the lineage leading to the angiosperms 300–350 million years ago, and at that time their light-harvesting apparatus must have looked the same as today since, in all cases studied so far, the proteins in Scots pine and *Arabidopsis* are orthologous. It seems as if no additional subtypes of LHC proteins have evolved after this split, since there are apparently no LHC proteins specific for certain angiosperm taxa, although there are many plant taxa not yet investigated. The separation of the newly identified Lhca5 protein apparently dates back to the time of the separation of the other Lhca subtypes (Ganeteg et al. 2004), whereas Lhca6 and Lhcb4.3 are perhaps of more recent origin. A schematic representation of the antenna systems of *Chlamydomonas* and angiosperms/conifers are shown in Fig. 1.

IV. The Evolution of LHC Proteins

A. One, Two, Four, Three

An outline for the early events in the evolution of the LHC proteins from the cyanobacterial one-helix proteins was put forward by Green and Pichersky in 1994 (Fig. 2). The HLIP proteins somehow gained a second, unrelated MSH and formed a two-helix protein that subsequently underwent internal gene duplication. The duplication presumably led to a four-helix configuration, typical for PsbS, and the forth helix was believed to have been lost in the evolution to the ELIPs and LHCs (Fig. 2). Although the two-helix proteins in *Arabidopsis* are clearly not the direct ancestors of PsbS, ELIPs, and LHCs, the identification of those proteins corroborated this very reasonable hypothesis. As mentioned above, it is interesting that all one-, two-, and four-helix proteins known today (and the three-helix ELIP protein) are, based on their expression characteristics,

A
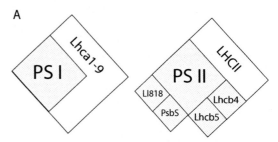
The light-harvesting antenna of Chlamydomonas

B
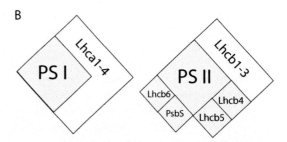
The light-harvesting antenna of angiosperms

Fig. 1. Schematic picture of antenna proteins associated with PS I and PS II in A) *Chlamydomonas* and B) angiosperms.

Chapter 10 A Protein Family Saga

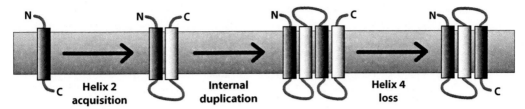

Fig. 2. Evolution of the LHC proteins. One-helix proteins, like HLIP, acquired a second helix resulting in a two-helix protein, like the higher plant SCPs, that underwent an internal gene duplication leading to a four-helix protein, like PsbS. After loss of the fourth helix, three helix proteins like ELIPs and LHCs evolved.

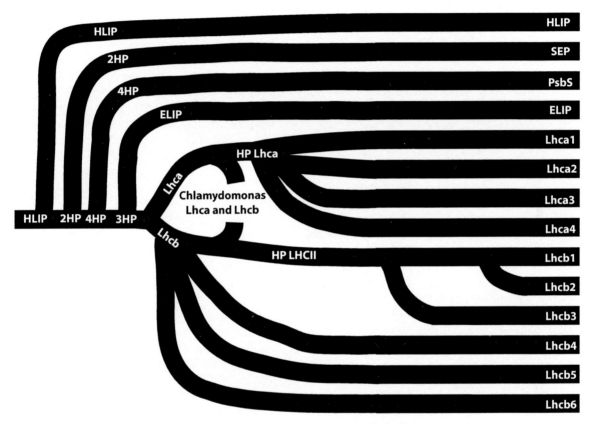

Fig. 3. Evolutionary tree of the LHC protein family.

more likely involved in high light protection than light harvesting. This makes it plausible that the gene family originally evolved to serve a function in photoprotection, and that the role in light harvesting is a derived function. The antenna function seems to have been a successful evolutionary step, so that once it appeared, perhaps in its earliest manifestation as an antenna of PSI as in today's rhodophytes, gene duplication and divergence created a set of LHC proteins with diversified functions in the antenna of PS I and PS II.

Based on today's knowledge, the following evolutionary scheme for the LHC family of proteins can be put forward (Figure 3). Starting with one-helix proteins, the early events resulting in two-, four-, and three-helix proteins must have taken place in the cyanobacterial lineage that was engulfed by a eukaryote and became the chloroplast. LHCI was probably the first LHC protein to appear, since in Rhodophytes (Red algae), proteins homologous to the LHCs are found associated with PS I (Wolfe et al., 1994), although the main light-harvesting antenna is made up by phycobilisomes, rather similar to those of cyanobacteria.

Before the lineages leading to *Chlamydomonas* and higher plants separated, the ELIP and LHC branches were formed and LHCI, LHCII, Lhcb4, and Lhcb5 subsequently diverged within the LHC branch, probably with LHCI separation as the first event. The split that created the Ll818 proteins probably happened before this separation, although the Ll818 function has been lost in the higher plant lineage. In the *Chlamydomas* lineage, diversification within the LHCI and LHCII branches led to the extant *Chlamydomonas* LHC polypeptide composition, and it is quite likely that other green algae will be found to have a different polypeptide composition.

About the time when plants started to inhabit the terrestrial ecosystem, further diversification took place. The individual Lhca proteins were formed in the LHCI-branch, and Lhcb6 and then Lhcb3 may have been the first to diverge from the LHCII branch. Alternatively, and perhaps more likely, Lhcb6 was present in the last common ancestor to *Chlamydomonas* and higher plants, but has been lost in the *Chlamydomonas* lineage. Two pieces of information support the latter alternative. First, Lhcb6 is more divergent from LHCII than Lhcb5, so the separation between Lhcb1/2/3 and Lhcb6 probably predated the split between Lhcb6 and Lhcb1/2/3. Secondly, there exist, in some respects, similarities between Lhcb6 and the Lhca proteins, even making it possible that Lhcb6 could have evolved from the LHCI-branch of the tree. The final step in the evolution, the separation of Lhcb1 and Lhcb2, took place after the separation of the liverwort and higher plant lineages, but before the separation of the conifers and angiosperms, i.e. about 400 million years ago. Apparently, the light-harvesting antenna developed at that time has proven to be very well suited for plants in the various terrestrial habitats, and no further evolution has been necessary.

B. From Photoprotection to Light Harvesting?

From the general picture outlined above, one obvious conclusion that could be drawn is that the proteins of this family originally had a photoprotective function. All one-, two-, and four-helix proteins as well as ELIPs studied so far have expression patterns consistent with a role in photoprotection rather than in optimizing light harvesting, since they are not expressed at all, or to low levels, under low light conditions (Jung and Niyogi, this volume). It is likely that pigment binding to this class of proteins first had a photoprotective function, either direct or indirect by involvement in pigment metabolism. Such pigment-binding proteins would be obvious candidates for recruitment as antenna proteins, as soon as mutations gave rise to pigment properties increasing the duration of excitation energy storage enough to allow energy transfer to neighboring pigments that are functionally connected to a reaction centrum. Whether this has happened once or several times in evolution is difficult to establish definitively. However, in the absence of evidence for several events, one can assume that it happened only once, and that this event conferred such an evolutionary advantage to the organism that the genes became fixed in the population. Later such genes became duplicated and diverged, resulting in the multitude of LHC antenna proteins that is present in photosynthetic organisms today.

However, a recent finding is complicating this picture slightly. As mentioned above, a few genes with different characteristics than the "standard" *Lhc* genes have been identified in *Arabidopsis*. The *Lhca5* gene has a regulation that is opposite to the norm (lower relative expression in leaves grown in low light), and this was also shown to be true for the *Lhca6*, *Lhcb4.3*, and an additional *Lhc* gene, most similar to *Lhcb5*, that has just been identified. Although low expression under conditions where a large antenna is needed is not very strong evidence of a photoprotective function, these four proteins (two potentially associating with PS I and two with PS II) are likely to have a distinct role in the photosynthetic antenna, suggesting that additional recruitments have occurred rather late in evolution. Tentatively, this set of LHC polypeptides can be assumed to be components of the light-harvesting antenna that partially replaces the "standard" LHC proteins under some conditions, and if the "standard" proteins are believed to be perfectly optimized for light harvesting, replacing these with other, potentially less efficient energy transfer proteins would probably be, in a wide sense, a photoprotective process. Thus, although the evolution has been from photoprotection to light harvesting, in some branches of the evolutionary tree there may have been a back-conversion into a protective function.

V. The Evolution of Feedback De-Excitation

How could the evolution of feedback de-excitation be put into this framework? A quenching mechanism that rapidly goes into operation when needed, but that does not decrease light harvesting under conditions where light is not present in excess, should represent a huge selective advantage to a plant that is exposed to rapid variations in light conditions, and could potentially represent a quantum leap in evolution allowing plants to occupy new habitats. Therefore it is important to

Chapter 10 A Protein Family Saga

consider how the mechanism could have evolved. In my opinion, the most likely sequence of evolutionary events leading to the development of feedback de-excitation is the following.

A. Cyanobacteria

Like all photosynthetic organisms, cyanobacteria need photoprotective systems (Nishiyama et al., this volume). The coupling of photosynthetic and respiratory electron transport, that in cyanobacteria take place in the same membrane, means that other possibilities to cope with fluctuations in excitation pressure exist, e.g. if electron transport driven by photosynthesis increases, electron transport from respiration could be decreased resulting in an unchanged total electron transport and proton pumping over the thylakoid membrane. Since no energy-dissipating mechanism seems to be present in the phycobilisome, a rapidly inducible energy dissipation system could be expected to be beneficial for the cyanobacteria and for this reason, such a role for the HLIPs/SCPs seem logical. Perhaps pigments bound to HLIPs/SCPs could be involved in energy transfer from excited singlet chlorophylls to e g zeaxanthin which accumulate in cyanobacteria grown in high light (Demmig-Adams and Adams, 1990) in a way not regulated by protonation. However, since the xanthophyll cycle is not operating in cyanobacteria, an inducible system cannot utilize the information of overcapacity in electron transport over ATP synthesis that a low pH-induced violaxanthin de-epoxidase will provide. Compared with the higher plant system, only one of the triggers for feedback de-excitation is therefore present. Carotenoid pigments serve a photoprotective role also in prokaryotes, but they probably do not directly influence the efficiency of light harvesting, and rather act as scavengers.

B. Xanthophyll Conversions are Old

Xanthophyll conversions date back to well before the process of feedback de-excitation. Cyclic xanthophyll conversions take place in Prochlorophytes (Clauster et al., 2002) and some Rhodophytes (Ursi et al., 2003). Although Prochlorophytes contain chlorophyll b, this pigment is bound to proteins unrelated to the higher plant LHC proteins. Thus, the feedback de-excitation mechanism is probably not present in this taxon. In Rhodophytes, proteins homologous to the LHCs are found associated with PS I, although the main light-harvesting antenna is made up by phycobilisomes and is rather similar to those of cyanobacteria. When cyanobacteria were engulfed by a eukaryotic host and became chloroplasts, the need for a respiratory chain creating a proton gradient over the thylakoid membrane disappeared (since energy and reducing power in the dark could be provided by the eukaryotic host), and photosynthetic electron transport became spatially separated from respiration. As a consequence, the proton gradient across the thylakoid became an indicator of excitation pressure and sensing of the lumenal pH became a direct way for an organism to detect whether electron transport or biochemistry limits photosynthesis. If biochemistry limits photosynthesis, excitation energy flow through PS I and PS II will be retarded and two adequate measures that a cell should take in order to avoid photodamage may occur. One is to increase the levels of photoprotectants in a wide sense and secondly, the efficient size of the light harvesting antenna can be decreased. A low pH-induced conversion of violaxanthin to zeaxanthin could serve the first purpose and e.g. prevent lipid peroxidation. Apparently this conversion appeared rather early in evolution, since it is found in both Rhodophytes and Prochlorophytes. The low pH will also induce protonation of proteins at the lumenal side of the thylakoid and, potentially, induce a conformational change that could lead to energy dissipation. Considering the extremely high concentration of pigments in light-harvesting systems, it is likely that the function of the antenna proteins is to keep the photosynthetic pigments apart in order to prevent quenching rather than to pull them together to enable energy transfer (Horton et al., 1996). Various conformational changes in the antenna will then bring pigment molecules together and enable them to quench each other. Since the violaxanthin and zeaxanthin molecules have different shapes, the low pH-dependent conversion of violaxanthin could also create a conformational change that in itself could lead to quenching. Together with protonation, this could have even bigger effects on the conformation resulting in today's feedback de-excitation. The mechanism for the quenching is outside the scope of this article, but several hypotheses are discussed elsewhere in this volume. However, since potentially any protein that has acidic residues in the lumen and binds xanthophylls could evolve to be subject to such a critical conformational change resulting in quenching, many proteins could be recruited for this function. Since PsbS has this function in higher plants (Li et al., 2000), and has its roots at about the correct time in evolution, it is likely that the PsbS-like precursors were the proteins where this function evolved.

C. What is the "Quenching Protein" in Chlamydomonas?

However, the situation in *Chlamydomonas* complicates the picture. The feedback de-excitation process seemed initially to share many features with that of higher plants, e.g. the xanthophyll conversions are similar and mutations of violaxanthin de-epoxidase, as well as zeaxanthin epoxidase, have similar effects (Niyogi et al., 1997). However, the molecular characterization of a *Chlamydomonas* mutant (*npq5*) with a similar phenotype as PsbS-less *Arabidopsis* (*npq4*) unexpectedly revealed that the gene involved coded for one of the proteins of LHCII (Lhcbm1) (Elrad et al., 2002). The recent identification of a PsbS-like protein in *Chlamydomonas* (Krishna Niyogi unpublished, cited in Elrad and Grossman, 2004), however, leaves open the possibility that PsbS is still the key protein in *Chlamydomonas* and that the *npq5* phenotype is caused by some other, yet unidentified, molecular change. Alternatively, one of the LHC proteins may have been recruited to fulfill this function in the *Chlamydomonas* lineage, whereas in the lineage leading to higher plants, PsbS was recruited. If this is so, and since low pH-dependent xanthophyll conversions arose deep in the evolutionary tree, it is possible that we will find other taxa where yet other pigment-binding proteins have a function to dissipate excess excitation energy. As mentioned above, a rapidly inducible energy dissipation system would give a very strong adaptive advantage to a photosynthetic organism. Since forward genetic screens have been successful in identifying the key protein for feedback de-excitation in *Arabidopsis* and *Chlamydomonas*, it is possible that, as the genome sequence of many other algal species will become available within the coming years, we may be able to find an unexpected heterogeneity in proteins responsible for energy dissipation.

VI. Conclusions

To conclude, the evolution of the LHC proteins illustrates how new protein functions are created by recruitment, and if there is a selective advantage conferred with the new function, it will be fixed in the gene pool. Subsequently, gene duplications could diversify the functions resulting in the development of gene families of a very complex kind. The events creating the recent antenna took place over 300 million years ago, after which the genes, for the most part, have been fixed in the population and little natural variation with adaptive significance may be found today. Nevertheless, due to the extensive knowledge in the area and high importance associated with changes in their function, genes coding for antenna proteins may be used as an illustration of converging and diverging gene evolution in plants.

Acknowledgments

This work was supported by grants from the Swedish Research Council and the Research Council for Environment, Agricultural Sciences and Spatial Planning.

References

Adamska I, Ohad I and Kloppstech K (1992) Synthesis of the early light-inducible protein is controlled by blue light and related to light stress. Proc Natl Acad Sci USA 89: 2610–2613

Andersson J, Wentworth M, Walters RG, Howard C, Ruban A, Horton P and Jansson S (2003a) Absence of the main light-harvesting complex of photosystem II affects photosynthesis and fitness but not grana stacking. Plant J 35: 350–361

Andersson U, Heddad M and Adamska I (2003b) Light stress-induced one-helix protein of the chlorophyll a/b-binding family associated with photosystem I. Plant Physiol 132: 811–820

Claustre H, Bricaud A, Babin M, Bruyant F, Guillou L, Le Gall F, Marie D and Partensky F (2002) Diel variations in Prochlorococcus optical properties. Limn Oceanogr 47: 1637–1647

Demmig-Adams B and Adams WW III (1990) The carotenoid zeaxanthin and 'high-energy-state quenching' of chlorophyll fluorescence. Photosynth Res 25: 187–197

Dolganov NAM, Bhaya D and Grossman AR (1995) Cyanobacterial protein with similarity to the chlorophyll a/b binding proteins of higher plants: Evolution and regulation. Proc Natl Acad Sci USA 92: 636–640

Elrad D and Grossman AR (2004) A genome's-eye view of the Lhc polypeptides of *Chlamydomonas reinhardtii*. Curr Genet 45: 61–75

Elrad D, Niyogi KK and Grossmann AR (2002) A major light-harvesting polypeptide of photosystem II functions in thermal dissipation. Plant Cell 14: 1801–1816

Funk C and Vermaas WFJ (1999) Expression of cyanobacterial genes coding for single-helix polypeptides resembling regions of light-harvesting proteins from higher plants. Biochem 38: 9397–9404

Ganeteg U, Klimmek F and Jansson S (2004) Lhca5 - an LHC-type protein associated with photosystem I. Plant Phys (in press)

Green BR (2001) Was "molecular opportunism" a factor in the evolution of different photosynthetic light-harvesting pigment systems? Proc Natl Acad Sci USA 98: 2119–2121

Green BR and Pichersky E (1994) Hypothesis for the evolution of three-helix Chl a/b and Chl a/c light-harvesting antenna proteins from two-helix and four-helix ancestors. Photosynth Res 39: 149–162

Havaux M, Guedeney G, He Q and Grossman AR (2003) Elimination of high-light-inducible polypeptides related to eukaryotic chlorophyll a/b-binding proteins results in aberrant photoacclimation in Synechocystis PCC680. Biochim Biophys Acta 1557: 21–33

He Q, Dolganov N, Bjorkman O, Grossman AR. (2001) The high light-inducible polypeptides in Synechocystis PCC6803. Expression and function in high light. J Biol Chem 276: 306–314

Heddad M and Adamska I (2000) Light stress-regulated two-helix proteins in *Arabidopsis thaliana* related to the chlorophyll a/b-binding gene family. Proc Natl Acad Sci USA 97: 3741–3746

Horton P, Ruban AV and Walters RG (1996) Regulation of light harvesting in green plants. Ann Rev Plant Physiol Plant Mol Biol 47: 644–684

Jansson S (1994) The light-harvesting chlorophyll a/b-binding proteins. Biochim Biophys Acta 1184: 1–19

Jansson S (1999) A guide to the Lhc genes and their relatives in Arabidopsis. Trends Plant Sci 4: 236–240

Jansson S, Pichersky E, Bassi R, Green BR, Ikeuchi M, Melis A, Simpson DJ, Spangfort M, Staehelin LA and Thornber JP (1992) A nomenclature for the genes encoding the chlorophyll *a/b*-binding proteins of higher plants. Plant Mol Biol Rep 10: 242–253

Jansson S, Andersson J, Kim SJ and Jackowski G (2000) An *Arabidopsis thaliana* protein homologous to cyanobacterial high light-inducible proteins. Plant Mol Biol 42: 345–351

Jung H-S and Niyogi KK (2005) Molecular analysis of photoprotection of photosynthesis. In: Demmig-Adams B, Adams WW III and Mattoo AK (eds) Photoprotection, Photoinhibition, Gene Regulation, and Environment, pp 127–143. Springer, Dordrecht

Kaneko T, Sato S, Kotani H, Tanaka A, Asamizu E, Nakamura Y, Miyajima N, Hirosawa M, Sugiura M, Sasamoto S, Kimura T, Hosouchi T, Matsuno A, Muraki A, Nakazaki N, Naruo K, Okumura S, Shimpo S, Takeuchi C, Wada T, Watanabe A, Yamada M, Yasuda M and Tabata S (1996) Sequence analysis of the genome of the unicellular cyanobacterium Synechocystis sp. strain PCC 6803. DNA Res 3: 109–136

Kaneko T, Nakamura Y, Wolk CP, Kuritz T, Sasamoto S, Watanabe A, Iriguchi M, Ishikawa A, Kawashima K, Kimura T, Kishida Y, Kohara M, Matsumoto M, Matsuno A, Muraki A, Nakazaki N, Shimpo S, Sugimoto M, Takazawa M, Yamada M, Yasuda M and Tabata S (2001) Complete genomic sequence of the filamentous nitrogen-fixing cyanobacterium Anabaena sp. strain PCC 7120. DNA Res 8: 205–213

Kühlbrandt W, Wang DN and Fujiyoshi Y (1994) Atomic model of plant light-harvesting complex. Nature 367: 614–621

Legen J, Misera S, Herrmann RG and Meurer J (2001) Map positions of 69 *Arabidopsis thaliana* genes of all known nuclear encoded constituent polypeptide and various regulatory factors of the photosynthetic membrane: a case study. DNA Res 8: 53–60

Leutweiler LS, Meyerowitz EM and Tobin EM (1986) Structure and expression of three light-harvesting chlorophyll a/b-binding protein genes in *Arabidopsis thaliana*. Nucleic Acids Res 14: 4051–4064

Li X-P, Björkman O, Shih C, Grossman AR, Rosenquist M, Jansson S and Niyogi KK (2000) A pigment binding protein essential for regulation of photosynthetic light harvesting. Nature 40: 391–395

McGrath JM, Terzaghi WB, Sridhar P, Cashmore AR and Pichersky E (1992) Sequence of the fourth and fifth photosystem II type I chlorophyll a/b-binding protein genes of *Arabidopsis thaliana* and evidence for the presence of a full complement of the extended CAB gene family. Plant Mol Biol 19: 725–733

Morosinotto T, Baroinio R and Bassi R (2002) Dynamics of chromophore binding to Lhc proteins in vivo and in vitro during operation of the xanthophyll cycle. J Biol Chem 277: 36913–36920

Nakamura Y, Kaneko T, Sato S, Ikeuchi M, Katoh H, Sasamoto S, Watanabe A, Iriguchi M, Kawashima K, Kimura T, Kishida Y, Kiyokawa C, Kohara M, Matsumoto M, Matsuno A, Nakazaki N, Shimpo S, Sugimoto M, Takeuchi C, Yamada M and Tabata S (2002) Complete genome structure of the thermophilic cyanobacterium *Thermosynechococcus elongatus* BP-1. DNA Res 9: 123–130

Nishiyama Y, Allakhverdiev SI and Murata N (2005) Regulation by environmental conditions of the repair of photosystem II in cyanobacteria. In: Demmig-Adams B, Adams WW III and Mattoo AK (eds) Photoprotection, Photoinhibition, Gene Regulation, and Environment, pp 193–203. Springer, Dordrecht

Niyogi KK, Björkman O and Grossman AR (1997) Chlamydomonas xanthophyll cycle mutants identified by video imaging of chlorophyll fluorescence quenching. Plant Cell 9: 1369–1380

Ohta N, Matsuzaki M, Misumi O, Miyagishima S, Nozaki H, Tanaka K, Shin-I T, Kohara Y and Kuroiwa T (2003) Complete sequence and analysis of the plastid genome of the unicellular red alga *Cyanidioschyuzon merolae*. DNA Res: 10: 67–77

Savard F, Richard C and Guertin M (1996) The *Chlamydomonas reinhardtii* LI818 gene represents a distant relative of the cabI/II genes that is regulated during the cell cycle and in response to illumination. Plant Mol Biol 32: 461–473

Ursi S, Pedersen M, Plastino E and Snoeijs P (2003) Red algae and xanthophyll cycle: Intraspecific variation of photosynthesis, respiration and photoprotective carotenoids in *Gracilaria birdiae* (Gracilariales: Rhodophyta). Mar Biol 142: 997–1007

Wistow G and Piatigorsky J (1987) Recruitment of enzymes as lens structural proteins. Science 236: 1554–1556

Wolfe GR, Cunningham FX, Durnford D, Green BR and Gantt E (1994) Evidence for a common origin of chloroplasts with light-harvesting complexes of different pigmentation. Nature 367: 566–568

Xu H, Vavilin D, Funk C and Vermaas W (2002) Small CAB-like proteins regulating tetrapyrrole biosynthesis. Plant Mol Biol 49: 149–160

Zhang H, Wang J and Goodman HM (1994) Differential expression in Arabidopsis of Lhca2, a PSI cab gene. Plant Mol Biol 25: 551–557

Chapter 11

Photoprotection of Photosystem II: Reaction Center Quenching Versus Antenna Quenching

Norman P. A. Huner*[1], Alexander G. Ivanov[1,2], Prafullachandra V. Sane[1,2], Tessa Pocock[1], Marianna Król[1], Andrius Balseris[1], Dominic Rosso[1], Leonid V. Savitch[3], Vaughan M. Hurry[2] and Gunnar Öquist[2]

[1]Department of Biology, University of Western Ontario, London, Ontario, Canada N6A 5B7; [2]Umeå Plant Science Center, Department of Plant Physiology, Umeå University, Umeå S-901 87, Sweden; [3]Agriculture and Agri-Food Canada, Eastern Cereal and Oilseed Research Centre, Ottawa, Canada K1A 0C6

Summary	155
I. Introduction	156
II. Antenna Quenching	157
III. Reaction Center Quenching	158
IV. Thermoluminescence	159
V. Photoprotection through Reaction Center Quenching	160
A. *Synechococcus* sp. PCC 7942	161
B. *Pinus sylvestris*	162
C. *Arabidopsis thaliana*	163
D. *Chlamydomonas reinhardtii*	164
VI. Bioenergetics of Reaction Center Quenching	165
VII. Molecular Mechanisms Regulating Reaction Center Quenching	167
A. D1 Exchange	167
B. Posttranslational Modifications of D1	168
C. Thylakoid Lipids and Fatty Acids	168
Acknowledgments	169
References	169

Summary

Understanding the role of the xanthophyll cycle and elucidating the mechanisms of antenna quenching through the non-photochemical dissipation of excess absorbed energy in the photoprotection of the photochemical apparatus continues to be a major focus of photosynthetic research. In addition to antenna quenching, there is evidence for the non-photochemical dissipation of excess energy through the PS II reaction center. Hence, this photoprotective mechanism is called reaction center quenching. One technique to assess reaction center quenching is photosynthetic thermoluminescence. This technique represents a simple but powerful probe of PS II photochemistry that measures the light emitted due to the reversal of PS II charge separation through the thermally-dependent recombination of the negative charges stabilized on Q_A^- and Q_B^- on the acceptor side of PS II with the positive charges accumulated in the S_2- and S_3-states of the oxygen evolving complex. Changes in the temperature maxima for photosynthetic thermoluminescence may reflect changes in redox potentials of recombining species within PS II reaction centers. Exposure of *Synechococcus* sp. PCC 7942, *Pinus sylvestris* L., *Arabidopsis thaliana*, and *Chlamydomonas reinhardtii* to either low temperatures or to high light induces a significant downshift in the temperature maxima for $S_2 Q_B^-$

---

*Author for correspondence, email: nhuner@uwo.ca

and $S_3Q_B^-$ recombinations relative to $S_2Q_A^-$ and $S_3Q_A^-$ recombinations. These shifts in recombination temperatures are indicative of lower activation energy for the $S_2Q_B^-$ redox pair recombination and a narrowing of the free energy gap between Q_A and Q_B electron acceptors. This, in turn, is associated with a decrease in the overall thermoluminescence emission. We propose that environmental factors such as high light and low temperature result in an increased population of reduced Q_A (Q_A^-), that is, increased excitation pressure, facilitating non-radiative $P680^+Q_A^-$ radical pair recombination within the PS II reaction center. The underlying molecular mechanisms regulating reaction center quenching appear to be species dependent. We conclude that reaction center quenching and antenna quenching are complementary mechanisms that may function to photoprotect PS II to different extents in vivo depending on the species as well as the environmental conditions to which the organism is exposed.

I. Introduction

Changes in irradiance, temperature, nutrient, and water availability result in imbalances between the light energy absorbed through photochemistry and energy utilization through photosynthetic electron transport coupled to carbon, nitrogen, and sulphur reduction. This leads to photoinhibition of photosynthesis under controlled laboratory conditions as well as natural field conditions (Powles, 1984; Krause, 1988; Aro et al., 1993; Long et al., 1994; Keren and Ohad, 1998). Recovery from photoinhibition in plants, green algae, and cyanobacteria is thought to involve a PS II repair cycle in which photodamaged D1 is degraded and the resynthesized D1 is re-inserted to form a functional PS II reaction center (Aro et al., 1993; Keren and Ohad, 1998; Melis, 1999). It has been shown in some chilling-sensitive plant species, green algae, and cyanobacteria that protection against photoinhibition may be accounted for, in part, by the rate of repair relative to the rate of photodamage to D1 (Nishida and Murata, 1996; Keren and Ohad, 1998; Melis, 1999). Alternatively, certain cold tolerant plant species such as winter wheat (*Triticum aestivum* L), rye (*Secale cereale* L), and *Arabidopsis thaliana*, exhibit a minimal dependence on D1 repair but exhibit increased photosynthetic capacity and reprogramming of photosynthetic carbon metabolism in response to cold acclimation (Huner et al., 1993; Hurry et al., 1995; Strand et al., 1997; Demmig-Adams et al., 1999; Adams et al., 2001; Stitt and Hurry, 2002; A. Strand et al., 2003). Although the Mehler reaction appears to contribute to photoprotection in cold tolerant cereals, cold acclimation of Monopol wheat results in the repression of photorespiration (Savitch et al., 2000). This reprogramming of metabolism results in an increased capacity to keep Q_A oxidized and PS II reaction centers open under high excitation pressure induced by either excessive irradiance or low temperatures (Huner et al., 1998; Huner et al., 2003; Öquist and Huner, 2003). Thus, photoprotection in these species is accomplished, in part, through an increase in photochemical quenching (q_P) (Krause and Jahns, 2003).

In contrast to the D1 repair cycle and photochemical quenching, the concept of radiationless dissipation of excess energy through antenna quenching was originally developed on the basis of the Butler model for energy transfer and used to account for Chl fluorescence quenching (Butler, 1978). Non-photochemical quenching (NPQ) of excess excitation energy in the antenna pigment bed of PS II is considered to be the major PS II photoprotective mechanism (Demmig-Adams and Adams, 1992; Horton et al., 1999; Demmig-Adams et al., 1999; Gilmore, 2000; Gilmore and Ball, 2000; Ort, 2001; Demmig-Adams and Adams, 2002). Recently, the term, feedback de-excitation, has been

Abbreviations: A-band – thermoluminescence band between −15° and −10°C; A – antheraxanthin; B_1-band – thermoluminescence band between +20°C and +30°C in the absence of DCMU; B_2-band – thermoluminescence band between +35° and +40°C in the absence of DCMU; C-band – thermoluminescence band between +50° and +60°C; CHB – cold hard band; Cyt b_{559} – cytochrome b_{559}; D1 – photosystem II reaction center polypeptide; D2 – photosystem II reaction center polypeptide; ELIPs – early light inducible proteins; F_o – minimum yield of chlorophyll fluorescence at open PS II centers in dark-adapted leaves; F_m – maximum yield of fluorescence at closed PS II reaction centers in dark adapted leaves; F_v – variable yield of fluorescence in dark adapted leaves; F_v/F_m – maximum PS II photochemical efficiency in dark adapted leaves; LHCII – the major Chl a/b pigment-protein complex associated with PSII; NPQ – non-photochemical quenching; OEC – oxygen evolving complex; Pheo – pheophytin; PI – photoinhibition; PS I – photosystem I; PS II – photosystem II; PS IIα – photosystem IIα centers; PS IIβ – photosystem II β centers; PsbS – PS II subunit and gene product of the *PsbS* gene; PsbZ – PS II subunit and gene product of *ycf9*; PQ – plastoquinone; Q-band – thermoluminescence band between 0° and +10°C in the presence of DCMU; Q_A – primary electron-accepting quinone in PS II reaction centers; Q_B – secondary electron-accepting quinone in PS II reaction centers; q_E – ΔpH-dependent high energy quenching; q_N – non-photochemical quenching coefficient; q_O – quenching coefficient for basal fluorescence; q_P – photochemical quenching coefficient; Q_y – chlorophyll a absorption band; TL – thermoluminescence; T_M – temperature of maximum thermoluminescence emission; V – violaxanthin; Z – zeaxanthin; Zv – thermoluminescence band between −80° and −30°C used to describe this protective mechanism (Kulheim et al., 2002). However, there is also evidence for the non-photochemical dissipation of excess energy through the PS II reaction center rather than through the antenna (Weis and Berry, 1987; Krause and Weis, 1991; Walters and Horton, 1993; Buhkov et al., 2001; Lee et al., 2001; Ivanov et al., 2001, 2002; Sane et al., 2002, 2003; Matsubara and Chow, 2004; Finazzi et al., 2004). Hence this photoprotective mechanism is called reaction center quenching.

We begin our discussion with a brief comparison of antenna quenching versus reaction center quenching and provide past and recent evidence that support a significant role for reaction center quenching in the photoprotection of PS II. This is followed by a discussion of thermoluminescence as a sensitive technique to detect reaction center quenching during photoinhibition of photosynthetic organisms as diverse as cyanobacteria, green algae, conifers, and herbaceous plants. We conclude that, as originally suggested by Krause and Weis (1991), it is probable that both reaction center and antenna quenching function in vivo to different extents depending on the environmental conditions to protect PS II from photodamage.

II. Antenna Quenching

Because antenna quenching is the focus of several other chapters in this volume, we will describe only its essential characteristics here. Although not always the case (Hurry et al., 1997), non-photochemical quenching (NPQ) of excess excitation energy is thought to occur through the interconversion of the light harvesting xanthophyll, violaxanthin (V), to the energy quenching xanthophylls, antheraxanthin (A) and zeaxanthin (Z), and is considered to be the major PS II photoprotective mechanism (Demmig-Adams and Adams, 1992; Horton et al., 1999; Demmig-Adams et al., 1999; Gilmore, 2000; Gilmore and Ball, 2000; Ort, 2001; Demmig-Adams and Adams, 2002). Two major mechanisms have been proposed to account for antenna quenching via the xanthophyll cycle. The direct mechanism proposes that the S_1 state of A and Z within LHCII is lower than that of Chl a within the antenna pigment bed. Thus, A and Z are not able to transfer energy to the S_1 state of antenna chlorophyll whereas V is able to transfer energy. Consequently, excited state A and Z decay to ground state with the release of heat (Frank et al., 1994). This light dependent, reversible interconversion of V to A and Z has been called a 'molecular gear shift' regulating energy transfer within LHCII (Frank et al., 1994). Although the data reported by Polivka et al. (1999; 2002) do not support the 'molecular gear shift' hypothesis, recent convincing evidence has been reported in support for the direct mechanism for non-photochemical quenching through zeaxanthin in the antenna (Ma et al., 2003; Dreuw et al., 2003a, 2003b; Holt et al., 2005). In contrast, the indirect mechanism proposes that the transthylakoid ΔpH gradient and the xanthophyll cycle pigments regulate the oligomerization state of LHCII that affects the rapidly relaxing energy-dependent component (q_E) of NPQ (Horton et al., 1999; Ruban et al., 2002; Aspinall-O'Dea et al., 2002; Wentworth et al., 2003). In support of the indirect mechanism, Elrad et al. (2002) reported that LHCII trimerization is required for antenna quenching in the *npq5* mutant of *Chlamydomonas reinhardtii*. Clearly, the underlying mechanism by which the xanthophyll cycle regulates antenna quenching remains controversial.

Regardless of the mechanism of antenna quenching, the persistent retention of Z and A in overwintering plants led to the development of the concept of sustained xanthophyll-dependent energy dissipation, which involved sustained thylakoid lumen acidification, even in the dark (Gilmore, 1997; Demmig-Adams et al., 1996; Adams et al., 2001). Sustained energy dissipation through the antenna has been suggested as an important protective mechanism enabling evergreen plants to maintain their leaves during the winter. Many mesophytic overwintering plants such as *Malva*, *Arabidopsis*, and winter cereals exhibit a 'cold-sustained', non-photochemical quenching which is rapidly reversible upon warming. In addition to this reversible form, schlerophytic evergreens also exhibit a sustained form of zeaxanthin-dependent non-photochemical quenching which predominates during the winter and is not rapidly reversible upon either warming or the presence of uncouplers. This persistent quenching appears to be associated with the reorganization of the LHCII into xanthophyll-containing aggregates induced by a combination of low temperature and high light (Gilmore and Ball, 2000; Öquist and Huner, 2003; Gilmore et al., 2003). However, this persistent type of sustained quenching has not been reported for mesophytic plant species (Adams et al., 2002; Adams et al., 2004). In addition to its role in non-photochemical quenching, Z also appears to act as an anti-oxidant to protect against photooxidative stress (Havaux and Niyogi, 1999; Baroli et al., 2003).

Recently, major insights into our understanding of the molecular mechanism(s) of NPQ have occurred as a consequence of the isolation of NPQ mutants of *Arabidopsis thaliana* and *Chlamydomonas reinhardtii*. The

PsbS deletion mutant, *npq4-1* (Li et al., 2000) and various *PsbS*-defective mutants of *Arabidopsis* (Niyogi, 1999; Havaux and Kloppstech, 2001; Peterson and Havir, 2001, 2003; Grasses et al., 2002) are impaired in the development of NPQ. However, despite the development of mutants specifically deficient in PsbS and NPQ, the precise function of the PsbS protein, and its specific role in NPQ remains equivocal. Based on the original observation that PsbS binds chlorophylls and xanthophylls (Funk, 2001), and its role in the development of q_E, Li et al. (2002) suggested that this protein is the site of ΔpH and xanthophyll-dependent NPQ. However, more detailed biochemical analyses suggest that the PsbS protein does not bind pigments (Dominici et al., 2002). This is in agreement with the fact that most of the highly conserved amino acids that form the ligands for chlorophyll in most of the LHC proteins (Kühlbrandt et al., 1994; Bassi et al., 1999) are not found in PsbS. In addition, the availability of the 3D map of the PS II supercomplex (Nield et al., 2000a) and the structure of the LHCII trimer (Kühlbrandt et al., 1994) indicate that there is not sufficient space to accommodate the PsbS protein within the LHCII-PSII supercomplex (Nield et al., 2000b). Through fluorescence analysis of the *npq4-1* mutant of *Arabidopsis thaliana* lacking PsbS, Peterson and Havir (2003) have suggested that the PsbS polypeptide may regulate exciton distribution within PS II. Wentworth et al. (2003) suggest that the role of PsbS is to regulate the oligomerization of antenna complexes involved in antenna quenching. However, Holt et al. (2005) reported that PsbS regulates NPQ through a chlorophyll-zeaxanthin heterodimer. A detailed summary of the structure and function of this intriguing protein is provided by Funk (2001) and by Niyogi et al. (2005).

Swiatek et al. (2001) have provided convincing evidence that the *ycf9* gene which encodes PsbZ, a core PS II subunit, plays a critical role in NPQ in tobacco and *Chlamydomonas reinhardtii*. PsbZ appears to stabilize the supramolecular organization of PS II core complexes with the peripheral antennae (Swiatek et al., 2001). Although NPQ was significantly inhibited in $\Delta ycf9$ tobacco plants (associated with a decrease in the level of PsbZ), PsbS accumulation was unaffected in this mutant. Furthermore, PsbZ is present in phycobilisome-containing eukaryotic and prokaryotic organisms that exhibit NPQ but no xanthophyll cycle. Swiatek et al. (2001) suggest that PsbZ is a critical component in the regulation of NPQ in these organisms.

Early light-inducible proteins (ELIPs) are a family of proteins related to the LHC gene family (Montane and Kloppstech, 2000). ELIPs have also been shown to accumulate under conditions of high light or low temperature stress in mature leaves under controlled as well as natural field conditions (Lindahl et al., 1997; Montane et al., 1997; Norén et al., 2003) and during chloroplast development (Meyer and Kloppstech, 1984; Krol et al., 1999). Although it is presumed that they are involved in the non-photochemical photoprotection of PS II, their precise role still remains to be elucidated.

III. Reaction Center Quenching

The concept of antenna quenching is based on Butler's model (Butler, 1978) and assumes that the rate constant for energy transfer to PS II reaction centers exceeds the rate constant for the back-transfer of energy from the reaction center to the antenna. This led to the concept of a PS II unit as an energy funnel. However, fluorescence lifetime measurements indicate that the equilibration of excitons between antennae and reaction centers is one order of magnitude faster than charge separation. Thus, PS II is trap limited and the reaction center appears to act not as a funnel but as a shallow trap (Schatz et al., 1988).

Although a major focus of recent research on photoprotection has been on the contribution of antenna quenching to NPQ, there is historical precedence for alternative mechanisms for the dissipation of excess light and photoprotection of PS II reaction centers (Krause, 1988; Krause and Weis, 1991; Walters and Horton, 1993). Considerable evidence for non-photochemical quenching through reaction center quenching has been provided in plants as well as cyanobacteria (Briantais et al., 1979; Weis and Berry, 1987; Krause, 1988; Vavilin and Vermaas, 2000; Bukhov et al., 2001; Sane et al., 2002). Exposure of spinach, wheat, and rye plants to photoinhibitory conditions is associated with quenching of F_v. This reflects the quenching of fluorescence within PS II reaction centers with minimal effects on antenna quenching as indicated by minimal changes in F_o fluorescence (Krause, 1988; Somersalo and Krause, 1990; Huner et al., 1993). It has been proposed that reaction center quenching is the result of the conversion of photochemically active, fluorescent, closed PS II reaction centers into photochemically inactive, non-fluorescent, PS II reaction centers (Krause, 1988; Krause and Weis, 1991). The photoinactivated PS II reaction centers act as PS II quenching centers and

$$(\text{PS II})_{\text{OPEN}} \overset{h\nu}{\leftrightarrow} (\text{PS II})_{\text{CLOSED}} \overset{[\text{H}^+]}{\leftrightarrow} (\text{PS II})_{\text{QUENCHER}} \quad (1)$$
$$\quad\quad\quad\quad\quad\text{ACTIVE}\quad\quad\quad\quad\text{INACTIVE}$$

Chapter 11 Reaction Center Quenching

dissipate energy as heat, preventing further damage not only to the photoinactivated reaction centers themselves but also neighboring active PS II reaction centers through their role as sinks for excitation energy (Öquist et al., 1992; Lee et al., 2001; Matsubara and Chow, 2004). The relative proportion of active PS II centers versus inactive centers is dependent both on the intrathylakoid ΔpH as well as proportion of closed reaction centers measured as the relative reduction state of Q_A (Weis and Berry, 1987; Krause and Weis, 1991; Krause and Jahns, 2003). Furthermore, the proportion of PS II quenching centers has been shown to be sensitive to the level of photoinhibition to which plants are exposed (Lee et al., 2001). Based on a theoretical assessment of alternative mechanisms for NPQ in Hordeum vulgare, Walters and Horton (1993) concluded that reaction center quenching is operative only when reaction centers are closed, that is, when Q_A is in the reduced state.

The presence of functionally distinct populations of PS II, that is PS IIα and PS IIβ centers, was used to explain the biphasic Chl fluorescence induction kinetics induced upon illumination (Melis and Homan,1976). In the PS II repair cycle, functionally active PS IIα centers are thought to be damaged by photoinhibition (PI), transformed into inactive PS IIβ centers which are subsequently repaired by the de novo synthesis of D1 (Melis, 1999). Thus, recovery from photoinhibition was presumed to be dependent upon chloroplastic protein synthesis (Greer et al., 1986; Aro et al., 1993; Keren and Ohad, 1998; Melis, 1999). However, photoinhibition of PS II in spinach (Somersalo and Krause, 1990; van Wijk and van Hasselt, 1993), wheat as well as rye (Huner et al., 1993) indicated the presence of a form of PS II that quenched Chl fluorescence through the reaction center that exhibited a rapid, temperature-independent recovery, with a $t_{1/2}$ of 15–30 min, the formation of which was light-dependent but independent of chloroplastic protein synthesis. This fast recovery component of PS II was suggested to represent PS IIα quenching centers that are rapidly and reversibly interconverted to active PS IIα centers (Krause and Weis, 1991; Huner et al., 1993).

$$\text{PS II}\alpha \text{ (active)} \underset{\text{fast}}{\overset{\text{PI}}{\leftrightarrow}} \text{PS II}\alpha \text{ (quencher)} \quad (2)$$

In a detailed flash-induced analysis of basal (F_o) and maximal (F_m) fluorescence, Delrieu (1998) demonstrated that fluorescence quenching might result from a conversion of PS IIα-centers (dimers) to PS IIβ–centers (monomers) in a low fluorescence state. The monomerization of PS II centers can be triggered by light (Kruse et al., 1997) that effectively decreases the absorption cross-section of PS II (Delrieu, 1998). Furthermore, it was shown that PS II quenching centers increase with the severity of photoinactivation (Lee et al., 2001) and dissipate excess excitation energy as heat (Krause, 1988). Krause and Weis (1991) suggested that both reaction center and antenna quenching play important roles in the photoprotection of PS II.

IV. Thermoluminescence

Photosynthetic thermoluminescence (TL) is the light emitted from a frozen (77°K), preilluminated leaf, algal, or cyanobacterial sample that is gradually heated in darkness. This technique represents a simple but powerful probe of PS II photochemistry (Inoue, 1996; Ke, 2001; Ducruet, 2003). Photosynthetic TL arises from the reversal of PS II charge separation through the thermally-dependent recombination of the negative charges stabilized on Q_A^- and Q_B^- on the acceptor side of PS II with the positive charges accumulated in the S_2 and S_3-states of the oxygen evolving complex (OEC) (Sane and Rutherford, 1986; Inoue, 1996; Ke, 2001; Ducruet, 2003). Light-induced PS II charge separation generates a singlet radical pair $^1[P680^+ \text{ Pheo}^-]$ within a few picoseconds that subsequently proceeds to stabilization of negative and positive equivalents at the acceptor (Q_A^-, Q_B^-) and donor sides (S_2^+, S_3^+-states) of PS II respectively and the regeneration of ground state P680. Although a major part of the light energy captured is stored as redox potential difference between the donor and acceptor sides of PS II, part of the captured energy is lost as 'stabilization energy' that results in the trapping of the separated donor-acceptor charge pair (Inoue, 1996; Ke, 2001; Ducruet, 2003). This free energy trap represents the activation energy barrier against charge recombination, and as a consequence, increases the probability of forward electron transfer relative to reverse electron transfer. However, when thermal activation energy is provided externally, charge recombination becomes possible with the re-excitation of P680 to P680* and the thermally-induced light emission from either P680* or core antenna Chl (Inoue 1996; Ducruet, 2003).

Thermally-induced PS II recombination events are distinguished by their characteristic emission peak temperatures (T_M). A deconvoluted TL glow curve of a photosynthetic sample pre-illuminated with continuous light (Fig. 1A) typically consists of six distinct emission peaks or bands (Vass and Govindjee, 1996; Inoue, 1996; Ke, 2001; Ivanov et al., 2002;

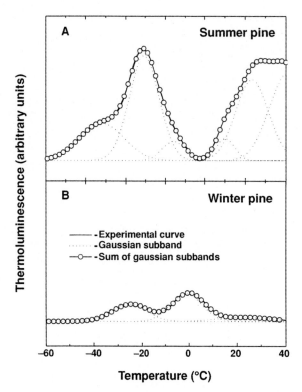

Fig. 1. Thermoluminescence glow curves and mathematical resolution of glow curves in sub-bands in Scots pine (*Pinus sylvestris* L.) needles collected during summer (A) and in winter (B). Experimental curves (-) represent averages of 3 to 5 scans. –o– Computer-generated sum of sub-bands; ··· – computer-fitted sub-band. (From Ivanov et al., 2002)

Ducruet, 2003). The Z_v-band occurs between $-80°$ and $-30°C$ and is thought to reflect $P680^+Q_A^-$ recombination whereas the A-band detected at about $-10°C$ has been assigned to $S_3Q_A^-$ recombinations. The B_1-band assigned to $S_3Q_B^-$ recombinations and the B_2-band associated with $S_2Q_B^-$ recombinations are typically detected between $+20°$ to $+30°C$ and between $+35°$ to $+40°C$ respectively. In the presence of DCMU, the B-bands are significantly reduced and replaced with a new emission band at between $0°$ and $+10°C$ called the Q-band which has been shown to be associated with $S_2Q_A^-$ recombinations (Inoue, 1996; Ke, 2001). The C-band assigned to $Y_D^+Q_A^-$ recombinations is detected at temperatures of about $+50°C$.

An upshift in the peak temperature for a particular recombination event indicates an increase in the activation energy required for that recombination event. This implies a change in the PS II reaction center that has increased the depth of the trap between the charged pairs making it energetically less favourable for charge recombination. Alternatively, a downshift in the peak temperature for a particular recombination event indicates a decrease in the activation energy required for that recombination event. This implies a shallower trap between the charged pairs of the PS II reaction center making it energetically more favourable for the recombination event.

Since the free energy of activation for recombination is related to the redox midpoint potential difference between the recombining species (Devault and Govindjee, 1990), an upward shift in the temperature for a TL peak emission maximum has been interpreted to indicate an increase in the redox potential difference between the recombining species. Conversely, a downward shift in the temperature for a TL emission maximum has been interpreted to reflect a decrease in the redox potential difference between recombining species (Devault and Govindjee, 1990). Thus, changes in the temperature maxima for TL emission have usually been discussed in terms of changes in redox potentials of recombining species within PS II reaction centers (Mayes et al., 1993; Nixon et al., 1995; Minagawa et al., 1999; Sane et al., 2002, 2003). However, through crystallographic analyses of reaction centers of *Rhodobacter sphaeroides*, Stowell et al. (1997) showed light-induced protein conformational changes within these reaction centers that altered the rate of electron transfer between Q_A and Q_B with no changes in redox potential. Thus, an alternative explanation for the shifts in the TL temperature maxima may be that they reflect alterations in the activation energy required to alter PS II reaction center protein conformation rather than changes in redox potential per se. Although changes in T_M have been primarily interpreted to indicate changes in the redox potential of PS II reaction center components, this alternative explanation should not be ignored.

V. Photoprotection through Reaction Center Quenching

The over-reduction of Q_A has been suggested as a major prerequisite for efficient dissipation of the excess light within the reaction center of PS II (Krause, 1988; Walters and Horton, 1993; Bukhov et al., 2001; Öquist and Huner, 2003). Non-radiative charge recombination between Q_A^- and the donor side of PS II has been suggested as a mechanism for the dissipation of excitation energy by PS II reaction center quenching (Briantais et al., 1979; Weis and Berry, 1987; Vavilin and Vermaas, 2000). This is consistent with the recent reports that the overwintering evergreens, snow gum and mistletoe, exhibit a distinctive 'cold-hard-band' (CHB)

Chapter 11 Reaction Center Quenching

in their 77K fluorescence emission spectrum that is associated with Chl aggregation and dissipates excess energy as heat from PS II while simultaneously decreasing the quantum yield of PS II (Gilmore and Ball, 2000; Gilmore et al., 2003). Below, we summarize recent data for the direct estimation of the redox properties of the acceptor side of PS II (Q_A and Q_B) using thermoluminescence and its implications for reaction center quenching as an alternative mechanism for the non-radiative dissipation of excess light energy during cold stress, cold acclimation, and high light stress of the cyanobacterium, *Synechococcus* sp. PCC 7942, the conifer, *Pinus sylvestris*, the model plant species, *Arabidopsis thaliana*, and the model green alga, *Chlamydomonas reinhardtii*. Since the effects of low temperature on the TL peak temperatures can be mimicked by high light, we suggest that the alterations in the activation energies for PS II charge recombination pairs reflect a response to excitation pressure, the relative reduction state of Q_A measured either as $1 - q_P$ (Huner et al., 1998; Bukhov et al., 2001) or $1 - q_L$ (Kramer et al., 2004). This is consistent with earlier theoretical considerations of reaction center quenching in barley (Walters and Horton, 1993).

A. Synechococcus sp. PCC 7942

Unlike eukaryotic photosynthetic organisms that contain a single chloroplastic *psbA* gene encoding the PS II reaction center polypeptide, D1, the cyanobacterium *Synechococcus* sp. PCC 7942 possesses three genes that are differentially regulated by light (Golden et al., 1986; Schaefer and Golden, 1989). Under normal growth light conditions, *Synechococcus* sp. PCC 7942 exhibits the presence of form one of the D1 reaction center polypeptide (D1:1). However, upon exposure to high light, D1:1 is exchanged for form two of the PS II reaction center polypeptide (D1:2). Cells expressing D1:2 exhibit decreased susceptibility to photoinhibition compared to those expressing D1:1 (Krupa et al., 1990, 1991). However, Campbell et al. (1995) reported that low temperature stress mimicked the effects of high light in inducing the exchange of D1:1 for D1:2 in *Synechococcus* sp. PCC 7942. Furthermore, they showed that this PS II reaction center polypeptide exchange was a transient phenomenon.

The effect of D1 replacement on the charge recombination events between the acceptor and donor sides of PS II were examined recently in *Synechococcus* sp. PCC 7942 cells exposed to short term low temperature stress (Sane et al., 2002). The TL data demonstrate that exposing *Synechococcus* cells grown at 36°C to 25°C

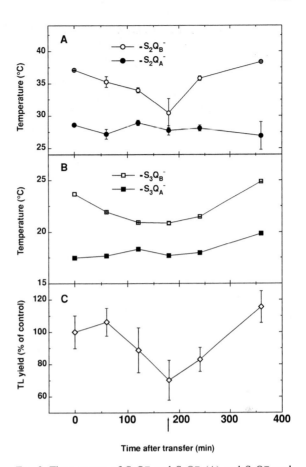

Fig. 2. Time course of $S_2Q_B^-$ and $S_2Q_A^-$ (A) and $S_3Q_B^-$ and $S_3Q_A^-$ (B) characteristic peaks in wild type *Synechococcus* sp. PCC 7942 cells during the temperature shift from the growth temperature of 36°C to 25°C for the first 180 min and back to 36°C for the second part of the curve. C – Relative TL yield measured as the total area under the experimental glow curves. The peak positions were estimated by decomposition analysis of the experimental TL curves after illumination with continuous white light. The presented mean values ± SE are calculated from 6–8 measurements in 3–5 independent experiments. ↑ – shift from 25°C back to 36°C. (From Sane et al., 2002)

shifted the recombination temperatures of $S_2Q_B^-$ and $S_3Q_B^-$ pairs closer to those of $S_2Q_A^-$ and $S_3Q_A^-$ pairs (Fig. 2A). The characteristic T_M of $S_2Q_B^-$ decreased gradually from about 40°C to 30°C after 180 min of low temperature stress. Transferring the cells from 25°C back to the normal growth temperature of 36°C caused a shift of the $S_2Q_B^-$ peak back to about 40°C, indicating that the shifts in T_M are completely reversible. A similar trend was observed for $S_3Q_B^-$ recombinations. In contrast, the T_M for $S_2Q_A^-$ and $S_3Q_A^-$ recombinations remained fairly constant during the temperature shifts indicating the changes in T_M are specific for Q_B^- recombinations (Fig. 2B). Furthermore, the overall TL yield also decreased by 30% after the 180 min exposure of the cells to low temperature, and this effect was also fully reversible after shifting the cultures back to 36°C (Fig. 2C; Sane et al., 2002). The reversible exchange of D1:1 for D1:2 followed the kinetics observed for the reversible changes in T_M (Fig. 3). These data indicate that in cold-stressed cells exhibiting D1:2, the redox potential of Q_B becomes lower approaching that of Q_A. A similar shift in the redox potential of Q_B was confirmed independently of growth temperature by examining the *Synechococcus* sp. PCC 7942 inactivation mutants R2S2C3 and R2K1 which possess either D1:1 or D1:2 respectively (Sane et al., 2002).

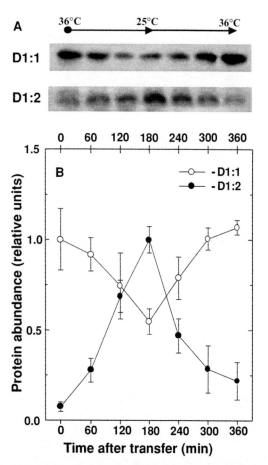

Fig. 3. Representative immunoblots (A) and densitometric analysis (B) of D1:1 and D1:2 polypeptides of PS II during the temperature shift of *Synechococcus* sp. PCC 7942 cells from 36°C to 25°C for 180 min and back to 36°C. Polypeptide abundance was detected by immunoblotting after SDS-PAGE with D1:1 and D1:2 specific antibodies. Mean values ± SE were calculated from 5–7 independent experiments. The data for relative abundance of D1:1 was normalized to its maximal values in control non-treated cells and for D1:2 to its maximal values after 180 min at 25°C. (From Sane et al., 2002)

We suggest that PS II reaction center protein exchange of D1:1 for D1:2 changes the redox properties of Q_B creating an altered charge equilibrium in favour of Q_A (Sane et al., 2002). This would increase the accumulation of Q_A^- and enhance the probability of non-radiative PS II reaction center quenching. Cyanobacteria synthesize zeaxanthin de novo from β-carotene in response to excess light (Adams et al., 1993; Ibelings et al., 1994; Masamoto and Furukawa, 1997), and although these prokaryotes lack xanthophyll cycle-dependent antenna quenching, they do possess the capability of utilizing zeaxanthin-dependent antenna quenching (Demmig-Adams et al., 1990). PS II reaction center quenching associated with the D1:1 / D1:2 exchange may, in part, contribute to the enhanced resistance to photoinhibition induced either by high light or low temperature in cyanobacteria (Krupa et al., 1990, 1991; Campbell et al., 1995).

B. Pinus sylvestris

The acceptor side of PS II has generally been considered to be a primary target for photoinhibition of photosynthesis (Powles, 1984; Krause, 1988; Öquist et al., 1992; Aro et al. 1993; Long et al., 1994; Keren and Ohad, 1998; Melis, 1999). The winter-induced inhibition of PS II photochemistry in Scots pine in vivo has been ascribed also to low temperature-induced photoinhibition of PS II (M. Strand and Öquist, 1985). Thus, the effects of photoinhibition on the charge recombination events between the acceptor and donor sides of PS II in Scots pine needles under both controlled environment as well as natural field conditions were assessed by TL (Ivanov et al., 2001, 2002).

As expected, TL glow emission curves of control, non-hardened pine needles pre-illuminated with continuous light were resolved into six distinct peaks with characteristic T_M corresponding to Z_V (P680$^+$Q$_A^-$ recombination), and A (S$_3$Q$_B^-$) bands below 0°C and to Q (S$_2$Q$_A^-$), B$_1$ (S$_2$Q$_B^-$), B$_2$ (S$_3$Q$_B^-$), and C (TyrD$^+$Q$_A^-$) bands above 0°C (Fig. 1A). In contrast, the cold hardened pine TL glow curves were best fitted with only three emission bands with T_M corresponding to the Z_V, Q, and B peaks with a concomitant decrease in the total TL emission (Fig. 1B; Ivanov et al., 2001). These effects on TL emission were reversible upon recovery of the pine needles from low temperature photoinhibition under laboratory conditions. In contrast to non-hardened needles, the treatment of cold hardened needles with DCMU to block the electron transfer from Q_A to Q_B, did not cause any significant changes in either the TL yield, the T_M, or the relative contribution of each peak to the overall glow curve. Furthermore, needles from cold hardened pine exhibited a significant inhibition of electron transfer from Q_A to Q_B relative to summer pine needles (Ivanov et al., 2001). This is consistent with the thesis that Q_A^- accumulates in PS II reaction centers of cold hardened pine needles to a greater extent than that in non-hardened pine needles due to an over-reduced PQ pool. These results are in agreement with earlier reports indicating that DCMU mimics the effects of low temperature photoinhibition of PS II in pine (Öquist and Martin, 1980), *Pisum sativum* (Farineau, 1993), and *Chlamydomonas reinhardtii* (Ohad et al., 1988). The lower total TL emission from winter pine needles and the relatively strong Z_v band accounting for almost 60% of the total luminescence is indicative of a preferred back reaction of Q_A with primary donors and a low probability of transfer of electrons from Q_A to Q_B (Ivanov et al., 2001). In addition, the $S_2Q_B^-$ charge recombinations were shifted to lower temperatures in cold hardened pine needles than non-hardened pine with little change in the T_M for $S_2/S_3Q_A^-$ recombinations (Ivanov et al., 2001; 2002). Thus, cold hardening conditions appear to cause major changes in the redox properties on the acceptor-side of PS II in *Pinus sylvestris*.

Seasonal dynamics of TL in Scots pine needles under natural field conditions showed that between November and April the contribution of the Q- and B-bands to the overall TL emission was less than 5%. During spring, the relative contribution of the Q- and B-bands, corresponding to charge recombination events between the acceptor and donor sides of PS II, rapidly increased, reaching maximal values in late July. Clearly, the reversible changes in the TL emission bands are observed both under controlled laboratory conditions as well as on a seasonal basis in *Pinus sylvestris* under natural conditions (Ivanov et al., 2002). Thus, the winter inhibition of photosynthesis in Scots pine is associated with major changes on the acceptor-side of PS II. We suggest that exposure to winter conditions narrows the redox potential gap between Q_B and Q_A causing the accumulation of Q_A^- and enhancing the probability for charge recombination within PS II reaction centers through a non-radiative pathway (Vavilin and Vermaas, 2000). This reaction center quenching may enhance the protection of PS II reaction centers through the dissipation of excess absorbed energy and complement the capacity for antenna quenching under conditions where the enzyme-dependent xanthophyll cycle is thermodynamically restricted resulting in a sustained non-photochemical quenching (Öquist and Huner, 2003).

C. Arabidopsis thaliana

Relative to winter wheat and winter rye, cold acclimation of *Arabidopsis thaliana* results in an incomplete recovery of photosynthetic capacity (Savitch et al., 2001). Thus, the possibility of cold induced alterations of PS II was also addressed by TL measurements for direct estimation and comparison of the redox properties of PSII in control, non-hardened, cold-stressed, and cold-acclimated *Arabidopsis* plants (Sane et al., 2003). As observed in *Synechococcus* and *Pinus sylvestris*, cold stress and cold acclimation of *Arabidopsis thaliana* resulted in significant shifts in the T_M for the flash induced TL-bands. However, in contrast to results for pine and *Synechococcus*, cold acclimated *Arabidopsis* exhibited a characteristic upshift of the T_M associated with $S_2Q_A^-$ recombination with a concomitant downshift in the T_M associated with $S_2Q_B^-$ recombinations relative to control plants (Fig. 4) (Sane et al., 2003). These data were confirmed by assessing the time course for the shift in the $S_2Q_A^-$ and $S_2Q_B^-$ peak temperatures when non-hardened control plants were transferred from 23°C to 5°C. The characteristic T_M of $S_2Q_B^-$ peak exhibited a gradual downshift while the T_M of $S_2Q_A^-$ exhibited a gradual upshift such that the initial gap of 55°C between $S_2Q_A^-$ and $S_2Q_B^-$ in non-hardened *Arabidopsis* was narrowed to about 36°C over a period of 4 weeks of cold acclimation (Sane et al., 2003). In addition, the relative TL yield measured as the integrated area under the glow curves also decreased in non-hardened plants transferred to 5°C such that after 24 days at low temperature, the non-hardened plants exhibited overall TL luminescence close to that in fully cold acclimated *Arabidopsis* (Sane et al., 2003). This has been interpreted to indicate major alterations in the redox properties of the acceptor-side of PS II during cold stress and cold acclimation in *Arabidopsis*, increasing the probability of non-radiative dissipation through PS II reaction center quenching and complementing the capacity for antenna quenching to protect PS II from over excitation.

To assess the relative contributions of antenna quenching to total non-photochemical quenching in non-acclimated and cold acclimated *Arabidopsis*, q_O, a measure of antenna quenching (Bukhov et al., 2001; Krause and Jahns, 2003), was plotted as a function of q_N (Fig. 5). If all of the non-photochemical quenching was due to antenna quenching, one would expect a positive, straight line relationship between q_O and q_N (Fig. 5). However, there is a clear curvilinear relationship between q_O and q_N in non-hardened *Arabidopsis* (Fig. 5), indicating significant contributions to q_N that

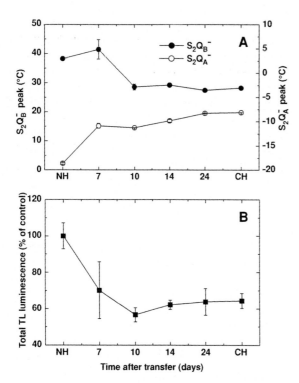

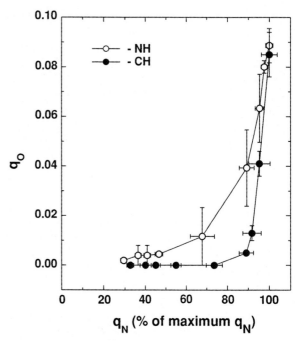

Fig. 4. A – Time course of the characteristic T_M of $S_2Q_B^-$ and $S_2Q_A^-$ peaks in *Arabidopsis* leaves during the temperature shift of control (NH) plants from the growth temperature of 23°C and 250 μmol photons m^{-2} s^{-1} to 5°C and the same irradiance. The peak positions were estimated by decomposition analysis of the experimental curves in control and in DCMU treated leaves. B – Relative TL yield measured as the total area under the experimental glow curves. The mean values ± SE were calculated from 6–8 measurements in 3 independent experiments. NH – control (non-hardened) plants grown at 23°C, 250 μmol photons m^{-2} s^{-1}, and a 16h photoperiod; CH – fully cold-acclimated plants grown at 5°C, 250 μmol photons m^{-2} s^{-1}, and a 16h photoperiod. (From Sane et al., 2003)

Fig. 5. Non-photochemical (q_N) versus q_O chlorophyll fluorescence quenching measured at different actinic light intensities in non-hardened (open symbols) and cold acclimated (closed symbols) *Arabidopsis* leaves. Mean values ± SE were calculated from 3–4 independent experiments. q_N values are presented as percentage of maximal q_N registered during illumination with 1600 μmol photons m^{-2} s^{-1} actinic white light. (From Sane et al., 2003)

do not originate from antenna quenching. This curvilinear relationship is accentuated in cold acclimated *Arabidopsis* where up to 75% of the non-photochemical quenching appears to be due to a quenching component(s) that is/are independent of q_O, that is, independent of the antenna. It has been suggested that PS II reaction center quenching represents the source of this additional quenching capacity (Buhkov et al., 2001; Sane et al., 2003).

D. Chlamydomonas reinhardtii

Photoinhibition and PS II photodamage has been studied extensively in green algae such as *Dunaliella salina* (Melis, 1999) and the model green alga, *Chlamydomonas reinhardtii* (Keren and Ohad, 1998). The consensus is that a continuous repair mechanism that requires the de novo synthesis of D1 is operative during photoinhibition and recovery in this green alga (Falk et al., 1990; Keren and Ohad, 1998; Melis, 1999). Thus, protection from photoinhibition is thought to reflect the relative rates of photodamage versus the rates of repair (Aro et al., 1993; Keren and Ohad, 1998; Melis, 1999). Similar to *Synechococcus* (Krupa et al., 1990, 1991), *Chlorella vulgaris* (Huner et al., 1998) and several plant species (Somersalo and Krause, 1990; Huner et al., 1993; Savitch et al., 2001), growth of *Chlamydomonas reinhardtii* at low temperature (12°C) increases the resistance of these cells to high light regardless of the temperature (Falk et al., 1990). It has been proposed that this increased resistance to photoinhibition in *Chlamydomonas reinhardtii* is, at least in part, due to an increased rate of repair of damaged D1 (Falk et al., 1990; Keren and Ohad, 1998).

Through the generation of NPQ mutants in *Chlamydomonas reinhardtii*, the important role of antenna quenching in photoprotection has been documented in this model green alga (Niyogi, 1999). Results on photoprotection in the *npq5* mutant of *Chlamydomonas reinhardtii* indicate that non-photochemical dissipation through antenna quenching is mediated by trimeric LHCII (Elrad et al., 2002). Furthermore, suppressors of the *npq1 lor1* double mutant in this green alga indicate that zeaxanthin acts as an antioxidant in the quenching of $^1O_2$ and free radicals in addition to its role in antenna quenching through the xanthophyll cycle (Baroli et al., 2003).

What role, if any, does reaction center quenching play in photoprotection of PS II of *Chlamydomonas reinhardtii*? The data in Table 1 illustrate that at a constant growth temperature of 29°C, an increase in growth irradiance from 20 to 500 μmol photons m^{-2} s^{-1} resulted in a decrease in the temperature gap between $S_2Q_B^-$ and $S_2Q_A^-$ from 21.5 to 7.5°C. This 14°C downward shift in the T_M for the $S_2Q_B^-$ recombination occurred with minimal changes in the $S_2Q_A^-$ recombinations. This indicates that exposure of *Chlamydomonas reinhardtii* to high light narrows the redox potential gap between Q_A and Q_B causing the accumulation of Q_A^- that increases the probability for charge recombination and PS II reaction center quenching. This apparent light-dependent effect on TL emission was reversed when high light-grown cells were shifted back to low light.

However, the downward shift in the T_M for $S_2Q_B^-$ can not be a simple high light effect since growth of *Chlamydomonas reinhardtii* at low temperature and moderate irradiance (15°C and 150 μmol photons m^{-2} s^{-1}) also caused a 15°C downward shift in the T_M for $S_2Q_B^-$, the extent of which was comparable to that observed for cells grown at high light with minimal changes in the T_M for $S_2Q_A^-$ recombinations (Table 1). Furthermore, cells grown at low temperature and low irradiance (15°C and 20 μmol photons m^{-2} s^{-1}) exhibited a comparable downward shift to the cells grown at 29°C and moderate irradiance (150 μmol

Table 1. The effects of growth irradiance and growth temperature on thermoluminescence T_M in *Chlamydomonas reinhardtii*.

Growth Regime (°C / μmol m^{-2} s^{-1})	Temperature Gap (°C)[a]
29 / 20	21.5
29 / 150	10.3
29 / 500	7.5
15 / 150	6.2
15 / 20	10.3

[a] The temperature gap was calculated as $T_M(S_2Q_B^-)-T_M(S_2Q_A^-)$. The light- and temperature-dependent changes in the temperature gap were due to downshifts in the T_M for $S_2Q_B^-$ since the T_M for $S_2Q_A^-$ remained fairly constant under all conditions tested.

photons m^{-2} s^{-1}). These results are consistent with the fact that either increased irradiance or low temperature does indeed increase the accumulation of Q_A^- in *Chlamydomonas reinhardtii*. Since low temperature can mimic high light effects due to comparable modulation of the relative redox state of Q_A (Huner et al., 1998), we conclude that reaction center quenching is modulated by excitation pressure in *Chlamydomonas reinhardtii*.

VI. Bioenergetics of Reaction Center Quenching

Summarizing the experimental data discussed above, it is evident that exposure to low temperatures causes major alterations in the redox properties of the acceptor side of PS II in various photosynthetic organisms. The shifts in the characteristic T_M of $S_2Q_A^-$ and $S_2Q_B^-$ recombinations in cold-acclimated *Arabidopsis thaliana* with the Q_A- and Q_B-associated peaks appearing at higher and lower temperatures, respectively, imply substantial changes in the activation energies associated with de-trapping of the electron from reduced Q_A and Q_B (Fig. 6A, B). Similar changes in the redox properties of PS II have been observed by Briantais et al. (1992), who reported a shift towards lower temperatures for the $S_2Q_B^-$ peak in cold acclimated spinach compared to non-hardened plants. More recently, a downshift in the T_M of the B-band ($S_2Q_B^-$) was observed in low temperature grown maize (Janda et al., 2000). Because the activation energies have been shown to be directly related to the redox potentials of the participating species (Devault and Govindjee, 1990), narrowing the temperature gap between the characteristic T_M for Q_A and Q_B may reflect a narrowing of the redox potential gap between Q_A and Q_B as a result of cold acclimation (Fig. 6).

The high temperature shift in the T_M of $S_2Q_A^-$ corresponding to increased activation energy of Q_A/Q_A^- in *Arabidopsis thaliana* would increase the free energy gap between Q_A^- and P680$^+$. This could cause stabilization of $S_2Q_A^-$ and decrease the probability for the back reaction through P680$^+$Pheo$^-$ (Minagawa et al., 1999; Vavilin and Vermaas, 2000). Moreover, the preferential localization of the electron on Q_A in cold acclimated *Arabidopsis* could also result from a change in the redox potential of Q_B. Lowering the redox potential of Q_B will narrow the gap between the redox potentials between Q_A and Q_B even further, and will decrease the probability for electron transfer between the two quinone acceptors by shifting the redox equilibrium

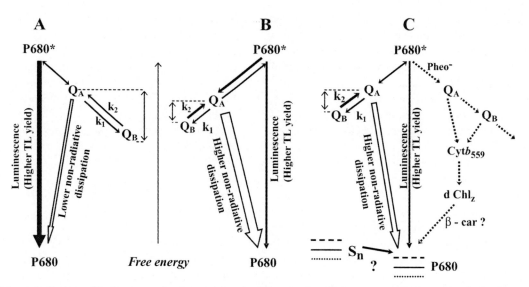

Fig. 6. Schematic diagram of the free energy levels explaining the differences in radiative vs non-radiative energy dissipation pathways in control (A) and cold acclimated (B), (C) plants. A - In control leaves, radiative energy dissipation pathway characterized by higher TL yield is predominant and probably involves the back-reaction *via* the P680$^+$Pheo$^-$ radical pair. B – In cold acclimated plants the increased free energy gap between P680$^+$ and Q_A^- would decrease the probability for a charge recombination pathway involving P680$^+$Pheo$^-$ and will cause stabilization of $S_2Q_A^-$ pair. In addition, shifting the redox potential of Q_B toward Q_A favors the k_2 rate constant and also results in an increased steady state proportion of reduced Q_A. It is proposed that this will increase the probability for direct recombination of Q_A^- with P680$^+$ *via* non-radiative interaction resulting in low TL yield without generating chlorophyll triplet. C – In some cases when there is no apparent shift of $S_2Q_A^-$ recombination to higher temperatures, but the shift of Q_B towards Q_A implies increased proportion of reduced Q_A and the low TL yield suggests increased non-radiative dissipation, the cyclic electron pathway: Cytb_{559} → dChl$_z$ → β-Car → P$_{680}^+$ might be involved. In this case, the redox properties of the donor side might also be modified during cold acclimation. In both types of plants the radiative charge recombination occurs, but is proportionally less in cold acclimated plants (Ivanov et al., 2003).

between $Q_A^-Q_B$ and $Q_AQ_B^-$ towards $Q_A^-Q_B$ (Minagawa et al., 1999). The retention of electrons preferentially on Q_A through a modification of the redox potentials of Q_A and Q_B in opposite directions would inhibit the reoxidation of Q_A^- (Mäenpää et al., 1995). The slower re-oxidation kinetics of Q_A^- in both overwintering Scots pine (Ivanov et al., 2001) and cold acclimated *Arabidopsis* leaves (Sane et al., 2003) are consistent with this interpretation. This would ensure that the Q_B site remains occupied by a quinone, which would protect PS II from photoinhibition and D1 degradation (Ohad and Hirschberg, 1992). Supporting evidence for this argument comes from experiments in which addition of DCMU had a protective effect on D1 turnover under photoinhibitory conditions (Komenda and Masojidek, 1998). When the Q_B site is occupied in the presence of DCMU and Q_A is in a reduced state, PS II shows increased resistance to photoinhibition.

A possible back reaction of the reduced Q_A with P680$^+$ has been suggested from earlier data (Prasil et al., 1996; Krieger-Liszkay and Rutherford, 1998), and this may be enhanced when Q_A remains reduced (Vavilin and Vermass, 2000). The accumulation of Q_A^- has been shown to inhibit the formation of the radical pair P680$^+$Pheo$^-$, thus preventing P680 triplet formation (Schatz et al., 1988; Vass et al., 1992). In addition, it has been suggested that there is a non-radiative pathway of charge recombination between Q_A^- and the donor side of PS II (Briantais et al., 1979; Weis and Berry, 1987; Vavilin and Vermaas, 2000). Such a pathway would increase the probability for non-radiative dissipation of excitation energy within the reaction center of PS II (Weis and Berry, 1987; Bukhov et al., 2001). The significantly lower total TL emission observed in winter pine (Ivanov et al., 2001, 2002), cold stressed *Synechococcus* (Sane et al., 2002), and in cold acclimated *Arabidopsis* (Sane et al., 2003) is consistent with such a non-radiative pathway within the PS II reaction center.

The reduction of Q_A has been suggested to be a major requirement for efficient reaction center quenching (Krause, 1988; Krause and Weis, 1991; Walters and Horton, 1993; Huner et al., 1993; Bukhov et al., 2001). In this regard, it is important to note that acclimation to low temperatures is strongly correlated with an increased proportion of reduced Q_A at the given growth temperature (Huner et al., 1993, 1998). Hence, it seems very likely that the increased population of Q_A^- due to the altered redox potential of Q_A and Q_B during the shift and acclimation to low temperature in *Arabidopsis* may enhance the dissipation of excess light within the reaction center of PS II via non-radiative $P680^+Q_A^-$ recombination, protecting the Q_A site from excessive excitation pressure (Huner et al., 1998; Öquist and Huner, 2003). Similar trends were reported for both winter Scots pine (Ivanov et al., 2001) and low temperature stressed *Synechococcus* (Sane et al., 2002), which are consistent with the results of Bukhov et al. (2001) and Grasses et al. (2002). The importance of excitation pressure and the relative redox state of Q_A in regulating reaction center quenching is further supported by the data for *Chlamydomonas reinhardtii* (Table 1). The low temperature-induced downshift in the T_M for $S_2Q_B^-$ recombination was mimicked by exposing cells to moderate temperatures but high light. Exposure to low temperature but moderate irradiance induces PS II closure because of a slower rate of Q_A^- oxidation relative to the rate of its reduction, whereas exposure to high light at moderate temperatures induces PS II closure due to a higher rate of Q_A reduction relative to the rate of its oxidation (Huner et al., 1998). This is consistent with the notion that the equilibrium $Q_A^- : Q_A$ / $Q_B^- : Q_B$ controls charge recombination within PS II (Keren and Ohad, 1998).

Cyclic electron transport around PS II (Fig. 6C) has been suggested as an alternative photoprotective mechanism operating within the PS II reaction centre (Telfer et al., 1991; Barber and De Las Rivas, 1993), and the role of the high potential form of Cyt b_{559} in this process has been discussed (Stewart and Brudwig, 1998). Allakhverdiev et al. (1997) provided direct evidence for the involvement of cyclic electron transport around PS II in protection against photoinhibitory damage. It was suggested that Cytb_{559} may protect PS II by acting as a secondary donor to $P680^+$ via the electron donation pathway: Cytb_{559} → dChl$_z$ → β-Car → $P680^+$, where Chl$_z$ is a chlorophyll molecule coordinated to His118 of the B trans-membrane helix of the D1 protein (Barber and De Las Rivas, 1993; Nield et al., 2000b). The accumulation of Chl$_z^+$ as a result of over-oxidation of P680 has been suggested as a site for photoprotection (Stewart and Brudvig, 1998). A simplified model illustrating the cyclic electron transport around PS II is presented in Figure 6. It seems reasonable to suggest that such a mechanism for photoprotection is a possible alternative to non-radiative reaction center quenching. Such a mechanism may take place in cases when there is no apparent upward shift of $S_2Q_A^-$ recombination to higher temperatures.

VII. Molecular Mechanisms Regulating Reaction Center Quenching

As discussed above, the prokaryotic and eukaryotic species examined exhibit significant downshifts in the T_M for the $S_2Q_B^-$ and $S_3Q_B^-$ recombinations whereas, in addition, *Arabidopsis thaliana* exhibits an upshift in the T_M for the $S_2Q_A^-$ and the $S_3Q_A^-$ recombinations (Fig. 4). Since these shifts in T_M can occur in response to the reduction state of Q_A induced either by high light or low temperature, we suggest that excitation pressure regulates reaction center quenching by altering the redox potentials of Q_A and Q_B in PS II reaction centers. What is/are the molecular mechanism(s) underlying these alterations in redox potentials of Q_A and Q_B in response to excitation pressure?

A. D1 Exchange

In the case of *Synechococcus* PCC 7942, excitation pressure and reaction center quenching are associated with D1 protein exchange; the D1:1 is exchanged for D1:2. We propose that, in *Synechococcus* sp. PCC 7942, it is primarily reaction center polypeptide exchange that results in a change in the microenvironment of the Q_B-binding site inducing a change in its redox properties. In support of this proposal, it has been shown that a single change in the crucial amino acid residue of D1 (Ohad and Hirschberg, 1992; Minagawa et al., 1999) or a deletion of the PEST-like sequence of D1 (Nixon et al., 1995) results in a shift of the $S_2Q_B^-$ TL peak towards lower temperatures.

However, growth temperature and growth irradiance also have a significant impact on the lipid and fatty acid composition of thylakoid membranes of cyanobacteria (Nishida and Murata, 1996; Los and Murata, 2002). The increased fatty acid unsaturation observed in cyanobacteria at low temperature appears to be a prerequisite for efficient D1 repair upon exposure to low temperature photoinhibition (Nishida and Murata, 1996). Recently, Sakurai et al. (2003) used the *pgsA* mutant of the cyanobacterium *Synechocystis* sp. PCC 6803 to show that the absence of the phospholipid, phosphatidylglycerol (PG), increased the susceptibility of the mutant cells to photoinhibition. Although *pgsA* cells exhibited comparable rates of D1 synthesis and degradation to those observed in the wild type cells, the mutant cells were impaired with respect to the dimerization of PS II core monomers and the reactivation of photoinhibited PS II core complexes (Sakurai et al., 2003). Thus, it is conceivable that changes in the thylakoid lipid and fatty acid composition could also alter the microenvironment of PS II reaction centers by altering lipid-protein interactions causing a shift in the T_M for the Q_B recombinations in *Synechococcus* sp. PCC 7942. However, since the R2S2C3 mutant of *Synechococcus* sp. PCC 7942 exhibiting D1:1 only and the R2K1 mutant exhibiting only D1:2 showed comparable changes in the T_M for $S_2Q_B^-$ and $S_3Q_B^-$ recombinations independent of any temperature change and presumably any changes in thylakoid lipid and fatty acid composition, we conclude that the observed changes in the redox properties of Q_B are most likely a consequence of D1 protein exchange rather than changes in lipid-protein interactions with PS II reaction centers of *Synechococcus* sp PCC 7942.

B. Posttranslational Modification of D1

In higher plants, the D1 polypeptide of PS II is subject to at least five post-translational modifications: C-terminal processing in the conversion of 34 kDa precursor polypeptide to the 32 kDa mature polypeptide; removal of the initiating methionine residue; N-acetylation of N-terminal threonine residue; covalent palmitoylation mapped to the N-terminal two thirds of the D1 polypeptide, and finally, reversible phosphorylation of the N-terminal threonine catalyzed by a light-dependent, redox-regulated kinase (Mattoo et al., 1993; Rintamäki and Aro, 2001). Although the functional role of D1 palmitoylation remains unknown, palmitoylation has been shown to regulate signal transduction through G-protein linked receptors by regulating protein-protein interactions (Milligan et al., 1995). There is no evidence for the role of D1 palmitoylation in altering the T_M for $S_2/S_3 - Q_A^-/Q_B^-$ recombinations. However, alterations in protein-protein interactions within PS II may be important since *Arabidopsis thaliana npq4-1* mutant lacking only the PsbS protein within PS II complexes exhibit significant downshifts in the T_M for $S_2/S_3 - Q_A^-/Q_B^-$ recombinations under normal growth conditions relative to wildtype. We hypothesize that this may be due to changes in protein-protein interactions within PS II reaction centers induced by the absence of PsbS. Further experimentation is ongoing to test this hypothesis.

In higher plants, both the D1 and D2 reaction center polypeptides undergo reversible phosphorylation during the PS II damage-repair cycle. The extent of D1 phosphorylation appears to be regulated by excitation pressure (Rintamäki and Aro, 2001) as well as by an endogenous circadian rhythm (Booij-James et al., 2002). Site-directed mutagenesis of *PsbA* in *Synechocystis* PCC 6803 indicates that alterations in a single amino acid can result in significant changes in the T_M for $S_2/S_3 - Q_A^-/Q_B^-$ recombinations (Minagawa et al., 1999; Vavilin and Vermaas, 2000). Thus, it is conceivable that post-translational modification of D1 and / or D2 PS II reaction center polypeptides by either palmitoylation or phosphorylation may alter the conformation of these polypeptides. This, in turn, may result in shifts in the T_M for $S_2/S_3 - Q_A^-/Q_B^-$ recombinations and hence the changes in the redox potentials of Q_A and Q_B. Consistent with the thesis that protein phosphorylation may convert PS II active centers into PS II quenching centers is the suggestion that the CHB observed in overwintering snow gum and mistletoe represents a Chl-protein complex containing PS II quenching centers formed as a result of PS II core protein phosphorylation (Gilmore et al. (2003).

Unlike in seed plants, no phosphorylation of the D1 polypeptide has been detected in *Chlamydomonas reinhardtii* (Keren and Ohad, 1998; Rintamäki and Aro, 2001). Clearly, this potential post-translational modification mechanism of the D1 polypeptide can not account for shifts in the T_M for $S_2/S_3 - Q_A^-/Q_B^-$ recombinations induced by excitation pressure in this green alga. However, D2 is phosphorylated in *Chlamydomonas reinhardtii* (Keren and Ohad, 1998) and may account for the shifts in the T_M for $S_2/S_3 - Q_A^-/Q_B^-$ recombinations and hence the changes in the redox potentials of Q_A and Q_B. In addition, the decrease in HCO_3^- concentrations in the chloroplast under saturating irradiance has also been shown to affect the redox potentials of Q_A and Q_B (Govindjee, 1993; Demeter et al., 1995). Thus, regulation of chloroplastic HCO_3^- concentrations may also contribute to modulating the redox potentials of Q_A and Q_B in this model green alga.

C. Thylakoid Lipids and Fatty Acids

Unlike *Synechococcus* sp PCC 7942, pine, *Arabidopsis thaliana*, and *Chlamydomonas reinhardtii* have only one *PsbA* gene coding for the D1 protein. Thus, a protein exchange mechanism similar to that observed for *Synechococcus* sp PCC 7942 cannot account for the modulation of the T_M for $S_2/S_3 - Q_A^-/Q_B^-$ recombinations and hence the changes in the redox potentials of Q_A and Q_B. However, the lipid and fatty acid compositions of thylakoid membranes of higher plants are sensitive to growth temperature and growth irradiance (Harwood, 1998; Vijayan et al., 1998; Selstam, 1998). PG and its fatty acid composition are important in regulating the oligomerization of LHCII in many higher plants as well as *Chlamydomonas reinhardtii* (Trémolières and Siegenthaler, 1998). Furthermore, Dobrikova et al. (1997) showed that the asymmetric surface charge distribution and electric polarizability of thylakoid membranes are significantly altered in the *fad*B and the *fad*C *Arabidopsis* mutants deficient in lipid fatty acid desaturases compared to wild type. Thus, it is possible that light- and temperature-induced changes in the thylakoid lipid / fatty acid composition may result in alterations in lipid-protein interactions within PS II reaction centers in pine, *Arabidopsis thaliana*, and *Chlamydomonas reinhardtii*. This, in turn, may result in the observed shifts in the T_M for $S_2/S_3 - Q_A^-/Q_B^-$ recombinations and hence changes in the redox potentials of Q_A and Q_B.

Although excitation pressure may be critical in regulating the shifts in the T_M for $S_2/S_3 - Q_A^-/Q_B^-$ recombinations and hence the redox potentials of Q_A and Q_B, the molecular alterations required to induce these shifts in T_M may be species dependent and vary with the environmental changes to which an organism is exposed. Whereas D1:1/D1:2 polypeptide exchange appears to be the primary molecular mechanism regulating the shifts in T_M for $S_2/S_3 - Q_A^-/Q_B^-$ recombinations in *Synechococcus* PCC 7942, other species such as *Pinus sylvestris*, *Arabidopsis thaliana*, and *Chlamydomonas reinhardtii* may use any one or a combination of the molecular mechanisms outlined above. We conclude that, as originally suggested by Krause and Weis (1991), both reaction center and antenna quenching function in vivo to different extents to protect PS II from photodamage depending on the species as well as the environmental conditions. However, further research is required not only to assess the contribution of any one of these mechanisms to the shifts in the T_M for $S_2/S_3 - Q_A^-/Q_B^-$ recombinations and hence reaction center quenching, but also to assess the timing for the onset of reaction center quenching versus antenna quenching associated with NPQ during exposure to increased excitation pressure.

Acknowledgments

This work was financially supported by the Natural Science and Engineering Research Council of Canada to NPAH, the Swedish Foundation for International Cooperation in Research and Higher Education (STINT) to GÖ and NPAH, and by the Swedish Research Council to GÖ and VMH.

References

Adams III WW, Demmig-Adams B and Lange OL (1993) Carotenoid composition and metabolism in green and blue algal lichens in the field. Oecologia 94: 576–584

Adams III WW, Demmig-Adams B, Rosenstiel TN and Ebbert V (2001) Dependence of photosynthesis and energy dissipation upon growth form and light environment during the winter. Photosynth Res 67: 51–62

Adams WW III, Demmig-Adams B, Rosenstiel TN, Brightwell AK and Ebbert V (2002) Photosynthesis and photoprotection in overwintering plants. Plant Biol 4: 545–557

Adams WW III, Zarter CR, Ebbert V and Demmig-Adams B (2004) Photoprotective strategies of overwintering evergreens. BioScience 54: 41–49

Allakhverdiev SI, Klimov VV and Carpentier R (1997) Evidence for the involvement of cyclic electron transport in the protection of photosystem II against photoinhibition: Influence of a new phenolic compound. Biochem 36: 4149–4154

Aspinall-O'Dea M, Wentworth M, Pascal A, Robert B, Ruban A and Horton P (2002) *In vitro* reconstitution of the activated zeaxanthin state associated with energy dissipation in plants. Proc Natl Acad Sci USA 99: 16331–16335

Aro E-M, Virgin I and Andersson B (1993) Photoinhibition of photosystem II: inactivation, protein damage and turnover. Biochim Biophys Acta 1143: 113–134

Barber J and De Las Rivas J (1993) A functional model for the role of cytochrome b_{559} in the protection against donor and acceptor side photoinhibition. Proc Natl Acad Sci USA 90:10942–10946

Baroli I, Do AD, Yamane T and Niyogi KK (2003) Zeaxanthin accumulation in the absence of a functional xanthophyll cycle protects *Chlamydomonas reinhardtii* from photooxidative stress. Plant Cell 15: 992–1008

Bassi R, Croce R, Cugini D and Sandona D (1999) Mutational analysis of a higher plant antenna protein provides identification of chromophores bound in multiple sites. Proc Natl Acad Sci USA 96: 10056–10061

Booij-James IS, Swegle WM, Edelman M and Mattoo AK (2002) Phosphorylation of the D1 Photosystem II reaction center protein is controlled by an endogenous circadian rhythm. Plant Physiol 130: 2069–2075

Briantais J-M, Vernotte C, Picaud M and Krause GH (1979) A quantitative study of the slow decline of chlorophyll *a* fluorescence in isolated chloroplasts. Biochim Biophys Acta 548: 128–138

Briantais J-M, Ducruet J-M, Hodges M and Krause GH (1992) The effects of low temperature acclimation and photoinhibitory treatments on photosystem 2 studied by thermoluminescence and fluorescence decay kinetics. Photosynth Res 31: 1–10

Bukhov NG, Heber U, Wiese C and Shuvalov VA (2001) Energy dissipation in photosynthesis: does the quenching of chlorophyll fluorescence originate from antenna complexes of photosystem II or from the reaction center? Planta 212: 749–758

Butler WG (1978) Energy distribution in the photochemical apparatus of photosynthesis. Ann Rev Plant Physiol 29: 345–378

Campbell D, Zhou G, Gustafsson P, Öquist G and Clarke AK (1995) Electron transport regulates exchange of two forms of photosystem II D1 protein in the cyanobacterium *Synechococcus*. EMBO J 14: 5457–5466

Delrieu MJ (1998) Regulation of thermal dissipation of absorbed excitation energy and violaxanthin deepoxidation in the thylakoids of *Lactuca sativa*. Photoprotective mechanism of a population of photosystem II centers. Biochim Biophys Acta 1363: 157–173

Demeter S, Janda T, Kovacs L, Mende D and Wiessner W (1995) Effects of in vivo CO_2-depletion on electron transport and photoinhibition in the green algae *Chlamydobotrys stellata* and *Chlamydomonas reinhardtii*. Biochim Biophys Acta 1229: 166–174

Demmig-Adams B and Adams WW III (1992) Photoprotection and other responses of plants to high light stress. Annu Rev Plant Physiol Plant Mol Biol 43: 599–626

Demmig-Adams B and Adams WW III (2002) Antioxidants in photosynthesis and human nutrition. Science 298: 2149–2153

Demmig-Adams B, Adams WW III, Czygan F-C, Schreiber U and Lange OL (1990) Differences in the capacity for radiationless energy dissipation in the photochemical apparatus of green and blue-green algal lichens associated with differences in carotenoid composition. Planta 180: 582–589

Demmig-Adams B, Gilmore AM and Adams WW III (1996) In vivo functions of carotenoids in higher plants. FASEB J 10: 403–412

Demmig-Adams B, Adams WW, Ebbert V and Logan BA (1999) Ecophysiology of the xanthophyll cycle. In: Frank HA, Young AJ, Britton G and Cogdell RJ (eds) The Photochemistry of Carotenoids, Advances in Photosynthesis, Vol 8, pp 245–269. Kluwer Academic Publishers, Dordrecht

Devault D and Govindjee (1990) Photosynthetic glow peaks and their relationship with the free energy changes. Photosynth Res 24: 175–181

Dobrikova A, Taneva SG, Busheva M, Apostolova E and Petkanchin I (1997) Surface electric properties of thylakoid membranes from *Arabidopsis thaliana* mutants. Biophys Chem 67: 239–244

Dominici P, Caffarri S, Armenante F, Ceoldo S, Crimi M and Bassi R (2002) Biochemical properties of the PsbS subunit of photosystem II either purified from chloroplast or recombinant. J Biol Chem 277: 22750–22758

Dreuw A, Fleming GR and Head-Gordon M (2003a) Charge-transfer state as a possible signature of a zeaxanthin-chlorophyll dimmer in the non-photochemical quenching process in green plants. J Phys Chem B 107: 6500–6503

Dreuw A, Fleming GR and Head-Gordon M (2003b) Chlorophyll fluorescence quenching by xanthophylls. Phys Chem Physics 5: 3247–3256

Ducruet J-M (2003) Chlorophyll thermoluminescence of leaf discs: simple instruments and progress in signal interpretation open the way to new ecophysiological indicators. J Exp Bot 54: 2419–2430

Elrad D, Niyogi KK and Grossman AR (2002) A major light harvesting polypeptide of photosystem II functions in thermal dissipation. Plant Cell 14: 1801–1816

Falk S, Samuelsson G and Öquist G (1990) Temperature-dependent photoinhibition and recovery of photosynthesis in the green alga *Chlamydomonas reinhardtii* acclimated to 12 and 27C. Physiol Plant 78: 173–180

Farineau J (1993) Compared thermoluminescence characteristics of pea thylakoids studied in vitro and in situ (in leaves). The effect of photoinhibitory treatments. Photosynth Res 36: 25–34

Finazzi G, Johnson GN, Dall'Osto L, Joliot P, Wollman F-A and Bassi R (2004) A zeaxanthin-independent nonphotochemical quenching mechanism localized in the photosystem II core complex. Proc Natl Acad Sci USA 101: 12375–12380.

Frank HA, Cua A, Chynwat V, Young A, Gosztola D and Wasielewski MR (1994) Photophysics of the carotenoids associated with the xanthophyll cycle in photosynthesis. Photosyn Res 41: 389–395

Funk C (2001) The PsbS protein: a Cab-protein with a function of its own. In: Aro E-M and Andersson B (eds) Regulation of Photosynthesis, Advances in Photosynthesis and Respiration, Vol 11, pp 453–467. Kluwer Academic Publishers, Dordrecht

Gilmore AM (1997) Mechanistic aspects of xanthophyll cycle-dependent photoprotection in higher plant chloroplasts and leaves. Physiol Plant 99: 197–209

Gilmore AM (2000) Mechanistic role of xanthophyll-dependent photoprotection in higher plant chloroplasts and leaves. Physiol Plant 99: 197–209

Gilmore AM and Ball MC (2000) Protection and storage of chlorophyll in overwintering evergreens. Proc Natl Acad Sci USA 97: 11098–11101

Gilmore AM, Matsubara S, Ball MC, Barker DH and Itoh S (2003) Excitation energy flow at 77K in the photosynthetic apparatus of overwintering evergreens. Plant Cell Environ 26: 1021–1034

Golden SS, Brussan J and Haselkorn R (1986) Expression of a family of *psbA* genes encoding a photosystem II polypeptide in the cyanobacterium *Anacystis nidulans* R2. EMBO J 5: 2789–2798

Govindjee (1993) Bicarbonate reversible inhibition of plastoquinone reduction in photosystem II. Z Naturforsch 48c: 251–258

Grasses, T, Pesaresi P, Schiavon F, Varotto C, Salamini F, Jahns P and Leister D (2002) The role of ΔpH-dependent dissipation of excitation energy in protecting photosystem II against light-induced damage in *Arabidopsis thaliana*. Plant Physiol Biochem 40: 41–49

Greer DH, Berry J and Björkman O (1986) Photoinhibition of photosynthesis in intact bean leaves: role of light and temperature and requirement of chloroplast-protein synthesis during recovery. Planta 168: 253–260

Harwood JL (1998) Involvement of chloroplast lipids in the reaction of plants submitted to stress. In: Siegenthaler P-A (ed) Lipids in Photosynthesis: Structure, Function and Genetics, Advances in Photosynthesis and Respiration, Vol 6, pp 287–302. Kluwer Academic Publishers, Dordrecht

Havaux M and Kloppstech K (2001) The protective functions of carotenoids and flavonoid pigments against excess visible radiation at chilling temperature investigated in *Arabidopsis npq* and *tt* mutants. Planta 213: 95–966

Havaux M and Niyogi KK (1999) The violaxanthin cycle protects plants from photooxidative damage by more than one mechanism. Proc Natl Acad Sci USA 96: 8762–8767

Holt NE, Zigmanta D, Valkunas L, Li X-P, Niyogi KK and Fleming GR (2005) Carotenoid cation formation and the regulation of photosynthetic light harvesting. Science 307: 433–436.

Horton P, Ruban AV and Young AJ (1999) Regulation of the structure and function of the light harvesting complexes of photosystem II by the xanthophyll cycle. In: Frank HA, Young AJ, Britton G and Cogdell RJ (eds) The Photochemistry of

Carotenoids, Advances in Photosynthesis, Vol 8, pp 271–291. Kluwer Academic Publishers, Dordrecht

Huner NPA, Öquist G, Hurry VM, Król M, Falk S and Griffith M (1993) Photosynthesis, photoinhibition and low temperature acclimation in cold tolerant plants. Photosynth Res 37: 19–39

Huner NPA, Öquist G and Sarhan F (1998) Energy balance and acclimation to light and cold. Trends Plant Sci 3: 224–230

Huner NPA, Öquist G and Melis A (2003) Photostasis in plants, green algae and cyanobacteria: the role of light harvesting antenna complexes. In: Green BR and Parson WW (eds) Light Harvesting Antennas in Photosynthesis, Advances in Photosynthesis and Respiration, Vol 13, pp 401–421. Kluwer Academic Publishers, Dordrecht

Hurry VM, Strand Å, Tobiæson M, Gardeström P and Öquist G (1995) Cold hardening of spring and winter-wheat and rape results in differential-effects on growth, carbon metabolism, and carbohydrate content. Plant Physiol 109: 697–706

Hurry VM, Anderson JM, Chow WS and Osmond CB (1997) Accumulation of zeaxanthin in abscisic acid-deficient mutants of *Arabidopsis* does not affect chlorophyll fluorescence quenching or sensitivity to photoinhibition *in vivo*. Plant Physiol 113: 639–648

Ibelings BW, Kroon BMA and Mur LR (1994) Acclimation of photosystem II in a cyanobacterial photosynthetic apparatus and a eukaryotic green alga to high and fluctuating photosynthetic photon flux densities simulating light regimes induced by mixing in lakes. New Phytol 128: 407–424

Inoue Y (1996) Photosynthetic luminescence as a simple probe of photosystem II electron transport. In: Amesz J and Hoff AJ (eds) Biophysical Techniques in Photosynthesis, Advances in Photosynthesis, Vol 3, pp 93–107. Kluwer Academic Publishers, Dordrecht

Ivanov AG, Sane PV, Zeinalov Y, Malmberg G, Gardeström P, Huner NPA and Öquist G (2001) Photosynthetic electron transport adjustments in overwintering Scots pine (*Pinus sylvestris* L.). Planta 213: 575–585

Ivanov AG, Sane PV, Zeinalov Y, Simidjiev I, Huner NPA and Öquist G (2002) Seasonal responses of photosynthetic electron transport in Scots pine (*Pinus sylvestris* L.) studied by thermoluminescence. Planta 215: 457–465

Ivanov AG, Sane PV, Hurry V, Krol M, Sveshnikov D, Hunter NPA, Öquist G (2003) Low temperature modulation of the redok properties of the acceptor side of photosystem II: Photoprotection through reaction center quenching of excess energy. Physiol Plant 119: 376–383.

Janda T, Szalai G and Páldi E (2000) Thermoluminescence investigation of low temperature stress in maize. Photosynthetica 38: 635–639

Ke B (2001) Charge recombination in photosystem II and thermoluminescence. In: Ke B. Photobiochemistry and Photobiophysics, Advances in Photosynthesis, Vol 10, pp 407–418. Kluwer Academic Publishers, Dordrecht

Keren N and Ohad I (1998) State transitions and photoinhibition. In: Rochaix J-D, Goldschmidt-Clermont M and Merchant S (eds) Molecular Biology of Chloroplasts and Mitochondria in *Chlamydomonas*, Advances in Photosynthesis, Vol 7, pp 569–598. Kluwer Academic Publishers, Dordrecht

Komenda J and Masojidek J (1998) The effect of photosystem II inhibitors DCMU and BNT on the light-induced D1 turnover in two cyanobacterial strains *Synechocystis* PCC 6803 and *Synechococcus* PCC 7942. Photosynth Res 57: 193–202

Kramer DM, Johnson G, Kiirats O and Edwards GE (2004) New fluorescence parameters for the determination of Q_A redox state and excitation energy fluxes. Photosyn Res 79: 209–218.

Krause GH (1988) Photoinhibition of photosynthesis. An evaluation of damaging and protective mechanisms. Physiol Plant 74: 566–574

Krause GH and Jahns P (2003) Pulse amplitude modulated chlorophyll fluorometry and its application in plant science. In: Green BR and Parson WW (eds) Light Harvesting Antennas in Photosynthesis, Advances in Photosynthesis and Respiration, Vol 13, pp 373–399. Kluwer Academic Publishers, Dordrecht

Krause GH and Weis E (1991) Chlorophyll fluorescence and photosynthesis: the basics. Ann Rev Plant Physiol Plant Mol Biol 42: 313–349

Krieger-Liszkay A and Rutherford AW (1998) Influence of herbicide binding on the redox potential of the quinone acceptor in photosystem II: relevance to photodamage and phototoxicity. Biochemistry 37: 17339–17344

Król M, Ivanov AG, Jansson S, Kloppstech K and Huner NPA (1999) Greening under high light and cold temperature affects the level of xanthophyll-cycle pigments, early light-inducible proteins, and light-harvesting polypeptides in wild-type barley and the *Chlorina f2* mutant. Plant Physiol 120: 193–203

Krupa Z, Öquist G and Gustafsson P (1990) Photoinhibition and recovery of photosynthesis in *psbA* gene-inactivated strains of the cyanobacterium *Anacystsis nidulans*. Plant Physiol 93: 1–6

Krupa Z, Öquist G and Gustafsson P (1991) Photoinhibition of photosynthesis and growth responses at different light levels in *psbA* gene mutants of the cyanobacterium *Synechococcus*. Physiol Plant 82: 1–8

Kruse O, Zheleva D and Barber J (1997) Stabilization of photosystem two dimers by phosphorylation: Implication for the regulation of the turnover of D1 protein. FEBS Lett 408: 276–280

Kühlbrandt W, Wang DN and Fujiyoshi Y (1994) Atomic model of plant light-harvesting complex by electron crystallography. Nature 367: 614–621

Kulheim C, Agren J and Jansson S (2002) Rapid regulation of light harvesting and plant fitness in the field. Science 297: 91–93.

Lee H-Y, Hong Y-N and Chow WS (2001) Photoinactivation of photosystem II complexes and photoprotection by non-functional neighbors in *Capsicum annuum* L. leaves. Planta 212: 332–342

Li X-P, Björkman O, Shin C, Grossman A, Rosenquist M, Jansson S and Niyogi KK (2000) A pigment-binding protein essential for regulation of photosynthetic light harvesting. Nature 403: 391–395

Li X-P, Phippard A, Pasari J and Niyogi KK (2002) Structure-function analysis of photosystem II subunit S (PsbS) in vivo. Func Plant Biol 29: 1131–1139

Lindahl M, Funk C, Webster J, Bingsmark S, Adamska I and Andersson B (1997) Expression of Elips and PsbS protein in spinach during acclimative reduction of the photosystem II antenna in response to increased light intensities. Photosynth Res 54: 227–236

Long SP, Humphries S and Falkowski PG (1994) Photoinhibition of photosynthesis in nature. Ann Rev Plant Physiol Plant Mol Biol 45: 633–662

Los DA and Murata N (2002) Sensing and responses to low temperature in cyanobacteria. In: Storey KB and Storey JM (eds) Sensing, Signalling and Cell Adaptation, pp 139–153. Elsevier Science BV, Amsterdam

Ma YZ, Holt NE, Li X-P, Niyogi KK and Fleming GR (2003) Evidence for direct carotenoid involvement in the regulation of photosynthetic light harvesting. Proc Natl Acad Sci USA 100: 4377–4382

Mäenpää P, Miranda T, Tyystjarvi E, Govindjee, Ducruet J-M and Kirilovsky D (1995) A mutation in the D-*de* loop of D1 modifies the stability of the $S_2Q_A^-$ and $S_2Q_B^-$ states in photosystem II. Plant Physiol 107: 187–197

Masamoto K and Furukawa K (1997) Accumulation of zeaxanthin in cells of the cyanobacterium, *Synechococcus* sp strain PCC 7942 under high irradiance. J Plant Physiol 151: 257–261

Matsubara S and Chow WS (2004) Populations of photoinactivated photosystem II reaction centers characterized by chlorophyll a fluroscence life time in vivo. Proc Natl Acad Sci USA 101: 18234–18239.

Mattoo AK, Elich TD, Ghirardi ML, Callahan FE and Edelman M (1993) Post-translational modification of chloroplast proteins and the regulation of protein turnover. In: Battey NH, Dickinson HG and Hetherington AM (eds) Post-translational Modification in Plants, Society for Experimental Biology Seminar Series, Vol 53, pp 65–78. Cambridge University Press, Cambridge UK

Mayes SR, Dubbs JM, Vass I, Nagy L and Barber J (1993) Further characterization of the *psbH* locus of *Synechocystis* sp PCC 6803: inactivation of *psbH* impairs Q_A to Q_B electron transport in photosystem 2. Biochemistry 32: 1454–1465

Melis A (1999) Photosystem-II damage and repair cycle in chloroplasts: what modulates the rate of photodamage *in vivo*? Trends Plant Sci 4: 130–135

Melis A and Homann PH (1976) Heterogeneity of the photochemical centers in system II chloroplasts. Photochem Photobiol 23: 343–350

Meyer G and Kloppstech K (1984) A rapidly light-induced chloroplast protein with high turnover coded for by pea nuclear DNA. Eur J Biochem 138: 201–207

Milligan G, Parenti M and Magee AI (1995) The dynamic role of palmitoylation in signal transduction. Trends Biochem Sci 20: 181–185

Minagawa J, Narusaka Y, Inoue Y and Satoh K (1999) Electron transfer between Q_A and Q_B in photosystem II is thermodynamically perturbed in phototolerant mutants of *Synechocystis sp*. PCC 6803. Biochemistry 38: 770–775

Montane M-H and Kloppstech K (2000) The family of light-harvesting-related proteins (LHCs, ELIPs, HLIPs): was the harvesting of light their primary function? Gene 258: 1–8

Nield J, Orlova E, Morris E, Cowen B, Van Heel M and Barber J (2000a) 3D map of the plant photosystem two supercomplex obtained by cryoelectron microscopy and single particle analysis. Nature Struct Biol 7: 44–47

Nield J, Funk C and Barber J (2000b) Supramolecular structure of photosystem II and location of the PsbS protein. Phil Trans Royal Soc B 355: 1337–1343

Nishida I and Murata N (1996) Chilling sensitivity in plants and cyanobacteria: the crucial contribution of membrane lipids. Ann Rev Plant Physiol Plant Mol Biol 47: 541–568

Nixon PJ, Komenda J, Barber J, Deak ZS, Vass I and Diner BA (1995) Deletion of the PEST-like region of photosystem two modifies the QB-binding pocket but does not prevent rapid turnover of D1. J Biol Chem 270: 14919–14927

Niyogi KK (1999) Photoprotection revisited: genetic and molecular approach. Annu Rev Plant Physiol Plant Mol Biol 50: 333–359

Niyogi KK, Li X-P, Rosenberg V and Jung H-S (2005) Is PsbS the site of non-photochemical quenching in photosynthesis? J Exp Bot 56: 375–382.

Norén H, Svensson P, Stegmark R, Funk C, Adamska I and Andersson B (2003) Expression of the early light-induced protein but not the PsbS protein is influenced by low temperature and depends on the developmental stage of the plant in field-grown cultivars. Plant Cell Environ 26: 245–253

Ohad I and Hirschberg J (1992) Mutations in the D1 subunit of photosystem II distinguish between quinone and herbicide binding sites. Plant Cell 4: 273–282

Ohad I, Koike H, Shochat S and Inoue Y (1988) Changes in the properties of reaction center II during the initial stages of photoinhibition as revealed by thermoluminescence measurements. Biochim Biophys Acta 933: 288–298

Öquist G and Huner N PA (2003) Photosynthesis of overwintering evergreen plants. Annu Rev Plant Biol 54: 329–355

Öquist G and Martin B (1980) Inhibition of photosynthetic electron transport and formation of inactive chlorophyll in winter-stressed *Pinus sylvestris*. Physiol Plant 48: 33–38

Öquist G, Chow WS and Anderson JM (1992) Photoinhibition of photosynthesis represents a mechanism for the long-term regulation of photosynthesis. Planta 186: 450–460

Ort DR (2001) When there is too much light. Plant Physiol 125: 29–32

Peterson RB and Havir EA (2001) Photosynthetic properties of an *Arabidopsis thaliana* mutant possessing a defective PsbS gene. Planta 214: 142–152

Peterson RB and Havir EA (2003) Contrasting modes of regulation of PSII light utilization with changing irradiance in normal and PsbS mutant leaves of *Arabidopsis thaliana*. Photosynth Res 75: 57–70

Polivka T, Herek JL, Zigmantis D and Åkerlund H-E (1999) Direct observation of the (forbidden) S1 state in carotenoids. Proc Natl Acad Sci USA 96: 4914–4917

Polivka Zigmantis D and Sundström V (2002) Carotenoid S1 state in a recombinant light-harvesting complex of photosystem II. Biochemistry 41: 439–450

Powles SB (1984) Photoinhibition of photosynthesis induced by visible light. Annu Rev Plant Physiol 35: 15–44

Prasil O, Kolber Z, Berry JA and Falkowski PG (1996) Cyclic electron flow around photosystem II *in vivo*. Photosynth Res 48: 395–410

Rintamäki E and Aro E M (2001) Phosphorylation of photosystem II proteins. In: Aro E-M and Andersson B (eds), Regulation of Photosynthesis, Advances in Photosynthesis and Respiration, Vol 11, pp 395–418. Kluwer Academic Publishers, Dordrecht

Ruban AV, Pascal AA, Robert B and Horton P (2002) Activation of zeaxanthin is an obligatory event in the regulation of photosynthetic light harvesting. J Biol Chem 277: 7785–7789

Sakurai I, Hagio M, Gombos Z, Tyystjarvi T, Paakkarinen V, Aro E-M and Wada H (2003) Requirement of phosphatidylglycerol for maintenance of photosynthetic machinery. Plant Physiol 133: 1376–1384

Sane PV and Rutherford AW (1986) Thermoluminescence from Photosynthetic Membranes. In: Govindjee, Amesz J and Fork DC (eds) Light Emission by Plants and Bacteria, pp 329–360. Academic Press, Orlando

Sane PV, Ivanov AG, Sveshnikov D, Huner NPA and Öquist G (2002) A transient exchange of the photosystem II reaction center protein D1:1 with D1:2 during low temperature stress of *Synechococcus* sp. PCC 7942 in the light lowers the redox potential of Q_B. J Biol Chem 277: 32739–32745

Sane PV, Ivanov AG, Hurry VM, Huner NPA and Öquist G (2003) Changes in the redox potential of Q_B confer increased resistance against photoinhibition in low temperature acclimated *Arabidopsis thaliana*. Plant Physiol 132: 2144–2151

Savitch LV, Massacci A, Gray GR and Huner NPA (2000) Acclimation to low temperature or high light mitigates sensitivity to photoinhibition: roles of the Calvin cycle and the Mehler reaction. Aust J Plant Physiol 27: 253–264

Savitch LV, Barker-Aström J, Ivanov AG, Hurry VM, Öquist G and Gardeström P (2001) Cold acclimation of *Arabidopsis thaliana* results in complete recovery of photosynthetic capacity and is associated with reduction of the chloroplast stroma. Planta 214: 295–303

Schaefer MR and Golden S (1989) Differential expression of members of a cyanobacterial *psbA* gene family in response to light. J Bacteriol 17: 3973–3981

Schatz GH, Brock H and Holzwarth AR (1988) Kinetic and energetic model for the primary processes in photosystem II. Biophys J 54: 397–405

Selstam E (1998) Development of thylakoid membranes with respect to lipids. In: Siegenthaler P-A and Murata N (eds) Lipids in Photosynthesis: Structure, Function and Genetics, Advances in Photosynthesis and Respiration, Vol 6, pp 209–224. Kluwer Academic Publishers, Dordrecht

Somersalo S and Krause GH (1990) Reversible photoinhibition of unhardened and cold-acclimated spinach leaves at chilling temperatures. Planta 180: 181–187

Stewart DH and Brudvig GW (1998) Cytochrome b559 of photosystem II. Biochim Biophys Acta 1367: 63–87

Stitt M and Hurry VM (2002) A plant for all seasons: alterations in photosynthetic carbon metabolism during cold acclimation in *Arabidopsis*. Curr Opinion Plant Biol 5: 199–206

Stowell MHB, McPhillips TM, Rees DC, Soltis SM, Abresch E and Feher G (1997) Light-induced structural changes in photosynthetic reaction center: implications for mechanism of electron-proton transfer. Science 276: 812–816

Strand Å, Hurry VM, Gustafsson P and Gardeström P (1997) Development of *Arabidopsis thaliana* leaves at low temperature releases the suppression of photosynthesis and photosynthetic gene expression despite the accumulation of soluble carbohydrates. Plant J 12: 605–614

Strand Å, Foyer CH, Gustafsson P, Gardeström P and Hurry VM (2003) Altering flux through the sucrose biosynthesis pathway in transgenic *Arabidopsis thaliana* modifies photosynthetic acclimation at low temperatures and the development of freezing tolerance. Plant Cell Environ 26: 523–535

Strand M and Öquist G (1985) Inhibition of photosynthesis by freezing temperatures and high light levels in cold-acclimated seedlings of Scots pine (*Pinus silvestris*). I. Effects of the light-limited and light saturated rates of CO_2 assimilation. Physiol Plant 64: 425–430

Swiatek M, Kuras R, Sokolenko A, Higgs D, Olive J, Cinque G, Müller B, Eichacker LA, Stern DB, Bassi R, Herrmann RG and Wollman F-A (2001) The chloroplast gene *ycf9* encodes a photosystem II (PSII) core subunit, PsbZ, that participates in PSII supramolecular architecture. Plant Cell 13: 1347–1367

Telfer A, De Las Rivas J and Barber J (1991) Beta-carotene within the isolated photosystem-II reaction center – photooxidation and irreversible bleaching of this chromophore by oxidized P680. Biochim Biophys Acta 1060: 106–114

Trémolières A and Siegenthaler P-A (1998) Reconstitution of photosynthetic structures and activities with lipids. In: Siegenthaler P-A (ed) Lipids in Photosynthesis: Structure, Function and Genetics, Advances in Photosynthesis and Respiration, Vol 6, pp 175–189. Kluwer Academic Publishers, Dordrecht

van Wijk KJ and van Hasselt PR (1993) Kinetic resolution of different recovery phases of photoinhibited photosystem II in cold-acclimated and non-acclimated spinach leaves. Physiol Plant 87: 187–198

Vass I and Govindjee (1996) Thermoluminescence from the photosynthetic apparatus. Photosynth Res 48: 117–126

Vass I, Styring S, Hundal T, Koivuniemi A, Aro E-M and Andersson B (1992) Reversible and irreversible intermediates during photoinhibition of photosystem II: Stable reduced Q_A species promote chlorophyll triplet formation. Proc Natl Acad Sci USA 89: 1408–1412

Vavilin DV and Vermass WFJ (2000) Mutations in the CD-loop region of the D2 protein in *Synechocystis* sp. PCC 6803 modify charge recombination pathways in photosystem II *in vivo*. Biochemistry 39: 14831–14838

Vijayan P, Routaboul JM and Browse J (1998) A genetic approach to investigating membrane lipid structure and photosynthetic function. In: Siegenthaler P-A (ed) Lipids in Photosynthesis: Structure, Function and Genetics, Advances in Photosynthesis and Respiration, Vol 6, pp 263–285. Kluwer Academic Publishers, Dordrecht

Walters RG and Horton P (1993) Theoretical assessment of alternative mechanisms for non-photochemical quenching of PSII fluorescence in barley leaves. Photosynth Res 36: 119–139

Weis E and Berry JA (1987) Quantum efficiency of photosystem II in relation to energy-dependent quenching of chlorophyll fluorescence. Biochim Biophys Acta 894: 198–208

Wentworth M, Ruban AV and Horton P (2003) Thermodynamic investigation into the mechanism of the chlorophyll fluorescence quenching in isolated photosystem II light harvesting complexes. J Biol Chem 278: 21845–21850

Chapter 12

Photoinhibition and Recovery in Oxygenic Photosynthesis: Mechanism of a Photosystem II Damage and Repair Cycle

Kittisak Yokthongwattana
Department of Biochemistry, Faculty of Science, Mahidol University, Rama 6 Road, Bangkok 10400 Thailand

Anastasios Melis*
Department of Plant and Microbial Biology, 111 Koshland Hall, University of California, Berkeley, CA 94720-3102, USA

Summary	175
I. Introduction	176
II. Photosystem II (PS II) Organization	176
III. PS II Heterogeneity	178
IV. PS II Damage and Repair Cycle in Chloroplasts	179
A. Excitation Energy Pressure at PS II Defines the Rate Constant of Photodamage	180
B. PS II Chlorophyll Antenna Size Modulates Rate of Photodamage	181
C. Electron Transport and Photosynthesis Mitigate Against Photodamage	182
D. Protein Phosphorylation and Disassembly of the PS II Holocomplex	182
E. Involvement of a Chloroplast-Localized Hsp70 in the PS II Repair Process	183
F. Role of Zeaxanthin and of the Cbr Protein in the PS II Damage and Repair Process	183
G. Sulfur-Deprivation Arrests the PS II Repair Process	184
H. A Novel Nuclear-Encoded and Chloroplast-targeted Sulfate Permease Regulates the PS II Repair Process in *Chlamydomonas reinhardtii*	184
V. DNA Insertional Mutagenesis for the Isolation and Functional Characterization of PS II Repair Aberrant Mutants	185
VI. Conclusions	186
Acknowledgments	187
References	187

Summary

This Chapter provides highlights on the mechanism of a photosystem II (PS II) damage and repair cycle in chloroplasts. Photo-oxidative damage to the PS II reaction center is a phenomenon that occurs in every organism of oxygenic photosynthesis. Through the process of evolution, an elaborate repair mechanism was devised, one that rectifies this presumably unavoidable and irreversible photoinhibition and restores the PS II charge separation activity. The repair process entails several enzymatic reactions for the selective removal and replacement of the inactivated D1/32 kD reaction center protein (the chloroplast-encoded *psbA* gene product) from the massive (>1,000 kD) H_2O-oxidizing and O_2-evolving PS II holocomplex. This repair process is unique in the annals of biology; nothing

---

*Author for correspondence, email: melis@nature.berkeley.edu

analogous in complexity and specificity has been reported in other biological systems. Elucidation of the repair mechanism may reveal the occurrence of hitherto unknown regulatory and catalytic reactions for the selective in situ replacement of specific proteins from within multi-protein complexes. This may not only have significant applications in photosynthesis and agriculture but also in medicine and other fields.

I. Introduction

Life on earth is sustained by oxygenic photosynthesis, a process that begins with the utilization of sunlight for the oxidation of water molecules. The chemical energy stored in this endergonic oxidation is processed through the electron-transport chain of the chloroplast thylakoids and is eventually delivered in the form of reductant (reduced ferredoxin) and high-energy phosphate bond (ATP). The absorption of light and the conversion of excitation energy to chemical energy take place in photosystem II (PS II) and photosystem I (PS I) in the thylakoid membrane (Hill and Bendall, 1960; Duysens et al., 1961). Light energy in PS II specifically facilitates the generation of a strong oxidant capable of oxidizing water molecules. The ability of PS II to utilize water molecules from which to extract electrons and protons was undoubtedly a significant event in the evolution of life on earth. It contributed to the gradual accumulation of oxygen in the atmosphere, thereby permitting the evolution of oxidative phosphorylation. Rightfully so, many scientists refer to PS II as '*the engine of life on earth*'.

From the biochemical point of view, PS II is a specialized H_2O-to-plastoquinone oxidoreductase. This enzyme features a rather sizable holocomplex consisting of more than 25 transmembrane and peripheral proteins. Most of the transmembrane proteins function as chlorophyll-protein light-harvesting complexes. The functional center of the holocomplex is the so-called D1/D2 heterodimer reaction center proteins that perform the light utilization, water oxidation, and primary electron transport reactions in PS II. These highly specialized functions of PS II take place in a protected and isolated microenvironment where oxygen abounds and where photons, in the form of excitation energy, are received by the photochemical reaction center at a rate of up to 5,000 per second. The transient formation of strong oxidants, the abundance of oxygen, and the presence of excitation energy are conditions that may lead to photo-oxidative damage (Barber, 1994). Indeed, such photodamage occurs frequently within the reaction center of PS II. It causes an irreversible inactivation in the PS II electron transport and stops photosynthesis (Powles, 1984).

Through the process of 2–3 billion years of evolution, organisms of oxygenic photosynthesis have not been able to either prevent or avoid this photo-oxidative adverse effect from occurring (Payton et al., 1998). Thus, to date, every oxygen-evolving photosynthetic organism known, from cyanobacteria to C4 plants, is subject to this irreversible photodamage. Nature, however, devised a repair mechanism that restores the functional status of PS II. The PS II damage and repair cycle, as the phenomenon has come to be known (Guenther and Melis, 1990), is of great importance for the maintenance and productivity of photosynthesis. In repair-aberrant mutants, oxygenic photosynthesis cannot be sustained (Zhang et al., 1997). Clearly, life on earth would have been quite different in the absence of the PS II repair process.

The objective of this article is to present highlights on the biochemical and molecular basis of the PS II damage and repair cycle. It examines specific reaction steps of the PS II repair process in chloroplasts and known mechanisms for the regulation of this phenomenon at the molecular and membrane levels. A more in-depth analysis of the PS II structure, function, photodamage, and repair is given below.

II. Photosystem II (PS II) Organization

All PS II electron transport intermediates are contained within the so-called D1/D2 32/34 kD heterodimer protein, coded for by chloroplast genes *psbA* and *psbD*, respectively (Barber et al., 1987; Nanba and Satoh, 1987; Seibert et al., 1988; Deisenhofer and Michel, 1989). Structural information on the architecture of the D1/D2 heterodimer complex at 3.8 Å resolution (Zouni et al., 2001) has shown how each protein binds one of the photochemical reaction center chlorophylls that form P680, one pheophytin, and one quinone binding site. D1 contains the plastoquinone (Q_B) binding site,

---

Abbreviations: Chl – chlorophyll; HL – High light; LHC – light-harvesting complex; LL – Low light; PS II – photosystem II; PS II-RC – PS II reaction center; D1 – the 32 kD PS II-RC protein; D2 – the 34 kD PSII-RC protein.

Chapter 12 Photosystem II Damage and Repair Cycle

which is also the site for herbicide-binding, whereas D2 contains a tightly bound plastoquinone molecule (Q_A) that can be reduced to the plastosemiquinone anion form. Each protein also contains an electron transport intermediate tyrosine residue (denoted as Y_Z and Y_D, respectively). Moreover, the X-ray crystal structure (Zouni et al., 2001; Ferreira et al., 2004) has shown the placement of the 4 Mn H_2O-oxidizing cluster (Cinco et al., 1999) close to Y_Z on the D1 side of the D1/D2 reaction center heterodimer. The PS II reaction center heterodimer complex (PS II-RC) may be noted for a structural symmetry (D1 and D2 are homologous proteins) and a functional asymmetry since electron transport is thought to proceed from Mn, Y_Z, Chl, pheophytin (in D1) to Q_A (in D2) to Q_B (in D1) (Fig. 1).

Earlier research revealed a highly unusual property for the D1/32 kD protein. The D1 accounts for less than 1% of the total thylakoid membrane protein, yet the rate of its synthesis was found to be comparable to or exceeding that of the abundant large subunit of Rubisco in the chloroplast (Bottomley et al., 1974; Eaglesham and Ellis, 1974; Edelman and Reisfeld, 1978). Since steady-state levels of D1 in thylakoids were low, it was inferred that rates of turnover must be high (Mattoo et al., 1984; Greenberg et al., 1987). The frequent light-dependent turnover of the D1/32 kD polypeptide was shown to be related to the phenomenon of "photoinhibition" in chloroplasts (Kyle et al., 1984; Ohad et al., 1984). Many of the mechanistic details for the selective degradation and specific replacement of the D1/32 kD PS II-RC protein are not yet known.

The photochemical apparatus of PS II contains, in addition to the reaction center proteins (PS II-RC), two "core antenna" chlorophyll-proteins CP47 and CP43 (the chloroplast *psb*B and *psb*C gene products, respectively) (Green and Camm, 1981; Bricker, 1990; Barber et al., 2000). In the structural organization of PS II (Rhee et al., 1998), CP43 is closer to D1 whereas CP47 is closer to D2 (Zouni et al., 2001; Ferreira et al., 2004). Collectively, PS II-RC, CP47, and CP43 define the PS II-core and they contain about 37 Chl *a* molecules (Glick and Melis, 1988). Light-harvesting by PS II is further aided by the auxiliary chlorophyll *a* − *b* light-harvesting complex (LHC II) (Thornber, 1986; Green, 1988; Bassi et al., 1990; Nield et al., 2000).

Present in PS II are a number of proteins that do not bind chlorophyll. Closely associated with the PS II-RC are two polypeptides (9 kD and 4 kD coded for by chloroplast genes *psb*E and *psb*F) that bind, via two

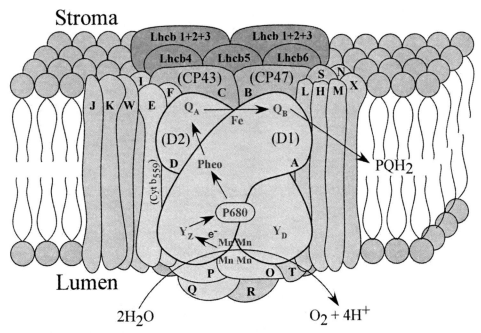

Fig. 1. A PS II organizational model illustrating the arrangement of the various proteins in the PS II complex and showing the pathway of electron transport within the D1/D2 32/34 kD reaction center proteins. Individual subunits are labeled by a single letter of the alphabet according to the gene product PsbA-PsbZ and Lhcb1-Lhcb6. The missing subunits (PsbG, U, V, Y, and Z) are proteins that either do not exist in higher plant PS II or their localization is presently unknown. The direction of electron transport within the reaction center is indicated by arrows.

histidine ligands, a b-type heme (thus forming a cytochrome b-559 molecule) (Cramer and Whitmarsh, 1977; Lam et al., 1983; Murata et al., 1984; Cramer et al., 1986; Miyazaki et al., 1989; Nedbal et al., 1992; Zouni et al., 2001; Ferreira et al., 2004). There are three extrinsic membrane proteins in the thylakoid lumen (33, 23, and 17 kD polypeptides are nuclear *psbO, psbP*, and *psbQ* gene products (Hallick, 1989)) believed to play a role in Mn stability and in the regulation of Ca^{++} and Cl^- concentrations near the water oxidizing complex (Ghanotakis and Yocum, 1990). A 10 kD phosphoprotein of unknown function (the chloroplast *psb*H gene product) is also a component of PS II (Farchaus and Dilley, 1986; Millner et al., 1986). The phosphorylation/dephosphorylation of this integral thylakoid membrane protein is believed to be subject to the redox control of the plastoquinone pool. There are additional polypeptides of unknown function (the *psbI-psbY* gene products) that are associated with PS II (Barber, 1989; Hallick, 1989; Green and Kühlbrandt, 1995; Lorkovic et al., 1995; Meetam et al., 1999; Shi et al., 1999; Ohnishi and Takahashi, 2001).

In summary, the PS II holocomplex is a massive (>1,000 kD) specialized H_2O-to-plastoquinone oxidoreductase, composed of more than 25 thylakoid membrane integral and peripheral proteins. It performs a unique function in nature, as it utilizes the energy of the sun to generate an intermediate reductant (reduced pheophytin), capable of reducing plastoquinone, and a strong oxidant ($P680^+$) capable of extracting electrons and releasing protons and oxygen from energy-poor but abundant H_2O molecules. This function of PS II sustains virtually all life on earth.

III. PS II Heterogeneity

In all higher plants and green algae examined to date, PS II shows heterogeneity in localization, structure, and function (Melis and Homann, 1976; Melis and Duysens, 1979; Black et al., 1986; Lavergne and Briantais, 1996). Approximately 70–80% of PS II units are localized in the tightly appressed membranes of the grana partition regions, they contain the full complement of the LHC II proteins with a total of 250 ± 40 Chl $a + b$ molecules in higher plants [up to 350 Chl $a + b$ molecules in green algae (Melis, 1996)], and are known as PS IIα (Melis and Homann, 1976; Melis and Anderson, 1983). The remaining PS II units (PS IIβ, about 20–30% of the total) are localized in stroma-exposed thylakoids (Anderson and Melis, 1983; Vallon et al., 1986, 1987), and contain a much smaller light-harvesting antenna with an estimated 130 ± 20 Chl $a + b$ molecules. In physical terms, this segregation is consistent with specific proteins localized in each thylakoid membrane domain (Wettern, 1986; Callahan et al., 1989).

In addition to the domain localization and antenna heterogeneity, PS IIα and PS IIβ centers have different properties with respect to electron transport from the reaction center to plastoquinone [$Q_A \rightarrow Q_B$ step (Crofts and Wraight, 1983; Van Gorkom, 1985)]. Early evidence in the literature suggested that PS IIβ centers, although photochemically competent, lacked the so-called two-electron gating mechanism operating between the primary quinone acceptor Q_A and the bound plastoquinone molecule Q_B (Thielen and Van Gorkom, 1981; Lavergne, 1982). Further evidence in the literature suggested that, in mature wild-type higher plants and green algae, PS IIβ in stroma-exposed thylakoids behave like DCMU-poisoned centers and cannot transfer electrons from Q_A to Q_B at physiologically significant rates (Melis, 1985; Graan and Ort, 1986; Chylla and Whitmarsh, 1989). In functional terms, PS IIβ centers are Q_B-nonreducing in contrast to PS IIα centers that are Q_B-reducing (Lavergne, 1982; Lavergne and Briantais, 1996).

The different LHC II antenna composition, localization in separate thylakoid regions, and the dissimilar properties of electron transport on the reducing side of PS IIα and PS IIβ are probably interrelated aspects of the PS II heterogeneity phenomenon in chloroplasts. Evidence suggested that these properties arise as a consequence of the operation of a PS II damage and repair cycle in chloroplasts (Guenther and Melis, 1990; Adir et al., 1990; Lavergne and Briantais, 1996). It is postulated that the PS II heterogeneity and separate thylakoid membrane localization of the two different types of PS II is a consequence of the spatial separation needed between the domain of functional centers (PS IIα) in the tightly appressed grana, and the stroma-exposed thylakoid membrane domain where the repair takes place. This arrangement provides for optimal functioning of the various processes associated with the repair, for instance, degradation of photodamaged D1, thylakoid membrane-association of ribosomes, biosynthesis and insertion of the nascent D1 in reaction center under repair, integration, and processing of the precursor of protein and re-assembly of PS II holocomplex prior to its translocation to the appressed grana thylakoid domain where functional PS II is localized (Mattoo and Edelman, 1987; Melis, 1991; Aro et al., 1993; see also below). In this respect, PS IIβ centers constitute a small reservoir of newly repaired centers that remain in stroma-exposed thylakoids (Anderson and Melis, 1983; Neale and Melis, 1990, 1991).

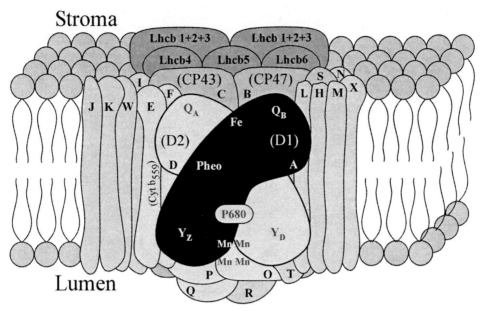

Fig. 2. The PS II complex with a photodamaged and, hence, inactive D1 reaction center protein. This irreversible configuration of PS II is unable to perform a stable charge separation and electron transport. For other details, see Fig. 1.

IV. PS II Damage and Repair Cycle in Chloroplasts

Photo-oxidative damage occurs frequently within the reaction center of PS II in a light-intensity-dependent manner (Barber, 1994; Melis, 1999). It causes an irreversible inactivation in the photochemical charge separation and electron transport within PS II (Cleland et al., 1986; Demeter et al., 1987) and irreversibly stops the function of the D1 reaction center protein (Fig. 2). This photodamage is followed by a partial disassembly of the PS II holocomplex, exposure of the photodamaged PS II-core to the chloroplast stroma, degradation of photodamaged D1, de novo D1 biosynthesis and insertion in the thylakoid membrane, re-assembly of the PS II reaction center complex, followed by activation of the electron-transport process through the reconstituted D1/D2 heterodimer. A temporal and spatial sequence of events concerning the PS II damage and repair cycle has been presented in the literature (Melis, 1991; Aro et al., 1993; Melis, 1998). A number of known biochemical reactions are associated with the mechanism of the PS II damage and repair cycle (Figs. 3 and 4):

a. A reversible phosphorylation of the D1 protein reportedly regulates the rate of inactive D1 degradation in higher plant chloroplasts (Callahan et al., 1990; Aro et al., 1992; Elich et al., 1992; Ebbert and Godde, 1996). However, phosphorylation of D1 is not observed in green algae and lack of D1 phosphorylation does not exert any adverse effect on the D1 turnover in these organisms (A.K. Mattoo, personal communication).

b. Following photodamage, there is a prompt disassembly of the PS II holocomplex and exclusion of a smaller PS II-core complex from the membrane of the grana partition regions (Adir et al., 1990; Guenther and Melis, 1990). This step serves to unfold the PS II holocomplex and to expose the D1/D2 heterodimer to the aqueous stroma-phase where removal of the photodamaged D1 and replacement by a de novo synthesized D1 can occur.

c. Under photoinhibition conditions, i.e., when the rate of photodamage exceeds that of the enzymatic repair (Greer et al., 1986), photodamaged PS II reaction centers accumulate in the chloroplast thylakoids. In the green alga *Dunaliella salina*, photodamaged reaction centers have been identified and isolated as distinct 160 kD complexes (Kim et al., 1993; Melis and Nemson, 1995; Baroli and Melis, 1996). The 160 kD complex appears to be a cross-linked derivative of D1, D2, and of repair-related proteins of unknown origin.

d. A chloroplast DegP2 protease performs the primary cleavage of the photodamaged D1 protein (Haußühl et al., 2001). The thylakoid FtsH protease plays a role in the further degradation of

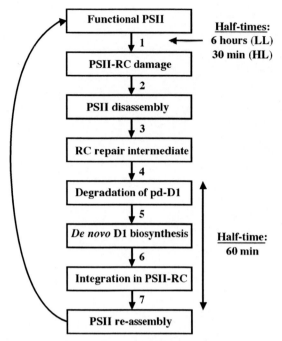

Fig. 3. Temporal sequence of events in the PS II damage, disassembly, D1 degradation, and replacement. The rate of photodamage to PS II (step 1) is directly proportional to the incident light intensity (Baroli and Melis 1996; Tyystjärvi and Aro, 1996). The rate of PS II disassembly (step 2) is not limiting under a broad range of incident intensities. Direct D1 degradation (step 3) and de novo D1 biosynthesis (step 4) become rate limiting under high-light intensity conditions. Under these conditions, PS II-RC intermediates in the form of 160 kD complexes accumulate in thylakoids (step 3). The rate of D1 degradation (slow) is the rate limiting step of the PS II repair process under high light intensities. It is estimated to occur with a half-time of 60 ± 15 min.

the cleaved D1 (Lindahl et al., 2000), thus contributing to the removal of the inactive D1 protein (Adam and Clark, 2002; Silva et al., 2002).

e. De novo biosynthesis (Kim et al., 1994) and a reversible D1 palmitoylation ostensibly facilitates insertion of the nascent D1 polypeptide in the PS II reaction center complex (Mattoo and Edelman, 1987).

f. Activation of electron transport through the Q_A–Q_B electron-gate converts a Q_B-nonreducing center to a Q_B-reducing form (Guenther et al., 1990; Neale and Melis, 1990, 1991).

g. Re-assembly of the PS II holocomplex and incorporation in the grana is the last step of the cycle.

This temporal (Fig. 3) and spatial (Fig. 4) sequence of events is consistent with the known properties of D1 turnover (Mattoo et al., 1984; Mattoo and Edelman, 1987) and also consistent with the heterogeneity in PS II configuration and function (Melis, 1991; Aro et al., 1993; Lavergne and Briantais, 1996). This knowledge, however, does not include information about many of the genes and proteins that are involved in the PS II repair process. Identification of the PS II repair genes and enzymes, and elucidation of the mechanistic and regulatory aspects of the PS II repair process is the goal of current research in the field.

A. Excitation Energy Pressure at PS II Defines the Rate Constant of Photodamage

The rate constant for photodamage in vivo was shown to be a linear function of incident irradiance, both in higher plants (Sundby et al., 1993; Tyystjärvi and Aro, 1996) and in green algae (Baroli and Melis, 1996). In green algae, linearity in the rate constant of photodamage was also observed as a function of 'growth irradiance' (meaning the higher the growth irradiance, the greater the rate constant of photodamage). At first, these observations were thought to indicate that PS II is a "photon counter" and that PS II photodamage will unavoidably occur after a fixed number of photons have been absorbed by the reaction center. However, this notion is no longer being entertained as the rate constant of photodamage was also shown to be modulated by physiological and metabolic parameters in the chloroplast.

A rigorous study on the modulation of PS II photodamage was completed and a unifying model for the in vivo modulation of photodamage was presented (Melis, 1999). The model postulates that the rate constant of photodamage depends primarily on the redox state of the primary quinone acceptor Q_A of PS II and on the rate of exciton arrival at the PS II reaction center (P680). It appears that the probability of photodamage is quite different in the two different redox states of PS II (the primary quinone acceptor Q_A being in the oxidized or reduced form during steady-state illumination). There is a low inherent probability for photodamage when Q_A is oxidized and excitation energy is dissipated by useful photochemistry, and a significantly higher probability for photodamage when Q_A is reduced and absorbed excitation is dissipated non-photochemically (Baroli and Melis, 1998; Melis, 1999). The two parameters, redox state of Q_A and rate of exciton arrival at the PS II reaction center, define the concept of PS II excitation pressure (Maxwell et al., 1995; Huner et al., 1996), which is the common denominator of many in vivo conditions that modulate the rate constant of PS II photodamage. This evidence was derived upon careful in vivo analysis

Chapter 12 Photosystem II Damage and Repair Cycle

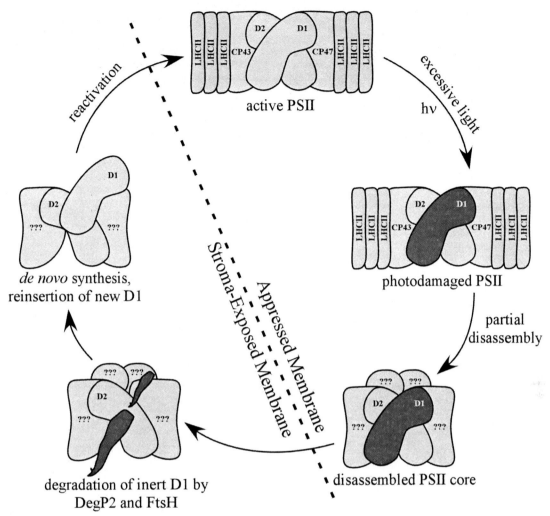

Fig. 4. Spatial sequence of events in the PS II photodamage, disassembly, D1 degradation, and replacement by a de novo synthesized protein. Disassembly of the PS II holocoplex is postulated to occur in the grana (following protein phosphorylation). The major portion of the LHCII remains in the grana, whereas the PS II core with the photochemically inert D1 is moved toward stroma-exposed thylakoid membranes where the repair is to take place. A selective degradation of the inert D1 by the chloroplast DegP2 and FtsH proteases then takes place. A de novo synthesized functional copy of the D1 protein replaces the inert D1 upon insertion into the PS II template, followed by a functional reactivation of the complex and insertion into the domain of the grana.

of the dependence of photodamage on light intensity, PS II chlorophyll antenna size, carbon dioxide availability, rate of photosynthetic electron transport, and effect of suboptimal temperature. Results from several related publications were summarized in a review article (Melis, 1999).

B. PS II Chlorophyll Antenna Size Modulates Rate of Photodamage

Early work with isolated thylakoid membranes indicated that the rate of photodamage was dependent on PS II absorption cross-section (Cleland et al., 1986; Cleland and Melis, 1987; Mäenpää et al., 1987). It has also been reported that photoinhibition could be totally independent of the PS II light-harvesting Chl antenna size (Tyystjärvi et al., 1991, 1994). More recent work has supported the notion that the rate of photodamage in vivo is modulated by the PS II Chl antenna size (Park et al., 1997; Baroli and Melis, 1998; Baroli et al., 2003). Accordingly, a PS II light-harvesting antenna size with fewer Chl molecules results in a lower rate constant of photodamage. Conversely, a large PS II Chl antenna size accentuates the rate constant of photodamage.

C. Electron Transport and Photosynthesis Mitigate Against Photodamage

Earlier studies suggested that a limitation in the rate of electron flow, e.g. caused by low CO_2 partial pressures, might accentuate photoinhibition in cyanobacteria (Kaplan, 1981) and higher plants (Demmig and Björkman, 1987; Gong et al., 1993). Furthermore, Park et al. (1996) reported that electron transport via the Mehler reaction to oxygen can protect against photoinhibition in pea leaves. However, antisense transgenic plants with a substantially lower cytochrome $b - f$ complex content, in which illumination produced slow rates of linear electron transport and in which Q_A accumulated in the reduced state, did not show the expected increase in their susceptibility to photoinhibition (Hurry et al., 1996). One possible reason for the confusion generated from the apparently contradictory results is that often *photoinhibition* is measured rather than *the rate of photodamage*. As discussed below, photoinhibition is a function of both photodamage and repair (Greer et al., 1986) and, therefore, measurements of photoinhibition are always more difficult to interpret. A more recent rigorous study was undertaken with *Dunaliella salina*, which was grown under high irradiance either with a limiting supply of inorganic carbon, provided by an initial concentration of 25 mM $NaHCO_3$ to the medium ($P_{max} = \sim 100$ pmol O_2 $(10^6$ cells$)^{-1}$ s^{-1}) or with supplemental CO_2 bubbled in the form of 3% CO_2 in air ($P_{max} = \sim 250$ pmol O_2 $(10^6$ cells$)^{-1}$ s^{-1}). There was a difference by a factor of ~ 2 in the rate constant of photodamage under these conditions, supporting the notion that electron transport and photosynthesis mitigate against this adverse effect (Baroli and Melis, 1998).

D. Protein Phosphorylation and Disassembly of the PS II Holocomplex

Photosystem II reaction centers occur as dimer complexes, which collectively contain 50, or more, transmembrane and peripheral proteins (Hankamer et al., 1997; Zouni et al., 2001; Ferreira et al., 2004). The repair process, which entails the selective removal and replacement of the affected D1 protein, requires a partial disassembly of the PS II holocomplex prior to D1 degradation. A prompt disassembly of the PS II holocomplex has been observed upon photodamage (Aro et al., 1993), however, the driving force for this disassembly is not well understood. It is possible that reversible PS II protein phosphorylation is responsible for the disassembly of photodamaged PS II holocomplexes.

In green plants and algae, several major PS II proteins including D1, D2, PsbH, CP43, and subunits of the LHC-II become reversibly phosphorylated upon exposure to strong illumination (Michel et al., 1988; Bennett, 1991). Although phosphorylation of the LHC-II is reported to serve in balancing the distribution of excitation energy between PS II and PS I (Bennett, 1979; Allen and Nilsson, 1997; Allen, 2003), the possibility cannot be excluded that such phosphorylation simply contributes to a negative charge density increase in the stroma-exposed regions of the NH_2-termini of these PS II proteins. Such a negative charge field on the PS II proteins might increase repulsive forces between them, leading to their electrostatic separation (unfolding of the PS II holocomplex).

Reversible phosphorylation of PS II proteins has been linked directly to the regulation of D1 protein turnover (Aro et al., 1992; Elich et al., 1992; Koivuniemi et al., 1995; Kruse et al., 1997). Under ambient physiological conditions, phosphorylation of the D1 reaction center protein appears to be controlled by an endogenous circadian rhythm (Booij-James et al., 2002). Phosphorylation of D1 does not alter its sensitivity to photodamage but rather prevents its degradation (Koivuniemi et al., 1995; Kruse et al., 1997). As a result, dephosphorylation of P $\sim$ D1 is required for the D1 protein degradation to occur (Rintamäki et al., 1996). In the presence of NaF, a protein phosphatase inhibitor, photoinhibited PS II was found to disassemble into monomers, while still in the phosphorylated state, and the monomer PS II complex was found to migrate from the appressed thylakoids of the grana to the stroma-exposed thylakoid membranes (Baena-González and Aro, 2002). It was proposed that phosphorylation of the PS II core proteins does not prevent monomerization of the PS II holocomplex but rather functions as a protective mechanism via inhibition of premature degradation of the damaged D1, i.e., before the latter reaches the stroma-exposed thylakoid region. Subsequent dephosphorylation of the PS II proteins in the stroma lamellae allows a coordinated D1 degradation and de novo D1 biosynthesis to take place. Phosphorylation of the D1 protein, however, has not been observed in cyanobacteria or red and green algae (Pursiheimo et al., 1998; A.K. Mattoo, personal communication). This suggests that D1 phosphorylation is not essential for D1 turnover in every photosynthetic organism. It also raises the prospect of other mechanisms employed by these organisms in the regulation of D1 degradation.

In a unicellular green alga, *Dunaliella salina*, photodamaged and disassembled PS II reaction centers have been identified and isolated as distinct 160 kD complexes on SDS-PAGE (Kim et al., 1993; Melis and Nemson, 1995). Kinetics of the 160 kD accumulation and decay matched those of photodamage and PS II repair (Kim et al., 1993; Baroli and Melis, 1996). The 160 kD complex was found to be a cross-linked derivative of D1, D2, CP47, and Hsp70B, the latter being a chloroplast-localized heat-shock protein (Yokthongwattana et al., 2001). Other investigators have also reported cross-linked products between D1 and proximal PS II proteins on SDS-PAGE during photoinhibition (Barbato et al., 1992; Ishikawa et al., 1999; Yamamoto, 2001). It was postulated that cross-linking of these proteins occurs specifically in photoinhibited thylakoids and is an artifact of the solubilization process. Such cross-linking is unlikely to occur in vivo as it is generally accepted that cross-linked proteins are subject to prompt degradation. In this case, other components of the cross-linked complex would also have a high turnover rate similar to that of D1. However, only the D1 protein is selectively degraded and replaced in the course of the PS II damage and repair process (Vasilikiotis and Melis, 1994). A frequent turnover of D2 and of other PS II subunits has not been observed under normal physiological conditions (Jansen et al., 1999).

E. Involvement of a Chloroplast-Localized Hsp70 in the PS II Repair Process

The PS II repair process is induced by irradiance and operates only in the light. As such, no photodamage or repair occurs in the dark (Polle and Melis, 1999). The repair involves a coordinated and light-regulated expression of several genes and their respective proteins. A clear example of this 'induction' was provided by the prompt (less than 1 h) and specific 70-fold increase in *Hsp70B* gene transcripts following a LL → HL transition of a green alga culture (Drzymalla et al., 1996; Schroda et al., 1999; Yokthongwattana et al., 2001).

Evidence suggested that a molecular chaperone, the Hsp70B protein, might play a critical role in the PS II damage and repair process. A full-length cDNA of the *D. salina Hsp70B* gene was cloned and sequenced (GenBank Accession No. AF420430/AJ271605). Expression patterns of the *Hsp70B* gene were investigated upon shifting a *D. salina* culture from low-light to high-light-growth conditions, designed to significantly accelerate the rate of PS II photodamage. Northern blot analyses and nuclear run-on transcription assays revealed a prompt and substantial irradiance-dependent induction of *Hsp70B* gene transcription, followed by a subsequent increase in Hsp70B protein synthesis and accumulation. Mild detergent solubilization of photoinhibited thylakoid membranes, in which photodamaged PS II centers had accumulated, followed by non-denaturing gel electrophoresis revealed formation of a 320 kD native protein complex that contained, in addition to the Hsp70B, the photodamaged but as yet undegraded D1 protein as well as D2 and CP47. Evidence suggested that the 320 kD complex is a transiently forming PS II repair intermediate. Denaturing solubilization of the 320 kD PS II repair intermediate by SDS-urea resulted in cross-linking of its constituent polypeptides, yielding a 160 kD protein complex. It was postulated that the Hsp70B protein plays a pivotal role in the repair process, e.g. in stabilizing the disassembled PS II-core complex and in facilitating the D1 removal and replacement (Fig. 5; see also Yokthongwattana et al., 2001).

Thus, the PS II damage and repair cycle lends itself to studies on the regulation of gene expression. Relevant and valid questions in this area range from signal (photodamage) perception to organelle-nucleus communication and coordination of gene expression for the repair.

F. Role of Zeaxanthin and of the Cbr Protein in the PS II Damage and Repair Process

The Cbr protein is a green alga homologue to the higher plant ELIP proteins (Banet et al., 2000), which are related to light stress (Adamska et al., 1992). In the

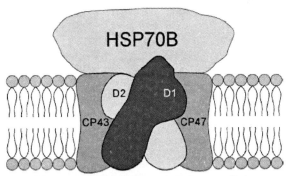

Photosystem II Repair Intermediate

Fig. 5. A schematic model depicting the interaction of the HSP70B with the disassembled PS II-core complex prior to the degradation and replacement of the photodamaged D1 reaction center protein. This association results in the formation of a PS II repair intermediate.

model green alga *Dunaliella salina*, synthesis of Cbr is induced by light stress and occurs in parallel with the accumulation of zeaxanthin (Levy et al., 1993; Andreasson and Melis, 1995; Jin et al., 2001, 2003). Both Cbr and zeaxanthin appear to be reversibly associated with PS II, in which zeaxanthin acts as a quencher of excited Chl* molecules (Frank et al., 2000; Baroli and Niyogi, 2000; Ma et al., 2003), thereby contributing to photoprotection. Recent work suggested a role for Cbr-zeaxanthin in the PS II repair process. This insight came as a result of irradiance-dependent studies, first with a photoinhibition-sensitive mutant of *Dunaliella salina* (denoted as *dcd1*; Jin et al., 2001). Photoinhibition in the *dcd1* was manifested by a lowering of the F_v/F_m ratio, inhibition in Q_A photoreduction, and accumulation of zeaxanthin and of the 320 kD protein complex in the thylakoid membranes, onset of which occurred at a lower threshold of irradiance than in the wild type. In addition to these accepted markers of photoinhibition, de-epoxidation of the xanthophyll cycle carotenoids, accumulation of zeaxanthin and enhanced levels of the Cbr protein were observed. Although the onset of these changes occurred at different levels of irradiance for the wild type and for the *dcd1* mutant, there appeared to be a strict correlation between xanthophyll de-epoxidation, amount of Cbr protein, and amount of photodamaged PS II centers. The notion of a relationship between PS II repair and Cbr-zeaxanthin was further strengthened in kinetic studies. These showed that zeaxanthin and the Cbr protein accumulate in parallel with the accumulation of photodamaged PS II centers following a LL → HL shift, and decay in tandem with a chloroplast recovery from photoinhibition (Jin et al., 2001, 2003).

Experimental evidence for the accumulation of zeaxanthin during photodamage, and possibly due to photodamage, was first presented by Trebst and coworkers (Depka et al., 1998). This body of evidence has been steadily growing in the literature (Smith et al., 1990; Baroli and Melis, 1996; Jahns and Miehe, 1996; Demmig-Adams et al., 1998; Xu et al., 1999; Jahns et al., 2000; Jin et al., 2001, 2003). Diverse observations, which cover both higher plants and green algae, raised the possibility that zeaxanthin and the Cbr protein accumulate not in response to irradiance per se but in proportion to photoinhibition. It was hypothesized that zeaxanthin and the Cbr protein might play a role in the protection of photodamaged and disassembled PS II reaction centers, apparently needed while PS II is in the process of degradation and replacement of the D1/32 kD reaction center protein (Jin et al., 2001, 2003). This notion is consistent with the recovery of pea chloroplasts from photoinhibition in which the kinetics of zeaxanthin epoxidation to violaxanthin resembled those of D1 degradation and replacement (Jahns and Miehe, 1996). The notion is also consistent with a study of an obligate shade species in which the de-epoxidation state of the xanthophyll-cycle carotenoids remained directly proportional to the level of photoinhibition in the leaves and independent of the light-intensity seen by the plant (Demmig-Adams et al., 1998). Further in this direction, of interest is the developing story of the underlying biochemistry in overwintering plant species, in which there appears to be interplay between photoprotection, zeaxanthin accumulation, and status of the PS II damage and repair cycle (Adams et al., 2002, 2004).

G. Sulfur-Deprivation Arrests the PS II Repair Process

In the absence of a sufficient supply of sulfur to the chloroplast, which is an essential component of cysteine and methionine (Hell, 1997), D1 protein biosynthesis is impeded and the repair cycle is arrested in the PS II Q_B-nonreducing configuration (Wykoff et al., 1998). In consequence, the rate of photosynthesis declines quasi-exponentially in the light as a function of time in S-deprivation with a half time of about 18 h (Wykoff et al., 1998; Melis et al., 2000; Cao et al., 2001). This effect is specific to PS II in the thylakoid membrane. Thus, the supply of inorganic sulfur to the chloroplast may determine the rate of D1 turnover and may thus represent a significant regulatory step in the PS II repair process.

H. A Novel Nuclear-Encoded and Chloroplast-targeted Sulfate Permease Regulates the PS II Repair Process in Chlamydomonas reinhardtii

Genomic, proteomic, phylogenetic, and evolutionary aspects of a novel gene encoding a putative chloroplast-targeted sulfate permease of prokaryotic origin in the green alga *Chlamydomonas reinhardtii* were described. This nuclear-encoded sulfate permease gene (*SulP*) contained four introns and five exons, whereas all other known chloroplast sulfate permease genes lack introns and are encoded by the chloroplast genome. The deduced amino acid sequence of the protein showed an extended N-terminus, which includes a putative chloroplast transit peptide. The mature protein contained 7 transmembrane domains and two large hydrophilic loops (Fig. 6). This novel prokaryotic-origin

Chapter 12 Photosystem II Damage and Repair Cycle

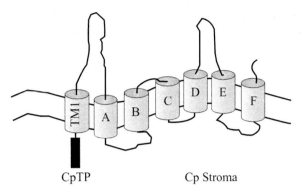

Fig. 6. Folding-model of the nuclear-encoded and chloroplast-targeted *Chlamydomonas reinhardtii* SulP protein. CpTP refers to the chloroplast transit peptide prior to cleavage by a stroma-localized peptidase. TM1 represents the first N-terminal transmembrane domain of the SulP protein, which is exclusive to *C. reinhardtii*. The other six conserved transmembrane domains of green alga chloroplast sulfate permease are shown as A through F. Note the two extended hydrophilic loops, occurring between transmembrane helices TM1-A and D-E, facing toward the exterior of the chloroplast. (From Chen et al., 2003.)

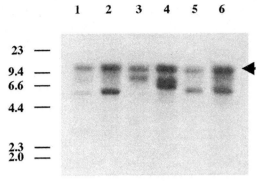

Fig. 7. Southern blot analysis to visualize the number of independent plasmid insertions in a group of plasmid-transformed *Chlamydomonas reinhardtii*. The arrow shows the position of the endogenous inactive *Arg7* gene. The other hybridization bands originate from the insertion and non-homologous recombination of plasmid DNA within the *Chlamydomonas reinhardtii* nuclear genome. An overall 2:1 single/double plasmid insertion ratio was found. Lane 1: *rep*53; Lane 2: *rep*55; Lane 3: *rep*27; Lane 4: *rep*66; Lane 5: *rep*16; Lane 6: *rep*18.

gene probably migrated from the chloroplast to the nuclear genome during evolution of *C. reinhardtii*. The *SulP* gene, or any of its homologues, has not been retained in vascular plants, e.g. *Arabidopsis thaliana*, although it is encountered in the chloroplast genome of a liverwort (*Marchantia polymorpha*). A comparative structural analysis and phylogenetic origin of chloroplast sulfate permeases in a variety of species was presented (Chen and Melis, 2002; Chen et al., 2002, 2003). Preliminary evidence suggested a dependence of the D1/32 kD protein turnover rate on the rate of sulfate uptake by the chloroplast. Thus, the *SulP* gene may directly or indirectly regulate the PS II repair process.

V. DNA Insertional Mutagenesis for the Isolation and Functional Characterization of PS II Repair Aberrant Mutants

Currently, DNA insertional mutagenesis appears to be the method of choice in efforts to unlock the "black box" of the PS II repair process. The successful isolation of many repair mutants will permit the identification and study of the respective genes and proteins, ultimately opening the way to a full elucidation of the repair process. A review of this technology and its application in this research is given below:

Mutagenesis and Screening Procedures: *Chlamydomonas reinhardtii* mutants are generated by transformation of an arginine auxotroph strain (CC-425) with plasmid DNA containing the complementing argininosuccinate lyase (arg+) gene (Gumpel and Purton, 1994; Davies et al., 1994, 1996). The integration of the transformant DNA occurs almost exclusively by nonhomologous recombination (Kindle, 1990; Tam and Lefebvre, 1993). Thus, transformants carrying integrated DNA at random locations in the *Chlamydomonas* nuclear genome are generated (Fig. 7). The following screening procedure is suitable and has been successfully employed to isolate PS II repair mutants:

- *Chlamydomonas reinhardtii* *Arg7* transformants are grown on TAP plates (medium lacking arginine).
- Replica plating on media lacking acetate is performed to identify and isolate acetate-requiring transformants. The latter are expected to include PS II repair mutants since they cannot grow photoautotrophically. This is an important initial screening step, as it eliminates about 90% of the *Arg7* transformants.
- Each acetate-requiring transformant is grown (in the presence of acetate) separately under low-light (LL: 10 μmol photons m^{-2} s^{-1}) or medium-light (ML: 150 μmol photons m^{-2} s^{-1}) conditions. The steady-state amount of functional PS II is measured in such replica colonies. In repair-aberrant *C. reinhardtii*, the steady-state amount of

functional PS II depends solely on the relationship between the rate of PS II biogenesis and photodamage. At 10 μmol photons m^{-2} s^{-1}, the rate of photodamage (about once every 48 h) is slower than the rate of de novo PS II biosynthesis (cell duplication time ~24 h), thus permitting accumulation and detection of functional PS II centers in the chloroplast thylakoids. At 150 μmol photons m^{-2} s^{-1}, the rate of photodamage (once every ~5 h) is faster than the rate of de novo PS II biosynthesis (cell duplication time of ~18 h), thus causing a nearly quantitative accumulation of photodamaged PS II centers in the chloroplast thylakoids.

- Chlorophyll fluorescence transient analysis of the LL and ML-grown mutants is applied to measure the activity of PS II from the yield of the non-variable (Fo), variable (Fv), and maximum (Fm) fluorescence emission (Guenther et al., 1990). Repair mutants are selected on the basis of differential fluorescence phenotype under LL versus ML conditions, as follows. LL-grown repair mutants display both Fo and Fv (Zhang et al., 1997), indicating the presence and functional integrity of PS II in the chloroplast thylakoids. ML-grown repair mutants display Fo but are mostly or entirely devoid of Fv. This irradiance-induced difference in the fluorescence induction characteristics indicates that chloroplasts synthesize, assemble, and retain functional D1 under low-light conditions (when the rate of photodamage is very slow) but could not repair photodamaged D1 under moderate light intensities.

- Transformants that meet the above criteria are isolated. Their functional and repair properties are investigated further by absorbance difference spectrophotometry, SDS-PAGE, Western blot analysis, and (^{35}S) sulfate pulse-chase labeling (Melis, 1989; Vasilikiotis and Melis, 1994; Zhang et al., 1997; Melis, 1999), leading to the isolation of repair-aberrant mutants.

In summary, DNA insertional mutagenesis efforts for the generation, isolation, and characterization of PS II repair aberrant mutants is a useful tool in the discovery of additional genes and proteins that are involved in the PS II repair process. Identification of such genes will permit a subsequent study of their functional and regulatory properties and will thus advance knowledge about the PS II repair process. This experimental approach, when carefully and persistently implemented, may thus contribute to the complete elucidation of the PS II repair process.

VI. Conclusions

The biological significance of the PS II repair cycle is the selective targeting, removal, and replacement of an individual subunit (the inactivated D1/32 kD protein) from the multi-protein PS II holocomplex (MW > 1,000 kD), which is localized in the inaccessible region of the grana membranes in chloroplasts. This phenomenon is unique in the annals of biology; nothing analogous in complexity and specificity has been reported in animals, fungi, or bacteria. Elucidation of the repair mechanism may reveal the occurrence of hitherto unknown regulatory and catalytic reactions for the selective in situ replacement of specific proteins in multi-protein complexes. This may have potentially significant applications in photosynthesis and plant biology but also in medicine and other fields. Thus, the phenomenon is both of fundamental and practical importance.

Fundamentally, elucidation of the PS II repair process will provide information about the genes and proteins, mechanisms and regulations that are involved in a unique process. Practically, the significance of the PS II repair to agriculture is that it maintains photosynthesis. Mutagenesis work in the laboratory of the senior author has shown that oxygenic photosynthesis cannot be sustained in repair-aberrant mutants (Zhang et al., 1997). However, the repair mechanism is not perfect. It is recognized that the PS II repair is efficient only under a narrow range of plant growth conditions. Under many field or laboratory conditions, the mechanism cannot fully cope with moderate and high rates of photodamage, resulting in photoinhibition of photosynthesis, i.e., lower yields and productivity of photosynthesis and potential losses in plant growth (Powles, 1984; Aro et al., 1993; Königer and Winter, 1993; Payton et al., 1998; Melis, 1999). Work in the laboratory of the senior author provided examples of the limitation in growth imposed by photoinhibition in the green alga *Dunaliella salina*. Under otherwise optimal growth conditions with an abundance of nutrients, the rate of photosynthesis and cell duplication in *Dunaliella salina* increased in the range of 100–800 μmol photons m^{-2} s^{-1}, reached a plateau in the range of 800–1,500 μmol photons m^{-2} s^{-1}, and declined at light intensities greater than 1,500 μmol photons m^{-2} s^{-1} (Smith et al., 1990; Baroli and Melis, 1996). At the intensity of full sunlight (2,500 μmol photons m^{-2} s^{-1}), the *actual rate* of cell growth was only ~60% of the *potential rate*. This loss in productivity was due to photoinhibition, i.e., to rates of photodamage that could not be compensated for by the repair mechanism (see also Smirnoff, 1995; McKersie et al., 1996). In summary, elucidation of the repair process, identification of its rate limiting step(s), and a genetic improvement of its performance may have positive implications in terms of improvement in the rate of photosynthesis under field conditions with an attendant increase in plant growth and productivity.

Acknowledgments

The work was financially supported by the United States Department of Agriculture National Research Initiative, Competitive Grants Office.

References

Adam Z and Clarke AK (2002) Cutting edge of chloroplast proteolysis. Trends Plant Sci 7: 451–456
Adams WW III, Demmig-Adams B, Rosenstiel TN, Brightwell AK and Ebbert V (2002) Photosynthesis and photoprotection in overwintering plants. Plant Biology 4: 545–557
Adams WW III, Zarter CR, Ebbert V and Demmig-Adams B (2004) Photoprotective strategies of overwintering evergreens. BioScience 54: 41–49
Adamska I, Ohad I and Kloppstech K (1992) Synthesis of the early light-inducible protein is controlled by blue light and related to light stress. Proc Natl Acad Sci USA 89: 2610–2613
Adir N, Shochat S and Ohad I (1990) Light-dependent D1 protein synthesis and translocation is regulated by reaction center II: Reaction center II serves as an acceptor for the D1 precursor. J Biol Chem 265: 12563–12568
Allen JF (2003) State transitions—a question of balance. Science 299: 1530–1532
Allen JF and Nilsson A (1997) Redox signaling and the structural basis of regulation of photosynthesis by protein phosphorylation. Physiol Plant 100: 863–868
Anderson JM and Melis A (1983) Localization of different photosystems in separate regions of chloroplast membranes. Proc Natl Acad Sci USA 80: 745–749
Andreasson E and Melis A (1995) Localization and characterization of a novel 20 kDa polypeptide in the chloroplast of the green alga *Dunaliella salina*. Plant Cell Physiol 136: 1483–1492
Aro E-M, Kettunen R and Tyystjärvi E (1992) ATP and light regulate D1 protein modification and degradation: role of D1 in photoinhibition. FEBS Lett 297: 29–33
Aro E-M, Virgin I and Andersson B (1993) Photoinhibition of photosystem II. Inactivation, protein damage and turnover. Biochim Biophys Acta 1143: 113–134
Baena-González E and Aro E-M (2002) Biogenesis, assembly and turnover of photosystem II units. Philos T Roy Soc B 357: 1451–1460
Banet G, Pick U and Zamir A (2000) Light-harvesting complex II pigments and proteins in association with Cbr, a homolog of higher-plant early light-inducible proteins in the unicellular green alga *Dunaliella*. Planta 210: 947–955
Barbato R, Friso G, Rigoni F, Frizzo A and Giacometti GM (1992) Characterization of a 41 kDa photoinhibition adduct in isolated photosystem II reaction centres. FEBS Lett 309: 165–169
Barber J (1989) Function and molecular biology of PS II. Oxford Surveys Plant Cell Biol 6: 115–162
Barber J (1994) Molecular basis of the vulnerability of PS II to damage by light. Aust J Plant Physiol 22: 201–208
Barber J, Chapman DJ and Telfer A (1987) Characterisation of a photosystem two reaction center isolated from *Pisum sativum*. FEBS Lett 220: 67–74
Barber J, Morris E and Buchel C (2000) Revealing the structure of the photosystem II chlorophyll binding proteins, CP43 and CP47. Biochim Biophys Acta 1459: 239–247
Baroli I and Melis A (1996) Photoinhibition and repair in *Dunaliella salina* acclimated to different growth irradiances. Planta 198: 640–646
Baroli I and Melis A (1998) Photoinhibitory damage is modulated by the rate of photosynthesis and by the photosystem II light-harvesting chlorophyll antenna size. Planta 205: 288–296
Baroli I and Niyogi KK (2000) Molecular genetics of xanthophyll-dependent photoprotection in green algae and plants. Philos T Roy Soc B 355: 1385–1394
Baroli I, Do AD, Yamane T and Niyogi KK (2003) Zeaxanthin accumulation in the absence of a functional xanthophyll cycle protects *Chlamydomonas reinhardtii* from photooxidative stress. Plant Cell 15: 992–1008
Bassi R, Rigoni F and Giacometti GM (1990) Chlorophyll binding proteins with antenna function in higher plants and green algae. Photochem Photobiol 52: 1187–1206
Bennett J (1979) The protein kinase of the thylakoid membrane is light-dependent. FEBS Lett 103: 342–344
Bennett J (1991) Protein phosphorylation in green plant chloroplasts. Annu Rev Plant Physiol Plant Mol Biol 42: 281–311
Black MT, Brearley TH and Horton P (1986) Heterogeneity in chloroplast PS II. Photosynth Res 8: 193–207
Booij-James IS, Swegle WM, Edelman M and Mattoo AK (2002) Phosphorylation of the D1 photosystem II reaction center protein is controlled by an endogenous circadian rhythm. Plant Physiol 130: 2069–2075
Bottomley W, Spencer D and Whitfeld PR (1974) Protein synthesis in isolated spinach chloroplasts: comparison of light-driven and ATP-driven synthesis. Arch Biochem Biophys 164: 106–117
Bricker TM (1990) The structure and function of CPa-1 and CPa-2 in photosystem-II. Photosynth Res 24: 1–13
Callahan FE, Wergin WP, Nelson N, Edelman M and Mattoo AK (1989) Distribution of thylakoid proteins between stromal and granal lamellae in *Spirodela*. Plant Physiol 91: 629–635
Callahan FE, Ghirardi ML, Sopory SK, Mehta AM, Edelman M and Mattoo AK (1990) A novel metabolic form of the 32 kDa-D1 protein in the grana-localized reaction center of PS II. J Biol Chem 265: 15357–15360
Cao H, Zhang L and Melis A (2001) Bioenergetic and metabolic processes for the survival of sulfur-deprived *Dunaliella salina* (Chlorophyta). J Appl Phycol 13: 25–34
Chen H-C and Melis A (2002) Complete genomic DNA (bases 1 through 3873) and protein sequence (amino acids 1 through 411) for a putative chloroplast-envelope localized sulfate permease (*CrcpSulP*) in the unicellular green alga *Chlamydomonas reinhardtii*. GenBank Accession Number AF467891

Chen H-C, Yolthongwattana K and Melis A (2002) *Chlamydomonas reinhardtii* chloroplast sulfate transport system permease (*SulP*) mRNA, complete cds; nuclear gene for chloroplast product. GenBank Accession Number AF481828

Chen H-C, Yokthongwattana K, Newton AJ and Melis A (2003) *SulP*, a nuclear gene encoding a putative chloroplast-targeted sulfate permease in *Chlamydomonas reinhardtii*. Planta 218: 98–106

Chylla RA and Whitmarsh J (1989) Inactive photosystem-II complexes in leaves—turnover rate and quantitation. Plant Physiol 90: 765–772

Cinco RM, Rompel A, Visser H, Aromi G, Klein M and Sauer K (1999) Comparison of the manganese cluster in oxygen-evolving photosystem II with distorted cubane manganese compounds through X-ray absorption spectroscopy. Inorg Chem 38: 5988–5998

Cleland RE and Melis A (1987) Probing the events of photoinhibition by altering electron-transport activity and light-harvesting capacity in chloroplast thylakoids. Plant Cell Environ 10: 747–752

Cleland RE, Melis A and Neale PJ (1986) Mechanism of photoinhibition: photochemical reaction center inactivation in system II of chloroplasts. Photosynth Res 9: 79–88

Cramer WA and Whitmarsh J (1977) Photosynthetic cytochromes. Annu Rev Plant Physiol 28: 133–172

Cramer WA, Theg SM and Widger WR (1986) On the structure and function of cytochrome *b*-559. Photosynth Res 10: 393–403

Crofts AR and Wraight CA (1983) The electrochemical domain of photosynthesis. Biochim Biophys Acta 726: 149–185

Davies JP, Yildiz F and Grossman AR (1994) Mutants of *Chlamydomonas* with aberrant responses to sulfur deprivation. Plant Cell 6: 53–63

Davies JP, Yildiz F and Grossman AR (1996) *Sac1*, a putative regulator that is critical for survival of *Chlamydomonas reinhardtii* during sulfur deprivation. EMBO J 15: 2150–2159

Deisenhofer J and Michel H (1989) The photosynthetic reaction center from the purple bacterium *Rhodopseudomonas viridis*. EMBO J 8: 2149–2170

Demeter S, Neale PJ and Melis A (1987) Photoinhibition: impairment of the primary charge separation between P680 and pheophytin in photosystem II of chloroplasts. FEBS Lett 214: 370–374

Demmig B and Björkman O (1987) Comparison of the effect of excessive light on chlorophyll fluorescence (77K) and photon yield of O_2 evolution in leaves of higher plants. Planta 171: 171–184

Demmig-Adams B, Moeller DL, Logan BA and Adams WW III (1998) Positive correlation between levels of retained zeaxanthin + antheraxanthin and degree of photoinhibition in shade leaves of *Schefflera arboricola* (Hayata) Merrill. Planta 205: 367–374

Depka B, Jahns P and Trebst A (1998) β-carotene to zeaxathin conversion in the rapid turnover of the D1 protein of photosytem II. FEBS Lett 424: 267–270

Drzymalla C, Schroda M and Beck CF (1996) Light-inducible gene *hsp70B* encodes a chloroplast-localized heat shock protein in *Chlamydomonas reinhardtii*. Plant Mol Biol 31: 1185–1194

Duysens LN, Amesz J and Kamp BM (1961) Two photochemical systems in photosynthesis. Nature 190: 510–511

Eaglesham ARJ and Ellis RJ (1974) Protein synthesis in chloroplasts: II. Light-driven synthesis of membrane protein by isolated pea chloroplasts. Biochim Biophys Acta 335: 396–407

Ebbert V and Godde D (1996) Phosphorylation of PS II polypeptides inhibits D1 protein-degradation and increases PS II stability. Photosynth Res 50: 257–269

Edelman M and Reisfeld A (1978) Characterization, translation and control of the 32,000 dalton chloroplast membrane protein in *Spirodela*. In: Akoyunoglou and Argyroudi-Akoyunoglou (eds) Chloroplast Development, pp 642–652. Elsevier/North Holland Biomedical Press, New York

Elich TD, Edelman M and Mattoo AK (1992) Identification, characterization, and resolution of the in vivo phosphorylated form of the D1 photosystem II reaction center protein. J Biol Chem 267: 3523–3529

Farchaus J and Dilley RA (1986) Purification and partial sequence of the Mr 10,000 phosphoprotein from spinach thylakoids. Arch Biochem Biophys 244: 94–101

Frank HA, Bautista JA, Josue JS and Young AJ (2000) Mechanism of non-photochemical quenching in green plants: energies of the lowest excited singlet states of violaxanthin and zeaxanthin. Biochemistry 39: 2831–2837

Ferreira KN, Tina M, Iverson TM, Maghlaoui K, Barber J and Iwata S (2004) Architecture of the photosynthetic oxygen-evolving center. Science 303: 1831–1838

Ghanotakis DF and Yocum CF (1990) Photosystem-II and the oxygen-evolving complex. Annu Rev Plant Physiol Plant Mol Biol 41: 255–276

Glick RE and Melis A (1988) Minimum photosynthetic unit size in system I and system II of barley chloroplasts. Biochim Biophys Acta 934: 151–155

Gong H, Nilsen S and Allen JF (1993) Photoinhibition of photosynthesis. In-vivo involvement of multiple sites in a photodamage process under carbon dioxide and oxygen-free conditions. Biochim Biophys Acta 1142: 115–122

Graan T and Ort DR (1986) Detection of oxygen-evolving photosystem-II centers inactive in plastoquinone reduction. Biochim Biophys Acta 852: 320–330

Green BR (1988) The chlorophyll-protein complexes of higher plant photosynthetic membranes or just what green band is that? Photosynth Res 15: 3–32

Green BR and Camm EL (1981) A model of the relationship of the chlorophyll-protein complexes associated with photosystem-II. In: Akoyunoglou G (ed) Photosynthesis, Proceedings of 5th International Congress, Vol III, pp 675–681. Balaban International Science Services, Philadelphia

Green BR and Kühlbrandt W (1995) Sequence conservation of light-harvesting and stress-response proteins in relation to the three-dimensional molecular structure of LHCII. Photosynth Res 44: 139–148

Greenberg BM, Gaba V, Mattoo AK and Edelman M (1987) Identification of a primary in vivo degradation product of the rapidly-turning-over 32 kD protein of photosystem II. EMBO J 6: 2865–2869

Greer DM, Berry JA and Björkman O (1986) Photoinhibition of photosynthesis in intact bean leaves: role of light and temperature, and requirement for chloroplast-encoded protein synthesis during recovery. Planta 168: 253–260

Guenther JE and Melis A (1990) The physiological significance of photosystem II heterogeneity in chloroplasts. Photosynth Res 23: 105–110

Guenther JE, Nemson JA and Melis A (1990) Development of PS II in dark grown *Chlamydomonas reinhardtii*. A light-dependent conversion of PS IIβ, Q_B-nonreducing centers to the PS IIα, Q_B-reducing form. Photosynth Res 24: 35–46

Gumpel NJ and Purton S (1994) Playing tag with *Chlamydomonas*. Trends Cell Biol 4: 299–301

Hallick RB (1989) Proposals for the naming of chloroplast genes. II. Update to the nomenclature of genes for thylakoid membrane polypeptides. Plant Mol Biol Rep 7: 266–275

Hankamer B, Barber J and Boekema EJ (1997) Structure and membrane organization of photosystem II in green plants. Annu Rev Plant Physiol Plant Mol Biol 48: 641–671

Haußühl K, Andersson B and Adamska I (2001) A chloroplast DegP2 protease performs the primary cleavage of the photodamaged D1 protein in plant photosystem II. EMBO J 20: 713–722

Hell R (1997) Molecular physiology of plant sulfur metabolism. Planta 202: 138–148

Hill R and Bendall F (1960) Function of the two cytochrome components in chloroplasts: a working hypothesis. Nature 186: 136–137

Huner NPA, Maxwell DP, Gray GR and Savitch LV (1996) Sensing environmental temperature change through imbalances between energy supply and energy consumption—redox state of photosystem II. Physiol Plant 98: 358–364

Hurry VM, Anderson JM, Badger MR and Price GD (1996) Reduced levels of cytochrome $b - f$ in transgenic tobacco increases the excitation pressure on photosystem II without increasing sensitivity to photoinhibition in vivo. Photosynth Res 50: 159–169

Ishikawa Y, Nakatani E, Henmi T, Ferjani A, Harada Y, Tamura N and Yamamoto Y (1999) Turnover of the aggregates and cross-linked products of the D1 protein generated by acceptor-side photoinhibition of photosystem II. Biochim Biophys Acta 1413: 147–158

Jahns P and Miehe B (1996) Kinetic correlation of recovery from photoinhibition and zeaxanthin epoxidation. Planta 198: 202–210

Jahns P, Depka B and Trebst A (2000) Xanthopyll cycle mutants from *Chlamydomonas reinhardtii* indicate a role for zeaxanthin in the D1 protein turn over. Plant Physiol Biochem 38: 371–376

Jansen MAK, Mattoo AK and Edelman M (1999) D1-D2 protein degradation in the chloroplast. Complex light saturation kinetics. Eur J Biochem 260: 527–532

Jin E, Polle J and Melis A (2001) Involvement of zeaxanthin and of the Cbr protein in the repair of photosystem-II from photoinhibition in the green alga *Dunaliella salina*. Biochim Biophys Acta 1506: 244–259

Jin E, Yokthongwattana K, Polle JEW and Melis A (2003) Role of the reversible xanthophyll cycle in the photosystem-II damage and repair cycle in *Dunaliella salina* (green alga). Plant Physiol 132: 352–364

Kaplan A (1981) Photoinhibition in *Spirulina platensis*: response of photosynthesis and HCO_3-uptake capability to CO_2-depleted conditions. J Exp Bot 32: 669–677

Kim J, Klein PG and Mullet JE (1994) Synthesis and turnover of photosystem II reaction center protein D1. Ribosome pausing increases during chloroplast development. J Biol Chem 269: 17918–17923

Kim JH, Nemson JA and Melis A (1993) Photosystem-II reaction center damage and repair in the green alga *Dunaliella salina*: analysis under physiological and adverse irradiance conditions. Plant Physiol 103: 181–189

Kindle KL (1990) High-frequency nuclear transformation of *Chlamydomonas reinhardtii*. Proc Natl Acad Sci USA 87: 1228–1232

Koivuniemi A, Aro E-M and Andersson B (1995) Degradation of the D1- and D2-proteins of photosystem II in higher plants is regulated by reversible phosphorylation. Biochemistry 34: 16022–16029

Königer M and Winter K (1993) Reduction of photosynthesis in sun leaves of *Gossypium hirsutum* L. under conditions of high light intensities and suboptimal leaf temperatures. Agronomie 13: 659–668

Kruse O, Zheleva D and Barber J (1997) Stabilization of photosystem two dimers by phosphorylation: implication for the regulation of the turnover of D1 protein. FEBS Lett 408: 276–280

Kyle DJ, Ohad I and Arntzen CJ (1984) Membrane protein damage and repair: selective loss of a quinone-protein function in chloroplast membranes. Proc Natl Acad Sci USA 81: 4070–4074

Lam E, Baltimore B, Ortiz W, Chollar S, Melis A and Malkin R (1983) Characterization of a resolved oxygen-evolving photosystem II preparation from spinach thylakoids. Biochim Biophys Acta 724: 201–211

Lavergne J (1982) Two types of primary acceptors in chloroplast PS II. Photobiochem Photobiophys 3: 257–285

Lavergne J and Briantais J-M (1996) Photosystem-II heterogeneity. In: Ort DR and Yocum CF (eds) Oxygenic Photosynthesis: The Light Reactions, pp 265–287. Kluwer Academic Publishers, Dordrecht, The Netherlands

Levy H, Tal T, Shaish A and Zamir A (1993) Cbr, an algal homolog of plant early light-induced proteins, is a putative zeaxanthin binding protein. J Biol Chem 268: 20892–20896

Lindahl M, Spetea C, Hundal T, Oppenheim AB and Andersson B (2000) The thylakoid FtsH protease plays a role in the light-induced turnover of the photosystem II D1 protein. Plant Cell 12: 419–431

Lorkovic ZJ, Schroder WP, Pakrasi HB, Irrgang K, Herrmann RG and Oelmuller R (1995) Molecular characterization of *psbW*, a nuclear-encoded component of the photosystem II reaction center complex in spinach. Proc Natl Acad Sci USA 92: 8930–8934

Ma Y-Z, Holt NE, Li X-P, Niyogi KK and Fleming GR (2003) Evidence for direct carotenoid involvement in the regulation of photosynthetic light harvesting. Proc Natl Acad Sci USA 100: 4377–4382

Mäenpää P, Andersson B and Sundby C (1987) Difference in sensitivity to photoinhibition between photosystem II in the appressed and non-appressed thylakoid regions. FEBS Lett 215: 31–36

Mattoo AK, Hoffman-Falk H, Marder J and Edelman M (1984) Regulation of protein metabolism: coupling of photosynthetic electron-transport to in vivo degradation of the rapidly metabolized 32-kDa protein of the chloroplast membranes. Proc Natl Acad Sci USA 81: 1380–1384

Mattoo AK and Edelman M (1987) Intramembrane translocation and posttranslational palmitoylation of the chloroplast 32-kDa herbicide-binding protein. Proc Natl Acad Sci USA 84: 1497–1501

Maxwell DP, Falk S and Huner NPA (1995) Photosystem II excitation pressure and development of resistance to photoinhibition. 1. Light harvesting complex II abundance and zeaxanthin content in *Chlorella vulgaris*. Plant Physiol 107: 687–694

McKersie BD, Bowley SR, Harjanto E and Leprince O (1996) Water-deficit tolerance and field performance of transgenic alfalfa overexpressing superoxide dismutase. Plant Physiol 111: 1177–1181

Meetam M, Keren N, Ohad I and Pakrasi HB (1999) The PsbY protein is not essential for oxygenic photosynthesis in the cyanobacterium Synechocystis sp PCC 6803. Plant Physiol 121: 1267–1272

Melis A (1985) Functional properties of PS IIβ in spinach chloroplasts. Biochim Biophys Acta 808: 334–342

Melis A (1989) Spectroscopic methods in photosynthesis: photosystem stoichiometry and chlorophyll antenna size. Philos T Roy Soc B 323: 397–409

Melis A (1991) Dynamics of photosynthetic membrane composition and function. Biochim Biophys Acta 1058: 87–106

Melis A (1996) Excitation energy transfer: functional and dynamic aspects of *Lhc (cab)* proteins. In: Ort DR and Yocum CF (eds), Oxygenic Photosynthesis: The Light Reactions, pp 523–538. Kluwer Academic Publishers, Dordrecht, The Netherlands

Melis A (1998) Photostasis in plants: mechanisms and regulation. In: Williams TP and Thistle A (eds) Photostasis and Related Phenomena, pp 207–221. Plenum Publishing Corporation, New York

Melis A (1999) Photosystem-II damage and repair cycle in chloroplasts: what modulates the rate of photodamage in vivo? Trends Plant Sci 4: 130–135

Melis A and Anderson JM (1983) Structural and functional organization of the photosystems in spinach chloroplasts: Antenna size, relative electron transport capacity, and chlorophyll composition. Biochim Biophys Acta 724: 473–484

Melis A and Duysens LNM (1979) Biphasic energy conversion kinetics and absorbance difference spectra of PS II of chloroplasts. Evidence for two different system II reaction centers. Photochem Photobiol 29: 373–382

Melis A and Homann PH (1976) Heterogeneity of the photochemical centers in system II of chloroplasts. Photochem Photobiol 23: 343–350

Melis A and Nemson JA (1995) Characterization of a 160 kD photosystem-II reaction center complex isolated from photoinhibited *Dunaliella salina* thylakoids. Photosynth Res 46: 207–211

Melis A, Zhang L, Forestier M, Ghirardi ML and Seibert M (2000) Sustained photobiological hydrogen gas production upon reversible inactivation of oxygen evolution in the green alga *Chlamydomonas reinhardtii*. Plant Physiol 122: 127–136

Michel H, Hunt DF, Shabanowitz J and Bennett J (1988) Tandem mass spectrometry reveals that three photosystem II protein of spinach chloroplasts contain N-acetyl-o-phosphothreonine at their NH$_2$ termini. J Biol Chem 263: 1123–1130

Millner PA, Marder JB, Gounaris K and Barber J (1986) Localization and identification of phosphoproteins within the photosystem-II core of higher-plant thylakoid membranes. Biochim Biophys Acta 852: 30–37

Miyazaki A, Shina T, Toyoshima Y, Gounaris K and Barber J (1989) Stoichiometry of cytochrome b-559 in photosystem-II. Biochim Biophys Acta 975: 142–147

Murata N, Miyao M, Omata T, Matsunami H and Kuwabara T (1984) Stoichiometry of components in the photosynthetic oxygen evolution system of photosystem-II particles prepared with Triton X-100 from spinach chloroplasts. Biochim Biophys Acta 765: 363–369

Nanba O and Satoh K (1987) Isolation of a photosystem-II reaction center containing D1 and D2 polypeptides and cytochrome b-559. Proc Natl Acad Sci USA 84: 109–112

Neale PJ and Melis A (1990) Activation of a reserve pool of photosystem II in *Chlamydomonas reinhardtii* counteracts photoinhibition. Plant Physiol 92: 1196–1204

Neale PJ and Melis A (1991) Dynamics of photosystem II heterogeneity during photoinhibition: depletion of PS IIβ from the non-appressed thylakoids during strong-irradiance exposure of *Chlamydomonas reinhardtii*. Biochim Biophys Acta 1056: 195–203

Nedbal L, Samson G and Whitmarsh J (1992) Redox state of a one-electron component controls the rate of photoinhibition of photosystem-II. Proc Natl Acad Sci USA 89: 7929–7933

Nield J, Kruse O, Ruprecht J, da Fonseca P and Barber J (2000) Three-dimensional structure of *Chlamydomonas reinhardtii* and *Synechococcus elongatus* photosystem II complexes allows for comparison of their oxygen-evolving complex organization. J Biol Chem 275: 27940–27946

Ohad I, Kyle DJ and Arntzen CJ (1984) Membrane protein damage and repair: removal and replacement of inactivated 32-kilodalton polypeptides in chloroplast membranes. J Cell Biol 99: 481–485

Ohnishi N and Takahashi Y (2001) PsbT polypeptide is required for efficient repair of photodamaged photosystem II reaction center. J Biol Chem 276: 33798–33804

Park YI, Chow WS, Osmond CB and Anderson JM (1996) Electron transport to oxygen mitigates against the photoinactivation of Photosystem II in vivo. Photosynth Res 50: 23–32

Park YI, Chow WS and Anderson JM (1997) Antenna size dependency of photoinactivation of photosystem II in light-acclimated pea leaves. Plant Physiol 115: 151–157

Payton P, Allen RD, Trolinder N and Holaday AS (1998) Overexpression of chloroplast-targeted Mn superoxide dismutase in cotton (*Gossypium hirsutum* L.) does not alter the reduction of photosynthesis after short exposures to low temperature and high light intensity. Photosynth Res 52: 233–244

Polle JEW and Melis A (1999) Recovery of the photosynthetic apparatus from photoinhibition during dark incubation of the green alga *Dunaliella salina*. Aust J Plant Physiol 26: 679–686

Powles SB (1984) Photoinhibition of photosynthesis induced by visible light. Annu Rev Plant Physiol 35: 15–44

Pursiheimo S, Rintamäki E, Baena-González E and Aro E-M (1998) Thylakoid protein phosphorylation in evolutionarily divergent species with oxygenic photosynthesis. FEBS Lett 423: 178–182

Rhee K-H, Morriss EP, Barber J and Kuhlbrandt W (1998) Three-dimensional structure of the plant photosystem II reaction centre at 8 angstrom resolution. Nature 396: 283–286

Rintamäki E, Kettunen R and Aro E-M (1996) Differential D1 phosphorylation in functional and photodamaged photosystem II centers: dephosphorylation is a prerequisite for degradation of damaged D1. J Biol Chem 271: 14870–14875

Schroda M, Vallon O, Wollman FA and Beck CF (1999) A chloroplast-targeted heat shock protein 70 (HSP70) contributes to the photoprotection and repair of PS II during and after photoinhibition. Plant Cell 11: 1165–1178

Seibert M, Picorel R, Rubin AB and Connolly JS (1988) Spectral, photophysical, and stability properties of isolated photosystem-II reaction center. Plant Physiol 87: 303–306

Shi LX, Kim SJ, Marchant A, Robinson C and others (1999) Characterisation of the *PsbX* protein from Photosystem II and light regulation of its gene expression in higher plants. Plant Mol Biol 40: 737–744

Silva P, Choi YJ, Hassan HAG and Nixon PJ (2002) Involvement of the HtrA family of proteases in the protection of the cyanobacterium *Synechocystis* PCC 6803 from light stress and in the repair of photosystem II. Philos T Roy Soc B 357: 1461–1467

Smirnoff N (1995) Antioxidant systems and plant responses to the environment. In: Smirnoff N (ed) Environment and Plant Metabolism, pp 217–242. BIOS Scientific Publishers Ltd, Oxford, UK

Smith BM, Morrissey PJ, Guenther JE, Nemson JA, Harrison MA, Allen JF and Melis A (1990) Response of the photosynthetic apparatus in *Dunaliella salina* (green algae) to irradiance stress. Plant Physiol 93: 1433–1440

Sundby C, McCaffery S and Anderson JM (1993) Turnover of the photosystem II D1 protein in higher plants under photoinhibitory and nonphotoinhibitory irradiance. J Biol Chem 268: 25476–25482

Tam L-W and Lefebvre PA (1993) Cloning of flagellar genes in *Chlamydomonas reinhardtii* by DNA insertional mutagenesis. Genetics 135: 375–384

Thielen APGM and VanGorkom HJ (1981) Electron transport properties of the photosystems IIα and IIβ. In: Akoyunoglou G (ed) Photosynthesis, Proceedings of 5th International Congress, Vol II, pp 57–64. Balaban International Science Services, Philadelphia

Thornber JP (1986) Biochemical characterization and structure of pigment-proteins of photosynthetic organisms. In: Staehelin LA and Arntzen CJ (eds) Encyclopedia of Plant Physiology, Vol 19, pp 98—115. Springer Verlag, New York

Tyystjärvi E and Aro E-M (1996) The rate constant of photoinhibition, measured in lincomycin-treated leaves, is directly proportional to light intensity. Proc Natl Acad Sci USA 93: 2213–2218

Tyystjärvi E, Koivuniemi A, Kettunen R and Aro E-M (1991) Small light-harvesting antenna does not protect from photoinhibition. Plant Physiol 97: 477–483

Tyystjärvi E, Kettunen R and Aro E-M (1994) The rate constant of photoinhibition in vitro is independent of the antenna size of photosystem II but depends on temperature. Biochim Biophys Acta 1186: 177–185

Vallon O, Wollman FA and Olive J (1986) Lateral distribution of the main protein complexes of the photosynthetic apparatus in *Chlamydomonas reinhardtii* and in spinach: an immunocytochemical study using intact thylakoid membranes and a PS II-enriched membrane preparation. Photobiochem Photobiophys 12: 203–220

Vallon O, Hoyer-Hansen G and Simpson DJ (1987) Photosystem-II and cytochrome *b*-559 in the stroma lamellae of barley chloroplasts. Carlsberg Res Commun 52: 405–421

Van Gorkom HJ (1985) Electron transfer in photosystem-II. Photosynth Res 6: 97–112

Vasilikiotis C and Melis A (1994) Photosystem-II reaction center damage and repair cycle—chloroplast acclimation strategy to irradiance stress. Proc Natl Acad Sci USA 91: 7222–7226

Wettern M (1986) Localization of 32,000 dalton chloroplast protein pools in thylakoids: significance in atrazine binding. Plant Sci 43: 173–177

Wykoff DD, Davies JP, Melis A and Grossman AR (1998) The regulation of photosynthetic electron transport during nutrient deprivation in *Chlamydomonas reinhardtii*. Plant Physiol 117: 129–139

Xu CC, Lee H-Y and Lee C-H (1999) Recovery from low temperature photoinhibition is not governed by changes in the level of zeaxanthin in rice (*Oryza sativa* L.) leaves. J Plant Physiol 155: 755–761

Yamamoto Y (2001) Quality control of photosystem II. Plant Cell Physiol 42: 121–128

Yokthongwattana K, Chrost B, Behrman S, Casper-Lindley C and Melis A (2001) Photosystem II damage and repair cycle in the green alga *Dunaliella salina*: involvement of a chloroplast-localized HSP70. Plant Cell Physiol 42: 1389–1397

Zhang L, Niyogi KK, Baroli I, Nemson JA, Grossman A and Melis A (1997) DNA insertional mutagenesis for the elucidation of a PS II repair process in the green alga *Chlamydomonas reinhardtii*. Photosynth Res 53: 173–184

Zouni A, Witt HT, Kern J, Fromme P, Krauss, N, Saenger W and Orth P (2001) Crystal structure of photosystem II from Synechococcus elongatus at 3.8 Å resolution. Nature 409: 739–743

Chapter 13

Regulation by Environmental Conditions of the Repair of Photosystem II in Cyanobacteria

Yoshitaka Nishiyama[1], Suleyman I. Allakhverdiev[2,3] and Norio Murata*[3]

[1] Cell-Free Science and Technology Research Center and Satellite Venture Business Laboratory, Ehime University, Bunkyo-cho, Matsuyama 790-8577, Japan; [2] Institute of Basic Biological Problems, Russian Academy of Sciences, Pushchino, Moscow Region 142290, Russia; [3] Department of Regulation Biology, National Institute for Basic Biology, Okazaki 444-8585, Japan

	Summary	193
I.	Introduction	194
II.	Effects of Light	194
	A. Photodamage and Repair of Photosystem II (PS II)	194
	B. Dissection of Photodamage and Repair	194
	C. Mechanisms of Photodamage	195
III.	Effects of Oxidative Stress	197
	A. Reactive Oxygen Species (ROS)	197
	B. ROS and Photodamage to PS II	197
	C. Inhibition of Repair by ROS	198
	D. Molecular Mechanisms of ROS-Induced Inhibition	198
IV.	Effects of Salt Stress	198
	A. Salt Stress and Photodamage to PS II	198
	B. Molecular Mechanisms of Salt-Induced Inhibition	199
	C. Overall Gene Expression under Salt Stress	199
V.	Effects of Low-Temperature Stress	199
	A. Low-Temperature Stress and Repair of PS II	199
	B. Molecular Mechanisms of Low-Temperature-Induced Inhibition	199
	C. Unsaturation of Fatty Acids	199
	D. Repair in Darkness	200
VI.	Conclusions and Future Perspectives	200
	Acknowledgments	200
	References	200

Summary

The activity of photosystem II (PS II) is severely restricted by a variety of environmental factors and, under environmental stress, is determined by the balance between the rate of damage to PS II and the rate of the repair of damaged PS II. The effects of environmental stress on damage and repair can be examined separately and it appears that, while light can damage PS II directly, most types of environmental stress act primarily by inhibiting the repair of PS II. Studies in cyanobacteria have demonstrated that repair-inhibiting conditions include oxidative stress, salt stress, and low-temperatures stress, each of which suppresses the de novo synthesis of proteins, in particular the D1 protein, which is required for the repair of PS II. The synergistic effects of combinations of different types of environmental stress suggest that it is the repair process that determines the sensitivity of PS II to specific environmental conditions.

*Author for correspondence, email: murata@nibb.ac.jp

I. Introduction

In natural environments, photosynthetic organisms are often exposed to unfavorable environmental conditions, such as strong light, high concentrations of salt, and low and high temperatures. Photosystem II (PS II) is very sensitive to changes in the environment and, under unfavorable or stressful environmental conditions, the activity of PS II declines more rapidly than many other physiological activities (Berry and Björkman, 1980; Demmig-Adams and Adams, 1992; Aro et al., 1993; Andersson and Aro, 2001; and references therein). Since the efficiency of photosynthesis is largely a reflection of the activity of PS II, considerable attention has been paid to the effects of environmental stress on PS II.

Initial studies directed toward an understanding of the mechanisms of the inhibition of PS II by environmental stress suggested that environmental stress might damage PS II directly (Jones and Kok, 1966a,b; Boyer and Bowen, 1970; Jones, 1973; Keck and Boyer, 1974; Björkmann and Powles, 1984; see also reviews by Powles, 1984; Aro et al., 1993; Ohad et al., 1994; Keren and Ohad, 1998; Melis, 1999; Andersson and Aro, 2001; Adir et al., 2003). This conclusion was based on the results of experiments in vitro in which isolated thylakoid membranes or PS II complexes were exposed to the conditions associated with environmental stress. Nonetheless, this approach revealed that damage to PS II was not repaired under the conditions of the experiments. Evidence for direct damage to PS II by environmental stress was also obtained from in vivo studies in which whole organisms were exposed to environmental stress. Again, conclusions were often derived from observations that failed to distinguish the process of damage to PS II from the repair of PS II.

In living photosynthetic cells, PS II is damaged by light and is repaired simultaneously (Kyle et al., 1984; Mattoo et al., 1984; Ohad et al., 1984). The rate of repair of PS II is coordinated with the rate of damage under non-stress conditions but the delicate balance is perturbed under stressful conditions. Environmental stress reduces the rate of repair and, as a result, the activity of PS II decreases. Thus, the activity of PS II that is detected under a given stress is determined by the balance between the rate of damage to PS II and the rate of repair. In order to clarify in full detail the nature of the inhibition of PS II, we must examine separately the effects of environmental stress on damage and on repair. Methods for monitoring the two processes separately have been established (Gombos et al., 1994; Wada et al., 1994) and their application has revealed that light can damage PS II directly, while a variety of other forms of environmental stress act primarily by inhibiting the repair of PS II. This review provides a summary of recent progress in this area and focuses on the elucidation of the mechanisms responsible for the regulation by environmental factors of the repair of PS II.

II. Effects of Light

A. Photodamage and Repair of PS II

Light is a prerequisite for photosynthesis but it is harmful to the photosynthetic machinery. Exposure of photosynthetic organisms to strong light results in severe inhibition of the activity of PS II (Powles, 1984; Aro et al., 1993; Ohad et al., 1994; Melis, 1999; Andersson and Aro, 2001; Adir et al., 2003). This phenomenon is referred to as photodamage to PS II or the photoinhibition of PS II. Although full details of mechanisms responsible for the photodamage to PS II remain unclear, there is general agreement that the primary target of photodamage is the photochemical reaction center. It is hypothesized that a primary event in photoinhibition causes damage to the D1 protein, which triggers the rapid degradation of the damaged D1 protein by several proteases (Prásil et al., 1992; Aro et al., 1993; Andersson and Aro, 2001).

In living photosynthetic cells, a system exists for the repair of photodamaged PS II (Aro et al., 1993; Andersson and Aro, 2001). The damaged D1 protein is replaced by a newly synthesized precursor to the D1 protein, which is encoded by the *psbA* gene (Ohad et al., 1984; Mattoo et al., 1984, 1989). The carboxy-terminal region of the precursor protein is removed by specific proteases (Anbudurai et al., 1994; Inagaki et al., 2001) and PS II is reactivated.

B. Dissection of Photodamage and Repair

Photosystem II normally undergoes photodamage and repair simultaneously in living cells. Thus, if we want to monitor the process of damage exclusively, it is necessary to block the repair process by exposure of cells to an inhibitor of protein synthesis, such as chloramphenicol or lincomycin. Figure 1A shows typical kinetics of the photodamage to PS II in the cyanobacterium *Synechocystis* sp. PCC 6803. Transformable cyanobacteria are useful model organisms for studies of the effects of environmental stress on photosynthesis. The incubation of *Synechocystis* cells for 2 h in light at

Chapter 13 Repair of Photosystem II and Environment

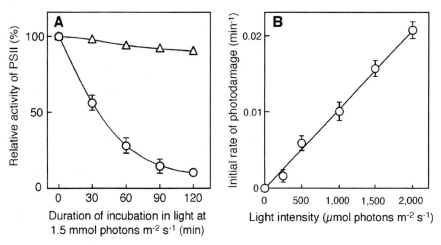

Fig. 1. Profiles of photodamage to PS II. (A) Photodamage to PS II was, apparently, not induced in cells of *Synechocystis* when they were exposed to strong light (1.5 mmol photons m^{-2} s^{-1}) for 2 h (open triangles), but the progress of actual photodamage was visualized when the repair of photodamaged PS II was blocked by the presence of 200 μg ml^{-1} chloramphenicol (open circles). (B) Relationship between photodamage to PS II and light intensity. Initial rates of photodamage to PS II under light at various intensities were determined in the presence of 200 μg ml^{-1} chloramphenicol in cells of *Synechocystis*. Reproduced from Nishiyama et al. (2004).

1.5 mmol photons m^{-2} s^{-1} had no significant effect on the activity of PS II. However, in the presence of chloramphenicol, similar incubation resulted in the inactivation of PS II, revealing the actual photodamage of PS II in the absence of repair. This strategy has been used to monitor the process of photodamage to PS II in cyanobacteria (Gombos et al., 1994; Wada et al., 1994; Nishiyama et al., 2001) and in plants (Moon et al., 1995; Alia et al., 1999). It appears, from Figure 1A, that light at 1.5 mmol photons m^{-2} s^{-1} for 2 h provides conditions under which the rate of photodamage is balanced by the rate of repair in *Synechocystis* (Fig. 1A). Stronger light promotes greater photodamage to PS II, and Figure 1B shows how the initial rate of photodamage to PS II varies with light intensity in the presence of chloramphenicol in *Synechocystis*. The initial rate of photodamage to PS II was proportional to the light intensity (Allakhverdiev and Murata, 2004; Nishiyama et al., 2004). Similar results were obtained in studies of plant leaves that had been exposed to light at various intensities in the presence of an inhibitor of protein synthesis (Park et al., 1995; Tyystjärvi and Aro, 1996; Lee et al., 2001). These findings suggest that photodamage to PS II depends solely on the intensity of incident light.

In experiments designed to monitor the repair of PS II, organisms are first exposed to very strong light, such as light at 3 mmol photons m^{-2} s^{-1}, to decrease the activity of PS II to about 10–20% of the original value. During subsequent exposure of the organisms to light at a moderate intensity, such as 70 μmol photons m^{-2} s^{-1}, the activity gradually returns to the original level (Fig. 2). Such experiments have often been used to examine the repair of PS II in cyanobacterial cells (Gombos et al., 1994; Wada et al., 1994; Nishiyama et al., 2001, 2004; Allakhverdiev et al., 2002, 2003; Allakhverdiev and Murata, 2004) and in leaves of higher plants (Moon et al., 1995; Alia et al., 1999). These studies have revealed, for example, that the rate of repair of PS II in *Synechocystis* depends on the intensity of light in light at intensities below approximately 300 μmol photons m^{-2} s^{-1} (Allakhverdiev and Murata, 2004). Thus, light acts not only by damaging PS II but also by inducing the repair of PS II.

C. Mechanisms of Photodamage

Intensive efforts have been made to define the molecular mechanisms of photodamage to PS II and many putative mechanisms have been proposed (Kyle et al., 1984; Mattoo et al., 1984; Arntz and Trebst, 1986; Callahan et al., 1986; Theg et al., 1986; Allakhverdiev et al., 1987; Demeter et al., 1987; Cleland, 1988; Greenberg et al., 1989; Klimov et al., 1990; Setlik et al., 1990; Vass et al., 1992; Keren et al., 1997). Among the many possible mechanisms, particular attention has been paid to three, namely, the "acceptor-side", "donor-side", and "low-light" mechanisms.

In the proposed "acceptor-side" mechanism, the double reduction of the quinone acceptor Q$_A$, as a result of excess light, facilitates the formation of the triplet state of the reaction-center chlorophyll, which transfers excitation energy to oxygen molecules. The

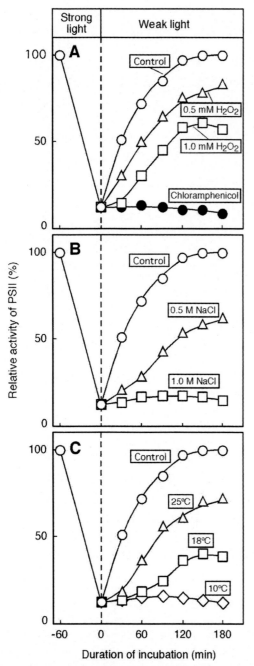

Fig. 2. Repair of photodamaged PS II and the effects of environmental stress. Cells of *Synechocystis* were exposed to strong light (3 mmol photons m^{-2} s^{-1}) for 1 h under normal conditions (20 mM NaCl, 34°C) to induce approximately 80% inactivation of PS II. Cells were then incubated in weak light (70 μmol photons m^{-2} s^{-1}) at 34°C to allow the activity of PS II to recover (Control). A, Repair of PS II under oxidative stress (0.5 and 1 mM H$_2$O$_2$). B, Repair of PS II under salt stress (0.5 and 1.0 M NaCl). C, Repair of PS II under low-temperature stress (10, 18, and 25°C).

singlet oxygen (^{1}O$_2$) that is produced damages to the D1 protein (Vass et al., 1992). By contrast, in the proposed "low-light" mechanism, charge recombination under low-intensity light between Q$_A^-$ or Q$_B^-$, another quinone acceptor, and the oxidized S$_{2,3}$ states of the donor side of PS II produces triplet chlorophyll, is responsible for the formation of ^{1}O$_2$ (Keren et al., 1997).

In the proposed "donor-side" mechanism, acidification of the lumen, due to the transfer of protons, inactivates the oxygen-evolving system and allows the formation of long-lived P680$^+$, the oxidized reaction-center chlorophyll, which damages the D1 protein (Callahan et al., 1986; Theg et al., 1986; Klimov et al., 1990; Chen et al., 1992). The evidence for these three possible mechanisms has been drawn mainly from studies of PS II complexes in vitro. However, none of the proposed mechanisms explains several aspects of the photodamage to PS II that have been observed in vivo. First, there is a distinct proportionality between the initial rate of photodamage to PS II and light intensity (Fig. 1B). The "acceptor-side" mechanism does not explain the proportionality of photodamage to light intensity under low-intensity light, and the "low-light" mechanism, in its turn, does not explain the proportionality of photodamage to light intensity under strong light.

Each of the three hypothetical mechanisms is based on the assumption that the rate of photodamage to PS II depends on the rate of transfer of electrons in PS II. However, interference with the transport of electrons by 3-(3,4-dichlorophenyl)-1,1-dimethylurea (DCMU) had no effect at all on the proportionality of photodamage to light intensity in *Synechocystis* (Nishiyama et al., 2004). Moreover, strict proportionality was even observed when the system for the transport of electrons in PS II was supersaturated by exposing the thylakoid membranes from pumpkin leaves to single-turnover flash conditions (Tyystjärvi et al., 2001). Thus, it appears that photodamage to PS II is independent of the transport of electrons.

The "acceptor-side" and "low-light" mechanisms also suggest that ^{1}O$_2$ might be responsible for photodamage to PS II. As we shall also see below, neither the overproduction of ^{1}O$_2$ nor the elimination of oxygen in *Synechocystis* cells has any effect on the proportionality of photodamage to light intensity (Nishiyama et al., 2004). Thus, it seems unlikely that ^{1}O$_2$ is responsible for photodamage to PS II in cyanobacteria.

The proportionality between light intensity and the photodamage to PS II indicates that photodamage depends on the number of photons absorbed and this proportionality also suggests the existence of a photon sensor in PS II. The action spectrum of photodamage to PS II has a peak in the ultraviolet and blue regions (Jones and Kok, 1966a; Greenberg et al., 1989; Jung and Kim, 1990; Tyystjärvi et al., 2002) and resembles the absorption spectra of model manganese compounds (Baffert et al., 2002; Carrell et al., 2003). Thus, Tyystjärvi et al. (2001) proposed that the primary event in photodamage might be the absorption of photons by the manganese cluster of the oxygen-evolving machinery, with subsequent dissociation of excited manganese ions and the formation of P680$^+$. Then P680$^+$, the strongest biological oxidant identified to date (1.12 volts; see Klimov et al., 1979), might damage the D1 protein (Anderson, 2001). Therefore, it is likely that the photodamage to PS II is a purely light-dependent event that occurs under light at all intensities and does not involve the action of oxygen. In this context, light might not be a stress factor and, furthermore, susceptibility to damage by light might be an intrinsic feature of PS II, as suggested by Tyystjärvi and Aro (1996).

III. Effects of Oxidative Stress

A. Reactive Oxygen Species (ROS)

Oxygen is essential for the viability of most organisms but it is also potentially toxic. In its ground state, oxygen is a triplet molecule that is generally unreactive as a result of spin restriction. The activation of oxygen in its ground state by various reactions overcomes the spin restriction with resultant formation of reactive oxygen species (ROS). Reduction of oxygen leads to the formation of the superoxide radical (O_2^-), hydrogen peroxide (H_2O_2), and the hydroxyl radical ($\cdot OH$), while electronic excitation leads to the formation of singlet-state oxygen (singlet oxygen; $^1O_2$). These various ROS can potentially damage many cellular components, such as proteins, lipids, and nucleic acids (Halliwell and Gutteridge, 1990).

In photosynthetic organisms, the major source of ROS is the photosynthetic machinery in thylakoid membranes (Asada, 1996, 1999). Both H_2O_2 and $\cdot OH$ are generated during the reduction of O_2^-, which is generated most abundantly on the acceptor side of photosystem I as a result of the photosynthetic transport of electrons. H_2O_2 and O_2^- are also generated in illuminated PS II (Ananyev et al., 1992; Chen et al., 1992). $^1O_2$ is generated by the transfer of energy from excited pigments, such as chlorophylls, in the light-harvesting complexes (Knox and Dodge, 1985; Zolla and Rinalducci, 2002); from excited Fe-S centers in photosystem I (Chung and Jung, 1995); and from photodamaged PS II (Anderson, 2001). The generation of various ROS is promoted when the photosynthetic machinery absorbs excess light and also when the availability of CO_2 or of NADP is limited (Asada, 1996, 1999).

B. ROS and Photodamage to PS II

As described above, various studies suggest that ROS and, in particular, $^1O_2$ are the primary cause of photodamage to PS II. According to the proposed "acceptor-side" and "low-light" mechanisms, the $^1O_2$ that is formed by the transfer of energy from triplet chlorophyll damages the D1 protein, and this damage triggers the enzymatic degradation of the D1 protein (Vass et al., 1992; Keren et al., 1997). The generation of $^1O_2$ can be detected when PS II complexes, thylakoid membranes, and plant leaves are illuminated (Telfer et al., 1994; Hideg et al., 1994, 1998). Exposure of thylakoid membranes to $^1O_2$ results in the selective and specific cleavage of the D1 protein (Okada et al., 1996). Other ROS, such as H_2O_2 and O_2^-, also induce the specific cleavage of the D1 protein in vitro (Miyao et al., 1995). Free-radical scavengers protect the D1 protein from degradation in *Spirodela* plants, suggesting that oxygen free radicals, such as $\cdot OH$, are involved in the light-dependent degradation of the D1 protein (Sopory et al., 1990). However, since effects of ROS on the repair system have not been investigated in these cited studies, the roles of ROS in photodamage in vivo remain to be fully clarified.

Details of the roles of ROS in the photodamage to PS II were examined in vivo in *Synechocystis* by separating photodamage from repair (Nishiyama et al., 2001, 2004). Increases in intracellular concentrations of $^1O_2$, caused by exposure of cells to photosensitizers, such as rose bengal and ethyl eosin at 10 μM, stimulated the apparent photodamage to PS II. However, the actual photodamage to PS II, as assessed in the presence of chloramphenicol, was unaffected by the production of $^1O_2$. These findings indicate that $^1O_2$ acts by inhibiting the repair of photodamaged PS II and not by accelerating damage to PS II. The proportionality between photodamage and light intensity, as shown in Figure 1B, was totally unaffected by the overproduction of $^1O_2$ in cells in the presence of photosensitizers and by the elimination of oxygen from cells in *Synechocystis* (Nishiyama et al., 2004). Thus, it appears that $^1O_2$ might not be responsible for photodamage to PS II.

Increases in the intracellular concentration of H_2O_2, caused by exposure of cells to 0.5 mM H_2O_2 or by disruption of genes for H_2O_2-scavenging enzymes, revealed that H_2O_2 also inhibits the repair of PS II but does not damage PS II directly (Nishiyama et al., 2001). Thus, it is likely that a variety of ROS act primarily by inhibiting the repair of photodamaged PS II.

C. Inhibition of Repair by ROS

Direct inhibition of the repair of PS II by ROS was demonstrated in studies that showed that the recovery of the activity of photodamaged PS II was severely inhibited by the intracellular production of $^1O_2$ and also by exposure of *Synechocystis* cells to 0.5 mM H_2O_2 (Fig. 2A; Nishiyama et al., 2001, 2004). Moreover, a mutant of *Synechocystis* that lacked two H_2O_2-scavenging enzymes was unable to repair photodamaged PS II in the presence of 0.5 mM H_2O_2 (Nishiyama et al., 2001).

D. Molecular Mechanisms of ROS-Induced Inhibition

The extent of the repair of PS II is determined by the rate of the synthesis of the D1 protein de novo (Aro et al., 1993). Suppression of the de novo synthesis of the D1 protein by $^1O_2$ and H_2O_2 was demonstrated by labeling proteins with [^{35}S]methionine in *Synechocystis* in vivo (Nishiyama et al., 2001, 2004; Allakhverdiev and Murata, 2004). These studies also revealed that $^1O_2$ and H_2O_2 not only suppressed the de novo synthesis of the D1 protein but also suppressed the synthesis of almost all of the other proteins in thylakoid membranes, suggesting that the target of suppression might be a process common to the synthesis of all proteins.

The synthesis of the mature D1 protein is accomplished by a sequence of events that includes transcription of *psbA* genes, translation of *psbA* mRNA, and processing of the precursor to the D1 protein (pre-D1). An attempt to identify the process that is suppressed by ROS was made by analyzing levels of *psbA* mRNA and pre-D1. Northern and immunoblot analysis revealed that the translation of *psbA* mRNA is suppressed by $^1O_2$ and H_2O_2 (Nishiyama et al., 2001, 2004). Furthermore, the subcellular localization of polysomes with bound *psbA* mRNA suggested that the primary target of $^1O_2$ and H_2O_2 might be the elongation step of translation (Nishiyama et al., 2001, 2004).

The sensitivity of elongation factors to oxidative stress has been reported in *Escherichia coli* and in mammals. Elongation factor G is sensitive to carbonylation by H_2O_2 in *Escherichia coli* (Tamarit et al., 1998), and this factor was also identified as one of the proteins that are most susceptible to oxidation in a mutant of *Escherichia coli* that lacked a superoxide dismutase (Dukan and Nyström, 1999). Moreover, elongation factor 2 is inactivated selectively by the oxidant cumene hydroperoxide in rat liver (Ayala et al., 1996). Thus, elongation factors appear to be the most probable candidates for primary targets of $^1O_2$ and H_2O_2.

In the chloroplast, the translation of *psbA* mRNA is regulated by redox components at the initiation step (Hirose and Sugiura, 1996; Yohn et al., 1996; Trebitsh and Danon, 2001) and at the elongation step (Zhang et al., 2000). It remains to be determined whether such redox components are affected by oxidative stress due to $^1O_2$ and H_2O_2. In contrast to cyanobacteria, the D1 protein of higher plants undergoes phosphorylation, which may regulate the degradation of the D1 protein (Rintamäki et al., 1995) and the metabolism of the D1 protein in a circadian-dependent manner (Booij-James et al., 2002). It is of interest to assess whether the phosphorylation of the D1 protein is affected by oxidative stress.

IV. Effects of Salt Stress

A. Salt Stress and Photodamage to PS II

Salt stress is an important environmental factor that limits the growth and productivity of plants (Boyer, 1982; Hagemann and Erdmann, 1997). In natural environments, salt stress often occurs in combination with light stress and there have been several studies of the effects of salt stress on PS II under strong light. Such studies suggest that salt stress might enhance photodamage to PS II in *Chlamydomonas reinhardtii* (Neale and Melis, 1989); leaves of barley (*Hordeum vulgare*; Sharma and Hall, 1991), sorghum (*Sorghum bicolor*; Sharma and Hall, 1991), and rye (*Secale cereale;* Hertwig et al., 1992); and in *Spirulina platensis* (Lu and Zhang, 1999). However, since the cited studies were performed under conditions that negatively impact synthesis and repair of PS II proteins, it remains to be determined whether salt stress induces damage to PS II directly.

The separate effects of salt stress on damage and repair have been examined in *Synechocystis*

Chapter 13 Repair of Photosystem II and Environment

(Allakhverdiev et al., 1999, 2002; Allakhverdiev and Murata, 2004). Salt stress, due to 0.5 M NaCl, inhibited the repair of photodamaged PS II (Fig. 2B) but did not accelerate damage to PS II directly. Thus, it appears that the enhanced photodamage to PS II that was observed in earlier studies might have been due to the synergistic effects of light and salt stress.

B. Molecular Mechanisms of Salt-Induced Inhibition

The labeling of proteins in *Synechocystis* in vivo revealed that the de novo synthesis of the D1 protein is inhibited by salt stress in the form of 0.5 M NaCl (Allakhverdiev et al., 2002; Allakhverdiev and Murata, 2004). Moreover, salt stress suppressed not only the synthesis of the D1 protein de novo but also the synthesis of almost all other proteins. In an analysis of the reactions that lead to the synthesis of the D1 protein de novo, northern and immunoblot analyses revealed that salt stress suppressed the synthesis of the D1 protein at both the transcriptional and the translational levels (Allakhverdiev et al., 2002).

C. Overall Gene Expression under Salt Stress

DNA microarrays have been used to examine the effects of environmental stress on the overall expression of genes in *Synechocystis* (Hihara et al., 2001; Kanesaki et al., 2002). Salt stress in the form of 0.5 M NaCl strongly diminished the inducibility by light of approximately 60% of light-inducible genes and moderately suppressed the inducibility by light of approximately 20% of light-inducible genes (Allakhverdiev et al. 2002). At 1.0 M NaCl, the expression of all the normally light-inducible genes was no longer inducible by light. Thus, salt stress depressed the expression of various light-inducible genes. It appears that direct interference with the transcriptional machinery by salt stress is partly responsible for the inhibition of the repair of PS II.

V. Effects of Low-Temperature Stress

A. Low-Temperature Stress and Repair of PS II

Low-temperature stress also has a significant negative effect on the growth and productivity of plants (Öquist et al., 1987; Adams et al. 2002, 2004; Öquist and Huner, 2003). It has often been noted that low-temperature stress apparently enhances photodamage to PS II under strong light (Öquist and Huner, 1991; Öquist et al., 1993). Studies of the effects of low-temperature stress on the repair of PS II demonstrated that repair is inhibited at low temperatures both in *Synechocystis* (Gombos et al., 1994; Wada et al., 1994) and in plants (Moon et al., 1995; Alia et al., 1996). This conclusion was confirmed in *Synechocystis* by monitoring damage and repair separately at low temperatures, such as 10°C and 18°C (Allakhverdiev and Murata, 2004). Low-temperature stress did not accelerate photodamage to PS II but inhibited the repair of photodamaged PS II (Fig. 2C). Thus, the combination of low-temperature stress and light stress has synergistic effects, namely, damage caused by light and inhibition of repair by low temperature.

B. Molecular Mechanisms of Low-Temperature-Induced Inhibition

Labeling of proteins in *Synechocystis* in vivo demonstrated that the de novo synthesis of the D1 protein was suppressed at low temperatures, such as 10°C and 18°C (Allakhverdiev and Murata, 2004). The extent of suppression depends on temperature and it seems likely that the suppression of the synthesis of proteins de novo might be a generalized phenomenon (Allakhverdiev and Murata, 2004). The step in the synthesis of the D1 protein that is suppressed by low temperature remains to be identified.

C. Unsaturation of Fatty Acids

The unsaturation of fatty acids in membrane lipids is an important factor in determining the tolerance of plants and cyanobacteria to low temperature (Wada et al., 1990; Murata et al., 1992). For example, the unsaturation of fatty acids protects PS II from the inhibition of the activity that is caused by strong light at low temperatures (Wada et al., 1990; Murata et al., 1992). Molecular biological studies in plants and cyanobacteria have demonstrated that the unsaturation of fatty acids protects PS II at low temperatures by accelerating the repair of photodamaged PS II (Gombos et al., 1994; Wada et al., 1994; Moon et al., 1995). Studies with a mutant of *Synechocystis* that was deficient in a fatty-acid desaturase revealed that the processing of pre-D1 was modulated by the unsaturation of fatty acids at low temperatures (Kanervo et al., 1997).

D. Repair in Darkness

Separation of photodamage from repair at low temperatures revealed the existence of an intermediate form of photodamaged PS II (Allakhverdiev et al., 2003). After photodamage to PS II in *Synechocystis* at low temperatures, which ranged from 0 to 10°C, the activity of PS II recovered to about 50% of the original level in darkness at moderate temperatures without the de novo synthesis of the D1 protein. This observation suggests that an intermediate form of photodamaged PS II might be repaired in a manner that is different from light-dependent repair. Full repair of photodamaged PS II requires the de novo synthesis of proteins in the light, which is accelerated with increases in temperature, mainly at the step that corresponded to the processing of pre-D1 (Allakhverdiev et al., 2003).

VI. Conclusions and Future Perspectives

The separation of photodamage from repair has helped to clarify the nature of the inhibition of the activity of PS II by environmental stress. Figure 3 summarizes the actions of environmental factors in the damage to and repair of PS II. Oxidative stress, salt stress, and low-temperature stress act primarily by inhibiting the repair of photodamaged PS II. By contrast, light acts by damaging PS II directly. The susceptibility of PS II to damage by light appears to be an intrinsic feature of PS II. However, full details of the mechanisms of photodamage to PS II remain to be elucidated. In addition to the proportionality between photodamage and light intensity, the action spectrum of photodamage to PS II, which has a peak in the ultraviolet and blue regions, forces us to reconsider the previously proposed mechanisms of photodamage that are based on the absorption of light by chlorophylls. Although the effects of other types of environmental stress, such as high-temperature stress, osmotic stress, and drought stress, remain to be examined, the similarities among the effects of oxidative stress, salt stress, and low-temperature stress suggest that it is the repair process that regulates the sensitivity of PS II to environmental conditions.

Investigations of molecular mechanisms have revealed that these repair-inhibiting stresses act similarly to suppress the de novo synthesis of proteins and, in particular, the synthesis of the D1 protein, which is required for the repair of photodamaged PS II. It seems likely that the suppression of the de novo synthesis of proteins is the rate-limiting step in the repair of PS II under environmental stress. The translational and transcriptional machineries appear to be particularly sensitive to environmental stress, for example, oxidative stress and salt stress. Future studies should be directed towards a full characterization of the molecular mechanisms whereby translation and transcription are inhibited by environmental stress.

Acknowledgments

This work was supported in part by a Grant-in-Aid for Scientific Research (no. 16013237 to Y.N.) from the Ministry of Education, Culture, Sports, Science, and Technology of Japan, and by the National Institute of Basic Biology Cooperative Research on Stress Tolerant Plants.

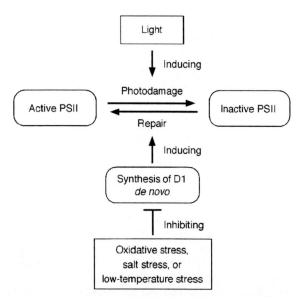

Fig. 3. Working model showing the effects of various types of environmental stress on damage and repair of PS II. Light damages PS II directly. Oxidative stress, salt stress, and low-temperature stress inhibit the repair of photodamaged PS II by suppressing the synthesis of the D1 protein *de novo*.

References

Adams WW III, Demmig-Adams B, Rosenstiel TN, Brightwell AK and Ebbert V (2002) Photosynthesis and photoprotection in overwintering plants. Plant Biol 4: 545–557

Adams WW III, Zarter CR, Ebbert V and Demmig-Adams B (2004) Photoprotective strategies of overwintering evergreens. Bioscience 54: 41–49

Adir N, Zer H, Shochat S and Ohad I (2003) Photoinhibition - a historical perspective. Photosynth Res 76: 343–370

Alia, Kondo Y, Sakamoto A, Nonaka H, Hayashi H, Pardha Saradhi P, Chen THH and Murata N (1999) Enhanced tolerance to light stress of transgenic *Arabidopsis* plants that express the *codA* gene for a bacterial choline oxidase. Plant Mol Biol 40: 279–288

Allakhverdiev SI and Murata N (2004) Environmental stress inhibits the synthesis de novo of D1 protein in the photodamage-repair cycle of photosystem II in *Synechocystis* sp. PCC 6803. Biochim Biophys Acta 1657: 23–32

Allakhverdiev SI, Setlikova E, Klimov VV and Setlik I (1987) In photoinhibited photosystem II particles pheophytin photoreduction remains unimpaired. FEBS Lett 226: 186–190

Allakhverdiev SI, Nishiyama Y, Suzuki I, Tasaka Y and Murata N (1999) Genetic engineering of the unsaturation of fatty acids in membrane lipids alters the tolerance of *Synechocystis* to salt stress. Proc Natl Acad Sci USA 96: 5862–5867

Allakhverdiev SI, Nishiyama Y, Miyairi S, Yamamoto H, Inagaki N, Kanesaki Y and Murata N (2002) Salt stress inhibits the repair of photodamaged photosystem II by suppressing the transcription and translation of *psbA* genes in *Synechocystis*. Plant Physiol 130: 1443–1453

Allakhverdiev SI, Mohanty P and Murata N (2003) Dissection of photodamage at low temperature and repair in darkness suggests the existence of an intermediate form of photodamaged photosystem II. Biochemistry 42: 14277–14283

Ananyev G, Wydrzynski T, Renger G and Klimov V (1992) Transient peroxide formation by the manganese-containing, redox-active donor side of photosystem II upon inhibition of O_2 evolution with lauroylcholine chloride. Biochim Biophys Acta 1100: 303–311

Anbudurai PR, Mor TS, Ohad I, Shestakov SV and Pakrasi HB (1994) The *ctpA* gene encodes the C-terminal processing protease for the D1 protein of the photosystem II reaction center. Proc Natl Acad Sci USA 91: 8082–8086

Anderson JM (2001) Does functional photosystem II complex have an oxygen channel? FEBS Lett 488: 1–4

Andersson B and Aro E-M (2001) Photodamage and D1 protein turnover in photosystem II. In: Aro E-M and Andersson B (eds) Regulation of Photosynthesis, pp 377–393. Kluwer Academic Publishers, Dordrecht, The Netherlands

Aro E-M, Virgin I and Andersson B (1993) Photoinhibition of photosystem II. Inactivation, protein damage and turnover. Biochim Biophys Acta 1143: 113–134

Arntz B and Trebst A (1986) On the role of the Q_B protein of photosystem II in photoinhibition. FEBS Lett 194: 43–49

Asada K (1996) Radical production and scavenging in the chloroplasts. In: Baker NR (ed) Photosynthesis and the Environment, pp 123–150. Kluwer Academic Publishers, Dordrecht, The Netherlands

Asada K (1999) The water-water cycle in chloroplasts: scavenging of active oxygens and dissipation of excess photons. Annu Rev Plant Physiol Plant Mol Biol 50: 601–639

Ayala A, Parrado J, Bougria M and Machado A (1996) Effect of oxidative stress, produced by cumene hydroperoxide, on the various steps of protein synthesis. J Biol Chem 271: 23105–23110

Baffert C, Collomb MN, Deronzier A, Pecaut J, Limburg J, Crabtree RH and Brudvig GW (2002) Two new terpyridine dimanganese complexes: a manganese(III,III) complex with a single unsupported oxo bridge and a manganese (III,IV) complex with a dioxo bridge. Synthesis, structure, and redox properties. Inorganic Chem 41: 1404–1411

Berry J and Björkman O (1980) Photosynthetic response and adaptation to temperature in higher plants. Ann Rev Plant Physiol 31: 491–543

Björkmann O and Powles SB (1984) Inhibition of photosynthetic reactions under water stress: interaction with light level. Planta 161: 490–504

Booij-James IS, Swegle WM, Edelman M and Mattoo AK (2002) Phosphorylation of the D1 photosystem II reaction center protein is controlled by an endogenous circadian rhythm. Plant Physiol 130: 2069–2075

Boyer JS (1982) Plant productivity and environment. Science 218: 443–448

Boyer JS and Bowen BL (1970) Inhibition of oxygen evolution in chloroplasts isolated from leaves with low water potentials. Plant Physiol 45: 612–615

Callahan FE, Becker DW and Cheniae GM (1986) Studies on the photoinactivation of the water-oxidizing enzyme. II. Characterization of weak-light photoinhibition of PSII and its light-induced recovery. Plant Physiol 82: 261–269

Carrell TG, Bourles E, Lin M and Dismukes GC (2003) Transition from hydrogen atom to hydride abstraction by Mn_4O_4 $(O_2PPh_2)_6$ versus $[Mn_4O_4(O_2PPh_2)_6]^+$: O-H bond dissociation energies and the formation of $Mn_4O_3(OH)(O_2PPh_2)_6$. Inorganic Chem 42: 2849–2858

Chen G-X, Kazimir J and Cheniae GM (1992) Photoinhibition of hydroxylamine-extracted photosystem II membranes: studies of the mechanism. Biochemistry 31: 11072–11083

Chung SK and Jung J (1995) Inactivation of the acceptor side and degradation of the D1 protein of photosystem II by singlet oxygen photogenerated from the outside. Photochem Photobiol 61: 383–389

Cleland RE (1988) Molecular events of photoinhibitory inactivation in the reaction centre of photosystem II. Aus J Plant Physiol 15: 135–150

Demeter S, Neale PJ and Melis A (1987) Photoinhibition: Impairment of the primary charge separation between P680 and pheophytin in PSII of chloroplasts. FEBS Lett 214: 370–374

Demmig-Adams B and Adams WW III (1992) Photoprotection and other responses of plants to high-light stress. Annu Rev Plant Physiol Plant Mol Biol 43: 599–626

Dukan S and Nyström T (1999) Oxidative stress defense and deterioration of growth-arrested *Escherichia coli* cells. J Biol Chem 274: 26027–26032

Gombos Z, Wada H and Murata N (1994) The recovery of photosynthesis from low-temperature photoinhibition is accelerated by the unsaturation of membrane lipids: a mechanism of chilling tolerance. Proc Natl Acad Sci USA 91: 8787–8791

Greenberg BM, Gaba V, Canaani O, Malkin S, Mattoo AK and Edelman M (1989) Separate photosensitizers mediate degradation of the 32-kDa photosystem II reaction center protein in the visible and UV spectral regions. Proc Natl Acad Sci USA 86: 6617–6620

Hagemann M and Erdmann N (1997) Environmental stresses. In: Rai AK (ed) Cyanobacterial Nitrogen Metabolism and Environmental Biotechnology, pp 156–221. Springer-Verlag, Heidelberg, Germany

Halliwell B and Gutteridge JMC (1990) Role of free radicals and catalytic metal ions in human disease: an overview. Methods Enzymol 186: 1–88

Hertwig B, Streb P and Feierabend J (1992) Light dependence of catalase synthesis and degradation in leaves and the influence of interfering stress conditions. Plant Physiol 100: 1547–1553

Hideg E, Spetea C and Vass I (1994) Singlet oxygen and free radical production during acceptor- and donor-side-induced photoinhibition. Studies with spin trapping EPR spectroscopy. Biochim Biophys Acta 1186: 143–152

Hideg E, Kálai T, Hideg K and Vass I (1998) Photoinhibition of photosynthesis *in vivo* results in singlet oxygen production: detection via nitroxide-induced fluorescence quenching in broad bean leaves. Biochemistry 37: 11405–11411

Hihara Y, Kamei A, Kanehisa M, Kaplan A and Ikeuchi M (2001) DNA microarray analysis of cyanobacterial gene expression during acclimation to high light. Plant Cell 13: 793–806

Hirose T and Sugiura M (1996) *Cis*-acting elements and *trans*-acting factors for accurate translation of chloroplast *psbA* mRNAs: development of an *in vitro* translation system from tobacco chloroplasts. EMBO J 15: 1687–1695

Inagaki N, Yamamoto Y and Satoh K (2001) A sequential two-step proteolytic process in the carboxy-terminal truncation of precursor D1 protein in *Synechocystis* sp. PCC 6803. FEBS Lett 509: 197–201

Jones HG (1973) Limiting factors in photosynthesis. New Phytol 72: 1089–1094

Jones LW and Kok B (1966a) Photoinhibition of chloroplast reactions. I. Kinetics and action spectra. Plant Physiol 41: 1037–1043

Jones LW and Kok B (1966b) Photoinhibition of chloroplast reactions. II. Multiple effects. Plant Physiol 41: 1044–1049

Jung J and Kim HS (1990) The chromophores as endogenous sensitizers involved in the photogeneration of singlet oxygen in spinach thylakoids. Photochem Photobiol 52: 1003–1009

Kanervo E, Tasaka Y, Murata N and Aro E-M (1997) Membrane lipid unsaturation modulates processing of the photosystem II reaction-center protein D1 at low temperatures. Plant Physiol 114: 841–849

Kanesaki Y, Suzuki I, Allakhverdiev SI, Mikami K and Murata N (2002) Salt stress and hyperosmotic stress regulate the expression of different sets of genes in *Synechocystis* sp. PCC 6803. Biochem Biophys Res Commun 290: 339–348

Keck RW and Boyer JS (1974) Chloroplast response to low leaf water potential. III. Differing inhibition of electron transport and photophosphorylation. Plant Physiol 53: 474–479

Keren N and Ohad I (1998) State transition and photoinhibition. In: Rochaix J-D, Goldschmidt-Clermont M and Merchant S (eds) The Molecular Biology of Chloroplast and Mitochondria in *Chlamydomonas*, Advances in Photosynthesis, Vol 7, pp 569–596. Kluwer Academic Publishers, Dordrecht, The Netherlands

Keren N, Berg A, van Kan PJM, Levanon H and Ohad I (1997) Mechanism of photosystem II photoinactivation and D1 protein degradation at low light: The role of back electron flow. Proc Natl Acad Sci USA 94: 1579–1584

Klimov VV, Allakhverdiev SI, Demeter S and Krasnovsky AA (1979) Photoreduction of pheophytin in the photosystem II of chloroplasts depending on the oxidation-reduction potential of the medium. Dokl Acad Nauk USSR 249: 227–230

Klimov VV, Shafiev MA and Allakhverdiev SI (1990) Photoinactivation of the reactivation capacity of photosystem II in pea subchloroplast particles after a complete removal of manganese. Photosynth Res 23: 59–65

Knox JP and Dodge AD (1985) Singlet oxygen and plants. Phytochemistry 24: 889–896

Kyle DJ, Ohad I and Arntzen CJ (1984) Membrane protein damage and repair: selective loss of quinone-protein function in chloroplast membranes. Proc Natl Acad Sci USA 181: 4070–4074

Lee HY, Hong YN and Chow WS (2001) Photoinactivation of photosystem II complex and photoprotection by non-functional neighbours in *Capsicum annuum* L. leaves. Planta 212: 332–342

Lu C-M and Zhang J-H (1999) Effects of salt stress on PSII function and photoinhibition in the cyanobacterium *Spirulina platensis*. J Plant Physiol 155: 740–745

Mattoo AK, Hoffman-Falk H, Marder JB and Edelman M (1984) Regulation of protein metabolism: Coupling of photosynthetic electron transport to *in vivo* degradation of the rapidly metabolized 32-kilodalton protein of the chloroplast membranes. Proc Natl Acad Sci USA 81: 1380–1384

Mattoo AK, Marder JB and Edelman M (1989) Dynamics of the photosystem II reaction center. Cell 56: 241–246

Melis A (1999) Photosystem-II damage and repair cycle in chloroplasts: what modulates the rate of photodamage *in vivo*? Trends Plant Sci 4: 130–135

Miyao M, Ikeuchi M, Yamamoto N and Ono T (1995) Specific degradation of the D1 protein of photosystem II by treatment with hydrogen peroxide in darkness: implication for the mechanism of degradation of the D1 protein under illumination. Biochemistry 34: 10019–10026

Moon BY, Higashi S, Gombos Z and Murata N (1995) Unsaturation of the membrane lipids of chloroplasts stabilizes the photosynthetic machinery against low-temperature photoinhibition in transgenic tobacco plants. Proc Natl Acad Sci USA 92: 6219–6223

Murata N, Ishizaki-Nishizawa O, Higashi S, Hayashi H, Tasaka Y and Nishida I (1992) Genetically engineered alteration in the chilling sensitivity of plants. Nature 356: 710–713

Neale PJ and Melis A (1989) Salinity-stress enhances photoinhibition of photosystem II in *Chlamydomonas reinhardtii*. J Plant Physiol 134: 619–622

Nishiyama Y, Allakhverdiev SI, Yamamoto H, Hayashi H and Murata N (2004) Singlet oxygen inhibits the repair of photosystem II by suppressing translation elongation of the D1 protein in *Synechocystis* sp. PCC 6803. Biochemistry 43: 11321–11330

Nishiyama Y, Yamamoto H, Allakhverdiev SI, Inaba M, Yokota A and Murata N (2001) Oxidative stress inhibits the repair of photodamage to the photosynthetic machinery. EMBO J 20: 5587–5594

Ohad I, Kyle DJ and Arntzen CJ (1984) Membrane protein damage and repair: removal and replacement of inactivated 32-kilodalton polypeptide in chloroplast membranes. J Cell Biol 99: 481–485

Ohad I, Keren N, Zer H, Gong HS, Mor TS, Gal A, Tal S and Eisenberg-Domovich Y (1994) Light-induced degradation of the photochemical reaction center II-D1 protein *in vivo*: An integrative approach. In: Baker NR, and Bowyer JR (eds) Photoinhibition of Photosynthesis: From Molecular Mechanisms to the Field, pp 161–177. Bios Scientific Publishers, Oxford

Okada K, Ikeuchi M, Yamamoto N, Ono T and Miyao M (1996) Selective and specific cleavage of the D1 and D2 proteins of photosystem II by exposure to singlet oxygen: factors responsible for the susceptibility to cleavage of the proteins. Biochim Biophys Acta 1274: 73–79

Öquist G and Huner NPA (1991) Effects of cold-acclimation on the susceptibility of photosynthesis to photoinhibition in Scots pine and in winter and spring cereals: A fluorescence analysis. Func Ecol 5: 91–100

Öquist G and Huner NPA (2003) Photosynthesis of overwintering evergreen plants. Annu Rev Plant Biol 54: 329–355

Öquist G, Greer DH and Ogren E (1987) Light stress at low temperature. In: Kyle DJ, Osmond CB and Arntzen CJ (eds) Topics in Photosynthesis: Photoinhibition, pp 67–87. Elsevier Science Publishers BV, Amsterdam, The Netherlands

Öquist G, Hurry VM and Huner NPA (1993) The temperature dependence of the redox state of Q_A and susceptibility of photosynthesis to photoinhibition. Plant Physiol Biochem 31: 683–691

Rintamäki E, Kettunen R, Tyystjärvi E and Aro E-M (1995) Light-dependent phosphorylation of D1 reaction center protein of photosystem II: hypothesis for the functional role in vivo. Physiol Plant 93: 191–195

Park Y-I, Chow WS and Anderson JM (1995) Light inactivation of functional photosystem II in leaves of peas grown in moderate light depends on photon exposure. Planta 196: 401–411

Powles SB (1984) Photoinhibition of photosynthesis induced by visible light. Annu Rev Plant Physiol 35: 15–44

Prásil O, Adir N and Ohad I (1992) Dynamics of photosystem II: mechanism of photoinhibition and recovery processes. In: Barber J (ed) Topics in Photosynthesis, Vol 11, The Photosystems: Structure, Function and Molecular Biology, pp 295–348. Elsevier Science Publishers, Amsterdam, The Netherlands

Setlik I, Allakhverdiev SI, Nedbal L, Setlikova E and Klimov VV (1990) Three types of photosystem II photoinactivation: I. Damaging processes on the acceptor side. Photosynth Res 23: 39–48

Sharma PK and Hall DO (1991) Interaction of salt stress and photoinhibition on photosynthesis in barley and sorghum. J Plant Physiol 138: 614–619

Sopory SK, Greenberg BM, Mehta RA, Edelman M and Mattoo AK (1990) Free radical scavengers inhibit light-dependent degradation of the 32 kDa photosystem II reaction center protein. Z Naturforsch 45c: 412–417

Tamarit J, Cabiscol E and Ros J (1998) Identification of the major oxidatively damaged proteins in *Escherichia coli* cells exposed to oxidative stress. J Biol Chem 273: 3027–3032

Telfer A, Bishop SM, Phillips D and Barber J (1994) The isolated photosynthetic reaction center of PS II as a sensitiser for the formation of singlet oxygen; detection and quantum yield determination using a chemical trapping technique. J Biol Chem 269: 13244–13253

Theg SM, Filar LJ and Dilley RA (1986) Photoinactivation of chloroplasts already inhibited on the oxidizing side of photosystem II. Biochim Biophys Acta 849: 104–111

Trebitsh T and Danon A (2001) Translation of chloroplast *psbA* mRNA is regulated by signals initiated by both photosystems II and I. Proc Natl Acad Sci USA 21: 12289–12294

Tyystjärvi E and Aro E-M (1996) The rate constant of photoinhibition, measured in lincomycin-treated leaves, is directly proportional to light intensity. Proc Natl Acad Sci USA 93: 2213–2218

Tyystjärvi E, Kairavuo M, Pätsikkä E, Keränen M, Khriachtchev L, Tuominen I, Guiamet JJ and Tyystjärvi T (2001) The quantum yield of photoinhibition is the same in flash light and under continuous illumination: implication for the mechanism. In: Critchley C (ed) Proceedings of the 12th International Congress of Photosynthesis, S8-P032. CSIRO Publishing, Melbourne, Australia

Tyystjärvi T, Tuominen I, Herranen M, Aro E-M and Tyystjärvi E (2002) Action spectrum of *psbA* gene transcription is similar to that of photoinhibition in *Synechocystis* sp. PCC 6803. FEBS Lett 516: 167–171

Vass I, Styring S, Hundal T, Koivuniemi A, Aro E-M and Andersson B (1992) The reversible and irreversible intermediates during photoinhibition of photosystem II: stable reduced Q_A species promote chlorophyll triplet formation. Proc Natl Acad Sci USA 89: 1408–1412

Wada H, Gombos Z and Murata N (1990) Enhancement of chilling tolerance of a cyanobacterium by genetic manipulation of fatty acid desaturation. Nature 347: 200–203

Wada H, Gombos Z and Murata N (1994) Contribution of membrane lipids to the ability of the photosynthetic machinery to tolerate temperature stress. Proc Natl Acad Sci USA 91: 4273–4277

Yohn CB, Cohen A, Danon A and Mayfield SP (1996) Altered mRNA binding activity and decreased translation initiation in a nuclear mutant lacking translation of the chloroplast *psbA* mRNA. Mol Cell Biol 16: 3560–3566

Zhang L, Paakkarinen V, van Wijk KJ and Aro E-M (2000) Biogenesis of the chloroplast-encoded D1 protein: regulation of translation elongation, insertion, and assembly into photosystem II. Plant Cell 12: 1769–1781

Zolla L and Rinalducci S (2002) Involvement of active oxygen species in degradation of light-harvesting proteins under light stresses. Biochemistry 41: 14391–14402

Chapter 14

Photosystem I and Photoprotection: Cyclic Electron Flow and Water-Water Cycle

Tsuyoshi Endo*
Division of Integrated Life Science, Graduate School of Biostudies, Kyoto University, Kyoto 606-8502, Japan

Kozi Asada*
Department of Biotechnology, Faculty of Life Science and Biotechnology, Fukuyama University, Gakuen-cho 1, Fukuyama, 729-0292, Japan

Summary 205
I. Introduction 206
II. Cyclic Electron Flow around Photosystem I (PS I) 207
    A. Cyclic Electron Flow Mediated by NAD(P)H Dehydrogenase (NDH) 207
        1. Function of NDH in Photoprotection 207
        2. Electron Donors of NDH 208
        3. Subunit Composition of NDH 208
    B. The Cyclic Electron Flow Mediated by Ferredoxin-Quinone Reductase (FQR) 210
    C. Cyclic Electron Flow and Chlororespiration in Algae and Plants 211
III. The Water-Water Cycle 212
    A. Reaction Sequence and Stoichiometry of the Water-Water Cycle 212
    B. Characteristics of the Water-Water Cycle 212
    C. Target Molecules of Reactive Oxygen Species in Chloroplasts 214
IV. Comparison between Cyclic Electron Flow and the Water-Water Cycle in Terms of Physiological Role 214
    A. Dissipation of Excess Photons and Electrons by Cyclic Electron Flow and the Water-Water Cycle 214
    B. Physiological Role of Cyclic Electron Flow 215
    C. Physiological Role of the Water-Water Cycle 216
V. Perspectives 217
Acknowledgments 217
References 217

Summary

Cyclic electron transport around photosystem I has been proposed to play dual roles in the regulation of photosynthetic electron transport: down-regulating PS II and adjusting the ATP/NADPH ratio. Recent molecular genetics revealed that cyclic electron flow is essential for normal photosynthesis and growth. The water-water-cycle would also play a role similar to cyclic electron transport, in addition to the effective scavenging of reactive oxygen species generated in PS I. Though their rates of electron flux are lower than that of linear electron transport at steady state,

*Author for correspondence, email: tuendo@kais.kyoto-u.ac.jp or asada@bt.fubt.fukuyama-u.ac.jp

these alternative electron flows are indispensable for acute responses to environmental changes and stress. Recent biochemical and molecular studies at the protein and gene level have clarified the components participating in the alternative electron transport. These new findings, including the dual functions of cyclic electron flow and the water-water cycle, and their respective roles in stress responses, are discussed in this chapter.

I. Introduction

The photosynthetic electron transport system is often regarded as a single chain of electron transfer. However, there are actually diverging and converging routes of electron flows around the central chain known as the 'Z-scheme' or 'linear electron transport chain'. These alternative routes of electron flow have been studied for some time, but their physiological functions are still a matter of debate. Among these alternative electron transport routes, cyclic electron transport around photosystem I (PS I) and the water-water cycle have been most intensively studied.

Cyclic electron flow around PS I was discovered by Arnon's group over 40 years ago (Tagawa et al., 1963). Reduced electron carriers in the stroma, such as reduced ferredoxin (redFd) and NADPH photoproduced in PS I, can donate electrons to the intersystem plastoquinone and thus move electrons in a cycle around PS I. Fd-dependent cyclic flow, which has been shown to be inhibited by antimycin A, is thought to be mediated by the putative enzyme Fd-quinone reductase (FQR). NAD(P)H-dependent cyclic flow has been shown to be mediated by NAD(P)H dehydrogenase (NDH), a homologue of the respiratory NADH dehydrogenase in mitochondria and bacteria, which has been referred to as Complex I in the respiratory electron transport. Recent molecular biological studies have characterized FQR and NDH, which are one of the main topics of this review. Advances in the study of cyclic electron transport as well as chlororespiration have been reviewed previously by Nixon (2000) and Peltier and Cournac (2002). Although transcriptional and post-transcriptional regulation of the expression of *ndh* genes have been used by many researchers as models for studying plastid gene expression, these studies are not included in this review. Heber and Walker (1992) have proposed that cyclic electron flow has dual functions, as an inducer of proton gradient-dependent down-regulation of PS II as well as a generator of additional ATP required for the CO_2-fixation cycle.

The very same dual function can also be applied to the water-water cycle, in addition to the effective scavenging of reactive oxygen species (ROS) in chloroplasts. The photoreduction of dioxygen was discovered by Mehler (1951), and has been referred to as the Mehler reaction. Subsequently, the primary photoreducing product of dioxygen was identified as superoxide anion radicals (O_2^-), that are photoproduced in PS I instead of reducing $NADP^+$ via ferredoxin (Asada et al., 1974). The hydrogen peroxide generated via SOD-catalyzed disproportionation of superoxide radicals is reduced to water via a series of enzymatic reactions in the vicinity of the PS I complex, by the electrons derived from the oxidation of water in PS II. Because the electrons derived from water reduce dioxygen to water, this effective scavenging system of reactive, reduced species of oxygen (superoxide and hydrogen peroxide) in PS I has been referred to as the water-water cycle. Molecular mechanisms and functions of the water-water cycle have been reviewed by Asada (1999, 2000) in plants and Miyake and Asada (2003) in algae.

The primary physiological function of the water-water cycle is the rapid scavenging of ROS generated in PS I prior to their interaction with target molecules in chloroplasts as a means for protection against photoinhibition. In addition, like the cyclic electron flow around PS I, the water-water cycle does not produce net reducing equivalents but only ATP through the generation of a proton gradient across the thylakoid membranes. This common property is a reason why both cyclic electron flow and the water-water cycle are thought to have similar physiological functions.

ATP production and generation of a proton gradient across the thylakoid membranes may both play roles in the protection against photoinhibition. ATP production is important for the fine-tuning of the ATP/NADPH ratio for effective CO_2-fixation, the chloroplastic and cellular requirements for which might vary from one condition to another. The ATP/NADPH ratio required for the Calvin cycle turnover is 1.5 in C_3 plants. Whether the linear or non-cyclic electron transport supports this molar ratio has been a matter of serious discussion.

Abbreviations: AsA – ascorbate; DHA – dehydroascorbate; Fd – ferredoxin; FNR – ferredoxin-NADPreductase; FQR – ferredoxin-quinone reductase; GSH – reduced glutathione; GSSG – oxidized glutathione; MDA – monodehydroascorbate; NDH – NAD(P)H dehydrogenase; PS I – photosystem I; PS II – photosystem II; redFd – reduced ferredoxin; ROS – reactive oxygen species; SOD – superoxide dismutase.

If the operation of the Q cycle is facultative and only occurs under specific light conditions, linear electron flow alone does not produce enough ATP for the Calvin cycle under full sunlight, resulting in over-production of NADPH. This over-reducing condition can often be observed when low light- or dark-adapted leaves are suddenly exposed to excessive, high light (Cornic et al., 2000; Barth et al. 2001). Additional ATP production, via either cyclic electron transport or the water-water cycle, might prevent the over-reduction in the intersystem electron carriers. In this way, the alternative flows of electrons play a role in protection from photoinhibition. The second possible protective role is related to proton gradient formation, which induces down regulation of PS II by increasing dissipation of photon energy as heat. This process is reviewed in detail in other chapters of this volume (Adams et al.; Demmig-Adams et al.; Jung and Niyogi). Recently, operation of cyclic electron transport around PS II has been shown (Miyake and Yokota, 2001; Miyake and Okamura, 2003), which may function to induce an extra proton gradient in a similar manner to cyclic electron flow around PS I.

II. Cyclic Electron Flow around Photosystem I (PS I)

A. Cyclic Electron Flow Mediated by NAD(P)H Dehydrogenase (NDH)

1. Function of NDH in Photoprotection

The presence of *ndh* (A-K) genes in the chloroplastic genomes of moss and tobacco was found independently by Ohyama et al. (1986) and Shinozaki et al. (1986). These genes have a high identity with the subunits of mitochondrial Complex I, that mediates electron transfer between NADH and the ubiquinones. Since those initial findings, similar encoding of *ndh* in the chloroplastic genome has been found in angiosperms, gymnosperms, and eukaryotic algae. Furthermore, similar *ndh* genes have also been found in cyanobacteria (T. Ogawa, 1991). These genes are, however, lacking in the plastid genomes of several symbiotic plants (dePamphilis and Palmer, 1990), suggesting that NDH plays an essential role in photosynthetic reactions. Even so, the function of NDH remained obscure until an *ndh*B-less cyanobacterium became available.

T. Ogawa (1991) isolated mutants of the cyanobacterium *Synechocystis* PCC6803 that could grow only in CO_2 at elevated concentrations, but not at air-concentration. Genetic analysis showed that these mutants have mutations in *ndh* genes, suggesting that NDH is essential for providing ATP to concentrate CO_2 in the cells from air. Cyclic electron transport around PS I, as observed by the reduction of $P700^+$ and an increase in chlorophyll fluorescence after actinic light illumination, was found in the wild-type cells, but not in a *Synechocystis* mutant in which the *ndh*B gene was inactivated (Mi et al., 1992a,b, 1994, 1995). This confirmed that NDH functions as the mediator between cytoplasmic electron donors photoproduced in PS I and plastoquinone in cyanobacteria. Further, NADPH functions as the electron donor to plastoquinone in the thylakoids of wild-type cells, but not those of the NDH-less mutant. Thus, NDH is the mediator between the cytoplasmic electron donor NADPH and plastoquinone in cyanobacteria, and it has been proposed that NDH-mediated cyclic electron transport supplies ATP required for the accumulation of CO_2 in cells. This proposed function of NDH is supported by the fact that the biosynthesis of NDH subunits is induced in *Synechocystis* cells under low CO_2 conditions (Deng et al., 2003).

Development of successful chloroplast transformation made it possible to analyze the function of NDH using gene disruption in higher plants. Four groups independently succeeded in the production of NDH-deficient transformants of tobacco (Burrows et al., 1998; Kofer et al., 1998; Joët et al., 1998; Shikanai et al., 1998). Unlike the *Synechocystis* mutants, the NDH-deficient tobacco plants can grow normally, at least under growth chamber conditions below 100 μmol photons m^{-2} s^{-1}. The common phenotype of all of these transformants is that they lack a transient increase in chlorophyll fluorescence in the dark after actinic illumination, which represents the transient reduction of the plastoquinone pool by reducing equivalents accumulated in the stroma during illumination (Asada et al., 1993; Mano et al., 1995). Thus, the lack of a post-illumination increase in chlorophyll fluorescence indicates that the NDH-deficient mutant does not have a pathway of electron transfer from the stromal reducing equivalents to plastoquinone. These results demonstrate the operation of an NDH-mediated cyclic flow around PS I in plants as well. The lack of a marked difference between the phenotypes of wild-type and NDH-less mutants under growth chamber conditions suggests that the ATP produced via NDH-dependent cyclic electron flow is either not necessary or is compensated by ATP produced via the Fd-dependent cyclic electron flow, as discussed below.

Upon illumination with saturating light for several minutes, NDH-deficient tobacco leaves accumulate photoreductants in the stroma and the intersystem chain to a greater degree than wild type, as judged from the redox changes of P700 (Endo et al., 1999) and chlorophyll fluorescence (Takabayashi et al., 2002). The accumulation of photoreductants, i.e., the over-reduction in the stroma and the intersystem carriers, is probably due to an inappropriate ATP/NADPH ratio for CO_2-fixation in the NDH-deficient mutants. Thus, the over-reduction found in NDH-deficient plants can be explained by assuming that NDH functions to produce supplemental ATP under strong light, similarly to what occurs under field conditions.

It should be noted that the phenotype of over-reduction under strong light is more pronounced in young leaves of an NDH-deficient tobacco mutant as compared to mature plants (Endo et al., 2001). The more distinct effects of the inactivation of NDH in young versus mature plants indicate that the requirement for NDH is more important in young leaves, probably due to their low CO_2-fixation capacity in the chloroplasts. In other words, NDH-dependent cyclic flow effectively dissipates excess photon energy when the photon energy-accepting capacity of the Calvin cycle is low in immature chloroplasts. Little effect of NDH-deficiency observed by Birth and Krause (2002) was probably because their study was limited to mature plants.

In conclusion, ATP production by NDH-mediated cyclic flow appears to be non-essential under low-light, growth chamber conditions, but is important under photo-oxidative stress and especially in young leaves. Under photo-oxidative conditions, NDH activity is enhanced in barley due to the phosphorylation of the NdhF subunit (Lascano et al., 2003). The growth retardation of NDH-deficient transformants under water stress conditions (Horváth et al., 2000) also supports this proposal. The phenotype of over-reduction in the tobacco NDH-less mutant is induced by chilling stress even under low irradiance (Li et al., 2004).

2. Electron Donors of NDH

The electron donor of chloroplastic NDH is a matter of debate. Partially purified NDH from the cyanobacterium *Synechocystis* PCC 6803 accepts electrons only from NADPH, but not from NADH (Mi et al., 1995; Matsuo et al., 1998). By contrast, purified NDH from plant thylakoids uses NADH rather than NADPH as electron donor (Sazanov et al., 1998; Elortza et al., 1999; Casano et al., 2000). In addition, NDH can be purified from plant thylakoids in the form of either an NADPH-specific complex (Guedeney, 1999) or a complex accepting both NADH and NADPH (Quiles and Cuello, 1998; Funk et al., 1999). Chlorophyll fluorescence analysis of spinach thylakoids indicates the dark reduction of plastoquinones by NADPH, but not by NADH. This NADPH-dependent reduction of plastoquinones was observed only when the thylakoids were prepared from intact chloroplasts in a medium containing 25 mM Mg^{2+} (Endo et al., 1997). Thus, the NADPH-dependent NDH appears to bind to the thylakoid membranes prepared only at this high concentration of Mg^{2+}, but not at 1 mM Mg^{2+}. By contrast, thylakoids from potato (Corneille et al., 1998) and barley (Teicher and Scheller, 1998) demonstrate electron donation to the plastoquinones from both NADPH and NADH. The activity in potato thylakoids is insensitive to rotenone, antimycin A, and piericidin A, whereas that in barley thylakoids is sensitive only to rotenone. Thylakoid membranes from wild-type tobacco demonstrate electron donation from redFd in an antimycin A-resistant manner, which is absent in the NDH-deficient mutant, suggesting that NDH also has Fd-plastoquinone reductase activity (Endo et al., 1998). The involvement of Fd-$NADP^+$ reductase (FNR) in NDH activity was suggested by immunological studies (Guedeney et al., 1996; Quiles and Cuello, 1998; Quiles et al., 2000). Thus, contradictory results have been presented regarding the electron donor of NDH, but it is clear that thylakoid membranes bind a distinct enzyme with NAD(P)H-plastoquinone oxidoreductase activity.

3. Subunit Composition of NDH

Chloroplastic NDH has homologous, well-characterized counterparts in mitochondria and heterotrophic bacteria. Since different nomenclatures have been used for *ndh* genes in these organisms, homologous subunits of NDH from various organisms are compared with one another in Table 1 according to their sequence alignments (Fearnley and Walker, 1992). The bacterial Complex I (NDH-1) is generally composed of 14 subunits. In *Escherichia coli*, however, NuoC and NuoD are fused to form NuoCD, resulting in 13 subunits (for a review see Friedrich, 1998). The genomes of tobacco chloroplasts and cyanobacteria lack the NADH-binding subunits found in *Bos taurus* (51, 24 and 75 kDa proteins) and *E. coli* (NuoE, F and G proteins). The 51 kDa (NuoF) subunit contains FMN and may have an NADH-binding domain. Thus, these three subunits have been referred to as the NADH-binding subcomplex (Friedrich, 1998). The other 11

Table 1. Subunit composition of mitochondrial Complex I of *Bos taurus*, NDH-1 of the bacteria *Escherichia coli* and *Rhodobacter capsulatus*, NDH of the cyanobacterium *Synechocystis* 6803, and the chloroplastic NDH complex of *Nicotiana tabacum*.

B. taurus[a] (Complex I)	*E. coli*[b] and *R. capsulatus* (NDH-1)	*Synechocystis* (NDH) and *N. tabacum* chloroplast (Chloroplastic NDH)
NAD(P)H-binding subcomplex		
51 kDa	NuoF	—[c]
24 kDa	NuoE	—[c]
75 kDa	NuoG	—[c]
Connecting subcomplex		
49 kDa	NuoD	NdhH
30 kDa	NuoC	NdhJ
20 kDa (PSST)	NuoB	NdhK
23 kDa (TYKY)	NuoI	NdhI
Hydrophobic subcomplex		
ND1	NuoH	NdhA
ND2	NuoN	NdhB
ND3	NuoA	NdhC
ND4	NuoM	NdhD
ND4L	NuoK	NdhE
ND5	NuoL	NdhF
ND6	NuoJ	NdhG

[a] Mitochondrial Complex 1 from mammals contains 27 additional subunits not shown here.
[b] NuoC and NuoD are fused to compose a single protein, NuoCD, in *E. coli*.
[c] Homologous subunits corresponding to those of NAD(P)H-binding subcomplex in bacteria are not encoded in chloroplastic DNA in plants and are not found in the cyanobacterium genome. For details, see the text.

subunits found in *E. coli* are all encoded in tobacco chloroplasts and cyanobacteria, which are referred to as NdhA-K. To avoid confusion in the nomenclature, we hereafter use the term 'NdhA-K', regardless of their origins.

Among these subunits, NdhH, I, J, and K may form a hydrophilic subcomplex that is easily dissociated from a membrane-embedded subcomplex (NdhA, B, C, D, E, and F). Since the four hydrophilic subunits (NdhH, I, J, and K) connect the NADH-binding subcomplex with the hydrophobic subcompex (NdhA-F), they are named the 'connecting subcomplex' in *E. coli* (Leif et al., 1995). Protease treatment of thylakoid membranes of spinach shows that the NdhI and J subunits are exposed to the stromal side of the thylakoid membranes (Lennon et al., 2003), suggesting that the subunit organization of chloroplastic NDH is similar to that of Complex I from bacteria and mitochondria. The NAD(P)H-binding and the connecting subcomplexes are together named the peripheral arm in *Neurospora crassa* and fragment I α in *E. coli*, while the subcomplex within the membranes is referred to as the 'membrane arm' in *N. crassa* and 'fragment I β' in *E. coli* (Videira, 1998). From the analogy with *E. coli*, the NdhA subunit might correspond to a quinone-binding protein, and NdhB, D, and F may be involved in proton transport through the thylakoid membranes (Friedrich, 1998). In mitochondrial Complex I from mammals, the subunits composing the hydrophobic subcomplex (NdhA-G) are encoded in the mitochondrial genome, while those of the NADH-binding and connecting subcomplexes are nuclear-encoded. In higher plants, two subunits (corresponding to NdhH and J) are encoded in the mitochondrial genome, but in mammals they are encoded in the nucleus (Rasmusson et al., 1998).

In *E.coli* NDH-1, the NADH-binding subunit is NuoF, but no corresponding subunit is encoded in the chloroplastic genome (Table 1). Thus, it is very likely that the NAD(P)H-binding subunit of chloroplastic NDH is encoded in the nuclear genome. However, no *nuoF* homologous gene has been found in the whole nuclear sequence of *Arabidopsis thaliana*, suggesting that NDH in chloroplasts has an NAD(P)H-binding subunit unique to plants. Proteomic analysis of active thylakoid-bound NDH complex should provide conclusive evidence of this point. It should also be noted that the NAD(P)H-binding subunit has not been isolated and characterized in cyanobacteria (Table 1), where NDH may function very similarly to the chloroplastic counterpart.

B. The Cyclic Electron Flow Mediated by Ferredoxin-Quinone Reductase (FQR)

The main reason why the NDH-deficient transformant of tobacco described above does not show a distinct phenotype with respect to its growth is probably due to the presence of an alternative pathway of cyclic electron flow mediated by ferredoxin around PS I. Since Arnon's group demonstrated ferredoxin-mediated cyclic electron flow (Tagawa et al., 1963), its molecular basis has remained to be clarified. A putative ferredoxin-quinone reductase (FQR), that is sensitive to antimycin A, has been thought to be the mediator of Fd-dependent cyclic electron transport around PS I (for review see Bendall and Manasse, 1995). Miyake et al. (1995) demonstrated the participation of a low potential cytochrome b-559 in the antimycin A-sensitive cyclic electron transport in mesophyll chloroplasts from maize. A spectrophotometric survey of the thylakoid membranes shows the far-red light-dependent reduction of a cytochrome with an α-peak at 559 nm. The reduction is dependent on ferredoxin and sensitive to antimycin A, suggesting cyclic flow of electrons around PS I, i.e., P700 → ferredoxin → Cyt b_{559} → plastoquinone. This Cyt b-559 is distinguished from Cyt b-559 in the PS II complex by its low redox potential. Recent X-ray structural analysis of the Cyt b_6/f complex reveals the presence of a novel c-type heme, which has been referred to as heme x or heme c_i. This novel heme is located at a position very close to the heme b_6 but oriented to the stroma side, which allows it to access the electrons from the stroma (Kurisu et al., 2003; Stroebel et al., 2003). This novel heme in the Cyt b_6/f complex could act as the electron carrier for the reduction of the plastoquinone in the cyclic electron flow around PS I.

Thus, chloroplasts have two pathways for the reduction of plastoquinone, ferredoxin-mediated and NAD(P)H-mediated cyclic flows. Furthermore, as stated above, chloroplastic NDH in tobacco shows FQR activity (Endo et al., 1998) that is not inhibited by antimycin A. Thus, this NDH-dependent FQR activity is distinguishable from the antimycin A-sensitive FQR. Based on these findings, the pathways for the reduction of plastoquinone around PS I are proposed to be as shown in Fig. 1.

The recent molecular identification of a cyclic factor PGR5 in *Arabidopsis* on a gene and protein basis (Munekage et al., 2002) may open a gate for the analysis of the long-unidentified FQR. PGR5 was originally identified as a factor that suppresses non-photochemical quenching of chlorophyll fluorescence.

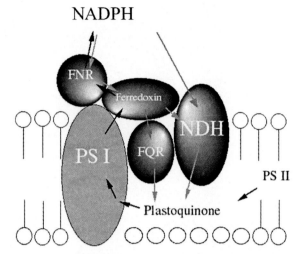

Fig. 1. Proposed pathways of cyclic electron transport around PS I. NAD(P)H dehydrogenase (NDH) may accept electrons from ferredoxin (Fd) as well as NADPH. The ferredoxin quinone reductase (FQR) pathway seems to be the main pathway, but the NDH pathway may generate extra ATP via a putative proton pump. If NADH is the substrate of NDH, the presence of a transhydrogenase, converting NADPH to NADH, can be assumed.

Further analysis of the *pgr5*-deficient mutant indicates that PGR5 is involved in antimycin A-sensitive electron transfer from redFd to the plastoquinones of the intersystem chain, suggesting that PGR5 is a subunit of FQR. Another important phenotype of the *pgr5* mutant is its susceptibility to photoinhibition in PS I. Under environmental conditions where the photoreductants accumulate in the stroma, photoinhibition of PS I is induced (Terashima et al., 1994; Sonoike, 1996). Thus, the deficiency of PGR5-associated cyclic flow probably results in the over-reduction of the stromal electron carriers such as $NADP^+$, ferredoxin, and thioredoxins. The lack of cyclic electron flow induces imbalance of ATP/NADPH rate due to insufficient production of ATP. Therefore, the clear phenotype of the *pgr5* mutant may settle the long-standing argument whether cyclic electron flow functions to produce sufficient ATP for the turnover of the CO_2-fixation cycle under physiological conditions. The role of PGR5 in the formation of non-photochemical quenching, as shown above, may also settle another argument about whether cyclic flow can down-regulate PS II via generation of a proton gradient across the thylakoid membranes. These two important phenotypic traits of the *pgr5* mutant provide the first clear evidence that supports the 'dual function' of cyclic electron transport proposed by Heber and Walker (1992). The remaining problem is to verify that PGR5 is really a subunit of FQR. The *pgr5* DNA sequence indicates the absence of redox reaction domains in PGR5, implying that unidentified subunits should participate in the electron transfer of FQR.

The relationship between the FQR- and NDH-mediated cyclic electron flows is another important point to be clarified. Recently, by crossing *pgr5* and *ccr* mutants, Munekage et al. (2004) isolated *Arabidopsis* mutants in which both the NDH- and FQR-mediated cyclic pathways are impaired. The *ccr* mutant lacked the expression of *ndh* genes (Hashimoto et al., 2003). The double mutants showed much slower growth than both parents did, suggesting that the NDH and FQR pathways can function in a complementary manner for the production of additional ATP required for CO_2 fixation. These results also strongly indicate that cyclic electron flow is an essential pathway for the normal operation of photosynthesis.

The growth rates of the mutants indicate that FQR represents the main pathway, while the NDH pathway compensates when the FQR pathway cannot operate. However, several other functional differences between the NDH and FQR pathway should be considered. The pool size of NADPH is larger than that of redFd in chloroplasts. Thus, the NDH pathway has a larger electron buffer for reducing equivalents when leaves are suddenly exposed to strong light. If NDH in chloroplasts has a proton-pump, as is the case for its counterpart in mitochondria, NDH can generate a larger proton gradient than FQR does. The study of the NDH-defective mutant indicated that the NDH pathway donates electrons to the plastoquinone pool for at least several minutes even after actinic illumination is turned off, while no such slow reaction has been indicated in the FQR pathway. This suggests a kinetic difference between the two cyclic pathways: the FQR pathway is very effective for acute responses and the NDH pathway for long-lasting responses. The physiological relevance of these differences must be further clarified to understand a reason why the two cyclic electron flows around PS I are necessary.

C. Cyclic Electron Flow and Chlororespiration in Algae and Plants

The term "chlororespiration" was introduced by Bennoun (1982) to refer to the electron transport chain oxidizing NAD(P)H at the expense of oxygen in chloroplasts of several eukaryotic algae. Chlororespiration is a chloroplastic analogue of the respiratory electron transport in mitochondria, and the oxygen uptake associated with this process is different from photorespiration and the Mehler reaction. Chlororespiration shares several electron carriers with those in the intersystem on the thylakoid membranes. Thus, the interaction between chlororespiration and photosynthetic linear electron transport in algal chloroplasts is very analogous to that between respiration and photosynthesis in cyanobacteria.

The NDH involved in chlororespiration has common properties with NDH involved in the mitochondrial respiratory chain. Godde and Trebst (1980) reported that NADH oxidation in *Chlamydomonas* thylakoids is inhibited by rotenone, and Godde (1982) purified membrane-associated NADH-plastoquinone oxidoreductase, which contains flavin and Fe-S. Ravenel et al. (1994) demonstrated NADPH-dependent cyclic flow around PS I in *C. reinhardtii*, as detected by the photoacoustic method. This activity is not inhibited by classical inhibitors of Complex I including rotenone, but the mediator of this cycle has not been identified. It was also demonstrated that Fd-dependent cyclic flow around PS I is inhibited by antimycin A and HQNO, both of which are inhibitors of FQR in plants. Cyclic electron flow in this alga, as measured by energy storage by PS I, is suppressed only when both of the alternative pathways are inhibited.

NDH in the thylakoid membranes of cyanobacteria has a similar role in dark respiration as in Complex I in respiratory electron transport, since the dark oxygen uptake of the NDH-defective mutant is very low compared to that of the wild type (Mi et al., 1994). In cyanobacteria, it appears that NDH-dependent electron flow is required for the reversible state transition between states 1 and 2, since the NDH-lacking mutant is locked in state 1. Thus, the mutant cannot respond to chromatic change (Schreiber et al., 1995). In addition, the chlororespiratory electron donation to the plastoquinone may have a function of down-regulating PS II via state transition under oxidative stress conditions, as demonstrated in *Chlamydomonas* (Endo et al., 1995; Endo and Asada, 1996).

NDH-defective tobacco indicates that NDH-mediated chlororespiration can operate in chloroplasts of higher plants. In the light, electrons in the intersystem chain may be used for $P700^+$ reduction, but in darkness the electrons may be transferred to dioxygen via a putative oxidase in the thylakoids. Dark reduction of plastoquinone in sunflower leaves is inhibited by rotenone and stimulated by carbon monoxide, an inhibitor of cytochrome *c* oxidase in mitochondria (Feild et al., 1998). Recently, the presence of a quinol oxidase, whose structure is similar to that of the alternative oxidase in mitochondria, has been shown to be present in thylakoid membranes of plants (for review, see Nixon, 2000; Peltier and Cournac, 2002). Protease treatment of thylakoid membranes degrades the IMMUTANS polypeptide, the alternative oxidase in thylakoids, suggesting that this oxidase faces the stromal side (Lennons et al., 2003). Thus, the oxidase cannot generate a proton gradient across the thylakoid membranes, but may function to dissipate excess electrons from the plastoquinone pool to oxygen in order to avoid over-reduction.

III. The Water-Water Cycle

A. Reaction Sequence and Stoichiometry of the Water-Water Cycle

The water-water cycle in the chloroplast is a process of photoreduction of one molecule of dioxygen to two molecules of water at the reducing side of PS I by four electrons generated in PS II from two molecules of water (Asada, 1996,1999). Therefore, no net change of oxygen apparently occurs, and this cycle can be directly observed only using $^{18}O_2$ or $H_2^{18}O_2$ in algal cells (Radmer and Kok, 1976; Miyake et al., 1991) or intact chloroplasts (Asada and Badger, 1984). Thus, the overall stoichiometry of the water-water cycle is as follows:

$$2H_2^{16}O \rightarrow {}^{16}O_2 + 4e^- + 4H^+ \quad \text{(PSII)}$$
$$^{18}O_2 + 4e^- + 4H^+ \rightarrow 2H_2^{18}O \quad \text{(PSI)}$$

where, e^- represents reducing equivalents. The whole sequence of reactions and participating enzymes involved in the water-water cycle are summarized in Table 2.

Allocation of the reducing equivalents generated in PS II to either the major pathway of the Calvin cycle or the water-water cycle occurs at PS I, where the electrons univalently reduce dioxygen in the water-water cycle (reactions 1 and 2) or $NADP^+$ via Fd in the major pathway. Superoxide anion radicals (O_2^-) produced by the univalent reduction of dioxygen are then disproportionated to oxygen and hydrogen peroxide via superoxide dismutase (SOD)-catalyzed reaction (reaction 3), and the hydrogen peroxide is reduced to water by ascorbate (AsA) via ascorbate-specific peroxidase (reaction 4). The primary oxidation product of this reaction, monodehydroascorbate radical (MDA), is reduced to ascorbate by either redFd or NADPH via MDA reductase (reaction 5). When MDA fails to be reduced to ascorbate via reaction 5, MDA radicals spontaneously disproportionate to dehydroascorbate (DHA) and AsA (reaction 6). Subsequently, DHA is reduced back to AsA by GSH via DHA reductase (reaction 7). Half of the electrons from two molecules of water (reaction 1) are used in the reduction of dioxygen to superoxide radicals (reaction 2), and the remaining half for the regeneration of AsA as the reducing equivalents (reaction 8). In either reducing routes via MDA and DHA, two electrons are required for the regeneration of one molecule of AsA using either redFd, NADPH, or GSH.

The water-water cycle, involving the reactions shown in Table 2, is present in chloroplasts of plants and eukaryotic algae. These photosythetic organisms are able to synthesize ascorbate, but cyanobacteria are not. $H_2^{18}O_2$-experiments in cyanobacteria, however, indicate the operation of the water-water cycle (Miyake et al., 1991; Miyake and Asada, 2003). In cyanobacteria, however, the role of ascorbate peroxidase in the water-water cycle of plants (reaction 4) is replaced by thioredoxin peroxidase, which reduces hydrogen peroxide with reduced thioredoxin (H. Yamamoto et al., 1999). Reduced thioredoxin is photoproduced in PS I by redFd via Fd-thioredoxin reductase.

The molecular and enzymatic properties of the enzymes participating in the water-water cycle have been characterized (Asada, 1999). For detailed physicochemical properties of ascorbate peroxidase as a family of heme peroxidases, see Raven (2003). Chloroplastic and cytosolic isoforms of ascorbate peroxidase, SOD, and MDA reductase have been found. In chloroplasts, ascorbate peroxidase occurs in the thylakoid-bound and stromal forms (Miyake et al., 1991), and these isoforms arise from a common pre-mRNA for the chloroplastic isoform by alternative splicing (Yoshimura et al., 1999). In *Arabidopsis*, the ascorbate peroxidase, monodehydroascorbate reductase, and glutathione reductase are also found in mitochondria, where they would function to scavenge ROS via a similar mechanism as in the water-water cycle of chloroplasts except for the electron donors. These enzymes are produced via differential transcripts from respective single genes and have been shown to be targeted to mitochondria and chloroplast stroma (Obara et al., 2002; Chew et al., 2003).

B. Characteristics of the Water-Water Cycle

The overall reaction of the water-water cycle is the reduction of oxygen by electrons derived from water. Through this pathway, electron flow from PS II to PS I and generation of the proton gradient across the thylakoid membranes are allowed to continue even in the absence of an electron acceptor of PS I, e.g. under CO_2-deficiency and/or excess photons. Thus, the

Table 2. Reaction sequence of the water-water cycle in chloroplasts and the participating enzymes in the cycle.

Reaction (enzyme)	Reaction
1) Photooxidation of water in PS II	$2 H_2O \rightarrow O_2 + 4 H^+ + 4e^-$
2) Photoreduction of oxygen in PS I	$2 O_2 + 2 e^- \rightarrow 2 O_2^-$
3) Disproportionation of superoxide (Superoxide dismutase)	$2 O_2^- + 2 H^+ \rightarrow O_2 + H_2O_2$
4) Reduction of hydrogen peroxide by ascorbate (Ascorbate peroxidase)	$H_2O_2 + 2 AsA \rightarrow 2 H_2O + 2 MDA$
5) Regeneration of AsA from MDA	
Spontaneous reduction of MDA by redFd	$2 MDA + 2 redFd \rightarrow 2 AsA + 2 Fd$
Reduction of MDA by NAD(P)H (MDA reductase)	$2 MDA + NAD(P)H \rightarrow 2 AsA + NAD(P)^+$
6) Spontaneous disproportionation of MDA	$2 MDA \rightarrow AsA + DHA$
7) Regeneration of AsA from DHA	
Reduction of DHA by GSH (DHA reductase)	$DHA + 2 GSH \rightarrow AsA + GSSG$
8) Generation of Reducing Eq. for reduction of MDA and DHA	
Photoreduction of Fd in PS I	$2 Fd + 2 e^- \rightarrow 2 redFd$
Photoreduction of NADP$^+$ in PS I via Fd (Fd-NADP$^+$ reductase)	$NADP^+ + H^+ + 2 redFd \rightarrow NADPH$
Reduction of GSSG by NADPH (Glutathione reductase)	$GSSG + NADPH \rightarrow 2 GSH + NADP^+$

AsA: ascorbate; MDA: monodehydroascorbate; DHA: dehydroascorbate; GSH: reduced glutathione; GSSG: oxidized glutathione; Fd: ferredoxin.

water-water cycle is able to produce ATP, but not NADPH, which is similar to the cyclic electron flow around PS I. Furthermore, the generation of a proton gradient across the thylakoid membranes makes it possible to dissipate excess photon energy as heat via the down regulation of PS II under excess light.

In contrast to cyclic electron flow, the intermediates of oxygen reduction are very reactive and, therefore, dangerous molecules for chloroplasts. To avoid damage by the reactive reduced species of oxygen, the reaction rates of the disproportionation of superoxide, the reduction of the resulting hydrogen peroxide to water, and the regeneration of AsA from MDA (reactions 3, 4, and 5) are several orders of magnitude higher than the rate of superoxide anion production in PS I (reaction 2, Table 2). Thus, the limiting step of the water-water cycle is the reduction of oxygen to form superoxide at PS I (reaction 2); its maximum half time is 5 ms^{-1}, even if all the electrons from PS II flow to dioxygen, but the half time for the SOD-catalyzed disproportionation of superoxide (reaction 3) is simulated to be 0.07 ms^{-1} (Asada 1996, 1999), as estimated from the reaction rate constant (2×10^9 M^{-1} s^{-1}) between superoxide and SOD and the local concentration of SOD on the PS I complex (1 mM). The simulation shows similar high rates for the reduction of hydrogen peroxide (reaction 4). Whole flux of the water-water cycle in chloroplasts has been simulated by Polle (2001).

These high rates of ROS scavenging are attained by nearly diffusion-controlled reaction rates of the respective enzymes with either superoxide or hydrogen proxide, and also by microcompartmentation of these enzymes at high concentrations at the PS I complex

of thylakoids as the site of the generation of superoxide. The chloroplastic isoform of SOD attaches on the PS I complex of the thylakoid membrane (K. Ogawa et al., 1995), and about half of the ascorbate peroxidase of chloroplasts occurs in the thylakoid-bound form in the vicinity of the PS I complex (Miyake et al., 1993). Actually, operation of the Water-water cycle in intact chloroplasts is largely retarded by disorganization of the microcompartmented system with an osmotic shock (Asada and Badger, 1984). Rapid disproportionation of superoxide and reduction of hydrogen peroxide shorten the life times of these intermediates, and do not allow diffusion of superoxide and hydrogen peroxide to target molecules in the chloroplasts, thus resulting in protection from damage by ROS. The very short life times of both superoxide and hydrogen peroxide give little chance to generate the highly reactive hydroxyl radical ($\cdot$ OH) via the transition metal (M^{n+})-dependent Harber-Weiss or Fenton reaction;

$$H_2O_2 + M^{n+} \rightarrow \cdot OH + OH^- + M^{(n+1)+}$$
$$M^{(n+1)+} + O_2^- \rightarrow M^{n+} + O_2,$$

where, M^{n+} represents either Fe^{2+} or Cu$^+$.

The occurrence of the water-water cycle even in cyanobacteria (H. Yamamoto et al., 1999) suggests that the effective scavenging system of ROS was acquired by photosynthetic organisms at very low concentrations of atmospheric oxygen, at least 3 billion years ago, when cyanobacteria evolved. Without this system, cyanobacteria may not have survived, because the cells would have been damaged by ROS derived from the oxygen they themselves produced (Asada, 2000).

C. Target Molecules of Reactive Oxygen Species in Chloroplasts

Under conditions where the water-water cycle cannot operate properly or light intensities are too high for CO_2-fixation, hydrogen peroxide would accumulate in chloroplasts. Several enzymes in chloroplasts are sensitive to hydrogen peroxide and would be target molecules of ROS. Ascorbate peroxidase is very sensitive to hydrogen peroxide at very low concentrations in the absence of AsA, because the reaction intermediate of the enzyme with H_2O_2, Complex I, is very labile (Miyake and Asada, 1996). When the concentration of hydrogen peroxide is increased and ascorbate is oxidized in chloroplasts by administration of paraquat in illuminated leaves, ascorbate peroxidase is primarily inactivated prior to other enzymes (Mano et al., 2001). Thus, the H_2O_2-scavenging enzyme ascorbate peroxidase itself is the primary target of H_2O_2, and would enhance a leak of ROS from the water-water cycle.

Several enzymes in the Calvin cycle, such as fructose 1, 6-bisphosphatase, glyceraldehyde 3-phosphate dehydrogenase, ribulose 5-phosphate kinase, and sedoheptulose 1, 7-bisphosphatase, are inactivated by hydrogen peroxide due to the oxidation of the functional sulfhydryl groups of the enzymes in intact chloroplasts (Kaiser, 1976). Inactive, disulfide forms of the above enzymes recover their activities via reduction to the sulfhydryl forms by reduced thioredoxin. The hydroxyl radical produced by the interaction of superoxide with hydrogen peroxide, catalyzed by transition metal ions, initiates the degradation of the large subunit of Rubisco (Mehta et al., 1992; Ishida et al., 1997, 1998). The inactivation of the Calvin cycle enzymes should lower the capacity of CO_2-fixation, resulting in a photon excess state even under weak light. Another important target of hydroxyl radial and singlet oxygen are polyunsaturated fatty acids and chlorophyll. The propagation of lipid oxidation initiated by ROS can be suppressed by tocopherol and also by thioredoxin peroxidase or glutathione peroxidase, which reduces lipid peroxide to its alcohol. The ferredoxin-dependent glutamate synthase in chloroplasts participating in photorespiration is inactivated by ROS. Inactivation of this enzyme suppresses dissipation of excess electrons via photorespiration, and eventually induces photobleaching of leaves (Kozaki and Takeba, 1996). In several chilling-sensitive plants like cucumber, the PSI reaction center is targeted by ROS at low temperatures (Sonoike, 1996; Tjus, 2001).

An electronically excited species of ROS, singlet oxygen, is thought to be generated in PS II under acceptor-limiting conditions via interaction of triplet excited reaction center chlorophyll ($^3P680^+$) formed via recombination with ground state triplet oxygen ($^3O_2$) (Hideg et al., 1998, 2000, 2001, 2002; Y. Yamamoto, 2001). Singlet oxygen is deduced to be involved in the light-induced breakdown of D_1(PsbA) polypeptides leading to photoinhibition of PS II (see Nishiyama et al., Yokthongwattana and Melis, this volume).

IV. Comparison between Cyclic Electron Flow and the Water-Water Cycle in Terms of Physiological Role

A. Dissipation of Excess Photons and Electrons by Cyclic Electron Flow and the Water-Water Cycle

Both the water-water cycle and cyclic electron flow around PS I are able to generate a proton gradient across the thylakoid membrane, and to produce ATP in association with this gradient. In the water-water cycle, the electrons from PS II are consumed just for the reduction of oxygen to water and are not stored as reducing equivalent in the form of NADPH, and therefore the quantum yield of ATP is lower than that of the cyclic electron flow around PS I. The energy dissipating nature of the water-water cycle is clearly understood when it is compared with the respiratory electron flow in which three coupling sites are involved in the oxidation of NADPH. In other words, the water-water cycle can dissipate excess electrons from PS II and can more effectively dissipate excess photon energy as compared with cyclic electron flow around PS I.

Both systems can trigger the proton gradient-dependent down-regulation of PS II quantum yield, and allow dissipation of excess photons by uncoupling the excitation energy transfer from antenna chlorophyll to the reaction center chlorophyll. Charge recombination within PS I under acceptor-limiting conditions has been proposed as a system for dissipation of excess photons (Barth et al., 2001), but its relation to the water-water cycle is unclear. However, both can function as potential energy-dissipating systems in PS I, which has long been neglected as compared with the well-studied energy dissipation in PS II.

In conclusion, although the cyclic electron flow around PS I and the water-water cycle may share similar physiological functions, it should be noted that the cyclic electron flow is an energy conserving system while the water-water cycle is an energy-dissipating system. The redox potentials of the Z scheme and alternative electron flow, including chlororespiration, are compared in Fig. 2.

Chapter 14 Cyclic Flow and Water-Water Cycle

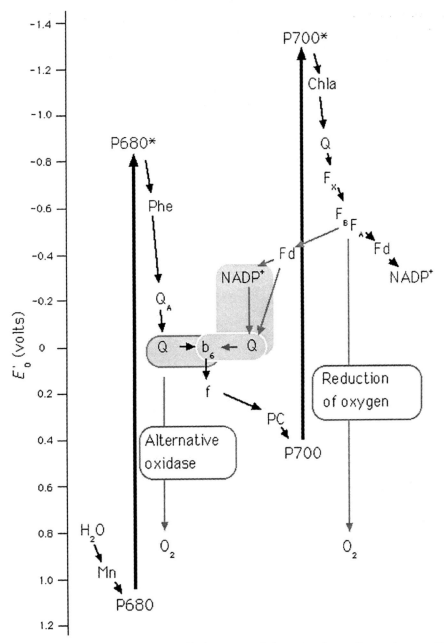

Fig. 2. Redox potentials of the Z scheme, represented by black arrows, and alternative electron flow routes around PS I, represented by gray arrows; cyclic electron flow, the water-water cycle, and alternative oxidase. The coupling sites are shaded gray. The two Qs, in the middle of the scheme, stand for plastoquinone molecules, whereas the third Q on the right side of the diagram stands for what others call A_1 (or phylloquinone); b_6 and f in the middle of the diagram are for cytochrome b_6 and cytochrome f, respectively.

B. Physiological Role of Cyclic Electron Flow

The very severe growth retardation of the double mutant of *pgr5* and *ccr* of *Arabidopsis* clearly indicates the requirement for NDH and FQR-dependent cyclic electron flows for photosynthesis, as presented in section IIB. Over-reduction of the intersystem electron carriers found in this mutant indicates that linear electron flow alone cannot generate an ATP/NADPH rate of 1.5 to drive the Calvin cycle, and does not allow the ΔpH-dependent down-regulation of PS II for suppression of photoinhibition. The rates of ATP/NADPH production as well as H^+/ATP production in linear electron transport (for a review, see Allen, 2002) must be reconsidered in light of this new finding. Cyclic activity has been shown to be controlled by the redox state of the stroma (Harbinson and Foyer, 1991; Cornic et al., 2000; Joët et al., 2002). A requirement of cyclic electron flow for the closure or opening of stomata has been proposed (Harbinson and Foyer, 1991; Heber, 2002). However, the clear difference in stromal redox level between wild-type and the *pgr*5 mutant, even under mild growth conditions where the stomata might remain open, suggests an obligatory operation of the FQR-dependent cyclic flow in the wild-type plants. Thus, cyclic flow may operate under a wide range of environmental fluctuations and may not require specific environmental conditions or a specific stromal redox state.

C. Physiological Role of the Water-Water Cycle

As stated above, there is no doubt that the primary function of the water-water cycle is the rapid and effective scavenging of superoxide and hydrogen peroxide, reactive intermediates of oxygen reduction, to suppress their interactions with target molecules in chloroplasts. Even when the electron flux rate for the reduction of oxygen is below 10% of that for the linear flux, it is estimated that ROS would inhibit CO_2 fixation by inactivation of the target molecules in chloroplasts within several minutes if the water-water cycle did not operate (Asada, 1999). As long as the water-water cycle operates properly and little ROS is released, this cycle is an effective system to dissipate excess photon energy in PS II and excess electrons in PS I. Actually, under anaerobic conditions where the water-water cycle cannot operate, photoinhibition is enhanced (Asada and Takahashi, 1987; Park et al., 1996).

The electron flux through the water-water cycle has been estimated by either photoreduction of $^{18}O_2$ (Radmer and Kok, 1976; Miyake et al., 1991; Asada and Badger, 1984) or by determination of the fate of electrons generated in PS II by measurement of chlorophyll fluorescence (Miyake and Yokota, 2000; Makino et al., 2002; Hirotsu et al., 2004). Electron flux through the water-water cycle is generally high before the onset of CO_2-fixation upon illumination in algae and plants, and then the flux is gradually decreased to its steady state level accompanied by an increase to a plateau level in CO_2-fixation. Early during the induction of CO_2-fixation, nearly all of the electrons from PS II pass through the water-water cycle and the rate of $^{18}O_2$ uptake is the same as that of the oxygen evolution rate in PS II, that is, O_2 is the electron acceptor in place of CO_2 during the induction period of CO_2-fixation (Radmer and Kok, 1976). During this period, the proton gradient through the thylakoid membranes is generally enhanced, as determined by the non-photochemical quenching of chlorophyll fluorescence (Miyake and Yokota, 2000; Makino et al. 2002). It has been supposed that oxygen reduction at PS I allows the oxidation of the intersystem electron carriers and supports efficient operation of the cyclic electron flow (Heber, 2002).

At an O_2 concentration below 0.5%, where the water-water cycle cannot operate, CO_2-fixation is not photoinduced and the stomata remain closed in maize (Owaki and Asada, 2003). Once CO_2-fixation is photoinduced under aerobic conditions, the lowering of oxygen concentration to a nearly anaerobic state does not affect the CO_2-fixation rate, indicating two possibilities: either the water-water cycle can operate even in an anaerobic atmosphere, using the oxygen generated by PS II, or the water-water cycle is not necessary after photoinduction of the Calvin cycle. Thus, a large flux of the electrons through the water-water cycle at the induction or activation period of the CO_2-fixation cycle is inferred from the three functions of the water-water cycle: 1) the dissipation of excess electrons from PS II in PS I via the transfer to oxygen, 2) an increase in the ratio of ATP/NADPH generated via the water-water cycle for the activation of the Calvin cycle, and 3) dissipation of excess photon in PS II via the generation of the proton gradient prior to the start of CO_2-fixation. Furthermore, at least in maize, the water-water cycle appears to be required for stomatal opening. The oxygen requirement for photochemical quenching in the chloroplasts of guard cells has been shown to produce ATP for stomatal opening (Goh et al., 1999).

When activation of the Calvin cycle and steady state CO_2-fixation have been attained, the flux through the water-water cycle is decreased to 10–30% (Asada, 2000), or below 10% of the total electron flux from PS II (Foyer and Noctor, 2000; Badger et al., 2000), although the level varies depending on the plant species and environmental conditions. Whether ATP production and PS II control by the water-water cycle is physiologically significant has been under debate (Osmond and Grace, 1995; Asada, 1999, 2000; Heber, 2002). Based on the generally low electron flux rates observed under steady state photosynthesis, Heber (2002) has proposed that the water-water cycle does not contribute much to controlling of PS II through the formation of a proton gradient across the thylakoid membranes. Even so, the water-water cycle is indispensable for the rapid scavenging of superoxide and hydrogen peroxide to protect plants from photoinhibition due to ROS. Tobacco plants in which expression of the key enzyme of the water-water cycle, thylakoid-bound ascorbate peroxidase, is suppressed, appears to be lethal, but its overexpression gives tolerance to photooxidative stress (Yabuta et al., 2002). In *Arabidopsis*, cytosolic ascorbate peroxidase-lacking plants causes accumulation of hydrogen peroxide indicating an importance to scavenge it in the cytosol for protection from ROS damage (Pnueli et al., 2003). Furthermore, the phenotype of *Arabidopsis* in which the expression of thylakoid-attached CuZn-SOD, another key enzyme of the water-water cycle, is suppressed and the water-water cycle cannot operate, is retarded growth and development, and such plants show typical photoinhibition such as low CO_2-fixation and low chlorophyll content, even under mild conditions (Rizhsky et al., 2003). These observations indicate that the water-water cycle is essential to protect photoinhibition even under mild growth conditions.

V. Perspectives

Genetic evidence for cyclic electron flows around PS I has promoted our understanding of physiological functions of the NDH- and FQR-dependent routes, and has established that cyclic electron flows are essential for photosynthesis and growth even under mild environmental conditions. However, the following key complexes participating in the mediation between the either Fd or NAD(P)H and the plastoquinones have still not been identified; 1) the subunits corresponding to bacterial and mitochondrial NAD(P)H-binding subcomplex of NDH in cyanobacteria, eukaryotic algae, and plants, and 2) the redox component of putative FQR, since PGR5 is supposed to be a component of the putative FQR complex, but does not have a redox domain as a mediator between Fd and the plastoquinones. It is also likely that the redox component of FQR is a component of the cytochrome *b/f* complex. Identification of these components should reveal the redox mechanism of cyclic electron flows, and possibly its regulation. For this goal, biochemical isolation and characterization of FQR- and NDH-complexes would be indispensable.

Requirement of two cyclic electron routes for normal photosynthesis has been well established, and their possible functions are the adjustment of the photoproduction ratio of ATP and NADPH for CO_2-fixation and down regulation of PS II to dissipate excess photons. However, the rates of NDH- and FQR-dependent cyclic flows and that of the linear electron flow with and without the Q-cycle are impossible to determine simultaneously under various environmental conditions. Kinetic contribution of each route to fine adjustment of the amounts of ATP and NADPH, their photoproduction ratio, and down regulation of the quantum yield in PS II via the proton gradient would provide basic data to elucidate the interrelation between cyclic and linear electron transports and their regulation.

The water-water cycle plays an essential role in scavenging of ROS immediately after its generation in PS I, and biochemical and genetic evidence has supported its role. In addition, the water-water cycle would play a similar function to that of cyclic electron transport around PS I as long as ROS is not leaked out from the cycle. Apparently the electron flux through the water-water cycle is high during induction period of CO_2-fixation and is lowered when CO_2-fixation has attained a steady state level. The maximum photoreduction rate of oxygen by isolated thylakoid membranes could not account for the electron flux through the water-water cycle at the maximum rate, where nearly all of the electron flux through PS I is passed through the cycle. Thus, the autooxidation of the thylakoid membrane-bound components is not enough for high operation of the water-water cycle, but stromal factors participate in the enhanced reduction of oxygen. Fd and/or MDA reductase (Miyake et al., 1998) has been assumed to participate in the univalent reduction of dioxygen in chloroplasts.

Acknowledgments

We thank Mr. A. Takabayashi for useful discussions.

References

Adams WW III, Zarter CR, Mueh KE, Amiard V and Demmig-Adams B (2005) Energy dissipation and photoinhibition: a continuum of photoprotection. In: Demmig-Adams B, Adams WW III and Mattoo AK (eds) Photoprotection, Photoinhibition, Gene Regulation, and Environment, pp 49–64. Springer, Dordrecht

Allen JF (2002) Photosynthesis of ATP – electrons, proton pumps, rotors, and poise. Cell 110: 273–276

Asada K (1996) Radical production and scavenging in the chloroplasts. In: Baker NR (ed) Photosynthesis and the Environment, pp 123–150. Kluwer Academic Publishers, Dordrecht

Asada K (1999) The water-water cycle in chloroplasts: Scavenging of active oxygens and dissipation of excess photons. Annu Rev Plant Physiol Plant Mol Biol 50: 601–639

Asada K (2000) The water-water cycle as alternative photon and electron sinks. Phil Trans Roy Soc Lond B 355: 1419–1431

Asada K and Badger MR (1984) Photoreduction of $^{18}O_2$ and $H_2^{18}O_2$ with concomitant evolution of $^{16}O_2$ in intact spinach chloroplasts; Evidence for scavenging of hydrogen peroxide by peroxidase. Plant Cell Physiol 25: 1169–1179

Asada K and Takahashi M (1987) Production and scavenging of active oxygen in photosynthesis. In: Kyle DJ, Osmond CB and Arntzen CJ (eds) Photoinhibition, pp 227–287. Elsevier, Amsterdam

Asada K, Kiso K and Yoshikawa K (1974) Univalent reduction of molecular oxygen by spinach chloroplasts on illumination. J Biol Chem 249: 2175–2181

Asada K, Heber U and Schreiber U (1993) Electron flow to the intersystem chain from stromal components and cyclic electron flow in maize chloroplasts, as detected in intact leaves by monitoring redox change of P700 and chlorophyll fluorescence. Plant Cell Physiol 34: 39–50

Badger MR, von Caemmerer R, Ruuska S and Nakano H (2000) Electron flow to oxygen in higher plants and algae: rates and control of direct photoreduction (Mehler reaction) and rubisco oxygenase. Phil Trans Roy Soc Lond B 355: 1433–1446

Barth C and Krause GH (2002) Study of tobacco transformants to assess the role of chloroplastic NAD(P)H dehydrogenase in photoprotection of photosystem I and II. Planta 216: 273–279

Barth C, Krause GH and Winter K (2001) Responses of photosystem I compared with photosystem II to high-light stress in tropical shade and sun leaves. Plant Cell Environ 24: 163–176

Bendall DS and Manasse RS (1995) Cyclic photophosphorylation and electron transport. Biochim Biophys Acta 1229: 23–38

Bennoun P (1982) Evidence for a respiratory chain in the chloroplast. Proc Natl Acad Sci USA 79: 4352–4356

Burrows PA, Sazanov A, Svab Z, Maliga P and Nixon PJ (1998) Identification of a functional respiratory complex in chloroplasts through analysis of tobacco mutants containing disrupted plastid *ndh* genes. EMBO J 17: 868–876

Casano LM, Zapata JM, Martín M and Sabater B (2000) Chlororespiration and poising of cyclic electron transport. Plastoquinone as electron transporter between thylakoid NADH dehydrogenase and peroxidase. J Biol Chem 275: 942–948

Chew O, Whelan J and Millar AH (2003) Molecular definition of the ascorbate-glutathione cycle in *Arabidopsis* mitochondria dual targeting of antioxidant defenses in plants. J Biol Chem 278: 46869–46877

Cornic G, Bukhov NG, Weise C, Bligny R and Heber U (2000) Flexible coupling between light-dependent electron and vectorial proton transport in illuminated leaves of C_3 plants. Role of photosystem I-dependent proton pumping. Planta 210: 468–477

Corneille S, Cournac L, Guedeney G, Havaux M and Peltier G (1998) Reduction of the plastoquinone pool by exogenous NADH and NADPH in higher plant chloroplasts. Characterization of a NAD(P)H-plastoquinone oxidoreductase activity. Biochim Biophys Acta 1363: 59–69

Demmig-Adams B, Ebbert V, Zarter CR and Adams WW III (2005) Characteristics and species-dependent employment of flexible versus sustained thermal dissipation and photoinhibition. In: Demmig-Adams B, Adams WW III and Mattoo AK (eds) Photoprotection, Photoinhibition, Gene Regulation, and Environment, pp 39–48. Springer, Dordrecht

Deng Y, Ye J and Mi H (2003) Effects of low CO_2 on NAD(P)H dehydrogenase, a mediator of cyclic electron transport around photosystem I in the cyanobacterium *Synechocystis* PCC6803. Plant Cell Physiol 44: 534–540

dePamphilis CW and Palmer JD (1990) Loss of photosynthetic and chlororespiratory genes from the plastid genome of a parasitic flowering plant. Nature 348: 337–339

Elortza F, Asturias JA and Arizmendi JM (1999) Chloroplast NADH dehydrogenase from *Pisum sativum*: characterization of its activity and cloning of *ndh*K gene. Plant Cell Physiol 40: 149–154

Endo T and Asada K (1996) Dark induction of non-photochemical quenching of chlorophyll fluorescence by acetate in *Chlamydomonus reinhardtii*. Plant Cell Physiol 37: 551–555

Endo T, Schreiber U and Asada K (1995) Suppression of quantum yield of photosystem II by hyperosmotic stress in *Chlamydomonus reinhardtii*. Plant Cell Physiol 36: 1253–1258

Endo T, Mi H, Shikanai T and Asada K (1997) Donation of electrons to plastoquinone by NAD(P)H dehydrogenase and by ferredoxin-quinone reductase in spinach chloroplasts. Plant Cell Physiol 38: 1272–1277

Endo T, Shikanai T, Sato F and Asada K (1998) NAD(P)H dehydrogenase-dependent, antimycin A-sensitive electron donation to plastoquinone in tobacco chloroplasts. Plant Cell Physiol 39: 1226–1231

Endo T, Shikanai T, Takabayashi A, Asada K and Sato F (1999) The role of chloroplastic NAD(P)H dehydrogenase in photoprotection. FEBS Lett 457: 5–8

Endo T, Takabayashi A, Shikanai T and Sato F (2001). Defect in chloroplastic NAD(P)H dehydrogenase complex resulted in stromal over-reduction after exposure to strong light. In: Procedings in International Congress of Photosynthesis. S11-018, CSIRO Publishing, Brisbane

Fearnley IM and Walker JE (1992) Conservation of sequences of subunits of mitochondrial complex I and their relationships with other proteins. Biochim Biophys Acta 1140: 105–134

Feild TS, Nedbal L and Ort DR (1998) Nonphotochemical reduction of the plastoquinone pool in sunflower leaves originated from chlororespiration. Plant Physiol 116: 1209–1218

Foyer CH and Noctor G (2000) Oxygen processing in photosynthesis: regulation and signaling. New Phytol 146: 359–388

Friedrich T (1998) The NADH: ubiquinone oxidoreductase (complex I) from *Escherichia coli*. Biochim Biophys Acta 1364: 134–146

Funk E, Schäfer E and Steinmüller K (1999) Characterization of the complex I-homologous NAD(P)H-plastoquinone-oxidoreductase (NDH-complex) of maize chloroplasts. J Plant Physiol 154: 16–23

Godde D (1982) Evidence for a membrane bound NADH-plastoquinone-oxidoreductase in *Chlamydomonas reinhardtii* CW-15. Arch Microbiol 131: 197–202

Godde D and Trebst A (1980) NADH as electron donorfor the photosynthetic membrane of *Chlamydomonas reinhardtii*. Arch Microbiol 127: 245–252

Goh CH, Schreiber U and Hedrich R (1999) New approach of monitoring changes in chlorophyll a fluorescence of single guard cells and protoplasts in response to physiological stimuli. Plant Cell Environ 22: 1057–1070

Guedeney G, Corneile S, Cuine S and Peltier G (1996) Evidence for an association of *ndh* B, *ndh* J gene products and ferredoxin-NADP reductase as components of a chloroplastic NAD(P)H dehydrogenase complex. FEBS Lett 378: 277–280

Harbinson J and Foyer CH (1991) Relationship between the efficiencies of photosytems I and II and stromal redox state in CO_2-free air. Evidence for cyclic electron flow in vivo. Plant Physiol 97: 41–49

Hashimoto H, Endo T, Peltier G, Tasaka M and Shikanai T (2003) A nucleus-encoded factor, CRR2, is essential for the expression of ndhB in *Arabidopsis*. Plant J 36: 541–549

Heber U (2002) Irrungen, Wirrungen? The Mehler reaction in relation to cyclic electron transport in C_3 plants. Photosynth Res 73: 223–231

Heber U and Walker DA (1992) Concerning a dual function of coupled cyclic electron transport in leaves. Plant Physiol 100: 1621–1626

Hideg E, Kálai A, Hideg K and Vass I (1998) Photoinhibition of photosytems in vivo results in singlet oxygen production. Detection via nitroxide-induced fluorescence quenching in broad bean leaves. Biochemistry 37: 11405–11411

Hideg E, Kalai A, Hideg K and Vass I (2000) Do oxidative stress conditions impairing photosynthesis in the light manifest as photoinhibition? Phil Trans Roy Soc London B 355: 1511–1516

Hideg E, Ogawa K, Kalai T and Hideg K (2001) Singlet oxygen imaging in *Arabidopsis thaliana* leaves under photoinhibition by excess photosynthetically active radiation. Physiol Plant 112: 10–14

Hideg E, Barta C, Kálai T, Vass I, Hideg K and Asada K (2002) Detection of singlet oxygen and superoxide with fluorescence sensors in leaves under stress by photoinhibition or UV radiation. Plant Cell Physiol 43: 1154–1164

Hirotsu N, Makino A, Ushio A and Mae T. (2004) Changes in the thermal dissipation and the electron flow in the water-water cycle in rice grown under conditions of physiologically low temperature. Plant Cell Physiol 45: 635–644

Horváth EM, Peter SO, Jöet T, Rumeau D, Cournac L, Horváth GV, Kavanagh TA, Schäfer C, Peltier G and Medgesy P (2000) Targeted inactivation of the plastid ndhB gene in tobacco results in an enhanced sensitivity of photosynthesis to moderate stomatal closure. Plant Physiol 123: 1337–1349

Ishida H, Nishimori Y, Sigisawa M, Makino M and Mae T (1997) The large subunit of ribulose-1,5-bisphosphate carboxylase/oxygenase is fragmented into 37-kDa and 16-kDa polypeptides by active oxygen in the lysates of chloroplasts from primary leaves of wheat. Plant Cell Physiol 38: 471–479

Ishida H, Shimizu S, Makino A and Mae T (1998) Light-dependent fragmentation of the large subunit of ribulose-1,5-bisphosphate carboxylase/oxygenase in chloroplasts isolated from wheat leaves. Planta 204: 305–309

Joët T, Cournac L, Guedeney G, Rumeau D, Peter SO, Schafer C, Horvath E, Medgyesy P and Peltier G (1998) Increased sensitivity of photosynthesis to anaerobic conditions induced by targeted inactivation of the chloroplast *ndhB* gene. In: Garab G (ed) Photosynthesis: Mechanisms and Effects, Vol 3, pp 1967–1970. Kluwer Academic Publishers, Dordrecht

Joët T, Cournac L, Peltier G and Havaux M (2002) Cyclic electron flow around photosytem I in C_3 plants. In vivo control by the redox state of chloroplasts and involvement of the NADH-dehydrogenase complex. Plant Physiol 128: 760–769

Jung H-S and Niyogi KK (2005) Molecular analysis of photoprotection and photosynthesis. In: Demmig-Adams B, Adams WW III and Mattoo AK (eds) Photoprotection, Photoinhibition, Gene Regulation, and Environment, pp 127–143. Springer, Dordrecht

Kaiser W (1976) Effect of hydrogen peroxide on CO_2-fixation of isolated chloroplasts. Biochim Biophys Acta 440: 476–482

Kofer W, Koop H-U, Wanner G and Steinmüller K (1998) Mutagenesis of the genes encoding subunits A, C, H, I, J and K of the plastid NAD(P)H-plastoquinone-oxidoreductase in tobacco by polyethylene glycol-mediated plastome transformation. Mol Gen Genet 258: 166–173

Kozaki A and Takeba G (1996) Photorespiration protects C_3 plants from photoinhibition. Nature 384: 557–560

Kurisu K, Zhang H, Smith JL and Cramer WA (2003) Structure of the cytochrome b_6f complex of oxygenic photosynthesis: tuning the cavity. Science 302: 1009–1014

Lascano HR, Casano LM, Martín M and Sabater B (2003) The activity of the chloroplastic Ndh complex is regulated by phosphorylation of NDH-F subunit. Plant Physiol 132: 256–262

Leif H, Sled VD, Ohnishi T, Weiss H and Friedrich T (1995) Isolation and characterization of the proton-translocating NADH:ubiquinone oxidoreductase form *Escherichia coli*. Eur J Biochem 230: 538–548

Lennon AM, Prommeenate P and Nixon PJ (2003) Location, expression and orientation of the putative chlororespiratory enzymes, Ndh and IMMUTANS, in higher-plant plastids. Planta 218: 254–260

Li X-G, Duan W, Meng Q-W, Zou Q and Zhao S-J (2004) The function of chloroplastic NAD(P)H dehydrogenase in tobacco during chilling stress under low irradiance. Plant Cell Physiol 45: 103–108

Makino A, Miyake C and Yokota A (2002) Physiological functions of the water-water cycle (Mehler reaction) and the cyclic electron flow around PSI in rice leaves. Plant Cell Physiol 43: 1017–1026

Mano J, Miyake C, Schreiber U and Asada K (1995) Photoactivation of the electron flow from NADPH to plastoquinone in spinach chloroplasts. Plant Cell Physiol 33: 12331237

Mano J, Ohno C, Domae Y and Asada K (2001) Chloroplastic ascorbate peroxidase is the primary target of methylviologen-induced photooxidative stress in spinach leaves: its relevance to monodehydroascorbate radical detected with in vivo ESR. Biochim Biophys Acta 1504: 275–287

Matsuo M, Endo T and Asada K (1998) Properties of the respiratory NAD(P)H dehydrogenase isolated from the cyanobacterium *Synechocystis* PCC6803. Plant Cell Physiol 39: 263–267

Mehler AH (1951) Studies on reactivities of illuminated chloroplasts. I. Mechanism of the reduction of oxygen and other Hill reagents. Arch Biochem Biophys 33: 65–77

Mehta RA, Fawcett TW, Porth D and Mattoo AK (1992) Oxidative stress causes rapid membrane translocation and in vivo degradation of riburose-1,5-bisphosphate carboxylase/oxygenase. J Biol Chem 267: 2810–2816

Mi H, Endo T, Schreiber U and Asada K (1992a) Donation of electrons to the intersystem chain in the cyanobacterium *Synechococcus* sp. PCC 7002 as determined by the reduction of P700$^+$. Plant Cell Physiol 33: 1099–1105

Mi H, Endo T, Schreiber U, Ogawa T and Asada K (1992b) Electron donation from cyclic and respiratory flows to the photosynthetic intersystem chain is mediated by pyridine nucleotide dehydrogenase in the cyanobacterium *Synechocystis* PCC 6803. Plant Cell Physiol 33: 1233–1237

Mi H, Endo T, Schreiber U, Ogawa T and Asada K (1994) NAD(P)H-dehydrogenase-dependent cyclic electron flow around photosystem I in the cyanobacterium *Synechocystis*

PCC 6803: a study of dark-starved cells and spheroplasts. Plant Cell Physiol 35: 163–173

Mi H, Endo T, Ogawa T and Asada K (1995) Thylakoid membrane-bound pyridine nucleotide dehydrogenase complex mediates cyclic electron transport in the cyanobacterium *Synechocystis* PCC 6803. Plant Cell Physiol 36: 661–668

Miyake C and Asada K (1996) Inactivation mechanism of ascorbate peroxidase at low concentrations of ascorbate: hydrogen peroxide decomposes compound I of ascorbate peroxidase. Plant Cell Physiol 37: 423–430

Miyake C and Asada K (2003) The water-water cycle in algae. In: Larkum AW, Dougles SE and Raven JA (eds) Photosynthesis in Algae, pp 183–204. Kluwer Academic Publishers, Dordrecht

Miyake C and Okamura M (2003) Cyclic electron flow within PSII protects PSII from its photoinhibition in thylakoid membranes from spinach chloroplasts. Plant Cell Physiol 44: 457–462

Miyake C and Yokota A (2000) Determination of the rate of photoreduction of O_2 in the water-water cycle in watermelon leaves and enhancement of the rate by limitation of photosynthesis. Plant Cell Physiol 41: 335–343

Miyake C and Yokota A (2001) Cyclic flow of electrons within PSII in thylakoid membranes. Plant Cell Physiol 45: 508–515

Miyake C, Michihata F and Asada K (1991) Scavenging of hydrogen peroxide in prokaryotic and eukaryotic algae. Plant Cell Physiol 32: 33–43

Miyake C, Cao WH and Asada K (1993) Purification and molecular properties of thylakoid-bound ascorbate peroxidase from spinach chloroplasts. Plant Cell Physiol 34: 881–889

Miyake C, Schreiber U and Asada K (1995) Ferredoxin-dependent and antimycin A-sensitive reduction of cytochrome *b*-559 by far-red light in maize thylakoids; participation of a menadiol-reducible cytochrome *b*-559 in cyclic electron flow. Plant Cell Physiol 36: 743–748

Miyake C, Schreiber U, Hormann H, Sano S and Asada K (1998) The FAD-enzyme monodehydroascorbate radical reductase mediates photoproduction of superoxide radicals in spinach thylakoid membranes. Plant Cell Physiol 39: 821–829

Munekage Y, Hojo M, Meurer, J, Endo T, Tasaka M and Shikanai T (2002) PGR5 is involved in cyclic electron flow around photosystem I and is essential for photoprotection in *Arabidopsis*. Cell 110: 361–371

Munekage Y, Hashimoto M, Miyake C, Tomizawa K, Endo T, Tasaka M and Shikanai T (2004) Cyclic electron flow around photosystem I is essential for photosynthesis. Nature 429: 579–582

Nishiyama Y, Allakhverdiev SI and Murata N (2005) Regulation by environmental conditions of the repair of photosystem II in cyanobacteria. In: Demmig-Adams B, Adams WW III and Mattoo AK (eds) Photoprotection, Photoinhibition, Gene Regulation, and Environment, pp 193–203. Springer, Dordrecht

Nixon PJ (2000) Chlororespiration. Phil Trans Roy Soc Lond B 355: 1541–1547

Obara K, Sumi K and Fukuda H (2002) The use of multiple transcription starts causes the dual targeting of *Arabidopsis* putative monodehydroascorbate reductase to both mitochondria and chloroplasts. Plant Cell Physiol 43: 697–705

Ogawa T (1991) A gene homologous to the subunit-2 gene of NADH dehydrogenase is essential to inorganic carbon transport of *Synechocystis* PCC6803. Proc Natl Acad Sci USA 88: 4275–4279

Ogawa K, Kanematsu S, Takabe K and Asada K (1995) Attachment of CuZn-superoxide dismutase to thylakoid membranes at the site of superoxide generation (PSI) in spinach chloroplasts: Detection by immuno-gold labeling after rapid freezing and substitution method. Plant Cell Physiol 36: 565–573

Ohyama K, Fukuzawa H, Kohchi T, Shirai H, Sano T, Sano S, Umesono K, Shiki Y, Takeuchi M, Chang Z, Aota S, Inokuchi H and Ozeki H (1986) Chloroplast gene organization deduced from complete sequence of liverwort *Marchantia polymorpha* chloroplast DNA. Nature 322: 572–574

Osmond CB and Grace SC (1995) Perspectives on photoinhibition and photorespiration in the field: quintessential inefficiencies of the light and dark reactions of photosynthesis? J Exp Bot 46: 1351–1363

Owaki T and Asada K (2003) Oxygen is indispensable to start photosynthetic CO_2-fixation in maize. Plant Biology 2003, Abstract Nr. 329

Park Y-I, Chow WS, Osmond CB and Anderson JM (1996) Electron transport to oxygen mitigates against the photoinactivation photosystem II in vivo. Photosynth Res 50: 23–32

Peltier G and Cournac L (2002) Chlororespiration. Annu Rev Plant Biol 53: 523–550

Pnueli L, Liang H, Rozenberg M and Mittler R (2003) Growth suppression, altered stomatal responses and augmented induction of heat shock proteins in cytosolic ascorbate (apx-1)-deficient arabidopsis plants. Plant J 34: 187–203

Polle A (2001) Dissecting the superoxide dismutase-ascorbate-glutathione pathway in chloroplasts by metabolic modeling. Computer simulation as a step towards flux analysis. Plant Physiol 126: 445–462

Quiles MJ and Cuello J (1998) Association of ferredoxin-NADP oxidoreductase with the chloroplastic pyridine nucleotide dehydrogenase complex in barley leaves. Plant Physiol 117: 235–244

Quiles MJ, Garcia A and Cuello J (2000) Separation by blue-native PAGE and identification of the whole NAD(P)H dehydrogenase complex from barley stroma thylakoids. Plant Physiol Biochem 38: 225–232

Radmer RJ and Kok B (1976) Photoreduction of O_2 primes and replaces CO_2 assimilation. Plant Physiol 58: 336–340

Rasmusson AG, Heiser V, Zabaleta E, Brennicke A and Grohmann L (1998) Physiological, biochemical and molecular aspects of mitochondrial complex I in plants. Biochim Biophys Acta 1364: 101–111

Raven EJ (2003) Understanding functional diversity and substrate specificity in heme peroxidase: What can we learn from ascorbate peroxidase? Nat Prod Rep 20: 367–381

Ravenel J, Peltier G and Havaux M (1994) The cyclic electron pathways around photosystem I in *Chlamydomonas reinhardtii* as determined in vivo by photoacoustic measurements of energy storage. Planta 193: 251–259

Rizhsky L, Liang H and Mittler R (2003) The water-water cycle is essential for chloroplast protection in the absence of stress. J Biol Chem 278: 38921–38925

Sazanov, LA, Burrows PA and Nixon PJ (1998) The plastid ndh genes code for an NADH-specific dehydrogenase: Isolation of a complex I analogue from pea thylakoid membranes. Proc Natl Acad Sci USA 95: 1319–1324

Schreiber U, Endo T, Mi H and Asada K (1995) Quenching analysis of chlorophyll fluorescence by the saturation pulse method:

particular aspects relating to the study of eukaryotic algae and cyanobacteria. Plant Cell Physiol 36: 873–882

Shikanai T, Endo T, Hashimoto T, Yamada Y, Asada K and Yokota A (1998) Directed disruption of the tobacco *ndhB* gene impairs cyclic electron flow around photosystem I. Proc Natl Acad Sci USA 95: 9705–9709

Shinozaki K, Ohme M, Tanaka M, Wakasugi T, Hayashida N, Matsubayashi T, Zaita N, Chunwongse J, Obokata J, Yamaguchi-Shinozaki K, Ohto C, Torazawa K, Meng BY, Sugita M, Deno H, Kampgashira T, Yamada K, Kusuda J, Takaiwa F, Kato A, Tohdoh N, Shimada H and Sugiura M (1986) The complete nucleotide sequence of the tobacco chloroplast genome: its gene organization and expression. EMBO J 5: 2043–2049

Sonoike K (1996) Photoinhibition of photosystem I: Its physiological significance in the chilling sensitivity of plants. Plant Cell Physiol 37: 239–247

Stroebel D, Choquet Y, Popot J-L and Picot D (2003) An atypical haem in the cytochorme $b_6 f$ complex. Nature 426: 413–418

Tagawa K, Tsujimoto HY and Arnon DI (1963) Role of chloroplast ferredoxin in the energy conversion process of photosynthesis. Proc Natl Acad Sci USA 49: 567–572

Takabayashi A, Endo T, Shikanai T and Sato F (2002) Post-illumination reduction of the plastoquinone pool in chloroplast transformants in which chloroplastic NAD(P)H dehydrogenase was inactivated. Biosci Biotech Biochem 66: 2107–2111

Teicher HB and Scheller HV (1998) The NAD(P)H dehydrogenase in barley thylakoids is photoactivatable and uses NADPH as well as NADH. Plant Physiol 117: 525–532

Terashima I, Funayama S and Sonoike K (1994) The site of photoinhibition in leaves of *Cucumis sativum* L. at low temperature is photosystem I, not photosystem II. Planta 193: 300–306

Tjus SE, Scheller HV, Andersson B and Moller BL (2001) Active oxygen produced during selective excitation of photosystem I is damaging not only photosystem I but also photosystem II. Plant Physiol 125: 2007–2015

Videira A (1998) Complex I from the fungus *Neurospora crassa*. Biochim Biophys Acta 1364: 89–100

Yabuta Y, Motoki T, Yoshimura K, Takeda T, Ishikawa T and Shigeoka S (2002) Thylakoid membrane-bound ascorbate peroxidase is a limiting factor of antioxidative systems under photo-oxidative stress. Plant J 32: 915–925

Yamamoto Y (2001) Quality control of photosystem II. Plant Cell Physiol 42: 121–128

Yamamoto H, Miyake C, Dietz K-J, Tomizawa KI, Murata N and Yokota A (1999) Thioredoxin peroxidase in the cyanobacterium *Synechocystis* sp. PCC6803. FEBS Lett 447: 269–273

Yokthongwattana K and Melis A (2005) Photoinhibition and recovery in oxygenic photosynthesis: Mechanism of a photosystem-II damage and repair cycle. In: Demmig-Adams B, Adams WW III and Mattoo AK (eds) Photoprotection, Photoinhibition, Gene Regulation, and Environment, this volume. Springer, Berlin

Yoshimura K, Yabuta Y, Tamoi M, Ishikawa M and Shigeoka S (1999) Alternatively spliced mRNA variants of chloroplast ascorbate peroxidase isoenzymes in spinach leaves. Biochem J 338: 41–48

Chapter 15

Integration of Signaling in Antioxidant Defenses

Philip M. Mullineaux*
Department of Biological Sciences, University of Essex, Wivenhoe Park, Colchester, Essex, CO4 3SQ, United Kingdom

Stanislaw Karpinski
Department of Botany, Stockholm University, Stockholm SE-106 91, Sweden

Gary P. Creissen
Department of Disease and Stress Biology, John Innes Centre, Colney, Norwich, Norfolk NR4 7UH, United Kingdom

Summary	224
I. Introduction	224
II. Signaling Networks and Cross-Talk	225
A. Antioxidant Networks	225
B. Coordinated Regulation of Antioxidant Defenses	225
1. Coordinated Up-Regulation of the Antioxidant Network	226
2. Down-Regulation of the Antioxidant Network; Evidence from Lesion Mimic Mutants	227
III. Reactive Oxygen Species (ROS)-Mediated Signaling	228
A. The Type of ROS	229
1. Oxylipin Signaling	229
B. The Source of ROS	229
1. Problems of Specificity	229
C. Dual Signaling with ROS	230
1. Signals from Photosynthetic Electron Transport	230
2. ROS and Calcium	230
3. ROS and Reactive Nitrogen Species	231
IV. Reconfiguration of the Antioxidant Network and the Regulatory Role of Glutathione	231
A. The Glutathione Paradox and Problems of Interpretation	232
V. ROS, Antioxidants, and Stress Hormones	232
A. Glutathione and the Salicylic Acid Signaling Pathway	232
B. Glutathione, Ascorbate, ROS, and the ABA Signaling Pathway	233
C. ROS, Antioxidants, Jasmonic Acid, and Ethylene	234
D. Growth, ROS, Antioxidants, and Auxins	234
E. Conclusions on Hormone Signaling and ROS/Antioxidant Metabolism	234
Acknowledgments	235
References	235

*Author for correspondence, email: mullin@essex.ac.uk

Summary

In the last few years, it has become apparent that reactive oxygen species (ROS) have important roles as signaling intermediaries in a large number of cellular processes, especially in relation to plants' interactions with their environment. A complex network of low molecular weight antioxidants, ROS scavenging enzymes, and enzymes that maintain antioxidant pools are required to control the levels of ROS in all subcellular compartments. The coordinated regulation of this network by ROS themselves and stress-associated hormones such as salicylic acid, abscisic acid, and jasmonic acid reveals that antioxidant metabolism is central to considerations of how signaling networks are regulated. Furthermore, it is becoming apparent that key antioxidants such as glutathione and ascorbate are involved in the regulation of stress hormone-directed signaling pathways without any interaction with ROS. Therefore ROS and antioxidants may be key points at which the coordination of different signaling pathways is achieved. These issues are considered in this chapter.

I. Introduction

Plant growth, development, and reproduction all require the continual involvement of reactive oxygen species (ROS; Pennel and Lamb, 1998). ROS is a collective term for the reduction intermediates or electronically excited forms of chemically relatively unreactive triplet state molecular oxygen ($^3O_2$). These are described in more detail in the chapters by Foyer et al. and Endo and Asada (this volume), but include both stable compounds (such as hydrogen peroxide [H_2O_2] and fatty acid hydroperoxides), free radicals of differing stabilities and reactivity (e.g. superoxide anion, the hydroxyl radical, lipid peroxy radicals), and singlet oxygen ($^1O_2$), a more reactive state of O_2. There is an extensive literature on how these various ROS can cause oxidative stress in plant cells, as in all aerobes, by the oxidation of cellular macromolecules and structures. However, it is now recognized that ROS in plants perform essential functions in diverse processes such as growth and development, acclimation to the environment, and establishing resistance to pathogens. The positive roles for ROS can be divided into either direct involvement in metabolic reactions or engagement in signaling mechanisms.

The involvement of ROS in metabolism is well established. Good examples are the numerous oxidases or peroxidases that are important in the interconversions of phenolic compounds, (poly)amines, ascorbate, and oxalate (Hiraga et al., 2001; Jansen et al., 2001; Sebala et al., 2001). Oxidative cross-linking of cells walls during normal growth and when challenged with disease agents is an important process in the life of plants (Bradley et al., 1992; Pennel and Lamb, 1998). However, it is the involvement of ROS as important components of signaling pathways that has attracted considerable attention in recent years. This is especially the case where such changes are associated with or initiate programmed cell death (PCD) and/or where other signaling molecules or phytohormones are involved.

This chapter focuses on those molecules and their associated signaling pathways that are affected by or influence ROS levels and antioxidant metabolism.

Abbreviations: ABA – abscisic acid; *ABI* – ABA insensitive gene (*ABI1*, *ABI2*, etc); APX – ascorbate peroxidase isoform (APX1, APX2, etc); *APX* – ascorbate peroxidase gene (*APX1*, *APX2*, etc); CAT1 – catalase, peroxisomal isoform; *CAT* – catalase gene (*CAT1*, *CAT2*, *CAT3*, etc); CBF1 – CRT/DRE binding factor 1; CHS – chalcone synthase; *CHS* – chalcone synthase gene; DHAR – dehydroascorbate reductase; γ-ECS – gamma-glutamylcysteine synthetase; *EDS1* – enhanced disease susceptibility 1 gene; *ELIP2* – early light induced protein 2 gene; ETR1 – ethylene resistant 1 protein; GR – glutathione reductase; GSH – reduced glutathione; GSSG – oxidised glutathione (glutathione disulphide); GST – glutathione-S-transferase (GST1, GST2, etc); *GST* – glutathione-S-transferase gene (*GST1*, *GST2*, etc); HR – hypersensitive response; JA – jasmonic acid; LSD1 – lesion simulating disease 1 protein; *LSD1* – lesion simulating disease 1 gene; *lsd1* – mutant form of *LSD1*; MAPK – mitogen-activated protein kinase; MDHR – monodehydroascorbate reductase; NADPH – nicotinamide adenine dinucleotide phosphate, reduced form; NPR1 – non-expressor of PR-1 1 protein; NO – nitric oxide; $^1O_2$ – singlet oxygen, reactive form of molecular oxygen (O_2); $^3O_2$ – triplet molecular oxygen, which is the form normally referred to as O_2; *PAD4* – phytoalexin deficient 4 gene; PAL – phenyl(alanine)ammonia lyase; *PAL* – phenyl(alanine)ammonia lyase gene; PCD – programmed cell death; PP2C – protein phosphatase 2C; *PR-1* – pathogenesis related protein-1 gene; RCD1 – runaway cell death protein (wild type); *RCD1* – runaway cell death gene (wild type); *rcd1* – mutant form of *RCD1*; RNS – reactive nitrogen species; ROS – reactive oxygen species; RuBisCo – ribulose bisphosphate carboxylase/oxygenase; SA – salicylic acid; SAR – systemic acquired resistance; SOD – superoxide dismutase; *SOD* – superoxide dismutase gene (*SOD1*, *SOD2*, etc); TGA-B-Zip – TGA class basic leucine zipper transcription factor; *vtc1* – vitamin C deficient 1 mutant;

Chapter 15 Signaling and Antioxidant Defenses

II. Signaling Networks and Cross-Talk

Inevitably, this chapter will consider the action of any one particular signaling pathway and the involvement of ROS in its function in controlling the expression of antioxidant defenses. However, the reader should be aware that it is best to think of a signaling network that controls a wide range and often common set of responses of the cell to diverse environmental challenges. The widespread analysis of signaling pathways has led to the realization that there is considerable "horizontal" interaction between them. Such "cross-talk" is now a subject of intense scrutiny, especially in those labs using *Arabidopsis* as their experimental species. The application of post-genomic techniques, such as the use of microarrays and mutants, is adding to this impression of a complex network of responses from the subcellular to whole plant level. Such integrated "systems biology" is seen by many as the way forward in studying plant-environment interactions (Rhaikel and Coruzzi, 2003). Added to this complexity is the problem of how plants respond to multiple environmental challenges, which is, of course, the norm for plants in their natural environment. We have barely begun to scratch the surface of how a response to multiple stresses is organized, but undoubtedly, it is why such signaling networks evolved. Thus the reality of how signaling pathways integrate their responses to environmental stimuli is akin to the complex behavior of neural networks and attempts to describe how such networks might operate are just beginning (Genoud and Metraux, 1999).

A. Antioxidant Networks

Wherever ROS are encountered in living cells, then antioxidant-driven reaction(s) must be available to prevent these reactive molecules from promoting oxidative stress or carrying out inappropriate signaling. A plethora of antioxidants is known, all of which are worthy of individual study. However, it is likely that the network of reactions is important in toto and provides comprehensive control of ROS in cells, rather than the action of any single reaction. In this chapter, it is this network of reactions, compounds, and enzymes that will be regarded as "antioxidant defenses". This requirement for an antioxidant network has come to be appreciated from the many attempts to genetically engineer resistance to oxidative stress using single genes encoding an enzyme of ROS or antioxidant metabolism. These approaches, for the most part, did not have any lasting value (Mullineaux and Creissen, 1999).

Such networks of reactions and interconversions have three features in common. First, they have one or more low molecular weight antioxidants (eg. glutathione, ascorbate, or α-tocopherol; Fig. 1). Second, they have enzymes that directly scavenge ROS, either by catalyzing disproportionation reactions (e.g. catalases and superoxide dismutases; Fig. 1) or through reduction reactions using low molecular weight antioxidants as electron donors (e.g. glutathione peroxidases or ascorbate peroxidases). Third, a network of linked enzyme-catalyzed reactions that maintain antioxidant compounds in their reduced form (e.g. glutathione reductase, violaxanthin deepoxidase).

The first antioxidant network to be proposed for plants was the ascorbate-glutathione cycle of the chloroplast stroma (Fig. 1; Foyer and Halliwell, 1976; Foyer and Mullineaux, 1994; Asada, 1999). While the individual enzyme components described in the ascorbate-glutathione cycle still receive a considerable amount of attention (e.g. superoxide dismutase and ascorbate peroxidase isoforms), it is now apparent that many more complex schemes exist that involve a wider range of enzymes and molecules. These networks would now include the movement of antioxidant compounds by specific transporters between and within cells and the use of membrane-spanning redox couples (for an example, see Fig. 2) (Polle, 1997; Jimenez et al., 1997; Jamai et al., 2000; Horemans et al., 2000; Foyer and Noctor, 2001). Such networks have been described for a range of tissues and organs, including photosynthetic tissues of leaves, developing embryos, roots, ripening fruits, and germinating seeds (Jimenez et al., 2002). In addition to the ascorbate glutathione cycle, further compounds, proteins, and enzymes have since been demonstrated to be important contributors to antioxidant defenses. These include potent antioxidant phenolic compounds (Gray et al., 1997; Grace and Logan, 2000; Horemans et al., 2000), small protein molecules such as (2-cys)peroxiredoxins, glutaredoxins, and thioredoxins (Meyer et al., 1999; Horling et al., 2003), enzymes that turn out to have potent ROS-scavenging potential such as glutathione-S-transferases/peroxidases (Marrs, 1996; Cummins et al., 1999), and the repair of oxidatively damaged proteins such as peptide methionine sulfoxide reductases (Bechtold et al., 2004).

B. Coordinated Regulation of Antioxidant Defenses

Whatever the precise configuration of antioxidant defenses in particular subcellular compartments or tissues

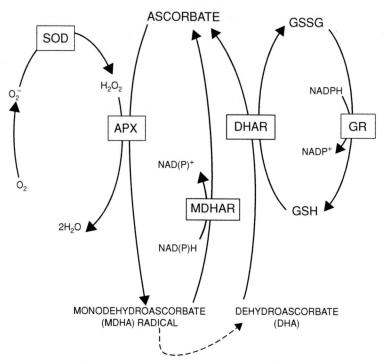

Fig. 1. An example of an antioxidant network. The ascorbate-glutathione cycle, which has been suggested to operate in several subcellular compartments (after Foyer and Halliwell, 1976; Foyer and Mullineaux, 1994; Jimenez et al., 1997). Reactive oxygen species, such as superoxide (O_2^-) and hydrogen peroxide (H_2O_2), are reduced to water by the concerted action of superoxide dismutases (SOD) and ascorbate peroxidase (APX). APX-catalysed reactions require ascorbate as the electron donor. Ascorbate is regenerated by reduction of either the monodehydroascorbate free radical, catalyzed by monodehydroascorbate reductase (MDHAR), or, after its dismutation to dehydroascorbate (dotted line), by dehydroascorbate reductase (DHAR). DHAR uses reduced glutathione (GSH) as its electron donor and the resulting oxidized glutathione (glutathione disulphide; GSSG) is re-reduced in a glutathione reductase (GR)-catalyzed reaction.

may be, the very existence of such networks suggests that a degree of coordinate regulation should occur. This would allow the adjustment of defenses throughout the life of the plant and in response to various environmental stimuli. Three reasons for modulating antioxidant defenses can be discerned:

- A requirement for a general increase in the activity of many components of the antioxidant network. This would be in response to an increase in the potential for cellular oxidative stress provoked by one or more environmental challenges.
- A down-regulation of antioxidant defenses under conditions where a localized increase in ROS concentrations is required.
- Reconfiguration of the antioxidant network as part of plant development or in acclimation to long term changes in the environment. This means that the activity of some components of the network will decrease while others increase.

1. Coordinated Up-Regulation of the Antioxidant Network

Much of the earlier literature focused on the response of multiple components of antioxidant defenses in plants suffering traumatic oxidative stress and bear out this notion of coordinated regulation of an antioxidant network (Foyer and Mullineaux, 1994; Noctor and Foyer, 1998; Asada, 1999). The activities of enzymes, levels of various antioxidants, and control at the levels of translation, RNA turnover, and de novo transcription have been described for the regulation of antioxidant defenses under these severe stresses. The above reviews provide many individual examples of the types of control on the expression of genes of the antioxidant network.

Investigations of global transcription (using microarrays) in response to oxidative stress, such as that caused by treatment of cells with hydrogen peroxide (Desikan et al., 2000), exposure of leaves to excess light (Rossel

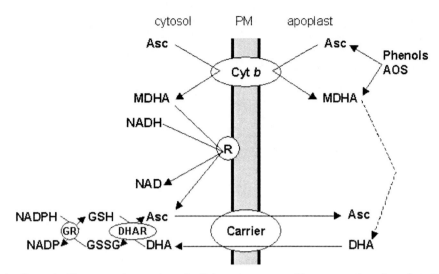

Fig. 2. An example of an antioxidant network spanning subcellular compartments. The regeneration of ascorbate (Asc) at the plasma membrane (PM; after Horemans et al., 2000). Asc is either directly re-reduced by a *b*-type cytochrome-mediated electron transfer (cyt *b*) or by the action of a cytoplasmic NADH-MDHA oxidoreductase (R). Alternatively, MDHA disproportionates to Asc and fully oxidized dehydroascorbate (DHA; broken line). The extracellular DHA is transported to the cytosol where it is re-reduced by DHA reductase (DHAR), using reduced glutathione (GSH) as the electron donor. The oxidized glutathione (GSSG) is, in turn, re-reduced by a glutathione reductase (GR) at the expense of NADPH. In addition, the plasma membrane ASC-DHA carrier is hypothesized to operate as an exchange carrier, thus keeping extracellular Asc levels constant. Both the figure scheme and part of the legend have been reproduced from Horemans et al. (2000) by kind permission of the publishers (Elsevier).

et al., 2002), or as a consequence of the deletion of a key enzyme of ROS metabolism (Pneuli et al., 2003), confirm that coordinated regulation of the antioxidant network is the norm. Nevertheless, it is also clear that such changes in the expression of genes encoding components of the antioxidant network occur alongside effects on the expression of often several hundred genes, only a proportion of which are implicated in defense responses in the broadest definition of the term.

2. Down-Regulation of the Antioxidant Network; Evidence from Lesion Mimic Mutants

The hypersensitive response (HR) leading to systemic acquired resistance (SAR) to certain pathogens is associated with an increase in the production of ROS. In this well-researched example, superoxide is generated via the action of plasma membrane-bound respiratory-burst oxidases at the site of lesion formation (Doke, 1985; Bestwick et al., 1997; Grant and Loake, 2000; Yoshika et al., 2003). Superoxide can then rapidly lead to production of hydrogen peroxide, via enzyme-catalyzed or spontaneous dismutation. In addition, cell wall peroxidases (Bolwell et al., 2002), and possibly oxalate and (poly)amine oxidases (Allan and Fluhr, 1997; Hu et al., 2003; Yoda et al., 2003), have also been named as sources of ROS for HR and establishment of resistance to certain pathogens.

In addition to ROS production, and of more relevance to this chapter, there could arguably be a down-regulation of expression of key enzymes of the antioxidant network which would point to coordinated control and therefore cross-talk between pathways. A good example of this concept may be the mode of action of LSD1. The prefix "LSD" means "lesion simulating disease" and refers to the way in which the gene was first identified in *Arabidopsis*. The mutant *lsd1* is one of a number of so-called lesion mimic mutants identified in several plant species that either show spontaneous but discrete HR-like lesions when not exposed to a pathogen, or lethal, spreading, uncontrolled lesion formation under non-permissive conditions. Such environmental conditions include a change in the light environment, exposure to an avirulent pathogen, or to ozone (Dietrich et al., 1994; Bowling et al., 1997; Gray et al., 1997; Takahashi et al., 1999; Overmyer et al., 2000; Devadas et al., 2002). LSD1 is a regulatory protein containing a zinc finger and likely a transcription factor (Dietrich et al., 1997). LSD1 may control superoxide levels by modulating the expression of SOD genes (Kleibenstein et al., 1999). Superoxide is a key ROS for the initiation of cell death processes that constitute the basis of the HR (Jabs et al., 1996). Thus LSD1 can be regarded as controlling superoxide levels by determining the activity of SOD gene expression and this, in turn, would determine whether superoxide attains levels that would trigger programmed cell death, which is associated with HR lesion formation (Kleibenstein et al., 1999). More recently, LSD1 has been shown to regulate the expression of *CAT1*, encoding peroxisomal catalase. This was revealed by studies on the inability of the *lsd1* mutant to acclimate to high light or long photoperiods that is caused by a reduced ability to deal with photorespiratory-derived H_2O_2. LSD1-mediated control of catalase activity would modulate increased ROS production arising from the photorespiratory cycle in the peroxisome. The light dependency of the propagation of HR lesions (Dietrich et al 1994), but not their initiation, may indicate that photorespiratory cycle-derived ROS can be recruited for cell death processes in foliar tissue. How LSD1 controls expression of SODs and catalase is not known, but one clue may be from tomato plants that ectopically express the *Arabidopsis* CBF1 transcription factor, which results in over-expression of *CAT1* (Hsieh et al., 2002). Perhaps, LSD1 exerts its control over ROS metabolism by influencing the expression of CBF transcription factors, which are known to be important in a range of abiotic stress responses (Thomashaw, 1999).

Similar conclusions can be drawn from the study of mutants and transgenic plants that are suppressed in the expression of components of antioxidant defenses. The best studied example is the suppression of *CAT1* expression in transgenic tobacco plants. These plants develop light intensity-dependent spreading chlorosis on their leaves (Changmonnpol et al., 1996; Willekens et al., 1997; Mittler et al., 1999; Dat et al., 2003). This phenotype is caused by failure of such plants to scavenge photorespiratory cycle-produced H_2O_2, which then triggers the expression of a wide range of defense genes associated with pathogen infection.

The ozone-hypersensitive mutant *rcd1* (*runaway cell death 1*; Overmyer et al., 2000) exhibits a clearly perturbed ROS metabolism in which, as in *lsd1*, the spreading lesion phenotype is again triggered by superoxide. The authors argue that the *rcd1* phenotype is not caused by a down-regulation of antioxidant defenses, but by a perturbation in pathogen defense gene expression. The hormones ethylene and jasmonic acid (JA) drive and inhibit superoxide-dependent lesion formation, respectively, and it is suggested that in wild type plants, at least some of the signaling actions of these hormones may be mediated by ROS. Furthermore, RCD1 (for runaway cell death protein) was shown to be a negative regulator of ethylene biosynthesis, leading to the notion of feedback loops for the regulation of cell death processes in which superoxide plays a prominent part (Overmyer et al., 2000; Mattoo and Handa, 2004). This negative regulation of ethylene biosynthesis by RCD1 could be an indirect effect of the latter's central role as a regulator of ROS homeostasis. This is because it has been suggested that ethylene can arise by the direct non-enzymatic action of some ROS on fatty acids and methionine. Therefore, an indirect effect of *rcd1* would be to influence this alternative route for the production of ethylene (Mattoo and Handa, 2004).

An inspection of the *rcd1*-sensitive defense genes, induced by ozone in the mutant and wild type plants, show that at least 4 genes (*CAT1*, *CAT3*, *GST1*, and *GST2*) encode products that can metabolize ROS and glutathione. This indicates that some of the processes related to superoxide function and hormone signaling may influence, and be influenced by, both ROS and glutathione metabolism and RCD1 plays a role in these processes. This is in keeping with more direct effects on antioxidant metabolism that can also be seen in other lesion mimic mutants, which again indicates how antioxidant metabolism must cut across several stress-associated processes. For example, the maize light-sensitive lesion mimic mutant *lethal spot 1* (*lls1*) encodes a defective oxygenase perhaps involved in maintaining phenolic antioxidant pools (Gray et al., 1997).

III. Reactive Oxygen Species (ROS)-Mediated Signaling

As considered above, ROS signaling may make a significant contribution towards bringing about wholesale changes to antioxidant defense gene expression. This raises questions about how specificity of regulation of antioxidant defenses is achieved in a ROS-driven signaling network. One can logically consider a number of means whereby such specificity could be achieved, none of which are mutually exclusive. These include:

- The type of ROS used for signaling
- The source of ROS
- Dual or multiple signaling in which ROS is at least one component

These will be the prime considerations here, but ROS-derived signaling may also be transduced by a complex

Chapter 15 Signaling and Antioxidant Defenses

cascade of protein kinases that elicit changes in the expression of a network of antioxidant defenses (Kovtun et al., 2000; Zhang and Klessig, 2001; Moon et al., 2003). It is not the intention in this chapter to describe this aspect of signaling.

A. The Type of ROS

H_2O_2 is well established as specifically inducing changes in antioxidant defense gene expression (Alvarez et al, 1998; Banzet et al., 1998; Karpinski et al., 1999; Wisniewski et al., 1999; Desikan et al., 2000; Kovtun et al., 2000; Dat et al., 2003; Fryer et al., 2003; Pneuli et al. 2003). Similarly, both superoxide and $^1O_2$ have been suggested to induce specific genes, including antioxidant defense genes such as GSTs (Jabs et al., 1996, Wisniewski et al., 1999; op den Camp et al., 2003).

1. Oxylipin Signaling

The ROS arising directly from O_2 give rise to a wide range of lipid peroxidation products through non-enzyme catalyzed oxidation reactions. These products are highly reactive and volatile electrophilic species, the most prominent of which are the cyclopentenone isoprostanes. These compounds are more reactive than other lipid peroxidation products and have been proposed to act as signaling molecules and to activate expression of a range of defense genes including those encoding GSTs (Vollenweider et al., 2000, Almeras et al., 2003, Thoma et al., 2003). In addition to the non-enzymatic-oxidation products of fatty acids, oxylipins are also synthesized which involves the action of lipoxygenase-catalyzed production of specific chiral forms of fatty acid hydroperoxides (Schaller, 2001). Oxylipins, specifically, have been convincingly implicated in signaling in cell death responses to the elicitor cryptogein, a process that might be initiated by H_2O_2 (Rusterucci et al., 1999). Furthermore, $^1O_2$ may elicit its signaling via the formation of a stereospecific isomer of hydroxyoctadecatrienoic acid, a peroxidation product of linolenic acid. This possibly implicates an enzyme-catalyzed reaction in the formation of this $^1O_2$-initiated signal (op den Camp, 2003).

B. The Source of ROS

The site in the cell at which ROS are generated might be responsible for initiating specific signaling pathways. The production of H_2O_2 or superoxide at the plasma membrane from NADPH oxidases, peroxidases, or (poly)amine oxidases (Doke, 1985; Jabs et al., 1996; Allan and Fluhr, 1997; Bestwick et al.; 1997; Bolwell et al., 2002; Yoda et al., 2003; Yoshioka et al., 2003) seems to be important for initiating defense responses against pathogens and is associated with HR-mediated cell death. Similarly, ozone fumigation causes production of a cocktail of ROS in the apoplast, initiating signaling responses very similar to those associated with the HR (Schraudner et al., 1998; Overmyer et al., 2000; Langebartels et al., 2002). The sources of ROS for ABA-mediated closure of stomata and lateral root hair development in *Arabidopsis* also have been shown to be plasma membrane-associated NADPH oxidases (Murata et al., 2001; Kwak et al., 2003; Foreman et al., 2003).

The chloroplast is a major source of cellular ROS from at least three distinct processes. One process is the light-mediated over-excitation of chlorophyll leading to the formation of $^1O_2$ at photosystem II (Fryer et al., 2003, op den Camp et al., 2003). A second process involves the Mehler reaction at photosystem I in which O_2 is reduced to ROS directly by photosynthetic electron transport (Asada, 1999; Badger et al., 2000; Ort and Baker, 2002). Finally, in photorespiration, the recycling of phosphoglycollate, formed by the oxygenase reaction of RuBisCo, leads to substantial production of H_2O_2 by a peroxisome-located glycollate oxidase (Willekens et al., 1997; Douce and Neuberger, 1999). The increases in foliar ROS associated with a large number of biotic and abiotic stresses may be a direct consequence of limiting photosynthetic carbon metabolism. This can be achieved by inhibiting the Calvin cycle through chilling or heat stress, or causing the closure of stomata by drought stress, wounding, or pathogens, thus limiting the supply of CO_2 for photosynthesis (Long et al., 1994; Asada, 1999; Ort and Baker 2002). Under these conditions, there will be increased activity of both the Mehler reaction and the RuBisCo oxygenase reaction, leading to increased ROS production. Hydrogen peroxide derived during the Mehler reaction has been strongly implicated in the regulation of *APX2* expression in bundle sheath cells of *Arabidopsis* leaves under excess light stress and in wounded leaves (Fryer et al., 2003; Chang et al., 2004), although H_2O_2 from photorespiration was ruled out, despite a closure of stomata in response to this stress.

1. Problems of Specificity

Whatever the subcellular source of ROS, the problem arises of how the specificity of a signal is maintained and can be distinguished from the same type of ROS coming from some other location in the cell. A good illustration of these questions comes from the signaling role of H_2O_2 produced in the chloroplast that drives induction of *APX2* (and presumably other antioxidant defense genes). The problem is how, once outside the chloroplast, can H_2O_2 be distinguished from that produced from any other source (Mullineaux and Karpinski, 2002)? The closure of stomatal guard cells involves the action of ROS (Pei et al., 2000) and, in at least one case, a photosynthetic electron transport-derived signal has been implicated (Zhang et al., 2000). These considerations are of broad relevance to understanding chloroplast-to-nucleus signaling (Rodermel, 2001; Mullineaux and Karpinski, 2002; Surpin et al., 2002). If the single ROS signal hypothesis in this case is to be accepted, then one has to propose that H_2O_2 produced from the Mehler reaction (for example) must be converted to some more specific signal either before or during its exit from the chloroplast. Such a signal is unknown but could involve trans-membrane electron transport chains (Jägger-Voterro et al., 1997; Murata and Takahashi, 1999).

Specificity for signaling has been suggested for ROS coming from the peroxisome of catalase-deficient plants (Willekens et al., 1997). The amount of H_2O_2 produced via photorespiration is substantial and the failure to scavenge it in the peroxisome leads to its diffusion into the cytosol. This was suggested to initiate signaling leading to induction of defense gene expression. Similarly, in response to fungal elicitors, ROS accumulates in the cytosol of tobacco epidermal cells, possibly originating from one of several subcellular locations and would lead to defense gene activation (Allan and Fluhr, 1997). ROS produced in mitochondria may be important in cell death responses in ozone-fumigated birch (aspen) and, therefore, it remains possible that ROS from this organelle are involved in signaling (Pellinen et al., 1997). Levels of transcripts encoding mitochondrial Mn-SOD were shown to rise substantially in tobacco suspension culture cells upon stimulation of their respiration by sucrose loading (Bowler et al., 1991). Such cells suffered oxidative stress, so it was suggested that ROS originating in mitochondria induced antioxidant defenses specifically for that organelle.

In conclusion, while ROS are generated in just about every subcellular compartment and enter the cell from external sources, it is difficult to see how specificity of signaling could be maintained, unless specific signaling is established very close to the point of production. There may be "receptors" for ROS that convert this signal into something more specific. Such a role has been suggested for the putative lipases encoded by the *Arabidopsis EDS1* and *PAD4* genes that may integrate ROS signals generated from the oxidative burst at the plasma membrane in response to attempted infection by certain pathogens (Rusterucci et al., 2001).

C. Dual Signaling with ROS

A further specificity could be imposed on ROS-mediated signaling if ROS did not act alone to control antioxidant defense gene expression. Possible examples of dual signaling are as follows.

1. Signals from Photosynthetic Electron Transport

In relation to antioxidant defense gene expression, signals originating directly from redox changes associated with photosynthetic electron transport under light stress conditions could provide a high degree of specificity (Mullineaux and Karpinski, 2002). Data from studies with inhibitors of photosynthetic electron transport led to the suggestion that expression of *APX2* and *APX1* of *Arabidopsis* could be regulated by signals initiated by changes in the redox state of the plastoquinone pool (Karpinski et al., 1997; 1999). A similar conclusion, i.e. that the redox state of the plastoquinone pool and other components of the photosynthetic electron transport chain may initiate signaling to the nucleus, comes from several studies. These include the expression of genes of carotenoid biosynthesis and turnover in the green alga *Haematococcus pluvialis*, expression of *ELIP2* in *Arabidopsis*, and expression of a nitrate reductase gene in duckweed and *Arabidopsis* (Kimura et al., 2003; Sherameti et al., 2003; Steinbrenner and Linden, 2003). Thus, a signal derived directly from photosynthetic electron transport may accompany an increase in ROS production from the same source, creating a dual signal. However, there are alternative explanations, and it is far from proven that plastoquinone is directly implicated in the regulation of nuclear gene expression (Fryer et al., 2003).

2. ROS and Calcium

Increases in intracellular Ca^{2+} have been observed for a wide range of biotic and abiotic stress responses that also bring about increases in ROS. There is considerable evidence to suggest that calcium, like ROS, can act as a major convergence point for signaling. Its association with hormones such as ABA and ethylene underscore this potential role. Many of the issues concerning specificity of ROS signaling are the same for calcium. The rate of mobilization from different internal stores, via a range of membrane-spanning ion channels, and the type of calcium "signature" produced (sustained, transient, single peak, waves, or oscillations) are important. Therefore, this controlled release and the spatial localization within cells can provide specificity (Bowler and Fluhr, 2000; Scrate-Field and Knight, 2003). Little is known about how calcium-mediated signal transduction is achieved, but it certainly involves a network of calcium-dependent protein kinases, calcium sensors (such as the calmodulins), and protein phosphatases such as the calcineurin B-like proteins (Bowler and Fluhr, 2000; Cheong et al., 2003; Gupta and Luan, 2003).

Oxidative stress responses in plants certainly involve mobilization of calcium (Price et al., 1994; Bowler and Fluhr, 2000; Scrate-Field and Knight, 2003), and calcium and ROS have been placed in at least one signal transduction pathway (Pei et al., 2000; Murata et al., 2001). The combined signal integration that could be achieved by ROS and Ca^{2+} may be much greater than either achieve on their own, but how such a mechanism might operate is still unknown (Bowler and Fluhr, 2000; Scrate-Field and Knight, 2003).

3. ROS and Reactive Nitrogen Species

Nitric oxide (NO) is a reactive nitrogen species (RNS) that has been strongly implicated in signaling for induction of defense responses against pathogens and closure of stomata in response to drought (Delledonne et al., 1998; Wendehenne et al., 2001; Neill et al., 2002). The action of NO was recognized early on to act in tandem with ROS, such that a combined NO/ROS signal has a synergistic effect on the induction of defense gene expression. It is not clear whether this dual requirement is a consequence of the formation of peroxynitrite from ROS and NO, but it nevertheless illustrates how further specificity can be imposed on ROS signaling.

IV. Reconfiguration of the Antioxidant Network and the Regulatory Role of Glutathione

There are numerous studies on a wide range of plant species, both in the natural environment and under laboratory conditions, showing that successful acclimation is associated with changes in antioxidant defenses (Rauser et al., 1991; De Vos et al., 1992; Walker and McKersie, 1993; Chaumont et al., 1995; Kampfenkel et al., 1995; Iturbe-Ormaetxe et al., 1998; Bruggemann et al., 1999; Hernandez et al., 2000). During the life of a plant, one may expect that antioxidant defenses will undergo subtle adjustments, with some components increasing their levels or activity, while others are decreasing. Recent evidence for this comes from the study of an *Arabidopsis APX1* null mutant, in which only a mild growth diminution was apparent that was accompanied by alterations in the expression of antioxidant defense genes as well as a range of genes involved in many other cellular functions (Pnueli et al., 2003). Similar observations have been made in transgenic tobacco with suppressed cytosolic APX expression, which again showed alterations in the expression of other antioxidant defense genes (Örvar and Ellis, 1997; Rizhsky et al., 2002).

An *Arabidopsis* mutant with altered glutathione levels and *APX2* expression, that displays no altered whole plant phenotypes under controlled environment conditions, shows specific alterations in a tight cluster of defense-associated genes, including genes encoding components of the antioxidant network (Ball et al., 2004). Under photo-oxidative stress conditions and in non-stressed situations, there was no evidence that this mutant had elevated levels of H_2O_2, which might indicate that a ROS signaling pathway had been affected. Rather, it has been concluded that glutathione levels and other aspects of glutathione metabolism, such as the flux through the biosynthetic pathway, have a direct impact on the configuration of antioxidant defenses. In support of these observations, *Arabidopsis* plants suffering prolonged sulfur depletion, which among many effects brings about diminished foliar glutathione levels, also display alteration in the expression of a wide profile of genes including components of antioxidant defenses (Hirai et al., 2003; Nikiforova et al., 2003). Changes in the configuration of defense gene expression may then influence the degree of response to environmental conditions that promote photo-oxidative stress. Interestingly, in contrast to glutathione-deficient mutants, the ascorbic acid deficient mutant, *vtc1*, showed no changes in antioxidant gene expression, which further indicates that glutathione has a particular role to play in regulation of cellular functions (Pastori et al., 2003).

In agreement with the observations that depressed levels of glutathione influence defense gene expression, elevated glutathione levels have a broadly opposite effect. Inhibition of *APX1* and *APX2* expression occurred when detached leaves, pre-treated with glutathione, were exposed to excess light. Under these conditions, the level of these transcripts rose dramatically in untreated leaves (Karpinski et al., 1997). Importantly, glutathione-treated leaves showed increased photoinhibition, strongly suggesting that they suffered a greater degree of photo-oxidative stress than controls. Similar results were obtained for rice leaves fed glutathione. The authors suggested that an inhibition of the antioxidant (carotenoid) xanthophyll cycle may have been the cause (Xu et al., 2000). Thus, paradoxically, treatment of leaves with one antioxidant, glutathione, has an inhibitory effect on the induction of other antioxidant defense components and compromises the ability of the plant to withstand this stress. However, this is not a universal response for all defense genes in all species. For example, treatment of Scots pine needles with glutathione (oxidized or reduced) caused an induction of Cu/Zn superoxide dismutase mRNA levels (Karpinski and Wingsle, 1996). Treatment of bean, tobacco, and parsley suspension cultures with glutathione also elicited increased expression of *CHS* and *PAL*, which encode enzymes of the phenylpropanoid pathway and whose products include potent antioxidants (Wingate et al., 1988; Herouart et al., 1995; Loyall et al., 2000). In the case of parsley cells, the induction of CHS by glutathione also required the expression of *GST1*, that encodes a glutathione-S-transferase, although its precise role in regulating *CHS* expression has not been elucidated (Loyall et al., 2000).

A. The Glutathione Paradox and Problems of Interpretation

While glutathione feeding experiments seem to give a coherent picture of how this antioxidant may influence defense gene expression, increasing the level of glutathione in transgenic plants by over-expression of the enzyme that catalyzes the first step in glutathione biosynthesis, γ-glutamylcysteine synthetase (γ-ECS), has given contradictory results. These include beneficial effects, such as increased heavy metal tolerance in Indian mustard (Zhu et al., 1999), no apparent effects in poplar and *Arabidopsis* (Noctor et al., 1998; Xiang et al., 2001), and deleterious effects in tobacco (Creissen et al., 1999). This may be due to possible differences in the targeting, expression, and controls exerted on the transgene-encoded γ-ECS, which will depend on the origins of the enzyme's coding sequence (bacterial or plant). For example, the plant enzyme may be subject to redox-dependent translational control that would act as a negative regulator of expression (Noctor et al., 2002), whereas a bacterial enzyme expressed in plants would escape such regulation. The negative effects on tobacco of over-expressing plastid-targeted bacterial γ-ECS included a light-sensitive spreading lesion phenotype reminiscent of those observed in catalase-deficient tobacco (see above; Creissen et al., 1999). Interestingly, the expression of the defense gene *PR-1* was readily detected in such plants, again underscoring the impact glutathione metabolism may have on defense gene expression. However, the effects were not due to increased levels of glutathione per se, but may have been caused by the redox state of the glutathione biosynthetic intermediate, γ-glutamylcysteine (Creissen et al., 1999). This illustrates that flux through the glutathione biosynthetic pathway, rather than the levels of this antioxidant, may be the key controlling factor. Despite these uncertainties and apparent contradictions, these various data point to glutathione playing an important regulatory role, at least in plants such as *Arabidopsis* and tobacco. Thus, glutathione may be a means by which various environmental stimuli and the action of signaling molecules integrate their effects on antioxidant defenses. As a consequence, the final section will address the relationship between various hormone-directed signaling pathways, ROS, and antioxidants.

V. ROS, Antioxidants, and Stress Hormones

A. Glutathione and the Salicylic Acid Signaling Pathway

The salicylic acid (SA) signaling pathway is a main route for the induction of defense-associated genes, as part of a plant immune response elicited by attempted infection via some avirulent pathogens (see Fig. 3). The immune response subsequently confers local and systemic resistance to the corresponding virulent pathogen. The key gene that regulates the induction of defense genes and resistance is called *NON EXPRESSOR OF PR1-1* (*NPR1*). The name of the gene comes from the way mutant alleles of the gene were first recognized (Cao et al., 1994). The NPR1 protein encoded by this gene contains ankyrin repeat domains that, from the outset, implicated it in protein-protein interactions. Impressive progress in our understanding of how this gene regulates expression of defense genes has been made in recent years and is outlined in Figure 3. This scheme shows that NPR1 can exist in an oxidized (and possibly aggregated) inactive form and a reduced active form, which permits its interaction with several members of the TGA class of B-Zip transcription factors, some of which also have to be in a reduced

Chapter 15 Signaling and Antioxidant Defenses

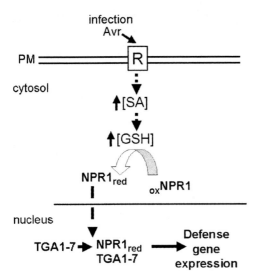

Fig. 3. The role of glutathione in the regulation of defense gene expression and establishment of immunity to pathogens (after Mou et al., 2003 and Deprés et al., 2003). A rise in intracellular concentration of salicylic acid (SA) caused by activation of a resistance (R) gene product at the plasma membrane (PM) by one or more factors from an avirulent pathogen (Avr), leads to a rise in reduced glutathione (GSH) levels. This may cause the reduction of the key defense gene regulator, NPR1, in the cytosol. Reduced NPR1 moves to the nucleus where it associates with one or more members of the TGA class of B-Zip transcription factors, some of which must themselves be in a reduced state. The NPR1 acts as a redox co-factor, activating this class of transcription factors. The NPR1:TGA complex then activates expression of defense genes as part of establishing whole plant immunity to subsequent infections by virulent pathogens.

state (Deprés et al., 2003; Mou et al., 2003). It is possible that it is these NPR1:TGA complexes that drive expression of defense genes. The activation of NPR1 is achieved by reduction of two cysteine residues, which have been shown to be important for NPR1 function in vivo. The reduction of these cysteine residues in vitro can be achieved by glutathione (as GSH). In cells, the reduction of NPR1 may well be mediated by specific factors such as glutaredoxins, since this is the case for functionally equivalent redox-sensitive regulators in bacteria and yeast (Delaunay et al., 2002; Paget and Buttner, 2003). Glutathione may, therefore, play a major role in the SA signaling pathway via its possible influence over NPR1 redox state, and could thus influence plant: pathogen interactions. SA levels vary, and in some instances, have been shown to protect against many other environmental challenges, such as UV irradiation, ozone fumigation, cadmium poisoning, heat stress, exposure to high light, drought, and salinity (Dat et al., 1998; Surplus et al., 1998; Rao and Davies 1999; Borsani et al., 2001; Karpinski et al., 2003; Metwally et al., 2003). These conditions are also known to influence glutathione levels, and thus it is conceivable that this antioxidant and its interaction with SA are important components of signaling for a wide range of environmental conditions.

B. Glutathione, Ascorbate, ROS, and the ABA Signaling Pathway

There has been considerable progress in recent years in understanding how ABA mediates stomatal closure. One of the most significant findings is that H_2O_2 mediates the ABA signal that regulates ion movement in guard cells (Pei et al., 2000; Murata et al., 2001; Zhang et al., 2001; Kwak et al., 2003). H_2O_2 has also been implicated in the many observations of ABA-mediated induction of antioxidant gene expression (Sakamoto et al., 1995; Guan et al., 2000; Fryer et al., 2003; Jiang and Zhang, 2003; Yoshida et al., 2003). Negative regulators of this pathway are two protein phosphatase 2C (PP2C) isoforms encoded by *ABI1* and *ABI2* (the prefix ABI stands for "ABA insensitive" and refers to the way the genes were first recognized as mutants with this phenotype; Koornneef et al., 1984; Leung et al., 1997). The protein substrates for these phosphatases are not known. In vitro, the ABI2 PP2C activity is substantially inhibited by physiological levels of H_2O_2 and can be protected, and the inhibition reversed, by dithiothreitol or glutathione (as GSH; Meinhard et al., 2002). This suggests that redox sensitive cysteines in the reduced form are required for functioning of the ABI2 phosphatase. There is good evidence that a mitogen-activated protein kinase (MAPK) cascade may mediate H_2O_2 induction of antioxidant defense genes (Kovtun et al., 2000; Zhang and Klessig, 2001). Therefore, it is tempting to speculate that the PP2C's inhibition by the same H_2O_2 may serve to amplify the action of a protein kinase-signaling pathway (Meinhard et al., 2002). The role of glutathione can be questioned on the grounds that the levels needed to block PP2C may not be physiologically relevant (Meinhard et al., 2002). However, one cannot exclude that the reduction of the *ABI2* PP2C cysteine residues in vivo may require the action of a glutaredoxin, thioredoxin, or some other mediator of cellular redox state. Furthermore, related protein tyrosine phosphatases in mammalian cells have been shown to be regulated by glutathionylation at the equivalent redox sensitive cysteines (Barrett et al., 1999).

Further links between antioxidant metabolism and ABA have come from the observation that the ascorbic acid-deficient mutant *vtc1* (Conklin et al., 1997) has 1.6 fold elevated levels of ABA (Pastori et al., 2003). It has been suggested that cellular responses to an alteration in ascorbate levels may be transduced by ABA (Pastori et al., 2003).

C. ROS, Antioxidants, Jasmonic Acid, and Ethylene

JA signaling for response to wounding of plants has been suggested to be mediated by H_2O_2 which, in turn, stimulates gene expression in the vicinity of the wound site that could help prevent opportunistic infections, deter herbivores, and repair wounds (Orozco-Cardenas et al., 2001). JA pre-treatment of plants can also protect tobacco and *Arabidopsis* against ozone damage, an effect that may, in part, be due to changes in expression of antioxidant defense genes such as *APX* (Orvar et al., 1997; Rao et al., 2000). Similarly, a coordinated regulation of the expression of genes of glutathione metabolism such as glutathione reductase (GR) and the enzymes of glutathione synthesis occurs upon treatment of *Arabidopsis* plants with JA (Xiang and Oliver, 1998). It is not clear that all these effects of JA on ROS/antioxidant metabolism are linked, but if they are, then those of the whole plant response to environmental challenges that are influenced by JA must involve a coordinated regulation of antioxidant defenses.

Under stress conditions, such as ozone fumigation or wounding, the relationship between signaling pathways orchestrated by both JA and ethylene may be important (Overmyer et al., 2000; Lorenzo et al., 2003). Plant cells challenged with ozone or undergoing carbon starvation produce ethylene that is implicated in directing a burst of ROS production mediated by membrane-bound NADPH oxidases (Overmyer et al., 2000; Chae and Lee, 2001; Langebartels et al., 2002). However, ROS can, in turn, impact on the production of ethylene, implying complex feedback regulation of these responses (Overmyer et al., 2000; Watanabe et al., 2001). In other situations, depression of antioxidant enzyme activities and levels of antioxidants may be regulated by ethylene, such as during the senescence of spinach leaves (Hodges and Forney 2000). The most direct and intriguing evidence that ethylene perception may have a redox-regulated component comes from a study of the ethylene response mediator, ETR1, in both *Arabidopsis* and yeast. ETR1 is located in membranes as a homodimer formed by an intermolecular disulphide bridge. The formation of this disulphide bridge was shown to be necessary for ETR1 function (Schaller et al., 1995). This raises the possibility that ETR1 could be sensitive to alterations in cellular redox state driven by changes in antioxidant and ROS metabolism.

D. Growth, ROS, Antioxidants, and Auxins

The induction of defense responses can also lead to a down-regulation of genes associated with the vegetative growth of the plant. For example, in the *vtc1* mutant, a slow growth rate has been attributed to elevated levels of ABA rather than partial depletion of its ascorbate content (Pastori et al., 2003). Furthermore, H_2O_2 treatment of *Arabidopsis* protoplasts led not only to MAPK-mediated up-regulation of antioxidant defense gene expression, but also to a down-regulation of auxin-regulated genes that might be implicated in meristem growth (Kovtun et al., 2000). This may reflect a recently described convergence of kinase-mediated signaling that would coordinate both defense and growth responses to H_2O_2 (Moon et al., 2003). Conversely, auxin treatment of plant tissues may induce ROS production and also induce expression of antioxidant defense genes such as GSTs (Chen and Singh, 1999; Joo et al., 2001; Pfeiffer and Höftberger, 2001). Again, these data indicate that there is a two-way relationship between a hormone and ROS.

E. Conclusions on Hormone Signaling and ROS/Antioxidant Metabolism

It is clear from the above considerations that antioxidant metabolism can be profoundly influenced by, and in turn influence, signaling pathways nominally directed by one or more hormones. It has been suggested by many authors that the overlap in the type of genes induced by chemically diverse molecules may be reconciled by invoking a common effect on ROS at the cellular level (reviewed in Garretón et al., 2002). However, it is also clear that, in all cases, complex feedback regulation occurs by which ROS and antioxidants can influence the potency of hormones, especially those associated with stress. Thus, circular pathways of regulation have to be invoked to link up these diverse observations (e.g. Overmyer et al., 2000).

Perhaps the most compelling part of the relationship between cellular redox state (influenced by antioxidants such as glutathione) and signaling pathways is that key players in the pathways studied to date have either been shown to be or may be redox regulated. These include NPR1, ABI2, and ETR1 of the SA, ABA, and ethylene signaling pathways, respectively (Schaller et al., 1995; Meinhard et al., 2002; Mou et al., 2003). Thus, it is now probably safe to state that antioxidants play a key role in the life of plants not simply as protectors against oxidative stress, but also in the regulation of many aspects of the way plants interact with their ever-changing environment.

Acknowledgments

PM and GC acknowledge the support of the Biotechnology and Biological Sciences Research Council (BBSRC) through both individual grants and the Core Strategic Grant to the John Innes Centre. SK acknowledges the financial support provided by the Department of Botany, Stockholm University, the Swedish National Research Councils (VR), the Swedish Research Council for Environment, Agricultural Sciences and Spatial Planning (FORMAS), the Swedish Council for International Cooperation in Research and Higher Education (STINT), and the Swedish Strategic Foundation (SSF).

References

Allan AC and Fluhr R (1997) Two distinct sources of elicited reactive oxygen species in tobacco epidermal cells. Plant Cell 9: 1559–1572

Alméras E, Stolz S, Vollenweider S, Reymond P, Mène-Saffrané and Farmer EE (2003) Reactive electrophile species activate defense gene expression in *Arabidopsis*. Plant J 34: 205–216

Alvarez ME, Pennell RI, Meijer PJ, Ishikawa A, Dixon RA and Lamb C (1998) Reactive oxygen intermediates mediate a systemic signal network in the establishment of plant immunity. Cell 92: 773–784

Asada K (1999) The water-water cycle in chloroplasts: Scavenging of active oxygen species and dissipation of excess photons. Annu Rev Plant Physiol Mol Biol 50: 601–639

Ball L, Accotto G-P, Bechtold U, Creissen GP, Funck D, Jimenez A, Kular B, Leyland N, Mejia-Carranza J, Reynolds H, Karpinski S and Mullineaux PM (2004) Evidence for a direct link between glutathione biosynthesis and stress defence gene expression in Arabidopsis. Plant Cell 16: 2448–2462.

Banzet N, Richaud C, Deveaux Y, Kazmaier M, Gagnon J and Triantaphylides C (1998) Accumulation of small heat shock proteins, including mitochondrial HSP22, induced by oxidative stress and adaptive responses in tomato cells. Plant J 13: 519–527

Barrett WC, DeGnore JP, König S, Fales HM, Keng Y-F, Zhang Z-Y, Yim MB and Chock PB (1999) Regulation of PTP1B via glutathionylation of the active site cysteine 215. Biochemistry 38: 6699–6705

Bechtold U, Murphy DJ and Mullineaux PM (2004) Arabidopsis peptide methionine sulfoxide reductase2 prevents cellular oxidative damage in long nights. Plant Cell: 16: 908–919

Bestwick CS, Brown IR, Bennett MHR and Mansfield JW (1997) Localization of hydrogen peroxide accumulation during the hypersensitive reaction of lettuce cells to *Pseudomonas syringae* pv *phaseolicola*. Plant Cell 9: 209–221

Bolwell GP, Bindschedler LV, Blee KA, Butt VS, Davies DR, Gardner SL, Gerrish C and Minibayeva F (2002) The apoplastic oxidative burst in response to biotic stress in plants: a three-component system. J Exp Bot 53: 1367–1376

Borsani O, Valpuesta V and Botella MA (2001) Evidence for a role of salicylic acid in the oxidative damage generated by NaCl and osmotic stress in Arabidopsis seedlings. Plant Physiol 126: 1024–1030

Bowler C and Fluhr R (2000) The role of calcium and activated oxygen as signals for controlling cross-tolerance. Trends Plant Sci. 5: 241–246

Bowler C, Alliotte T, De Loose M, Van Montagu M and Inzé D (1989). The induction of manganese superoxide dismutase in response to stress in *Nicotiana plumbaginifolia*. EMBO J 8: 31–38

Bowling SA, Clarke JD, Liu YD, Klessig DF and Dong XN (1997) The *cpr5* mutant of *Arabidopsis* expresses both NPR1-dependent and NPR1-independent resistance. Plant Cell 9: 1573–1584

Bradley DJ, Kjellbom P and Lamb CJ (1992) Elicitor-induced and wound-induced oxidative cross-linking of a proline-rich plant-cell wall protein – a novel, rapid defense response. Cell 70: 21–30

Brüggemann W, Beyel V, Brodka M, Poth H, Weil M and Stockhaus J (1999) Antioxidants and antioxidative enzymes in wild-type and transgenic *Lycopersicon* genotypes of different chilling tolerance. J Plant Physiol 140: 145–154

Cao H, Bowling SA, Gordon AS and Dong XN (1994) Characterization of an *Arabidopsis* mutant that is nonresponsive to inducers of systemic acquired-resistance. Plant Cell 6: 1583–1592

Chae HS and Lee WS (2001) Ethylene- and enzyme-mediated superoxide production and cell death in carrot cells grown under carbon starvation. Plant Cell Rep 20: 256–261

Chamnongpol S, Willekens H, Langebartels C, Van Montagu M, Inzé D and Van Camp W (1996) Transgenic tobacco with a reduced catalase activity develops necrotic lesions and induces pathogenesis-related expression under high light. Plant J 10: 491–503

Chaumont M, Morot-Gaudry JF and Foyer CH (1995) Effects of photoinhibitory treatment on CO_2 assimilation, D_1 protein, ascorbate, glutathione and xanthophyll contents and the electron transport rate in vine leaves. Plant Cell Environ 18: 1358–1366

Chen W and Singh KS (1999) The auxin, hydrogen peroxide and salicylic acid induced expression of the Arabidopsis *GST6* promoter is mediated in part by an ocs element. Plant J 19: 667–677

Cheong YH, Kim K-N, Pandey GK, Gupta R, Grant JJ and Luan S (2003) CBL1, a calcium sensor that differentially regulates salt, drought and cold responses in Arabidopsis. Plant Cell 15: 1833–1845

Conklin PL, Williams EH and Last RL (1996) Environmental stress sensitivity of an ascorbic acid-deficient *Arabidopsis* mutant. Proc Natl Acad Sci USA 93: 9970–9974

Creissen GP, Firmin J, Fryer M, Kular B, Leyland N, Reynolds H, Pastori G, Wellburn F, Baker N, Wellburn A and Mullineaux PM (1999) Elevated glutathione biosynthetic capacity in the chloroplasts of transgenic tobacco plants paradoxically causes increased oxidative stress. Plant Cell 11: 1277–1292

Cummins I, Cole DJ and Edwards R (1999) A role for glutathione transferases functioning as glutathione peroxidases in resistance to multiple herbicides in black-grass. Plant J 18: 285–292

Dat J, Foyer CH and Scott IM (1998) Changes in salicylic acid and antioxidants during induced thermotolerance in mustard seedlings. Plant Physiol 118: 1455–1461

Dat JF, Pellinen R, Beeckman T, Van De Cotte B, Langebartels C, Kangasjärvi J, Inzé D and Van Breusegem F (2003) Changes in hydrogen peroxide homeostasis trigger an active cell death process in tobacco. Plant J 33: 621–632

De Vos RHC, Vonk MJ, Vooijs R and Schat H (1992) Glutathione depletion due to copper-induced phytochelatin synthesis causes oxidative stress in Silene cucubalus. Plant Physiol 98: 853–858

Delaunay A, Pflieger D, Barrault M-B, Vinh J and Toledano MB (2002) A thiol peroxidase is an H_2O_2 receptor and a redox-transducer in gene activation. Cell 111: 471–481

Delledonne M, Xia Y, Dixon RA and Lamb C (1998) Nitric oxide functions as a signal in plant disease resistance. Nature 394: 585–588

Després C, Chubak C, Rochon A, Clark R, Bethune T, Desveaux D and Fobert PR (2003) The Arabidopsis NPR1 disease resistance protein is a novel cofactor that confers redox regulation of DNA binding activity to the basic domain/leucine zipper transcription factor TGA1. Plant Cell 15: 2181–2191

Desikan R, A-H Mackerness S, Hancock JT and Neill SJ (2000) Regulation of the Arabidopsis transcriptome by oxidative stress. Plant Physiol 127: 159–172

Devadas SK, Enyedi A and Raina R (2002) The Arabidopsis hrl1 mutation reveals novel overlapping roles for salicylic acid, jasmonic acid and ethylene signaling in cell death and defense against pathogens. Plant J 30: 467–480

Dietrich RA, Delaney TP, Uknes SJ, Ward ER, Ryals JA and Dangl JL (1994) Arabidopsis mutants simulating disease resistance response. Cell 77: 565–577

Dietrich RA, Richberg MH, Schmidt R, Dean C and Dangl JL (1997) A novel zinc finger protein is encoded by the Arabidopsis LSD1 gene and functions as a negative regulator of plant cell death. Cell 88: 685–694

Doke N (1985) NADPH-dependent O_2^- generation in membrane fractions isolated from wounded potato tubers inoculated with Phytophthora infestans. Physiol Plant Pathol 27: 311–322

Douce R and Neuberger M (1999) Biochemical dissection of photorespiration. Curr Opin Plant Biol 2: 214–222

Foreman J, Demidchik V, Bothwell JHF, Mylona P, Miedema H, Torres MA, Linstead P, Costa S, Brownlee C, Jones JDG, Davies JM and Dolan L (2003) Reactive oxygen species produced by NADPH oxidase regulate plant cell growth. Nature 422: 442–445

Foyer CH and Halliwell B (1976) Presence of glutathione and glutathione reductase in chloroplasts: A proposed role in ascorbic acid metabolism. Planta 133: 21–25

Foyer CH and Mullineaux PM (eds) (1994) Causes of Photooxidative Stress and Amelioration of Defense Systems in Plants. CRC Press, Boca Raton, Florida

Fryer MJ, Ball L, Oxborough K, Karpinski S, Mullineaux PM and Baker NR (2003) Control of ascorbate peroxidase 2 expression by hydrogen peroxide and leaf water status during excess light stress reveals a functional organisation of Arabidopsis leaves. Plant J 33: 691–705

Garretón V, Carpinelli J, Jordana X and Holuigue L (2002) The as-1 promoter element is an oxidative stress-responsive element and salicylic acid activates it via oxidative species. Plant Physiol 130: 1516–1526

Genoud T and Metraux J-P (1999) Crosstalk in plant cell signaling: structure and function of the genetic network. Trends Plant Sci 4: 503–507

Grace SC and Logan BA (2000) Energy dissipation and radical scavenging by the plant phenylpropanoid pathway. Phil Trans Royal Soc Lond Ser B - Biol Sci 355: 1499–1510

Grant JJ and Loake GJ (2000) Role of reactive oxygen intermediates and cognate redox signaling in disease resistance. Plant Physiol 124: 21–29

Gray J, Close PS, Briggs SP and Johal GS (1997) A novel suppressor of cell death in plants encoded by the Lls1 gene of maize. Cell 89: 25–31

Guan LM, Zhao J and Scandalios JG (2000) Cis-element and trans-factors that regulate expression of the maize Cat1 antioxidant gene in response to ABA and osmotic stress: H_2O_2 is the likely intermediary signaling molecule for the response. Plant J 22: 87–95

Gupta R and Luan S (2003) Redox control of protein tyrosine phosphatases and mitogen-activated protein kinases in plants. Plant Physiol 132: 1149–1152

Hernandez JA, Jimenez A, Mullineaux PM and Sevilla F (2000) Tolerance of pea (Pisum sativum L.) to long-term salt stress is associated with induction of antioxidant defenses. Plant Cell Environ 23: 853–862

Hérouart D, Van Montagu M and Inzé D (1993) Redox activated expression of a cytosolic copper/zinc superoxide dismutase gene in Nicotiana. Proc Natl Acad Sci USA 90: 3018–3112

Hiraga S, Sasaki K, Ito H, Ohashi Y and Matsui H (2001) A large family of class III plant peroxidases. Plant Cell Physiol 42: 462–468

Hirai MY, Fujiwara T, Awazuhara M, Kimura T, Noji M and Saito K (2003) Global gene expression profiling of sulfur-starved Arabidopsis by DNA macroarray reveals the role of O-acetyl-L-serine as a general regulator of gene expression in response to sulfur nutrition. Plant J 33: 651–663

Hodges DM and Forney CF (2000) The effects of ethylene, depressed oxygen and elevated carbon dioxide on antioxidant profiles of senescing spinach leaves. J Exp Bot 51: 645–655

Horemans N, Foyer CH and Asard H (2000) Transport and action of ascorbate at the plant plasma membrane. Trends Plant Sci 5: 263–267

Horling F, Lamkemeyer P, König J, Finkemeier I, Kandlbinder A, Baier M and Dietz K-J (2003) Divergent light-, ascorbate-, and oxidative stress-dependent regulation of expression of the peroxiredoxin gene family in Arabidopsis. Plant Physiol 131: 317–325

Hsieh TH, Lee JT, Charng YY and Chan MT (2002) Tomato plants ectopically expressing Arabidopsis CBF1 show enhanced resistance to water deficit stress. Plant Physiol 130: 618–626

Hu X, Bidney DL, Yalpani N, Duvick JP, Crasta O, Folkerts O and Lu G (2003) Overexpression of a gene encoding hydrogen peroxide-generating oxalate oxidase evokes defense responses in sunflower. Plant Physiol 133: 170–181

Iturbe-Ormataexe I, Escuredo PR, Arresse-Igor C and Becana M (1998) Oxidative damage in pea plants exposed to water deficit or paraquat. Plant Physiol 116: 173–181

Jabs T, Dietrich RA and Dangl JL (1996) Initiation of runaway cell death in an Arabidopsis mutant by extracellular superoxide. Science 27: 1853–1856

Jamaï A, Tommasini R, Martinoia E and Delrot S (1996) Characterization of glutathione uptake in broad bean leaf protoplasts. Plant Physiol 111: 1145–1152

Jansen MAK, van den Noort RE, Tan MYA, Prinsen E, Lagrimini LM and Thorneley RNF (2001) Phenol-oxidizing peroxidases contribute to the protection of plants from ultraviolet radiation stress. Plant Physiol 126: 1012–1023

Jiang M and Zhang J (2003) Cross-talk between calcium and reactive oxygen species originated from NADPH oxidase in abscisic acid-induced antioxidant defense in leaves of maize seedlings. Plant Cell Environ 26: 929–939

Jimenez A, Hernandez JA, del Rio JA and Sevilla F (1997) Evidence for the presence of the ascorbate-glutathione cycle in mitochondria and peroxisomes of pea leaves. Plant Physiol 114: 275–284

Jimenez A, Creissen GP, Kular B, Firmin J, Robinson S, Verhoeyen M and Mullineaux PM (2002): Changes in oxidative processes and components of the antioxidant system during tomato fruit ripening. Planta 214: 751–758

Joo HJ, Bae YS and Lee JS (2001) Role of auxin-induced reactive oxygen species in root gravitropism. Plant Physiol 126: 1055–1060

Kampfenkel K, Van Montagu M and Inzé D (1995) Effects of iron excess on *Nicotiana plumbaginifolia* plants. Plant Physiol 107: 725–735

Karpinski S, Escobar C, Karpinska B, Creissen GP and Mullineaux PM (1997) Photosynthetic electron transport regulates the expression of cytosolic ascorbate peroxidase genes in Arabidopsis during excess light stress. Plant Cell 9: 627–640

Karpinski S, Reynolds H, Karpinska B, Wingsle G, Creissen GP and Mullineaux PM (1999) Systemic signaling and acclimation in response to excess excitation energy in Arabidopsis. Science 284: 654–657

Karpinski S, Gabrys H, Mateo A, Karpinska B and Mullineaux PM (2003) Light perception in plant disease defense signaling. Curr Opin Plant Biol 6: 390–396

Kimura M, Manabe K, Abe T, Yoshida S, Matsui M and Yamamoto YY (2003) Analysis of hydrogen peroxide-independent expression of the high-light-inducible ELIP2 gene with the aid of the ELIP2 promoter-luciferase fusion. Photochem Photobiol 77: 668–674

Kleibenstein DJ, Dietrich RA, Martin AC, Last RL and Dangl JL (1999) LSD1 regulates salicylic acid induction of copper zinc superoxide dismutase in *Arabidopsis thaliana*. Mol Plant-Microbe Interact 12: 1022–1026

Koornneef M, Reuling G and Karssen CM (1984) The isolation and characterization of abscisic acid-insensitive mutants of *Arabidopsis thaliana*. Physiol Plant 61: 377–383

Kovtun Y, Chiu W-L, Tena G and Sheen J (2000) Functional analysis of oxidative stress-activated mitogen-activated protein kinase cascade in plants. Proc Natl Acad Sci USA 97: 2940–2945

Kwak JM, Mori IC, Pei ZM, Leonhardt N, Torres MA, Dangl JL, Bloom RE, Bodde S, Jones JD and Schroeder JI (2003) NADPH oxidase AtrbohD and AtrbohF genes function in ROS-dependent ABA signaling in Arabidopsis. EMBO J 22: 2623–2633

Langebartels C, Wohlgemuth H, Kschieschan S, Grun S and Sandermann H (2002) Oxidative burst and cell death in ozone-exposed plants. Plant Physiol Biochem 40: 567–575

Leung J, Merlot S and Giraudat J (1997) The Arabidopsis *ABSCISIC ACID-INSENSITIVE2* (*ABI2*) and *ABI1* genes encode homologous protein phosphatases 2C involved in abscisic acid signal transduction. Plant Cell 9: 759–771

Long SP, Humphries S and Falkowski PG (1994) Photoinhibition of photosynthesis in nature. Annu Rev Plant Physiol Plant Mol Biol 45: 633–662

Lorenzo O, Piqueras R, Sànchez-Serrano JJ and Solano R (2003) ETHYLENE RESPONSE FACTOR1 integrates signals from ethylene and jasmonate pathways in plant defense. Plant Cell 15: 165–178

Loyall L, Uchida K, Braun S, Furuya M and Frohnmeyer H (2000) Glutathione and a UV light-induced glutathione-S-transferase are involved in signaling to chalcone synthase in cell cultures. Plant Cell 12: 1939–1950

Marrs KA (1996) The functions and regulation of glutathione-S-transferases in plants. Annu Rev Plant Physiol Plant Mol Biol 47: 127–148

Mattoo AK and Handa AK (2004) Ethylene signaling in plant cell death. In: Noodén, LD (ed) Plant Cell Death Processes. pp 125–142. Elsevier, Amsterdam

Meinhard M, Rodriguez PL and Grill E (2002) The sensitivity of ABI2 to hydrogen peroxide links the abscisic acid-response regulator to redox signaling. Planta 214: 775–782

Metwally A, Finkemeier I, Georgi M and Dietz K-J (2003) Salicylic acid alleviates the cadmium toxicity in barley seedlings. Plant Physiol 132: 272–281

Meyer Y, Verdoucq L and Vignols F (1999) Plant thioredoxins and glutaredoxins: identity and putative roles. Trends Plant Sci 4: 388–391

Mittler R, Herr EH, Orvar BL, van Camp W, Willekens H, Inzé D and Ellis BE (1999) Transgenic tobacco plants with reduced capability to detoxify reactive oxygen intermediates are hyperresponsive to pathogen infection. Proc Natl Acad Sci USA 96: 14165–14170

Moon H, Lee B, Choi G, Shin D, Prasad DT, Lee O, Kwak S-S, Kim DH, Nam J, Bahk J, Hong, JC, Lee SY, Cho MJ, Lim CO and Yun D-J (2003) NDP kinase 2 interacts with two oxidative stress-activated MAPKs to regulate cellular redox state and enhances multiple stress tolerance in transgenic plants. Proc Natl Acad Sci USA 100: 358–363

Mou Z, Fan W and Dong X (2003) Inducers of plant systemic acquired resistance regulate NPR1 function through redox changes. Cell 113: 935–944

Mullineaux PM and Creissen GP (1999) Manipulating oxidative stress responses using transgenic plants: successes and dangers. In: Altman A, Ziv M, Izhar S (eds) Plant Biotechnology and In Vitro Biology in the 21st Century, pp 525–532. Kluwer Academic Publishers, Dordrecht

Mullineaux PM and Karpinski S (2002) Signal transduction in response to excess light: getting out of the chloroplast. Curr Opin Plant Biol 5: 43–48

Murata Y and Takahashi M (1999) An alternative electron transfer pathway mediated by chloroplast envelope. Plant Cell Physiol 40: 1007–1013

Murata Y, Pei Z-M, Mori IC and Schroeder J (2001) Abscisic acid activation of plasma membrane Ca^{2+} channels in guard cells requires cytosolic NAD(P)H and is differentially disrupted upstream and downstream of reactive oxygen species production in *abi1-1* and *abi2-1* protein phosphatase 2C mutants. Plant Cell 13: 2513–2523

Neill SJ, Desikan R, Clarke A and Hancock JT (2002) Nitric oxide is a novel component of abscisic acid signaling in stomatal guard cells. Plant Physiol 128: 13–16

Nikiforova V, Freitag J, Kempa S, Adamik M, Hesse H and Hoefgen R (2003) Transcriptome analysis of sulfur depletion in *Arabidopsis thaliana*: interlacing of biosynthetic pathways provides response specificity. Plant J 33: 633–650

Noctor G and Foyer CH (1998) Ascorbate and glutathione: Keeping active oxygen under control. Annu Rev Plant Physiol Plant Mol Biol 49: 249–279

Noctor G, Arisi A-C, Joanin L, Kunert KJ, Rennenberg H and Foyer CH (1998) Glutathione: biosynthesis, metabolism and relationship to stress tolerance explored in transformed plants. J Exp Bot 49: 623–647

Noctor G, Gornez L, Vanacker H and Foyer CH (2002) Interactions between biosynthesis, compartmentation and transport in the control of glutathione homeostasis and signaling. J Exp Bot 53: 1283–1304

Op den Camp RGL, Przbyla D, Ochsenbein C, Laloi C, Kim C, Danon A, Wagner D, Hideg E, Göbel C, Feussner I, Nater M and Apel K. (2003) Rapid induction of distinct stress responses after the release of singlet oxygen in Arabidopsis. Plant Cell 15: 2320–2332

Orozco-Cárdenas ML, Narváez-Vásquez J and Ryan CA (2001) Hydrogen peroxide acts as a secondary messenger for the induction of defense genes in tomato plants in response to wounding, systemin and methyl jasmonate. Plant Cell 13: 179–191

Ort DR and Baker NR (2002) A photoprotective role for O_2 as an alternative electron sink in photosynthesis. Curr Opin Plant Biol 5: 193–198

Örvar BL and Ellis BE (1997) Transgenic tobacco plants expressing antisense RNA for cytosolic ascorbate peroxidase show increased susceptibility to ozone injury. Plant J 11: 1297–1305

Örvar BL, McPherson J and Ellis BE (1997) Pre-activating wounding response in tobacco prior to high-level ozone exposure prevents necrotic injury. Plant J 11: 203–212

Overmyer K, Tuominen H, Kettunene R, Betz C, Langebartels C, Sandermann Jr. H and Kangasjärvi J (2000) Ozone-sensitive Arabidopsis rcd1 mutant reveals opposite roles for ethylene and jasmonate signaling pathways in regulating superoxide-dependent cell death. Plant Cell 12: 1849–1862

Paget MSB and Buttner MJ (2003) Thiol-based regulatory switches. Annu Rev. Genet. 37: 91–121

Pastori GM, Kiddle G, Antoniw J, Bernard S, Veljovic-Jovanovic S, Verrier PJ, Noctor G and Foyer CH (2003) Leaf vitamin C contents modulate plant defense transcripts and regulate genes that control development through hormone signaling. Plant Cell 15: 939–951

Pei ZM, Murata Y, Benning G, Thomine S, Klüsener B, Allen GJ, Grill E and Schroeder JI (2000) Calcium channels activated by hydrogen peroxide mediate abscisic acid signaling in guard cells. Nature 406: 731–734

Pellinen R, Palva T and Kangasjärvi J (1999) Subcellular localization of ozone-induced hydrogen peroxide production in birch (Betula pendula) leaf cells. Plant J 20: 349–356

Pennell RI and Lamb C (1997) Programmed cell death in plants. Plant Cell 9: 1157–1168

Pfeiffer W and Höftberger M (2001) Oxidative burst in Chenopodium rubrum suspension culture cells: Induction by auxin and osmotic changes. Physiol Plant 111: 144–150

Pneuli L, Liang H, Rozenberg M and Mittler R (2003) Growth suppression, altered stomatal responses, and augmented induction of heat shock proteins in cytosolic ascorbate peroxidase (Apx1)-deficient Arabidopsis plants. Plant J 34: 187–203

Polle A (1997) Defense against photooxidative damage in plants. In: Scandalios, JG (ed) Oxidative Stress and the Molecular Biology of Antioxidant Defenses, pp 623–666. Cold Spring Harbor Laboratory Press, Cold Spring Harbor, NY

Price AH, Taylor A, Ripley SJ, Griffiths A, Trewavas AJ and Knight MR (1994) Oxidative signals in tobacco increase cytosolic calcium. Plant Cell 6: 1301–1310

Rao MV and Davies KR (1999) Ozone-induced cell death occurs via two distinct mechanisms in Arabidopsis: the role of salicylic acid. Plant J 17: 603–614

Rao MV, Lee H-I, Creelman RA, Mullet JE and Davies KR (2000) Jasmonic acid signaling modulates ozone-induced hypersensitive cell death. Plant Cell 12: 1633–1646

Rauser WE. Schupp R and Rennenberg H (1991) Cysteine, γ-glutamylcysteine, and glutathione levels in maize seedlings: Distribution and translocation in normal and cadmium-exposed plants. Plant Physiol 97: 128–138

Rhaikel NV and Coruzzi GM (2003) Plant systems biology. Plant Physiol 132: 403

Rizhsky L, Hallak-Herr E, Van Breusegem F, Rachmilevitch S, Barr JE, Rodermel S, Inzé D and Mittler R (2002) Double antisense plants lacking ascorbate peroxidase and catalase are less sensitive to oxidative stress than single antisense plants lacking ascorbate peroxidase or catalase. Plant J 32: 329–342

Rodermel S (2001) Pathways of plastid-to-nucleus signaling. Trends Plant Sci 6: 471–474

Rossel JB, Wilson IM and Pogson BJ (2002) Global changes in gene expression in response to high light in Arabidopsis. Plant Physiol 130: 1109–1120

Rustérucci C, Montillet, J-L, Agnel, J-P, Battesti C, Alonso B, Knoll A, Bessoule J-J, Etienne P, Suty L, Blein J-P and Triantaphyllides C (1999) Involvement of lipoxygenase-dependent production of fatty acid hydroperoxides in the development of the hypersensitive cell death by cryptogein on tobacco leaves. J Biol Chem 274: 36446–36455

Rustérucci C, Aviv DH, Holt BF, Dangl JL and Parker JE (2001) The disease resistance signaling components EDS1 and PAD4 are essential regulators of the cell death pathway controlled by LSD1 in Arabidopsis. Plant Cell 13: 2211–2224

Sakamoto A, Okumura T, Kaminaka H, Sumi K and Tan K (1995) Structure and differential response to abscisic acid of two promoters for the cytosolic copper/zinc-superoxide dismutase genes, SodCc1 and SodcC2, in rice protoplasts. FEBS Lett. 358: 62–66

Schaller F (2001) Enzymes of the biosynthesis of octadecanoid-derived signaling molecules. J Exp Bot 52: 11–23

Schaller GE, Ladd AN, Lanahan MB, Spanbauer JM and Bleecker AB (1995) The ethylene response mediator ETR1 from Arabidopsis forms a disulfide-linked dimer. J Biol Chem 270: 12526–12530

Schraudner M, Moeder W, Wiese C, Van Camp W, Inzé D, Langebartels C and Sandermann H Jr. (1998) Ozone-induced oxidative burst in the ozone biomonitor plant, tobacco Bel W3. Plant J 16: 235–245

Scrase-Field SAMG and Knight MR (2003) Calcium: just a chemical switch? Curr Opin Plant Biol 6: 500–506

Sebala M, Radova A, Angelini R, Tavladoraki P, Frebort I and Pec P (2001) FAD-containing polyamine oxidases: a timely challenge for researchers in biochemistry and physiology of plants. Plant Sci. 160: 197–207

Sherameti I, Sopory SK, Trebicka A, Pfannschmidt T and Oelmüller R (2002) Photosynthetic electron transport determines nitrate reductase gene expression and activity in higher plants. J Biol Chem 277: 46594–46600

Steinbrenner J and Linden H (2003) Light induction of carotenoid biosynthesis genes in the green alga *Haematococcus pluvialis*. Plant Mol Biol 52: 343–356

Surpin M, Larkin, RM and Chory J (2002) Signal transduction between the chloroplast and the nucleus. Plant Cell Supplement: S327–S338

Surplus SL, Jordan BR, Murphy AM, Carr JP, Thomas B and MacKerness SA-H (1998) Ultraviolet-B-induced responses in *Arabidopsis thaliana*: role of salicylic acid and reactive oxygen species in the regulation of transcripts encoding photosynthetic and acidic pathogenesis-related proteins. Plant Cell Environ 21: 685–694

Takahashi A, Kawasaki T, Henmi K, Shii K, Kodama O, Satoh H and Shimamoto K (1999) Lesion mimic mutants of rice with alterations in early signaling events of defense. Plant J 17: 535–545

Thoma I, Loeffler C, Sinha AK, Gupta M, Krischke M, Steffan B, Roitsch T and Mueller MJ (2003) Cyclopentenone isoprostanes induced by reactive oxygen species trigger defense gene activation and phytoalexin accumulation in plants. Plant J 34: 363–375

Thomashow MF (1999) Plant cold acclimation: Freezing tolerance genes and regulatory mechanisms. Ann Rev Plant Physiol Plant Mol Biol 50: 571–599

Vollenweider S, Weber H, Stolz S, Chételat A and Farmer EE (2000) Fatty acid ketodienes and fatty acid ketotrienes: Michael addition acceptors that accumulate in wounded and diseased Arabidopsis leaves. Plant J 24: 467–476

Walker MA and McKersie BD (1993) Role of the ascorbate-glutathione antioxidant system in chilling resistance in tomato. J Plant Physiol 141: 234–239

Watanabe T, Seo S and Sakai S (2001) Wound-induced expression of a gene for 1-aminocyclopropane-1-carboxylate synthase and ethylene production are regulated by both reactive oxygen species and jasmonic acid in *Cucurbita maxima*. Plant Physiol Biochem 39: 121–127

Wendehenne D, Pugin A, Klessig DF and Durner J (2001) Nitric oxide: comparative synthesis and signaling in animal and plant cells. Trends Plant Sci 6: 177–180

Willekens H, Chamnongpol S, Davey M, Schraudner M, Langebartels C, Van Montagu M, Inzé D and Van Camp W (1997) Catalase is a sink for H_2O_2 and is indispensable for stress defense in C3 plants. EMBO J 16: 4806–16

Wingate VMP, Lawton MA and Lamb CJ (1988) Glutathione causes a massive and selective induction of plant defense genes. Plant Physiol 87: 206–210

Wingsle G and Karpinski S (1996) Differential redox regulation of glutathione reductase and Cu/Zn superoxide dismutase genes expression in *Pinus sylvestris* (L) needles. Planta 198: 151–157

Wisniewski J-P, Cornille P, Agnel J-P and Montillet J-L (1999) The extensin multigene family responds differentially to superoxide or hydrogen peroxide in tomato cell cultures. FEBS Lett 447: 264–268

Xiang C and Oliver DJ (1998) Glutathione metabolic genes coordinately respond to heavy metals and jasmonic acid in Arabidopsis. Plant Cell 10: 1539–1550

Xiang C, Werner BL, Christensen E.M and Oliver DJ (2001) The biological functions of glutathione revisited in Arabidopsis transgenic plants with altered glutathione levels. Plant Physiol 126: 564–574

Xu C-C, Li L and Kuang, T. (2000) The inhibited xanthophyll cycle is responsible for the increase in sensitivity to low temperature photoinhibition in rice leaves fed with glutathione. Photosynth Res 65: 107–114

Yoda H, Yamaguchi Y and Sano H (2003) Induction of hypersensitive cell death by hydrogen peroxide produced through polyamine degradation in tobacco plants. Plant Physiol 132: 1973–1981

Yoshida K, Igarashi E, Mukai M, Hirata K and Miyamoto K (2003) Induction of tolerance to oxidative stress in the green alga, *Chlamydamonas reinhardtii*, by abscisic acid. Plant Cell Environ 26: 451–457

Yoshioka H, Numata N, Nakajima K, Katou S, Kawakita K, Rowland O, Jones JDG and Doke N (2003) *Nicotiana benthamiana* gp91^{phox} homologs NbrbohA and NbrbohB participate in H_2O_2 accumulation and resistance to *Phytophthora infestans*. Plant Cell 15: 706–718

Zhang S and Klessig DF (2001) MAPK cascades in plant defense signaling. Trends Plant Sci 6: 520–527

Zhang X, Zhang L, Dong F, Gao J, Galbraith DW and Song CP (2001) Hydrogen peroxide is involved in abscisic acid-induced stomatal closure in *Vicia faba*. Plant Physiol 126: 1438–1448

Zhu YL, Pilon-Smits EAH, Tarun AS, Weber SU, Jouanin L and Terry N (1999) Cadmium tolerance and accumulation in Indian mustard is enhanced by overexpressing γ-glutamylcysteine synthetase. Plant Physiol 121: 1169–1177

Chapter 16

Signaling and Integration of Defense Functions of Tocopherol, Ascorbate and Glutathione

Christine H. Foyer*
Crop Performance and Improvement Division, Rothamsted Research, Harpenden, Hertfordshire AL5 2JQ, UK

Achim Trebst
Plant Biochemistry, Ruhr University, 44780 Bochum, Germany

Graham Noctor
Institut de Biotechnologie des Plantes, UMR 8618 CNRS, Université Paris XI, Bâtiment 630, 91405 Orsay cedex, France

Summary	242
I. Introduction	242
II. Singlet Oxygen and Tocopherol Function in PS II	243
A. Sources of Singlet Oxygen	243
B. Tocopherol	248
C. Function of Tocopherol in PS II	248
1. Blocking Tocopherol Biosynthesis with Inhibitors	248
2. Tocopherol Synthesis Mutants	250
III. Ascorbate: A Key Player in Leaf Development and Responses to the Environment	250
A. Ascorbate in the Chloroplast	250
B. Ascorbate-Mediated Regulation of Gene Expression	251
C. Ascorbate, Abscisic acid, and Stomatal Regulation	252
D. Ascorbate Oxidase and the Regeneration of Apoplastic Ascorbate	253
E. Respiration, Reactive Oxygen and Ascorbate Biosynthesis	254
1. Plant Mitochondria and Reactive Oxygen	254
2. Respiration and Ascorbate Synthesis	255
IV. Glutathione and the Importance of Cellular Thiol/Disulfide Status	256
A. Glutathione and Antioxidant Defense	256
B. Glutathione and Gene Expression in Plants	256
C. The Glutathione Redox Couple and Sensing of Reactive Oxygen	257
D. Glutathione Synthesis and the Redox State of the Cell	258
E. Coupling of the Ascorbate and Glutathione Pools	259
V. Conclusions and Perspectives: All for One and One for All?	259
Acknowledgments	260
References	260

*Author for correspondence, email: christine.foyer@bbsrc.ac.uk

B. Demmig-Adams, William W. Adams III and A.K. Mattoo (eds), Photoprotection, Photoinhibition, Gene Regulation, and Environment, 241–268.
© Springer Science+Business Media B.V. 2008

Summary

Ascorbate, glutathione, and tocopherol are the three major low molecular weight antioxidants of plant cells. While tocopherol is hydrophobic and is found only in lipid membranes, ascorbate and glutathione are hydrophilic, accumulating to high concentrations in the chloroplast stroma and other compartments of the plant cell. Ascorbate and glutathione not only limit photo-oxidative damage but can also act independently as signal-transducing molecules regulating defense gene expression. Both metabolites transmit information concerning oxidative load and redox-buffering capacity. Ascorbate modifies the expression of chloroplast genes. Net glutathione synthesis during stress restores the cellular redox state and allows orchestration of systemic acquired resistance. The degree of redox coupling between these antioxidants has profound implications for regulation, function, and signaling associated with the two major energy-generating systems, i.e. photosynthesis and respiration. Tocopherol fulfills an essential protective function, counter-acting the harmful effects of singlet oxygen production at photosystem II. Ascorbate reduces and thus regenerates oxidized tocopherol, but flux through this reaction is not sufficient to maintain the reduced tocopherol pool under high light stress. This may be because tocopherol regeneration draws on the ascorbate pool of the chloroplast lumen, which may be depleted under stress. Moreover, while glutathione always reduces oxidized ascorbate (dehydroascorbate), the degree of coupling between the ascorbate and glutathione redox couples is variable. The flexibility of coupling between these antioxidant pools is crucial to differential redox signaling, particularly by ascorbate and glutathione.

I. Introduction

Plants are autotrophic organisms fueled by light-driven redox chemistry (Noctor et al., 2000). Through the necessity of harnessing light energy to drive photosynthetic metabolism, plants have optimized strategies for redox control, including ways of minimizing the generation of reactive oxygen species (ROS) and employing a network of pathways of ROS detoxification. Moreover, redox signals exert extensive control on gene expression, adapting plant growth and development to environmental inputs and cues.

Oxygenic photosynthesis and aerobic respiration are both able to undertake the concerted, four-electron exchange between water and oxygen, at photosystem II and terminal oxidases, respectively. In addition, electron transport reactions associated with both processes are a major source of ROS in plants, generating superoxide, hydrogen peroxide, and singlet oxygen (Endo and Asada, this volume). Large amounts of hydrogen peroxide are also formed by the photorespiratory pathway (Foyer and Noctor, 2003), and many other metabolic processes in plants catalyze only partial reduction of oxygen, thus generating superoxide and hydrogen peroxide (Foyer and Noctor, 2000; Mittler, 2002). For example, superoxide and/or H_2O_2 are generated at significant rates by oxidative phosphorylation, fatty acid β-oxidation, and also by many types of oxidase activity. One of the most intensively studied of these pro-oxidant events is the oxidative burst that occurs in the apoplast in response to pathogen attack (Lamb and Dixon, 1997).

There is no doubt that oxygen is potentially toxic and that excessive ROS production is incompatible with cell functions. However, aerobic organisms have evolved means of using the strong oxidizing potential of the O_2/H_2O redox couple ($Em_7 = +815$ mV) in a controlled fashion and, furthermore, also exploit the reactivity of ROS in plant metabolism, signaling, and defense. The roles of ROS in stress-induced damage have long been recognized, but it is now also generally accepted that ROS are an integral component of cellular signaling in both animals (Hancock et al., 2001) and plants (Vranova et al., 2002a, b; Mittler, 2002; Foyer and Noctor, 2003). ROS are involved in innate immune responses in plants as well as in acquired resistance to biotic and abiotic stresses (Alvarez et al., 1998; Dat

---

Abbreviations: ABA – abscisic acid; AO – ascorbate oxidase; APX – ascorbate peroxidase; DCMU – 3-(3, 4-dichlorphenyl)-1, 1-dimethylurea; DHA – dehydroascorbate; DHAR – dehydroascorbate reductase; FBPase – fructose-1, 6-*bis*phosphatase; G6PDH – glucose-6-phosphate dehydrogenase; GLDH – galactono-1, 4-lactone dehydrogenase; GPX – glutathione peroxidase; GR – glutathione reductase; GRX – glutaredoxin; GSH – reduced glutathione; GSSG – glutathione disulfide; GST – glutathione *S*-transferase; HPP – hydroxymethylphenyl pyruvate; MDHA – monodehydroascorbate; MDHAR – monodehydroascorbate reductase; NAT – nucleobase L-ascorbic acid transporters; NPR1 – nonexpressor of PR genes 1; NCED – 9-*cis*-epoxycarotenoid dioxygenase; PHGPX – phospholipid hydroperoxide glutathione peroxidase; PR – pathogenesis-related; PRK – phosphoribulokinase; PRX – peroxiredoxin; PS I – photosystem I; PS II – photosystem II; ROS – reactive oxygen species; SA – salicylic acid; SBPase – sedoheptulose-1, 7-*bis*phosphatase; TRX – thioredoxin.

et al., 2000). H_2O_2 was first shown to be a second messenger in plant-pathogen interactions (Green and Fluhr, 1995), wounding (Orozco-Cardenas et al., 2001), and abiotic stress (Pastori and Foyer, 2002). More recently, H_2O_2 and other ROS have been reported to be intimately involved in a wide range of hormone-dependent developmental signaling processes, as well as in cell wall cleavage and associated cell wall growth in a diverse range of developing and expanding organs such as embryonic axes (Puntarulo et al., 1988), roots (Joo et al., 2001), germinating seeds (Schopfer, 2001), expanding leaves (Rodriguez et al., 2002), and coleoptiles (Schopfer et al., 2002). In particular, Foreman et al. (2003) showed that ROS generated by a plasma membrane NADPH oxidase regulate root hair growth. In this case, ROS were found to regulate cell expansion through activation of calcium channels. ROS can be positioned both up- and down-stream of plant hormone action. For example, application of auxin promotes ROS production and gravitropic responses in roots (Joo et al., 2001) while abscisic acid (ABA) mediates H_2O_2 accumulation during stomatal closure (Pei et al., 2000), and gibberellic acid alters the sensitivity of barley aleurone protoplasts to H_2O_2-induced programmed cell death (Fath et al., 2001). On the other hand, iron-induced oxidative stress was shown to modulate auxin levels in protoplast-derived cells (Pasternak et al., 2002) and photorespiratory H_2O_2 accumulation caused by catalase deficiency induced accumulation of stress hormones such as salicylic acid (Chamnongpol et al., 1998). During ozone exposure, ROS action is situated both upstream and downstream of ethylene synthesis (Moeder et al., 2002). ROS are also associated with stimulation of homologous recombination (Kovalchuk et al., 2003).

Like nitric oxide (NO), ROS have typical features of signaling molecules in that they are metabolically active and have a short life-time. Signal intensity is therefore easily controllable by variations in rates of production and/or removal. The duration of the ROS signal is determined by the effective antioxidant network in which tocopherol, ascorbate, and glutathione are key players. These highly abundant metabolites are ubiquitous in the photosynthetic cells of higher plants and many algae, where they are present predominantly in the reduced state, except under extreme stress conditions. Ascorbate and glutathione fulfill many essential roles in plants (May et al., 1998; Noctor et al., 1998; Arrigoni and De Tullio, 2000). For example, ascorbate is involved in the regulation of quiescence, mitosis, and cell growth (Potters et al., 2000), while glutathione is important in root formation and senescence (Vernoux et al., 2000; Ogawa et al., 2004). The antioxidant functions of tocopherol, ascorbate, and glutathione, and their participation in cellular defense against oxidative stress, is well known and has been discussed many times previously (Noctor and Foyer, 1998; Foyer and Noctor, 2000). We consider the roles of the ascorbate and glutathione pools as major redox buffers of plant cells (Figure 1) to be equally important to their functions in metabolism and defense. The membrane tocopherol pool is coupled to the glutathione pool through the mediation of ascorbate (Figure 2) and the levels of these two antioxidants tend to vary in a similar manner. While levels of tocopherol, ascorbate, and glutathione are fairly constant throughout the day/night cycle in the leaves of some species, they show marked diurnal variation in other species. For example, the ascorbate, glutathione, and tocopherol contents are diurnally modulated in *Arabidopsis* rosette leaves and in spruce needles (Schupp and Rennenberg, 1990; Tamaoki et al., 2003). Net synthesis of these antioxidants increases in the light and maximum accumulation often corresponds with the maximum day light intensities in field grown plants. In tobacco leaves, ascorbate contents fall in darkness and are high throughout the duration of the light period (Figure 3). Similarly, glutathione synthesis can be faster in the light than the dark (Noctor et al., 1998). Here, we discuss the importance of the degree of interaction between the low molecular weight antioxidant pools in relation to function. A major theme of this overview is that redox coupling is efficient enough to provide long term buffering of cellular redox state while allowing short-term perturbations in each redox couple to provide specific signaling information.

II. Singlet Oxygen and Tocopherol Function in PS II

A. Sources of Singlet Oxygen

Photosynthesis is the major source of singlet oxygen in plants. Singlet oxygen is mainly formed through energy transfer from the triplet state of excited pigments to triplet oxygen, but is also an intermediate in chemi-excitation, such as occurs in lipid peroxidation. Of course, since photosynthesis involves constant pigment excitation, the potential for singlet oxygen production is always present. Chlorophyll triplets are formed both in the antenna as well as in the photosystems. In the antenna, the triplets are formed by inter-system crossing from the excited singlet state. Their yield can be very

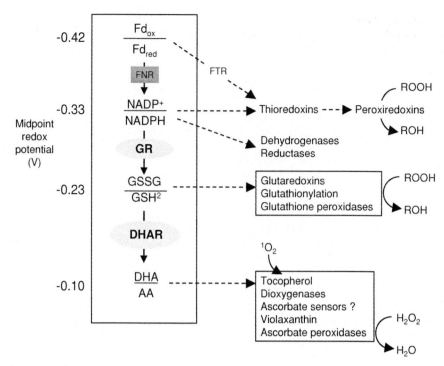

Fig. 1. Soluble reductant couples arranged according to their midpoint potentials (left, linked by solid arrows) and principal components and processes with which they interact (right, linked by dashed arrows). DHAR, dehydroascorbate reductase; Fd, ferredoxin; FTR, ferredoxin-thioredoxin reductase; GR, glutathione reductase. ROOH denotes H_2O_2 or organic peroxides. Chloroplast-specific components (Fd, FTR) are shown in rectangles. For simplicity, only the principal interactions are shown. For example, certain peroxiredoxins may also be regenerated by glutaredoxins (Rouhier et al., 2002); glutaredoxins and thioredoxins, as well as other enzymes, may also be involved in DHA reduction (Nishizawa and Buchanan, 1981; Wells et al., 1990); de-epoxidation of both violaxanthin and antheraxanthin requires ascorbate (Demmig-Adams and Adams, 1992); and other reactive oxygen species can act as oxidants of the couples shown (Polle, 2001). Irrespective of these details, and although in non-photosynthetic plastids net electron transfer may be from NADPH to Fd (Bowsher et al., 1989), net electron flow will generally be from Fd (in the chloroplast) or from NAD(P)H (in other compartments) to ascorbate. Specificity in signaling linked to production of reactive oxygen species is likely to be determined by the extent to which the reductant redox couples are kinetically linked, and the relative importance of peroxiredoxins, glutathione peroxidases and ascorbate peroxidases in peroxide scavenging. For further discussion, see text.

high (Witt, 1996; Telfer, 2002), but they are effectively quenched by the abundant carotenoid pigments (Cogdell and Frank, 1996), mainly the xanthophylls in the distal and proximal antenna and β-carotene in the core antenna, and therefore little harm is done. Triplets of the PS II reaction center chlorophylls are not formed by intersystem crossing, as the excited singlet state immediately leads to charge separation. Here the triplets are formed in recombination reactions (Diner and Rappoport, 2002). Their yield depends not only on light input but also on the redox state of the primary stable electron acceptor quinone of PS II, Q_A.

Figure 4 illustrates the energy levels of the primary reactants in the reaction center of purple bacteria under anaerobic conditions according to Angerhofer (1991). In forward electron flow, charge separation occurs when Q_A becomes oxidised (Figure 4, left). In a situation where Q_A is already reduced (Figure 4, right), the energy levels of the radical pair are in the singlet and triplet state and are very close because of spin interaction with Q_A minus. Recombination under these conditions leads to the triplet state of the reaction center chlorophyll. An equivalent scheme is used by Krieger and Rutherford for PS II under aerobic conditions (Krieger et al., 1998; Rutherford and Krieger-Liszkay, 2001; Fufezan et al., 2002). As oxygen is present in this latter case, decay of the P680 triplet state competes with energy transfer to oxygen and singlet oxygen may be formed. Because of the physical arrangement of the PS II reaction center, where the β-carotenes are bound to the D2 protein while P_{680} is localised on the D1 protein, the quenching of chlorophyll triplets as well as of the short-lived and localized singlet oxygen by carotene in the reaction center of aerobic photosynthetic organisms may not play the same role as in the antenna. There is no orbital overlap between the carotenes and the

Chapter 16 Tocopherol, Ascorbate and Glutathione

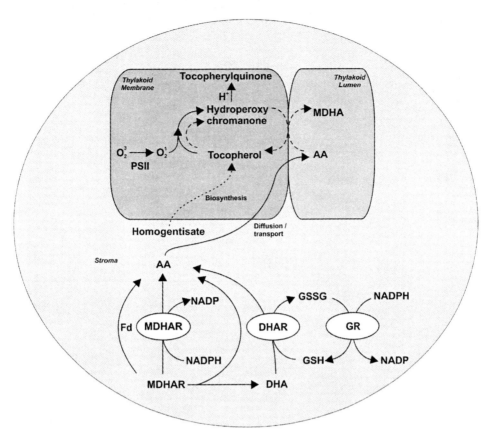

Fig. 2. Antioxidant cycling in the chloroplast. Singlet oxygen arising in PS II oxidizes tocopherol to an intermediate hydroperoxide, which may be hydrolysed to tocoquinone. The tocopherol is regenerated by biosynthesis from homogentisate (broken arrows) or the intermediate may be re-reduced by ascorbate (solid arrows). Coupling of tocopherol regeneration to ascorbate oxidation potentially engages the three reductant couples shown in Fig.1: Fd through direct photochemical reduction of MDHA, NADPH through MDHAR (and GR), and glutathione through DHA reduction following dismutation of MDHA. For abbreviations, see list.

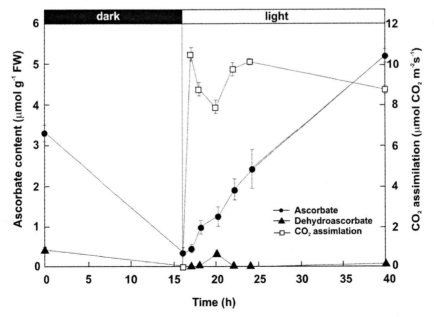

Fig. 3. The effect of dark and light on tobacco leaf ascorbate and dehydroascorbate contents. The leaf ascorbate and dehydroascorbate contents of 6-week-old tobacco plants was measured before and after a dark period of 16h and then during a period of 24 h continuous light (350 μmol m^{-2} s^{-1}). Photosynthetic CO_2 assimilation rates in the light are also shown. For further information see Pignocchi (2000).

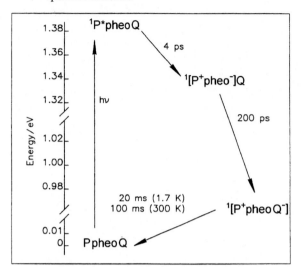 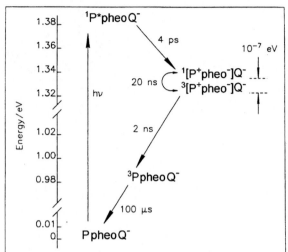

Fig. 4. Energy levels and energy states of the purple bacteria radical pair under anaerobic conditions showing the path of recombination following charge separation when Q_A is either oxidized (left) or reduced (right). The figure is adapted from Angerhofer (1991).

primary P_{680} triplets (Van Gorkom and Schelvi, 1993; Telfer et al., 1994a; Telfer and Barber, 1995) as shown by recent X-ray structure analysis of PS II by Kamiya and Shen (2003) and Ferreira et al. (2004). Indeed, at low temperatures no carotene quenching occurs at all in PS II (Mathis et al., 1979; Satoh and Mathis, 1981; Durrant et al., 1990). It is important to note that the yield of P680 triplets is not only a function of light intensity, but also of the redox state of the electron flow system. It follows therefore that P680 triplets and singlet oxygen are formed in greatest abundance when the demand for reducing power from the photosynthetic electron flow system is small, but light excitation remains high, i.e. the basic conditions for photoinhibition (Keren et al., 1997; Osmond and Förster, this volume), a process defined by Osmond (1994) as the light-dependent inhibition of the light-dependent reactions of photosynthesis. The amount of singlet oxygen formed in enriched PS II preparations can be measured by spectroscopic and electron spin resonance techniques (Macpherson et al., 1993; Telfer et al., 1994a; Hideg et al., 1994, 1998; Krieger et al., 1998; op den Camp, 2003). To date, measurements and detection of $^1O_2$ in vivo have failed, but this should not be taken as an argument that singlet oxygen is not formed in vivo, as is discussed below in relation to the turnover of the D1 protein. Production of $^1O_2$ in vivo is discussed in more detail in relation to tocopherol by Trebst (1999, 2003).

With this caveat in mind, singlet oxygen generated in the PS II reaction center (Hideg et al., 2002) has several options for further reactions:

a. It may react with the D1 protein and induce its degradation (Keren et al., 1995, 1997). This is the onset of the well-known rapid turnover of the D1 protein (Mattoo et al., 1989), which is an integral part of photoinhibition (reviewed in Adir et al., 2003). As long as the rapid degradation rate of the D1 protein is compensated by an equally rapid re-synthesis of the protein and re-assembly into a new functional PS II, photosynthetic flux continues unabated. If not, the D1 protein is not replaced and functional PS II units disappear. There can then be no PS II activity and the whole chlorophyll/pigment system eventually bleaches (Aro et al., 1993; Melis, 1999; Adir et al., 2003). In the process of disassembly of PS II, chlorophyll and carotenes are released from their binding sites on the D1 protein. This is also likely to occur on the D2 protein but, at least at first, not for the 43 and 47 kDa core antenna proteins. For reassembly, newly synthesized β-carotene (Trebst and Depka, 1997) and probably also new chlorophyll (Feierabend and Dehne, 1996) are required.

b. Singlet oxygen generated at the P680 chlorophyll binding site on the D1 protein may, in spite of its short life time of about 10 μsec, diffuse to the carotene binding site on the D2 protein to be quenched there (Telfer et al., 1994b).

c. Singlet oxygen could be scavenged by tocopherol, as discussed below in detail.

d. In view of possibilities a–c, singlet oxygen, or a stable product or reaction with singlet oxygen, may act as a signal transducer.

Another source of singlet oxygen in the thylakoid membrane is the light-induced excited state of protoporphyrin IX, an intermediate in the biosynthesis of cytochromes and chlorophylls. The amount of protoporphyrin IX can be greatly enhanced by inhibition of protoporphyrinogen IX oxidase (PPO). This inhibition is achieved by herbicides that target PPO (Duke et al., 1991; Hock et al., 1995) and cause bleaching via exacerbated rates of ROS generation through light excitation of protoporphyrin. Protoporphyrin IX and protochlorophyllide are also formed via this reaction in a mutant (*flu*) defective in the regulation of porphyrin synthesis (Meskauskiene et al., 2001). This mutant is viable only when grown in continuous light where the intermediate is consumed. In the dark, the photosensitizers accumulate. Exposure to intermittent light leads to singlet oxygen production. In this situation, mature *flu* mutants stop growing, while seedlings bleach and die (op den Camp et al., 2003). These responses do not merely result from oxidative damage but rather result from the activation of distinct cell sucide programs. Evidence in support of this view comes from molecular genetic screens that have shown that inactivation of a single gene, EXECUTER 1 (EX1), is sufficient to prevent singlet oxygen-induced *flu* seedling death and growth inhibition in mature plants (Wagner et al. 2004). Inactivation of the EX1 protein not only suppresses the induction of death in *flu* seedlings but it also prevents death in wild type plants treated with 3-(3, 4-dichlorphenyl)-1, 1-dimethylurea (DCMU). However, this observation is more difficult to explain, as DCMU prevents singlet oxygen formation in PS II and prevents D1 turnover.

It remains to be demonstrated conclusively whether singlet oxygen is a unique signal per se or whether it is indistinguishable from other ROS-induction phenomena. Of course, singlet oxygen is the primary reactant, and quenching by pyridoxamine prevents the initiation of the cell suicide program. Similarly, the adverse effects of PPO herbicides can be abolished by adding diphenylamine as a singlet oxygen quencher. There is no doubt that singlet oxygen generated at random by photosensitizers quickly induces a cascade of ROS that orchestrate a genetically determined suicide program. This is true for both the singlet oxygen formed as a result of PPO herbicide action as well as in the *flu* mutants. In vivo, singlet oxygen generated at a specific "natural" site, i.e. within the PS II reaction center, fufills this role. However, signal transduction will only occur if this singlet oxygen escapes destruction (i.e. specific scavenging by D1 or by tocopherol) or is otherwise consumed in a ROS cascade.

The expression patterns of approximately 5% of the total *Arabidopsis* genes are modified in the *flu* mutant within the first 30 minutes after release of singlet oxygen (op den Camp et al., 2003). However, the pattern of differential gene expression observed upon singlet oxygen generation is practically identical to that obtained by applying H_2O_2 and light (Kimura et al., 2001, 2003). The majority of the singlet-oxygen induced changes in gene expression observed in the *flu* mutant are not suppressed in the *ex1* mutant. Thus, singlet oxygen production activates at least two distinct signal transduction pathways, one that requires *EX1* and one that is *EX1* independent (K.P. Lee et al., 2003; op den Camp et al., 2003, Wagner et al. 2004). Moreover, singlet oxygen-mediated signaling has similarities to that associated with another recently proposed plastid signaling molecule, the tetrapyrrole intermediate Mg^{2+} protoporphyrin IX (Strand et al., 2003; Mg^{2+} proto IX). This molecule was identifed via analysis of a group of nuclear recessive mutants called *gun* (*genomes uncoupled*) that were generated by mutating plants that contain a fusion of the *light-harvesting chlorophyll a/b binding protein* (*Lhcb*) promoter to the B-glucuronidase (*GUS*) and screening for plants that express *GUS* when treated with norflurazon, an inhibitor of phytoene desaturase that leads to carotenoid deficiency. The *gun* mutants de-repress transcription of nuclear genes involved in photosynthesis (Rodermel, 2001). Moreover, like *flu* (op den Camp et al., 2003), the large changes in gene expression (affecting about 4% of the total *Arabidopsis* genome) brought about by release of the plastid signaling factor include many known stress responsive genes (Strand et al. 2003). The plastid signaling factor and singlet oxygen may therefore use similar signaling pathways.

As mentioned above, chlorophyll is released during the degradation of the D1 protein. Free chlorophyll is a very effective photosensitizing agent, producing singlet oxygen first and then, by a chain reaction, numerous other ROS that eventually bleach the chlorophyll. The latter, rather than singlet oxygen per se, are the probable source of ozone damage to the D1 protein (Sandermann, 2000). Cyanobacteria employ largely transcriptional and translational regulation of gene expression in response to exposure to high light intensities or other photoinhibitory conditions. Flowering plants also employ circadian controls and post-translational regulation including redox regulation of the phosphorylation of the D1 protein that may add an extra level of protection at PS II (Booij-James et al. 2002) and prevent the orchestration of the signal transduction pathways described above that ultimately lead to death.

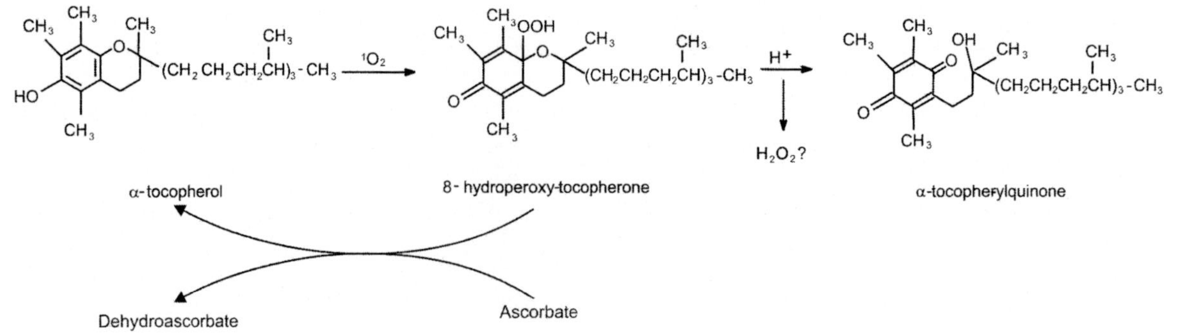

Fig. 5. The reaction of tocopherol with singlet oxygen (according to Neely et al., 1998).

B. Tocopherol

Since the occurrence and properties of tocopherols in plants have been extensively reviewed previously (Fryer, 1992; Elstner et al., 1998; Foyer and Harbinson, 1999; Bramley et al., 2000; Munné-Bosch and Alegre, 2002a,b; Trebst et al., 2002; Munné-Bosch and Falk, 2003), the present discussion will concentrate largely on the specific roles of tocopherol in relation to photosynthesis and its functions in PS II in particular. Indeed, tocopherol is restricted to the chloroplast where it is synthesized from homogentisic acid. Chloroplasts contain all of the tocopherol biosynthetic enzymes. The genes encoding tocopherol-synthesizing proteins are located in the nucleus and their respective proteins are imported into the chloroplast. Tocopherol is localized largely to the thylakoid membranes and the plastoglobuli, with perhaps a small amount present in the chloroplast envelope as well (Lichtenthaler, 1968, 1998; Tevini and Lichtenthaler, 1970; Soll et al., 1985). Leaf tocopherol contents, like those of ascorbate, are light dependent, i.e. increase as growth light intensity increases. An extensive literature exists on the chemistry of tocopherol and its functions because of the importance of tocopherol (vitamin E) for human health (Fryer, 1992; Kaiser et al., 1990). Tocopherol is slowly oxidized to radical forms in air, but reacts rapidly with singlet oxygen to form a hydroperoxychromanone (Neely et al., 1988). This intermediate can then be hydrolyzed to tocopherylquinone—shown to occur in the chloroplast (Kruk and Strzalka, 1995)—in which the chromane ring is irreversibly opened (Figure 5). However, the hydroperoxychromanone intermediate can be re-reduced to tocopherol by ascorbate (Figure 5) (Neely et al., 1988).

C. Function of Tocopherol in PS II

As stated above, the major potential source of singlet oxygen in the chloroplast is the triplet state of the

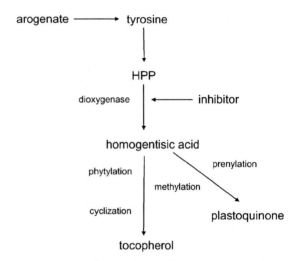

Fig. 6. Intermediates in the biosynthesis of plastoquinone and tocopherol. HPP stands for *h*ydroxymethyl *p*henyl *p*yruvate.

reaction center chlorophyll, P680. Conditions required for singlet oxygen production are 1) high light and 2) a reduced plastoquinone pool. It is also suggested above that tocopherol functions as a singlet oxygen scavenger in vivo. This hypothesis has been tested with inhibitors that impair tocopherol biosynthesis and in mutants defective in this biosynthesis.

1. Blocking Tocopherol Biosynthesis with Inhibitors

Commercial herbicides (Schulz et al., 1993; D.P. Lee et al., 1997; Pallett et al., 1998; Pallett, 2000; Gressel, 2002; Matsomoto et al., 2002) have been developed that block hydroxyphenylpyruvate (HPP) dioxygenase, the first enzyme in the pathway from tyrosine that produces homogentisic acid for the subsequent synthesis of plastoquinone and tocopherol (Figure 6). Their secondary action as bleaching herbicides is attributed to a deficiency in plastoquinone that is a substrate for phytoene desaturase (Norris et al., 1995; Pallett, 2000;

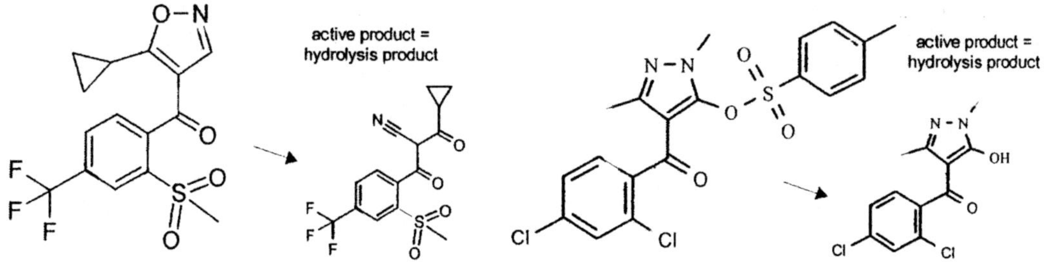

Fig. 7. Two commercial herbicides that block the HPP dioxygenase. Left, isoxaflutole. Right, pyrazolate.

Matsomoto, 2002). Plants treated with such herbicides have decreased levels of carotenes and plastoquinone as well as greatly reduced tocopherol levels, suggesting that the dioxygenase is important for both plastoquinone and tocopherol biosynthesis. Similar studies with *Chlamydomonas reinhardtii* have shown that tocopherol is an essential component in the maintenance of PS II structure and function (Trebst et al., 2002, Trebst, 2003):

Algae grown on pyrazolate or isoxaflutole, that form active breakdown products in the plant (Fig. 7) at concentrations of about half those required for maximal inhibition, synthesize adequate plastoquinone to maintain maximal rates of photosynthetic electron flow. Under such conditions, algal growth and photosynthesis rates were similar to controls, even though tocopherol contents were decreased by the herbicide. However, in algae exposed to strong light (of about twenty times growth light), photosynthesis was rapidly inhibited (within one to two hours), with PS II activity and D1 protein being lost quickly. The tocopherol content of the algae grown with the low herbicide concentrations rapidly fell to below detectable levels in high light. The turnover rate for the tocopherol pool increased as light intensity increased. Thus, it would appear that the tocopherol pool size depends on continuous synthesis with both degradation and synthesis being enhanced at high light. If synthesis is restricted, as occurs in the presence of herbicides, the algae rapidly become tocopherol-deficient in high light, and as a consequence the D1 protein is degraded and PS II activity (though not PS I activity) is lost. These results are in good agreement with tocopherol acting as a scavenger of singlet oxygen generated in PS II and thus protecting the D1 protein from degradation. In these experiments, tocopherol is assumed to be largely oxidized to tocopherylquinone (Fig. 5) via hydroperoxy-tocopherone. This would indicate that re-reduction of the intermediate by ascorbate was too slow to maintain the tocopherol pool in these conditions. Very little information is available on the ascorbate contents of algae or on the factors that govern ascorbate synthesis. In particular, we do not know the extent of ascorbate accumulation in the lumen, which is probably the primary consideration assuming that tocopherol acts on the donor side of PS II and its reaction center.

The interpretation of the function of tocopherol derived from experiments with herbicides may be flawed as these compounds also cause a degree of inhibition of plastoquinone formation and, consequently, of carotene biosynthesis. Even though the experiments were designed to allow optimal photosynthesis rates at growth light intensities, the plastoquinone pool may have become limiting in high light. To test whether PS II breakdown in these conditions was caused by effects on carotene biosynthesis rather than tocopherol deficiency, attempts were made to rescue the algae in high light by adding plastoquinone and/or tocopherol. The high light-induced disappearance of PS II activity and the D1 protein could indeed be delayed by adding decylplastoquinone (PQ-45 will not permeate into the cell). However, within two hours of exposure to high light added plastoquinone was no longer able to protect PS II from inhibition. In contrast, the addition of a cell-permeable tocopherol (a short chain tocopheryl-derivative synthesized with a bromoethyl sidechain according to Scott et al. [1974] instead of the normal 16 C-atom sidechain of natural tocopherols) gave complete protection of PS II function in high light (Trebst et al., 2004). In the short term, the increase in endogenous tocopherol turnover rate was sufficient to give full protection in high light. This continued until the endogenous tocopherol pool was exhausted and then the added cell-permeable tocopherol facilitated protection. Additional support of the view that singlet oxygen is the cause of PS II inactivation in high light in tocopherol-deficient algae is derived from experiments with singlet oxygen quenchers such as diphenylamine or its derivatives. In the presence of singlet oxygen quenchers, herbicide-treated algae showed no loss of photosynthesis activity in high light and the tocopherol pool remained high even when synthesis was completely blocked (Trebst et al., 2002).

2. Tocopherol Synthesis Mutants

A bleaching mutant deficient in HPP-dioxygenase was first identified via a phenotype related to plastoquinone deficiency in the phytoene desaturase reaction (Norris et al., 1995; see also Dähnhardt et al. (2002) and Falk et al. (2002)). This mutant was important as it enabled resolution of the dispute as to whether plastoquinone had redox roles in addition to those associated with electron transport (Mayer et al., 1990; Breitenbach et al., 2001). Enhancing HPP-dioxygenase or homogentisate phytytransferase activity in transgenic *Arabidopsis* led to higher tocopherol levels in leaves and seeds (Tsegaye et al. 2002; Collakova et al., 2003) and allowing the conclusions that 1) these enzymes provide major sites of flux control over the tocopherol synthetic pathway and 2) that later steps, such as that catalysed by tocopherol cyclase, are not limiting for tocopherol accumulation (Collakova et al., 2003a; Kanwischer et al., 2005). HPP-dioxygenase or homogentisate phytyltransferase are induced upon exposure to abiotic stress (Collakova et al., 2003b). A deficiency in carotene biosynthesis has serious consequences under all growth conditions since it diminishes the protective capacity of growing photosynthetic tissues, and more relevant here, prevents the re-assembly of functional PS II reaction centers during D1 protein turnover by preventing the obligatory reattachment of newly synthesized β-carotene (Trebst and Depka, 1997).

The roles of tocopherol in photosynthesis can also be elucidated by analysis of biosynthesis mutants (Bishop and Wong, 1974; Graßes et al., 2001; Schledz et al., 2001; Porfirova et al., 2002; Collakova and DellaPenna, 2001, 2003; Havaux et al., 2003; Hofius and Sonnewald, 2003; Sattler et al., 2003; Bergmüller et al., 2003; Kanwischer et al., 2005). Very early on, a tocopherol-deficient *Scenedesmus* mutant that showed bleaching in high light was identified, but the precise location of the mutated locus was not found and the mechanism of bleaching was not resolved (Bishop and Wong, 1974). More recently, a more targeted approach using RNAi technology has yielded much more information (Graßes et al., 2001). Antisense tobacco lines were produced by targeting the gene responsible for the geranylgeraniol conversion to the phytyl side chain. The leaves of the antisense lines have very little tocopherol and the plants are very sensitive to light, bleaching rapidly upon exposure to high light. This phenotype was attributed to the accumulation of singlet oxygen and the resultant effects of oxidative stress on the thylakoid membranes, as opposed to the high membrane stability maintained by tocopherol in the wild type leaves (Havaux et al., 2003).

Other cyanobacteria and *Arabidopsis* mutants deficient in tocopherol synthesis cited above have thus far given less convincing results with regard to tocopherol function in photosynthesis. The *Arabidopsis vte1* mutant is a knockout in tocopherol cyclase and is hence tocopherol-deficient (Porfirova et al., 2002). The mutant has a very similar phenotype to wild type under most growth conditions. However, in contrast to wild type, *vte1* leaves develop necrotic lesions when grown under high light. Although cyanobacterial mutants with lesions in phytylation, cyclization, or methylation steps of tocopherol synthesis have been identified, there have thus far been no reports of phenotypic effects. Recently, Kanwischer et al. (2005) have characterised Arapidopsis plants where tocopherol cyclase has either been deleted or overexpressed. They also examined double mutants with either low ascorbate and low tocopherol, or low glutathione and low tocopherol. These studies with double mutants clearly show the compensatory effects of these three ROS scavengers (tocopherol, ascorbate and glutathione). However, the maize mutant *sxd*, which is orthogogous to *vte1*, shows a much more severe phenotype with accumulation of metabolites such as anthocyanins and sugars and also starch in the leaves (Sattler et al., 2003).

Taken together, these results indicate that tocopherol deficiency becomes critical for photosynthetic function only under high light growth conditions. High light alone may not, however, be sufficient to generate P680 triplets and singlet oxygen at concentrations exceeding D1 protein turnover rates. The role of the redox state of PQ in the recombination reactions and the triplet yield has already been stressed above. If D1 protein turnover rates are high, this alone could be viewed as a singlet oxygen scavenger system.

III. Ascorbate: A Key Player in Leaf Development and Responses to the Environment

It is tempting to categorize ascorbate primarily as a low molecular weight antioxidant. However, considerable data reflect the central importance of this metabolite in plant biology, with roles in hormone synthesis, gene expression, cell division, and growth. Ascorbate is closely linked to photosynthesis, and current data suggest that ascorbate metabolism can influence photosynthesis at several levels, including antioxidant defense in the chloroplast, the control of nuclear gene expression, and the regulation of stomatal opening.

A. Ascorbate in the Chloroplast

The production of superoxide and H_2O_2 as a result of processes associated with photosynthesis has been described many times previously (Asada, 1999; Foyer and Noctor, 2000, 2003) and will not therefore be discussed here in detail. Chloroplasts contain a hierarchy of H_2O_2-detoxification mechanisms in which ascorbate plays a central role (Noctor and Foyer, 1998; Foyer and Harbinson, 1999; Foyer and Noctor, 2000). Ascorbate also functions in the systems that reduce the probability of ROS generation in the chloroplast, particularly the xanthophyll cycle, in which ascorbate provides the reducing power necessary to convert violaxanthin to zeaxanthin in the de-epoxidation sequence (Demmig-Adams and Adams, 1992; Foyer and Harbinson, 1999). Thus, like the regeneration of tocopherol discussed above, the violaxanthin de-epoxidation reaction oxidizes ascorbic acid in the lumen. Hence, two intrinsic defense processes protecting PS II draw on the lumen ascorbate pool. Since no kinetic evidence has been obtained to indicate the presence of a thylakoid ascorbate transport or re-reduction system, we must assume that the high stromal ascorbate pool allows sufficient passive diffusion to meet these demands. Similarly, a putative lumenal ascorbate peroxidase (APX) form could help draw ascorbate from the thylakoid lumen.

The abundant stromal ascorbate pool supports the Mehler-peroxidase (water-water) cycle and the ascorbate-glutathione cycle (Noctor and Foyer, 1998; Asada, 1999), both of which use reducing power from the photosynthetic electron transport chain via APX. Chloroplasts contain high concentrations of all components associated with H_2O_2 detoxification. The thylakoid contains superoxide dismutase and APX, and isoforms of these enzymes also exist in the stroma together with glutathione, ascorbate, MDHAR, DHAR, and GR (Foyer and Noctor, 2000). APX was first reported in chloroplasts by Groden and Beck (1979) and Kelly and Latzko (1979), but cytosolic, mitochondrial, glyoxysomal, and leaf peroxisomal isoforms have since been characterized (Chen and Asada, 1989; Mittler and Zilinskas, 1991; Amako et al., 1994; Yamaguchi et al., 1995; Bunkelmann and Trelease, 1996; Jimenez et al., 1997). Stromal and thylakoid-bound forms of chloroplast APX are rapidly inactivated at ascorbate concentrations below 20 μM (Hossain and Asada, 1984). As the ascorbate concentrations in the chloroplast and cytosol are likely to be 100–1000 times higher than that necessary to prevent APX inactivation (Foyer et al., 1983; Foyer and Lelandais, 1996), the physiological significance of this process is not clear. The sensitivity of the lumen APX to inactivation when ascorbate availability is restricted is unknown. However, the non-chloroplastic isoforms are stable in the absence of ascorbate (Chen and Asada, 1989; Yoshimura et al., 1998).

Chloroplasts contain other proteins that may also function in reductive detoxification of H_2O_2. Such enzymes include peroxiredoxins (PRX; Dietz et al., this volume), glutathione-peroxidases (GPX; Eshdat et al., 1997), and glutathione S-transferases (GST; Bartling et al., 1993). It has been suggested that the chloroplast 2-cys peroxiredoxins (PRX) / thioredoxin (TRX) system operate in a similar manner to the Mehler-peroxidase cycle in the removal of H_2O_2 (Dietz, 2003). Together, the chloroplast 2-Cys PRX/TRX and Mehler-peroxidase reaction would allow very high rates of H_2O_2 turnover. PRXs are clearly a key part of the intricate web of chloroplast antioxidant defenses (Baier et al., 2000; Dietz, 2003; Collin et al., 2003). Four PRXs in *Arabidopsis* are predicted to be targeted to the chloroplasts (Horling et al., 2003). To date, only the physiological role of the 2-Cys PRX has been studied in detail in photosynthesizing cells (Baier and Dietz, 1999, Baier et al., 2000). 2-Cys PRX protects photosynthesis from photo-damage and loss of this antioxidant could not be fully compensated by other components of the antioxidant network (Baier et al., 2000). Ascorbate-regenerating enzymes were induced in antisense *Arabidopsis* plants with decreased 2-Cys PRX contents and the ascorbate pool became more oxidized indicating that loss of 2-Cys PRX placed an increased oxidative load on the ascorbate system. The antisense plants had greater amounts of mRNAs encoding thylakoid APX, stroma-soluble APX, and stromal MDHAR, and this was reflected in markedly enhanced activity of the latter enzyme in leaf extracts (Baier et al., 2000). These changes in expression were correlated with increased oxidation of leaf ascorbate, but not glutathione (Baier et al., 2000). The expression of the gene encoding 2-Cys PRX (*prx*) is markedly inhibited by ascorbate (Horling et al., 2003). This would suggest that in many growth conditions where leaf ascorbate is very high, 2-Cys PRX expression would be largely suppressed. This is an example of how cytoplasmic ascorbate content can modulate gene expression. Antioxidant-mediated regulation of gene expression appears to be a widespread phenomenon with many repercussions for plant growth and defense, as discussed in the next two sections.

B. Ascorbate-Mediated Regulation of Gene Expression

Just as *Arabidopsis vte1* mutants have been useful in elucidating the role of tocopherol in PS II (Porfirova et al., 2002; Bergmüller et al., 2003; Kanwischer et al., 2005), the ascorbate-deficient *Arabidopsis* mutant, *vtc1*, has been an invaluable tool in elucidating the roles of ascorbate in relation to the control of photosynthesis. Unlike the *vte1* mutant, however, which has a very similar phenotype to the respective Col 2 wild type under most conditions, the *vtc1* mutant grows more slowly than the Col 0 wild type when grown under short days (Velojovic-Jovanovic et al., 2001). We have recently shown that the smaller leaf size observed in the *vtc1* mutant is caused by a much reduced cell size, presumably because limiting ascorbate prevents full cell expansion (authors unpublished data). The *vtc1* mutant, which was isolated via its sensitivity to ozone, shows increased sensitivity to other abiotic stresses such as freezing and UV-B irradiation (Conklin et al., 1996). Microarray analysis was used to identify genes that were differentially expressed in *vtc1* compared to Col 0 and also following feeding *vtc1* leaf discs with ascorbate (Kiddle et al., 2003; Pastori et al., 2003). In addition to dramatic changes in the abundance of transcripts encoding pathogenesis related (PR) proteins and similar proteins associated with plant defense responses against biotic stress, effects of ascorbate on the abundance of transcripts encoding chloroplast proteins were observed (Kiddle et al., 2003). This occurs despite the fact that the rate of photosynthesis was similar in *vtc1* and Col 0 leaves and rates of energy dissipation and electron transport to alternative sinks such as oxygen were unchanged in the mutant compared to the wild type (Veljovic-Jovanovic et al., 2001; Pastori et al., 2003). This would suggest that gene expression has a higher threshold for sensing ascorbate depletion than the thermal dissipation reactions associated with photosynthesis. Thus, we can envisage that changes in photosynthetic gene expression precede changes in photosynthetic metabolism as ascorbate becomes depleted.

The concept that antioxidants could be signal transducing molecules leading to specific changes in gene expression has only been established relatively recently (Karpinski et al., 1997, 1999; Foyer and Noctor, 2000; Noctor et al., 2000; Pastori et al., 2003). Low ascorbate in *vtc 1* leads to the activation of a suite of genes that might be considered to provide a molecular signature of ascorbate deficiency in plants (Pastori et al, 2003), while high ascorbate not only leads to repression of these transcripts but can result also in changes in other transcripts, at least in the short term (Kiddle et al., 2003; Pastori et al., 2003). Transcripts responsive to leaf ascorbate content include electron transport chain components such as those of the light harvesting system, PS I, PS II, and the oxygen-evolving complex (Kiddle et al., 2003). Transcripts are also affected for stromal enzymes such as glucose 6-phosphate dehydrogenase (G6PDH), phosphoribulokinase (PRK), fructose-1,6-*bis*phosphatase (FBPase), and sedoheptulose-1, 7-*bis*phosphatase (SBPase). The abundance of G6PDH and CP12 transcripts are increased by high ascorbate while mRNAs encoding PRK, FBPase, and SBPase are decreased (Kiddle et al., 2003). This suggests an inverse relationship between regulation of enzymes whose activity is modulated by TRX (G6PDH is inactivated by reduced TRX while FBPase, SBPase, and PRK activities are activated) and the high ascorbate-mediated changes in transcript abundance (G6PDH induced while FBPase and SBPase transcripts were repressed). This is a further example of crosstalk between the two sources of reductant for the major chloroplast peroxidases.

C. Ascorbate, Abscisic acid, and Stomatal Regulation

Ascorbate may exert effects on stomatal opening in at least two ways. First, ascorbate is a co-factor in the synthesis of ABA, the phytohormone that regulates stomatal pore size and hence entry of CO_2 and loss of water from the leaf. Stomatal pores close in response to water stress through the action of ABA, which causes an increase in cytosolic calcium concentrations via H_2O_2–activated channels and release from the vacuole and other intracellular stores (Price et al., 1994; Köhler and Blatt, 2002). Ascorbate-dependent dioxygenases are involved in the pathway of ABA biosynthesis. In particular, ascorbate is required for activity of 9-*cis*-epoxycarotenoid dioxygenase (NCED), an enzyme catalyzing the formation of xanthoxin, a precursor of ABA. It is noteworthy that NCED expression is modulated by ascorbate, such that transcripts are increased when ascorbate is low and decreased when ascorbate is high (Pastori et al., 2003).

The second interaction between ascorbate and stomata involves a role for ascorbate in regulating the redox state of the guard cell apoplast. As mentioned in section I, H_2O_2 interacts strongly with (ABA) signaling. In particular, H_2O_2 is thought to fulfill a signaling role in guard cells, activating plasma membrane-localized anion channels, leading to guard cell depolarisation, potassium efflux, and loss of turgor and volume concluding in stomatal closure (Schroeder et al., 2001a,b). Although the significance of H_2O_2 in stomatal regulation has been questioned (Köhler et al., 2003), two genes encoding subunits of NADPH oxidase (*AtrbohD* and *AtrbohF*) appear to be involved in stomatal closure, as well as in other ABA responses such as seed dormancy (Kwak et al., 2003). Double mutants in which both genes are inactivated show impaired closure of stomata, an effect that can be reversed by exogenous H_2O_2 (Kwak et al., 2003). These findings clearly implicate oxidative processes in ABA signaling and suggest that antioxidant status may be a factor that regulates this signaling. Indeed, H_2O_2-induced stomatal closure was reversed by the application of exogenous ascorbate, presumably reflecting H_2O_2-scavenging (Zhang et al., 2001). Moreover, high constitutive expression of DHAR in transgenic plants increased the amount of ascorbate relative to DHA in leaves and guard cells (Chen et al., 2003) and significantly affected guard cell signaling and stomatal movement. The leaves of the DHAR-overexpressors contained less H_2O_2 in the guard cells and had a higher percentage of open stomata and increased stomatal conductance (Chen and Gallie, 2004). These results would suggest that plants with higher cellular ascorbate are better able to maintain open stomata and hence photosynthesis, at least under optimal conditions of water supply. The pH of the apoplast is about pH 5.5–6.3 in well-watered plants as opposed to about pH 7.2 in drought-stressed plants, favoring a preferential accumulation of ABA and ascorbic acid in stomata under drought stress. The above evidence suggests that the distribution of ascorbate between the cytoplasm and the apoplast could also be influential in the regulation of stomatal opening. However, high apoplastic ascorbate concentrations would act antagonistically to ABA by limiting the extent of ABA-induced H_2O_2 accumulation.

ABA elicits H_2O_2 production on the apoplastic face of the plasma membrane as part of the signaling process required to promote stomatal closure (Pei et al., 2000; Murata et al., 2001; Schroeder et al., 2001a,b; Zhang et al., 2001). The influence of apoplastic ascorbate in ABA signaling may be complex, and a key factor could be the activity of peroxidases. Like other antioxidants, ascorbate is capable of chemically reducing superoxide to H_2O_2 (Allen and Hall, 1973; Polle, 2001). As ascorbate reduces H_2O_2 more slowly than it does superoxide, ascorbate could *increase* rates of H_2O_2 production, if both superoxide dismutation rates and peroxidase activities were limiting. However, it is clear that the apoplast and cell wall contain numerous peroxidase activities able to use ascorbate as a reductant (Vanacker et al., 1998). Furthermore, the exact nature of the oxidizing signal during the 'oxidative burst' is not yet fully elucidated and, as is also the case for oxidative signals associated with plant-pathogen interactions, the receptor(s) immediately downstream remain(s) to be identified. Thus, it is likely that, even in the absence of NADPH oxidase induction, the apoplastic ascorbate pool could be important in regulating stomatal opening by contributing to peroxidase-catalyzed detoxification of H_2O_2.

D. Ascorbate Oxidase and the Regeneration of Apoplastic Ascorbate

Apoplastic ascorbate is regarded as the first defense against ozone (Burkey et al., 2003) and also ROS produced via the action of the plasma-membrane bound H_2O_2-producing systems such as the NADPH oxidase that is triggered, among other things, by ABA action (Conklin and Barth, 2004). About 4 to 10% of the leaf ascorbate pool resides in the apoplast (Noctor and Foyer, 1998; Veljovic et al., 2001), giving an ascorbate concentration in the low millimolar range. This fraction depends on the overall ascorbate content of the leaf since, for example, feeding the ascorbate precursor L-galactono-1, 4-lactone enhanced apoplastic ascorbate as well as total leaf ascorbate (Maddison et al., 2002) while no apoplastic ascorbate was detectable in the ascorbate–deficient *vtc1 Arabidopsis* mutant, which has only 30% of the total leaf ascorbate of the wild-type (Veljovic-Jovanovic et al., 2001).

The redox state of the apoplastic ascorbate pool, which can be defined as % total ascorbate in the reduced form, i.e. 100 [ascorbate] / ([MDHA] + [DHA] + [ascorbate]), may be a key regulator of apoplastic defense and redox-linked signaling. The apoplast contains ascorbate oxidases (AO) whose function is to produce MDHA and DHA, thus removing ascorbate in such a way as to regulate growth and signal transduction processes (Pignocchi and Foyer, 2003; Pignocchi et al., 2003). The effect of AO on the redox state of the apoplastic ascorbate pool and ozone tolerance has been clearly demonstrated in transformed plants with an AO transgene expressed in the sense or antisense orientation. High constitutive over-expression of AO in tobacco decreased the apoplastic ascorbate redox state to only 3% without any appreciable effect on the total leaf ascorbate accumulation or the redox state of the whole leaf ascorbate pool (Pignocchi et al., 2003). The decrease in apoplastic ascorbate greatly enhanced leaf injury upon chronic ozone exposure (Sanmartin et al.,

2003). These results with AO transgenic plants also show that a steep ascorbate gradient can be maintained across the plasma membrane such that the ascorbate redox state of the apoplast can be very different (i.e., much more oxidized) from that of the cytosol, which is largely reduced under most conditions. This gradient arises because of the very low capacity of the apoplast to reduce MDHA and DHA and is linked to transport across the plasma membrane. An ascorbate-reducible transmembrane b-type cytochrome c (PM cyt b_{561}) has been shown to reduce apoplastic MDHA to ascorbate (Asard et al., 2001). Also, an NADH-MDHAR is associated with the cytoplasmic face of the plasma membrane (Berczi and Møller, 1998). Moreover, apoplastic DHA is actively transported across the plasma membrane to be reduced by DHAR in the cytosol (Horemans et al., 2000). In animals, the major ascorbate carriers are nucleobase sodium-dependent L-ascorbic acid transporters (NATs; Tsukaguchi et al., 1999). Although a number of plant NATs have been identified, none to date has been shown to facilitate transport of ascorbic acid. In plants, ascorbate is taken up in both its reduced and oxidized (DHA) forms (Horemans et al., 2000). In animals, DHA transport is facilitated by GLUT-type glucose transporters (Liang et al., 2001). Intracellular transport of ascorbate is another question requiring resolution. Ascorbate may be transported from its place of synthesis by mitochondrial proteins that transport both DHA and glucose (possibly by a closely related transport system) as well as ascorbic acid. Chloroplasts also contain a glucose transporter that is considered to be closely related to the mammalian GLUT transporter family but differs from other known plant hexose transporters (Weber et al., 2000). DHA and ascorbic acid are transported into the chloroplast stroma through the envelope membrane *via* carrier-mediated facilitated diffusion (Foyer and Lelandais, 1996; Horemans et al., 1998). The low affinity of the envelope carrier for ascorbate indicates that pathways of ascorbate regeneration are vital in maintaining high concentrations of this antioxidant in the chloroplasts.

E. Respiration, Reactive Oxygen and Ascorbate Biosynthesis

1. Plant Mitochondria and Reactive Oxygen

The potential of plant mitochondria as an important source of ROS had not been appreciated until relatively recently (Møller, 2001), because reactions associated with photosynthesis have a much greater capacity for superoxide and hydrogen peroxide generation than respiration (Vanlerberghe and Ordog 2002; Foyer and Noctor, 2003). In darkness, however, mitochondrial respiration is likely the major source of ROS, even in green tissues. Similarly, in non-photosynthetic cells that are not actively degrading lipid, the mitochondrion is likely to be the major source of ROS (Puntarulo et al., 1991). Although it is clear that the respiratory chains of both animal and plant mitochondria are capable of producing ROS, there is no clear consensus in the literature as to the quantity of ROS produced (Møller, 2001). Reported rates of either superoxide or hydrogen peroxide production in isolated mitochondria from a range of plant species vary by almost two orders of magnitude (summarised in Møller, 2001).

Although the exact site of this ROS production in plant mitochondria has not been formally established, the transfer of single electrons to oxygen to form superoxide can occur via electron transport carriers with an appropriate (i.e. sufficiently low) redox potential. The high degree of conservation of electron transport chain components between plant and animal mitochondria suggests common sites of superoxide production. In animals, there are two defined sites in the respiratory chain where single electrons can be transferred to molecular oxygen resulting in the formation of the superoxide anion. These are complex I (the major NADH-ubiquinone oxidoreductase) and complex III (ubiquinol-cytochrome c oxidoreductase) (Boveris et al., 1972; Beyer, 1990). Both of these complexes are involved in ubiquinone-ubiquinol interconversions and it is the presence of the reaction-intermediate radical, ubisemiquinone, that is thought to be responsible for electron leakage from both complexes (Raha and Robinson, 2000). However, the exact mechanism and site of ubisemiquinone production within complex I has not been fully elucidated (Beyer, 1990; Du et al., 1998). Iron-sulphur centres and/or the flavin active site are also possible sources of superoxide (Genova et al., 2001; Liu et al., 2002). In contrast, the topology of the Q cycle within complex III is well established and it is known that ubisemiquinone occurs twice in the cycle, either close to the intermembrane-space side of complex III, or close to the matrix side (Trumpower, 1990). Thus, the activity of complex III can lead to the generation of superoxide both in the mitochondrial matrix and in the intermembrane space (Han et al., 2001). Compared to animals, where mitochondrial ROS generation has been shown to cause ageing, degenerative diseases, cancer, and programmed cell death (Liu et al., 2002), relatively little is known about the role of mitochondria in related processes in plants. However, PCD in plants appears to involve ROS generated in mitochondria and

Chapter 16 Tocopherol, Ascorbate and Glutathione

release of cytochrome c as it does in animals (Sweetlove and Foyer, 2004). Moreover, studies of plant complex I mutants have provided evidence that mitochondrial electron transport partitioning is a key factor in cellular NADH homeostasis, and is important in controlling oxidative load and setting the cellular redox-stat (Dutilleul et al., 2003a,b; Noctor et al., 2004).

2. Respiration and Ascorbate Synthesis

Mitochondria, like other compartments of the plant cell, house both enzymic and non-enzymic antioxidants to prevent ROS accumulation (Møller, 2001; Sweetlove and Foyer, 2004). In addition, plant mitochondria fulfill a unique role in plant redox metabolism as they are the site of synthesis of ascorbic acid, the major redox buffer of plant cells. Although ascorbic acid is a high abundance metabolite, with many important functions in plant biology, relatively little is known about the factors controlling its accumulation in leaves (Noctor and Foyer, 1998; Smirnoff and Wheeler, 2000). The biosynthetic pathway proposed by Wheeler et al. (1998) probably represents the major pathway of ascorbate production in plants. However, other routes have been demonstrated and several biosynthetic pathways may coexist (Davey et al., 1999; Agius et al., 2003; Smirnoff et al., 2004). Regardless of the early steps of ascorbate biosynthesis, the final step in all schemes proposed to date is catalyzed by L-galactono-1, 4-lactone dehydrogenase (GLDH; Ôba et al., 1995; Østergaard et al., 1997; Pallanca and Smirnoff, 1999; Siendones et al., 1999; Bartoli et al., 2000), an enzyme located in the inner mitochondrial membrane. Isolated mitochondria produce ascorbate when supplied with the substrate L-galactono-1, 4-lactone, using the respiratory chain cytochrome *c* as the electron acceptor (Siendones et al., 1999; Bartoli et al., 2000). Thus, ascorbate synthesis in isolated mitochondria is inhibited by KCN and stimulated by antimycin A, confirming that GLDH activity is restricted by the availability of oxidized cytochrome *c* (Bartoli et al., 2000). GLDH is encoded by a single gene in *Arabidopsis*, cauliflower, sweet potato, and tobacco (Østergaard et al., 1997; Imai et al., 1998; Yabuta et al., 2000). Antisense suppression of GLDH in synchronous BY-2 tobacco cell cultures resulted in a decline in enzyme activity and a 30% reduction in ascorbate levels (Tabata et al., 2001). The decrease in cellular ascorbate had a pronounced negative effect on cell division, growth, and cell structure.

Little is known about the factors that regulate the expression of GLDH and other enzymes of ascorbate synthesis. Light appeared to be an essential trigger for expression of GLDH transcripts over a five day period in tobacco leaves (Tabata et al., 2002) and conversion of exogenously-supplied L-galactono-1, 4-lactone to ascorbate over 24 h was increased by light treatment of barley leaf slices (Smirnoff, 2000). However, there is some indication of interspecific variation with regard to diurnal regulation of GLDH transcript abundance, since this was observed in *Arabidopsis* (Tamaoki et al., 2003), but not in tobacco (Pignocchi et al., 2003).

GLDH is bound to mitochondrial complex I in *Arabidopsis* and regulated by the activity of the electron transport chain (Heazlewood et al., 2003; Millar et al., 2003). Inhibition of complex I electron transport by rotenone also inhibits ascorbate biosynthesis in intact mitochondria. Respiration and ascorbate synthesis appear to be intricately co-ordinated and mitochondria might therefore have a profound influence on the ability of the plant to produce ascorbate during stress. We have recently shown that there is no relationship between GLDH protein/maximal activity and ascorbate content in the leaves of a range of plant species (Bartoli et al., 2004). Since GLDH catalyzes the final step of ascorbate synthesis, and is regulated by the rate of respiration in *Arabidopsis* (Millar et al., 2003), it is perhaps not surprising that the absolute amount of GLDH protein or activity is not always indicative of the capacity for ascorbate accumulation.

Ascorbate synthesis in leaves is light-dependent (Smirnoff and Pallanca, 1996). Leaves from high light grown plants tend to have more ascorbate than those grown at lower light (Gillham and Dodge, 1987). Ascorbate accumulates in plants exposed to continuous light (Morimura et al., 1999) and transfer from low light to high light can prompt rapid increases in leaf ascorbate contents (Eskling and Åkerlund, 1998). Moreover, a positive correlation between leaf ascorbate contents and light has been found in natural environments (Logan et al., 1996). It would appear that light-dependent ascorbate accumulation is an indirect effect of the increased substrate availability in plant tissues (Smirnoff and Pallanca, 1996), but this remains to be proven. However, ascorbate appears to be the major sink for photoassimilate in the leaves of certain alpine plants (Streb et al., 2003). Since high ascorbate contents tend to be associated with stress tolerance, it is tempting to suggest that its biosynthesis may be regulated by redox triggers, and particularly ROS. Perturbation of mitochondrial redox processes in a complex I-deficient mutant is associated with altered expression of several antioxidant components, both within and outside the mitochondrion (Dutilleul et al., 2003b). Though ascorbate contents are not increased in this mutant, and much remains to be discovered concerning the mitochondrial signals and sensors that perceive stress and orchestrate appropriate responses, it is clear that mitochondrial signals linked to redox reactions influence stress resistance.

Taken together, the results discussed above present a complex picture with regard to the regulation of ascorbate synthesis and accumulation. Influential factors include metabolic limitations and environmental cues, where light, photosynthesis, respiration, and stress act synchronously to determine the extent of ascorbate accumulation in leaves. This is perhaps not surprising given the multiple roles of ascorbate in plants.

IV. Glutathione and the Importance of Cellular Thiol/Disulfide Status

Like ascorbate, glutathione is a multifunctional compound with functions that extend beyond the antioxidative system (Alscher, 1989; Rennenberg, 1997; May et al., 1998; Noctor and Foyer, 1998). It is indisputable, however, that a major role of glutathione is as a thiol/disulfide buffer, i.e. to allow appropriate conditions for protein function. We have recently reviewed the regulation of GSH biosynthesis and function (Foyer et al., 2005) and thus we will describe only certain key features below. A major aim of the following discussion is to situate the GSH/GSSG couple within the context of redox signaling in the soluble phase of photosynthetic cells.

A. Glutathione and Antioxidant Defense

Even in the absence of an enzyme, glutathione is able to interact rapidly with free radicals such as superoxide and the hydroxyl radical (Polle, 2001). In addition, glutathione plays an important role in peroxide detoxification through several possible enzyme systems. First, GSH can regenerate ascorbate (Fig. 1), either directly or through the action of various enzymes able to catalyze DHA reduction, including glutaredoxins (GRX) and GSTs (Wells et al., 1990; Trumper et al., 1994; Shimaoka et al., 2000; Urano et al., 2000; Dixon et al., 2002). Secondly, both GRX and GSTs can also reduce hydroperoxides to the corresponding alcohol or H_2O (Bartling et al., 1993; Collinson et al., 2002). Third, certain PRXs are regenerated by a GR/GRX system (Rouhier et al., 2002). Fourth, specific glutathione peroxidases (GPXs) exist in plants (Eshdat et al., 1997). Genes encoding plant GPXs have been cloned from a wide variety of plant species, and include genes encoding proteins targeted to the chloroplast (Mullineaux et al., 1998). Along with certain GSTs, some GPXs are among genes strongly induced by oxidative stress (Levine et al., 1994; Willekens et al., 1997). As for PRX, the reaction mechanism of plant-type GPXs involves cysteine residues, contrasting with mammalian H_2O_2-detoxifying GPXs (which involve selenocysteine) and with other plant peroxidases (APX, guaiacol-type peroxidases), which contain heme. All plant GPX cDNAs isolated to date show high homology to the mammalian phospholipid hydroperoxide glutathione peroxidases (PHGPX), which have a higher affinity to lipid hydroperoxides than to H_2O_2. This is explained by the absence of selenocysteine at the active site, which diminishes the nucleophilic properties of the enzyme and probably accounts for its much lower activity with H_2O_2. It has also been demonstrated that at least two plant PHGPXs probably represent a novel isoform of TRX peroxidase, exhibiting much higher oxidation activity for TRX than for glutathione (Herbette et al., 2002; Jung et al., 2002). H_2O_2 is the preferred substrate for TRX peroxidase. PHGPXs are induced by a wide range of biotic and abiotic stresses, suggesting that they play a specific role in limiting the extent of lipid peroxidation during stress by removing phospholipid hydroperoxides and H_2O_2.

B. Glutathione and Gene Expression in Plants

GSH is a physiological regulator of many thiol-disulphide exchange reactions, including chloroplast transcription (Liere and Link, 1997). GSH can be a strong inducer of defense gene expression (Wingate et al., 1988; Dron et al., 1988). GSH induces pathogenesis-related (PR) gene expression in a similar manner to salicylic acid (SA; Mou et al., 2003). Of the transcriptome changes caused by ascorbate deficiency in *vtc1*, the specific induction of PR proteins in the absence of any major changes in expression of antioxidant defense transcripts was perhaps the most unexpected (Pastori et al., 2003; Kiddle et al., 2003). The pathway of ascorbate-mediated induction of PR genes has been suggested to involve both enhanced ABA (Pastori et al., 2003) and enhanced SA (Barth and Conklin, 2004). It has long been known that during systemic acquired resistance (SAR), SA induces the concerted expression of PR genes. These encode small proteins that are either secreted from the cell or targeted to the vacuole (Sticher et al., 1997). Mutants at a single locus, *npr1*, nonexpressor of PR genes (also known as *nim1* and *asi1*), block SA signaling in *Arabidopsis*. It has recently been shown that thiol-disulphide exchange reactions involving glutathione are crucial for PR gene expression (Mou et al., 2003). Thus, induction of PR gene expression in *vtc1* may at least partly be due to the small but significant increase in GSH and total glutathione present in *vtc 1* leaves (Veljovic-Jovanovic et al., 2001).

Thiol/disulfide exchange is crucial in SAR, particularly due to the mechanism of activation of NPR1. The *NPR1* gene encodes a protein containing a bipartite localization sequence and accumulates in the nucleus in response to both chemical and biotic inducers of SAR (Kinkema et al., 2000). The nuclear localization of the NPR1 protein is a prerequisite for PR gene expression, and is determined by oligomerization. NPR1 is localized in the cytosol as an oligomer or in the nucleus as a monomer. The balance between these two forms is modulated by thiol-disulphide exchange reactions that can be influenced by GSH and GSSG (Mou et al., 2003). Inducers of SAR, such as SA, regulate *PR* gene expression through redox changes in NPR1. In order to be active, NPR1 has to be reduced to the monomeric form that is then able to move to the nucleus to induce PR gene expression.

The NPR1 signal transduction pathway is highly conserved between plant species. Any biotic or abiotic stimulus that can perturb the cellular redox state may serve to upregulate the same set of defense genes via the NPR1 pathway (Mou et al., 2003). This may also explain PR gene expression in catalase-deficient mutants (Willekens et al., 1997). Catalase deficiency is associated with extensive oxidation of the glutathione pool and greatly enhanced glutathione accumulation (Smith et al., 1984; Willekens et al., 1997; Noctor et al., 2000), reminiscent of that observed in plant-pathogen interactions (Vanacker et al., 2000; Mou et al., 2003). In this situation, SAR induction involves an early burst of ROS and a transient or more sustained increase in oxidative signals followed by a sharp decrease in cellular reduction potential as a result of accumulation of antioxidants such as GSH.

The enzymes of GSH synthesis and metabolism are induced together in response to stress (Mittova et al., 2003). This suggests that there is considerable overlap in the signal transduction cascades that induce the genes encoding the enzymes of GSH synthesis and those involved in the induction of GPX homologous genes by singlet oxygen (Op den Camp et al., 2003) and GST1 by H_2O_2 (Rentel and Knight, 2004).

Recent evidence based on molecular genetics suggests that gamma-glutamylcysteine synthetase (γ-ECS) may fulfill a signaling function as well as a catalytic role (Ball et al., 2004; Foyer at al. 2005). Moreover, the intra-cellular distribution of the GSH biosynthetic enzymes between the chloroplast and cytosol has been questioned, suggesting that not only GSH synthesis but also GSH sensing may be more important in the cytosol than the chloroplasts of photosynthetic cells (Foyer at al., 2005). Similarly, many questions with regard to the transport of gamma glutamylcysteine (γ-EC), GSH, and GSSG remain to be answered, but it appears likely that the chloroplast transport system is rather different from that of other membranes (Foyer at al., 2005).

C. The Glutathione Redox Couple and Sensing of Reactive Oxygen

Thiol/disulfide exchange is likely a key mechanism in sensing and relaying redox changes, including those linked to increases in ROS production. The TRX system is well known in plants as a redox signal transducer that mediates light/dark signaling through post-translational modification of target proteins (Buchanan, 1991; Schürmann and Jacquot, 2000). Thiol/disulfide exchange involves changes in protein conformation that are both reversible and stable, and key roles for such processes in ROS sensing probably await discovery in plants. In *E.coli*, the oxyR transcription factor senses changes in redox state through reversible thiol/disulfide exchange that affects its DNA-binding activity (Bauer et al., 1999). In yeast, the transcription factor YAP-1 is a redox sensor that induces a regulon of several genes in response to peroxides. Like oxyR, YAP-1 operates through thiol/disulfide exchange, and like NPR1 its localization is dependent on thiol status. In the case of YAP-1, oxidation of two Cys residues to form an intramolecular disulfide bond acts to trap the protein in the nucleus, thereby increasing its activity (Kuge et al., 1997). Recently, it has been shown that YAP-1 is oxidized via H_2O_2 reduction by a GPX-like protein (Delauney et al., 2002). Re-reduction of YAP-1 may occur by a TRX-dependent pathway (Delauney et al., 2002). Although no sequences homologous to YAP-1 have as yet been identified in plants, there could be components that function in a similar way.

Many stresses cause net oxidation of the glutathione pool, often accompanied or followed by increases in total glutathione (Smith et al., 1984; Sen Gupta et al., 1991; Willekens et al., 1997; Noctor et al., 2002b; Gomez et al., 2004). An important mechanism in perceiving this redox perturbation could be glutathionylation of proteins. In this process, glutathione forms a mixed disulfide with a target protein, with possible important consequences for activity of enzymes and components such as transcription factors. In yeast and animals, glutathionylation is considered likely to play an important role in redox signaling, as well as in the protection of protein structure and function (Klatt and Lamas, 2000). Glutathionylation can occur as a result of either increases in GSSG concentration or through ROS-mediated oxidation of thiols to thiyl radicals (Starke et al., 2003). The process is likely reversed by specific GRXs and/or TRXs. Both these types of redox-active proteins are encoded by quite large gene families in plants (Lemaire, 2004) and considerable work will be required to elucidate the functions of specific members. A recent report has shown that two Calvin cycle enzymes (aldolase and triose phosphate isomerase) undergo glutathionylation (Ito et al., 2003), and that glutathionylation of recombinant triose phosphate isomerase in vitro caused inactivation of the enzyme (Ito et al., 2003).

D. Glutathione Synthesis and the Redox State of the Cell

In animal cells, substantial evidence implicates redox potential as an important factor determining cell fate, and the glutathione redox couple is considered the key player (Schäfer and Buettner, 2001). Unlike many other redox couples such as ascorbate/DHA and NADPH/NADP, the glutathione redox potential depends on both GSH/GSSG and absolute glutathione concentration. Because the concentration-dependent term of the Nernst equation is second-order with respect to GSH but first-order with respect to GSSG, accumulation of GSH can offset the change in redox potential caused by decreases in the GSH/GSSG ratio.

As noted above, various stress conditions cause not only net oxidation of the glutathione redox couple, but also increases in the total glutathione pool. The simplest explanation of this effect is that oxidation of glutathione causes activation of glutathione synthesis. This would represent a homeostatic mechanism in which increases in [GSH] act to offset decreases in [GSH]/[GSSG]. This point can be illustrated by data produced from maize leaves exposed to a 2 day chilling period followed by two days recovery at optimal growth temperatures (Gomez et al., 2004). Table 1 shows how chilling and recovery affect the overall glutathione redox potential, which is calculated from the measured % reduction of leaf glutathione and the fold increase in the total pool. Because we do not know the absolute concentration of glutathione in the different cell compartments, we have calculated data for the likely range in the chloroplast and cytosol (2-10 mM; Noctor et al., 2002a). Neither the GSH/GSSG ratio nor the total glutathione pool was modified during the chilling period. Thus, the leaf was able to maintain an unchanged glutathione redox potential at low temperature. Upon

Table 1. Calculated changes in the redox potential of the glutathione pool in response to chilling in maize: the compensatory effect of up-regulated glutathione synthesis.

Experimentally obtained values for glutathione (Gomez et al., 2004) were converted to redox potentials of the GSH/GSSG couple for different assumed total pool sizes, ranging from 2 to 10 mM (Noctor et al., 2002). Midpoint redox potential at pH 7 (E_{m7}) of glutathione was taken as −240 mV (Foyer and Noctor, 2000; Schafer and Buettner, 2001), and the actual redox potential at pH 7 was calculated as $E_{m7} - 29.6 \log ([GSH]^2/[GSSG])$, where concentrations were expressed as M (Schafer and Buettner, 2001). Control: redox potential calculated from measured % reduction for leaves maintained at 25/19°C. Recovery: redox potential calculated from measured % reduction and measured increase in total glutathione (1.7-fold) following transfer to 10/8°C day/night for two days and then transfer back to 25/19°C. Constant pool size: redox potential calculated from measured % reduction in Recovery leaves but assuming total glutathione as in Control leaves; ie, these values show the hypothetical redox potential that would result from the observed decreases in GSH/GSSG without the observed increased accumulation of total glutathione.

Condition	Total glutathione (pool size)	% GSH	Redox potential (mV)
Control	Minimum (2 mM)	95.9	−209
	Intermediate (5 mM)	95.9	−221
	Maximum (10 mM)	95.9	−230
Recovery	Minimum × 1.7	89.5	−202
	Intermediate × 1.7	89.5	−214
	Maximum × 1.7	89.5	−223
Constant pool size	Minimum (2 mM)	89.5	−195
	Intermediate (5 mM)	89.5	−207
	Maximum (10 mM)	89.5	−216

return to optimal temperatures, however, GSSG increased from 34 (control leaves) to 146 (recovery) nmol g^{-1}FW, causing a 3-fold decrease in GSH/GSSG. Table 1 shows that this change means that the redox potential becomes more positive by 7 mV (Table 1; compare 'recovery' and 'control'). However, the increase in GSH associated with the 1.7-fold increase in total glutathione means that this change is only half of that (14 mV) which would occur if GSH/GSSG decreased 3-fold but [GSH] + [GSSG] remained constant (Table 1; compare 'recovery' and 'constant pool size'). Thus, induction of glutathione accumulation acts as a mechanism to offset chilling-initiated oxidation of glutathione, though it is clear that the maize leaf is unable to activate glutathione synthesis to fully compensate for this sustained oxidation. The nature of the link between redox state perturbation and enhanced glutathione accumulation is unresolved to date. Available data suggest at least two links. First, the activity of adenosine phosphosulfate reductase, a key enzyme in sulfate assimilation, may be activated by decreases in GSH/GSSG (Bick et al., 2001). Second, H$_2$O$_2$ or GSH/GSSG ratio dominates control of the translation of γ-glutamylcysteine synthetase, the first and limiting enzyme in the synthesis of glutathione (Xiang and Oliver, 1998).

E. Coupling of the Ascorbate and Glutathione Pools

Redox coupling between ascorbate and glutathione is a universal phenomenon in cells that contain both compounds (Fig. 1). This coupling is a necessary part of a H$_2$O$_2$ detoxification pathway in chloroplasts, linked to ascorbate peroxidation (Foyer and Halliwell, 1976; Anderson et al., 1983a, b). GSH can regenerate ascorbate from DHA both chemically and by acting as the reductant for several enzymes with DHAR activity. There remains some uncertainty regarding the importance of DHARs in chloroplasts. The relative redox potentials of the ascorbate and glutathione couples strongly favor net electron flow from GSH to DHA (Fig. 1; Foyer and Noctor, 2000), a reaction that can occur at significant rates even in the absence of DHAR, particularly at alkaline pH values (Winkler, 1992; Polle, 2001). Under conditions that cause high fluxes of ascorbate peroxidation, however, DHAR activity is necessary to ensure effective maintenance of the reduced ascorbate pool (Polle, 2001). Glutathione disulphide (GSSG) generated as a result of this reaction is subsequently reduced by GR. Evidence in support of the roles of GR and DHAR in this coupling has come from analysis of transformed plants over-expressing these enzymes (Foyer et al., 1995; Chen et al., 2003). The significance of the degree of coupling of the ascorbate and glutathione pools in different leaf compartments, particularly in relation to signal transduction, has been discussed previously (Noctor et al., 2000). A detailed mathematical model has provided an excellent means of analyzing the relationship between chloroplastic superoxide production and changes in ascorbate and glutathione (Polle, 2001). Recently, evidence has been provided for the presence of a glutathione-independent pathway of ascorbate regeneration from DHA in tobacco BY-2 cell cultures (Potters et al., 2004). The glutathione-independent pathway of ascorbate regeneration remains to be elucidated, but other antioxidants such as lipoic acid can reduce oxidized ascorbate. Moreover, other thiol-containing proteins, particularly those with a dicysteinyl motif, are capable of fulfilling a similar function as DHAR. In animals, this activity can be catalyzed by proteins such as GRXs and protein disulphide isomerases (Wells et al., 1990). Such proteins, as well as the TRX/TRX reductase system (Wells et al., 1990; Hou et al., 1999; Potters et al., 2002) and certain types of trypsin inhibitor (Trumper et al., 1994), might also catalyze the reaction in plants. The TRX/TRX reductase system has established links to the ascorbate/glutathione cycle since accumulation of DHA and GSSG will tend to inactivate TRX f–activated enzymes (Nishizawa and Buchanan, 1981) and ascorbate modulates the transcription of TRX-modulated proteins (Kiddle et al., 2003). Taken together, these observations suggest that various pathways of DHA reduction exist in plant cells. While the importance of mechanisms identified in vitro remains to be elucidated, the presence of numerous components that allow ascorbate to be regenerated from DHA in a GSH-independent manner suggests a high degree of flexibility in regulation and signaling.

V. Conclusions and Perspectives: All for One and One for All?

Tocopherol, ascorbate, and glutathione are central components of plant antioxidant defenses acting together to limit ROS life-time and accumulation (Mullineaux et al., this volume). Like Dumas' *Three Musketeers*, and together with D'Artagnan (NADPH), these compounds fight the potentially villainous ROS in a unified fashion, with a high degree of coupling that provides inter-dependency in the regenerative cycle and some compensation of function. Nevertheless, there is clear distinctiveness to be found in their specificity of defense function and, we suggest, specificity of signaling. The antioxidant role of tocopherol is the elimination of singlet oxygen while ascorbate and glutathione control hydrogen peroxide accumulation, but also other species such as superoxide and hydroxyl radicals (Polle, 2001).

The cycling of antioxidants is not tightly coupled under all conditions, allowing a high degree of independent function for each pool. The amounts and extent of coupling between these antioxidants can provide precise information on redox state and buffering capacity in specific locations, e.g. tocopherol on membrane redox buffering capacity, ascorbate on the overall availability of reductant in the hydrophilic phase, and glutathione on the thiol/disulphide status of the cell, particularly in relation to the thiol/disulphide status of proteins. Ascorbate and glutathione differentially modulate gene expression. The effects of ascorbate deficiency and feeding on gene expression demonstrate the extensive metabolic cross-talk between the different defense processes in plants. Glutathione has comparable roles in orchestration of defense gene expression particularly during SAR. Oxidation caused by triggered ROS production plays a key part in determining leaf responses to stress. This inherently involves a change in cellular redox potential and, if the ascorbate and glutathione pools are well coupled, an increase in GSSG. An accumulation of GSSG precedes a net increase in the glutathione pool and this appears to be a salient feature triggering SAR. A net oxidation of the glutathione pool, followed by an increase in total leaf GSH, for example, occurs in barley leaves upon attack by *Blumeria gramis,* the fungus that causes powdery mildew (Vanacker et al., 2000) and in catalase-deficient mutants upon exposure to air after growth in CO_2 enriched conditions (Noctor et al., 2002b). Net GSH synthesis not only buffers changes in cellular redox state but it also causes changes in the intracellular localisation of NPR1, as discussed above.

Moreover, a clear mechanism is beginning to be defined through which increased ROS production and/or increases of GSSG could be signaled through posttranslational modification of proteins by glutathionylation. In animals and yeast, targets for glutathionylation include signal transduction components such as transcription factors (Klatt and Lamas, 2000) and recent data demonstrate that this process also occurs in plants (Ito et al., 2003). Signal transduction processes associated with tocopherol have not yet been described. However, singlet oxygen overproduction clearly impacts on gene expression (Op den Camp et al., 2003) and, we consider that as for ascorbate and glutathione, it may soon be shown that tocopherol exerts a significant influence on cellular control mechanisms linked to changes in redox state.

Acknowledgments

Achim Trebst is grateful to Deutsche Forschungsgemeinschaft (SFB 480-Germany) for finacial support. Rothamsted Research receives grant-aided support from the Biotechnology and Biological Sciences Research council of the U.K.

References

Adir N, Zer H, Shochat S and Ohad I (2003) Photoinhibition—a historical perspective. Photosynth Res 76: 343–370

Agius F, González-Lamothe R, Caballero JL, Muñoz-Blanco J, Botella MA and Valpuesta V (2003) Engineering increased vitamin C levels in plants by overexpression of a D-galacturonic acid reductase. Nature Biotechnol 21: 177–181

Allen JF and Hall DO (1973) Superoxide reduction as a mechanism of ascorbate-stimulated oxygen uptake by isolated chloroplasts. Biochem Biophys Res Commun 52: 856–862

Alscher RG (1989) Biosynthesis and antioxidant function of glutathione in plants. Physiol Plant 77: 457–64

Alvarez ME, Penell RI, Meijer PJ, Ishikawa A, Dixon RA and Lamb C (1998) Reactive oxygen intermediates mediate a systemic signal network in the establishment of plant immunity. Cell 92: 773–784

Amako K, Chen G-X and Asada K (1994) Separate assays specific for ascorbate peroxidase and guaiacol peroxidase and for the chloroplastic and cytosolic isozymes of ascorbate peroxidase in plants. Plant Cell Physiol 35: 497–504

Anderson JW, Foyer CH and Walker DA (1983a) Light-dependent reduction of hydrogen peroxide by intact spinach chloroplasts. Biochim Biophys Acta 724: 69–74

Anderson JW, Foyer CH and Walker DA (1983b) Light-dependent reduction of dehydroascorbate and uptake of exogenous ascorbate by spinach chloroplasts. Planta 158: 442–450

Angerhofer A (1991) Chlorophyll triplets and radical pairs. In: Scheer H (ed), Chlorophylls, pp 945–992. CRC Press, Boca Raton

Aro E-M, Virgin I and Andersson B (1993) Photoinhibition of photosystem II: inactivation, protein damage and turnover. Biochem Biophys Acta 1143: 113–134

Arrigoni O and de Tullio MC (2000) The role of ascorbic acid in cell metabolism: between gene-directed functions and unpredictable chemical reactions. J Plant Physiol 157: 481–488

Asada K (1999) The water-water cycle in chloroplasts: Scavenging of active oxygens and dissipation of excess photons. Annu Rev Plant Physiol Plant Mol Biol 50: 601–639

Asard H, Kapila J, Verelst W and Bérczi A (2001) Higher-plant plasma membrane cytochrome b_{561}: a protein in search of a function. Protoplasma 217: 77–93

Baier M and Dietz KJ (1997) The plant 2-cys peroxiredoxin BAS1 is a nuclear-encoded chloroplast protein: its expressional regulation, phylogenetic origin, and implications for its specific physiological function in plants. Plant J 12: 179–190

Baier M and Dietz K-J (1999a) Alkyl hydroperoxide reductases: the way out of the oxidative breakdown of lipids in chloroplasts. Trends Plant Sci 4: 166–168

Baier M and Dietz K-J (1999b) Protective function of chloroplast 2-cysteine peroxiredoxin in photosynthesis. Evidence from transgenic *Arabidopsis*. Plant Physiol 119: 1407–1414

Baier M, Noctor G, Foyer CH and Dietz KJ (2000) Antisense suppression of 2-cysteine peroxiredoxin in *Arabidopsis* specifically enhances the activities and expression of enzymes associated with ascorbate metabolism but not glutathione metabolism. Plant Physiol 124: 823–832

Ball L, Accotto G, Bechtold U, Creissen G, Funck D, Jimenez A, Kular B, Leyland N, Mejia-Carranza J, Reynolds H, Karpinski S and Mullineaux PM (2004) Evidence for a direct link between glutathione biosynthesis and stress defense gene expression in *Arabidopsis*. Plant Cell 16: 2448–2462

Bartling D, Radzio R, Steiner U and Weiler EW (1993) A glutathione-S-transferase with glutathione peroxidase activity from *Arabidopsis thaliana*. Molecular cloning and functional characterization. Eur J Biochem 216: 579–586

Bartoli CG, Pastori GM and Foyer CH (2000) Ascorbate biosynthesis in mitochondria is linked to the electron transport chain between complexes III and IV. Plant Physiol 123: 335–343

Bartoli CG, Guiamet JJ, Kiddle G, Pastori G, Di Cagno R, Theodoulou FL and Foyer CH (2004) The relationship between L-galactono-1, 4-lactone dehydrogenase (GalLDH) and ascorbate content in leaves under optimal and stress conditions. Plant Cell Environ, In press

Bauer CE, Elsen S and Bird TH (1999) Mechanisms for redox control of gene expression. Annu Rev Microbiol 53: 495–523

Berczi A and Møller IM (1998) NADH-monodehydroascorbate oxidoreductase is one of the redox enzymes in spinach leaf plasma membranes. Plant Physiol 116: 1029–1036

Bergmüller E, Porfirova S and Dörmann P (2003) Characterization of an *Arapidopsis* mutant deficient in γ-tocopherol methyltransferase. Plant Mol Biol 52: 1181–1190

Beyer RE (1990) The participation of coenzyme Q in free radical production and antioxidation. Free Radic Biol Med 8: 545–565

Bick JA, Setterdahl AT, Knaff DB, Chen Y, Pitcher LH, Zilinskas BA and Leustek T (2001) Regulation of the plant-type 5'-adenylylsulfate reductase by oxidative stress. Biochemistry 40: 9040–9048

Bishop NI and Wong J (1974) Photochemical characteristics of a vitamin E deficient mutant of *Scenedesmus obliquus*. Ber dtsch bot Ges 87: 359–371

Boveris A, Oshino N and Chance B (1972) The cellular production of hydrogen peroxide. Biochem J 128: 617–630

Bramley PM, Elmadfá I, Kafatos A, Kelly FJ, Manios Y, Roxborough HE, Schuch W, Sheehy PJA and Wagner KH (2000) Vitamin E. J Sci Food Agric 80: 913–938

Booij-James I, Swegle W M, Edelman M and Mattoo A (2002) Phosphorylation of the D1 photosystem II reaction center protein is controlled by an endogenous circadian rhythm. Plant Physiol 130: 2069–2075

Bowsher CG, Hucklesby DP and Emes MJ (1989) Nitrite reduction and carbohydrate metabolism in plastids purified from roots of *Pisum sativum* L. Planta 177: 359–366

Breitenbach J, Zhu C and Sandmann G (2001) The bleaching herbicice norflurazon inhibits phytoene desaturase by competetion with the cofactors. J Agric Food Chem 49: 5270–5272

Buchanan BB (1991) Regulation of CO_2 assimilation in oxygenic photosynthesis: the ferredoxin/thioredoxin system. Arch Biochem Biophys 228: 1–9

Bunkelmann JR and Trelease RN (1996) Ascorbate peroxidase. A prominent membrane protein in oilseed glyoxysomes. Plant Physiol 110: 589–598

Burkey KO, Eason G and Fiscus EL (2003) Factors that affect leaf extracellular ascorbic acid content and redox status. Physiol Plant 117: 51–57

Chamnongpol S, Willekens H, Moeder W, Langebartels C, Sandermann H Jr, Van Montagu M, Inze D and Van Camp W (1998) Defense activation and enhanced pathogen tolerance induced by H_2O_2 in transgenic tobacco. Proc Natl Acad Sci USA 95: 5818–5823

Chen G and Asada K (1989) Ascorbate peroxidase in tea leaves: Occurrence of two isozymes and the differences in their enzymatic and molecular properties. Plant Cell Physiol 30: 987–998

Chen Z and Gallie DR (2004). The ascorbic acid redox state controls guard cell signaling and stomatal movement. Plant Cell 16: 1143–1162

Chen Z, Young TE, Ling J, Chang SC and Gallie DR (2003) Increasing vitamin C content of plants through enhanced ascorbate recycling. Proc Natl Acad Sci USA 100: 3525–3530

Cogdell R and Frank HA (1996) Carotenoids in photosynthesis. Photochem Photobiol 63: 257–264

Collakova E and DellaPenna D (2001) Isolation and functional analysis of homogentisate phytyltransferase from *Synechocystis sp.* PCC 6803 and *Arabidopsis*. Plant Physiol 127: 1113–1124

Collakova E and DellaPenna D (2003a) Homogentisate phytyltransferase activity is limiting for tocopherol biosynthesis in *Arabidopsis*. Plant Physiol 131: 632–642

Collakova E and DellaPenna D (2003b) The role of homogentisate phytyltransferase and other tocopherol pathway enzymes in the regulation of tocopherol synthesis during abiotic stress. Plant Physiol 131: 930–940

Collin V, Issakidis-Bourguet E, Marchand C, Hirasawa M, Lancelin JM, Knaff DB and Miginiac-Maslow M (2003) The *Arabidopsis* plastidial thioredoxins: new functions and new insights into specificity. J Biol Chem 278: 23747–23752

Collinson EJ, Wheeler GL, Garrido EO, Avery AM, Avery SV and Grant CM (2002) The yeast glutaredoxins are active as glutathione peroxidases. J Biol Chem 277: 16712–16717

Conklin PL and Barth C (2004) Ascorbic acid, a familiar small molecule intertwined in the response of plants to ozone, pathogens and the onset of senescence. Plant Cell Environ 27: 959–970

Conklin PL, Williams EH and Last RL (1996) Environmental stress sensitivity of an ascorbic acid-deficient *Arabidopsis* mutant. Proc Natl Acad Sci USA 93: 9970–9974

Dähnhardt D, Falk J, Appel J, van der Kooij TAW, Schulz-Friedrich R and Krupinska K (2002) The hydroxyphenylpyruvate dioxygenase from *Synechocystis sp.* PCC 6803 is not required for plastoquinone biosynthesis. FEBS Lett 523: 177–181

Dat J, Vandenabeele S, Vranová E, Van Montagu M, Inzé D and Van Breusegem F (2000) Dual action of AOS during plant stress responses. Cell Mol Life Sci 57: 779–795

Davey MW, Gilot C, Persiau G, Ostergaard J, Han Y, Bauw GC and Van Montagu MC (1999) Ascorbate biosynthesis in *Arabidopsis* cell suspension culture. Plant Physiol 121: 535–543

Delauney A, Pflieger D, Barrault MB, Vinh J and Toledano MB (2002) A thiol peroxidase is an H_2O_2 receptor and redox-transducer in gene activation. Cell 111: 1–11

Demmig-Adams B and Adams WW III (1992) Photoprotection and other responses of plants to high light stress. Annu Rev Plant Physiol Plant Mol Biol 43: 599–626

Dietz KJ (2003) Plant peroxiredoxins. Annu Rev Plant Biol 54: 93–107

Dietz K-J, Stork T, Finkemeier I, Lamkemeyer P, Li W-X, El-Tayeb MA, Michel K-P, Pistorius E and Baier M (2005) The role of peroxiredoxins in oxygenic photosynthesis of cyanobacteria and higher plants: peroxide detoxification or redox sensing? In: Demmig-Adams B, Adams WW III and Mattoo AK (eds) Photoprotection, Photoinhibition, Gene Regulation, and Environment, pp 303–319. Springer, Dordrecht

Diner BA and Rappoport F (2002) Structure, dynamics, and energetics of the primary photochemistry of photosystem II of oxygenic photosynthesis. Annu Rev Plant Biol 53: 552–580

Dixon DP, Davis BG and Edwards R (2002) Functional divergence in the glutathione transferase superfamily in plants. Identification of two classes with putative functions in redox homeostasis in *Arabidopsis thaliana*. J Biol Chem 277: 30859–30869

Dron M, Clouse SD, Dixon RA, Lawton MA and Lamb CJ (1988) Glutathione and fungal elicitor regulation of a plant defense gene promoter in electroporated protoplasts. Proc Natl Acad Sci USA 85: 6738–6742

Du G, Mouithys-Mickalad A and Sluse FE (1998) Generation of superoxide anion by mitochondria and impairment of their functions during anoxia and reoxygenation in vitro. Free Radic Biol Med 25: 1066–1074

Duke SO, Lydon JM, Becerril TD, Sherman L and Matsumoto H (1991) Protoporphyrinogen oxidase inhibiting herbicides. Weed Sci 39: 465–473

Durrant JB, Giorgi LB, Barber J, Klug DR and Porter G (1990) Characterization of triplet states in isolated photosystem II reaction centers: oxygen quenching as a mechanism for photodamage. Biochim Biophys Acta 1017: 167–175

Dutilleul C, Driscoll S, Cornic G, De Paepe R, Foyer CH and Noctor G (2003a) Functional mitochondrial complex I is required by tobacco leaves for optimal photosynthetic performance in photorespiratory conditions and during transients. Plant Physiol 313: 264–275

Dutilleul C, Garmier M, Noctor G, Mathieu CD, Chétrit P, Foyer CH and De Paepe R (2003b) Leaf mitochondria modulate whole cell redox homeostasis, set antioxidant capacity and determine stress resistance through altered signaling and diurnal regulation. Plant Cell 15: 1212–1226

Elstner EF, Wagner GA and Schultz W (1998) Activated oxygen in green plants in relation to stress situation. Curr Top Plant Biochem Physiol 7: 159–187

Endo T and Asada K (2005) Photosystem I and photoprotection: cyclic electron flow and water-water cycle. In: Demmig-Adams B, Adams WW III and Mattoo AK (eds) Photoprotection, Photoinhibition, Gene Regulation, and Environment, pp 205–221. Springer, Dordrecht

Eshdat Y, Holland D, Faltin Z and Ben-Hayyim G (1997) Plant glutathione peroxidases. Physiol Plant 100: 234–240

Eskling M and Åkerlund HE (1998) Changes in the quantities of violaxanthin de-epoxidase, xanthophylls and ascorbate in spinach upon shift from low to high light. Photosynth Res 57: 41–50

Falk J, Krauß N, Dähnhardt D and Krupinska K (2002) The senescence associated gene of barley encoding 4-hydroxyphenylpyruvate dioxygenase is expressed during oxidative stress. J Plant Physiol 159: 1245–1253

Fath A, Bethke PC and Jones RL (2001) Enzymes that scavenge reactive oxygen species are down-regulated prior to gibberellic acid-induced programmed cell death in barley aleurone. Plant Physiol 126: 156–166

Feierabend J and Dehne S (1996) Fate of the porphyrin cofactors during the light-dependent turnover of catalase and of the photosystem II reaction center protein D1 in mature rye leaves. Planta 198: 413–422

Ferreira KN, Iverson TM, Maghlaoui K, Barber J and Iwata S (2004) Architecture of the Photosynthetic Oxygen-Evolving Center Science 303: 1831–1838.

Foreman J, Demidchik V, Bothwell JHF, Mylona P, Miedema H, Torres MA, Linstead P, Costa S, Brownlee C, Jones JDG, Davies JM and Dolan L (2003) Reactive oxygen species produced by NADPH oxidase regulate plant cell growth. Nature 422: 442–446

Foyer CH and Halliwell B (1976) The presence of glutathione and glutathione reductase in chloroplasts: a proposed role in ascorbic acid metabolism. Planta 133: 21–25

Foyer CH and Harbinson J (1999) Relationships between antioxidant metabolism and carotenoids in the regulation of photosynthesis. In: Frank HA, Young AJ, Britton G and Cogdell RJ (eds) The Photochemistry of Carotenoids, pp 305–325. Kluwer Academic Publishers, Dordrecht

Foyer CH and Lelandais M (1996) A comparison of the relative rates of transport of ascorbate and glucose across the thylakoid, chloroplast and plasma membranes of pea leaf mesophyll cells. J Plant Physiol 148: 391–398

Foyer CH and Noctor G (2000) Oxygen processing in photosynthesis: regulation and signalling. New Phytol 146: 359–388

Foyer CH and Noctor G (2003) Redox sensing and signaling associated with reactive oxygen in chloroplasts, peroxisomes and mitochondria. Physiol Plant 119: 355–364

Foyer CH, Rowell J and Walker D (1983) Measurements of the ascorbate content of spinach leaf protoplasts and chloroplasts during illumination. Planta 157: 239–244

Foyer CH, Souriau N, Perret S, Lelandais M, Kunert KJ, Pruvost C and Jouanin L (1995) Overexpression of glutathione reductase but not glutathione synthetase leads to increases in antioxidant capacity and resistance to photoinhibition in poplar trees. Plant Physiol 109: 1047–1057

Foyer CH, Gomez LD and van Heerden PDR. (2005) Glutathione. In: Smirnoff N (ed) Antioxidants and Reactive Species in Plants, pp 000–000. Blackwell Publishing, London

Fryer MJ (1992) The antioxidant effect of thylakoid vitamin E (α-tocopherol). Plant Cell Environ 15: 381–392

Fufezan C, Rutherford AW and Krieger-Liszkay A (2002) Singlet oxygen production in herbicide-treated photosystem II. FEBS Lett 532: 407–410

Genova ML, Ventura B, Giuliano G, Bovina C, Formiggini G, Castelli GP and Lenaz G (2001) The site of production of superoxide radical in mitochondrial Complex I is not a bound ubisemiquinone but presumably iron-sulfur cluster N2. FEBS Lett 505: 364–368

Gillham DJ and Dodge AD (1986) Hydrogen peroxide scavenging systems within pea chloroplasts. A quantitative study. Planta 167: 246–251

Gomez L, Vanacker H, Buchner P, Noctor G and Foyer CH (2004) Regulation of glutathione metabolism during the short-term chilling response of maize leaves. Plant Physiol 134: 1662–1671

Graßes T, Grimm B, Koroleva O and Jahns P (2001) Loss of α-tocopherol in tobacco plants with decreased geranylgeranyl reductase activity does not modify photosynthesis in optimal growth conditions but increases sensitivity to high-light stress. Planta 213: 620–628

Green R and Fluhr R (1995) UV-B-induced PR-1 accumulation is mediated by active oxygen species. Plant Cell: 203–212

Gressel J (2002) Molecular Biology of Weed Control. Taylor & Francis, London

Groden D and Beck E (1979) H_2O_2 destruction by ascorbate-dependent systems from chloroplasts. Biochim Biophys Acta 546: 426–435

Han D, Canali R, Rettori D and Cadenas E (2001) Production of superoxide into the intermembrane space and cytoplasm by heart mitochondria. Free Radic Biol Med 31: 45

Hancock JT, Desikan R and Neill SJ (2001) Does the redox status of cytochrome C act as a fail-safe mechanism in the regulation of programmed cell death? Free Rad Biol Medic 31: 697–703

Havaux M, Lütz C and Grimm B (2003) Chloroplast membrane photostability in *chlP* transgenic tobacco plants deficient in tocopherols. Plant Physiol 132: 300–310

Heazlewood JL, Howell KA and Millar AH (2003) Mitochondrial complex I from *Arabidopsis* and rice: orthologs of mammalian and fungal components coupled with plant-specific subunits. Biochim Biophys Acta 1604: 159–169

Herbette S, Lenne C, Leblanc N, Julien JL, Drevet JR and Roeckel-Drevet P (2002) Two GPX-like proteins from *Lycopersicon esculentum* and *Helianthus annuus* are antioxidant enzymes with phospholipid hydroperoxide glutathione peroxidase and thioredoxin peroxidase activities. Eur J Biochem 269: 2414–2420

Hideg E, Spetea C and Vass I (1994) Singlet oxygen production in thylakoid membranes during photoinhibition as detected by ESR spectroscopy. Photosynth Res 39: 191–199

Hideg E, Kalai T, Hideg K and Vass I (1998) Photoinhibition of photosynthesis in vivo results in singlet oxygen production. Detection via nitroxide-induced fluorescence quenching in broad bean leaves. Biochemistry 37: 11405–11411

Hideg E, Barta C, Kalai T, Vass I, Hideg K and Asada K (2002) Detection of singlet oxygen and superoxide with fluorescent sensors in leaves under stress by photoinhibition or UV radiation. Plant Cell Physiol 43: 1154–1164

Hofius D and Sonnewald U (2003) Vitamin E biosynthesis: biochemistry meets cell biology. Trends Plant Sci 8: 6–8

Horemans N, Asard H, Van Gestelen P and Caubergs RJ (1998) Facilitated diffusion drives transport of oxidized molecules into purified plasma membrane vesicles of *Phaseolus vulgaris*. Physiol Plant 104: 783–789

Horemans N, Foyer CH and Asard H (2000) Transport and action of ascorbate at the plant plasma membrane. Trends Plant Sci 5: 263–267

Horling F, Lamkemeyer P, Konig J, Finkemeier I, Kandlbinder A, Baier M and Dietz KJ (2003) Divergent light-, ascorbate-, and oxidative stress-dependent regulation of expression of the peroxiredoxin gene family in *Arabidopsis*. Plant Physiol 131: 317–325

Hossain MA and Asada K (1984) Inactivation of ascorbate peroxidase in spinach chloroplasts on dark addition of hydrogen peroxide: its protection by ascorbate. Plant Cell Physiol 25: 1285–1295

Hou WC, Chen HJ and Lin YH (1999) Dioscorins, the major tuber storage proteins of yam (*Dioscorea batatas* Decne), with dehydroascorbate reductase and monodehydroascorbate reductase activities. Plant Sci 149: 151–156

Imai T, Karita S, Shiratori G, Hattori M, Nunome T, Oba K and Hirai M (1998) L-galactono-γ-lactone dehydrogenase from sweet potato: Purification and cDNA sequence analysis. Plant Cell Physiol 39: 1350–1358

Ito H, Iwabuchi M and Ogawa K (2003) The sugar-metabolic enzymes aldolase and triose-phosphate isomerase are targets of glutathionylation in *Arabidopsis thaliana*: Detection using biotinylated glutathione. Plant Cell Physiol 44: 655–660

Jiménez A, Hernández JA, del Río L and Sevilla F (1997) Evidence for the presence of the ascorbate-glutathione cycle in mitochondria and peroxisomes of pea leaves. Plant Physiol 114: 275–284

Joo BH, Bae YS and Lee JS (2001) Role of auxin-induced reactive oxygen species in root gravitropism. 126: 1055–1060

Jung BG, Lee KO, Lee SS, Chi YH, Jang HH, Kang SS, Lee K, Lim D, Yoon SC, Yun DJ, Inoue Y, Cho MJ and Lee SY (2002) A Chinese cabbage cDNA with high sequence identity to phospholipid hydroperoxide glutathione peroxidases encodes a novel isoform of thioredoxin-dependent peroxidases. J Biol Chem 277: 12572–12578

Kaiser S, DiMascio P, Murphy ME and Sies H (1990), Physical and chemical scavenging of singlet molecular oxygen by tocopherols. Arch Biochem Biophys 277: 101–108

Kamiya N and Shen J-R (2003) Crystal structure of oxygen-evolving photosystem II from *Thermosynechoccus vulcanus* at 3.7 Å resolution. Proc Natl Acad Sci USA 100: 98–103

Kanwischer M, Porfirova S, Bergmüller E and Dörmann P (2005) Alterations in tocopherol cyclase activity in transgenic and mutant plants of Arapidosis affect tocopherol content, tocopherol composition, and oxidative stress. Plant Physiol 137: 713–723.

Karpinski S, Escobar C, Karprinska B, Creissen G and Mullineaux PM (1997) Photosynthetic electron transport regulates the expression of cytosolic ascorbate peroxidase genes in *Arabidopsis* during excess light stress. Plant Cell 9: 627–640

Karpinski S, Reynolds H, Karpinksa B, Wingsle G, Creissen G and Mullineaux PM (1999) Systemic signaling and acclimation in response to excess excitation energy in Arabidopsis. Science 284: 654–657

Kelly GJ and Latzko E (1979) Soluble ascorbate peroxidase. Naturwissenschaften 66: 377–382

Keren N, Gong.H and Ohad I (1995) Oscillations of reaction center II-D1 protein degradation in vivo induced by repetitive light flashes. J Biol Chem 270: 806–814

Keren N, Berg A, van Kann PJM, Levanon H and Ohad I (1997) Mechanism of photosystem II inactivation and D1 protein degradation at low light: the role of back electron flow. Proc Natl Acad Sci USA 94: 1579–1584

Kiddle G, Pastori GM, Bernard S, Pignocchi C, Antoniw J, Verrier PJ and Foyer CH (2003) Effects of ascorbate signaling on defense and photosynthesis genes. Antioxidant Redox Signaling 5: 23–32

Kimura M., Yoshizumi T, Manabe K, Yamamoto YY and Matsui M (2001) Arabidopsis transcriptional regulation by light stress via hydrogen peroxide-dependent and -independent pathways. Genes Cells 6: 607–617

Kimura M, Yamamoto YY, Seki M, Sato M, Abe T, Yoshida S, Manabe K, Shinozaki K and Matsui M (2003) Identification of Arabidopsis genes regulated by light stress using cDNA microarray. Photochem. Photobiol 77: 226–233

Kinkema M, Fan WH and Dong XN (2000) Nuclear localization of NPR1 is required for activation of PR gene expression. Plant Cell 12: 2339–2350

Klatt P and Lamas S (2000) Regulation of protein function by S-glutathiolation in response to oxidative and nitrosative stress. Eur J Biochem 267: 4928–4944

Köhler B and Blatt MR (2002) Protein phosphorylation activates the guard cell Ca^{2+} channel and is a prerequisite for gating by abscisic acid. Plant J 32: 185–194

Köhler B, Hills A and Blatt MR (2003) Control of guard cell ion channels by hydrogen peroxide and abscisic acid indicates their action through alternate signaling pathways. Plant Physiol 131: 385–388

Kovalchuk I, Bojko V, Kovalchuk O, Gloeckler V, Filkowski J, Heinlein M and Hohn B (2003) Pathogen induced systemic plant signal triggers genome instability. Nature 423: 760–762

Krieger A, Rutherford AW, Vass I and Hideg E (1998) Relationship between activity, D1 loss, and Mn binding in photoinhibition of photosystem II. Biochemistry 37: 16262–16269

Kruk J and Strzalka K (1995) Occurrence and function of alpha-tocopherol quinone in plants. J Plant Physiol 145: 405–409

Kuge S, Jones N and Nomoto A (1997) Regulation of YAP-1 nuclear localization in response to oxidative stress. EMBO J 16: 1710–1720

Kwak JM, Mori IC, Pei ZM, Leonhardt N, Torres MA, Dangl JL, Bloom RE, Bodde S, Jones JD and Schroeder JI (2003) NADPH oxidase *AtrbohD* and *AtrbohF* genes function in ROS-dependent ABA signaling in *Arabidopsis*. EMBO J 22: 2623–2633

Lamb C and Dixon RA (1997) The oxidative burst in plant disease resistance. Annu Rev Plant Physiol Plant Mol Biol 48: 251–275

Lee DL, Prisbylla MP, McLean Provan W, Frase T and Mutter LC (1997) The discovery and structural requirements of inhibitors of *p*-hydroxyphenylpyruvate dioxygenase. Weed Sci 45: 601–609

Lee KP, Kim C, Lee DW and Apel K (2003) TIGRINA d, required for regulating the biosynthesis of tetrapyrroles in barley, is an ortholog of the FLU gene of *Arabidopsis thaliana*. FEBS Lett 553: 119–124

Lemaire SD (2004) The glutaredoxin family in oxygenic photosynthetic organisms. Photosynth Res 79: 305–318

Lichtenthaler HK (1968) Plastoglobuli and the fine structure of plastids. Endeavour XXVII: 144–149

Lichtenthaler HK (1998) The plants 1-deoxy-d-xylulose-5-phosphate pathway for biosynthesis of isoprenoids. Fett Lipid 100: 128–138

Liere K and Link G (1997) Chloroplast endoribonuclease p54 involved in RNA 3'-end processing is regulated by phosphorylation and redox state. Nucleic Acids Res 25: 2403–2408

Liu YB, Fiskum G and Schubert D (2002) Generation of reactive oxygen species by the mitochondrial electron transport chain. J Neurochem 80: 780–787

Logan BA, Barker DH, Demmig-Adams B and Adams WW III (1996) Acclimation of leaf carotenoid composition and ascorbate levels to gradients in the light environment within an Australian rainforest. Plant Cell Environ 19: 1083–1090

Macpherson AN, Telfer A, Barber J and Truscott TG (1993) Direct detection of singlet oxygen from isolated photosystem II reaction centers. Biochim Biophys Acta 1143: 301–309

Maddison J, Lyons TM, Plöchl M and Barnes JD (2002) Hydroponically-cultivated radish fed L-galactono-1, 4-lactone exhibit increased tolerance to ozone. Planta 214: 383–391

Mathis P, Butler WL and Satoh K (1979) Carotenoid triplet state and chlorophyll fluorescence quenching in chloroplasts and subchloroplast particles. Photochem Photobiol 30: 603–614

Matsumoto H, Mizutani M, Yamaguchi T and Kadotani J (2002) Herbicide pyrazolate causes cessation of carotenoids synthesis in early watergrass by inhibiting 4-hydroxyphenylpyruvate dioxygenase. Weed Biol Manag 2: 39–45

Mattoo AK, Marder JB and Edelman M (1989) Dynamics of the photosystem II reaction center. Cell 56: 241–246

May MJ, Vernoux T, Leaver C, Van Montagu M and Inzé D (1998) Glutathione homeostasis in plants: implications for environmental sensing and plant development. J Exp Bot 49: 649–667

Mayer M, Beyer P and Kleinig H (1990) Quinone compounds are able to replace molecular oxygen as terminal electron acceptor in phytoene desaturation in chromoplasts of *Narcissus pseudonarcissus* L.Eur J Biochem 191: 359–363

Melis A (1999) Photosystem II damage and repair cycle in chloroplasts: what modulates the rate of photodamage in vivo? Trends Plant Sci 4: 130–135

Meskauskine R, Nater M, Goslings D, Kessler F, op den Camp R and Apel K (2001) FLU: A negative regulator of chlorophyll biosynthesis in *Arabidopsis thaliana*. Proc Natl Acad Sci USA 98: 12826–12831

Millar AH, Mittova V, Kiddle G, Heazlewood JL, Bartoli CG, Theodoulou FL and Foyer CH (2003) Control of ascorbate synthesis by respiration and its implications for stress responses. Plant Physiol 133: 443–447

Mittler R (2002) Oxidative stress, antioxidants and stress tolerance. Trends Plant Sci 7: 405–410

Mittler R and Zilinskas BA (1992) Molecular cloning and characterization of a gene encoding pea cytosolic ascorbate peroxidase. J Biol Chem 267: 21802–21807

Mittova V, Kiddle G, Theodoulou FL, Gomez L, Volokita M, Tal M, Foyer CH and Guy M (2003) Co-ordinate induction of glutathione biosynthesis and glutathione-metabolising enzymes is correlated with salt tolerance in tomato. FEBS Lett 554: 417–421

Moeder W, Barry CS, Tauriainen AA, Betz C, Tuomainen J, Utriainen M, Grierson D, Sandermann H, Langebartels C and Kangasjärvi J (2002) Ethylene sunthesis regulated by biphasic induction of 1-aminocyclopropane-1-carboxylic acid synthase and 1-aminocyclopropane-1-carboxylic acid oxidase is required for hydrogen peroxide accumulation and cell death in ozone-exposed tomato. Plant Physiol 130: 1918–1926

Møller IM (2001) Plant mitochondria and oxidative stress: electron transport, NADPH turnover, and metabolism of reactive oxygen species. Annu Rev Plant Physiol Plant Mol Biol 52: 561–591

Morimura Y, Iwamoto K, Ohya T, Igarashi T, Nakamura Y, Kubo A, Tanaka K and Ikawa T (1999) Light-enhanced induction of ascorbate peroxidase in Japanese radish roots during postgerminative growth. Plant Sci 142: 123–132

Mou Z, Fan WH and Dong XN (2003) Inducers of plant systemic acquired resistance regulate NPR1 function through redox changes. Cell 113: 935–944

Mullineaux PM, Karpinski S, Jimenez A, Cleary SP, Robinson C and Creissen GP (1998) Identification of cDNAS encoding plastid-targeted glutathione peroxidases. Plant J 13: 375–379

Mullineaux PM, Karpinski S and Creissen GP (2005) Integration of signaling in antioxidant defenses. In: Demmig-Adams B, Adams WW III and Mattoo AK (eds) Photoprotection, Photoinhibition, Gene Regulation, and Environment, pp 223–239. Springer, Dordrecht

Munné-Bosch S and Alegre L (2002a) Plant aging increases oxidative stress in chloroplasts. Planta 214: 608–615

Munné-Bosch S and Alegre L (2002b) The function of tocopherols and tocotrienols in plants. Crit Rev Plant Sci 21: 31–57

Munné-Bosch S and Falk J (2003) New insights into the function of tocopherols in plants. Planta 218: 323–326

Murata Y, Pei ZM, Mori IC and Schroeder JI (2001) Abscisic acid activation of plasma membrane Ca^{2+} channels in guard cells requires cytosolic NAD(P)H and is differentially disrupted upstream and downstream of reactive oxygen species production in *abi1-1* and *abi2-1* protein phosphatase 2C mutants. Plant Cell 13: 2513–2523

Neely WC, Martin JM and Barker SA (1988) Products and relative reaction rates of the oxidation of tocopherols with singlet molecular oxygen. Photochem Photobiol 48: 423–428

Nishizawa AN and Buchanan BB (1981) Enzyme regulation in C_4 photosynthesis. Purification and properties of thioredoxin-linked fructose bisphosphatase and sedoheptulose bisphosphatase from corn leaves. J Biol Chem 256: 6119–6126

Noctor G and Foyer CH (1998) Ascorbate and glutathione: keeping active oxygen under control. Annu Rev Plant Physiol Plant Mol Biol 49: 249–279

Noctor G, Arisi ACM, Jouanin L, Kunert KJ, Rennenberg H and Foyer CH (1998) Glutathione: biosynthesis, metabolism and relationship to stress resistance explored in transformed plants. J Exp Bot 49: 623–647

Noctor G, Veljovic-Jovanovic S and Foyer CH (2000) Peroxide processing in photosynthesis: antioxidant coupling and redox signalling. Proc Roy Soc Lond B 355: 1465–1475

Noctor G, Gomez L, Vanacker H and Foyer CH (2002a) Interactions between biosynthesis, compartmentation and transport in the control of glutathione homeostasis and signalling. J Exp Bot 53: 1283–1304

Noctor G, Veljovic-Jovanovic SD, Driscoll S, Novitskaya L and Foyer CH (2002b) Drought and oxidative load in the leaves of C_3 plants: a predominant role for photorespiration? Ann Bot 89: 841–850

Noctor G, Dutilleul C, De Paepe R and Foyer CH (2004) The use of mitochondrial mutants to evaluate the effects of redox state on photosynthesis, stress tolerance, and the integration of carbon and nitrogen metabolism. J Exp Bot 55: 49–57

Norris SR, Barette TR and DellaPenna D (1995) Genetic dissection of carotenoid synthesis in *Arabidopsis* defines plastoquinone as an essential component of phytoene desaturation. Plant Cell 7: 2139–2148

Oba K, Ishikawa S, Nishikawa M, Mizuno H and Yamamoto T (1995) Purification and properties of L-galactono-γ-lactone dehydrogenase, a key enzyme for ascorbic acid biosynthesis, from sweet potato roots. J Biochem 117: 120–124

Ogawa K, Hatano-Iwasaki A, Yanagida M and Iwabuchi M (2004) Level of glutathione is regulated by ATP-dependent ligation of glutamate and cysteine through photosynthesis in *Arabidopsis thaliana*: Mechanism of strong interaction of light intensity with flowering. Plant Cell Physiol 45: 1–8

Op den Camp RG, Przybyla D, Ochsenbein C, Laloi C, Kim C, Danon A, Wagner D, Hideg E, Gobel C, Feussner I, Nater M and Apel K (2003) Rapid induction of distinct stress responses after the release of singlet oxygen in *Arabidopsis*. Plant Cell 15: 2320–2332

Orozco-Cardenas ML, Narvaez-Vazquez J and Ryan CA (2001) Hydrogen peroxide acts as a second messenger for the induction of defense genes in tomato plants in response to wounding, systemin, and methyl jasmonate. Plant Cell 13: 179–191

Ostergaard J, Persiau G, Davey MW, Bauw G and VanMontagu M (1997) Isolation of a cDNA coding for L-galactono-γ-lactone dehydrogenase, an enzyme involved in the biosynthesis of ascorbic acid in plants. Purification, characterization, cDNA cloning, and expression in yeast. J Biol Chem 272: 30009–30016

Osmond CB (1994) What is photoinhibition? Some insights from comparisons of shade and sun plants. In: Baker NR and Bowyer JR (eds) Photoinhibition of Photosynthesis from Molecular Mechanisms to the Field, pp 1–24. Bios Scientific Publishers, Oxford, UK

Osmond CB and Förster B (2005) Photoinhibition: then and now. In: Demmig-Adams B, Adams WW III and Mattoo AK (eds) Photoprotection, Photoinhibition, Gene Regulation, and Environment, pp 11–22. Springer, Dordrecht

Pallanca JE and Smirnoff N (1999) Ascorbic acid metabolism in pea seedlings. A comparison of D-glucosone, L-sorbosone, and L-galactono-1, 4-lactone as ascorbate precursors. Plant Physiol 120: 453–461

Pallett K (2000) The mode of action of isoxaflutole: a case study of an emerging target site. In: Cobb AH and Kirkwood RC (eds) Herbicides and Their Mechanism of Action, pp 215–238. CRC Press, Boca Raton, FL

Pallett KE, Little JP, Sheekey M and Veeasekaran P (1998) The mode of action of isoxaflutole. Pestic Biochem Physiol 62: 113–124

Pasternak TP, Prinsen E, Ayaydin F, Miskolczi P, Potters G, Asard H, Van Onckelen HA, Dudits D and Fehér A (2002) The role of auxin, pH, and stress in the activation of embryogenic cell division in leaf protoplast-derived cells of alfalfa. Plant Physiol 129: 1807–1819

Pastori GM and Foyer CH (2002) Common components, pathways and networks of cross tolerance to stress: the central role of "redox" and hormone-mediated controls. Plant Physiol 129: 460–468

Pastori GM, Kiddle G, Antoniw J, Bernard S, Veljovic-Jovanovic S, Verrier PJ, Noctor G and Foyer CH (2003) Leaf vitamin C contents modulate plant defense transcripts and regulate genes controlling development through hormone signaling. Plant Cell 15: 939–951

Pei ZM, Murata Y, Benning G, Thomine S, Klüsener B, Allen GJ, Grill E and Schroeder JI (2000) Calcium channels activated by hydrogen peroxide mediate abscisic acid signalling in guard cells. Nature 406: 731–734

Pignocchi C (2000) Establishing the functional significance of ascorbate oxidase *in Planta*. PhD Thesis. University of Newcastle, U.K.

Pignocchi C and Foyer CH (2003) Apoplastic ascorbate metabolism and its role in the regulation of cell signalling. Curr Opin Plant Biol 6: 379–389

Pignocchi C, Fletcher JM, Wilkinson JE, Barnes JD and Foyer CH (2003) The function of ascorbate oxidase in tobacco. Plant Physiol 132: 1631–1641

Polle A (2001) Dissecting the superoxide-dismutase-ascorbate-glutathione pathway in chloroplasts by metabolic modeling. Computer simulations as a step towards flux analysis. Plant Physiol 126: 445–462

Porfirova S, Bergmüller E, Tropf S, Lemke R and Dörmann P (2002) Isolation of an *Arabidopsis* mutant lacking vitamin E and identification of a cyclase essential for all tocopherol biosynthesis. Proc Natl Acad Sci USA 99: 12495–12500

Potters G, Horemans N, Caubergs RJ and Asard H (2000) Ascorbate and dehydroascorbate influence cell cycle progression in a tobacco cell suspension. Plant Physiol 124: 17–20

Potters G, De Gara L, Asard H and Horemans N (2002) Ascorbate and glutathione: guardians of the cell cycle, partners in crime? Plant Physiol Biochem 40: 537–548

Potters G, Horemans N, Bellone S, Caubergs R J, Trost P, Guisez Y and Asard H (2004) Dehydroascorbate influences the plant cell cycle through a glutathione-independent reduction mechanism. Plant Physiol 134: 1479–1487

Price AH, Taylor A, Ripley SJ, Griffiths A, Trewavas AJ and Knight M (1994) Oxidative signals in tobacco increase cytosolic calcium. Plant Cell 6: 1301–1310

Puntarulo S, Sanchez RA and Boveris A (1988) Hydrogen peroxide metabolism in soybean embryonic axes at the onset of germination. Plant Physiol 86: 626–630

Puntarulo S, Galleano M, Sanchez RA and Boveris A (1991) Superoxide anion and hydrogen peroxide metabolism in soybean embryonic axes during germination. Biochim Biophys Acta 1074: 277–283

Raha S and Robinson BH (2000) Mitochondria, oxygen free radicals, disease and ageing. Trends Biochem Sci 25: 502–508

Rennenberg H (1997) Molecular approaches to glutathione biosynthesis. In: Cram WJ, DeKok LJ, Stulem I, Brunnold C and Rennenberg H (eds) Sulphur Metabolism in Higher Plants, pp 59–70. Backhuys Publishers, Leiden, The Netherlands

Rentel MC and Knight MR (2004) Oxidative stress-induced calcium signaling in *Arabidopsis*. Plant Physiol 135: 1471–1479

Rodermel S (2001) Pathways of plastid-to-nucleus signalling. Trends Plant Sci 6: 471–478

Rodriguez AA, Grunberg KA and Taleisnik EL (2002) Reactive oxygen species in the elongation zone of maize leaves are necessary for leaf extension. Plant Physiol 129: 1627–1632

Rouhier N, Gelhaye E and Jacquot JP (2002) Glutaredoxin-dependent peroxiredoxin from poplar. Protein-protein interaction and catalytic mechanism. J Biol Chem 277: 13609–13614

Rutherford AW and Krieger-Liszkay A (2001) Herbicide-induced oxidative stress in photosystem II. Trends Biochem Sci 26: 648–653

Sandermann H (2001) Active oxygen species as mediators of plant immunity: three case studies. Biol. Chem 381: 649–653

Sanmartin M, Drogouti PD, Lyons T, Barnes J and Kanellis AK (2003) Overexpression of ascorbate oxidase in the apoplast of transgenic tobacco results in altered ascorbate and glutathione redox states and increased sensitivity to ozone. Planta 216: 918–928

Satoh K and Mathis P (1981) Photosystem II chlorophyll α-protein complex: a study by flash absorption spectroscopy. Photobiochem Photobiophys 2: 189–198

Sattler SE, Cahoon EB, Coughlan SJ and DellaPenna D (2003) Characterisation of tocopherol cyclases from higher plants and cyanobacteria. Evolutionary implications for tocopherol synthesis and function. Plant Physiol 132: 2184–2195

Schafer FQ and Buettner GR (2001) Redox environment of the cell as viewed through the redox state of the glutathione disulfide/glutathione couple. Free Rad Biol Medic 30: 1191–1212

Schledz M, Seidler A, Beyer P and Neuhaus G (2001) A novel phytyltransferase from *Synechocystis sp.* PCC 6803 involved in tocopherol biosynthesis. FEBS Lett 499: 15–20

Schopfer P (2001) Hydroxyl radical-induced cell-wall loosening *in vitro* and *in vivo*: Implications for the control of elongation growth. Plant J 28: 679–688

Schopfer P, Liszkay A, Bechtold M, Frahry G andWagner A (2002) Evidence that hydroxyl radicals mediate auxin-induced extension growth. Planta 214: 821–828

Schroeder JI, Allen GJ, Hugouvieux V, Kwak JM and Waner D (2001a) Guard cell signal transduction. Annu Rev Plant Physiol Plant Mol Biol 52: 627–658

Schroeder JI, Kwak JM and Allen GJ (2001b) Guard cell abscisic acid signalling and engineering drought hardiness in plants. Nature 410: 327–330

Schulz A, Ot O, Beyer P and Kleinig H (1993) SC-0051, a benzoyl-cyclohexane-1, 3-dione bleaching herbicide, is a potent inhibitor of the enzyme *p*-hydroxyphenylpyruvate dioxygenase. FEBS Lett 318: 162–166

Schupp R and Rennenberg H (1990) Diurnal changes in the thiol composition of spruce needles. In: Rennenberg H, Brunold CH, de Kok LJ and Stulen I (eds) Sulfur Nutrition and Sulfur Assimilation in Higher Plants, pp 249–254. SPB Acad Publ, The Hague, Netherlands

Schürmann P and Jacquot J-P (2000) Plant thioredoxin systems revisited. Annu Rev Plant Physiol Plant Mol Biol 51: 371–400

Scott JW, Cort WM, Harley H, Parrish DR and Saucy G (1974) 6-Hydroxychroman-2-carboxylic acid: Novel antioxidants. J Am Oil Chem Soc 51: 200–203

Sen Gupta A, Alscher RG and McCune D (1991) Response of photosynthesis and cellular antioxidants to ozone in *Populus* leaves. Plant Physiol 96: 650–655

Shimaoka T, Yokota A and Miyake C (2000) Purification and characterization of chloroplast dehydroascorbate reductase from spinach leaves. Plant Cell Physiol 41: 1110–1118

Siendones E, Gonzalez-Reyes JA, Santos-Ocana C, Navas P and Cordoba F (1999) Biosynthesis of ascorbic acid in kidney bean. L-galactono-γ-lactone dehydrogenase is an intrinsic protein located at the mitochondrial inner membrane. Plant Physiol 120: 907–912

Smirnoff N (2000) Ascorbic acid: metabolism and functions of a multi-facetted molecule. Curr Opin Plant Biol 3: 229–235

Smirnoff N and Pallanca JE (1996) Ascorbate metabolism in relation to oxidative stress. Biochem Soc Trans 24: 472–478

Smirnoff N and Wheeler GL (2000) Ascorbic acid in plants. Biosynthesis and function. Crit Rev Biochem Mol Biol 35: 291–314

Smirnoff N, Running JA and Gatzek S (2004) Ascorbate metabolism in relation to oxidative stress. In: Asrad H, May JM and Smirnoff N (eds) Vitamin C Functions and Biochemistry in Animals and Plants, pp 1–30. Bios Scientific Publishers. London, UK

Smith IK, Kendall AC, Keys AJ, Turner JC and Lea PJ (1984) Increased levels of glutathione in a catalase-deficient mutant of barley (*Hordeum vulgare* L.). Plant Sci Lett 37: 29–33

Soll J, Schultz G, Joyard J, Douce R and Block MA (1985) Localization and synthesis of prenylquinones in isolated outer and inner envelope membranes from spinach chloroplasts. Arch Biochem Biophys 238: 290–299

Starke DW, Chock PB and Mieyal JJ (2003) Glutathione thiyl radical scavenging and transferase properties of human glutaredoxin (thioltransferase). Potential role in redox signal transduction. J Biol Chem 278: 14607–14613

Sticher L, MauchMani B and Metraux JP (1997) Systemic acquired resistance. Annu Rev Phytopath 35: 235–270

Streb P, Aubert S, Gout E and Bligny R (2003) Cold- and light-induced changes of metabolite and antioxidant levels in two high mountain plant species *Soldanella alpina* and *Ranunculus glacialis* and a lowland species *Pisum sativum*. Physiol Plant 118: 96–104

Strand A, Asami T, Alonso J, Ecker JR and Chory J (2003) Chloroplast to nucleus communication triggered by accumulation of Mg-protoporphyrin IX. Nature 421: 79–83

Sweetlove L and Foyer CH (2004). Roles for reactive oxygen species and antioxidants in plant mitochondria. In: Day DA, Millar AH and Whelan J (eds) Plant Mitochondria: From Genome to Function. Advances in Photosynthesis and Respiration, Vol 17, pp 307–320. Kluwer Academic Publishers, Dordrecht, The Netherlands

Tabata K, Oba K, Suzuki K and Esaka M (2001) Generation and properties of ascorbic acid-deficient transgenic tobacco cells expressing antisense RNA for L-galactono-1, 4-lactone dehydrogenase. Plant J 27: 139–148

Tabata K, Takaoka T and Esaka M (2002) Gene expression of ascorbic acid-related enzymes in tobacco. Phytochemistry 61: 631–635

Tamaoki M, Mukai F, Asai N, Nakajima N, Kubo A, Aono M and Saji H (2003) Light-controlled expression of a gene encoding L-galactono-γ-lactone dehydrogenase which affects ascorbate pool size in *Arabidopsis thaliana*. Plant Sci 164: 1111–1117

Telfer A (2002) What is β-carotene doing in the photosystem II reaction centre? Phil Trans Roy Soc Lond 357: 1431–1440

Telfer A and Barber J (1995) Role of carotenoid bound to the photosystem II reaction centre. In: Mathis P (ed) Photosynthesis: from Light to Biosphere, Vol IV, pp 15–20. Kluwer Academic Publishers, Dordrecht

Telfer A, Bishop SM, Philipps D and Barber J (1994a) Isolated photosynthetic reaction center of photosystem II as a sensitizer for the formation of singlet oxygen. Detection and quantum yield determination using a chemical trapping technique. J Biol Chem 269: 13244–13253

Telfer A, Dhami S, Bishop SM, Phillips D and Barber J (1994b) β-carotene quenches singlet oxygen formed in isolated photosystem II reaction center. Biochemistry 33: 14469–14474

Tevini M and Lichtenthaler HK (1970) Untersuchungen über die Pigment- und Lipochinonausstattung der zwei photosynthetischen Pigmentsysteme. Z Pflanzenphysiol 62: 17–32

Trebst A (1999) Singlet oxygen in photosynthesis. In: Denke A and Dornisch K (eds) Different Pathways through Life, pp 125–142. Lincom Europe, München, Germany

Trebst A (2003) Function of β-carotene and tocopherol in photosystem II. Z Naturforsch 58c: 609–620

Trebst A and Depka B (1997) Role of carotene in the rapid turnover and assembly of photosystem II in *Chlamydomonas reinhardtii*. FEBS Lett 400: 359–362

Trebst A, Depka B and Holländer-Czytko H (2002) A specific role for tocopherol and of chemical singlet oxygen quenchers in the maintenance of photosystem II structure and function in *Chlamydomonas reinhardtii*. FEBS Lett 516: 156–160

Trebst A, Depka B, Jäger J and Oettmeier W (2004) Reversal of the inhibition of photosynthesis by herbicides affecting hydroxyphenylpyruvate dioxygenase by plastoquinone- and tocopheryl-derivatives in *Chlamydomonas reinhardtii*. Pest Manag Sci publication 60: 669–674.

Trumper S, Follmann H and Haberlein I (1994) A novel dehydroascorbate reductase from spinach chloroplasts homologous to plant trypsin inhibitor. FEBS Lett 352: 159–162

Trumpower BL (1990) The protomotive Q cycle. Energy transduction by coupling of proton translocation to electron transfer by the cytochrome bc1 complex. J Biol Chem 265: 11409–11412

Tsegaye Y, Shintani DK and DellaPenna D (2002) Overexpression of the enzyme *p*-hydroxyphenolpyruvate dioxygenase in *Arabidopsis* and its relation to tocopherol biosynthesis. Plant Physiol Biochem 40: 913–920.

Tsukaguchi H, Tokui T, Mackenzie B, Berger UV, Chen XZ, Wang Y, Brubaker RF and Hediger MA (1999) A family of mammalian Na$^+$-dependent L-ascorbic acid transporters. Nature 399: 70–75

Urano J, Nakagawa T, Maki Y, Masumura T, Tanaka K, Murata N and Ushimaru T (2000) Molecular cloning and characterization of a rice dehydroascorbate reductase. FEBS Lett 466: 107–111

Vanacker H, Carver TL and Foyer CH (1998) Pathogen-induced changes in the antioxidant status of the apoplast in barley leaves. Plant Physiol 117: 1103–1114

Vanlerberghe G C and Ordog SH (2002) Alternative oxidase: integrating carbon metabolism and electron transport in plant respiration. In: Foyer CH and Noctor G (eds) Photosynthetic Nitrogen Assimilation and Associated Carbon and Respiratory Metabolism, Advances in Photosynthesis and Respiration, Vol 12, pp 173–191. Kluwer Academic Publishers, Dordrecht

Van Gorkom HJ and Schelvi JPM (1993) Kok's oxygen clock: what makes it tick? The structure of P680 and consequences of its oxidizing power. Photosynth Res 38: 297–301

Veljovic-Jovanovic SD, Pignocchi C, Noctor G and Foyer CH (2001) Low ascorbic acid in the vtc-1 mutant of *Arabidopsis* is associated with decreased growth and intracellular redistribution of the antioxidant system. Plant Physiol 127: 426–435

Vernoux T, Wilson RC, Seeley KA, Reichheld JP, Muroy S, Brown S, Maughan SC, Cobbett CS, Van Montagu M, Inzé D, May MJ and Sung ZR (2000) The ROOTMERISTEMLESS1/CADMIUM SENSITIVE2 gene defines a glutathione-dependent pathway involved in initiation and maintenance of cell division during post-embryonic root development. Plant Cell 12: 97–100

Vranova E, Inze D and Van Breusegem F (2002a) Signal transduction during oxidative stress. J Exp Bot 53: 1227–1236

Vranova E, Atichartpongkul S, Villarroel R., Van Montagu M., Inzé D and Van Camp W (2002b) Comprehensive analysis of gene expression in *Nicotiana tabacum* leaves acclimated to oxidative stress. Proc Natl Acad Sci USA 99: 10870–10875

Wagner D, Przybyla D, op den Camp R, Kim C, Landgraf F, Lee K P, Wursch M, Laloi C, Nater M, Higee E and Apel K (2004)

The genetic basis of singlet oxygen-induced stress reactions of *Arabidopsis thaliana*. Science 306: 1183–1185.

Wells WW, Xu DP, Yang Y and Rocque PA (1990) Mammalian thioltransferase (glutaredoxin) and protein disulfide isomerase have dehydroascorbate reductase activity. J Biol Chem 265: 15361–15364

Weber A, Servaites JC, Geiger DR, Kofler H, Hille D, Groner F, Hebbeker U and Flugge UI (2000) Identification, purification, and molecular cloning of a putative plastidic glucose translocator. Plant Cell 12: 787–802

Wheeler GL, Jones MA and Smirnoff N (1998) The biosynthetic pathway of vitamin C in higher plants. Nature 393: 365–369

Willekens H, Chamnongpol S, Davey M, Schraudner M, Langebartels C, Van Montagu M, Inzé D and Van Camp W (1997) Catalase is a sink for H_2O_2 and is indispensable for stress defense in C_3 plants. EMBO J 16: 4806–4816

Wingate VPM, Lawton MA and Lamb CJ (1988) Glutathione causes a massive and selective induction of plant defense genes. Plant Physiol 87: 206–210

Winkler BS (1992) Unequivocal evidence in support of the non-enzymatic redox coupling between glutathione/glutathione disulfide and ascorbic acid/dehydroascorbic acid. Biochim Biophys Acta 1117: 287–290

Witt HT (1996) Photosynthesis. Ber Bunsenge Phys Chem 100: 1923–1927

Xiang C and Oliver DJ (1998) Glutathione metabolic genes co-ordinately respond to heavy metals and jasmonic acid in *Arabidopsis*. Plant Cell 10: 1539–1550

Yabuta Y, Yoshimura K, Takeda T and Shigeoka S (2000) Molecular characterization of tobacco mitochondrial L-galactono-γ-lactone dehydrogenase and its expression in *Escherichia coli*. Plant Cell Physiol. 41: 666–675

Yamaguchi K, Mori H and Nishimura M (1995) A novel isozyme of ascorbate peroxidase localized on glyoxysomal and leaf peroxisomal membranes in pumpkin. Plant Cell Physiol 36: 1157–1162

Yoshimura K, Ishikawa T, Nakamura Y, Tamoi M, Takeda T, Tada T, Nishimura K and Shigeoka S (1998) Comparative study on recombinant chloroplastic and cytosolic ascorbate peroxidase isozymes of spinach. Arch Biochem Biophys 353: 55–63

Zhang W, Zhang L, Dong F, Gao J, Galbraith DW and Song CP (2001) Hydrogen peroxide is involved in abscisic acid-induced stomatal closure in *Vicia faba*. Plant Physiol 126: 1438–1448

Chapter 17

Redox Regulation of Chloroplast Gene Expression[#]

Sacha Baginsky*
*Plant Science Institute, ETH Center, Swiss Federal Institute of Technology,
Universitätsstr. 2, CH-8092 Zurich, Switzerland*

Gerhard Link*
*University of Bochum, Plant Cell Physiology & Molecular Biology,
Universitätsstr. 150, D-44780 Bochum, Germany*

Summary	269
I. Introduction	270
A. The Chloroplast – a Center for Sensing, Transmission, and Expression of Responses to Environmental Signals	270
B. Redox Regulation of Gene Expression – a Universal Theme in Biology	270
C. Transcriptional versus Posttranscriptional Regulation of Chloroplast Gene Expression	272
II. Posttranscriptional Processes	274
A. Redox Regulation of Translation: A Paradigm	275
B. Redox Regulation of RNA Stability and Degradation: An Emerging Field	277
C. Chloroplast RNA Splicing and Translation Elongation: Examples for Mixed or Unknown Light-Driven Signal Transduction Chains	279
III. Transcription	279
A. The Complexity of the Chloroplast Transcription Apparatus	279
B. Players and Mechanisms of Redox-Regulated Chloroplast Transcription	280
IV. Connections, Outlook and Perspectives	282
Acknowledgments	283
References	283

Summary

The chloroplast is the most important biosynthetic compartment of a green plant cell, being the site of photosynthesis and aspects of carbon, sulfur, and nitrogen assimilation as well as other pathways. At the same time, the complex enzymatic machinery of the organelle is a key target for photooxidative stress. The chloroplast contains an evolutionarily conserved set of genes and a specially adaptable gene expression machinery that is in close physical proximity to the photosynthetic apparatus, i.e. the primary source of reactive oxygen species. This adaptability somehow links the rapid gene expression response to the activity status of photosynthetic electron transport and accompanying redox reactions. In this chapter, we address the following questions: (i) which plastid gene products are subject to redox control? (ii) which stage(s) of organellar gene expression are redox-controlled? and (iii) what are the mechanisms and mediators involved?

---

*Author for correspondence, email: sbaginsky@ethz.ch or gerhard.link@ruhr-uni-bochum.de
[#]This chapter is dedicated to Professor Achim Trebst on the occasion of his 75th birthday

I. Introduction

A. The Chloroplast – a Center for Sensing, Transmission, and Expression of Responses to Environmental Signals

In "green" eukaryotes, from unicellular algae to higher plants, the chloroplast is the intracellular site of photosynthesis (Buchanan et al., 2000; Aro and Andersson, 2001). It is a member of the plastid family, i.e. a cell-specific group of organelles comprising a number of differentiated forms, each with a specialized function. In a given plant species, all plastid types contain closed circular plasmid-type DNA molecules with essentially the same organization, but with different copy numbers. A typical higher plant plastid DNA molecule harbors the genes for complete sets of rRNAs and tRNAs for protein synthesis on the organellar ribosomes as well as genes for more than a hundred proteins (Bogorad and Vasil, 1991; Sugita and Sugiura, 1996). Considering the large number of more than 3,000 (mostly nuclear-encoded and imported) putative chloroplast proteins, as suggested by proteomics and genomics database analyses (Kleffmann et al., 2004), the coding capacity of the organellar DNA seems surprisingly small. One may ask why the plant cell invests at all in the effort of establishing a complete gene expression system inside the plastid compartment. According to current views and consistent with a considerable body of compelling data, at least part of the answer lies in the integration of the plastid into the regulatory network that determines the differentiation and activity state of the entire cell. This complex network is based on the integration of different genetic compartments within the plant cell, i.e. the plastid, the mitochondrion, and the nucleus, necessitating a concerted regulation of gene expression. This is immediately obvious for the nuclear-encoded plastid proteins, which—in addition to many proteins involved in photosynthesis and other metabolic reactions—include proteins and regulatory factors involved in organellar gene expression itself. Conversely, it has become increasingly clear that the physiological status of the chloroplast is signaled back to the nucleus in a retrograde fashion. Although the biochemical nature of such (a) plastid signal(s) is only beginning to emerge, it is easily envisaged that both the forward and retrograde communication mechanisms, that have evolved together during evolution, have overcome the need to retain a full extra genome with thousands of genes initially brought in by (an) ancient endosymbiont(s).

Despite the small (less than 3%) proportion of chloroplast proteins that are encoded by organellar genes, these proteins comprise a group of essential components in photosynthesis and carbon assimilation as well as in plastid gene expression. In fact, the very first proteins to be identified as organellar gene products were the large subunit of RubisCO and the D1 protein of the PS II reaction center – both functionally important plant proteins that rank amongst the most extensively studied chloroplast gene products (Bogorad and Vasil, 1991; Sugita and Sugiura, 1996). It is interesting to note that, despite the overall small proportion of organelle-encoded (vs. imported) plastid proteins, the ratio is much higher (almost 1:1) in the case of the photosynthetic proteins. This emphasizes the importance of having genes for photosynthetic proteins—and their expression – inside the organelle, where their expression can be tightly regulated by the activity status of the photosynthetic electron transport chain. This is an obvious advantage in a situation such as photooxidative stress where there is a need for rapid replenishment of reaction center proteins (and possibly other photosynthetic proteins as well) (Aro and Andersson, 2001; Demmig-Adams and Adams, 2002). The question is which of the organelle-encoded proteins are affected by photosynthetic signals, and to what extent, and at which stage(s) of gene expression does the response to photostress operate? Before we focus on these specific points, we will review the general principles of reduction-oxidation (redox) regulation of gene expression.

B. Redox Regulation of Gene Expression – a Universal Theme in Biology

Redox chemistry is a universal aspect of living organisms. As defined by the Nernst equation, the redox potential of an electron carrier depends on the relative concentrations of its reduced versus oxidized forms. Redox reactions involve the transfer of electrons

---

Abbreviations: APX – ascorbate peroxidase; bromanil – tetrabromo-1,4-benzoquinone; CCCP – carbonal cyanide-m-chlorophenylhydrazone; CDK – cyclin-dependent kinase; CK2 – casein kinase 2; cpCK2 – chloroplast casein kinase 2; DB-MIB – 2,5-dibromo-3-methyl-6-isopropyl-1,4-benzoquinone; DCMU – 3-(3,4-dichlorophenyl)-1,1-dimethyl urea; DCPIP – 2,6-dichlorophenolindophenol; DTT – dithiothreitol; FNR – fumarate and nitrate reduction regulator; GSSG – oxidized disulfide form of glutathione; GSH – reduced dithiol from of glutathione; GSK – glycogen synthase kinase; MAPK – mitogen-activated protein kinase; IBZ – iodosobenzoic acid; NEM – N-ethyl maleimide; NEP – nuclear-encoded plastid RNA polymerase; PDI – protein disulfide isomerase; PEP – plastid-encoded RNA polymerase; PTK – plastid transcription kinase; ROS – reactive oxygen species; RRM – RNA recognition motif; SLF – sigma-like factor; SOD – superoxide dismutase

Table 1. Redox signals and sensing modules, mediators within signaling chains and at the interface with gene expression, and organisms that typically use these mechanisms. This table is far from complete, listing a representative set of well-characterized systems. One and the same component can act both as a signal and a mediator, possibly at more than one level. For further details, see text.

Redox signals, sensing modules	Mediators	Organisms
Quinones	plastoquinone/plastoquinol	plants, bacteria
Molecular oxygen	H_2O_2, O_2^-, NO	plants, bacteria, animals
Thiol/disulfide	thioredoxin	algae, bacteria, animals
	glutathione	plants, animals
	protein disulfide isomerase	algae, mammals, plants
	OxyR	bacteria
	AP-1, NF-kappaB	mammals
	Yap1	yeast
Metal ions		
[2Fe/2S] cluster	SoxR	bacteria
[4Fe/4S] cluster	FNR	bacteria
Heme	FixL	bacteria
Flavines as co-factors	NifL	bacteria
Other	ascorbate	plants

from one compound to another, resulting in the oxidation of one reactant and the reduction of the other reactant. Among the "classical" biological electron transfer chains is the photosynthetic apparatus. Its multisubunit pigment-protein complexes together produce energy (ATP) and reducing equivalents – mostly NADPH – for subsequent biochemical reactions (for a general overview, see e.g. Buchanan et al., 2000). Since oxygen is produced during photosynthesis, this also results in the generation of reactive oxygen species (ROS) that are potentially harmful to the plastid (Noctor and Foyer, 1998).

In addition to light energy capture, a second important function of photosynthetic electrochemistry is its role as a redox sensor for downstream signaling responses. For instance, changes in the steady-state levels of reducing equivalents can affect the properties of many chloroplast proteins, including e.g. several Calvin cycle enzymes (Schürmann and Jacquot, 2000). Furthermore, gene expression responses – both inside and outside the organelle – are apparently mediated by the energy state of the photosynthetic electron transport chain (Foyer and Noctor, 1999). In the mature functional chloroplast, the efficiency of photosynthetic electron transport can vary considerably depending on the species, age, cell position, and environmental conditions. Moreover, there is a constant requirement for a balanced supply of newly-synthesized photosynthetic proteins, due to the inherent turnover of e.g. reaction center proteins during active photosynthesis. It is therefore necessary that both the chloroplast and nucleo-cytoplasmic genetic systems be in communication to monitor the state of photosynthesis in an integrated fashion. Signaling chains based on redox chemistry would seem to be an ideal means to serve this role.

Redox regulation of gene expression has emerged as a common theme in biology and several examples are available from a large variety of organisms (Table 1). In most of these cases, redox regulation is mediated by a signal transduction chain involving low molecular weight molecules that change their redox state in response to a changing cellular redox potential. Prominent examples for the latter group of molecules are the ROS (see also Mullineaux et al., this volume). Although every cell can potentially become a source of ROS, for example by the activity of NAD(P)H oxidases, the chloroplast photosynthetic machinery is especially prone to generating these molecules. Although a number of ROS have been implied in redox signaling, the signal transduction chain and its constituents remain largely elusive. Signaling functions have been described for superoxide radicals (O_2^-) that can be converted into H_2O_2 by superoxide dismutases. Both molecules trigger gene expression responses in the bacterial system (for a review see Bauer et al., 1999). Recently, nitric oxide (NO) has received widespread attention for its essential role in the redox modulation of a wide variety of transcription factors in mammalian cells (reviewed in Stamler et al., 2001). Mechanistically, all ROS can modulate the activity of transcription factors directly or by acting further upstream in signal transduction cascades that mediate gene expression responses via redox-sensitive proteins. Several examples for signaling molecules and redox sensitive proteins are summarized in Table 1. Although a large variety of redox-responsive proteins has been described, two molecular mechanisms of redox regulation dominate in biological systems. One of them involves the oxidation of thiol groups to disulfides. The other mechanism utilizes the redox modulation of iron sulfur clusters (Table 1). In addition to these common redox mechanisms, the reduction/oxidation of protein-bound cofactors such as flavines can regulate the activity of a protein (Table 1).

There is considerable evidence from many organisms for redox regulation of proteins by oxidation of sulfhydryl (SH) groups and/or reduction of disulfide (S-S) bonds, both of which can influence enzyme activity, conformation, and interactive properties. Thiol redox reactions, involving mostly cysteine residues, are important for the regulation of single proteins and their immediate interaction partners. The activity of several transcription factors is known to be directly regulated by dithiol/disulfide exchange. Among the most prominent examples for the latter are the mammalian factor NF-kappaB (Hirota et al., 1999), the bacterial factor OxyR (S.O. Kim et al., 2002), and the yeast factor Yap1 (Delaunay et al., 2002). As the thiol redox signal can be transmitted also to third, fourth, or more redox components, it is not unexpected that this mechanism can play a role in signaling pathways across compartmental boundaries (Hogg, 2003). The principal mediators of this type of regulation include both specialized proteins, such as the thioredoxins (Nishiyama et al., 2001; Collin et al., 2003), and low-molecular weight peptides such as glutathione (Meister, 1995; Schafer and Buettner, 2001). As is the case in other organisms, glutathione is the most abundant non-protein thiol compound in plant cells (Bergmann and Rennenberg, 1993) and has a multi-compartment role as a key antioxidant and redox buffer (May et al., 1998; Foyer et al., 2001). The role of glutathione in ROS scavenging can result in drastic short-term changes of the cellular glutathione redox state, i.e. the ratio between oxidized (GSSG) and reduced glutathione (GSH). In the chloroplast, for example, ROS-scavenging is carried out by the collective action of superoxide dismutases (SOD), ascorbate peroxidase (APX), and other enzymes of the ascorbate-glutathione cycle (Bowler et al., 1992; Noctor and Foyer, 1998). Glutathione feeds electrons into this cycle and becomes oxidized when the capacity of antioxidant defense mechanisms is exceeded, particularly under exposure to stress factors such as high or low temperature, drought, or salinity. The resulting shift in the glutathione redox state is accompanied by changes in gene expression both inside and outside the organelle (Karpinski et al., 1999; Baena-González et al., 2001; see also Foyer et al., this volume).

Redox modulation of proteins by reduction/oxidation of iron sulfur clusters is a common mechanism in bacteria. Perhaps the best characterized example for this type of regulation is the SoxR-SoxS system that helps protect the bacterial cell against oxidative damage caused by superoxide radicals (Hidalgo et al., 1997). Under normal physiological conditions, SoxR contains a reduced [2Fe-2S] cluster that becomes oxidized by superoxide radicals or H_2O_2 under conditions of oxidative stress. Oxidized SoxR activates the transcription of SoxS, most likely via interaction with RNA polymerase. SoxS induces several proteins with functions in redox homeostasis, resulting in a concerted expression of antioxidant defense systems. Similarly, an oxidation-induced disassembly of [4Fe-4S] clusters has been reported for FNR (fumarate and nitrate reduction regulator) (Table 1), a global transcription regulator responsible for the anaerobic/aerobic regulation of several target genes in *E. coli* (Ralph et al., 1998).

The possible existence of redox-regulated gene expression in chloroplasts (see Fig. 1) was initially raised by the above-described reports on redox-responsive gene regulation in bacteria, including those capable of photosynthesis (Campbell et al., 1993; Bauer and Bird, 1996; Demple, 1998; Zeilstra-Ryalls et al., 1998; Zheng et al., 1998). Likewise, there have been an increasing number of reports on redox regulation in eukaryotic systems (Abate et al., 1990; Toledano and Leonard, 1991; Pahl and Baeuerle, 1994; Sen and Packer, 1996; Tell et al., 1998; Zheng et al., 1998; for recent reviews, see e.g. Kim et al., 2002; Georgiou, 2002). A direct analysis of similar processes in the chloroplast has been hampered by the complex nature of the dual-origin gene expression system in the organelle and by a lack of techniques to rigorously define the mechanisms and players involved. Progress in both aspects has been made over the past several years, resulting in improved knowledge of the architecture and expression dynamics of the major multi-protein complexes involved.

C. Transcriptional versus Posttranscriptional Regulation of Chloroplast Gene Expression

It had been a matter of debate whether or not chloroplast gene regulation is restricted to the nuclear-encoded genes responsible for proteins localized to the plastids. This question has now been clearly answered by numerous studies demonstrating changes in proteins and RNA corresponding to plastid-encoded genes in response to cellular and environmental cues, suggesting a tight regulation of chloroplast gene expression.

Chapter 17 Plastid Gene Regulation

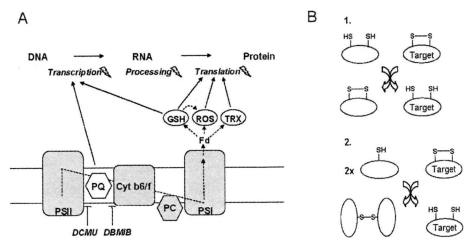

Fig. 1. (A) Regulation of chloroplast gene expression by photosynthetic electron flow. This scheme depicts the various layers of gene expression that are regulated by light (indicated by a flash) and the different known signal mediators involved in this regulation. Electron flow from PSII to PSI and to the soluble electron carriers is depicted as a dashed line. Glutathione (GSH), reactive oxygen species (ROS), and thioredoxin (TRX) are known signal mediators that are reduced by electrons from the photosynthetic electron transport chain via ferredoxin (Fd). Arrows indicate a known regulation of gene expression by redox reactions involving the soluble redox carrier. DCMU and DBMIB inhibit the reduction/oxidation of PQ as indicated. Abbreviations: PSII and I: photosystem II and I; PQ: plastoquinone/plastoquinol; Cyt: cytochrome; PC: plastocyanine; Fd: ferredoxin; GSH: glutathione; ROS: reactive oxygen species; TRX: thioredoxin. (B) Mechanism of dithiol/disulfide exchange at target proteins (Target) by (1) dithiol reductants, such as thioredoxin or protein disulfide isomerases, and (2) monothiol reductants, such as glutathione. In the case of glutathione, a mixed disulfide can be formed with a single thiol group (glutathionylation) (Danon, 2002).

This has led to the next question, i.e. whether plastid gene expression is regulated predominantly at a "transcriptional" or "posttranscriptional" level (terms often used synonymously for changes in the steady-state concentration of specific RNAs or proteins under investigation). It should be noted that the process of transcription strictly means de novo synthesis (and not accumulation) of RNA and that post-transcriptional processes encompass everything beyond the transcription stage. At the RNA level, this includes processing, modification, and degradation of the newly-synthesized transcripts. At the protein level, the stages involved range from the initiation of translation, i.e. the mRNA-directed de novo synthesis of (precursor) polypeptides, via elongation and termination, to the proteolytic processing and suite of posttranslational modification(s) that can have dramatic effects on the activity and stability of the mature functional gene products.

In view of the complex gene expression machinery of the chloroplast, it has become clear that more than one single regulatory event is often, if not always, involved. As an example, an RNA molecule that is not properly processed or edited (Sugiura et al., 1998) may no longer be an efficient template for protein synthesis. Such "non-functional" transcripts may be present in amounts and sizes that are similar to those of their functional counterparts (Hirose et al., 1996), and are therefore not easily distinguished from the latter by the common northern and dot blot techniques. As a consequence, one might be misled in the interpretation of the transcript population and the extent of regulation therein. Furthermore, even if the steady-state concentrations of a specific transcript remains virtually constant, the expression of that particular gene may be highly regulated if changes in transcription rates are compensated e.g. by changes in RNA stability (Shiina et al. 1998). However, in many cases it does seem appropriate to assign changes in RNA (e.g. based on northern analysis) or protein concentration (based on immunoblots) to bona fide post-transcriptional or translational control, as long as it is kept in mind that mechanisms at other control levels are not investigated rigorously by such techniques.

Furthermore, it is dangerous to generalize findings obtained for one particular organism, tissue, developmental stage, and environmental condition. Many conflicting conclusions reached by different groups regarding the control level(s) of light-regulated plastid gene expression fall into this category. For instance, it is not unexpected to find regulatory details in studies on a single-celled green alga like *Chlamydomonas reinhardtii* to be different from those found in tissues of multi-cellular higher plants. The same is true for a comparison of light/dark regulation in young seedlings versus fully mature, dark-adapted, and re-illuminated plants. Furthermore, the gene expression program in a dicotyledonous seedling will not necessarily be the same as that along the axis of a monocotyledonous leaf with its basal meristem. In view of these differences, it is now widely accepted that gene regulation in the plastid and elsewhere is quite flexible in a specific cellular and environmental context. Hence, it is important to keep in mind the extreme complexity and interdependence of the gene regulatory systems, within a single cell and across cellular boundaries, and as affected by the environmental context. In the next section recent work related to redox regulation at various levels of plastid gene expression is discussed (summarized in Table 2).

II. Posttranscriptional Processes

The aim of this section is to summarize our current knowledge about the mechanisms and key players of redox regulation of posttranscriptional events in chloroplast gene expression. Since it is our goal to provide mechanistic details of redox-control we will first focus on well-characterized signal transduction chains, such as the control of translation initiation in *Chlamydomonas reinhardtii*, followed by some interesting but indirect observations that infer a light-mediated redox regulation of posttranscriptional processes, and finally provide an outlook on strategies to define novel targets for redox-control by soluble electron carriers.

Table 2. Mechanisms and players of redox regulation of plastid gene expression, a summary. PET: photosynthetic electron transport; ROS: reactive oxygen species; PDI: protein disulfide isomerase; ?: unknown signal mediator.

Level of redox regulation	Experimental evidence	Mediator	Reference
Transcription			
transcription factor phosphorylation	in vitro kinase assays/phosphorylation studies/high-light illumination experiments	GSH	Baginsky et al., 1997, 1999; Baena Gonzalez et al., 2001
gene transcription of PSI and II components	illumination with different light qualities, DCMU/DBMIB inhibitor treatments	PET, ?	Deng and Gruissem, 1987; Pfannschmidt et al., 1999
RNA synthesis	illumination, DCMU/DBMIB inhibitor treatments	PET, ?	Pearsson et al., 1993
RNA splicing			
psbA splicing	illumination, DCMU/DBMIB inhibitor treatments, electron transport mutants	PET, ?	Deshpande et al., 1997
RNA stability			
Chimeric model RNA	illumination DCMU/DBMIB inhibitor treatments, in vitro reduction/oxidation	PET, ?	Salvador and Klein, 1999
fed-mRNA	illumination DCMU/DBMIB inhibitor treatments	PET, ?	Petracek et al., 1997, 1998; Dickey et al., 1998
mRNA processing and degradation	illumination and in vitro RNA degradation	?	Baginsky and Gruissem, 2002
endonuclease activity	in vitro RNA processing analyses	GSH	Liere and Link, 1997
polyadenylation activity	in vitro assays following illumination	?	Baginsky and Gruissem, 2002, Kudla et al., 1996
Translation initiation			
rbcL translation (*Chlamydomonas*)	high-light illumination	ROS, GSH	Shapira et al., 1997; Irhimovitch and Shapira, 2000
psbA translation initiation (higher plant)	high-light illumination	?	Kuroda et al., 1996; Kettunen et al., 1997
psbA translation initiation (*Chlamydomonas*)	illumination experiments DCMU/DBMIB inhibitor treatments, in vitro reduction/oxidation, electron transport mutants	TRX, PDI, PET	Danon and Mayfield, 1994a,b; Kim and Mayfield, 2002; Trebitsh et al., 2000; Trebitsh and Danon, 2001; Fong et al., 2000
Translation elongation			
photosynthetic proteins	illumination experiments, DCMU/DBMIB treatments, in vitro reduction/oxidation	PET, ?	Mühlbauer and Eichacker, 1998; Edhofer et al., 1998; Zhang et al., 2000

A. Redox Regulation of Translation: A Paradigm

Probably the best characterized redox-controlled system of chloroplast gene expression is the regulation of translation initiation in *Chlamydomonas reinhardtii*. Danon and Mayfield (1994a) identified a protein complex that regulates translation initiation of the D1 protein by specific binding to the 5'-untranslated region (5'-UTR) of *psbA* mRNA as described below. Although the RNA-binding complex is equally abundant in light- and dark-grown cells, its RNA-binding activity is substantially increased in the light. A mutant strain deficient in the assembly of photosystem I (cc703) lacked the light activation of *psbA* 5'-UTR binding, suggesting a direct involvement of photosynthetic electron transport in the activation of the protein complex. In vivo pulse labeling experiments with this mutant demonstrated a deficiency in light-activated translation of photosynthetic proteins (D1, D2, CP47, and CP43), confirming the important function of the RNA binding complex and its activation in the regulation of translation. The lack of light activation in a photosystem I-deficient mutant suggests that light signaling is mediated by signals "downstream" of photosystem I. Since photosystem I is the primary reducer of ferredoxin and thus of thioredoxin (Fig. 1A), the involvement of these soluble redox carrier in the regulation of translation initiation appears reasonable. In vitro experiments supported the view that thioredoxin is the critical redox-reactive component for the light-activation of translation. In vitro oxidation of the complex from light-grown cells with the oxidant dithionitrobenzoic acid (DTNB) completely abolishes its RNA-binding activity. The dithiol reductant DTT (with two SH-groups) but not the monothiol reductant β-mercaptoethanol (with one SH group) is able to restore the RNA-binding activity of the oxidized complex, suggesting a redox control mechanism that involves dithiol-disulfide exchange at two vicinal cysteine residues (Fig. 1B). Consistent with the proposed mechanism of reduction, thioredoxin, that contains two vicinal SH groups, restored the RNA-binding activity of the DTNB-oxidized complex to levels beyond those achieved by reduction with DTT alone.

A more detailed dissection of the components of the RNA-binding complex provided deeper insights into the molecular mechanism of its regulation and supported the role of vicinal cysteine residues. The complex consists of four subunits, RB38, RB55, RB47, and RB60. The functions of RB38 and RB55 remain unknown, although the gene for RB38 has been cloned recently (reviewed in Barnes and Mayfield, 2003).

RB47 is a chloroplast poly A-binding protein that constitutes the RNA-binding activity of the complex. Two mutants defective in RB47 are impaired in *psbA* translation supporting the critical role of this protein in the regulation of translation. RB60 is a chloroplast protein disulfide isomerase (cPDI) and appears to be the key component in the signal transduction chain. The active site of cPDI consists of two adjacent cysteine residues that are separated by two amino acids (Fig. 1B). Since this active site is similar to the redox-active site of thioredoxin, RB60 is an ideal candidate for thioredoxin-mediated redox control of its activity. According to the model proposed by Mayfield and colleagues (Fong et al., 2000), the thioredoxin-mediated reduction of cPDI (RB60) at its regulatory disulfide bond delivers the electrons for the RB60-mediated reduction of RB47, which results in the activation of its RNA-binding activity. RB47 contains several RNA-recognition motifs (RRMs) that are highly conserved among all known poly A binding proteins but, unlike the other poly A-binding proteins, RB47 contains a cysteine residue in each of the four RRMs (Fong et al., 2000). Substitutions of single cysteine residues do not alter the redox regulation of the RNA-binding activity of RB47. When both cysteines of RRM2 and RRM3 are substituted simultaneously, however, RB47 loses its redox responsiveness, suggesting that a regulatory disulfide bridge is formed between two different RNA-recognition motifs. Oxidation of the cysteine residues results in the formation of a disulfide bridge that connects RRM2 with RRM3 covalently and inactivates the RNA-binding activity of RB47 (Fong et al., 2000). Reduction of the disulfide bridge by RB60 separates the two RNA recognition motifs, opens the protein for its RNA substrate, and activates the RNA-binding activity of RB47.

This simple mechanism of dithiol/disulfide exchange, controlling the RNA-binding activity of RB47, suggests that alterations in the cellular redox-state are a major signal for the regulation of RB47 activity. Recent data indicated, however, that this view is too simplistic. When the RNA-binding complex is isolated from dark-grown cells, in vitro reduction is not sufficient to activate its RNA-binding activity (Trebitsh et al., 2000). It appears that redox-signaling in vivo is specifically mediated by RB60, rather than by global changes of the cellular redox conditions. It was demonstrated that RB60 needs to be primed to mediate this redox-control, and two distinct mechanisms for this priming have been suggested. One mechanism involves redox-dependent priming by light-induced oxidation of RB60, while the other is a redox-independent mechanism triggered by ADP-dependent phosphorylation of RB60. Light-dependent oxidation of RB60 was inferred from labeling experiments, with RB60. In these experiments the reactive thiols of RB60 were labeled with N-iodoacetyl-[^{125}I]-3-iodotyrosine. A decrease in labeling efficiency was observed upon illumination, suggesting a decrease of accessible thiol groups in the light (Trebitsh et al., 2000).

Light-induced oxidation contradicts the current understanding of light-mediated regulatory processes that are thought to be mediated by reduction rather than oxidation. The active photosynthetic electron transport reduces soluble redox carriers at photosystem I, resulting in a reducing cellular environment. Light-induced oxidation therefore requires a specific, oxidizing protein factor that is activated by reducing signals derived from the photosynthetic electron transport chain and is capable of oxidizing RB60 against a decreasing cellular redox potential (Trebitsh et al., 2000; Trebitsh and Danon, 2001). Although such a light-activated RB60 oxidoreductase is unknown to date, a recent study suggested that the redox-dependent priming of RB60 is mediated by the redox status of the plastoquinol/plastoquinone pool (Fig. 1A) (Trebitsh and Danon, 2001). Since the reductive signals for the activation of RB47 are produced by the soluble redox carrier at photosystem I (Fig. 1A), this separation of light signals provides a greater variability for the regulation of complex processes of chloroplast gene expression (Trebitsh and Danon, 2001).

An even greater variability is achieved with the redox-independent priming of RB60 activity by ADP-dependent phosphorylation (Danon and Mayfield, 1994b). The control of RB60 activity by an ADP-dependent kinase expands the possibilities for an integration of the control of translation initiation with cellular processes and signal transduction chains. ADP-dependent phosphorylation is thought to be regulated by the chloroplastic ATP/ADP ratio. It has been proposed that the ADP levels are higher in the dark, resulting in phosphorylation of RB60 in the dark (Danon and Mayfield, 1994b). ADP-dependent phosphorylation and oxidizing conditions both diminish the RNA-binding activity of the complex and, by inference, downregulate translation of photosynthetic proteins in a synergistic fashion (Danon and Mayfield, 1994a, b). To date, however, no data are available that suggest light-dependent variations of the phosphorylation status of RB60. Until this information is available, the regulatory role of RB60 phosphorylation and its function in the regulation of translation initiation remains elusive.

Translation initiation in higher plant chloroplasts is also regulated by light, but the mechanistic details of the signal transduction and the key player of this regulation are not defined to date (Kuroda et al., 1996; Kettunen et al., 1997). For example, the number of translation initiation complexes on *psbA* mRNA increases upon transfer of pea plants (*Pisum sativum* L.) from standard growth light to high light (Kettunen et al., 1997). The mechanism by which high light triggers the regulation of translation initiation is unknown, but recent data suggested that high light mediates signals via the glutathione redox system (Baginsky et al., 1999; Baena-González et al., 2001). It is conceivable that a redox-regulatory mechanism similar to that found at the level of transcription also plays a role in the regulation of translation initiation. This would thus be an effective mechanism for signal cross-talk between different stages of the plastid gene expression pathway (see also part III of this review).

Support for a possible involvement of the glutathione redox system in the regulation of translation initiation came from studies with *Chlamydomonas reinhardtii*. Here, an arrest of RBCL translation initiation correlates with ROS-induced changes in the redox status of the glutathione pool (Shapira et al., 1997; Irihimovitch and Shapira, 2000). The enhanced production of ROS upon high light treatment shifts the glutathione redox system towards its oxidized form (GSSG), and a return to the normal GSH/GSSG ratio is observed after 6 hours (Irihimovitch and Shapira, 2000). The increase in ROS and the parallel shift in the glutathione redox status upon high light treatment can be prevented by ascorbate that also prevents the arrest of RBCL translation. These observations suggest a direct regulatory role for the glutathione redox system in the translation machinery in *Chlamydomonas reinhardtii*. A possible scenario is that oxidized glutathione (GSSG) prevents RBCL translation by oxidation of a specific RBCL mRNA-binding protein. To date, these protein factors have not been identified, but they may be specific for RBCL since the translation of other photosynthetic proteins is unaffected by high light (Irihimovitch and Shapira, 2000). Utilization of the glutathione redox system for the high light regulation of gene expression appears reasonable, assuming that high light requires cellular adaptations that are different from normal light-adaptation processes. Therefore, high light- and growth light-mediated processes must be separated and are likely to act via different signal transduction chains using distinct mediators.

Although the previous examples document the growing information about redox regulation of translation initiation, they represent only a snapshot of the research in this field. Several further examples are available that report light and redox regulation of RNA-binding activities in the 5'-UTR of chloroplast mRNAs and the light regulation of translation initiation (Eibl et al., 1999; Shen et al., 2001; Zerges et al., 2002). The molecular function of 5'-UTR binding, as well as the regulation of 5'-UTR binding activities, will be elucidated in greater detail below.

B. Redox Regulation of RNA Stability and Degradation: An Emerging Field

The stability of an mRNA is reflected by its half life that has a significant impact on its translation rate and thus the level of its protein product. Indirect evidence for light regulation of mRNA stability has come from studies with *Chlamydomonas reinhardtii*. In this organism, light induces the degradation of several photosynthetic mRNAs, most likely via crucial sequence elements in the 5'-UTR (Salvador and Klein, 1999). First, insights into the mechanistic details of light-regulation were obtained with a chimeric model RNA (rbcL promoter: β-glucuronidase:*psaB* 3'-end) that has a half life of 20 minutes in the light and is stabilized to a half life of 5 hours in the dark. DCMU prevents the light-induced destabilization of this mRNA, suggesting a regulatory role of redox signals originating from the photosynthetic electron transport chain (Fig. 1). This view is substantiated by the observation that reduced DTT restores the RNA destabilization in DCMU-blocked cells, while the oxidant diamide prevents RNA degradation even when photosynthetic electron flow is not inhibited. These data suggest that the RNA-degradation machinery (endonucleases and/or exonucleases) is activated by reducing conditions and inactivated by oxidation, most likely via dithiol/disulfide exchange. Alternatively, RNA stabilizing protein factors such as RNA-binding proteins (Schuster and Gruissem, 1991; Nickelsen et al., 1994; Hayes et al., 1996; Nakamura et al., 2001; Zerges et al., 2002) could be inactivated by reducing conditions and activated by oxidation.

Several reports describe light-regulated binding of proteins to the 5'-UTR of *Chlamydomonas* mRNAs (Baginsky and Gruissem, 2001, 2002; Zerges et al., 2002). Since the dithiol reductant DTT restores RNA degradation in DCMU-blocked cells more efficiently than the monothiol reductant β-mercaptoethanol, thioredoxin has been proposed as the physiological redox mediator. To date, however, no data are available to support this hypothesis (Salvador and Klein, 1999).

In higher plant plastids, light triggers the stabilization of mRNAs rather than their degradation. Light-dependent stabilization of ferredoxin-1 *(fed-1)* mRNA has been observed in tobacco (*Nicotiana tabacum*) (Hansen et al., 2001). Transfer of tobacco plants to darkness resulted in a rapid decrease of *fed-1* mRNA half life from 2.4 hours in the light to 1.2 hours in the dark. DCMU-treatment decreased mRNA half life in the light, suggesting involvement of the photosynthetic electron transport chain in the regulation of mRNA half life (Petracek et al., 1997, 1998). Stabilization of *fed-1* mRNA is accompanied by its increased polysome association in the light, which might be the reason for its increased stability (Dickey et al., 1998). The stabilization of mRNA in the light is in line with in vitro data obtained with chloroplasts from spinach. Initial observations indicated that the transcription rates of *psbA*, that encodes the D1 protein, and other photosynthetic genes are not significantly different between etioplasts and chloroplasts, although the steady-state levels of their mRNAs, as well as their protein products, accumulate to higher levels in the light (Deng and Gruissem, 1987; Deng et al., 1987, 1989). This suggested that a significant proportion of light regulation of photosynthetic mRNAs in higher plant chloroplasts occurs at the posttranscriptional level. With the *petD* 3'-UTR as a model transcript and a soluble protein extract that correctly reproduces mRNA processing (i.e. the in vitro and in vivo 3'-ends of the mature *petD* mRNA are identical), the regulatory mechanism of posttranscriptional control were elucidated in further detail by a series of in vitro experiments.

The in vitro activities of protein extracts from light- and dark-grown plants reflect the light-mediated change in mRNA half lives in vivo, i.e. processing and stabilization in the light versus degradation in the dark (Baginsky and Gruissem, 2002). Dark-induced *petD* mRNA degradation is initiated by endonucleolytic cleavage within the coding region, most likely followed by polyadenylation of the cleavage products and their rapid exonucleolytic degradation (Lisitsky et al., 1996; Kudla et al., 1996; Baginsky and Gruissem, 2002). An RNA-binding protein in the extract from the light-grown plants prevented RNA degradation induced by the addition of extract (Baginsky and Gruissem, 2002), offering several possible scenarios for the mechanism of dark-induced mRNA degradation. Light may activate the RNA-binding protein that masks the endonucleolytic cleavage site utilized by a constitutive endonuclease. Alternatively, an endonuclease could be activated in the dark, either by posttranslational modification or by an elevated expression of its respective gene.

Endonucleases are key players in the regulation of mRNA stability not only in chloroplasts but also in prokaryotes (reviewed in Carpousis, 1999;

Grunberg-Manago, 1999; Hayes et al., 1999; Rauhut and Klug, 1999). The initial cleavage in the coding region of plastid mRNAs removes the secondary structure barrier at the 3'-end of most chloroplast mRNAs and offers an unprotected 3'-end to 3' to 5' processing exonucleases (Stern and Gruissem, 1987; Kudla et al., 1996; Baginsky and Gruissem, 2002). Several chloroplast endonucleolytic activities have been described (reviewed by Monde et al., 2000). One well-characterized endonuclease, that might play a key role in the regulation of mRNA stability, is p54 that has been purified and characterized from mustard seedlings (Nickelsen and Link, 1993; Liere and Link, 1997). P54 is an endonuclease that specifically binds to a conserved U-rich sequence element (UUUAUCU) in chloroplast precursor transcripts (Nickelsen and Link, 1993). In vitro experiments demonstrated that its endonucleolytic activity is regulated by phosphorylation as well as redox state acting synergistically (Liere and Link, 1997; Liere et al., 2001). Redox control is specifically mediated by the glutathione redox system (GSH/GSSG), and other reductants have no appreciable effect on p54 activity (Liere and Link, 1997). Reduced glutathione (GSH) inactivates the endonucleolytic activity while oxidized glutathione (GSSG) enhances it (Liere and Link, 1997). Consistent with a regulatory mechanism that involves reversible dithiol/disulfide exchange, oxidized GSSG can re-activate GSH-inactivated p54 activity. Phosphorylation of p54 with a partially purified CKII-like plastid kinase increases its endonucleolytic activity while dephosphorylation with calf intestinal alkaline phosphatase decreases it (Liere and Link, 1997). This suggests that p54 is a phospho-protein in vivo (Liere and Link, 1997). Phosphorylation and redox-control of p54 activity make it an ideal candidate for the integration of cellular signal transduction cascades and the regulation of mRNA stability in the chloroplast in response to intra- and extra-cellular signals (Liere et al., 2001). The gene for p54 has not been cloned and its homologies to known endonucleases from the chloroplast or the prokaryotic system remain to be established.

Initiation of RNA degradation by endonucleolytic cleavage is required to remove protective secondary structures from the 3'-end and to make the RNA accessible to 3' to 5' degradation (Stern and Gruissem, 1987). To date, polynucleotide phosphorylase (PNPase) is the only exonuclease identified from chloroplasts (Hayes et al., 1996; Baginsky et al., 2001). Recent data suggested that PNPase functions not only as an exonuclease but also as a 3'-polyadenylating enzyme (Yehudai-Resheff et al., 2001). Unlike in eukaryotes, polyadenylation in chloroplasts and prokaryotes serves as an RNA-degradation signal (Kudla et al., 1996; Lisitsky et al., 1997). Consistent with its function in mRNA degradation, PNPase-catalyzed polyadenylation activity is enhanced in the dark whereas its exonucleolytic activity is not altered (Kudla et al., 1996; Baginsky and Gruissem, 2002). The mechanism underlying the control of PNPase activity is unknown.

RNA-binding proteins play a crucial role in the regulation of mRNA stability. A conserved family of RNA-binding proteins, with an N-terminal acidic region and two RNA recognition motifs (RRMs) that are separated by a spacer region, has been identified from chloroplasts of higher plants. A recent study confirmed that these RNA-binding proteins stabilize ribosome free mRNA in the chloroplast stroma (Nakamura et al., 2001). A 28 kDa RNA-binding protein (28RNP) from spinach chloroplasts has been analyzed and characterized in greater detail. It was demonstrated that the RNA-binding activity of 28RNP is regulated by phosphorylation of the N-terminal acidic region (Schuster and Gruissem, 1991; Lisitsky and Schuster, 1995). In addition to being subject to control by phosphorylation, 28RNP has been identified as a possible novel thioredoxin target, suggesting additional redox control of its activity (Balmer et al., 2003). Thioredoxin was immobilized to a column and one of its two catalytic cysteines was exchanged for a serine, which prevented complete reduction and trapping of the substrate on the column. This suggests that 28RNP in this screen interacts directly with thioredoxin via one of its two cysteine residues (Schuster and Gruissem, 1991; Balmer et al., 2003).

A possible regulation of RNA-binding activities by thioredoxin provides an efficient mechanism to connect RNA-stability regulation to the redox state of thioredoxin and thus to photosynthetic electron transport. A direct light control of the RNA-binding activity of 28RNP, however, has not been demonstrated to date. Reconstitution of a protein extract from dark-adapted plants with 28RNP, that was isolated from light-grown chloroplasts, or with reduced recombinant 28RNP did not prevent the degradation of the *petD* precursor probe in vitro, not even when 28RNP was reduced with thioredoxin m or f in the presence of DTT prior to the reconstitution (Baginsky and Gruissem, 2002). It is possible that 28RNP participates in the redox-dependent assembly of an RNA-binding complex that is required for the regulation of RNA stability. A conceivable scenario is the formation of disulfide bridges either between different RRMs, as has been reasoned for the chloroplast poly A binding protein (RB47) from

Chapter 17 Plastid Gene Regulation

Chlamydomonas (Fong et al., 2000), and possibly also across different RNA-binding proteins. Further experiments are necessary to confirm the light regulation of RNA-binding activities in higher plant chloroplasts and to elucidate the mechanistic details of light-regulated mRNA stability regulation.

C. Chloroplast RNA Splicing and Translation Elongation: Examples for Mixed or Unknown Light-Driven Signal Transduction Chains

Apart from the initial indirect observations, light-regulated RNA splicing is not a well-established mechanism, and no data about the nature of the signals or the mediators are available. Splicing of *psbA* RNA plays a crucial role in the assembly of the D1 protein, that contains four large group-I introns in *Chlamydomonas reinhardtii*. Control of D1 protein assembly is tightly linked to the activity status of the photosynthetic electron transport chain. Light markedly stimulates the splicing process and supports the assembly of a functional D1 protein that can replace degraded D1 in the reaction center of photosystem II (Deshpande et al., 1997). This activation is abolished by inhibitors of the photosynthetic electron transport chain, such as DCMU and DBMIB (Fig. 1), but not by the ATP-synthesis inhibitor CCCP (Deshpande et al., 1997). These data strongly suggest that *psbA* splicing is regulated by the activity status of the photosynthetic electron transport chain. Support for this view comes from the observation that a mutant defective in the cytochrome b_6/f complex lacks the light induction of *psbA* splicing (Deshpande et al., 1997). To date, it is not clear, however, whether the signal transduction occurs via redox signals, or other light-driven signals such as changes in the trans-thylakoid pH gradient. A regulatory role of the redox status of the plastoquinone/plastoquinol pool is unlikely since DCMU and DBMIB both prevent the light induction of mRNA splicing. Redox regulation by thioredoxin has been suggested, but no data are available to support this view or to exclude a role of other redox carriers or the trans-thylakoid pH gradient in this regulatory system (Deshpande et al., 1997). Genetic screens for nuclear genes involved in *psbA* splicing of *Chlamydomonas* chloroplast genes could help to identify possible targets for this regulation and provide clues about its mechanistic basis. However, only one single nuclear gene locus with a function in *psbA* splicing has been identified to date (Li et al., 2002).

The light-regulation of translation elongation is an example that demonstrates that light regulates enzymatic activities not only via redox signals, but also via the trans-thylakoid pH gradient that is generated during photosynthetic electron transport (Mühlbauer and Eichacker, 1998; Edhofer et al., 1998; Zhang et al., 2000). Light-regulation of translation elongation was initially deduced from experiments with inhibitors of the photosynthetic electron transport chain, such as DCMU and DBMIB that were both capable of diminishing the rate of D1 translation elongation. Feeding of electron donors such as DCPIP and duroquinone partially restores the productive translation elongation in DCMU-blocked chloroplasts in the light but not in the dark (Edhofer et al., 1998; Mühlbauer and Eichacker, 1998). In a series of experiments, Mühlbauer and Eichacker presented compelling evidence that the trans-thylakoid pH gradient is responsible for the regulation of translation elongation. Consistent with this, soluble redox carriers that accept electrons from photosystem I do not restore translation elongation activity in DCMU-blocked cells (Mühlbauer and Eichacker, 1998). Similarly, Zhang et al. (2000) reported that oxidizing reagents such as NEM or IBZ were capable of inhibiting translation elongation. These data suggest that, in addition to the trans-thylakoid pH gradient, dithiol/disulfide exchange is an important regulatory event in translation elongation, although the extent of contribution and a possible signal hierarchy remain to be addressed (Zhang et al., 2000). It was further demonstrated that oxidizing conditions slowed down the co-translational assembly of nascent polypeptide chains into photosystem II. The redox-responsiveness of the assembly of multi-subunit protein complexes opens up a completely new level of regulation via redox regulation of protein/protein interactions. Redox regulation at this level significantly broadens and diversifies the impact of photosynthetic electron transport on regulatory processes.

III. Transcription

A. The Complexity of the Chloroplast Transcription Apparatus

As in bacteria and eukaryotic nuclear systems, de novo synthesis of RNA in chloroplasts is a complex enzymatic process with a number of aspects to consider, including multiple RNA polymerases, transcription factors, and protein modifications (Stern et al., 1997; Maliga, 1998; Hess and Börner, 1999). Light and other environmental cues can affect the plastid transcription apparatus and, despite initial arguments against it (e.g. Deng and Gruissem, 1987), clear-cut evidence based on in organello run-on transcription data suggests the participation of transcriptional regulation in light- versus dark-grown tissues (Klein and Mullet, 1990; Schrubar et al., 1991; Isono et al., 1997; Sakai et al., 1998).

The notion that plastid transcription can indeed be subject to regulation was further supported by the findings of multiple RNA polymerases in the organelle, each with a specific role in a developmental context (Mullet, 1993; Maliga, 1998). Based on current nomenclature (Maliga, 1998), at least two principally different transcription enzymes can be distinguished. One of them, the "Plastid-Encoded Polymerase" (PEP) according to the organellar coding of its multi-subunit core, resembles bacterial RNA polymerase (Cramer, 2002). The "Nuclear-Encoded Polymerase" (NEP), on the other hand, has a single catalytic subunit like those of the "uneven" (T3/T7) bacteriophages and mitochondria, and this polypeptide is encoded by a nuclear gene (Stern et al., 1997; Maliga, 1998; Hess and Börner, 1999). Even more intriguing, the PEP enzyme can exist in two different forms, "A" and "B" (Pfannschmidt and Link, 1997). While the "B" enzyme closely resembles bacterial RNA polymerase, the "A" enzyme has accessory polypeptides that are nuclear-encoded and provide "eukaryotic" features to this enzyme. The "A" form is the major RNA polymerase of functional chloroplasts (Pfannschmidt and Link, 1997) and is subject to redox regulation as will be outlined below (for a review, see Link 2001, 2003).

B. Players and Mechanisms of Redox-regulated Chloroplast Transcription

The possibility of redox regulation of chloroplast transcription was theoretically addressed by Allen (1993). Studies by Pearson et al. (1993) on RNA synthesis in isolated lettuce chloroplasts lent initial support to this idea. It was shown that dark-incubated chloroplasts were more active in RNA synthesis than those incubated in the light, and inhibitors of electron flow acting after PS II (DCMU, DBMIB) or on the cytochrome $b_6 f$ complex (bromanil) resulted in enhanced RNA synthesis in the light but not in the dark. The authors therefore concluded that the incorporation of radiolabel under the experimental conditions was favored if the cytochrome $b_6 f$ complex was in the oxidized state. This statement was clearly justified, even though these early experiments were carried out in a way that did not rigorously distinguish transcription and other enzymatic reactions leading to the incorporation of radioactive label into RNA. Although the ^{3}H-NAD used as the radioactive label was found to be rapidly converted to 5'-AMP and adenosine, and then incorporated into RNA, RNA synthesis was only partially inhibited by actinomycin D. This suggested that only a fraction of the labeled products represented the result of DNA template-dependent RNA synthesis (transcription). The most likely interpretation of these results is that the incorporation represented a partial terminal addition of NAD-derived adenine residues to preexisting RNA molecules, i.e. polyadenylation (see part II above).

Using the established in organello run-on transcription technique (Deng and Gruissem, 1987; Mullet and Klein, 1987), Pfannschmidt et al. (1999) provided further evidence for control of de novo template-dependent RNA synthesis in chloroplasts by photosynthetic electron flow. They preincubated plastids isolated from mustard (*Sinapis alba*) cotyledons in the absence or presence of DCMU or DBMIB, followed by in organello run-on transcription reactions. The formation of newly-synthesized transcripts from single chloroplast genes was furthermore assessed using gene-specific hybridization probes (Link, 1994) (Fig. 2). The mustard seedlings from which chloroplasts

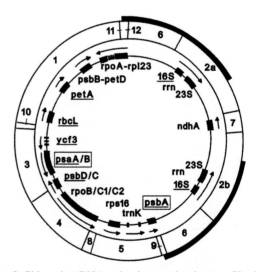

Fig. 2. Chloroplast DNA and redox regulated genes. Physical map of the cloned DNA molecule from the crucifer *Sinapis alba* (mustard), showing the architecture and arrangement of typical genes. For more complete details of higher-plant chloroplast DNA features, see e.g. Sugita and Sugiura (1996). The gene map shown here is based on that summarized by Link (1994) with added information listed more recently (e.g. Homann and Link, 2003). Genes that have been shown to be activated by high-intensity light (Baena-González et al., 2001) are underlined, and those selectively affected by light quality changes (Pfannschmidt et al, 1999a, b) are boxed.

were isolated were grown under two spectrally different light sources known to favor either PS I or PS II, thus resulting in an imbalance of electron flow between the two photosystems (Glick et al., 1986). Using a *psaA/B* probe for the genes encoding the two major reaction center proteins of PSI, less transcriptional activity was detected in plastids from seedlings grown under red light primarily absorbed by PS I (PS I light) than those grown under yellow light preferentially absorbed by PS II (PS II light). In addition, the negative effect of PS I light could be mimicked by DCMU and, conversely, the positive effect of PS II light was partially mimicked by DBMIB. This led the authors to conclude that *psaA/B* transcription decreases when the plastoquinone pool is oxidized (PS I light or DCMU), and increases when the pool is reduced (PS II light or DBMIB).

Rather than using spectral light quality changes, Baena-González at al. (2001) investigated the effect of different light intensity on the in organello rates of specific transcription of chloroplast genes. These experiments showed that chloroplasts from mustard seedlings grown under high light (1,000 μmol photons m^{-2} s^{-1}) are transcriptionally more active than those from seedlings grown under standard growth light (50 μmol photons m^{-2} s^{-1}). High-light conditions affect photosynthetic electron flow as well as turnover of reaction center proteins (photoinhibition) and result in changes in chloroplast redox state (Aro et al., 1993; Karpinski et al., 1997). The run-on transcription data by Baena-González et al. (2001) suggest that this change in redox balance results in a global, rather than differential, activation of chloroplast transcription. This is borne out by the observation that not only were genes for photosynthetic proteins activated by high light, but the same was true for genes for non-photosynthetic products (mostly tRNA and rRNA genes) that were included in this analysis. This is different from the results obtained for chloroplast transcription following light quality adaptation (Pfannschmidt et al., 1999a, b). It is thus possible that plastids use more than a single strategy in regulating gene expression (Figs. 1 and 2).

Hence, the question emerging from both the light quality and light quantity experiments described was: what are the mechanism(s) transducing the redox signal to the chloroplast transcription apparatus? Some experimental evidence suggested that phosphorylation/dephosphorylation of regulatory proteins might be involved. As suggested by Allen (1993) on the basis of two-component regulation in photosynthetic prokaryotes, it was expected that chloroplasts might have retained this highly successful regulatory mechanism that can be found in a wide range of eukaryotic organisms including plants. Despite intense research in this direction, however, the only evidence pointing to organellar two-component regulation stems from a chrysophyte alga (Jacobs et al., 1999), whereas other photosynthetic eukaryotes, from green algae to higher plants, have thus far proven negative in this respect.

This does not necessarily imply that phosphorylation control of plastid transcription is absent. On the contrary, evidence for an involvement of this mechanism has been available for more than a decade. Targets for this type of reversible protein modification in chloroplasts are the sigma-like factors (SLFs), i.e. proteins that, like their bacterial counterparts, confer promoter selection and transcription initiation specificity on the core RNA polymerase (Burgess and Anthony, 2001). Three different SLFs, named SLF67, SLF52, and SLF29, were purified from mustard chloroplasts (Tiller et al., 1991) and same-sized SLFs were also detected in etioplasts (Tiller and Link, 1993a). Interestingly, however, the equivalent factors from each plastid-type behaved differently with respect to their enzymatic characteristics, and explained by a different extent of phosphorylation (Tiller and Link, 1993b), suggesting that plastid transcription might indeed be regulated via reversible phosphorylation of sigma factors, and perhaps other regulatory proteins (for reviews see e.g. Link 1996; Link et al., 1997).

Support for this idea came from the purification of the chloroplast protein kinase responsible for SLF phosphorylation. This serine-specific enzyme, termed Plastid Transcription Kinase (PTK), was found to be loosely associated with chloroplast RNA polymerase PEP-A (Baginsky et al., 1997). Furthermore, PTK activity itself was found to respond to phosphorylation, suggesting that it might be the terminal component in a signaling cascade that controls chloroplast transcription in vivo and contains other member(s) that still await identification.

Subsequent work showed that PTK uses not only chloroplast sigma factors as substrates, but also several others of the 15 or so polypeptides that comprise the PEP-A chloroplast RNA polymerase (Baginsky et al., 1999; Pfannschmidt et al., 2000; Baena-González at al., 2001). The PTK activity state was found to be a critical factor that determines the extent of faithful transcription from the *psbA* promoter in an in vitro homologous plastid system. Furthermore, the kinase activity in vitro was affected both by phosphorylation and reduced glutathione (GSH), albeit in an antagonistic manner. The unphosphorylated (active) enzyme was inhibited by GSH, whereas the phosphorylated (inactive) PTK form was re-activated by the redox reagent (Baginsky et al., 1999). This suggested that PTK might indeed be a key player in both phosphorylation and redox control in vivo, perhaps connecting a sensor of photosynthetic electron flow to chloroplast gene expression directly at the transcriptional level or in a joint manner with other member(s) of a signaling chain (for review see Link et al., 1997; Link 2001).

The purified PTK enzyme from mustard chloroplasts (Baginsky et al., 1997, 1999) was classified as a member of the CMGC group of protein kinases, a name that is based on the first letters of the principal families representing this group: CDK (cyclin-dependent kinases), MAPK (mitogen-activated kinases), GSK-3 (glycogen synthase kinase 3), and CK2 (casein kinase 2) (Stone and Walker, 1995). Its (mostly nucleo-cytosolic) members include terminal components of signaling chains that phosphorylate nuclear transcription factors (Pinna, 1997). A prototype kinase of this type is CK2 that is a known transcriptional regulator in animal and yeast (Ghavidel and Schultz, 2001) as well as in plant systems (Klimczak et al., 1995).

In an effort to further characterize the chloroplast transcription kinase (PTK), Ogrzewalla et al. (2002) set out to clone the gene for the catalytic component. This work provided evidence that PTK is a nuclear-encoded chloroplast transcription factor closely related to nucleo-cytosolic CK2, with high sequence similarity to the (catalytic) alpha subunit from a wide range of organisms (Ogrzewalla et al., 2002). The gene product synthesized in vitro by coupled transcription-translation was found to be imported into isolated chloroplasts as a precursor and processed correctly. Following bacterial overexpression of the full-length cDNA for the mature product, the purified recombinant protein was found to have biochemical characteristics typical of CK2-alpha. Furthermore, using either antibodies against the bacterially expressed protein or mass spectrometry, the authentic chloroplast protein was detected as a component of the organellar transcription apparatus. Based on these criteria, the catalytic component of chloroplast PTK was named cpCK2-alpha (Ogrzewalla et al., 2002). A comparison of the substrate specificity and glutathione sensitivity of the recombinant and authentic chloroplast enzymes revealed very similar properties. Both enzyme preparations are capable of using plastid sigma factor(s) and several other RNA polymerase-associated proteins as phosphorylation substrates, and both are inhibited by GSH (Baginsky et al., 1997; Baena-González at al., 2001; Ogrzewalla et al., 2002). Hence, current evidence suggests that it is this CK2-type activity that is responsible

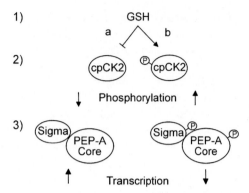

Fig. 3. SH Redox control of chloroplast transcription by glutathione mediated by the plastid transcription kinase now identified as a CK2-class protein kinase. The scheme depicts the situation borne out by studies of the mustard chloroplast transcription apparatus (Baginsky et al., 1997, 1999) and the functional characterization of the cloned kinase named cpCK2 (Ogrzewalla et al., 2002). PEP-A is the multi-subunit RNA polymerase that is responsible for transcription of most organellar genes in mature chloroplasts (Pfannschmidt et al., 2000; Baena-González et al., 2001). The cloned mustard sigma transcription factors have been recently described (Homann and Link, 2003). In essence, the scheme indicates that GSH inhibits the (unphosphorylated) kinase form and hence activates transcription. Conversely, the phosphorylated kinase form is re-activated under these conditions, which leads to transcriptional down-regulation. Both sigma factors and several other polypeptides of the transcription complex are substrates for cpCK2 (for review, see e.g. Link, 2001, 2003).

for phosphorylation and redox control of the PEP-A transcription apparatus. The cloned gene product will be useful, e.g. to define interaction partners, and mutants can be expected to provide further functional insights into the role of this kinase in the (redox) regulation of chloroplast gene transcription (Fig. 3).

IV. Connections, Outlook and Perspectives

How can the redox signaling mechanisms that probably connect photosynthetic electron transport to gene expression at various levels (Fig. 1 and Table 1) be further pinpointed? The answers will likely be obtained from a combination of molecular genetics and biochemistry efforts. Both naturally occurring mutants (e.g. Barkan et al., 1994; Lopez-Juez et al., 1998; Meurer et al., 1998) and transgenic systems (Allison and Maliga, 1995; Rochaix, 1997; Shiina et al., 1998) will be important approaches to carry out functional analyses. The cloning of most, if not all genes for photosynthesis proteins has been achieved, and so has that of the genes encoding the core components of the organellar gene expression machinery (reviewed by Sugita and Sugiura, 1996; Maliga, 1998). What remains to be a challenging task for the years to come is the definition of the various "layers" of accessory and regulatory proteins that surround the basic core. These components can be expected to reveal a dynamic composition, including factors that act in a highly stage- and/or stimulus-specific manner, those that are only transiently involved, and those that connect large multi-protein complexes such as those involved in the various steps of gene expression. At the level of transcription, cloning and functional analysis of the genes for plastid sigma factors is in rapid progress (for a review, see e.g. Allison, 2000; Privat et al., 2003; Homann and Link, 2003), as is the cloning of genes for other accessory factors involved in chloroplast transcription (PTK, SOD) and possible connectors to post-transcriptional processes (RNA-binding proteins of the PEP-A transcription complex) (Pfannschmidt et al., 2000; Ogrzewalla et al., 2002).

Light-dependent redox control of posttranscriptional processes and, more recently discovered, of transcription (see details above) plays a key role in the regulation of chloroplast gene expression (Fig. 1). An emerging theme is the signal transduction by soluble redox carrier such as thioredoxin and glutathione. Thioredoxin is reduced at photosystem I and well known to mediate redox-control of enzymatic activities by reversible dithiol/disulfide exchange (Buchanan, 1991). This type of enzymatic control adjusts the activities of metabolic enzymes directly to the levels of photosynthetic electron transport. Recently, the potential for thioredoxin-mediated redox control of proteins from regulatory pathways has been recognized. This dual role of thioredoxin has been elegantly summarized in a search for novel thioredoxin targets combined with a proteomics approach (Balmer et al., 2003). In this screen, Buchanan and colleagues identified several known thioredoxin targets confirming the validity of this type of experiment. Furthermore, they uncovered a number of novel thioredoxin targets, among them proteins with a function in chloroplast mRNA stability and translation (Balmer et al., 2003). It is notable that no proteins involved in transcription or transcription regulation were identified suggesting mechanistic differences in the light-mediated redox regulation of gene expression.

A systematic search for novel targets of redox regulation will help complete the picture of redox-regulated processes at the posttranscriptional level. Similar to the screen for thioredoxin targets, other potential redox carrier such as glutathione can be used as an affinity matrix for column chromatography. Combined with proteomics approaches, these experiments could offer deeper insights into redox-regulated processes, uncover a novel set of putative targets for redox regulation, and help to elucidate mechanisms of signal cross-talk under a variety of environmental conditions.

Acknowledgments

We apologize that, because of space constraints, not all relevant work in the field could be cited. Particular thanks are due to our lab members for valuable comments and discussion. Work from our groups was funded by the Deutsche Forschungsgemeinschaft and the Swiss National Science Foundation.

References

Abate C, Patel L, Rauscher FJ III and Curran T (1990) Redox regulation of Fos and Jun DNA-binding activity in vitro. Science 249: 1157–1161

Allen JF (1993) Control of gene expression by redox potential and the requirement for chloroplast and mitochondrial genomes. J Theor Biol 165: 609–631

Allison LA (2000) The role of sigma factors in plastid transcription. Biochimie 82: 537–548

Allison LA and Maliga P (1995) Light-responsive and transcription-enhancing elements regulate the plastid *psbD* core promoter. EMBO J 14: 3721–3730

Aro E-M, Virgin I and Andersson B (1993) Photoinhibition of Photosystem II. Inactivation, protein damage and turnover. Biochim Biophys Acta Bioenerg 1143: 113–134

Aro E-M and Andersson B (eds) (2001) Regulation of Photosynthesis. Kluwer Academic Publishers, Dordrecht

Baena-González E, Baginsky S, Mulo P, Summer H, Aro E-M and Link G (2001) Chloroplast transcription at different light intensities. Glutathione-mediated phosphorylation of the major RNA polymerase involved in redox-regulated organellar gene expression. Plant Physiol 127: 1044–1052

Baginsky S and Gruissem W (2001) Chloroplast mRNA 3'-end nuclease complex. Methods Enzymol 342: 408–419

Baginsky S and Gruissem W (2002) Endonucleolytic activation directs dark-induced chloroplast mRNA degradation. Nucleic Acids Res 30: 4527–4533

Baginsky S, Tiller K and Link G (1997) Transcription factor phosphorylation by a protein kinase associated with chloroplast RNA polymerase from mustard (*Sinapis alba*). Plant Mol Biol 34: 181–189

Baginsky S, Tiller K, Pfannschmidt T and Link G (1999) PTK, the chloroplast RNA polymerase-associated protein kinase from mustard (*Sinapis alba*), mediates redox control of plastid *in vitro* transcription. Plant Mol Biol 39: 1013–1023

Baginsky S, Shteiman-Kotler A, Liveanu V, Yehudai-Resheff S, Bellaoui M, Settlage RE, Shabanowitz J, Hunt DF, Schuster G and Gruissem W (2001) Chloroplast PNPase exists as a homomultimer enzyme complex that is distinct from the *Escherichia coli* degradosome. RNA 10: 1464–1475

Balmer Y, Koller A, Del Val G, Manieri W, Schürmann P and Buchanan BB (2003) Proteomics gives insight into the regulatory function of chloroplast thioredoxins. Proc Natl Acad Sci USA 100: 370–375

Barkan A, Walker M, Nolasco M and Johnson D (1994) A nuclear mutation in maize blocks the processing and translation of several chloroplast mRNAs and provides evidence for the differential translation of alternative mRNA forms. EMBO J 13: 3170–3181

Barnes D and Mayfield SP (2003) Redox control of posttranscriptional processes in the chloroplast. Antiox Redox Signal 5: 89–94

Bauer CE and Bird TH (1996) Regulatory circuits controlling photosynthesis gene expression. Cell 85: 5–8

Bergmann L and Rennenberg H (1993) Glutathione metabolism in plants. In: De Kok LJ, Stulen I, Rennenberg H, Brunold C and Rauser WE (eds) Sulfur Nutrition and Assimilation in Higher Plants, pp 109–123. SPB Academic Publishers, The Hague

Bogorad L and Vasil IK (eds.) (1991) The Molecular Biology of Plastids. Academic Press, San Diego

Bowler C, Van Montagu M and Inzé D (1992) Superoxide dismutase and stress tolerance. Annu Rev Plant Physiol Plant Mol Biol 43: 83–116

Buchanan BB (1991) Regulation of CO_2 assimilation in oxygenic photosynthesis: The ferredoxin/thioredoxin system: Perspective on its discovery, present status, and future development. Arch Biochem Biophys 288: 1–9

Buchanan BB, Gruissem W and Jones RL (2000) Biochemistry & Molecular Biology of Plants. American Society of Plant Biologists, Rockville, Maryland

Burgess RR and Anthony L (2001) How sigma docks RNA polymerase and what sigma does. Curr Opin Microbiol 4: 126–131

Campbell D, Houmard J and Tandeau de Marsac N (1993) Electron transport regulates cellular differentiation in the filamentous cyanobacterium *Calothrix*. Plant Cell 5: 451–463

Carpousis AJ, Vanzo NF and Raynal LC (1999) mRNA degradation. A tale of poly(A) and multiprotein machines. Trends Genet 15: 24–28

Collin V, Issakidis-Bourguet E, Marchand C, Hirasawa M, Lancelin JM, Knaff DB and Miginiac-Maslow M (2003) The *Arabidopsis* plastidial thioredoxins - New functions and new insights into specificity. J Biol Chem 278: 23747–23752

Cramer P (2002) Multisubunit RNA polymerases. Curr Opin Struct Biol 12: 89–97

Danon A (2002) Redox reactions of regulatory proteins: do kinetics promote specificity? Trends Biochem Sci 27: 197–203

Danon A and Mayfield SP (1994a) Light-regulated translation of chloroplast messenger RNAs through redox potential. Science 266: 1717–1719

Danon A and Mayfield SP (1994b) ADP-dependent phosphorylation regulates RNA-binding in vitro: Implications in light-modulated translation. EMBO J 13: 2227–2235

Delaunay A, Pflieger D, Barrault MB, Vinh J and Toledano MB (2002) A thiol peroxidase is an H_2O_2 receptor and redox-transducer in gene activation. Cell 111: 471–481

Demmig-Adams B and Adams WW III (2002) Antioxidants in photosynthesis and human nutrition. Science 298: 2149–2153

Demple B (1998) Signal transduction – A bridge to control. Science 279: 1655–1656

Deng X-W and Gruissem W (1987) Control of plastid gene expression during development: the limited role of transcriptional regulation. Cell 49: 379–387

Deng X-W, Stern DB, Tonkyn JC and Gruissem W (1987) Plastid run-on transcription. Application to determine the transcriptional regulation of spinach plastid genes. J Biol Chem 262: 9641–9648

Deng X-W, Tonkyn JC, Peter GF, Thornber JP and Gruissem W (1989) Post-transcriptional control of plastid mRNA accumulation during adaptation of chloroplasts to different light quality environments. Plant Cell 1: 645–654

Deshpande NN, Bao Y and Herrin DL (1997) Evidence for light/redox-regulated splicing of psbA pre-RNAs in *Chlamydomonas* chloroplasts. RNA 3: 37–48

Dickey LF, Petracek ME, Nguyen TT, Hansen ER and Thompson WF (1998) Light regulation of *Fed-1* mRNA requires an element in the 5' untranslated region and correlates with differential polyribosome association. Plant Cell 10: 475–484

Edhofer I, Mühlbauer SK and Eichacker LA (1998) Light regulates the rate of translation elongation of chloroplast reaction center protein D1. Eur J Biochem 257: 78–84

Eibl C, Zou Z, Beck A, Kim M, Mullet J and Koop HU (1999) In vivo analysis of plastid psbA, rbcL and rpl32 UTR elements by chloroplast transformation: tobacco plastid gene expression is controlled by modulation of transcript level and translation efficiency. Plant J 19: 333–346

Fong CL, Lentz A and Mayfield SP (2000) Disulfide bond formation between RNA binding domains is used to regulate mRNA binding activity of the chloroplast poly(A)-binding protein. J Biol Chem 275: 8275–8278

Foyer CH and Noctor G (1999) Leaves in the dark see the light. Science 284: 599–601

Foyer CH, Theodoulou FL and Delrot S (2001) The functions of inter- and intracellular glutathione transport systems in plants. Trends Plant Sci 6: 486–492

Foyer CH, Trebst A and Noctor G (2005) Signaling and integration of defense functions of tocopherol, ascorbate, and glutathione. In: Demmig-Adams B, Adams WW III and Mattoo AK (eds) Photoprotection, Photoinhibition, Gene Regulation, and Environment, pp 241–268. Springer, Dordrecht

Georgiou G (2002) How to flip the (redox) switch. Cell 111: 607–610

Ghavidel A and Schultz MC (2001) TATA binding protein-associated CK2 transduces DNA damage, signals to the RNA polymerase III transcriptional machinery. Cell 106: 575–584

Glick RE, McCauly SW, Gruissem W and Melis A (1986) Light quality regulates expression of chloroplast genes and assembly of photosynthetic membrane complexes. Proc Natl Acad Sci USA 83: 4287–4291

Grunberg-Manago M (1999) Messenger RNA stability and its role in control of gene expression in bacteria and phages. Annu Rev Genet 33: 193–227

Hansen ER, Petracek ME, Dickey LF and Thompson WF (2001) The 5' end of the pea ferredoxin-1 mRNA mediates rapid and reversible light-directed changes in translation in tobacco. Plant Physiol 125: 770–778

Hayes R, Kudla J, Schuster G, Gabay L, Maliga P and Gruissem W (1996) Chloroplast mRNA 3'-end processing by a high molecular weight protein complex is regulated by nuclear encoded RNA binding proteins. EMBO J 15: 1132–1141

Hayes R, Kudla J and Gruissem W (1999) Degrading chloroplast mRNA: the role of polyadenylation. Trends Biochem Sci 24: 199–202

Hess WR and Börner T (1999) Organellar RNA polymerases of higher plants. Int Rev Cytol 190: 1–59

Hidalgo E, Ding HG and Demple B (1997) Redox signal transduction via iron-sulfur clusters in the SoxR transcription activator. Trends Biochem Sci 22: 207–210

Hirose T, Fan H, Suzuki JY, Wakasugi T, Tsudzuki T, Kössel H and Sugiura M (1996) Occurrence of silent RNA editing in chloroplasts: Its species specificity and the influence of environmental and developmental conditions. Plant Mol Biol 30: 667–672

Hirota K, Murata M, Sachi Y, Nakamura H, Takeuchi J, Mori K and Yodoi J (1999) Distinct roles of thioredoxin in the cytoplasm and in the nucleus - A two-step mechanism of redox regulation of transcription factor NF-kappaB. J Biol Chem 274: 27891–27897

Hogg PJ (2003) Disulfide bonds as switches for protein function. Trends Biochem Sci 28: 210–214

Homann A and Link G (2003) DNA-binding and transcription characteristics of three cloned sigma factors from mustard (*Sinapis alba* L.) suggest overlapping and distinct roles in plastid gene expression. Eur J Biochem 270: 1288–1300

Irihimovitch V and Shapira M (2000) Glutathione redox potential modulated by reactive oxygen species regulates translation of Rubisco large subunit in the chloroplast. J Biol Chem 275: 16289–16295

Isono K, Shimizu M, Yoshimoto K, Niwa Y, Satoh K, Yokota A and Kobayashi H (1997) Leaf-specifically expressed genes for polypeptides destined for chloroplasts with domains of SIGMA[70] factors of bacterial RNA polymerases in *Arabidopsis thaliana*. Proc Natl Acad Sci USA 94: 14948–14953

Jacobs MA, Connell L and Cattolico RA (1999) A conserved His-Asp signal response regulator-like gene in *Heterosigma akashiwo* chloroplasts. Plant Mol Biol 41: 645–655

Karpinski S, Escobar C, Karpinska B, Creissen G and Mullineaux PM (1997) Photosynthetic electron transport regulates the expression of cytosolic ascorbate peroxidase genes in Arabidopsis during excess light stress. Plant Cell 9: 627–640

Kettunen R, Pursiheimo S, Rintamäki E, Van Wijk KJ and Aro E-M (1997) Transcriptional and translational adjustments of *psbA* gene expression in mature chloroplasts during photoinhibition and subsequent repair of photosystem II. Eur J Biochem 247: 441–448

Kim J and Mayfield SP (2002) The active site of the thioredoxin-like domain of chloroplast protein disulfide isomerase, RB60, catalyzes the redox-regulated binding of chloroplast poly(A)-binding protein, RB47, to the 5' untranslated region of *psbA* mRNA. Plant Cell Physiol 43: 1238–1243

Kim SO, Merchant K, Nudelman R, Beyer WF Jr., Keng T, DeAngelo J, Hausladen A and Stamler JS (2002) OxyR: A molecular code for redox-related signaling. Cell 109: 383–396

Kleffmann T, Russenberger D, von Zychlinski A, Christopher W, Sjölander K, Gruissem W and Baginsky S (2004) The *Arabidopsis thaliana* chloroplast proteome reveals pathway abundance and novel protein functions. Curr Biol 14: 354–362

Klein RR and Mullet JE (1990) Light-induced transcription of chloroplast genes. *psbA* transcription is differentially enhanced in illuminated barley. J Biol Chem 265: 1895–1902

Klimczak LJ, Collinge MA, Farini D, Giuliano G, Walker JC and Cashmore AR (1995) Reconstitution of Arabidopsis casein kinase II from recombinant subunits and phosphorylation of transcription factor GBF1. Plant Cell 7: 105–115

Kudla J, Hayes R and Gruissem W (1996) Polyadenylation accelerates degradation of chloroplast mRNA. EMBO J 15: 7137–7146

Kuroda H, Kobashi K, Kaseyama H and Satoh K (1996) Possible involvement of a low redox potential component(s) downstream of photosystem I in the translational regulation of the D1 subunit of the photosystem II reaction center in isolated pea chloroplasts. Plant Cell Physiol 37: 754–761

Li X-P, Gilmore AM and Niyogi KK (2002) Molecular and global time-resolved analysis of a *psbS* gene dosage effect on pH- and xanthophyll cycle-dependent nonphotochemical quenching in photosystem II. J Biol Chem 277: 33590–33597

Liere K and Link G (1997) Chloroplast endoribonuclease p54 involved in RNA 3'-end processing is regulated by phosphorylation and redox state. Nucleic Acids Res 25: 2403–2408

Liere K, Nickelsen J and Link G (2001) Chloroplast p54 endoribonuclease. Methods Enzymol 342: 420–428

Link G (1994) Plastid differentiation: organelle promoters and transcription factors. In: Nover L (ed) Plant Promoters and Transcription Factors – Results and Problems in Cell Differentiation, Vol. 20, pp 65–85. Springer-Verlag, Berlin-Heidelberg

Link G (1996) Green life: control of chloroplast gene transcription. BioEssays 18: 465–471

Link G (2001) Redox regulation of photosynthetic genes. In: Aro EM and Andersson B (eds) Regulation of Photosynthesis, pp 85–107. Kluwer Academic Publishers, Dordrecht

Link G (2003) Redox regulation of chloroplast transcription. Antiox Redox Signal 5: 79–88

Link G, Tiller K and Baginsky S (1997) Glutathione, a regulator of chloroplast transcription. In: Hatzios KK (ed) Regulation of Enzymatic Systems Detoxifying Xenobiotics in Plants, pp 125–137. Kluwer Academic Publishers, Dordrecht

Lisitsky I and Schuster G (1995) Phosphorylation of a chloroplast RNA-binding protein changes its affinity to RNA. Nucleic Acids Res 23: 2506–2511

Lisitsky I, Klaff P and Schuster G (1996) Addition of destabilizing poly(A)-rich sequences to endonuclease cleavage sites during the degradation of chloroplast mRNA. Proc Natl Acad Sci USA 93: 13398–13403

Lisitsky I, Kotler A and Schuster G (1997) The mechanism of preferential degradation of polyadenylated RNA in the chloroplast. The exoribonuclease 100RNP/Polynucleotide phosphorylase displays high binding affinity for poly(A) sequence. J Biol Chem 272: 17648–17653

López-Juez E, Jarvis RP, Takeuchi A, Page AM and Chory J (1998) New Arabidopsis *cue* mutants suggest a close connection between plastid- and phytochrome regulation of nuclear gene expression. Plant Physiol 118: 803–815

Maliga P (1998) Two plastid RNA polymerases of higher plants: an evolving story. Trends Plant Sci 3: 4–6

May MJ, Vernoux T, Leaver CJ, Van Montagu M and Inzé D (1998) Glutathione homeostasis in plants: implications for environmental sensing and plant development. J Exp Bot 49: 649–667

Meister A (1995) Glutathione metabolism. Methods Enzymol 251: 3–7

Meurer J, Grevelding C, Westhoff P and Reiss B (1998) The PAC protein affects the maturation of specific chloroplast mRNAs in *Arabidopsis thaliana*. Mol Gen Genet 258: 342–351

Monde RA, Schuster G and Stern DB (2000) Processing and degradation of chloroplast mRNA. Biochimie 82: 573–582

Mühlbauer SK and Eichacker LA (1998) Light-dependent formation of the photosynthetic proton gradient regulates translation elongation in chloroplasts. J Biol Chem 273: 20935–20940

Mullet JE (1993) Dynamic regulation of chloroplast transcription. Plant Physiol 103: 309–313

Mullet JE and Klein RR (1987) Transcription and RNA stability are important determinants of higher plant chloroplast RNA levels. EMBO J 6: 1571–1579

Mullineuax PM, Karpinski S and Creissen GP (2005) Integration of signaling in antioxidant defenses. In: Demmig-Adams B, Adams WW III and Mattoo AK (eds) Photoprotection, Photoinhibition, Gene Regulation, and Environment, pp 223–239. Springer, Dordrecht

Nakamura T, Ohta M, Sugiura M and Sugita M (2001) Chloroplast ribonucleoproteins function as a stabilizing factor of ribosome-free mRNAs in the stroma. J Biol Chem 276: 147–152

Nickelsen J and Link G (1993) The 54 kDa RNA-binding protein from mustard chloroplasts mediates endonucleolytic transcript 3' end formation in vitro. Plant J 3: 537–544

Nickelsen J, Van Dillewijn J, Rahire M and Rochaix J-D (1994) Determinants for stability of the chloroplast *psbD* RNA are located within its short leader region in *Chlamydomonas reinhardtii*. EMBO J 13: 3182–3191

Nishiyama Y, Yamamoto H, Allakhverdiev SI, Inaba M, Yokota A and Murata N (2001) Oxidative stress inhibits the repair of photodamage to the photosynthetic machinery. EMBO J 20: 5587–5594

Noctor G and Foyer CH (1998) Ascorbate and glutathione: Keeping active oxygen under control. Annu Rev Plant Physiol Plant Mol Biol 49: 249–279

Ogrzewalla K, Piotrowski M, Reinbothe S and Link G (2002) The plastid transcription kinase from mustard (*Sinapis alba* L.) – A nuclear-encoded CK2-type chloroplast enzyme with redox-sensitive function. Eur J Biochem 269: 3329–3337

Pahl HL and Baeuerle PA (1994) Oxygen and the control of gene expression. BioEssays 16: 497–502

Pearson CK, Wilson SB, Schaffer R and Ross AW (1993) NAD turnover and utilisation of metabolites for RNA synthesis in a reaction sensing the redox state of the cytochrome $b_6 f$ complex in isolated chloroplasts. Eur J Biochem 218: 397–404

Petracek ME, Dickey LF, Huber SC and Thompson WF (1997) Light-regulated changes in abundance and polyribosome association of ferredoxin mRNA are dependent on photosynthesis. Plant Cell 9: 2291–2300

Petracek ME, Dickey LF, Nguyen TT, Gatz C, Sowinski DA, Allen GC and Thompson WF (1998) Ferredoxin-1 mRNA is destabilized by changes in photosynthetic electron transport. Proc Natl Acad Sci USA 95: 9009–9013

Pfannschmidt T and Link G (1997) The A and B forms of plastid DNA-dependent RNA polymerase from mustard (*Sinapis alba* L.) transcribe the same genes in a different developmental context. Mol Gen Genet 257: 35–44

Pfannschmidt T, Nilsson A and Allen JF (1999a) Photosynthetic control of chloroplast gene expression. Nature 397: 625–628

Pfannschmidt T, Nilsson A, Tullberg A, Link G and Allen JF (1999b) Direct transcriptional control of the chloroplast genes *psbA* and *psaAB* adjusts photosynthesis to light energy distribution in plants. IUBMB Life 48: 271–276

Pfannschmidt T, Ogrzewalla K, Baginsky S, Sickmann A, Meyer HE and Link G (2000) The multisubunit chloroplast RNA polymerase A from mustard (*Sinapis alba* L.): Integration of a prokaryotic core into a larger complex with organelle-specific functions. Eur J Biochem 267: 253–261

Pinna LA (1997) Molecules in focus: protein kinase CK2. Int J Biochem Cell Biol 29: 551–554

Privat I, Hakimi MA, Buhot L, Favory JJ and Lerbs-Mache S (2003) Characterization of *Arabidopsis* plastid sigma-like transcription factors SIG1, SIG2 and SIG3. Plant Mol Biol 51: 385–399

Ralph ET, Guest JR and Green J (1998) Altering the anaerobic transcription factor FNR confers a hemolytic phenotype on *Escherichia coli* K12. Proc Natl Acad Sci USA 95: 10449–10452

Rauhut R and Klug G (1999) mRNA degradation in bacteria. FEMS Microbiol Lett 23: 353–370

Rochaix J-D (1997) Chloroplast reverse genetics: new insights into the function of plastid genes. Trends Plant Sci 2: 419–425

Sakai A, Suzuki T, Miyazawa Y, Kawano S, Nagata T and Kuroiwa T (1998) Comparative analysis of plastid gene expression in tobacco chloroplasts and proplastids: Relationship between transcription and transcript accumulation. Plant Cell Physiol 39: 581–589

Salvador ML and Klein U (1999) The redox state regulates RNA degradation in the chloroplast of *Chlamydomonas reinhardtii*. Plant Physiol 121: 1367–1374

Schafer FQ and Buettner GR (2001) Redox environment of the cell as viewed through the redox state of the glutathione disulfide/glutathione couple. Free Radic Biol Med 30: 1191–1212

Schrubar H, Wanner G and Westhoff P (1991) Transcriptional control of plastid gene expression in greening *Sorghum* seedlings. Planta 183: 101–111

Schuster G and Gruissem W (1991) Chloroplast mRNA 3' end processing requires a nuclear-encoded RNA-binding protein. EMBO J 10: 1493–1502

Schürmann P and Jacquot JP (2000) Plant thioredoxin systems revisited. Annu Rev Plant Physiol Plant Mol Biol 51: 371–400

Shapira M, Lers A, Heifetz PB, Irihimovitz V, Osmond CB, Gillham NW and Boynton JE (1997) Differential regulation of chloroplast gene expression in *Chlamydomonas reinhardtii* during photoacclimation: Light stress transiently suppresses synthesis of the Rubisco LSU protein while enhancing synthesis of the PS II D1 protein. Plant Mol Biol 33: 1001–1011

Shen YX, Danon A and Christopher DA (2001) RNA bindingproteins interact specifically with the Arabidopsis chloroplast psbA mRNA 5' untranslated region in a redox-dependent manner. Plant Cell Physiol 42: 1071–1078

Shiina T, Allison LA and Maliga P (1998) rbcL transcript levels in tobacco plastids are independent of light: Reduced dark transcription rate is compensated by increased mRNA stability. Plant Cell 10: 1713–1722

Stamler JS, Lamas S and Fang FC (2001) Nitrosylation: The prototypic redox-based signaling mechanism. Cell 106: 675–683

Stern DB and Gruissem W (1987) Control of plastid gene expression: 3' inverted repeats act as mRNA processing and stabilizing elements, but do not terminate transcription. Cell 51: 1145–1157

Stern DB, Higgs DC and Yang J (1997) Transcription and translation in chloroplasts. Trends Plant Sci 2: 308–315

Stone JM and Walker JC (1995) Plant protein kinase families and signal transduction. Plant Physiol 108: 451–457

Sugita M and Sugiura M (1996) Regulation of gene expression in chloroplasts of higher plants. Plant Mol Biol 32: 315–326

Sugiura M, Hirose T and Sugita M (1998) Evolution and mechanism of translation in chloroplasts. Annu Rev Genet 32: 437–459

Tell G, Scaloni A, Pellizzari L, Formisano S, Pucillo C and Damante G (1998) Redox potential controls the structure and DNA binding activity of the paired domain. J Biol Chem 273: 25062–25072

Tiller K and Link G (1993a) Sigma-like transcription factors from mustard (*Sinapis alba* L.) etioplast are similar in size to, but functionally distinct from, their chloroplast counterparts. Plant Mol Biol 21: 503–513

Tiller K and Link G (1993b) Phosphorylation and dephosphorylation affect functional characteristics of chloroplast and etioplast transcription systems from mustard (*Sinapis alba* L.). EMBO J 12: 1745–1753

Tiller K, Eisermann A and Link G (1991) The chloroplast transcription apparatus from mustard (*Sinapis alba* L.)—Evidence for three different transcription factors which resemble bacterial SIGMA factors. Eur J Biochem 198: 93–99

Toledano MB and Leonard WJ (1991) Modulation of transcription factor NF-kappaB binding activity by oxidation-reduction in vitro. Proc Natl Acad Sci USA 88: 4328–4332

Trebitsh T and Danon A (2001) Translation of chloroplast *psbA* mRNA is regulated by signals initiated by both photosystems II and I. Proc Natl Acad Sci USA 98: 12289–12294

Trebitsh T, Levitan A, Sofer A and Danon A (2000) Translation of chloroplast *psbA* mRNA is modulated in the light by counteracting oxidizing and reducing activities. Mol Cell Biol 20: 1116–1123

Trebitsh T, Meiri E, Ostersetzer O, Adam Z and Danon A (2001) The protein disulfide isomerase-like RB60 is partitioned between stroma and thylakoids in *Chlamydomonas reinhardtii* chloroplasts. J Biol Chem 276: 4564–4569

Yehudai-Resheff S, Hirsh M and Schuster G (2001) Polynucleotide phosphorylase functions as both an exonuclease and a poly(A) polymerase in spinach chloroplasts. Mol Cell Biol 21: 5408–5416

Zeilstra-Ryalls J, Gomelsky M, Eraso JM, Yeliseev A, O'Gara J and Kaplan S (1998) Control of photosystem formation in *Rhodobacter sphaeroides*. J Bacteriol 180: 2801–2809

Zerges W, Wang SW and Rochaix JD (2002) Light activates binding of membrane proteins to chloroplast RNAs in *Chlamydomonas reinhardtii*. Plant Mol Biol 50: 573–585

Zhang LX, Paakkarinen V, Van Wijk KJ and Aro E-M (2000) Biogenesis of the chloroplast-encoded D1 protein: regulation of translation elongation, insertion, and assembly into photosystem II. Plant Cell 12: 1769–1782

Zheng M, Åslund F and Storz G (1998) Activation of the OxyR transcription factor by reversible disulfide bond formation. Science 279: 1718–1721

Chapter 18

Intracellular Signaling and Chlorophyll Synthesis

Robert M. Larkin*
MSU-DOE Plant Research Laboratory and Department of Biochemistry and Molecular Biology, Michigan State University, East Lansing, MI 48824, USA

Summary		289
I.	Introduction	289
II.	Chlorophyll Biosynthetic Mutant, Inhibitor, and Feeding Studies	290
III.	Plastid-to-Nucleus Signaling Mutants Inhibit Mg-Porphyrin Accumulation	293
IV.	Mechanism of Mg-Proto/Mg-ProtoMe Signaling	294
V.	Plastid and Light Signaling Pathways Appear to Interact	297
VI.	Conclusions and Perspectives	298
Acknowledgments		296
References		298

Summary

The chloroplast proteome is encoded by genes that reside in both the chloroplast and the nucleus. This separation of genetic material necessitates a system for coordinating the expression of genes that reside in each compartment. Because the overwhelming majority of genes that encode chloroplast proteins reside in the nucleus, the regulation of nuclear genes by developmental and environmental cues plays a dominant role in chloroplast development and function. However, the chloroplast is not indifferent to its own protein composition. In fact, the chloroplast generates signals that have dramatic effects on the expression of nuclear genes that encode particular chloroplast proteins. Currently it is known that plastids produce at least a few distinct signals during chloroplast development that are required for proper expression of particular nuclear genes that encode components of the photosynthetic machinery. In response to certain environmental signals, mature chloroplasts send additional signals that regulate nuclear gene expression. The molecular nature of most of these plastid-to-nucleus signaling pathways is not well established. However, a number of studies have suggested that accumulation of certain chlorophyll precursors within plastids is a signal that regulates nuclear gene expression during chloroplast development and during the diurnal cycle. Future work in this area should provide detailed molecular information on the influence of chlorophyll synthesis and other plastid-localized metabolism on nuclear gene expression and how plants utilize this form of interorganellar communication during their lifecycles.

I. Introduction

The photoautotrophic lifestyle of plants is absolutely dependent on chloroplasts, which are the products of an endosymbiotic relationship between an ancient eukaryote and a predecessor of modern cyanobacteria. As the endosymbiotic relationship between eukaryotic and cyanobacterial cells gave rise to modern photosynthetic organisms, the overwhelming majority of the ancient cyanobacterial genome was lost or transferred to the nucleus of the eukaryotic host (Herrmann et al., 2003). Currently chloroplast genomes of higher plants are known to encode 60 to 80 proteins, and more than 3500 nuclear genes are predicted to encode chloroplast proteins in *Arabidopsis thaliana* (Martin

---

*Author for correspondence, email: larkinr@msu.edu

Abbreviations: GUN – Genomes uncoupled; Lhcb – light-harvesting chlorophyll *a/b*-binding protein of photosystem II; Mg-Proto – magnesium protoporphyrin IX; Mg-ProtoMe – magnesium protoporphyrin IX 6-monomethyl ester; Pchlide – protochlorophyllide; Proto – protoporphyrin IX; RBCS – ribulose 1,5-bisphosphate carboxylase/oxygenase small subunit.

and Hermann, 1998; Arabidopsis Genome Initiative, 2001). As expected, the nucleus plays a dominant role in chloroplast development, but the expression of nuclear genes that encode chloroplast proteins is also dependent on the functional and developmental state of the chloroplast. Chloroplasts send signals to the nucleus that are essential for proper expression of nuclear genes that encode proteins with functions related to photosynthesis, coordinating expression of chloroplast and nuclear genomes, and proper leaf morphogenesis (Rodermel, 2001; Surpin et al., 2002; Gray et al., 2003; Rodermel and Park, 2003; Reinbothe and Reinbothe, this volume). Because chloroplasts perform essential metabolic functions, it is not surprising that the functional state of the chloroplast has a dramatic influence on gene expression and developmental decisions, but little is known about the molecular mechanisms that chloroplasts use to communicate with other cellular compartments.

Mayfield and Taylor (1984) used maize seedlings in which chloroplast development was arrested at an early developmental stage to provide the first evidence that the developmental state of the chloroplast has a powerful effect on the expression of nuclear genes that encode particular chloroplast proteins. Subsequently, a number of studies have indicated that proper expression of nuclear genes that encode proteins with functions related to photosynthesis is dependent on normal chloroplast development in diverse monocotyledonous and dicotyledonous plants (Oelmüller, 1989; Gray et al., 2003; Strand et al., 2003). More recent experiments indicate that there are a number of distinct plastid-to-nucleus signaling pathways. Each of these signaling pathways is likely essential for proper metabolism under particular developmental and environmental conditions (Mochizuki et al., 2001; Rodermel et al., 2001; Mullineaux and Karpinski, 2002; Surpin et al., 2002; Gray et al. 2003; Pfannschmidt, 2003; Rodermel and Park, 2003).

The physiological significance of some chloroplast signaling pathways seems clear. For example, some pathways fine tune the expression of nuclear genes that encode components of the photosynthetic machinery to particular light environments (Pfannschmidt, 2003); other pathways induce the expression of nuclear-encoded antioxidant defense proteins as reactive oxygen species accumulate within the chloroplast (Mullineaux and Karpinski, 2002; Mullineaux et al., this volume). It is likely that plastid signaling pathways linking chloroplast development to other cellular processes are important as seedlings emerge from underneath the soil and/or ground cover and begin the transition from heterotrophic to photoautotrophic growth.

Examples of plastid-to-nucleus signaling pathways that might be important during photomorphogenesis include pathways that link leaf morphogenesis and the expression of nuclear genes that encode proteins active in photosynthesis to chloroplast development (reviewed in Oelmüller, 1989; Rodermel, 2001; Surpin et al., 2002; Gray et al., 2003; Rodermel and Park, 2003; Reinbothe and Reinbothe, this volume). Plastids appear to use different pathways for influencing leaf development and coordinating the expression of nuclear genes encoding proteins that are active in photosynthesis with chloroplast development (Rodermel, 2001). There is now substantial evidence that the cell uses more than one signal to monitor chloroplast development, and one of these signals appears to be the accumulation of particular chlorophyll precursors. In this chapter, the data supporting a role for chlorophyll precursors in intracellular communication are reviewed and current models for chlorophyll precursor signaling as well as future directions of this field are discussed.

II. Chlorophyll Biosynthetic Mutant, Inhibitor, and Feeding Studies

Tetrapyrroles are the intermediates and end products of the chlorophyll, heme, and phytochromobilin biosynthetic pathway (Fig. 1a). Tetrapyrroles are best known for their importance in metabolism, but there is precedence for heme (Fig. 1b) also functioning as a ligand for factors that regulate gene expression in yeast, animal, and bacterial cells (Chen and London, 1995; Sassa and Nagai, 1996; Zhang and Hach, 1999; O'Brian and Thony-Meyer, 2002). Mg-protoporphyrin IX monomethyl ester (Mg-ProtoMe, Fig. 1c), a chlorophyll precursor that bears a striking resemblance to heme, was first suggested to act as a regulator of nuclear gene expression in *Chlamydomonas reinhardtii* (Johanningmeier and Howell, 1984). Using mutations and inhibitors that were previously shown to affect accumulation of the chlorophyll precursor Mg-ProtoMe, these researchers reported an inverse correlation between Mg-ProtoMe accumulation and the levels of the light-harvesting chlorophyll a/b-binding protein of photosystem II (Lhcb) mRNA and ribulose 1, 5-bisphosphate carboxylase/oxygenase small subunit RBCS mRNA (Johanningmeier and Howell, 1984; Johanningmeier, 1988; Jasper et al., 1991). Protoporphyrin IX (Proto) and protochlorophyllide (Pchlide) appeared to be less effective regulators of nuclear transcription than Mg-ProtoMe (Johanningmeier and Howell, 1984).

Chapter 18 Mg-Porphyrin Signaling

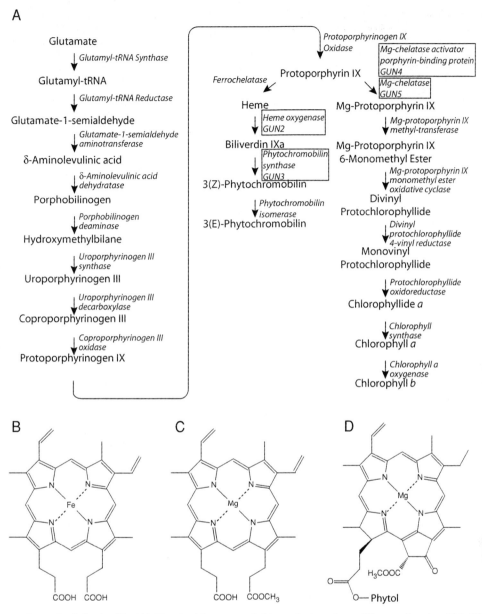

Fig. 1. Tetrapyrrole synthesis in plastids. (A) The plastid tetrapyrrole biosynthetic pathway. The reactions in which *GUN2*, *GUN3*, *GUN4*, and *GUN5* gene products participate (discussed below) are boxed. Synthesis of common precursors (i.e., glutamate through protoporphyrin IX) and the chlorophyll biosynthetic pathway are adapted from Beale (1999). Synthesis of 3(E)-phytochromobilin from protoporphyrin IX is adapted from Terry et al. (1993). (B) The structure of heme. (C) Structure of Mg-protoporphyrin IX monomethyl ester. (D) The structure of chlorophyll. Phytol is a long (C_{20}) hydrocarbon side chain.

Considering the precedence for heme-regulated gene expression, it is reasonable to propose that some chlorophyll precursors, which bear a striking resemblance to heme, might regulate the expression of particular nuclear genes in algal and plant cells. Unfortunately, serious technical difficulties have hampered efforts to examine the influence of Mg-ProtoMe and other related tetrapyrroles on the regulation of nuclear gene expression in plants and algae. For example,

Mg-ProtoMe is extremely hydrophobic and is not taken up well by plants. Also, Mg-ProtoMe and other chlorophyll precursors are excellent photosensitizing agents. Excited triplet states of these porphyrins generate reactive oxygen species via collisions with molecular oxygen, and misregulation of the chlorophyll biosynthetic pathway intermediates can cause lethal photooxidative damage. The photosensitizing properties of Mg-ProtoMe and other chlorophyll precursors are especially troubling because some of the most dramatic effects of plastid-to-nucleus signaling pathways on nuclear gene expression are directed at light-induced nuclear genes that encode proteins with functions related to photosynthesis (Oelmüller, 1989; Rodermel, 2001; Surpin et al., 2002; Gray et al., 2003; Rodermel and Park 2003; Strand et al., 2003). These genes are expressed at low levels under conditions where chlorophyll precursors are nonphotosensitizing (e.g., constant darkness; Terzaghi and Cashmore, 1995). Nonetheless, a number of researchers overcame these formidable technical barriers and subsequently reported additional experimental support for a role for particular chlorophyll precursors in the regulation of nuclear gene expression.

Kropat et al. (1997) found that chlorophyll precursor feeding does not appear to have toxic effects on *Chlamydomonas* cells in the dark. In contrast, feeding chlorophyll precursors to cells in the light causes lethal photooxidative damage. Kropat et al. (1997) further reported that feeding Mg-protoporphyrin IX (Mg-Proto) and Mg-ProtoMe to *Chlamydomonas* cells induced nuclear *HSP70* genes in the dark (Kropat et al., 1997, 2000). More detailed analysis of one *Chlamydomonas HSP70* promoter implied that Mg-Proto and Mg-ProtoMe mediated-induction of HSP70 genes required a light-responsive promoter element. Low concentrations of Mg-Proto induced *HSP70* genes and induction kinetics were rapid (Kropat et al., 1997). Interestingly, *Chlamydomonas* cells take up Proto and convert it to Mg-Proto in the chloroplast. Because Proto feeding does not induce nuclear *HSP70* gene expression in the dark, Kropat et al. (2000) proposed that Mg-Proto and Mg-ProtoMe are exported from the plastid in the light and that export of Mg-Proto and Mg-ProtoMe is required for activation of *HSP70* gene transcription.

Plants do not take up Mg-Proto and Mg-ProtoMe, but it is possible to perturb endogenous levels of these Mg-porphyrins by feeding plants δ-aminolevulinic acid (ALA), which is a small hydrophilic tetrapyrrole precursor. ALA feeding can flood the plastid tetrapyrrole biosynthetic pathway with intermediates and cause accumulation of chlorophyll precursors (Granick, 1959; Gough, 1972; Mascia, 1978). ALA feeding was reported to repress Lhcb mRNA levels in cress and in *Arabidopsis* seedlings (Kittsteiner et al., 1991; Vinti et al., 2000). In these experiments, Lhcb mRNA levels were monitored after a brief, nonphotosensitizing pulse of red light, during continuous illumination with nonphotosensitizing far-red light, or in complete darkness. Because the accumulation of plastid tetrapyrroles correlated with reduced levels of Lhcb mRNA under all of these nonphotosensitizing conditions, these experiments suggest that accumulation of Mg-porphyrins and/or other plastid-derived tetrapyrroles may regulate nuclear transcription in plants.

Some inhibitor treatments appear to more specifically promote Mg-Proto and Mg-ProtoMe accumulation compared to ALA feeding. For example, the Rüdiger laboratory found that thajaplicin and amitrole treatments that cause Mg-ProtoMe and Mg-Proto accumulation, respectively, specifically inhibit *Lhcb* and/or *RBCS* expression (Oster et al., 1996; La Rocca et al., 2001). In these experiments, Lhcb and RBCS mRNA levels were reduced after Mg-Proto or Mg-ProtoMe levels were induced in a nonphotosensitizing white light fluence, nonphotosensitizing far-red light, or complete darkness. These results further support the hypothesis that Mg-Proto/Mg-ProtoMe accumulation might affect nuclear gene expression. Because the ALA-feeding and inhibitor studies were carried out under nonphotosensitizing conditions, this work suggests that tetrapyrrole accumulation and not tetrapyrrole-induced photooxidative stress affected nuclear gene expression in these experiments. These results also imply that Mg-Proto/Mg-ProtoMe signaling may differ between plants and *Chlamydomonas* because buildup of Mg-Proto/Mg-ProtoMe in *Chlamydomonas* chloroplasts does not affect nuclear gene expression in the dark (Kropat et al., 2000), but Mg-porphyrin accumulation in plant plastids has dramatic affects on nuclear gene expression in dark (Vinti et al., 2000; La Rocca et al., 2001).

The plastid signaling studies carried out under nonphotosensitizing conditions are important to consider when interpreting other experiments that utilized the photosensitizing properties of tetrapyrroles to evaluate the influence of chloroplast development on nuclear transcription. For example, many plastid-to-nucleus signaling studies have employed norflurazon, a photobleaching herbicide. Norflurazon inhibits the synthesis of carotenoids, which quench excited triplet states of chlorophyll. Therefore, when norflurazon-treated seedlings are grown in bright light, substantial amounts of reactive oxygen species are produced, and the resulting photooxidative damage blocks chloroplast development at an early stage (Oelmüller, 1989). Cells containing these developmentally arrested plastids also exhibit severe repression of photosynthesis-related genes that reside in the nucleus. Strand et al. (2003) showed that Mg-Proto accumulation in norflurazon-treated seedlings and feeding Mg-Proto, but not porphobilinogen, protoporphyrin IX, or heme to *Arabidopsis* protoplasts results in a decrease in *Lhcb* gene expression.

The results described above provide compelling evidence that Mg-Proto and Mg-ProtoMe accumulation in plastids regulates nuclear transcription in plants and algae. But it is also important to consider whether these particular Mg-porphyrin-induced effects on gene expression might occur during normal growth and development or whether buildup of these Mg-porphyrins is specific to the unphysiological precursor feeding and inhibitor treatments. Actually, Mg-Proto and Mg-ProtoMe levels change dramatically during the diurnal cycle. Mg-Proto and Mg-ProtoMe have been reported to transiently increase 50- to 100-fold at dawn (Pöpperl et al., 1998; Papenbrock et al., 1999). When analyzed, the ALA feeding and the inhibitor treatments described above caused only a 5- to 15-fold increase in Mg-Proto levels (La Rocca et al., 2001; Strand et al., 2003). Therefore, the experimental conditions commonly used to perturb Mg-Proto and Mg-ProtoMe levels do not induce these porphyrins to accumulate above levels routinely occurring in plants during normal growth.

Mg-Proto transiently increases 5-fold and Mg-ProtoMe accumulates to a lesser degree after *Chlamydomonas* cells are transferred from the dark to the light. Therefore, a transient burst of Mg-Proto and Mg-ProtoMe at dawn appears to be conserved in both plants and algae. Moreover, this transient increase in *Chlamydomonas* Mg-Proto and Mg-ProtoMe levels appears to be essential for induction of a nuclear *HSP70* gene after the dark-to-light shift (Kropat et al., 2000). By analogy, the Mg-Proto/Mg-ProtoMe buildup at dawn might also affect nuclear transcription in plants. The *Arabidopsis cs* mutant, which contains a mutation in a gene encoding a subunit of the Mg-Proto-producing enzyme Mg-chelatase, exhibits reduced induction of *HSP70* expression after a dark-to-light shift. Surprisingly, however, the *Arabidopsis cch1* mutant, which contains a mutation in a different Mg-chelatase subunit gene, does not exhibit reduced induction kinetics of *HSP70* expression after the dark-to-light shift (Brusslan and Peterson, 2002). These results suggest that Mg-Proto might participate in *HSP70* expression in plants. However, more work will be required to confirm a role for Mg-Proto in the induction of *HSP70* gene expression in plants and to determine why *HSP70* expression is impaired in *cs* mutants but not in *cch1* mutants. Exactly how the transient peak of Mg-Proto/Mg-ProtoMe affects the plant transcriptome and whether additional bursts of Mg-Proto/Mg-ProtoMe regulate gene expression during particular phases of plant development remains to be determined.

III. Plastid-to-Nucleus Signaling Mutants Inhibit Mg-Porphyrin Accumulation

Susek and Chory (1992) developed a reporter gene-based screen in *Arabidopsis thaliana* to isolate mutants in which *Lhcb* expression is uncoupled from chloroplast development. Because the normal coordinated expression of the nuclear and chloroplast genomes is uncoupled in these mutants, they are referred to as *gun* for *genomes uncoupled*. Mutations defining 5 loci were isolated from this screen, *gun1-gun5* (Susek et al., 1993; Mochizuki et al., 2001). The recessive nature of these mutations suggests that they affect repressive signaling pathways. Double mutant studies indicate that these five *gun* mutants define two partially redundant signaling pathways: one pathway is defined by *gun1*, the other by *gun2*, *gun3*, *gun4*, and *gun5* (Mochizuki et al., 2001).

GUN2, *GUN3*, *GUN4*, and *GUN5* have been cloned, and all four of these genes are essential for proper tetrapyrrole biosynthesis in the plastid (Mochizuki et al., 2001; Larkin et al., 2003). The identity of the *GUN1* gene is not known. Besides the intracellular signaling defect, the only phenotype reported for *gun1* mutants is enhanced light sensitivity during deetiolation (Mochizuki et al., 1996). *gun1* mutants do not exhibit defects in tetrapyrrole metabolism (Vinti et al., 2000; Mochizuki et al., 2001), which is consistent with *GUN1* and *GUN2-GUN5* encoding components of two distinct signaling pathways. Microarray studies and experiments utilizing plastid translation inhibitors provide additional support for the conclusion that these five *gun* mutants define two different pathways (Gray et al., 2003; Strand et al., 2003).

gun2 and *gun3* are alleles of previously described genes, *HY1* and *HY2* (Mochizuki et al., 2001). *HY1/GUN2* encodes heme oxygenase (Davis et al., 1999; Muramoto et al., 1999), and *HY2/GUN3* encodes phytochromobilin synthase (Kohchi et al., 2001), both of which participate in phytochromobilin synthesis. Heme oxygenase and phytochromobilin synthase gene mutants in *Arabidopsis*, tomato, and pea produce lower levels of chlorophyll and chlorophyll precursors than do wild type. The reduced chlorophyll accumulation in these mutants is currently believed to result largely from feedback inhibition aimed at glutamyl-tRNA reductase, the second enzyme in the plastid tetrapyrrole biosynthetic pathway. Heme inhibits glutamyl-tRNA reductase *in vitro*, and heme appears to accumulate in heme oxygenase and phytochromobilin synthase gene mutants (Cornah et al., 2003; Franklin et al., 2003; Reinbothe and Reinbothe, this volume). Because of feedback inhibition from the heme/phytochromobilin branch of the plastid tetrapyrrole pathway, *gun2* and *gun3* were predicted to contain lower levels of Mg-Proto and Mg-ProtoMe compared to wild type (Mochizuki et al., 2001), and these predictions have been confirmed for *gun2*. *gun2* mutants accumulate lower levels of Mg-Proto than wild type when grown in standard conditions or after a Mg-Proto-inducing norflurazon treatment (Mochizuki et al., 2001; Strand et al., 2003). *GUN5* encodes the Proto/Mg-Proto-binding subunit of Mg-chelatase (Mochizuki et al., 2001), and *GUN4* is a previously uncharacterized gene that encodes a Proto/Mg-Proto-binding protein that stimulates Mg-chelatase (Larkin et al., 2003). Therefore, *gun4* and *gun5* mutants contain lower levels of chlorophyll than wild type because of reduced Mg-chelatase activity (Mochizuki et al., 2001; Larkin et al., 2003). Like *gun2* mutants, *gun5* mutants accumulate lower levels of Mg-Proto than do wild type in the presence or absence of norflurazon (Strand et al., 2003). *gun4* alleles are pale green or do not accumulate chlorophyll (Vinti et al., 2000; Mochizuki et al., 2001; Larkin et al., 2003). The paleness of these phenotypes is at least partially due to the substantial reduction or loss of the dramatic stimulatory effect of GUN4 on Mg-chelatase (Larkin et al., 2003). Therefore, it is very likely that norflurazon-treated *gun4* seedlings also contain lower Mg-Proto and Mg-ProtoMe levels than norflurazon-treated wild-type seedlings. Thus, *gun2-gun5* mutants appear to uncouple *Lhcb* expression from chloroplast development in photobleached seedlings by partially inhibiting the accumulation of Mg-Proto. Mutations affecting four additional genes that encode plastid tetrapyrrole biosynthetic enzymes or enzyme subunits were recently reported to cause *gun* phenotypes (Strand et al., 2003). However, not all chlorophyll deficient mutants are *gun* mutants and among chlorophyll deficient *gun* mutants there is no correlation between the severity of chlorophyll deficient phenotypes and *gun* phenotypes (Vinti et al., 2000; Mochizuki et al., 2001). In fact, Vinti et al. (2000) reported that although *hy1* and *gun5* mutants have strikingly different pale phenotypes and similar *gun* phenotypes, the *hy1gun5* double mutant has a severe pale phenotype but no *gun* phenotype. Also, McCormac and Terry (2002) showed that plants overexpressing protochlorophyllide oxidoreductase (POR), the penultimate enzyme in chlorophyll biosynthesis, have *gun* phenotypes. Mg-Proto/Mg-ProtoMe levels have not been determined in POR overexpressing plants, but POR overexpression is expected to reduce Mg-Proto/Mg-ProtoMe levels. It is likely that only a particular class of chlorophyll deficient mutants that prevent Mg-Proto from accumulating to a critical threshold can affect the plastid-to-nucleus signaling pathway defined by the *gun2-gun5* mutants.

The *GUN2-GUN5* pathway probably influences nuclear gene expression only during periods of Mg-Proto and Mg-ProtoMe accumulation, and it might be possible to block chloroplast development without causing Mg-Proto/Mg-ProtoMe buildup. For example, after blocking plastid development with a plastid translation inhibitor, the *gun* phenotype was observed in the *gun1* but not in the *gun2*, *gun4*, or *gun5* mutants (Gray et al., 2003). This result may be explained by the inability of plastid translation inhibitors to induce Mg-Proto/Mg-ProtoMe accumulation. Thus, this observation provides additional support for the *GUN1* gene participating in a signaling pathway that does not monitor the accumulation of specific Mg-porphyrins. Instead, the *GUN1* pathway probably senses some other molecular indicator of plastid development that is triggered by both photobleaching herbicides and plastid translation inhibitors. In fact, *gun1* mutants have previously been shown to uncouple *Lhcb* transcription from chloroplast development after chloroplast development was blocked with three mechanistically distinct approaches, including inhibition of plastid translation (Susek et al., 1993).

The *long after far-red 6* (*laf6*) mutant was identified because of defects in phytochrome A signal transduction, and *laf6* also contains elevated levels of protoporphyrin IX (Proto). Interestingly, *LAF6* encodes a soluble ATP-binding cassette protein that localizes in the chloroplast periphery. LAF6 was proposed to help retain Proto within the chloroplast, and accumulation of Proto in the cytoplasm was proposed to affect phytochrome A signal transduction in *laf6* mutants (Møller et al., 2001). Although *laf6* mutants exhibit defects in plastid tetrapyrrole metabolism, *laf6*, *gun1*, and *gun2-gun5* have been reported to affect different signaling pathways (Surpin et al., 2002).

IV. Mechanism of Mg-Proto/Mg-ProtoMe Signaling

The mechanism by which Mg-Proto/Mg-ProtoMe regulates nuclear gene expression is not understood. A popular model suggests that Mg-Proto and/or Mg-ProtoMe are exported from the plastid where they participate in a signaling pathway localized in the cytoplasm that affects gene expression in the nucleus (Kropat et al., 1997, 2000; La Rocca et al., 2001; Mochizuki et al., 2001; Møller et al., 2001; Larkin et al., 2003; Strand et al., 2003; Fig. 2). This model is

Chapter 18 Mg-Porphyrin Signaling

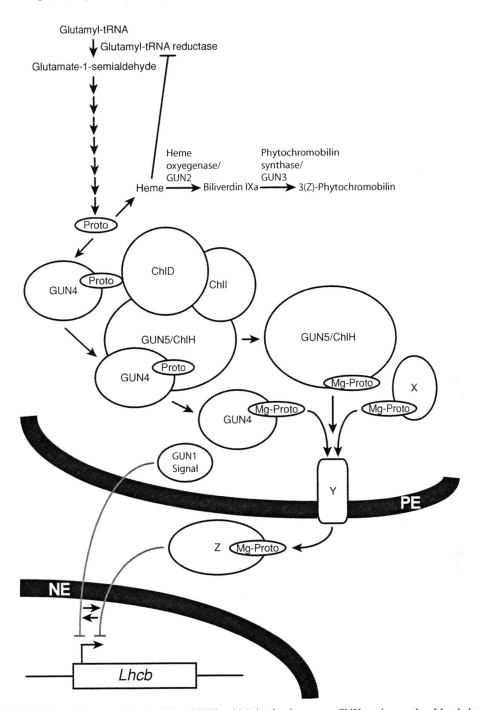

Fig. 2. Model for Mg-Proto/Mg-ProtoMe signaling. GUN5, which is also known as ChlH, and two other Mg-chelatase subunits named ChlI and ChlD insert Mg^{2+} into the Proto ring with assistance from GUN4. GUN4 binds the substrate and the product of the Mg-chelatase reaction, Proto and Mg-Proto. Heme accumulates in *gun2* and *gun3* mutants. High levels of heme repress Mg-Proto/Mg-ProtoMe synthesis by inhibiting glutamyl-tRNA reductase. Under conditions of Mg-Proto accumulation, Mg-Proto may be guided to a tetrapyrrole transporter (Y) by GUN4, GUN5, or by another Mg-Proto-binding protein (X). After export from the plastid, a cytoplasmic factor (Z) binds Mg-Proto. Factor Z affects a signaling pathway that represses *Lhcb* transcription when chloroplast development is blocked; Mg-Proto regulates the activity of factor Z. Mg-ProtoMe signaling may function in a similar manner. This Proto and Mg-Proto trafficking probably takes place in the plastid envelope, but for the sake of clarity, Proto and Mg-Proto trafficking is not shown to be associated with the envelope. The molecular nature of the pathway defined by the *gun1* mutant is not known. However, the pathways defined by the *gun1* and the other *gun* mutants, *gun2*, *gun3*, *gun4*, and *gun5*, are partially redundant. The molecular details of the interaction(s) between these two plastid-to-nucleus signaling pathways are not understood. The interaction between these two pathways is shown in the plastid, but the interaction between these two pathways may occur in another cellular compartment (e.g., the cytoplasm or the nucleus). The details of the model are explained in the text. NE, nuclear envelope; PE, plastid envelope.

attractive because the plastid appears to be the major site of cellular tetrapyrrole synthesis and other tetrapyrroles are routinely exported from plastids. For example, heme is exported from plastids (Thomas and Weinstein, 1990), and plastids are thought to be the major source of cellular heme synthesis (Cornah et al., 2002). The limited heme biosynthesis in plant mitochondria has been proposed to rely entirely on plastid-derived protoporphyrinogen IX (Beale, 1999; Brusslan and Peterson, 2002). Phytochromobilin and chlorophyll degradation products are also exported from plastids (Terry et al., 1993; Takamiya et al., 2000). This model is also attractive because, as mentioned above, there is precedence for heme functioning as a ligand for regulators of gene expression. An alternative model suggests that the cell monitors flux through the chlorophyll biosynthetic pathway and somehow, possibly via a signaling function of GUN5, uses this information to regulate nuclear gene expression (Mochizuki et al., 2001; Brusslan and Peterson, 2002). This sort of scenario might resemble the situation in *Bradyrhizobium japonicum* in which the iron response regulator Irr represses heme biosynthesis in response to catalytic activity of ferrochelatase (Qi and O'Brian, 2002). However, results from several experiments suggest the first model is more likely. For example, Kropat et al. (1997, 2000) reported that feeding Proto to *Chlamydomonas* cells in the dark causes Mg-Proto/Mg-ProtoMe buildup, presumably in chloroplasts, but feeding Proto to *Chlamydomonas* cells in the dark does not affect nuclear gene expression. Additionally, Kropat et al. (1997, 2000) reported that feeding Mg-Proto to *Chlamydomonas* cells in the dark has a dramatic affect on nuclear transcription. These results suggest that Mg-Proto must get out of the plastid to regulate gene expression in *Chlamydomonas*. Moreover, Strand et al. (2003) reported that feeding Mg-Proto to *Arabidopsis* protoplasts represses *Lhcb* expression, which indicates that Mg-Proto buildup does not necessarily have to occur within a plastid to affect gene expression in the nucleus.

The only proteins known to bind Mg-Proto in vivo are GUN4 and GUN5. Dissociation constants (K_d) for binding to Mg-deuteroporphyrin IX, a more water-soluble derivative of Mg-Proto, have been determined for *Synechocystis* relatives of GUN4 and GUN5. A K_d value of 0.26 ± 0.029 μM was estimated for a *Synechocystis* GUN4 relative (Larkin et al., 2003), and a K_d value of 2.43 ± 0.46 μM was estimated for *Synechocystis* GUN5 (Karger et al., 2001). Thus, GUN4 binds Mg-deuteroporphyrin IX and probably Mg-Proto with a higher affinity than GUN5 in *Synechocystis*. These two components of the chlorophyll biosynthetic pathway have been localized to each part of the chloroplast: stroma, thylakoids, and envelope (Gibson, et al., 1996; Nakayama et al., 1998; Larkin et al., 2003). GUN4 is monomeric in the stroma but associated with large complexes in thylakoid and envelope membranes. The level of GUN4 is extremely low in the thylakoids compared to the levels in the stroma and envelope (Larkin et al., 2003), but there is precedence for localization of chlorophyll biosynthetic enzymes in both thylakoid and envelope membranes (Block et al., 2002; Tottey et al., 2003). A small GUN4 complex was purified from *Arabidopsis* thylakoids using ion exchange and immunoaffinity chromatography; the thylakoid complex was found to contain GUN4 and GUN5 (Larkin, et al., 2003). The envelope GUN4 complex may also contain GUN5, but this has not been demonstrated. The chloroplast envelope is the expected location for proteins that participate in Mg-Proto/Mg-ProtoMe export; however, a role for GUN4 and GUN5 in Mg-Proto/Mg-ProtoMe trafficking remains to be established even though it is difficult to imagine that extremely lipophilic and photosensitizing molecules like Mg-Proto/Mg-ProtoMe diffuse freely within the cell. Because of the potentially toxic nature of these chlorophyll precursors, particular proteins should be required to target Mg-Proto and Mg-ProtoMe to particular enzymes in the chlorophyll biosynthetic pathway and prevent collisions between these Mg-porphyrins and molecular oxygen. Mg-Proto/Mg-ProtoMe binding proteins might also be very important for inhibiting unproductive degradation of these intermediates by catabolic enzymes in the plastid (Whyte and Castelfranco, 1993). Therefore, it is possible that, in addition to catalytic functions, GUN5 and/or GUN4 perform Mg-Proto/Mg-ProtoMe scavenging/photoprotective functions during Mg-Proto/Mg-ProtoMe buildup that are somewhat analogous to the scavenging/photoprotective functions performed by protochlorophyllide oxidoreductase during Pchlide accumulation (Reinbothe et al., 1996; Reinbothe and Reinbothe, this volume). In fact, a few observations indicate that GUN4 and GUN5 perform protective functions. For example, Willows et al. (2003) reported that in *Rhodobacter capsulatus*, which carries out anoxygenic photosynthesis, GUN5 forms covalent adducts with bound Proto in the presence of bright light and oxygen. This reaction inactivates Mg-chelatase and appears to play a significant role in rapidly shutting down bacteriochlorophyll synthesis in *Rhodobacter capsulatus* when oxygen and bright light are present (Willows et al., 2003). Interestingly, *Rhodobacter sphaeroides,* another species that performs anoxygenic photosynthesis, does not appear to contain relatives of GUN4 (Larkin et al., 2003). Whether *Rhodobacter capsulatus* and other species that perform anoxygenic photosynthesis contain relatives of GUN4 remains to be determined, but all species that carry out oxygenic photosynthesis appear to contain at least one GUN4 relative (Larkin et al., 2003). These observations along with the observation that *gun4-2*, a null mutant, does not accumulate chlorophyll under normal light conditions (e.g., 75 μmol m^{-2} s^{-1}) but does accumulate chlorophyll under dim light conditions (e.g., 16 μmol m^{-2} s^{-1}) are consistent with a model in which GUN4 by itself or in conjunction with other proteins (e.g., GUN5) shield Mg-Proto and possibly other porphyrins from collisions with molecular oxygen in species that carry out oxygenic photosynthesis (Larkin et al., 2003). If GUN5 and/or GUN4 turn out to participate in Mg-Proto/Mg-ProtoMe trafficking for the chlorophyll biosynthetic pathway, it seems reasonable to hypothesize that GUN5 and/or GUN4 might also participate in Mg-Proto/Mg-ProtoMe trafficking for a plastid-to-nucleus signaling pathway.

V. Plastid and Light Signaling Pathways Appear to Interact

As mentioned earlier, plastid-to-nucleus signaling pathways that couple nuclear gene expression to chloroplast development severely inhibit the expression of nuclear genes that encode proteins with functions related to photosynthesis when chloroplast development is blocked. These repressive signaling pathways mediate their effects at the transcriptional level (Oelmüller, 1989; Gray et al., 2003). Several laboratories have identified plastid-responsive promoter elements by carrying out promoter deletion studies in transgenic plants that were treated with an herbicide that blocks chloroplast development at an early developmental stage. All plastid-responsive promoters analyzed to date are also induced by light, and in many plastid-responsive promoters, plastid and light signaling pathways utilize common promoter elements (Bolle et al., 1994, 1996; Lübberstedt et al., 1994; Kusnetsov et al., 1996; Puente et al., 1996; McCormac et al., 2001; Martinez-Hernandez et al., 2002). Light and plastid signals also appear to utilize common promoter elements in *Chlamydomonas* (Kropat et al., 1997; Hahn and Kück, 1999). However, Strand et al. (2003) reported that the mutations in the CUF-1 element from the plastid-responsive *Lhcb1*1* promoter cause a *gun* phenotype strikingly similar to *gun5* (e.g., 2–4% of the promoter activity observed in green tissue; Mochizuki et al., 2002; Strand et al., 2003) in wild type seedlings treated with an herbicide that blocks chloroplast development. Moreover, CUF-1 element mutations affected *Lhcb1*1* expression similarly in wild type and *gun5* backgrounds Therefore, Strand et al. (2003) concluded that the CUF-1 element in the *Lhcb1*1* promoter responds to the Mg-Proto signal. CUF-1 (CACGTA) is a binding site for the CAB upstream factor-1 (Anderson et al., 1994) and is related to the G-box element (CACGTG), which is a binding site for bZIP transcription factors (Foster et al., 1994). CUF-1 is not essential for light regulation of *Lhcb1*1* transcription, and CUF-1 appears to only contribute to the *Lhcb1*1* promoter strength (Anderson et al., 1994; Anderson and Kay, 1995). However, G-box elements are components of promoter elements that are both light- and plastid-responsive (Martinez-Hernandez et al., 2002), and the expression of a basal promoter fragment from the nopaline synthase gene becomes dependent on chloroplast development when fused to a G-box tetramer (Puente et al., 1996). Thus, G-box elements likely contribute to the plastid-responsiveness of other promoters. On the other hand, G-box elements do not appear to perceive plastid signals for the spinach *AtpC* gene, which encodes a chloroplast ATP synthase subunit. Bolle et al. (1996) reported that site-directed mutagenesis of five nucleotides (AAAAT) near the transcription start site (-59/-55) of the *AtpC* promoter completely inhibits the transcriptional repression of the spinach *AtpC* gene that occurs when chloroplast development is blocked with a photobleaching herbicide treatment. Changing these five bases also causes the spinach *AtpC* promoter to become fully induced and unresponsive to light and tissue-specific signals. Thus, plastid-response elements may vary among plastid-responsive nuclear genes.

Although light and plastid signals utilize common promoter elements in many genes, two groups have reported that plastid signaling pathways are active in darkness, which indicates that plastid and light signals are clearly distinct. La Rocca et al. (2001) showed that plastid-derived Mg-Proto can reduce the low levels of the light-inducible RBCS mRNA that remains in darkness. Also, Sullivan and Gray (1999) showed that inhibitors of plastid development block the overexpression of light-inducible genes in the dark-grown photomorphogenic mutants *cop1* and *lip1*. Together, these data and results from plastid-responsive promoter element studies suggest that plastid signaling pathways regulate light-inducible nuclear genes that encode chloroplast proteins active in photosynthesis by interacting with light signaling pathways upstream of transcription. Plastid-derived signals have been suggested to gate the light signaling pathway (McCormack and Terry, 2001). Because the plastid signal is required for the overexpression of nuclear genes in dark-grown *cop1* and *lip1* mutants, it was suggested that the light and plastid signaling pathways might interact downstream of the COP1 and LIP1 proteins (Sullivan and Gray, 1999), which are related proteins from *Arabidopsis* and pea (Sullivan and Gray, 2000). COP1 is an ubiquitin-protein ligase that targets components of light signal transduction pathways for degradation via the ubiquitin-proteasome pathway within the nucleus (Osterlund et al., 2000; Seo et al., 2003, 2004).

VI. Conclusions and Perspectives

A number of ALA feeding, inhibitor, and genetic studies suggest that Mg-Proto and Mg-ProtoMe accumulation within the plastid represses a number of nuclear genes with functions related to photosynthesis. How Mg-Proto/Mg-ProtoMe accumulation regulates nuclear gene expression is an open question. The simplest and most attractive model is that Mg-Proto/Mg-ProtoMe exits the plastid and interacts with cytosolic signaling pathways. Whether Mg-porphyrins exit the plastid using a plastid envelope ABC transporter, which would be similar to the heme transporters in bacteria (Köster, 2001) and animal mitochondria (Shirihai et al., 2000), or some other export mechanism remains to be established. Because of the lipophilic and photosensitizing nature of Mg-Proto and Mg-ProtoMe, these molecules probably move within the plastid via carrier proteins (e.g., GUN4 and GUN5). Additional porphyrin-binding proteins are likely involved if extraplastidic transport occurs. The nature of interactions between these Mg-porphyrins and cytosolic signaling pathways is also an open question. Whether factors similar to the heme-binding transcription described in bacteria and yeast (Zhang and Hach, 1999; Qi and O'Brian, 2002) participate in the *GUN2-GUN5* plastid-to-nucleus signaling pathway remains to be determined. Proto appears to influence phytochrome A signaling, but the interaction between Proto and the phytochrome A signaling pathway does not appear to involve the *GUN* pathways. Thus, plastid tetrapyrroles may affect multiple signaling pathways. Additional plastid signaling pathways that may be important during photomorphogenesis include the pathway defined by *GUN1* and the pathway(s) that link chloroplast development to leaf morphogenesis. The molecular nature of these pathways is not clear, although the leaf morphogenesis pathways and the *GUN* pathways appear to be distinct (Rodermel, 2001). In contrast, the *GUN1* and the *GUN2-GUN5* pathways are partially redundant (Mochizuki et al., 2001), although this redundancy is not understood in molecular terms. Identification of additional genes that participate in *GUN* pathways will be critical for understanding both the molecular nature of these pathways and the interactions between them.

Probably the most important issue to resolve is whether plastid-to-nucleus signaling pathways participate in photomorphogenesis. Furthermore, do these signaling pathways influence gene expression and development at other stages of the plant life cycle and during environmental stress? Recently, Richly et al. (2003) reported that the *GUN1* and *GUN5* pathways appear to be embedded in a larger network that regulates the nuclear chloroplast transcriptome under a large number of different conditions. Thus, the *GUN* pathways may make important contributions to transcription regulation decisions throughout the plant life cycle.

Acknowledgments

Research in the laboratory of R.M.L. is supported by funds from the Department of Energy (Energy Biosciences Program).

References

Anderson SL and Kay SA (1995) Functional dissection of circadian clock- and phytochrome-regulated transcription of the *Arabidopsis CAB2* gene. Proc Natl Acad Sci USA 92: 1500–1504

Anderson SL, Teakle GR, Martino-Catt SJ and Kay SA (1994) Circadian clock- and phytochrome-regulated transcription is conferred by a 78 bp cis-acting domain of the *Arabidopsis CAB2* promoter. Plant J 6: 457–470

Arabidopsis Genome Initiative (2000) Analysis of the genome sequence of the flowering plant *Arabidopsis thaliana*. Nature 408: 796–815

Beale SI (1999) Enzymes of chlorophyll synthesis. Photosynth Res 60: 43–73

Block MA, Tewari AK, Albrieux C, Marechal E and Joyard J (2002) The plant S-adenosyl-L-methionine:Mg-protoporphyrin IX methyltransferase is located in both envelope and thylakoid chloroplast membranes. Eur J Biochem 269: 240–248

Bolle C, Sopory S, Lubberstedt T, Klosgen RB, Herrmann RG and Oelmüller R (1994) The role of plastids in the expression of nuclear genes for thylakoid proteins studied with chimeric β-glucuronidase gene fusions. Plant Physiol 105: 1355–1364

Bolle C, Kusnetsov VV, Herrmann RG and Oelmüller R (1996) The spinach *AtpC* and *AtpD* genes contain elements for light-regulated, plastid-dependent and organ-specific expression in the vicinity of the transcription start sites. Plant J 9: 21–30

Brusslan JA and Peterson MP (2002) Tetrapyrrole regulation of nuclear gene expression. Photosynth Res 71: 185–194

Chen JJ and London IM (1995) Regulation of protein synthesis by heme-regulated eIF-2 alpha kinase. Trends Biochem Sci 20: 105–108

Cornah JE, Roper JM, Pal Singh D and Smith AG (2002) Measurement of ferrochelatase activity using a novel assay suggests that plastids are the major site of haem biosynthesis in both photosynthetic and non-photosynthetic cells of pea (*Pisum sativum* L.). Biochem J 362: 423–432

Cornah JE, Terry MJ and Smith AG (2003) Green or red: what stops the traffic in the tetrapyrrole pathway? Trends Plant Sci 8: 224–230

Davis SJ, Kurepa J and Vierstra RD (1999) The *Arabidopsis thaliana HY1* locus, required for phytochrome-chromophore biosynthesis, encodes a protein related to heme oxygenases. Proc Natl Acad Sci USA 96: 6541–6546

Foster R, Izawa T and Chua N-H (1994) Plant bZIP proteins gather at ACGT elements. FASEB J 8: 192–200

Franklin KA, Linley PJ, Montgomery BL, Lagarias JC, Thomas B, Jackson SD and Terry MJ (2003) Misregulation of tetrapyrrole biosynthesis in transgenic tobacco seedlings expressing mammalian biliverdin reductase. Plant J 35: 717–728

Gibson LC, Marrison JL, Leech RM, Jensen PE, Bassham DC, Gibson M and Hunter CN (1996) A putative Mg chelatase subunit from *Arabidopsis thaliana* cv C24. Sequence and transcript analysis of the gene, import of the protein into chloroplasts, and in situ localization of the transcript and protein. Plant Physiol 111: 61–71

Gough S (1972) Defective synthesis of porphyrins in barley plastids caused by mutation in nuclear genes. Biochem Biophys Acta 286: 36–54

Granick S (1959) Magnesium porphyrins formed by barley seedlings treated with δ-aminolevulinic acid. Plant Physiol 34: XVIII

Gray JC, Sullivan JA, Wang JH, Jerome CA and MacLean D (2003) Coordination of plastid and nuclear gene expression. Philos Trans R Soc Lond B Biol Sci 358: 135–144

Hahn D and Kück U (1999) Identification of DNA sequences controlling light- and chloroplast-dependent expression of the *lhcb1* gene from *Chlamydomonas reinhardtii*. Curr Genet 34: 459–466

Herrmann RG, Maier RM and Schmitz-Linneweber C (2003) Eukaryotic genome evolution: rearrangement and coevolution of compartmentalized genetic information. Philos Trans R Soc Lond B Biol Sci 358: 87–97

Jasper F, Quednau B, Kortenjann M and Johanningmeier U (1991) Control of *cab* gene expression in synchronized *Chlamydomonas reinhardtii* cells. Photochem Photobiol 11: 139–150

Johanningmeier U (1988) Possible control of transcript levels by chlorophyll precursors in *Chlamydomonas*. Eur J Biochem 177: 417–424

Johanningmeier U and Howell SH (1984) Regulation of light-harvesting chlorophyll-binding protein mRNA accumulation in *Chlamydomonas reinhardtii*. Possible involvement of chlorophyll synthesis precursors. J Biol Chem 259: 13541–13549

Karger GA, Reid JD and Hunter CN (2001) Characterization of the binding of deuteroporphyrin IX to the magnesium chelatase H subunit and spectroscopic properties of the complex. Biochemistry 40: 9291–9299

Kittsteiner U, Brunner H and Rüdiger W (1991) The greening process in cress seedlings. II. Complexing agents and 5-aminolevulinate inhibit accumulation of cab messenger RNA coding for the light-harvesting chlorophyll a/b protein. Physiol Plant 81: 190–196

Kohchi T, Mukougawa K, Frankenberg N, Masuda M, Yokota A and Lagarias JC (2001) The *Arabidopsis HY2* gene encodes phytochromobilin synthase, a ferredoxin-dependent biliverdin reductase. Plant Cell 13: 425–436

Köster W (2001) ABC transporter-mediated uptake of iron, siderophores, heme and vitamin B12. Res Microbiol 152: 291–301

Kropat J, Oster U, Rüdiger W and Beck CF (1997) Chlorophyll precursors are signals of chloroplast origin involved in light induction of nuclear heat-shock genes. Proc Natl Acad Sci USA 94: 14168–14172

Kropat J, Oster U, Rüdiger W and Beck CF (2000) Chloroplast signalling in the light induction of nuclear *HSP70* genes requires the accumulation of chlorophyll precursors and their accessibility to cytoplasm/nucleus. Plant J 24: 523–531

Kusnetsov V, Bolle C, Lubberstedt T, Sopory S, Herrmann RG and Oelmüller R (1996) Evidence that the plastid signal and light operate via the same cis-acting elements in the promoters of nuclear genes for plastid proteins. Mol Gen Genet 252: 631–639

Larkin RM, Alonso JM, Ecker JR and Chory J (2003) GUN4, a regulator of chlorophyll synthesis and intracellular signaling. Science 299: 902–906

La Rocca N, Rascio N, Oster U and Rüdiger W (2001) Amitrole treatment of etiolated barley seedlings leads to deregulation of tetrapyrrole synthesis and to reduced expression of *Lhc* and *RbcS* genes. Planta 213: 101–108

Lübberstedt T, Oelmüller R, Wanner G and Herrmann RG (1994) Interacting cis-elements in the plastocyanin promoter from spinach ensure regulated high-level expression. Mol Gen Genet 242: 602–613

Martin W and Herrmann RG (1998) Gene transfer from organelles to the nucleus: how much, what happens, and why? Plant Physiol 118: 9–17

Martinez-Hernandez A, Lopez-Ochoa L, Arguello-Astorga G and Herrera-Estrella L (2002) Functional properties and regulatory complexity of a minimal *RBCS* light-responsive unit activated by phytochrome, cryptochrome, and plastid signals. Plant Physiol 128: 1223–1233

Mascia P (1978) An analysis of precursors accumulated by several chlorophyll biosynthetic mutants of maize. Mol Gen Genet 161: 237–244

Mayfield SP and Taylor WC (1984) Carotenoid-deficient maize seedlings fail to accumulate light-harvesting chlorophyll a/b binding protein (LHCP) mRNA. Eur J Biochem 144: 79–84

McCormac AC and Terry MJ (2002) Loss of nuclear gene expression during the phytochrome A-mediated far-red block of greening response. Plant Physiol 130: 402–414

McCormac AC, Fischer A, Kumar AM, Soll D and Terry MJ (2001) Regulation of *HEMA1* expression by phytochrome and a plastid signal during de-etiolation in *Arabidopsis thaliana*. Plant J 25: 549–561

Mochizuki N, Susek R and Chory J (1996) An intracellular signal transduction pathway between the chloroplast and nucleus is involved in de-etiolation. Plant Physiol 112: 1465–1469

Mochizuki N, Brusslan JA, Larkin R, Nagatani A and Chory J (2001) *Arabidopsis genomes uncoupled 5* (*GUN5*) mutant reveals the involvement of Mg-chelatase H subunit in plastid-to-nucleus signal transduction. Proc Natl Acad Sci USA 98: 2053–2058

Møller SG, Kunkel T and Chua N-H (2001) A plastidic ABC protein involved in intercompartmental communication of light signaling. Genes Dev 15: 90–103

Mullineaux P and Karpinski S (2002) Signal transduction in response to excess light: getting out of the chloroplast. Curr Opin Plant Biol 5: 43–48

Mullineaux PM, Karpinski S and Creissen GP (2005) Integration of signaling in antioxidant defenses. In: Demmig-Adams B, Adams WW III and Mattoo AK (eds) Photoprotection, Photoinhibition, Gene Regulation, and Environment, pp 223–239. Springer, Dordrecht

Muramoto T, Kohchi T, Yokota A, Hwang I and Goodman, HM (1999) The *Arabidopsis* photomorphogenic mutant *hy1* is deficient in phytochrome chromophore biosynthesis as a result of a mutation in a plastid heme oxygenase. Plant Cell 11: 335–348

Nakayama M, Masuda T, Bando T, Yamagata H, Ohta H and Takamiya K (1998) Cloning and expression of the soybean chlH gene encoding a subunit of Mg-chelatase and localization of the Mg^{2+} concentration-dependent ChlH protein within the chloroplast. Plant Cell Physiol 39: 275–284

O'Brian MR and Thony-Meyer L (2002) Biochemistry, regulation and genomics of heme biosynthesis in prokaryotes. Adv Microb Physiol 46: 257–318

Oelmüller, R (1989) Photooxidative destruction of chloroplasts and its effect on nuclear gene expression and extraplastidic enzyme levels. Photochem Photobiol 49: 229–239

Oster U, Brunner H and Rüdiger W (1996) The greening process in cress seedlings. V. Possible interference of chlorophyll precursors, accumulated after thujaplicin treatment, with light-regulated expression of*Lhc* genes. Photochem Photobiol 36: 255–261

Osterlund MT, Hardtke CS, Wei N and Deng XW (2000) Targeted destabilization of HY5 during light-regulated development of *Arabidopsis*. Nature 405: 462–466

Papenbrock J, Mock H-P, Kruse E and Grimm B (1999) Expression studies in tetrapyrrole biosynthesis: inverse maxima of magnesium chelatase and ferrochelatase activity during cyclic photoperiods. Planta 208: 264–273

Pfannschmidt, T (2003) Chloroplast redox signals: how photosynthesis controls its own genes. Trends Plant Sci 8: 33–41

Pöpperl G, Oster U and Rüdiger W (1998) Light-dependent increase in chlorophyll precursors during the day-night cycle in tobacco and barley seedlings. J Plant Physiol 153: 40–45

Puente P, Wei N and Deng X-W (1996) Combinatorial interplay of promoter elements constitutes the minimal determinants for light and developmental control of gene expression in *Arabidopsis*. EMBO J 15: 3732–3743

Qi Z and O'Brian MR (2002) Interaction between the bacterial iron response regulator and ferrochelatase mediates genetic control of heme biosynthesis. Mol Cell 9: 155–162

Reinbothe C and Reinbothe S (2005) Regulation of photosynthetic gene expression by the environment: from seedling de-etiolation to leaf senescence. In: Demmig-Adams B, Adams WW III and Mattoo AK (eds) Photoprotection, Photoinhibition, Gene Regulation, and Environment, pp 333–365. Springer, Dordrecht

Reinbothe S, Reinbothe C, Apel K and Lebedev N (1996) Evolution of chlorophyll biosynthesis-the challenge to survive photooxidation. Cell 86: 703–705

Richly E, Dietzmann A, Biehl A, Kurth J, Laloi C, Apel K, Salamini F and Leister D (2003) Covariations in the nuclear chloroplast transcriptome reveal a regulatory master-switch. EMBO Rep 4: 491–498

Rodermel S (2001) Pathways of plastid-to-nucleus signaling. Trends Plant Sci 6: 471–478

Rodermel S and Park S (2003) Pathways of intracellular communication: Tetrapyrroles and plastid-to-nucleus signaling. Bioessays 25: 631–636

Sassa S and Nagai T (1996) The role of heme in gene expression. Int J Hematol 63: 167–178

Seo HS, Yang JY, Ishikawa M, Bolle C, Ballesteros ML and Chua N-H (2003) LAF1 ubiquitination by COP1 controls photomorphogenesis and is stimulated by SPA1. Nature 424: 995–999

Seo HS, Watanabe E, Tokutomi S, Nagatani A and Chua N-H (2004) Photoreceptor ubiquitination by COP1 E3 ligase desensitizes phytochrome A signaling. Genes Dev In press

Shirihai OS, Gregory T, Yu C, Orkin SH and Weiss MJ (2000) ABC-me: a novel mitochondrial transporter induced by GATA-1 during erythroid differentiation. EMBO J 19: 2492–2502

Strand Å, Asami T, Alonso J, Ecker JR and Chory J (2003) Chloroplast to nucleus communication triggered by accumulation of Mg-protoporphyrinIX. Nature 421: 79–83

Sullivan JA and Gray JC (1999) Plastid translation is required for the expression of nuclear photosynthesis genes in the dark and in roots of the pea *lip1* mutant. Plant Cell 11: 901–910

Sullivan JA and Gray JC (2000) The pea *light-independent photomorphogenesis1* mutant results from partial duplication of *COP1* generating an internal promoter and producing two distinct transcripts. Plant Cell 12: 1927–1938

Surpin M, Larkin RM and Chory J (2002) Signal transduction between the chloroplast and the nucleus. Plant Cell 14: S327–S338

Susek R and Chory J (1992) A tale of two genomes: role of a chloroplast signal in coordinating nuclear and plastid genome expression. Aust J Plant Physiol 19: 387–399

Susek RE, Ausubel FM and Chory J (1993) Signal transduction mutants of *Arabidopsis* uncouple nuclear *CAB* and *RBCS* gene expression from chloroplast development. Cell 74: 787–799

Takamiya KI, Tsuchiya T, and Ohta H (2000) Degradation pathway(s) of chlorophyll: what has gene cloning revealed? Trends Plant Sci 5: 426–431

Terry MJ, Wahleithner JA and Lagarias JC (1993) Biosynthesis of the plant photoreceptor phytochrome. Arch Biochem Biophys 306: 1–15

Terzaghi WB and Cashmore AR (1995) Light-regulated transcription. Annu Rev Plant Physiol Plant Mol Biol 46: 445–474

Thomas J and Weinstein JD (1990) Measurement of heme efflux and heme content in isolated developing cotyledons. Plant Physiol 94: 1414–1423

Tottey S, Block MA, Allen M, Westergren T, Albrieux C, Scheller HV, Merchant S and Jensen PE (2003) *Arabidopsis* CHL27, located in both envelope and thylakoid membranes, is required for the synthesis of protochlorophyllide. Proc Natl Acad Sci USA 100: 16119–16124.

Vinti G, Hills A, Campbell S, Bowyer JR, Mochizuki N, Chory J and Lopez-Juez E (2000) Interactions between *hy1* and *gun* mutants of *Arabidopsis*, and their implications for plastid/nuclear signaling. Plant J 24: 883–894

Whyte BJ and Castelfranco PA (1993) Breakdown of thylakoid pigments by soluble proteins of developing chloroplasts. Biochem J 290: 361–367

Zhang L and Hach A (1999) Molecular mechanism of heme signaling in yeast: the transcriptional activator Hap1 serves as the key mediator. Cell Mol Life Sci 56: 415–426

Chapter 19

The Role of Peroxiredoxins in Oxygenic Photosynthesis of Cyanobacteria and Higher Plants: Peroxide Detoxification or Redox Sensing?

Karl-Josef Dietz*[1], Tina Stork[1], Iris Finkemeier[1], Petra Lamkemeyer[1],
Wen-Xue Li[1], Mohamed A. El-Tayeb[1,3], Klaus-Peter Michel[2],
Elfriede Pistorius[2], and Margarete Baier[1]

[1]*Biochemistry and Physiology of Plants, W5, University of Bielefeld, 33501 Bielefeld, Germany;* [2]*Molecular Cell Physiology, W6, University of Bielefeld, 33501 Bielefeld, Germany;* [3]*Present address: South Valley University, Faculty of Science, Qena, Egypt*

Summary		303
I.	Oxidative Stress	304
II.	Cyanobacteria as Model Organisms to Study Oxygenic Photosynthesis	304
	A. Detoxification of Superoxide Anion (O_2^-) in Cyanobacteria	305
	B. Detoxification of Hydrogen Peroxide (H_2O_2) in Cyanobacteria	305
	C. Peroxiredoxins in Cyanobacteria	306
III.	Peroxiredoxins in Eukaryotes and Their Subcellular Compartmentation	307
	A. Chloroplast Peroxiredoxins	308
	B. Expressional Regulation of Chloroplast Peroxiredoxin by Photosynthetic Signals	308
	C. Function of Peroxiredoxins in Photosynthesis	312
IV.	The Reaction Mechanisms of Peroxiredoxins	312
V.	Involvement of Organellar Peroxiredoxins in Stress Response	313
VI.	Peroxiredoxins in Redox Signaling	314
VII.	Conclusions	316
Acknowledgments		316
References		316

Summary

Peroxiredoxins (Prx) constitute a group of recently identified peroxidases that detoxify a broad range of peroxides in distinct subcellular compartments, including chloroplasts. They are ubiquitously expressed in all organisms, i.e. bacteria, fungi, and animals, as well as in cyanobacteria and plants, in which they frequently represent a considerable fraction of total cellular and organellar protein. At least seven *prx* genes are expressed in leaves of *Arabidopsis*. The gene products of four of them are targeted to chloroplasts. Five genes encoding (putative) Prx are found in *Synechocystis* sp. PCC 6803. Based on such circumstantial evidence, as well as biochemical analysis and observations on photosynthetic organisms with modified levels of Prx, it has been established that a subset of Prx plays a role in the context of photosynthesis. The conclusion is further strengthened by studies that showed a modulation of *prx* gene expression in response to photosynthetic activity. This chapter describes the properties of peroxiredoxins in general and focuses on the role of Prx in protecting the photosynthetic apparatus from oxidative damage and, possibly, in redox signaling in photooxygenic cells.

---

*Author for correspondence, email: karl-josef.dietz@uni-bielefeld.de

I. Oxidative Stress

Many non-enzymatic as well as enzymatic reactions in cell metabolism involve the formation of radicals, reactive oxygen species (ROS), and reactive nitrogen species (RNS) (Patel et al., 1999; Foyer and Noctor, 2000; Janssen-Heininger et al., 2002). These highly reactive compounds are also implicated in signaling (Van Breusegem et al., 2001; Vranova et al., 2002; Neill et al., 2003). To prevent damage to cell structures, ROS and RNS levels need to be tightly regulated. A delicate antioxidant network, consisting of low molecular mass antioxidants and enzymes, suppresses the accumulation of these reactive and harmful intermediates under normal growth conditions. Defense against oxidative and radical damage is particularly crucial for photosynthetic organisms. Generation of singlet oxygen and H_2O_2 in PS II (Melis, 1999), of superoxide anion radicals by PS I, and of hydrogen peroxide during photorespiration (Foyer and Noctor, 2000) are examples for ROS production associated with photosynthesis. ROS production may proceed at high rates under certain unfavorable environmental conditions, such as excess light, low temperature, drought, and limited electron acceptor availability. High levels of ROS production are frequently associated with lipid peroxidation and initiation of radical chain reactions that may lead to membrane damage and destruction (Smirnoff, 1993; Rawyler et al., 2002).

A set of antioxidant metabolites and enzymes detoxify ROS, RNS, and lipid peroxides in the various cell compartments such as chloroplasts, cytosol, mitochondria, and peroxisomes (Noctor and Foyer, 1998). Prominent antioxidant enzymes are superoxide dismutases, ascorbate peroxidases, glutathione peroxidases, and catalases (Foyer et al., 1994; Baier and Dietz, 1998). In 1996, 1-Cys and 2-Cys peroxiredoxins were identified as new players within the antioxidant network of barley (Baier and Dietz, 1996; Stacy et al., 1996). Since then, peroxiredoxins have been cloned from various photosynthetic organisms, identified in proteomes, and characterized *in vitro* and *in vivo*. Many of these findings have been summarized in recent reviews (Dietz et al., 2002; Rouhier et al., 2002; Dietz, 2003a). Recent advancements concerning function, regeneration, and regulation of peroxiredoxins and their role in photosynthetic cells, including cyanobacteria, are the focus of this review.

II. Cyanobacteria as Model Organisms to Study Oxygenic Photosynthesis

Cyanobacteria are a remarkable group of phylogenetically old prokaryotic organisms that probably evolved up to 3.5 billions years ago (Schopf, 2000) and contribute about 40 % of the present-day global photosynthetic primary biomass production (Paerl, 2000). In evolutionary terms, cyanobacteria represent the link between heterotrophic bacteria and algae or higher plants (Fay, 1983). Cyanobacteria are characterized by their ability to synthesize chlorophyll *a* (Whitton and Potts, 2000) and perform oxygenic photosynthesis (Carr and Whitton, 1982), thus driving the switch from an anoxygenic to an oxygenic atmosphere and the development of complex eukaryotic life forms. Due to their considerable morphological diversity (Whitton and Potts, 2000) and metabolic flexibility (Vermaas, 2001), cyanobacteria are able to colonize virtually all terrestrial and aquatic habitats. Cyanobacteria lack the eukaryotic type of compartmentation. Thus, oxygenic photosynthesis, respiration, and nitrogen assimilation take place in the same compartment (Schmetterer, 1994; Vermaas, 2001). Moreover, most cyanobacteria contain two distinct and fully functional respiratory chains, one located in the cytoplasmic membrane and the other in the thylakoid membrane. The latter uses, in part, the same components as the photosynthetic electron transport chain implying a complex interrelationship between respiratory and photosynthetic electron transport.

Cyanobacteria were the first photosynthetic organisms with an oxygenic type of photosynthesis, utilizing water as electron donor and producing dioxygen as a by-product. Therefore, they are the most ancient organisms to have coped with the production of reactive oxygen species (ROS) as a result of incomplete reduction of molecular oxygen via electron transport processes. Cyanobacteria have evolved effective mechanisms to prevent large-scale ROS production, such as modification of their light harvesting antenna, evolution of state transitions, changes in the stoichiometry of PS II to PS I, excitation energy dissipation by carotenoids, and, of course, enzymes for the detoxification of inevitably formed ROS (Nishiyama et al., this volume). Since the redox state of the cell and the electron transport

---

Abbreviations: aa – amino acid; ABA – abscisic acid; *Apx2* – cytosolic ascorbate peroxidase gene; Cys – cysteine; Fd – ferredoxin *katG* – catalase-peroxidase; ORF – open reading frame; *PetE* – plastocyanin gene; Prx – peroxiredoxin (protein); *prx* – peroxiredoxin (gene or transcript); PS – photosystem; RNS – reactive nitrogen species; ROS – reactive oxygen species; SOD – superoxide dismutase; Trx – thioredoxin

chain modulates many essential cellular processes, and ROS are an integral part of some signaling networks, the cellular level of the various ROS species has to be tightly controlled. This metabolic control is especially complex in an oxygenic photosynthetic prokaryote lacking the eukaryotic type of compartmentation. Moreover, it has recently been shown that the acclimatory response to oxidative stress is closely intertwined with iron homeostasis in cyanobacteria (Michel and Pistorius, 2004). It was suggested that H_2O_2 may be a master trigger that coordinates the adaptive response to oxidative stress as well as to iron starvation (Li et al., 2003; Yousef et al., 2003; Singh et al., 2004) causing a modification of PS II by the IdiA protein that leads to protection of the acceptor side of PS II in a yet unknown way, and causing a modification of PS I by IsiA protein that results in a new chlorophyll a-containing antenna around PS I trimers (see e.g. Michel and Pistorius, 2004).

Cyanobacteria may represent an ideal system to study the function of peroxiredoxins in the context of photosynthesis. In order to detoxify potentially lethal ROS originating from various sources of metabolism, cyanobacteria have evolved several enzymatic mechanisms, which in part are identical to plant enzymes, but also show some significant differences. The specific detoxification reactions and detoxifying enzyme systems in *Synechocystis sp.* PCC 6803, *Synechococcus elongatus* PCC 7942, and the closely related species *Synechococcus elongatus* PCC 6301 will be discussed.

A. Detoxification of Superoxide Anion (O_2^-) in Cyanobacteria

Superoxide anions can be generated at the acceptor side of PS I and, to a minor extent, at the acceptor side of PS II. This reactive oxygen species is efficiently scavenged by a family of enzymes, the superoxide dismutases (SOD). Superoxide dismutases are metalloenzymes catalyzing the dismutation of O_2^- to H_2O_2 and O_2 (Frausto da Silva and Williams, 1993; Kaim and Schwederski, 1995). There are three types of SODs that are distinguished by the metal iron cofactor present at the active site of the enzyme, iron (FeSOD), manganese (MnSOD), and copper/zinc (Cu/ZnSOD). All cyanobacteria so far investigated contain an FeSOD (Herbert et al., 1992; Campbell and Laudenbach, 1995) in the cytosol and some also contain a MnSOD that is associated with the thylakoid membranes (Campbell and Laudenbach, 1995; Thomas et al., 1998). In addition, Cu/ZnSOD activity has also been reported in a cyanobacterium (Chadd et al., 1996), even though Cu/ZnSOD was initially thought to be specific for eukaryotes.

The variability of SOD expression in cyanobacteria is high, indicating evolution of specific superoxide scavenging or avoiding systems. In *Synechocystis sp.* PCC 6803, few superoxide anions are generated at photosystem I (Helmann et al., 2003). In the Mehler reaction, two flavoproteins, Flv1 and Flv3, directly reduce oxygen to water (Helmann et al., 2003). Therefore, it is not surprising that bioinformatic evaluation of the genomic sequence of *Synechocystis sp.* PCC 6803 (Kashino et al., 2002) reveals the presence of a single gene encoding an SOD, *sodB* (*slr1516*) (http://www.kazusa.or.jp/cyano/cyano.html). *SodB* encodes an FeSOD, suggesting that this FeSOD is likely the sole superoxide anion-scavenging enzyme in *Synechocystis sp.* PCC 6803 (Tichy and Vermaas, 1999; Kaneko et al., 1996).

B. Detoxification of Hydrogen Peroxide (H_2O_2) in Cyanobacteria

H_2O_2, that can be formed via the disproportionation of O_2^- catalyzed by SOD (Frausto da Silva and Williams, 1993; Kaim and Schwederski, 1995) or by incomplete water oxidation in PS II (Wydrzynski et al., 1989; Samuilov, 1997), is the most stable one among the different types of ROS. However, in combination with metal ions, such as Fe^{2+}, it catalyzes the formation of highly reactive hydroxyl radicals (OH·) (Elstner, 1990; Guerinot and Yi, 1994). Early reports indicated the presence of several types of hydrogen peroxide-degrading enzyme activities. The common classification of hydrogen peroxidases distinguishes between two groups: catalases and peroxidases. Enzymes belonging to the former group catalyze the breakdown of H_2O_2 to H_2O and O_2, while peroxidases reduce H_2O_2 to H_2O with the use of a wide variety of substances as electron donors (Frausto da Silva and Williams, 1993; Kaim and Schwederski, 1995). In general, specific peroxidases are named after their reducing compound. Prokaryotes have been shown to possess a unique type of H_2O_2-detoxifying enzyme, termed catalase-peroxidase. This enzyme that has both catalase and peroxidase activity, oxidizes a broad range of electron donors, such as o-dianisidine (Regelsberger et al., 2002). Moreover, dehydroascorbate peroxidase and monodehydroascorbate peroxidase have been reported to be major H_2O_2 scavengers in chloroplasts. However, cyanobacteria contain a 250-fold lower concentration of ascorbate compared to higher plant chloroplasts (Tel-Or et al., 1985, 1986), and evaluation of the *Synechocystis* sp. PCC 6803 genome sequence did not identify homologues to those genes encoding ascorbate peroxidases (Kaneko et al., 1996). Thus, ascorbate peroxidase might not be a major player in the H_2O_2 detoxification of cyanobacteria. Nevertheless, ascorbate peroxidase activity has been reported for several cyanobacteria, including *Synechocystis* sp. PCC 6803 (Tel-Or et al., 1985, 1986). It has been reported that cyanobacteria contain millimolar concentrations of glutathione, but no glutathione peroxidase activity has been detected so far (Tel-Or et al., 1985), although two ORFs (*slr1171* and *slr1992*) with significant similarity to glutathione peroxidase genes are present in the genome of *Synechocystis* sp. PCC 6803 (Kaneko et al., 1996).

In cyanobacteria, catalase-peroxidases, encoded by the *katG* genes, were detected in *Synechococcus elongatus* PCC 7942, *Synechococcus elongatus* PCC 6301, and *Synechocystis sp.* PCC 6803, and identified as cytosolic hydrogenperoxide-scavenging hemoproteins (Mutsuda et al., 1996; Obinger et al., 1997; Jakopitsch et al., 1999). These catalase-peroxidases have little sequence homology to typical heme-containing monofunctional catalases, but exhibit similarity to yeast cytochrome *c* peroxidase and plant ascorbate peroxidase, and thus belong to the class I of the super family of peroxidases (Welinder, 1992). Characterization of a catalase-peroxidase-free *Synechocystis* sp. PCC 6803 mutant revealed a normal resistance of this mutant to H_2O_2 and methyl viologen despite an H_2O_2 decomposition rate that was 30 times lower in the mutant compared to wild type (Tichy and Vermaas, 1999). It has been suggested that the main function of the catalase-peroxidase, with its low H_2O_2 affinity and high V_{max}, is to break down external H_2O_2 entering the cell, thus providing a protective role against environmental H_2O_2 generated by algae or bacteria in the ecosystem (Tichy and Vermaas, 1999). Furthermore, it was concluded by this group that, at least under laboratory conditions, there are only two enzymatic mechanisms for H_2O_2 decomposition present in *Synechocystis* sp. PCC 6803. One is catalyzed by a catalase-peroxidase and the other is by a thiol-specific peroxidase belonging to the family of peroxiredoxin proteins.

C. Peroxiredoxins in Cyanobacteria

The involvement of peroxiredoxins in antioxidant defense in cyanobacteria has recently been verified using a *Synechocystis* sp. PCC 6803 double mutant (Nishiyama et al., 2001) lacking catalase-peroxidase and 2-Cys peroxiredoxin. This is consistent with previous suggestions by Klughammer et al. (1998) and Yamamoto et al. (1999) who analyzed 2-Cys peroxiredoxin loss-of-function mutants. When treated with H_2O_2, this double mutant showed accelerated photodamage and inhibition of photorepair, as is typical when H_2O_2-scavenging enzyme activity is absent (Nishiyama et al., 2001).

Peroxiredoxins are ubiquitously found among prokaryotes and eukaryotes (Chae et al., 1994; Baier and Dietz, 1997). They were initially termed thiol-specific antioxidant proteins (Chae et al., 1993) and later shown to be peroxidases reducing H_2O_2, alkyl hydroperoxides, and peroxinitrite (Chae et al., 1994; Bryk et al., 2000). These enzymes, which use sulfhydryl compounds for regeneration of the active site (Chae et al., 1994), are now termed peroxiredoxins. Peroxiredoxins are members of the thioredoxin-fold super family (Schröder and Ponting, 1998).

Peroxiredoxins can be divided into four different subgroups: 1-Cys Prx, 2-Cys Prx, PrxQ, and Type II Prx (Horling et al., 2002) (see below). The bioinformatic evaluation of the genome sequence of *Synechocystis* sp. PCC 6803 revealed that at least five genes encoding putative Prx are present in this cyanobacterium. Table 1 summarizes the different subgroups, molecular masses, values for the isoelectric point, and identity/similarity values for peroxiredoxins.

Only one of these five Prx has been investigated in greater detail, Slr0755 (Klughammer et al., 1998; Yamamoto et al., 1999; Nishiyama et al., 2001). This protein represents a Prx of the 2-Cys Prx type, having the two cysteine residues in highly conserved motifs (FTFVCPTEI and EVCP). Characterization of a *Synechocystis* sp. PCC 6803 mutant lacking Slr0755 revealed that growth was not perturbed under low light intensity, but was much reduced under high light conditions in comparison to the wild type (Klughammer et al., 1998). Results with a KatG-Sll0755 double mutant support these findings (Nishiyama et al., 2001). Moreover, addition of peroxides to the wild type resulted in a peroxide-dependent oxygen evolution in the light, indicating the coupling of the thioredoxin-dependent Prx activity to the photosynthetic electron transport chain (Yamamoto et al., 1999). This peroxide-dependent O_2 evolution was not observed in the mutant. This result is consistent with the function of PS I in supplying electrons to the 2-Cys Prx via ferredoxin and thioredoxin (Yamamoto et al., 1999).

A phylogenetic tree of the gene sequences of a number of cyanobacterial Prx has been published (Jin and Jeang, 2000; Regelsberger et al., 2002). Studies with a *Synechocystis* sp. PCC 6803 mutant lacking catalase-peroxidase indicate that under low H_2O_2 levels the peroxide is mainly detoxified by one or several of the five

Chapter 19 Peroxiredoxins in Photosynthesis

Table 1. Genes encoding peroxiredoxins identified in the genomes of *Arabidopsis thaliana* and *Synechocystis sp.* PCC 6803. The table gives the MIPS accession numbers and the gene numbers from Cyanobase, the lengths of the deduced gene products in number of aa residues, their molecular masses, and isoelectric points, respectively. The characteristics of gene products predicted to be targeted to organelles following processing at the predicted cleavage site are given in brackets. Identity and similarity between amino acid sequences of homologous peroxiredoxins of *Arabidopsis thaliana* and *Synechocystis* were calculated for sequence stretches of core proteins. *expression of PrxIIA has not yet been shown.

	Arabidopsis thaliana						*Synechocystis sp.* PCC 6803			
MIPS#	Length [aa-residues]	Molecular mass MM [kDa]	pI	Homologue	Identity/ Similarity [%]	Number of compared aa-residues	Protein	Length [aa-residues]	MM [kDa]	pI
At1g48130	216	24.1	6.13	1-Cys Prx	45/64	212	Slr1198	211	23.6	5.08
A3tg11630	266 (183)	29.1 (20.5)	6.91 (5.22)	2-Cys Prx A	63/82	192	Sll0755	200	22.5	4.83
At5g06290	271 (183)	29.6 (20.5)	5.55 (4.90)	2-Cys Prx B	64/82	192				
At1g65990*	553	62.7	6.06	Type II Prx A	34/61	124				
At1g65980	162	17.4	5.17	Type II Prx B	44/66	118				
At1g65970	162	17.4	5.33	Type II Prx C	43/66	118	Sll1621	189	21.2	4.94
At1g60740	162	17.4	5.33	Type II Prx D	43/66	118				
At3g52960	234 (164)	24.7 (17.3)	9.12 (5.02)	Type II Prx E	45/67	120				
At3g06050	199 (171)	21.2 (18.3)	8.98 (6.29)	Type II Prx F	35/57	162				
At3g26060	215 (159)	23.6 (17.5)	9.53 (9.36)	PrxQ	39/55	139	Sll0221	184	23.3	4.76
					33/52	147	Slr0242	160	17.6	5.18

Prx present in PCC 6803 (Tichy and Vermaas, 1999). At high H_2O_2 concentrations, the catalase-peroxidase plays a major role. This suggests that Prx have a very important function in hydrogen peroxide scavenging in *Synechocystis* sp. PCC 6803 under oxidative stress conditions involving an imbalance in the excitation of the photosynthetic electron transport chain versus energy dissipation. This also implies that cyanobacteria, at least those so far investigated in greater detail, employ a strategy for scavenging H_2O_2 generated by photosynthesis that is different from the ascorbate peroxidase pathway mainly utilized in plant chloroplasts. It should also be pointed out that cyanobacteria are a very diverse group of oxygenic photosynthetic organisms and, therefore, this might not be the case for all of them.

III. Peroxiredoxins in Eukaryotes and Their Subcellular Compartmentation

As has been well characterized in humans and yeast, peroxiredoxins are found in the cytosol, nucleus, peroxisome, and mitochondrion of animals and fungi (Table 2). The yeast genome encodes five *prx* genes (Rhee et al., 2001; Nguyen-Nhu 2003), and 6 *prx* genes have been identified in mammals (Wood et al., 2003; Table 2). These figures compare to 10 *prx* genes found in the *Arabidopsis* genome (Dietz, 2003a). Based on structural and biochemical properties, Prx are assigned to four functional groups (Hofmann et al., 2002, Horling et al., 2002), i.e. the groups of 2-Cys Prx, 1-Cys Prx, PrxQ, and type II Prx (Dietz et al., 2002).

Table 2. Compilation of peroxiredoxins found in humans, yeast, and *Arabidopsis*. The data were taken from Wood et al. (2003), Nguyên-Nhu (2003), Rhee et al. (2001), Dietz et al. (2002), and Dietz (2003a). Expression of *PrxIIA* gene has not been established.

	Yeast		Humans		Arabidopsis	
Cytosol	cTPxI	2-Cys Prx	PRDX1	2-Cys Prx	PrxIIA (?)	type II Prx
	cTPxII	2-Cys Prx	PRDX2	2-Cys Prx	PrxIIB	type II Prx
	cTPxIII	2-Cys Prx	PRDX4	2-Cys Prx	PrxIIC	type II Prx
			PRDX5	type II Prx	PrxIID	type II Prx
			PRDX6	1-Cys Prx	1-Cys Prx	1-Cys Prx
Mitochondrion	mTPx	1-Cys Prx	PRDX3	2-Cys Prx	PrxIIF	type II Prx
			PRDX5	type II Prx		
Nucleus	nTPx	PrxQ	PRDX1	2-Cys Prx	1-Cys Prx	1-Cys Prx
			PRDX5	type II Prx		
Peroxisome	cTPxIII (?)	type II Prx	PRDX5	type II Prx	– (?)	–
Chloroplast	–	–	–	–	PrxIIE	type II Prx
					2-Cys PrxA	2-Cys Prx
					2-Cys PrxB	2-Cys Prx
					PrxQ	PrxQ

Through comparison of their subcellular distribution, the assignment to the different subgroups and the differences between the organisms revealed: (1) In all species, the highest number of different Prx is present in the cytosol. (2) 2-Cys Prx constitute the predominant type of Prx in animals and yeast, whereas type II Prx represent the largest and most diverse group of Prx in *Arabidopsis*. (3) Various Prx are targeted to several subcellular compartments. For example, 1-Cys Prx of plants (Stacy et al., 1999) and the human PRDX1 (Neumann et al., 2003) have been detected in both cytosol and nucleus, respectively, and the yeast cTPxIII is suggested to be located in the cytosol as well as in peroxisomes (Lee et al., 1999). The most peculiar distribution is described for the type II Prx PRDX5 of humans that has been detected in the cytosol, mitochondria, peroxisomes, and the nucleus (Nguyên-Nhu, 2003). (4) Localization of Prx to the peroxisomes has been proven in humans. However, no unequivocal evidence for peroxisomal compartmentation exists in yeast and plants. (5) Diverse Prx are targeted to mitochondria and the nucleus. (6) The most striking feature of plant Prx is that four are targeted to chloroplasts, while the homologous proteins are cytosolic in yeast (Chae et al., 1994) and humans (Cha and Kim, 1996; Chae et al., 1999). This implies specific functions of the plant Prx in the context of photosynthesis. In fact, the increased Prx-number in *Arabidopsis* as compared to yeast and humans is fully accounted for by the four extra genes encoding chloroplast Prx.

A. Chloroplast Peroxiredoxins

All plant 2-Cys Prx analyzed so far are nuclear-encoded chloroplast proteins, as shown by immunocytochemistry, post-translational import into isolated chloroplasts, life imaging of cells expressing chimeric in-frame fusions of 2-Cys Prx with green fluorescence protein (GFP), and chloroplast proteome analysis (Dietz and Baier, 1997; Peltier et al., 2000; Genot et al., 2001; König et al., 2002). In barley chloroplasts, they have been estimated to account for 0.6% of total protein (König et al., 2003). Genomic sequence analysis of *Arabidopsis* revealed two 2-Cys Prx genes (*2-Cys prx A* and *2-Cys prx B*). The two isogenes are considered to be functionally equivalent, but differ slightly in transcript regulation (Horling et al., 2003) and, as shown in preliminary *in vitro* experiments, in regeneration efficiency seen with different thioredoxins.

Type II Prx, with N-terminal extension typical for plastid targeting, have been identified for *Arabidopsis* and rice (Bréhelin et al., 2003; Horling et al., 2003). The predicted chloroplast localization of Prx IIE was confirmed by immunochemical analysis. The sequence variation of the core protein separates the chloroplast type II Prx from other Prx (Fig. 1). On the basis of phylogenetic analysis (Fig. 2), the chloroplast type II Prx cannot be unequivocally linked to a specific group. Therefore, no suggestion on the evolutionary origin can be made. Interestingly, the highest PrxIIE promoter activity was observed in the stamen of young flowers as well as the embryo sac and albumen of older seeds (Bréhélin et al., 2003), suggesting that PrxIIE is not specifically involved in photosynthesis but in the antioxidant metabolism of plastids in general.

For plant PrxQ, as for 2-Cys Prx (Baier and Dietz, 1997), an endosymbiotic origin was suggested. Amino acid sequence alignment of PrxQ of higher plants (Kong et al., 2000), cyanobacteria, *E. coli*, and *Saccharomyces cerevisiae* (Fig. 3) demonstrates that plant and cyanobacterial homologues have several amino acid residues in common, which are not conserved in *E. coli* and *S. cerevisiae* PrxQ. In contrast to *Arabidopsis*, *Synechocystis sp.* PCC 6803 encodes two PrxQ-like proteins that are more distantly related to higher plant PrxQ than to *E. coli* PrxQ (Fig. 2).

B. Expressional Regulation of Chloroplast Peroxiredoxin by Photosynthetic Signals

Consistent with the proposed function in protection against photooxidative stress (Baier and Dietz, 1999a; Broin et al. 2002; Dietz, 2003a), transcript levels of the nuclear-encoded chloroplast Prx increase following exposure to high light or a pro-oxidant treatment in *Arabidopsis thaliana* (Horling et al., 2003), but decrease following transfer to low light or application of antioxidants (Horling et al., 2003). Direct comparison of transcript level modulation indicated a slightly greater response of *2-Cys prx A* and *prxQ* to oxidants, while *2-Cys prx B* and *prxIIE* transcript amounts decreased more strongly following antioxidant treatment (Horling et al., 2003).

Transcriptional regulation of *2-Cys prx A* was recently studied in more detail using a reporter gene approach (Baier et al., 2004). Photosynthesis was found to be a key regulator of the promoter activity. In contrast to regulation of *PetE* (Pfannschmidt et al., 2001) and *Apx2* expression (Karpinski et al., 1997), the photosynthetic redox signal involved in 2-Cys Prx A regulation is independent of the redox state of the plastoquinone pool in *Chlamydomonas* and *Arabidopsis* (Goyer et al.,

Chapter 19 Peroxiredoxins in Photosynthesis

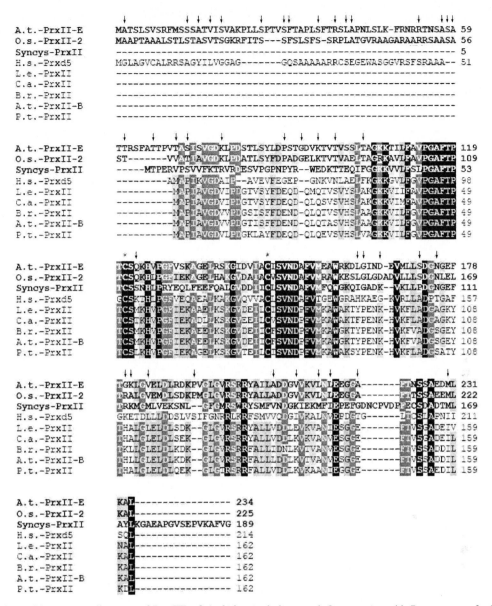

Fig. 1. Amino acid sequence alignment of Prx IIE of *Arabidopsis thaliana* and *Oryza sativa* with 7 sequences of other type II peroxiredoxins. Identical amino acid residues are shaded black, similar aa from dark to light grey. The catalytic Cys is highlighted with an asterisk. Identical amino acid residues of plastidic Prx are marked with an arrow. A.t.-PrxII-E/-B = *Arabidopsis thaliana* (At3g5290, At1g65980); O.s.-PrxII-2 = *Oryza sativa* (BAA82377); Syncys-PrxII = *Synechocystis* PCC6803 (sll 1621); H.s.-Prxd5 = *Homo sapiens* (AP001453); L.e.-PrxII = *Lycopersicon esculentum* (AAP34571.1); C.a.-PrxII = *Capsicum annuum* (AAL35363); B.r.-PrxII = *Brassica rapa* (AAD33602); P.t.-PrxII = *Populus tremula*Populus tremuloides* (AAL90751).

2002, Baier et al., 2004). The response to altered electron sink capacity of chloroplast metabolism points towards a signal associated with acceptor availability downstream of PS I (Baier et al., 2004). *2-Cys prx A*-promoter activity is strongly reduced by high CO_2 that induces a decrease in the NADPH/NADP$^+$-ratio (Dietz and Heber, 1984). The promoter shows little response to low CO_2 concentrations that do not change the redox state of NADPH/NADP$^+$ in the chloroplasts due to activation of the malate valve (Fridlyand and Scheibe, 1999). ROS play a secondary role in *2-Cys prx A* promoter regulation. Promoter activity increases only under severe (photo)-oxidative conditions, such as illumination in the cold or wounding, pathogen attack, and treatment with the plant hormone ethylene or H_2O_2 (Baier et al., 2004).

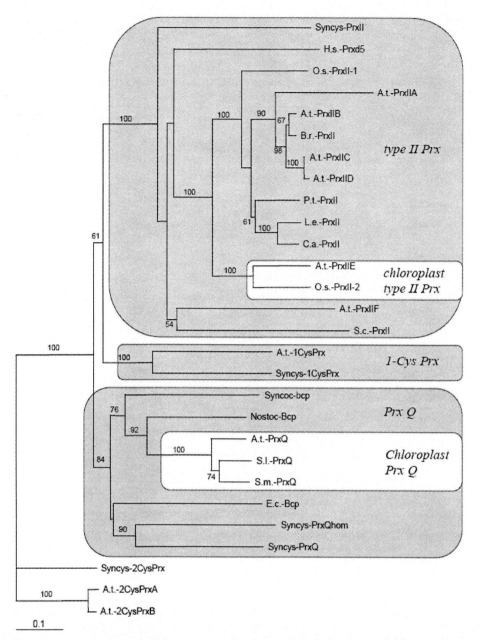

Fig. 2. Phylogenetic distance tree of type II PRX, Prx Q, and 1-Cys Prx inferred with neighbor joining algorithm of uncorrected distance matrix based on an amino acid data set (PAUP* 4.0b version 10 [Swofford, 2002]). Bootstrap values of 1000 replicates above 50 % are shown. 2-Cys Prx from *Arabidopsis* and *Synechocystis* were taken as external reference sequence. **Type II PRX:** A.t.-PrxII-A/-B/-C/-C/-D/-E/-F = *Arabidopsis thaliana* (At1g65990, At1g65980, At1g65970, At1g60740, At3g5290, At3g06050); O.s.-PrxII-1/-2 = *Oryza sativa* (AAG40130, BAA82377); B.r.-PrxII = *Brassica rapa* (AAD33602); C.a.-PrxII = *Capsicum annuum* (AAL35363); H.s.-Prxd5 = *Homo sapiens* (AP001453); L.e.-PrxII = *Lycopersicon esculentum* (AAP34571); P.t.-PrxII = *Populus tremula*Populus tremuloides* (AAL90751); S.c.-PrxII = *Saccharomyces cerevisiae* (P38013); Syncys-PrxII = *Synechocystis* PCC 6803 (sll 1621). **PRX Q:** A.t.-PrxQ = *Arabidopsis thaliana* (At3g26060); E.c.-Bcp = *Escherichia coli* (NP289033); Nostoc-Bcp = *Nostoc sp.* PCC 7120 (NP487223); S.c.-nTPx = *Saccharomyces cerevisiae* (P40553); S.l.-PrxQ = *Sedum lineare* (BAA90524); S.m.-PrxQ = *Suaeda maritima subsp. salsa* (AAQ67661); Syncoc-Bcp = *Synechococcus sp.* WH 8102 (NP897108); Syncys-PrxQ = *Synechocystis* PCC 6803 (sll0221); Syncys-PrxQhom = *Synechocystis spec* (sll0242). **1-Cys PRX:** A.t.-1-CysPrx = *Arabidopsis thaliana* (At1g48130); Syncys-1CysPrx = *Synechocystis sp.* (slr1198). **2-Cys PRX:** A.t.-2CysPrxA/-B = *Arabidopsis thaliana* (At3g11630, At5g06290); Syncys-2CysPrx = *Synechocystis* PCC 6803 (sll0755).

Chapter 19 Peroxiredoxins in Photosynthesis

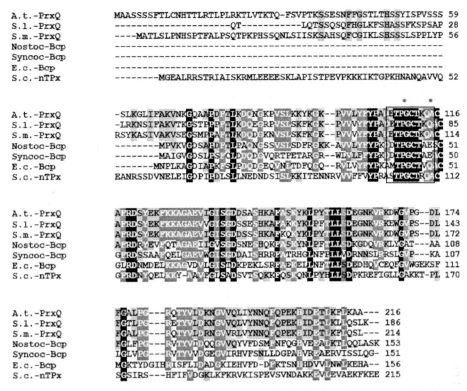

```
A.t.-PrxQ      MAASSSSFTLCNHTTLRTLPLRKTLVTKTQ-FSVPTKSSESNFFGSTLTHSSYISPVSSS  59
S.l.-PrxQ      ---------------------QT------------LQTSSQSQFHGLKFSHASSFKSPSAP  28
S.m.-PrxQ      ----MATLSLPNHSPTFALPSQTPKPHSSQNLSIISKSAHSQFCGIKLSHSSSLSPPLYP  56
Nostoc-Bcp     ------------------------------------------------------------
Syncoc-Bcp     ------------------------------------------------------------
E.c.-Bcp       ------------------------------------------------------------
S.c.-nTPx      --------MGEALRRSTRIAISKRMLEEEESKLAPISTPEVPKKKIKTGPKHNANQAVVQ  52

A.t.-PrxQ      -SLKGLIFAKVNKGQAAEDTIKDONCKPVSLKKYKGK--PVVLYFYPAIETPGCTKQAC  116
S.l.-PrxQ      -LRKNSIFAKVTKGSTPLPTIKDOEGRPVSLSKFKGK--PVVVYFYPAIETPGCTKQAC   85
S.m.-PrxQ      RSYKASIVAKVSEGSMPEATIKDODKNVSLSKFKGK--PVVVYFYPAIETPGCTKQAC  114
Nostoc-Bcp     -------MPVKVGDSAEDTIPANSSSVSISDFRGKK-AVVLYFYPKDTPGCTAESC   51
Syncoc-Bcp     --------MAIGVGDSLESCIDDODVQRTPETARGR--WLVLFYPKDTPGCTAEIC   50
E.c.-Bcp       -------MNPLKAGDIAPKISTPDODEQONITDFQGQ--RLVIYFYPKAMTPGCTVQAC   51
S.c.-nTPx      EANRSSDVNELEIGDPIEDLSFLNEDNDSISLKKITENNRVVFFVYPRASTPGCTIRQAC  112

A.t.-PrxQ      AFRDSYEKFKKAGAEVIGISGDDSAHKAFASKYKIPYIALIADEGNKVRKDWGVPG--DL  174
S.l.-PrxQ      AFRDSYEKFKKAGAEVVGISGDSSEHKAFAKKYKIPFILLIADEGNKVRKEWGVPS--DL  143
S.m.-PrxQ      AFRDSYEKFKKAGAEVIGISGDDSSHKAFAKQYKIPYIALIADEGNKVRKDWGVPS--DL  172
Nostoc-Bcp     AFRDRYEVFQTIAGABIIGVSGDSNEHQKFASKYNIPFSILLIADKGDQVLKLYGAT--AA  108
Syncoc-Bcp     GFRDSSAAFQEIGAEVIGISGDDAILHRRFITRHGINFPILAVDRNNSIIRSLGVP---KA  107
E.c.-Bcp       GLRDNMDEILKKAGVDLVGISTDKPEKLSRFAEKELNFTILSDEDHQVCEQFGVWGEKSF  111
S.c.-nTPx      GRDNIYQEIKKYAAVFGISADSVTIQKKFQSIQNIPYHILSDPKREFIGLLGAKKT--PL  170

A.t.-PrxQ      FGALPG----RQTYVILDKNGVVQLIYNNQFQPEKIDETIKFIKAA---  216
S.l.-PrxQ      FGTLPG----RETYVLDKNGVVQLVYNNQFQPEKIDETIKLIQSLK--  186
S.m.-PrxQ      FGALPG----RQTYVLDRNGVVRLVYNNQFQPEKIDETIKFIQSL---  214
Nostoc-Bcp     FGLFPG----RVTYVIDQQGVVQYVFDSMFNFQGIVEPAIKTIQQLASK  153
Syncoc-Bcp     LGLVPG----RVTYVVDGEVIRHVFSNLLDGPAIVRIAERVISSLQG-  151
E.c.-Bcp       MGKTYDGIHRISFLIDADGKIEHVFD-DKTSNIHDVVINWIKEHA--  156
S.c.-nTPx      SGSIRS----HFIFVDGKLKFKRVKISPEVSVNDAKKEVIEVAEKFKEE  215
```

Fig. 3. Amino acid sequence alignment of PrxQ. Three cDNA- or genome-derived amino acid sequences of PrxQ type peroxiredoxins of higher plants were aligned with PrxQ sequences of two cyanobacteria, yeast, and *E. coli*. AA residues conserved throughout all sequences are shaded black, aa residues conserved in some sequences are shaded grey. A.t.-PrxQ = *Arabidopsis thaliana* (At3g26060); S.l.-PrxQ = *Sedum lineare* (BAA90524); S.m.-PrxQ = *Suaeda maritima subsp. salsa* (AAQ67661); Nostoc-Bcp = *Nostoc sp.* PCC 7120 (NP487223); Syncoc-Bcp = *Synechococcus sp.* WH 8102(NP897108); E.c.-Bcp = *Escherichia coli* (NP289033); S.c.-nTPx = *Saccharomyces cerevisiae* (P40553).

Abscisic acid (ABA) further controls *2-Cys prx A* promoter activity in *Arabidopsis* (Fig. 4), and this is distinct from the ABA-dependent regulation of other antioxidant genes such as superoxide dismutase (Sakamoto et al., 1995). Studies with the ABA-biosynthetic mutants *aba2* and *aba3* (Schwartz et al., 1997) and the abscisic acid-insensitive mutants *abi1* and *abi2* (Armstrong et al., 1995) demonstrated that ABA is a strong suppressor of *2-Cys prx A* transcription. Redox signaling of *2-Cys prx A* expression depends on ABA, and it has been proposed that redox signaling is integrated into the ABA signaling cascade (Baier et al., 2004). The promoter regulation by ABA on one hand and the peroxidase function of peroxiredoxin on the other hand (König et al., 2002) suggest that the ABA-mediated suppression of *2-Cys prx A* might be important for the stabilization of the ABA-induced oxidative burst. Enzyme inactivation by overoxidation (König et al., 2003; Wood et al., 2003) and transcriptional suppression by ABA (Baier et al., 2004) could simultaneously decrease the capacity of the peroxiredoxin and thus allow peroxides to accumulate.

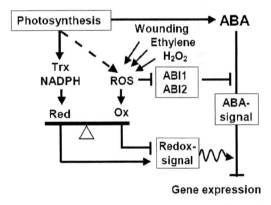

Fig. 4. Simplified scheme of promoter regulation of 2-Cys Prx by redox and ABA-signals. ABA-biosynthesis and the redox signal are under control of photosynthesis. Outside the plastids, the redox signal integrates into the ABA-signaling cascade by lowering or strengthening the suppressive ABA-signal. In addition, by ROS-mediated inhibition of ABA-antagonistic phosphatases like ABI1 and ABI2 (Meinhard and Grill, 2001; Meinhard et al., 2002) the ABA-signal is strengthened under stress conditions, which makes 2CPA-expression strongly responsive to antioxidants, but less inducible by oxidants (Baier and Dietz, 1996; Horling et al., 2003).

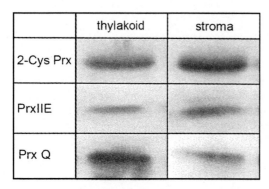

Fig. 5. Distribution of chloroplast Prx between soluble and membrane fraction. Intact chloroplasts were isolated from barley leaves and incubated for 5 min in 10 mM ascorbic acid. The chloroplasts were lysed by hypoosmotic shock. Thylakoids were separated from the soluble stroma by centrifugation. Proteins were separated by SDS-PAGE under reducing conditions and analyzed by immunoblotting with antiserum against 2-Cys Prx, PrxIIE, and PrxQ.

C. Function of Peroxiredoxins in Photosynthesis

A first indication of the function of chloroplast peroxiredoxins may be derived from their suborganellar compartmentation. Figure 5 shows that a considerable fraction of chloroplast 2-Cys Prx and PrxQ is associated with thylakoids, similar to other antioxidants like superoxide dismutase and ascorbate peroxidase (Asada, 2000). Conversely, PrxIIE is mostly recovered in the soluble (stroma) fraction of chloroplasts. As pointed out above, expression of PrxIIE is not related to photosynthetic activity.

That 2-Cys Prx protects photosynthesis was revealed by studies with transgenic plants expressing an antisense construct (Baier and Dietz, 1999b, Baier et al., 2000) and *Synechocystis* mutants with an inactivated *2-Cys prx* gene (Klughammer et al., 1998; Yamamoto et al., 1999). *Arabidopsis* expressing the barley *2-Cys prx* gene in antisense orientation under the control of the 35S-promoter showed decreased levels of 2-Cys Prx protein and retarded growth during the early phase of development. The high stability of the Prx protein allowed accumulation of Prx in older tissues despite the low level of expression. In young, 2-6 week-old seedlings, the decreased *2-Cys prx* expression was related to lower levels of photosynthetic activity, chlorophyll, and photosynthetic proteins (Baier and Dietz, 1999b). At the same time, glutathione status was unchanged, whereas ascorbate levels were increased and considerably more oxidized (Baier et al., 2000). Ascorbate is involved in H_2O_2 detoxification by ascorbate peroxidase in the Halliwell-Foyer cycle, reduction of the tocopheroxyl radical, and in the violaxanthin cycle. Decreased 2-Cys Prx apparently shifts the antioxidant burden to the ascorbate system, either by stimulation of ascorbate peroxidase or by non-enzymatic hydroperoxide detoxification by ascorbate. The Prx-based reducing mechanism is alternative to the ascorbate peroxidase-dependent pathway (Asada, 2000). The broad substrate specificity of Prx, that includes activities towards hydrogen peroxide, alkyl hydroperoxides, and peroxinitrite (Cheong et al., 1999; Choi et al., 1999; König et al., 2002, 2003; Sakamoto et al., 2003; Rouhier et al., 2004;), implies additional functions of Prx in chloroplasts. Prx are likely to be involved in reducing peroxidized thylakoid lipids. This activity may either be part of a detoxification mechanism or it may modulate liberation of signaling molecules such as oxylipins (Baier and Dietz, 1999a).

IV. The Reaction Mechanisms of Peroxiredoxins

All Prx possess a cysteinyl residue in their catalytic center within a highly conserved sequence environment. The sulfhydryl group of this Cys residue, i.e. Cys-64 in *Hordeum vulgare* 2-Cys Prx, is the primary target for the nucleophilic attack of the peroxide substrate (Fig. 6). Sequence comparisons, site directed mutagenesis, and crystal structure analysis have revealed that a catalytic triad composed of threonine, cysteine, and arginine is essential for Prx function (Alphey et al., 2000; Schröder et al., 2000). The corresponding amino acid residues in barley 2-Cys Prx are Thr-62, Cys-64, and Arg-140. The Cys64-Ser variant is inactive when assayed for enzymatic activity with either hydrogen peroxide or alkylhydroperoxide as substrates and either dithiothreitol or thioredoxin as regenerants, respectively (König et al., 2003). An additional series of residues in the catalytic center, Arg-140, Arg-163, and Trp-99, are critically important for efficient peroxide reduction (König et al., 2003).

Rouhier et al. (2004) showed that the catalytic triad in plant type II Prx from poplar consists of Thr-48, Cys-51, and Arg-129. These studies with site-directed mutagenized variants of plant Prx showed that substrate specificity can be altered by exchange of critical amino acid residues. For example, wild type poplar type II Prx showed an activity sequence with H_2O_2, t-butyl hydroperoxide, cumene hydroperoxide, and phospholipid hydroperoxide of 2.6 s^{-1}, 1.5 s^{-1}, 0.9 s^{-1}, and 2.1 s^{-1}, respectively. The Thr-48-Val variant revealed residual activities of 2.3, 0.4, 16.3, and 2.2 %, respectively, whereas Arg-129-Glu had activities of 0.3,

Chapter 19 Peroxiredoxins in Photosynthesis

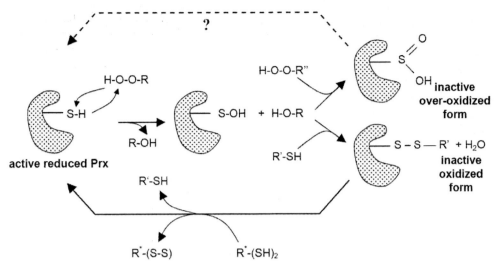

Fig. 6. Mechanism of peroxide detoxification by Prx. A peroxide substrate ROOH reacts with the catalytic sulfhydryl group that is converted to a sulfenic acid intermediate. The corresponding alcohol is liberated. The sulfenic acid is either converted to a disulfide by a 'resolving' sulfhydryl group or further oxidized to sulfinic or sulfonic acid by interaction with another oxidizing agent R"OOH. The resolving SH-group is either present on the same polypeptide chain (PrxQ, type II Prx), on another Prx subunit (2-Cys Prx), or is yet unknown (1-Cys Prx). Regeneration of reduced active catalytic sulfhydryl group is achieved by electron donors such as thioredoxin, glutaredoxin, cyclophilin, or other thiols. It appears likely that the overoxidized form can be re-reduced (question mark), however the mechanism has not yet been identified.

3.6, 7.8, and 0% of the wild type protein, respectively (Rouhier et al., 2004). There appears to be a potential of engineering differential substrate specificity in Prx. Lipid peroxides like cumene hydroperoxide rapidly inactivate type II Prx; however, the inactivation was not observed in the Thr-48 and Arg-129 variants. A similar propensity to inactivation has been described before for eukaryotic 2-Cys Prx, including the chloroplast 2-Cys Prx of barley (König et al., 2002, 2003), and will be discussed below in the context of a role of 2-Cys Prx in cellular redox signaling. The chemical basis of inactivation involves further oxidation of the intermediate sulfenic acid form of the catalytic Cys generated during the catalytic cycle to sulfinic or sulfonic acid (Wood et al., 2003). The overoxidation reaction is stimulated when peroxide concentrations are high (König et al., 2002) or when peroxide substrates with bulky alkyl groups bind to the 2-Cys Prx *in vitro* (König et al., 2003). From these observations, it is concluded that the sulfenic acid intermediate has to be rapidly converted to a disulfide, by reacting with another sulfhydryl group, in order to avoid attack of another peroxide resulting in sulfinic acid formation. Depending on the Prx group, the interacting sulfhydryl group is either within the same protein subunit (type II Prx, PrxQ), on the second homomeric subunit of the functional dimer (2-Cys Prx), or, in case of 1-Cys Prx, presented by a yet unknown donor (Rouhier et al., 2002; Dietz, 2003a). The disulfide bridge formed is reduced to dithiol via other thiol-disulfide exchange proteins such as thioredoxins, glutaredoxins, or cyclophilins (Chae et al., 1994; Lee et al., 2001; Rouhier and Jacquot, 2002). Reductive regeneration of Prx by thioredoxins was first demonstrated for yeast TpxI by Chae et al. (1994), and later thioredoxin-dependent regeneration of oxidized plant Prx was described for 2-Cys Prx from Chinese cabbage (Choi et al., 1999). In addition to Trx from *E. coli* and yeast, isoforms of chloroplast Trx-m and Trx-f regenerated 2-Cys Prx *in vitro* (König et al., 2002). In a more detailed study by Collin et al. (2003), five chloroplast Trx were compared for their efficiency in regenerating 2-Cys Prx. Trx-x reduces oxidized 2-Cys Prx most efficiently with a $k_{cat}/k_M = 7.33 \pm 0.28$ l·s/mol, whereas the k_{cat}/k_M of Trx-m2 was as low as 0.33 l·s/mol. These results demonstrate specificity in the redox network of chloroplasts. Similar differences in the regeneration efficiency of diverse chloroplast thioredoxins were observed for 2-Cys Prx A, 2-Cys Prx B, and PrxQ from *Arabidopsis thaliana*.

V. Involvement of Organellar Peroxiredoxins in Stress Response

Peroxiredoxins have been implicated in stress adaptation mainly on the basis of stress-specific stimulation of gene expression (Horling et al., 2002, 2003). Interestingly, among all Prx isoforms, the cytosolic

PrxIIC was most responsive to a set of different stresses. For example, in fully expanded leaves of *Arabidopsis*, phosphorous deficiency led to a 5-fold increase in *catalase-1* levels and an approximately 15-fold increase in *prxIIC* transcript level (Kandlbinder et al., 2004). The transcript levels of organellar peroxiredoxins also responded to redox-active effectors, metabolic constraints, and abiotic stresses, albeit mostly in a more subtle way.

Peroxiredoxins are a part of the redox network of cells, which consists of many different thiol-disulfide exchange proteins such as thioredoxins and glutaredoxins. In a screen for drought- and oxidative stress-induced proteins, Pruvot et al. (1996) identified a thioredoxin-like protein denominated CDSP32 with two thioredoxin domains and an active disulfide center in the C-terminus (Rey et al., 1998). Sensitivity towards oxidative stress increased in transgenic potato with decreased amounts of the CDSP32 (Broin et al., 2002). 2-Cys Prx was identified as an interacting partner of CDSP32 (Broin et al., 2002). Leaf discs of transgenic potato plants with decreased CDSP32 levels showed twice the amount of heat-induced chlorophyll thermoluminescence (TL) in the presence of the oxidizing reagent methyl viologen. Chlorophyll TL reflects energy transfer from excited forms of lipid peroxides to chlorophyll and heat leads to de-excitation of chlorophyll through photon emission (Broin et al., 2003). Simultaneously, the fraction of over-oxidized 2-Cys Prx was increased. The authors concluded that CDSP32 is the electron donor to 2-Cys Prx, and that this system is a critical component in the defense against lipid peroxidation in photosynthetic membranes. In the same study, the authors observed no change in 2-Cys Prx amounts under drought, whereas 2-Cys Prx levels decreased in plants with lowered CDSP32 levels. In another study, plants differing in drought sensitivity were maintained under water-limited conditions for 14 days. The field capacity of the soil was maintained at 40%, which is close to permanent wilting. 2-Cys Prx levels were unchanged, and no differences were observed between highly sensitive and less sensitive *Vicia* and *Triticum* varieties (Fig. 7). PrxQ was slightly up-regulated under drought and showed some correlation with drought sensitivity. The fraction of monomeric 2-Cys Prx increased in wilted plants, suggesting overoxidation of the catalytic Cys-residue to sulfinic or sulfonic acid (Broin et al., 2003). Likewise, König et al. (2002) reported on the appearance of oligomerized 2-Cys Prx under salt stress, severe drought, and aging. Overoxidized 2-Cys Prx are known to oligomerize similar to the reduced form (Wood et al., 2003). The decamer attaches to the thylakoid membrane, where it could act as a signal as proposed below (König et al., 2003). The fate of overoxidized 2-Cys Prx is not clear at present. It may be degraded, or a yet unknown reduction mechanism could enzymatically regenerate active thiols in the active sites.

VI. Peroxiredoxins in Redox Signaling

Reduction of toxic peroxides appears to be the principal activity and direct function of Prx. Either through this activity or by other mechanisms, Prx are suggested to play a role in ROS and RNS signaling. Three scenarios are discussed below:

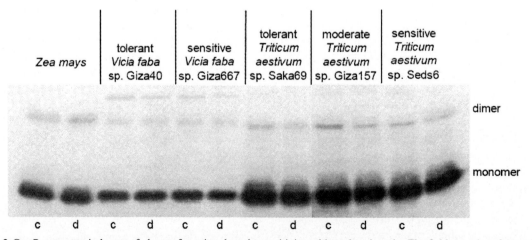

Fig. 7. 2-Cys Prx amount in leaves of plants of varying drought sensitivity subjected to drought. The field capacity of 4 weeks old plants was adjusted to 40% for 2 weeks. The 2-Cys Prx amounts in control (c) and droughted (d) leaves of *Zea maize*, tolerant and sensitive *Vicia faba* lines, and tolerant, moderate, and sensitive *Triticum aestivum* lines were analyzed, respectively. The figure shows consistent 2-Cys Prx content in response to drought independent of the level of tolerance.

Chapter 19 Peroxiredoxins in Photosynthesis

(i) Prx modulate ROS and RNS levels

Catalytic activities have only been studied for some plant Prx. Barley 2-Cys Prx has a k_{cat} of 0.23 [s^{-1}] and a catalytic efficiency k_{cat}/K_m of $1.1 \cdot 10^5 M^{-1}s^{-1}$ (König et al., 2003). k_{cat} of type II Prx varied between 0.9 s^{-1} for cumene hydroperoxide and 2.6 s^{-1} for H$_2$O$_2$ with catalytic efficiency in the range of $8 \cdot 10^4 M^{-1}s^{-1}$ for phosphatidylcholine hydroperoxide (Rouhier et al., 2004). Thus, Prx are classified as good catalysts with intermediate activities that are slightly lower than the activities of enzymes like ascorbate peroxidase and considerably lower than those of superoxide dismutases and catalases (Asada, 1999). It should be noted that it is not clear whether the most efficient and natural regenerants have been identified thus far. Different thioredoxins vary in their efficiency to regenerate oxidized 2-Cys Prx (Collin et al., 2003). However, since 2-Cys Prx is an abundant protein in the chloroplast, even an intermediate catalytic efficiency will affect the steady state concentration of ROS and RNS in the chloroplast and, thereby, modulate ROS- and RNS-dependent signaling (Dietz, 2003b; Neill et al., 2003; Pfannschmidt, 2003). Protein levels of PrxIIE and PrxQ appear to be lower than those of 2-Cys Prx. However, the catalytic efficiency of PrxQ is almost 100-fold higher than that of 2-Cys Prx. This fact supports the view that PrxQ also significantly contributes to antioxidant defense in chloroplasts. Regulation of Prx activities will affect redox signaling. The functional unit of 2-Cys Prx is the dimer that oligomerizes to decamers, i.e. five dimers form a ring-like structure (Wood et al., 2003). Oligomerization has also been observed for plant 2-Cys Prx under high salt and reducing conditions (König et al., 2002, 2003). At elevated peroxide concentration, 2-Cys Prx are over-oxidized and also oligomerize (Fig. 8). Thus, there is a cycling between membrane-associated reduced Prx and soluble oxidized Prx. Recently, Wood et al. (2002) suggested that 2-Cys Prx might function as a floodgate in redox signaling. Beyond a certain threshold of peroxide concentration, the 2-Cys Prx is over-oxidized. Due to this inactivation, Prx cannot detoxify ROS or RNS any longer, and ROS- and RNS-mediated signal transduction may proceed at maximum efficiency (Wood et al., 2002). The scenario was developed for 2-Cys Prx from non-plant

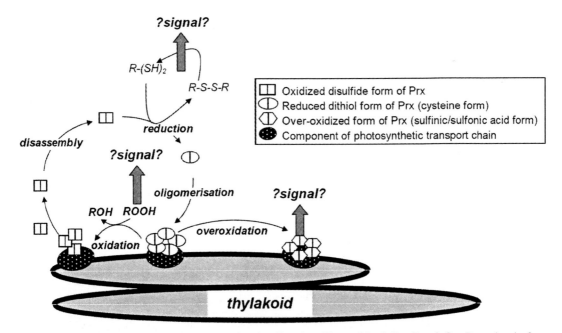

Fig. 8. Regulatory cycle of 2-Cys Prx and possible signals affected or liberated by 2-Cys Prx. 2-Cys Prx exists in four states, i.e. oxidized dimer, reduced dimer, reduced oligomer, and over-oxidized oligomer. Redox-dependent signaling may be initiated by affecting the free peroxide concentration, by kinetically tuning the redox state of another dithiol/disulfide component that in turn modulates downstream events, or by the distribution of Prx between thylakoid and stroma. The overoxidized form may act as a static redox sensor (Dietz, 2003b). See text for further details.

eukaryotes and may also apply to plant 2-Cys Prx.

(ii) Prx sense ROS and may kinetically transmit information to other regulatory elements
Following peroxide reduction, 2-Cys Prx need to be regenerated prior to the next catalytic cycle. Various regenerants have been shown to function in 2-Cys Prx reduction, namely thioredoxins, glutaredoxins, and cyclophilins (Chae et al., 1994; Lee et al., 2001; Rouhier et al., 2002). Assuming a bifunctional role of the regenerant, i.e. in reducing both Prx and another possibly signal transduction element, the rate of electron drainage from the regenerant by Prx reduction may transmit information to other signaling pathways. In this case, Prx may act as a kinetic peroxide sensor without necessarily functioning in substantial peroxide detoxification (Fig. 8) (Hofmann et al., 2002; Dietz, 2003a, b).

(iii) Prx function as static ROS sensors
A third role in redox sensing may be hypothesized on the basis of the dimer-oligomer transition (Fig. 8). Both reduced and over-oxidized 2-Cys Prx bind to the thylakoid membrane (König et al., 2003). The nature of the binding site or sites is unknown. The reduction versus over-oxidation state affects complex assembly and thereby could trigger signaling events.

VII. Conclusions

Taken together, the comparison of plant and cyanobacterial peroxiredoxin gene families and the more advanced knowledge on Prx function in higher plants indicate a significant role of peroxiredoxins in photosynthesis. It is hypothesized that 2-Cys Prx and PrxQ evolved in the functional context of oxygenic photosynthesis and were brought into the plant cell by endosymbiosis. In addition to their role in detoxifying hydrogen peroxide, alkylhydroperoxide, and peroxinitrite, a function in redox sensing and signaling appears likely. Experiments with plants and cyanobacteria expressing either modified levels of Prx or site-directed mutagenized variants of Prx should provide answers to the question of redundancy versus specificity of physiological function. Furthermore, proteins interacting with Prx, for example the binding sites for 2-Cys Prx und PrxQ at the thylakoid membrane, need to be identified to develop a model of Prx involvement in signal transduction and site-specific detoxification. Another aspect for further investigation relates to the role of peroxiredoxins in the development of antioxidant defense during evolution. Towards this end, interesting clues can be expected from the increasing number of completed genomic sequences of various cyanobacteria and green algae.

Acknowledgments

Our work reported in this review was funded by the Deutsche Forschungsgemeinschaft (Di 346/6 and FOR 387, TP3, TP7, and TP10). We are grateful to Ingo Busse (University of Bielefeld) for help in constructing the phylogenetic tree of selected Prx. W.X. Li acknowledges support by the DAAD, and M. El-Tayeb by the DFG.

References

Alphey MS, Bond CS, Tedaud E, Fairlamb AH and Hunter WN (2000) The structure of reduced tryparedoxin peroxidase reveals a decamer and insight into reactivity of 2-Cys peroxiredoxins. J Mol Biol 300: 903–916

Armstrong F, Leung J, Grabov A, Brearley J, Giraudat J and Blatt MR (1995) Sensitivity of abscisic acid of guard cell K^+ channels is suppressed by abi1-1, a mutant Arabidopsis gene encoding a putative protein phosphatase. Proc Natl Acad Sci USA 92: 9520–9524

Asada K (1999) The water-water cycle in chloroplasts: Scavenging of active oxygen and dissipation of excess photons. Annu Rev Plant Physiol Plant Mol Biol 50: 601–639

Asada K (2000) The water-water cycle as alternative photon and electron sinks. Philos Trans Roy Soc Lond 355: 1419–1432

Baier M and Dietz K-J (1996) Primary structure and expression of plant homologues of animal and fungal thioredoxin-dependent peroxide reductases and bacterial alkyl hydroperoxide reductases. Plant Mol Biol 31: 553–564

Baier M and Dietz K-J (1997) The plant 2-Cys peroxiredoxin BAS1 is a nuclear-encoded chloroplast protein: its expressional regulation, phylogenetic origin, and implications for its specific physiological function in plants. Plant J 12: 179–190

Baier M and Dietz K-J (1998) The costs and benefits of oxygen for photosynthesizing cells. Prog Bot 60: 282–314

Baier M and Dietz K-J (1999a) Alkylhydroperoxide reductases: the way out of oxidative breakdown of lipids in chloroplasts. Trends Plant Sci 4: 166–168

Baier M and Dietz K-J (1999b) Protective function of chloroplast 2-Cysteine peroxiredoxin in photosynthesis. Evidence from transgenic Arabidopsis. Plant Physiol 119: 1407–1414

Baier M, Noctor G, Foyer CH and Dietz K-J (2000) Antisense suppression of 2-Cys peroxiredoxin in Arabidopsis specifically enhances the activities and expression of enzymes associated with ascorbate metabolism but not glutathione metabolism. Plant Physiol 124: 823–832

Baier M, Ströher E and Dietz K-J (2004) The acceptor availability at photosystem I and ABA control nuclear expression of 2-Cys peroxiredoxin-A in *Arabidopsis thaliana*. Plant Cell Physiol 45: 997–1006

Bréhélin C, Meyer EH, de Souris JP, Bonnard G and Meyer Y (2003) Resemblance and dissemblance of Arabidopsis type II peroxiredoxins: Similar sequences for divergent gene expression, protein localization, and activity. Plant Physiol 132: 2045–2057

Broin M and Rey P (2003) Potato plants lacking the CDSP 32 plastidic thioredoxin exhibit overoxidation of the BAS1 2-Cysteine peroxiredoxin and increased lipid peroxidation in thylakoid under photooxidative stress. Plant Physiol 132: 1335–1343

Broin M, Cuiné S, Eymery F and Rey P (2002) The plastidic 2-Cysteine peroxiredoxin is a target for a thioredoxin involved in the protection of the photosynthetic apparatus against oxidative damage. Plant Cell 14: 1417–1432

Bryk R, Griffin P and Nathan C (2000) Peroxinitrite reductase activity of bacterial peroxiredoxins. Nature 407: 211–215

Campbell WS and Laudenbach DE (1995) Characterization of four superoxide dismutase genes from a filamentous cyanobacterium. J Bacteriol 177: 964–972

Carr NG and Whitton BA (1982) The Biology of Cyanobacteria. Blackwell Scientific Publishers, Oxford, UK

Cha M-K and Kim I-H (1996) Thioredoxin-linked peroxidase from human red blood cell: evidence for the existence of thioredoxin and thioredoxin reductases in human red blood cell. Biochem Biophys Res Commun 217: 900–907

Chadd HE, Newman J, Mann NH and Carr NG (1996) Identification of iron superoxide dismutase and a copper/zinc superoxide dismutase enzyme activity within the marine cyanobacterium *Synechococcus sp.* WH 7803. FEMS Microbiol Lett 138: 161–165

Chae HZ, Kim IH, Kim K and Rhee SG (1993) Cloning, sequencing, and mutation of thiol-specific antioxidant gene of *Saccharomyces cerevisiae*. J Biol Chem 268: 16815–16821

Chae HZ, Chung SJ and Rhee SG (1994) Thioredoxin-dependent peroxide reductase from yeast. J Biol Chem 269: 27670–27678

Chae HZ, Kang SW and Rhee SG (1999) Isoforms of mammalian peroxiredoxin that reduce peroxides in presence of thioredoxin. Methods Enzymol 300: 219–226

Cheong NE, Choi YO, Lee KO, Kim WY, Jung BG, Chi YH, Jeong JS, Kim K, Cho MJ and Lee SY (1999) Molecular cloning, expression, and functional characterization of a 2-Cys-peroxiredoxin in Chinese cabbage. Plant Mol Biol 40: 825–834

Choi YO, Cheong NE, Lee KO, Jung BG, Hong CH, Jeong JH, Chi YH, Kim K, Cho MJ and Lee SY (1999) Cloning and expression of a new isotype of the peroxiredoxin gene in Chinese cabbage and its comparison to 2-Cys peroxiredoxin isolated from the same plant. Biochem Biophys Res Commun 258: 768–771

Collin V, Issakidis-Bourguet E, Marchand C, Hirasawa M, Lancelin J-M, Knaff DB and Miginiac-Maslow M (2003) The Arabidopsis plastidial thioredoxins: New functions and new insights into specificity. J Biol Chem 278: 23747–23752

Dietz K-J (2003a) Plant peroxiredoxins. Annu Rev Plant Biol 54: 93–107

Dietz K-J (2003b) Redox regulation, redox signalling and redox homeostasis in plants. Int Rev Cytol 228, 111–193

Dietz K-J and Heber U (1984) Rate-limiting factors in leaf photosynthesis. I. Carbon fluxes in the Calvin cycle. Biochim Biophys Acta 767: 432–443.

Dietz K-J, Horling F, Konig J and Baier M (2002) The function of the chloroplast 2-cysteine peroxiredoxin in peroxide detoxification and its regulation. J Exp Bot 53: 1321–1329

Elstner EF (1990) Der Sauerstoff. Biochemie, Biologie, Medizin. Spektrum Akademischer Verlag, Heidelberg, Germany

Fay P (1983) The Blue-Greens. Edward Arnold, Baltimore, USA

Foyer CH and Noctor G (2000) Oxygen processing in photosynthesis: regulation and signalling. New Phytol 146: 359–388

Foyer CH, Lelandais M and Kunert K-J (1994) Photooxidative stress in plants. Physiol Plant 92: 696–717

Frausto da Silva JJR and Williams RJP (1993) The Biological Chemistry of the Elements. The Inorganic Chemistry of Life. Clarendon Press, Oxford

Fridlyand LE and Scheibe R (1999) Controlled distribution of electrons between acceptors in chloroplasts: a theoretical consideration. Biochim Biophys Acta 1413: 31–42

Genot G, Wintz H, Houlné G and Jamet E (2001) Molecular characterization of a bean chloroplastic 2-Cys peroxiredoxin. Plant Physiol Biochem 39: 449–459

Goyer A, Haslekas C, Miginiac-Maslow M, Klein U, Le Marechal P, Jacquot JP and Decottignies P (2002) Isolation and characterization of a thioredoxin-dependent peroxidase from *Chlamydomonas reinhardtii*. Eur J Biochem 269: 272–282

Guerinot ML and Yi Y (1994) Iron: Nutritious, noxious, and not readily available. Plant Physiol 104: 815–820

Helman Y, Tchernov D, Reinhold L, Shibata M, Ogawa T and Kaplan A (2003) Genes encoding A-type flavoproteins are essential for photoreduction of O_2 in cyanobacteria. Curr Biol 13: 230–235

Herbert SK, Samson G, Fork DC and Laudenbach DE (1992) Characterization of damage to photosystems I and II in a cyanobacterium lacking detectable iron superoxide dismutase activity. Proc Natl Acad Sci USA 89: 8716–8720

Hofmann B, Hecht HJ and Flohé L (2002) Peroxiredoxins. Biol Chem 383: 347–364

Horling F, König J and Dietz K-J (2002) Type II peroxiredoxin C, a member of the peroxiredoxin family of *Arabidopsis thaliana*: Its expression and activity in comparison with other peroxiredoxins. Plant Physiol Biochem 40: 491–499

Horling F, Lamkemeyer P, König J, Finkemeier I, Kandlbinder A, Baier M and Dietz K-J (2003) Divergent light-, ascorbate- and oxidative stress-dependent regulation of expression of the peroxiredoxin gene family in *Arabidopsis thaliana*. Plant Physiol 131: 317–325

Jakopitsch C, Ruker F, Regelsberger G, Dockal M, Peschek GA and Obinger C (1999) Catalase-peroxidase from the cyanobacterium *Synechocystis* PCC 6803: cloning, overexpression in *Escherichia coli*, and kinetic characterization. Biol Chem 380: 1087–1096

Janssen-Heininger YMW, Persinger RL, Korn SH, Pantano C, McElhinney B, Reynaert NL, Langen RCJ, Ckless K, Shrivastava P and Poynter ME (2002) Reactive nitrogen species and cell signaling – Implications for death or survival of lung epithelium. Am J Resp Crit Care Med 166: 9–16

Jin D-Y and Jeang K-T (2000) Peroxiredoxins in cell signalling and HIV infection. In: Sen CK, Sies H and Baeuerle PA (eds) Antioxidants and Redox Regulation of Genes, pp 382–409. Academic Press, San Diego

Kaim W and Schwederski B (1995) Bioanorganische Chemie. Teubner, Stuttgart

Kandlbinder A, Finkemeier I, Wormuth D, Hanitzsch M and Dietz K-J (2004) The antioxidant status of photosynthesising leaves under nutrient deficiency: redox regulation, gene expression and antioxidant activity in *Arabidopsis thaliana*. Physiol Plant 120: 63–73

Kaneko T, Sato S, Kotani H, Tanaka A, Asamizu E, Nakamura Y, Miyajima N, Hirosawa M, Sugiura M, Sasamoto S, Kimura T, Hosouchi T, Matsuno A, Muraki A, Nakazaki N, Naruo K, Okumura S, Shimpo S, Takeuchi C, Wada T, Watanabe A, Yamada M, Yasuda M and Tabata S (1996) Sequence analysis of the genome of the unicellular cyanobacterium *Synechocystis sp.* strain PCC 6803. II. Sequence determination of the entire genome and assignment of potential protein-coding regions (supplement). DNA Res 3: 185–209

Karpinski S, Escoubar C, Karpinska B, Creissen G and Mullineaux PM (1997) Photosynthetic electron transport regulates the expression of cytosolic ascorbate peroxidase genes in Arabidopsis during excess light stress. Plant Cell 9: 627–640

Kashino Y, Koike H, Yoshio M, Egashira H, Ikeuchi M, Pakrasi HB and Satoh K (2002) Low-molecular-mass polypeptide components of a photosystem II preparation from the thermophilic cyanobacterium *Thermosynechococcus vulcanus*. Plant Cell Physiol 43: 1366–1373

Klughammer B, Baier M and Dietz K-J (1998) Inactivation by gene disruption of 2-Cysteine-peroxiredoxin in *Synechocystis sp.* PCC 6803 leads to increased stress sensitivity. Physiol Plant 104: 699–706

Kong W, Shiota S, Shi YX, Nakayama H and Nakayama K (2000). A novel peroxiredoxin of the plant *Sedum lineare* is a homologue of the *Escherichia coli* bacterioferritin co-migratory protein (Bcp). Biochem J 351: 107–114

König J, Baier M, Horling F, Kahmann U, Harris G, Schürmann P and Dietz K-J (2002) The plant-specific function of 2-Cys peroxiredoxin-mediated detoxification of peroxides in the redox-hierarchy of photosynthetic electron flux. Proc Natl Acad Sci USA 99: 5738–5743

König J, Lotte K, Plessow R, Brockhinke A, Baier M and Dietz K-J (2003) Reaction mechanism of plant 2-Cys peroxredoxin: Role of the C terminus and the quaternary structure. J Biol Chem 278: 24409–24420

Lee J, Spector D, Godon C, Labarre J and Toledano MB (1999) A new antioxidant with alkyl hydroperoxide defense properties in yeast. J Biol Chem 274: 4537–4544

Lee SP, Hwang YS, KimYJ, Kwon KS, Kim HJ, Kim K and Chae HZ (2001) Cyclophylin-A binds to peroxiredoxins and activates its peroxidase activity. J Biol Chem 276: 29826–29832

Li H, Singh AK and Sherman LA (2003) Differential gene expression and the regulation of oxidative stress in the cyanobacterium *Synechocystis sp.* PCC 6803. In: Hagemann M, Herrmann RG, Omata O and Tabata S (eds) Functional Genomics in Cyanobacteria, pp 54–55. Benediktbeuren, Germany

Meinhard M and Grill E (2001) Hydrogen peroxide is a regulator of ABI1, a protein phosphatase 2C from Arabidopsis. FEBS Lett 508: 443–446

Meinhard M, Rodriguez PL and Grill E (2002) The sensitivity of ABI2 to hydrogen peroxide links the abscisic acid-response regulator to redox signaling. Planta 214: 775–782

Melis A (1999) Photosystem-II damage and repair cycle in chloroplasts: what modulates the rate of photodamage in vivo? Trends Plant Sci 4: 130–135

Michel KP and Pistorius EK (2004) Adaptation of the photosynthetic electron transport chain in cyanobacteria to iron deficiency: The function of IdiA and IsiA. Physiol Plant 120: 36–50

Mutsuda M, Ishikawa T, Takeda T and Shigeoka S (1996) The catalase-peroxidase of *Synechococcus* PCC 7942: purification, nucleotide sequence analysis and expression in *Escherichia coli*. Biochem J 316: 251–257

Neill SJ, Desikan R and Hancock JT (2003) Nitric oxide signalling in plants. New Phytol 159: 11–35

Neumann CA, Krause DS, Carman CV, Das S, Dubey DP, Abraham JL, Bronson RT, Fujiwara Y, Orkin SH and Van Etten RA (2003) Essential role for the peroxiredoxin Prdx1 in erythrocyte antioxidant defence and tumour suppression. Nature 424: 561–565

Nguyên-Nhu NT (2003) Characterization of human peroxiredoxin 5 and homologous *Saccharomyces cerevisiae* alkyl hydroperoxide reductase 1. PhD-thesis, Université Catholique de Louvain, Faculté des Sciences

Nishiyama Y, Yamamoto H, Allakhverdiev SI, Inaba M, Yokota A and Murata N (2001) Oxidative stress inhibits the repair of photodamage to the photosynthetic machinery. EMBO J 20: 5587–5594

Nishiyama Y, Allakhverdiev SI and Murata N (2005) Regulation by environmental conditions of the repair of photosystem II in cyanobacteria. In: Demmig-Adams B, Adams WW III and Mattoo AK (eds) Photoprotection, Photoinhibition, Gene Regulation, and Environment, pp 193–203. Springer, Dordrecht

Noctor G and Foyer CH (1998) Ascorbate and glutathione: Keeping active oxygen under control. Annu Rev Plant Physiol Plant Mol Biol 49: 249–279

Obinger C, Regelsberger G, Strasser G, Burner U and Peschek GA (1997) Purification and characterization of a homodimeric catalase-peroxidase from the cyanobacterium *Anacystis nidulans*. Biochem Biophys Res Commun 235: 545–552

Paerl HW (2000) Marine plankton. In: Whitton BA and Potts M (eds) The Ecology of Cyanobacteria: Their Diversity in Time and Space, pp 121–148. Kluwer Academic Publishers, Dordrecht, London, Boston

Patel RP, McAndrew J, Sellak H, White CR, Jo HJ, Freeman BA and Darley-Usmar VM (1999) Biological aspects of reactive nitrogen species. Biochim Biophys Acta 1411: 385–400

Peltier JB, Friso G, Kalume DE, Roepstorff P, Nilsson F, Adamska I and van Wijk KJ (2000) Proteomics of the chloroplast: Systematic identification and targeting analysis of lumenal and peripheral thylakoid proteins. Plant Cell 12: 319–341

Pfannschmidt T (2003) Chloroplast redox signals: how photosynthesis controls its own genes. Trends Plant Sci 8: 33–41

Pfannschmidt T, Allen JE and Oelmüller R (2001) Principles of redox control in photosynthesis gene expression. Physiol Plant 112: 1–9

Pruvot G, Massimino J, Peltier G and Rey P (1996) Effects of low temperature, high salinity and endogenous ABA on the synthesis of two chloroplastic drought-induced proteins in *Solanum tuberosum*. Physiol Plant 97: 123–131

Rawyler A, Arpagaus S and Braendle R (2002) Impact of oxygen stress and energy availability on membrane stability of plant cells. Ann Bot 90: 499–507

Regelsberger G, Jakopitsch C, Plasser L, Schwaiger H, Furtmüller G, Peschek G, Zamocky M and Obinger C (2002) Occurrence and biochemistry of hydroperoxidases in oxygenic phototrophic prokaryotes (cyanobacteria). Plant Physiol Biochem 40: 479–490

Rey P, Pruvot G, Becuwe N, Eymery F, Rumeau D and Peltier G (1998) A novel-thioredoxin-like protein located in the chloroplast is induced by water deficit in *Solanum tuberosum* L. Plant J 13: 97–107

Rhee SG, Kang SW, Chang TS, Jeong W, Kim K (2001) Peroxiredoxin, a novel family of peroxidases. IUBMB Life 52: 35–41

Rouhier N and Jacquot JP (2002) Plant peroxiredoxins: alternative hydroperoxide scavenging enzymes. Photosynth Res 74: 259–268

Rouhier N, Gelhaye E, Corbier C and Jacquot JP (2004) Active site mutagenesis and phospholipid hydroperoxide reductase activity of poplar type II peroxiredoxin. Physiol Plant 120: 57–62

Sakamoto A, Okumura T, Kaminaka H, Sumi K and Tanaka K (1995) Structure and differential response to abscisic acid of two promoters for the cytosolic copper/zinc-superoxide dismutase genes, SpdCc1 and SodCc2, in rice protoplasts. FEBS Lett 358: 62–66

Sakamoto A, Tsukamoto S, Yamamoto H, Ueda-Hashimoto M, Takahashi M, Suzuki H and Morikawa H (2003) Functional complementation in yeast reveals a protective role of chloroplast 2-Cys peroxiredoxin against reactive nitrogen species. Plant J 33: 841–851

Samuilov VD (1997) Photosynthetic oxygen: the role of H_2O_2. A review. Biochemistry (Moscow) 62: 451–454

Schmetterer G (1994) Cyanobacterial respiration. In: Bryant DA (ed) The Molecular Biology of Cyanobacteria, pp 409–435. Kluwer Academic Publishers, The Netherlands, Dordrecht, Boston, London

Schopf JW (2000) The fossil record: Tracing the roots of the cyanobacterial lineage. In: Whitton BA, Potts M (eds) The Ecology of Cyanobacteria: Their Diversity in Time and Space. Kluwer Academic Publishers, Dordrecht, London, Boston, pp 13–35

Schröder E and Ponting CP (1998) Evidence that peroxiredoxins are novel members of the thioredoxin fold superfamily. Protein Sci 7: 2465–2468

Schröder E, Littlechild JA, Lebedev AA, Errington N, Vagin AA and Isupov MN (2000) Crystal structure of decameric 2-Cys peroxiredoxin from human erythrocytes at 1.7 A resolution. Structure 8: 605–615

Schwartz SH, Léon-Kloosterziel KM, Koornneef M and Zeevaart JAD. (1997) Biochemical characterization of the aba2 and aba3 mutant in *Arabidopsis thaliana*. Plant Physiol 114: 161–166

Singh AK, Li H and Sherman LA (2004) Microarray analysis and redox control of gene expression in the cyanobacterium *Synechocystis sp.* PCC 6803. Physiol Plant 120:27–35

Smirnoff N (1993) Tansley review 52: The role of active oxygen in the response of plants to water deficit and desiccation. New Phytol 125: 27–58

Stacy RAP, Munthe E, Steinum T, Sharma B and Aalen RB (1996) A peroxiredoxin antioxidant is encoded by a dormancy-related gene, Per1, expressed during late development in the aleurone and embryo of barley grains. Plant Mol Biol 31: 1205–1216

Stacy RAP, Nordeng TW, Culianez-Macia FA and Aalen RB (1999) The dormancy-related peroxiredoxin antioxidant, PER1, is localized to the nucleus of barley embryo and aleurone cells. Plant J 19: 1–8

Swofford D L (2002) PAUP*. Phylogenetic Analysis Using Parsimony (*and other methods). Version 4. Sinauer Associates, Sunderland, Massachusetts

Tel-Or E, Huflejt M and Packer L (1985) The role of glutathione and ascorbate in hydroperoxide removal in cyanobacteria. Biochem Biophys Res Commun 132: 533–539

Tel-Or E, Huflejt ME and Packer L (1986) Hydroperoxide metabolism in cyanobacteria. Arch Biochem Biophys 246: 396–402

Thomas DJ, Avenson TJ, Thomas JB and Herbert SK (1998) A cyanobacterium lacking iron superoxide dismutase is sensitized to oxidative stress induced with methyl viologen but is not sensitized to oxidative stress induced with norflurazon. Plant Physiol 116: 1593–1602

Tichy M and Vermaas W (1999) *In vivo* role of catalase-peroxidase in *Synechocystis sp.* strain PCC 6803. J Bacteriol 181: 1875–1882

Van Breusegem F, Vranova E, Dat JF and Inzé D (2001) The role of active oxygen species in plant signal transduction. Plant Sci 161: 405–414

Vermaas WFJ (2001) Photosynthesis and respiration in cyanobacteria. In: Encyclopedia of Life Sciences, pp 1–7. Nature Publishing Group, London

Vranova E, van Breusegem F, Dat J, Belles-Boix E and Inzé D (2002) The role of active oxygen species in plant signal transduction. In: Scheel D and Wasternack C (eds) Plant Signal Transduction, pp 45–73. Oxford University Press, Oxford

Welinder KG (1992) Superfamily of plant, fungal and bacterial peroxidases. Curr Opin Struct Biol 2: 388–393

Whitton BA and Potts M (2000) Introduction to the cyanobacteria. In: Whitton BA and Potts M (eds) The Ecology of Cyanobacteria: Their Diversity in Time and Space, pp 1–11. Kluwer Academic Publishers, Dordrecht, London, Boston

Wood ZA, Poole LB, Hantgan RR and Karplus PA (2002) Dimers to oligomers: Redox-sensitive oligomerization of 2-Cysteine peroxiredoxins. Biochemistry 41: 5493–5504

Wood ZA, Schröder E, Harris JR and Poole LB (2003) Structure, mechanism and regulation of peroxiredoxins. Trends Biochem Sci 28: 32–40

Wydrzynski T, Angström J and Vännegard T (1989) H_2O_2 formation by photosystem II. Biochim Biophys Acta 973: 23–28

Yamamoto H, Miyake C, Dietz K-J, Tomizawa K, Murata N and Yokota A (1999) Thioredoxin peroxidase in the cyanobacterium *Synechocystis sp.* PCC 6803. FEBS Lett 447: 269–273

Yousef N, Pistorius EK and Michel KP (2003) Comparative analysis of idiA and isiA transcription under iron starvation and oxidative stress in *Synechococcus elongatus* PCC 7942 wild type and selected mutants. Arch Microbiol 180: 471–483

Chapter 20

Lipoxygenases, Apoptosis, and the Role of Antioxidants

Mauro Maccarrone*
Department of Biomedical Sciences, University of Teramo, Piazza A. Moro 45, 64100 Teramo, Italy, and the IRCCS C. Mondino, Mondino-Tor Vergata- Santa Lucia Center for Experimental Neurobiology, Via Ardeatina 306, 00179 Rome, Italy

Summary	321
I. Introduction	321
II. Involvement of Lipoxygenases in Apoptosis	323
III. Role of Antioxidants in Lipoxygenase-mediated Apoptosis	326
IV. Concluding Remarks	328
Acknowledgments	328
References	328

Summary

Lipoxygenases are a family of enzymes that dioxygenate unsaturated fatty acids, thus initiating lipoperoxidation of membranes and the synthesis of signaling molecules, or inducing structural and metabolic changes in the cell. This activity is the basis for the critical role of lipoxygenases in a number of pathophysiological conditions, in both animals and plants. In the past few years, a pro-apoptotic effect of lipoxygenase has been reported in different cells and tissues, leading to cell death along unrelated apoptotic pathways. However, anti-apoptotic effects of lipoxygenases have also been reported, often based on the use of enzyme inhibitors. In the present review, the characteristics of the lipoxygenase family and the role of lipoxygenase activation in apoptosis of animal and plant cells are discussed, suggesting a common signal transduction pathway in cell death conserved through the evolution of both kingdoms. In addition, the inhibition of lipoxygenases by antioxidants and its consequences on apoptosis will be presented.

I. Introduction

Lipoxygenases (linoleate:oxygen oxidoreductase, EC 1.13.11.12; LOXs) are a family of monomeric non-heme, non-sulfur iron dioxygenases that catalyze the conversion of polyunsaturated fatty acids into conjugated hydroperoxides. The unsaturated fatty acids, which are essential in humans, are absent in most bacteria and thus LOXs are also absent in typical prokaryotes. LOXs are widely expressed in animal and plant cells, sometimes at high level, and their activity may initiate the synthesis of a signaling molecule or may induce structural or metabolic changes in the cell. Mammalian lipoxygenases have been implicated in the pathogenesis of several inflammatory conditions such as arthritis, psoriasis, and bronchial asthma (Kühn and Borngraber, 1999). They are also thought to have a role in atherosclerosis (Cathcart and Folcik, 2000), brain aging (Manev et al., 2000), HIV infection (Maccarrone et al., 2000a), kidney disease (Maccarrone et al., 1999a; Montero and Badr, 2000), and terminal differentiation of keratinocytes (Heidt et al., 2000). In plants, lipoxygenases play a role in germination, and participate in the synthesis of traumatin and jasmonic acid and in the response to abiotic stress (Grechkin, 1998; Feussner and Wasternack, 2002; Weichert et al., 2002). Remarkably, several of the above-mentioned conditions are associated with apoptosis (programmed cell death, PCD) in both animals (Han et al., 1995; Lizard et al., 1996; Martín-Malo et al., 2000; Fumelli et al., 2000)

---

*Author for correspondence, email:maccarrone@unite.it

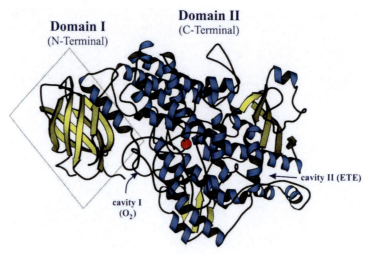

Fig. 1. Schematic diagram of the three-dimensional structure of soybean (*Glycine max*) lipoxygenase-1, showing the small N-terminal domain I (rectangular box) and the large C-terminal domain II. The iron-containing active site is located in domain II, and can be reached by molecular oxygen (O_2) through cavity I and by arachidonic acid (eicosatetraenoic acid, ETE) through cavity II. β-Sandwiches are represented in yellow, α-helices in blue, random coils in gray, and iron is the red sphere. The three-dimensional structure was modelled through the RASMOL program, using the lipoxygenase-1 sequence (PDB accession number: 2SBL).

and plants (Greenberg, 1996; Jones and Dangl, 1996; Wang H. et al., 1996; Wang M. et al., 1996; Koukalová et al., 1997).

LOXs from animal and plant tissues have been sequenced, purified, and characterized, and have been shown to form a closely related family with no similarities to other known sequences. The phylogenetic tree shows that plant and animal enzymes are separate branches, each forming several subgroups (Brash, 1999). When arachidonic (eicosatetraenoic, C20:4; ETE) acid is the substrate, different LOX isozymes can add a hydroperoxy group at carbons 5, 12, or 15, and are therefore designated 5-, 12-, or 15-lipoxygenases. Linoleic (octadecadienoic, C18:2; OD) acid and linolenic (octadecatrienoic, C18:3; OT) acid are also substrates of LOXs. Soybean (*Glycine max* (L.) Merrill) lipoxygenase-1 (LOX-1) is a 15-lipoxygenase widely used as a prototype for studying the homologous family of lipoxygenases from tissues of different species, both in structural (Boyington et al., 1993; Minor et al., 1996; Gan et al.; 1996; Sudharshan and Appu Rao, 1999; Sudharshan et al., 2000) and kinetic (Glickman and Klinman, 1995; Jonsson et al., 1996; Maccarrone et al., 2001a, 2001b; Di Venere et al., 2003) investigations. The primary sequence (Shibata et al., 1987) and three dimensional structure (Boyington et al., 1993; Minor et al., 1996) of LOX-1 have been determined, showing that it is a prolate ellipsoid of 90 by 65 by 60 Å, with 839 amino acid residues and a molecular mass of 93840 Da. LOX-1 is made of two domains: a 146-residue β-barrel at the N-terminal (domain I) and a 693-residue helical bundle at the C-terminal (domain II). The iron-containing active site is in the center of domain II, liganded to four conserved histidines and to the carboxyl group of the C-terminal conserved isoleucine. It can be reached through the two cavities (I and II) shown in Fig. 1.

Cavity I presents an ideal path for the access of molecular oxygen to iron, whereas cavity II can accommodate arachidonic acid or even slightly larger fatty acids (Boyington et al., 1993). Mammalian lipoxygenases lack the N-terminal domain present in LOX-1 and related plant lipoxygenases, thus showing smaller molecular mass (75 - 80 kDa compared to 94 -104 kDa in plants). It has been suggested that the N-terminal domain in LOX-1 makes only a loose contact with the C-terminal domain (Boyington et al., 1993), and that it may be dispensable for plant lipoxygenases (Minor et al., 1996), because all of the amino acid side chains responsible for catalysis are located in

---

Abbreviations: ERK – extracellular regulated kinase; ETE – eicosatetraenoic (arachidonic) acid; FLAP – 5-lipoxygenase activating protein; HR – hypersensitive response; LOX – lipoxygenase; LRP – lentil root protoplast; MAPK – mitogen-activated protein kinase; NDGA – nordihydroguaiaretic acid; NF-kB – nuclear factor-kB; NSAID, – nonsteroidal anti-inflammatory drug; OD – octadecadienoic (linoleic) acid; OT – octadecatrienoic (linolenic) acid; PCD – programmed cell death; PLA$_2$ – phospholipase A$_2$; PPAR – peroxisomal proliferator-activated receptor; ROS – reactive oxygen species; SPD – spermidine; SPN – spermine; TGF – transforming growth factor; TNF – tumor necrosis factor.

the C-terminal domain. However, limited proteolysis experiments indicated that the two domains are instead tightly associated (Ramachandran et al., 1992), and that domain-domain interactions play a role in the reversible unfolding of LOX-1 (Sudharshan and Appu Rao, 1999) through ionic interactions (Sudharshan and Appu Rao, 2000). We have recently shown that the N-terminal domain of LOX-1 is a built-in inhibitor of the activity and membrane binding ability of the enzyme (Maccarrone et al., 2001a), conferring to LOX-1 lower flexibility and lower hydration and, hence, higher stability than domain II alone (Di Venere et al., 2003). LOX-1 is usually in the inactive Fe^{2+} form (Brash et al., 1998). Oxidation to the active Fe^{3+} enzyme is required for catalysis, and cleavage of the C-H bond of the substrate is the rate-limiting step of the overall dioxygenation reaction (Glickman and Klinman, 1995; Jonsson et al., 1996).

II. Involvement of Lipoxygenases in Apoptosis

The role of lipoxygenases in programmed cell death of animal and plant cells has been recently reviewed (Maccarrone et al., 2001b), and here the main findings will be summarized. Cellular membranes are the primary site of several PCD inducers (Fadok et al., 1998, 2000; Ren and Savill, 1998; Kagan et al., 2000) and lipid messengers have long been known to act as regulators of apoptosis (McGahon et al., 1995). Indeed, mobilization of esterified fatty acids from membranes represents a key regulatory step in cellular response to various stimuli such as growth factors, cytokines, chemokines, and circulating hormones (Locati et al., 1994). The role of fatty acids in signal transduction through interactions with protein kinases, lipases, or G proteins (Los et al., 1995), and gene transcription through interactions with nuclear receptors of the peroxisomal proliferator-activated receptor (PPAR) family (Devchand et al., 1996; Nolte et al., 1997), is increasingly apparent. More recently, attention has been drawn to the role of the lipoxygenase-triggered "arachidonate cascade" (Funk, 2001) in the execution of mammalian cell apoptosis. In fact LOXs, by introducing a peroxide into the fatty acid moieties of (phospho)lipids (Kühn et al., 1990; Feussner et al., 1995, 1997; Schnurr et al., 1996; Pérez-Gilabert et al., 1998), modify the fluidity and permeability of biomembranes (Wratten et al., 1992; Maccarrone et al., 1995a). These membrane modifications may play a regulatory role in the physiological clearance of apoptotic cells (Mic et al., 1999; Godson et al., 2000). Moreover, lipid oxidation products can damage proteins by reacting with lysine amino groups, cysteine sulfhydryl groups, and histidine imidazole groups (Refsgaard et al., 2000). Besides the direct lipid membrane and indirect protein modifications, LOXs can also generate reactive oxygen species (ROS), which are well known inducers of PCD (Eu et al., 2000; Hensley et al., 2000; Miller et al., 2000; Schaul, 2000; Sekharam et al., 2000). Furthermore, the hydroperoxides generated by LOXs from fatty acids released from membranes by phospholipase activity (Cummings et al., 2000) can act as lipid messengers along different apoptotic pathways, all leading to nuclear changes like chromatin condensation and DNA fragmentation that are typical of PCD (Maccarrone et al., 2001b). Lipid peroxides have long been considered critical in apoptosis (Hockenbery et al., 1993; Torres-Roca et al., 1995), and LOX products have been shown to induce PCD in human T cells (Sandstrom et al., 1994, 1995), neutrophils (Hébert et al., 1996), PC12h cells (Aoshima et al., 1997), and Jurkat cells (Liberles and Schreiber, 2000). Therefore, it is not surprising that activation of lipoxygenases has been associated with programmed death of different cells and tissues induced by different, unrelated stimuli (Table 1 and references therein). Several studies have shown the pro-apoptotic activity of 5-LOX (Maccarrone et al., 1997, 1998a), of leukocyte-type 12-LOX (Zhou et al., 1998; Kamitani et al., 1999), and of 15-LOX (Zhou et al., 1998; Ikawa et al., 1999; Kasahara et al., 2000), and have identified the molecular targets for lipoxygenase interaction that contribute to the induction of apoptosis (Table 1 and references therein).

In the context of the induction of apoptosis by activation of lipoxygenases, it is noteworthy that these enzymes are also able to uncouple mitochondria within hours (Maccarrone et al., 2000c). LOX activity can dioxygenate mitochondrial membranes (Kühn et al., 1990; Feussner et al., 1995), leading to formation of pore-like structures in the lipid bilayer which initiate programmed organelle disruption (Van Leyen et al., 1998). It can be proposed that LOX products perturb the mitochondrial membrane by increasing fluidity (Maccarrone et al., 2000c), thus allowing calcium release. In turn, cytosolic calcium stimulates LOX activity (Hammarberg and Rådmark, 1999), which can further uncouple mitochondria and trigger a cascade of events leading to cell death. In this line, it can be recalled that lipoxygenase activation also affects protein phosphorylation (Rice et al., 1998) and cytoskeleton organization (Provost et al., 1999), both of which can be instrumental in promoting PCD. In fact, we have

Table 1. Pro-apoptotic stimuli that upregulate lipoxygenase activity and cellular targets for lipoxygenase interaction.

Stimulus	Reference	Molecular target	Reference
TNFα[a]	O'Donnell et al., 1995	Membrane lipids	Maccarrone et al., 1998a
			Kamitani et al., 1998
TGFβ1 ± cisplatin	Maccarrone et al., 1996	Phospholipase A_2	Tang et al., 1997
Fas/APO-1 ligand	Wagenknecht et al., 1998	Phospholipase C	Szekeres et al., 2000
Retinoic acid	Maccarrone et al., 1997	Mitochondria	Biswal et al., 2000
			Maccarrone et al., 2000c
H_2O_2	Croft et al., 1990	Free radicals	Maccarrone et al., 1997
	Levine et al., 1994		
	Buonaurio and Servili, 1999		
	Rusterucci et al., 1999		
	Maccarrone et al., 2000b		
Thapsigargin	Zhou et al., 1998	Calcium stores	Buyn et al., 1997
			Metzler et al., 1998
Bleomycin	Vernole et al., 1998	Caspases	Wagenknecht et al., 1998
Tamoxifen	Maccarrone et al., 1998b	MAPK phosphatase1	Metzler et al., 1998
	Kamitani et al., 1998		
Sodium butyrate	Ikawa et al., 1999	p38-MAPK	Aoshiba et al., 1999
	Kamitani et al., 1999		
X-ray irradiation	Matyshevskaia et al., 1999	Protein kinase C	Szekeres et al., 2000
NSAIDs	Shureiqi et al., 2000	ERK1/2, Ras	Szekeres et al., 2000
			Maccarrone et al., 2000c

[a] TNFα, tumor necrosis factor α; TGFβ1, transforming growth factor β1; NSAIDs, nonsteroidal anti-inflammatory drugs; MAPK, mitogen-activated protein kinase; ERK, extracellular regulated kinase.

previously shown that LOX activation is an early event along different apoptotic pathways, and occurs several hours before any typical sign of apoptosis at the nuclear level (Maccarrone et al., 1999b). Furthermore, early LOX activation was not observed in necrotic cells (Maccarrone et al., 1999b), thus representing the first biochemical difference between the initiation of apoptosis and that of necrosis so far described. Altogether, the available experimental evidence supports the scheme shown in Fig. 2. Physical (UV light) or chemical (drugs, oxidative stress, serum starvation) induction of PCD, as well as apoptotic pathways mediated by receptors, like CD28, interleukin receptor or Fas (Maccarrone et al., 2001b), all involve LOX activation, leading to membrane lipid peroxidation and decreased mitochondrial potential. These events are accompanied by increased release of cytochrome c, followed by the activation of caspases as executioners of apoptosis (Fadok et al., 2000). In addition, LOX-mediated programmed cell death implies modifications of: *i*) membrane properties (exposure of phosphatidylserine, increased levels of cholesterol and consequent altered Ras expression, generation of free hydroperoxides); *ii*) cytoskeleton (at the level of coactosin-like protein, lamins, Gas 2, and α-fodrin), and *iii*) gene transcription (through nuclear factor (NF)-kB, poly(ADP-ribose) polymerase, and ROS).

In plants, PCD has been associated with germination (Wang M. et al., 1996), various phases of development and senescence (Greenberg, 1996; Jones and Dangl, 1996), and response to salt (Katsuhara and Kawasaki, 1996) or cold (Koukalová et al., 1997) stress. A type of PCD of particular interest, termed "hypersensitive response (HR)", has been observed during the plant response to pathogen attack (Jones and Dangl, 1996; Wang H. et al., 1996). Signal transduction pathways are activated during HR, leading to biosynthesis or release of potential antimicrobial effector molecules that are thought to contribute to both host and pathogen cell death (Levine et al., 1994; Jones and Dangl, 1996; Wang H. et al., 1996). Among other signals, rapid generation of ROS has been implicated in HR of plants against pathogens (Jabs et al., 1996). In particular, hydrogen peroxide has been shown to be a crucial component of the HR control circuit, since a short pulse of exogenous H_2O_2 is sufficient to activate the hypersensitive cell death programme in soybean cells (Levine et al., 1994). A similar HR has been observed in bean (Croft et al., 1990), pepper (Buonaurio and Servili, 1999), and tobacco (Lacomme and Santa Cruz, 1999; Rusterucci et al., 1999) leaves infected with different pathogens. Interestingly, LOX activation has been associated with HR (Buonaurio and Servili, 1999; Rusterucci et al., 1999), and H_2O_2 oxidative stress has been shown to induce animal cell apoptosis (P.F. Li et al., 1997; Maccarrone et al., 1997; Quillet-Mary et al., 1997) through activation of 5-LOX (Maccarrone et al., 1997).

Chapter 20 Antioxidants and Lipoxygenase-Mediated Apoptosis

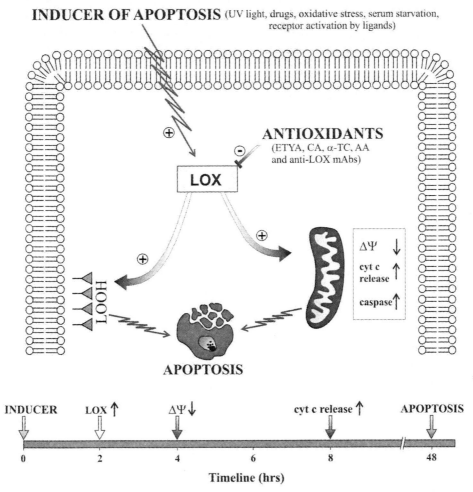

Fig. 2. Model of lipoxygenase-dependent apoptosis. In different cell types, pro-apoptotic inducers like UV light, oxidative stress, or receptor activation by ligands upregulate various lipoxygenase (LOX) isozymes that are directly associated with decreased mitochondrial potential ($\Delta\Psi$), increased cytochrome c release, and caspase activation, ultimately leading to apoptosis. LOX activation also causes the formation of membrane lipoperoxides (LOOH) that contribute to the induction of apoptosis. Antioxidants like eicosatetraynoic acid (ETYA), caffeic acid (CA), α-tocopherol (α-TC), and ascorbic acid (AA), or specific anti-LOX monoclonal antibodies (mAbs), are able to inhibit LOX activity, thus preventing membrane peroxidation and mitochondrial damage responsible for LOX-induced apoptosis.

We have recently used oxidative stress by exogenous H_2O_2 to investigate whether LOX activation was a cause rather than an effect of plant cell apoptosis. Lentil (*Lens culinaris*) lipoxygenase has been characterized (Maccarrone et al., 1992), cloned (Hilbers et al., 1994), and expressed in *Escherichia coli* (Hilbers et al., 1996), and therefore lentil root protoplasts (LRP) were chosen as a model. It was shown that the early phase of hydrogen peroxide-induced apoptosis is characterized by enhancement of lipoxygenase activity, attributable to up-regulation of gene expression at both the transcriptional and translational level. The increase of lipoxygenase was paralleled by an enhancement of ultraweak luminescence, a marker of LOX-related membrane lipid peroxidation (Nakano et al., 1994; Maccarrone et al., 2000b). These findings suggest that lipoxygenase activation might contribute to membrane peroxidation during hydrogen peroxide-induced PCD, resembling previous observations from animal cells (Maccarrone et al., 1997). Nordihydroguaiaretic acid (NDGA) protected LRPs against H_2O_2-induced DNA fragmentation and ultraweak luminescence. However, in view of the lack of specificity of this LOX inhibitor (see later in this review), we also used inhibitory anti-LOX monoclonal antibodies that are capable of blocking LOX activity in LRPs without affecting the redox state of the cell (Maccarrone et al., 2000b). These antibodies too protected LRPs against H_2O_2-induced

PCD, whilst the specific products of LRP lipoxygenase, 9-hydroperoxylinoleic acid (9-HPOD) and 13-HPOD (Hilbers et al., 1995), were able *per se* to induce DNA fragmentation (Maccarrone et al., 2000b). Furthermore, 9-hydroxylinoleic acid (9-HOD) and 13-HOD, the reduced forms of the LOX products in LRPs, were able to induce cell death and DNA fragmentation to almost the same extent as the corresponding HPODs, again matching our previous observations on human cells (Maccarrone et al., 2000b). Collectively, the results obtained with LRPs show that activation of lipoxygenase with the consequent membrane peroxidation is a critical step in the initiation of apoptosis in plants. Since similar findings were reported in H_2O_2-induced PCD of human cells, these observations suggest that animals and plants share a common signal transduction pathway triggering apoptosis after oxidative stress. Whether or not LOX products regulate the physiological clearance of apoptotic cells in plants, as they do in animals, remains to be elucidated.

III. Role of Antioxidants in Lipoxygenase-mediated Apoptosis

Anti-apoptotic effects of LOXs have also been reported, mainly based on the observation that LOX inhibitors, most often NDGA and MK886, had pro-apoptotic activity (Tang et al., 1996, 1997; Ghosh and Myers, 1998; Korystov et al., 1998; Ding et al., 1999; L. Li et al., 1999). Interestingly, in other cellular types NDGA or MK886 protected against apoptosis (O'Donnell et al., 1995; Maccarrone et al., 1996, 1997; Wagenknecht et al., 1998; Zhou et al., 1998; Shureiqi et al., 2000) or induced PCD even in cells completely devoid of LOX activity (Vanags et al., 1997; Datta et al., 1999; Biswal et al., 2000). In this context, it should be recalled that most of the LOX inhibitors used in PCD studies, including NDGA, act by reducing the active Fe^{3+} enzyme to the inactive Fe^{2+} form (Fig. 3).

Yet, NDGA also blocks voltage-activated calcium currents, inhibits P450 monooxygenase activity, and acts as a general radical scavenger (O'Donnell et al., 1995). On the other hand, MK886, which inhibits 5-LOX by blocking the 5-LOX activating protein (FLAP) (Datta et al., 1999), can also induce PCD via a caspase-3-dependent pathway that is related to bcl-xL but unrelated to 5-LOX (Datta et al., 1998). Recently, MK886 has been found to be effective as a PCD inducer at doses 100-fold higher than those required for 5-LOX inhibition in a human cell line which lacks FLAP and its mRNA (Datta et al., 1999). Resveratrol, a natural cancer preventive agent present in red wine (Jang et al., 1997; Soleas et al., 1997; Rotondo et al., 1998), has been shown to inhibit PCD induced by unrelated stimuli in human erythroleukemia K562 cells through competitive inhibition of 5-LOX and 15-LOX activity (Maccarrone et al., 1999c; Pinto et al., 1999). However, resveratrol is a general antioxidant that also inhibits nonenzymatic lipoperoxidation (Belguendouz et al., 1997) and cyclooxygenase activity (Subbaramaiah et al., 1998; Maccarrone et al., 1999c), and acts as an agonist for the estrogen receptor (Gehm et al., 1997). Finally, caffeic acid, which prevents apoptosis linked to 5-LOX activation (Maccarrone et al., 1997), is a LOX inhibitor with general antioxidant properties (Sud'ina et al., 1993), much like the LOX inhibitors α-tocopherol (Lomnitski et al., 1993) and γ-tocopherol (Jiang and Ames, 2003). In addition to antioxidants able to reduce active Fe^{3+} to inactive Fe^{2+} in the LOX active site (Fig. 3), a number of other antioxidants have been shown to inhibit LOX-1 activity via different mechanisms (Maccarrone et al., 1995b, 1998b; Lomnitski et al., 1997; Trono et al., 1999; Pastore et al., 2000; Jiang and Ames, 2003). Here we summarize the effect of some of the antioxidants that might be more relevant for apoptosis and for human nutrition (Demmig-Adams and Adams, 2002).

Ascorbic acid (Fig. 3) was found to be a competitive inhibitor of LOX-1 (Maccarrone et al., 1995b), showing an inhibition constant (Ki) of 27 μM (Table 2).

Inhibition by ascorbic acid was reversible, with LOX-1 regaining full activity after 4 min of dialysis of enzyme/inhibitor mixtures (15 nM/100 μM) at 4 °C, and was dependent on the pH of the reaction (Maccarrone et al., 1995b). Interestingly, dehydroascorbic acid inhibited LOX-1 activity to a lesser extent than ascorbic acid. In fact, 150 μM dehydroascorbic acid reduced linoleic acid oxygenation to ∼70% of the control value, compared to ∼40% obtained with the same concentration of ascorbic acid. Similar inhibition studies were performed with the chain-breaking antioxidants 6-palmitoylascorbic acid and trolox, a water-soluble analogue of α-tocopherol. Both compounds were found to be reversible and competitive inhibitors of LOX-1 activity (Maccarrone et al., 1995b), Ki values being 3 μM and 18 μM, respectively (Table 2). Unlike other antioxidant inhibitors of LOX-1, ascorbic acid, 6-palmitoylascorbic acid, and trolox did not reduce Fe^{3+} to inactive Fe^{2+} in the active site, as demonstrated by the reaction progress curves showing that the lag phase of the reaction was not prolonged (Maccarrone et al., 1995b). EPR spectroscopy confirmed this

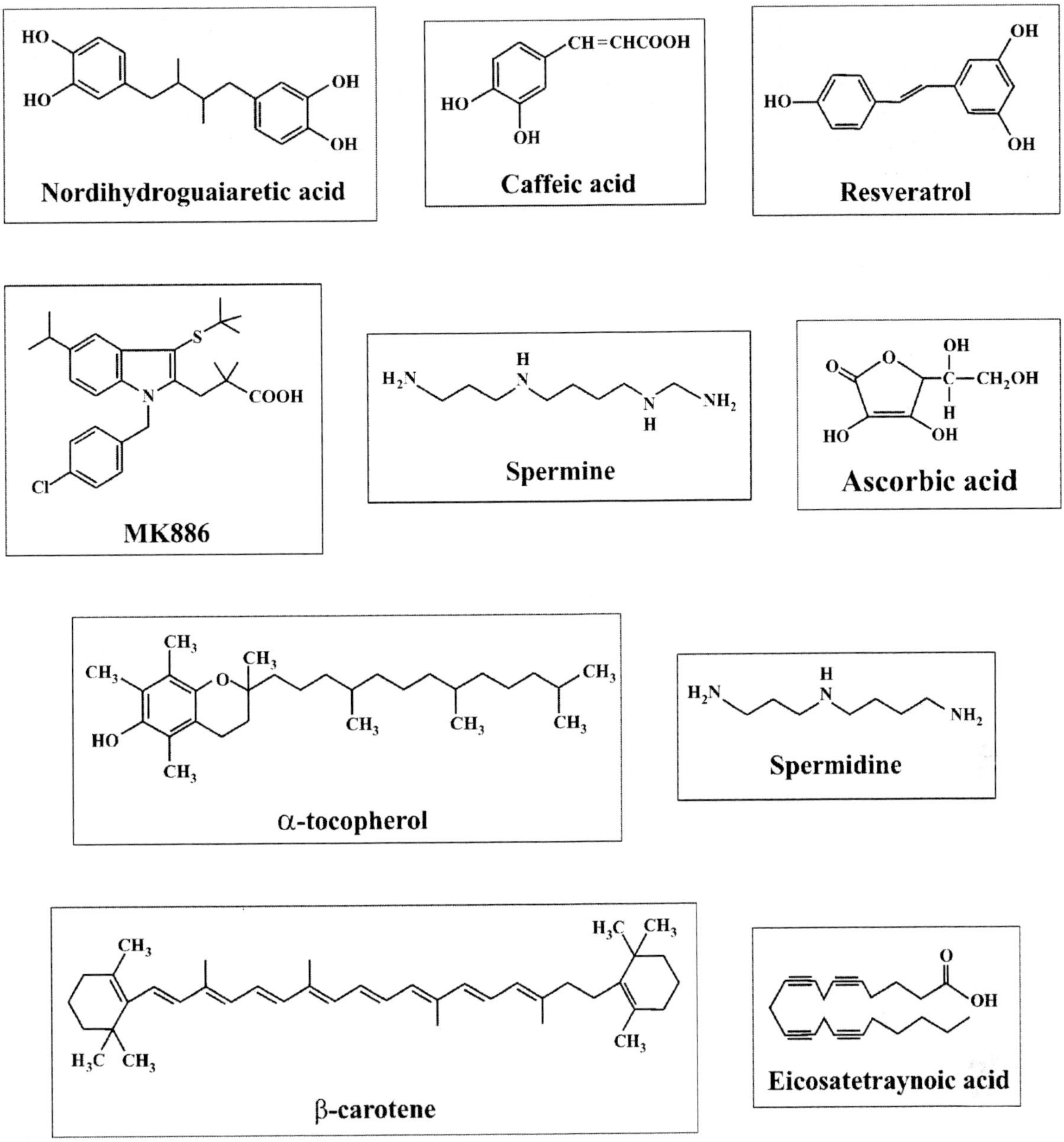

Fig. 3. Antioxidant inhibitors of lipoxygenase activity, some of which are widely used in investigations on apoptosis of animal and plant cells.

Table 2. In vitro inhibition of lipoxygenase activity by antioxidants.

Inhibitor	Type of inhibition	Inhibition constant (K_i, µM)	Reference
Ascorbic acid[a]	competitive	27	Maccarrone et al., 1995b
6-Palmitoylascorbic acid[a]	competitive	3	Maccarrone et al., 1995b
Trolox[a]	competitive	18	Maccarrone et al., 1995b
α-Tocopherol[b]	noncompetitive	30	Maccarrone et al., 1999a
β-Carotene[c]	mixed	2.4	Lomnitski et al., 1993
Retinol[c]	competitive	1.7	Lomnitski et al., 1993
Spermine[a]	competitive	800	Maccarrone et al., 1998b
Spermidine[a]	uncompetitive	2,700	Maccarrone et al., 1998b

[a] Assays were performed with soybean (*Glycine max*) lipoxygenase-1.
[b] Assays were performed with barley (*Hordeum vulgare*) 5-lipoxygenase.
[c] Assays were performed with soybean (*Glycine max*) lipoxygenase-2.

finding, clearly indicating the lack of radical formation from the inhibitor molecules during LOX-1 inhibition (Maccarrone et al., 1995b). It can be speculated that the chain-breaking antioxidants might interact with the interior of the "funnel" leading to the active site (cavity II in Fig. 1), because enhancing the lipophilicity of the inhibitor made it more effective (compare ascorbate and 6-palmitoylascorbate in Table 2). Similar results have been reported for soybean LOX-2 inhibition by β-carotene (Lomnitski et al., 1993; Trono et al., 1999) and by α-tocopherol (Lomnitski et al., 1993; Pastore et al., 2000).

Interestingly, the activity of LOX-1 was also inhibited by polyamines in a dose-dependent manner (Maccarrone et al., 1998b). The inhibitory power followed the order: spermine > spermidinespermidine > cadaverine • putrescine. Therefore, spermine (SPN) and spermidine (SPD) were chosen to further characterize the inhibition mechanism. Lineweaver-Burk plots of the reaction catalyzed by LOX-1, in the presence or absence of SPN or SPD, indicated that both compounds are uncompetitive inhibitors of LOX-1 with respect to linoleic acid, the inhibition constants being 800 μM and 2700 μM for SPN and SPD respectively (Table 2). Cadaverine (1,5-diaminopentane) was approximately as effective as putrescine (1,4-diaminobutane), yielding a maximum inhibition of LOX-1 activity of $\sim$20% (Maccarrone et al., 1998b). Therefore, the spacing between the amino groups does not seem to influence the inhibitory ability of polyamines. Moreover, the uncompetitive inhibition of LOX-1 with respect to linoleic acid indicated that SPN and SPD bind to the enzyme-substrate complex only. Also, the presence of a negative charge on the substrate molecule does not seem to be involved in the interaction with polyamines, because dioxygenation of linoleic acid (negatively charged) and methyl-linoleate (uncharged) was inhibited to the same extent (Maccarrone et al., 1998b). Neither SPN nor SPD inhibited LOX-1 by reducing active Fe^{3+} to inactive Fe^{2+} in the catalytic site, because the duration of the lag phase of the reaction, which is a function of the iron redox state, was not affected (Maccarrone et al., 1998b). It seems noteworthy that the radical trapping ability of polyamines increases from putrescine to spermine, in keeping with their inhibitory power towards LOX-1. Of further interest is the fact that polyamines, in a concentration range close to the physiological values, were shown to efficiently inhibit the activity of LOX-1 in intact cells, suggesting that they can rapidly cross the plasma membranes, possibly through specific polyamine channels. How this transport can contribute to the modulation of intracellular LOX activity remains to be elucidated.

IV. Concluding Remarks

Activation of different LOX isozymes has been shown to be associated with an early phase of apoptosis triggered by several, unrelated stimuli. Consistently, the hydro(pero)xides generated by LOX activity have been shown to induce programmed cell death in different cellular models through a series of events including membrane lipid peroxidation and mitochondrial uncoupling, followed by cytochrome c release and caspase activation. Of major interest is the observation that animal and plant cells share a common signal transduction pathway triggering apoptosis after oxidative stress, based upon lipoxygenase activation, that has been conserved through evolution. In addition, evidence has been reviewed that the chain-breaking antioxidants ascorbic acid, 6-palmitoylascorbic acid, and trolox act as competitive inhibitors of soybean LOX-1. Also natural polyamines have been shown to bind to the enzyme-substrate complex only, leading to uncompetitive inhibition of LOX-1. Taken together, these findings support the hypothesis that lipid peroxidation might be prevented by antioxidants or polyamines through a direct inhibition of LOX-1. These results also seem to have physiological significance in the light of the nutritional value of many of the LOX inhibitors (Demmig-Adams and Adams, 2002), which may retard cell ageing by preventing cell death.

Acknowledgments

I we would like to thank Prof. Alessandro Finazzi Agrò (University of Rome Tor Vergata) for stimulating discussions and continuing support, and all colleagues at the University of Rome Tor Vergata and at Utrecht University who took part over the years in the study of the role of lipoxygenases in programmed cell death. This study was partly supported by Istituto Superiore di Sanità (II AIDS project) and by Ministero dell'Istruzione, dell'Università e della Ricerca (COFIN 2002), Rome.

References

Aoshiba K, Yasui S, Nishimura K and Nagai A (1999) Thiol depletion induces apoptosis in cultured lung fibroblasts. Am J Respir Cell Mol Biol 21: 54–64

Aoshima H, Satoh T, Sakai N, Yamada M, Enokido Y, Ikeuchi T and Hatanaka H (1997) Generation of free radicals during lipid hydroperoxide-triggered apoptosis in PC12h cells. Biochim Biophys Acta 1345: 35–42

Belguendouz L, Fremont L and Linard A (1997) Resveratrol inhibits metal ion-dependent and independent peroxidation

of porcine low-density lipoproteins. Biochem Pharmacol 53:1347–1355
Biswal SS, Datta K, Shaw SD, Feng X, Robertson JD and Kehrer JP (2000) Glutathione oxidation and mitochondrial depolarization as mechanisms of nordihydroguaiaretic acid-induced apoptosis in lipoxygenase-deficient FL5.12 cells. Toxicol Sci 53: 77–83
Boyington JC, Gaffney BJ and Amzel LM (1993) The three-dimensional structure of an arachidonic acid 15-lipoxygenase. Science 260: 1482–1486
Brash AR (1999) Lipoxygenases: occurrence, functions, catalysis, and acquisition of substrate. J Biol Chem 274: 23679–23682
Buonaurio R and Servili M (1999) Involvement of lipoxygenase, lipoxygenase pathway volatiles, and lipid peroxidation during the hypersensitive reaction of pepper leaves to *Xanthomonas campestris* pv. *vesicatoria*. Physiol Mol Plant Pathol 54: 155–169
Buyn T, Dudeja P, Harris JE, Ou D, Seed T, Sawlani D, Meng J, Bonomi P and Anderson KM (1997) A 5-lipoxygenase inhibitor at micromolar concentration raises intracellular calcium in U937 cells prior to their physiologic cell death. Prostaglandins Leukot Essent Fatty Acids 56: 69–77
Cathcart MK and Folcik VA (2000) Lipoxygenases and atherosclerosis: protection versus pathogenesis. Free Radic Biol Med 28: 1726–1734
Croft KPC, Voisey CR and Slusarenko AJ (1990) Mechanism of hypersensitive cell collapse: correlation of increased lipoxygenase with membrane damage in leaves of *Phaseolus vulgaris* (L.) cv. Red Mexican inoculated with avirulent race 1 of *Pseudomonas syringae* pv. *phaseolicola*. Physiol Mol Plant Pathol 36: 49–62
Cummings BS, McHowat J and Schnellmann RG (2000) Phospholipase A_2s in cell injury and death. J Pharmacol Exp Ther 294: 793–799
Datta K, Biswal SS, Xu J, Towndrow KM, Feng X and Kehrer JP (1998) A relationship between 5-lipoxygenase-activating protein and bcl-xL expression in murine pro-B lymphocytic FL5.12 cells. J Biol Chem 273: 28163–28169
Datta K, Biswal SS and Kehrer JP (1999) The 5-lipoxygenase-activating protein (FLAP) inhibitor, MK886, induces apoptosis independently of FLAP. Biochem J 340: 371–375
Demmig-Adams B and Adams WW III (2002) Antioxidants in photosynthesis and human nutrition. Science 298: 2149–2153
Devchand PR, Keller H, Peters JM, Vazquez M, Gonzalez P and Whali W (1996) The PPARα-leukotriene B_4 pathway to inflammation control. Nature 384: 39–43
Di Venere A, Salucci ML, Van Zadelhoff G, Veldink G, Mei G, Rosato N, Finazzi-Agrò A and Maccarrone M (2003) Structure-to-function relationship of mini-lipoxygenase, a 60 kDa fragment of soybean lipoxygenase-1 with lower stability but higher enzymatic activity. J Biol Chem 278: 18281–18288
Ding XZ, Kuszynski CA, El-Metwally TH and Adrian TE (1999) Lipoxygenase inhibition induced apoptosis, morphological changes, and carbonic anhydrase expression in human pancreatic cancer cells. Biochem Biophys Res Commun 266: 392–399
Eu JP, Liu L, Zeng M and Stamler JS (2000) An apoptotic model for nitrosative stress. Biochemistry 39: 1040–1047
Fadok VA, Bratton DL, Frasch SC, Warner ML and Henson PM (1998) The role of phosphatidylserine in recognition of apoptotic cells by phagocytes. Cell Death Differ 5: 551–562

Fadok VA, Bratton DL, Rose DM, Pearson A, Ezekewitz RA and Henson PM (2000) A receptor for phosphatidylserine-specific clearance of apoptotic cells. Nature 405: 85–90
Feussner I and Wasternack C (2002) The lipoxygenase pathway. Annu Rev Plant Biol 53: 275–297
Feussner I, Wasternack C, Kindl H and Kühn H (1995) Lipoxygenase-catalyzed oxygenation of storage lipids is implicated in lipid mobilization during germination. Proc Natl Acad Sci USA 92: 11849–11853
Feussner I, Balkenhohl TJ, Porzel A, Kühn H and Wasternack C (1997) Structural elucidation of oxygenated storage lipids in cucumber cotyledons. J Biol Chem 272: 21635–21641
Fumelli C, Marconi A, Salvioli S, Straface E, Malorni W, Offidani AM, Pellicciari R, Schettini G, Giannetti A, Monti D, Franceschi C and Pincelli C (2000) Carboxyfullerenes protect human keratinocytes from ultraviolet B-induced apoptosis. J Invest Dermatol 115: 835–841
Funk CD (2001) Prostaglandins and leukotrienes: advances in eicosanoid biology. Science 294: 1871–1875
Gan Q-F, Browner MF, Sloane DL and Sigal E (1996) Defining the arachidonic acid binding site of human 15-lipoxygenase. J Biol Chem 271: 25412–25418
Gehm BD, McAndrews JM, Chien PY and Jameson JL (1997) Resveratrol, a polyphenolic compound found in grapes and wine, is an agonist for the estrogen receptor. Proc Natl Acad Sci USA 94: 14138–14143
Ghosh J and Myers CE (1998) Inhibition of arachidonate 5-lipoxygenase triggers massive apoptosis in human prostate cancer cells. Proc Natl Acad Sci USA 95: 13182–13187
Glickman MH and Klinman JP (1995) Nature of the rate-limiting steps in the soybean lipoxygenase-1 reaction. Biochemistry 34: 14077–14092
Godson C, Mitchell S, Harvey K, Petasis NA, Hogg N and Brady HR (2000) Cutting edge: lipoxins rapidly stimulate nonphlogistic phagocytosis of apoptotic neutrophils by monocyte-derived macrophages. J Immunol 164: 1663–1667
Grechkin A (1998) Recent developments in biochemistry of the plant lipoxygenase pathway. Prog Lipid Res 37: 317–352
Greenberg JT (1996) Programmed cell death: a way of life for plants. Proc Natl Acad Sci USA 93: 12094–12097
Hammarberg T and Rådmark O (1999) 5-Lipoxygenase binds calcium. Biochemistry 38: 4441–4447
Hammarberg T, Provost P, Persson B and Rådmark O (2000) The N-terminal domain of 5-lipoxygenase binds calcium and mediates calcium stimulation of enzyme activity. J Biol Chem 275: 38787–38793
Han DKM, Haudenschild CC, Hong MK, Tinkle BT, Leon MB and Liau G (1995) Evidence for apoptosis in human atherogenesis and in a rat vascular injury model. Am J Pathol 147: 267–277
Hébert M-J, Takano T, Holthöfer H and Brady HR (1996) Sequential morphologic events during apoptosis of human neutrophils induced by lipoxygenase-derived eicosanoids. J Immunol 157: 3105–3115
Heidt M, Furstenberger G, Vogel S, Marks F and Krieg P (2000) Diversity of mouse lipoxygenases: identification of a subfamily of epidermal isozymes exhibiting a differentiation-dependent mRNA expression pattern. Lipids 35: 701–707
Hensley K, Robinson KA, Gabbita SP, Salsman S and Floyd RA (2000) Reactive oxygen species, cell signaling, and cell injury. Free Radic Biol Med 28: 1456–1462
Hilbers MP, Rossi A, Finazzi-Agrò A, Veldink GA and Vliegenthart JFG (1994) The primary structure of a lipoxygenase from

the shoots of etiolated lentil seedlings derived from its cDNA. Biochim Biophys Acta 1211: 239–242

Hilbers MP, Kerkhoff B, Finazzi-Agrò A, Veldink GA and Vliegenthart JFG (1995) Heterogeneity and developmental changes of lipoxygenase in etiolated lentil seedlings. Plant Sci 111: 169–180

Hilbers MP, Finazzi-Agrò A, Veldink GA and Vliegenthart JFG (1996) Purification and characterization of a lentil seedling lipoxygenase expressed in *E. coli*. Int J Biochem Cell Biol 28: 751–760

Hockenbery DM, Oltvai ZN, Yin X-M, Milliman CL and Korsmeyer SJ (1993) Bcl-2 functions in an antioxidant pathway to prevent apoptosis. Cell 75: 241–251

Ikawa H, Kamitani H, Calvo BF, Foley JF and Eling TE (1999) Expression of 15-lipoxygenase-1 in human colorectal cancer. Cancer Res 59: 360–366

Jabs T, Dietrich RA and Dangl JL (1996) Initiation of a runaway cell death in an *Arabidopsis* mutant by extracellular superoxide. Science 273: 1853–1856

Jang M, Cai L, Udeani GO, Slowing KV, Thomas CF, Beecher CWW, Fong HHS, Farnsworth NR, Kinghorn AD, Metha RG, Moon RC and Pezzuto JM (1997) Cancer chemopreventive activity of resveratrol, a natural product derived from grapes. Science 275: 218–220

Jiang Q and Ames BN (2003) γ-Tocopherol, but not α-tocopherol, decreases proinflammatory eicosanoids and inflammation damage in rats. FASEB J 17: 816–822

Jones AM and Dangl JL (1996) Logjam at the Styx: programmed cell death in plants. Trends Plant Sci 1: 114–119

Jonsson T, Glickman MH, Sun S and Klinman JP (1996) Experimental evidence for extensive tunneling of hydrogen in the lipoxygenase reaction: implications for enzyme catalysis. J Am Chem Soc 118: 10319–10320

Kagan VE, Fabisiak JP, Shvedova AA, Tyurina YY, Tyurin VA, Schor NF and Kawai K (2000) Oxidative signaling pathway for externalization of plasma membrane phosphatidylserine during apoptosis. FEBS Lett 477: 1–7

Kamitani H, Geller M and Eling T (1998) Expression of 15-lipoxygenase by human colorectal carcinoma Caco-2 cells during apoptosis and cell differentiation. J Biol Chem 273: 21569–21577

Kamitani H, Ikawa H, Hsi LC, Watanabe T, DuBois RN and Eling TE (1999) Regulation of 12-lipoxygenase in rat intestinal epithelial cells during differentiation and apoptosis induced by sodium butyrate. Arch Biochem Biophys 368: 45–55

Kasahara Y, Tuder RM, Cool CD and Voelkel NF (2000) Expression of 15-lipoxygenase and evidence for apoptosis in the lungs from patients with COPD. Chest 117: 260S

Katsuhara M and Kawasaki T (1996) Salt stress induced nuclear and DNA degradation in meristematic cells of barley roots. Plant Cell Physiol 37: 169–173

Korystov YuN, Shaposhnikova VV, Levitman MKh, Kudryavtsev AA, Kublik LN, Narimanov AA and Orlova OE (1998) The effect of inhibitors of arachidonic acid metabolism on proliferation and death of tumor cells. FEBS Lett 431: 224–226

Koukalová B, Kovarik A, Fajkus J and Siroky J (1997) Chromatin fragmentation associated with apoptotic changes in tobacco cells exposed to cold stress. FEBS Lett 414: 289–292

Kühn H and Borngraber S (1999) Mammalian 15-lipoxygenases. Enzymatic properties and biological implications. Adv Exp Med Biol 447: 5–28

Kühn H, Belkner J, Wiesner R and Brash AR (1990) Oxygenation of biological membranes by the pure reticulocyte lipoxygenase. J Biol Chem 265: 18351–18361

Lacomme C and Santa Cruz S (1999) Bax-induced cell death in tobacco is similar to the hypersensitive response. Proc Natl Acad Sci USA 96: 7956–7961

Levine A, Tenhaken R, Dixon R and Lamb C (1994) H_2O_2 from the oxidative burst orchestrates the plant hypersensitive disease resistance response. Cell 79: 583–593

Li L, Zhu Z, Joshi B, Porter AT and Tang DG (1999) A novel hydroxamic acid compound, BMD188, demonstrates antiprostate cancer effects by inducing apoptosis. I: In vitro studies. Anticancer Res 19: 51–60

Li PF, Dietz R and von Harsdorf R (1997) Differential effect of hydrogen peroxide and superoxide anion on apoptosis and proliferation of vascular smooth muscle cells. Circulation 96: 3602–3609

Liberles SD and Schreiber SL (2000) Apoptosis-inducing natural products found in utero during murine pregnancy. Chem Biol 7: 365–372

Lizard G, Deckert V, Dubrez L, Moisant M, Gambert P and Lagrost L (1996) Induction of apoptosis in endothelial cells treated with cholesterol oxides. Am J Pathol 148: 1625–1638

Locati M, Zhou D, Luini W, Evangelista V, Mantovani A and Sozzani S (1994) Rapid induction of arachidonic acid release by monocyte chemotactic protein-1 and related chemokines. Role of Ca^{2+} influx, synergism with platelet-activating factor and significance for chemotaxis. J Biol Chem 269: 4746–4753

Lomnitski L, Bar-Natan R, Sklan D and Grossman S (1993) The interaction between β-carotene and lipoxygenase in plant and animal systems. Biochim Biophys Acta 1167: 331–338

Lomnitski L, Grossman S, Bergman M, Sofer Y and Sklan D (1997) *In vitro* and *in vivo* effects of β-carotene on rat epidermal lipoxygenases. Int J Vit Nutr Res 67: 407–414

Los M, Schenk H, Hexel K, Baeuerle PA, Droge W and Schulze-Osthoff K (1995) IL-2 gene expression and NF-kB activation through CD28 requires reactive oxygen production by 5-lipoxygenase. EMBO J 14: 3731–3740

Maccarrone M, Veldink GA and Vliegenthart JFG (1992) Inhibition of lipoxygenase activity in lentil protoplasts by monoclonal antibodies introduced into the cells via electroporation. Eur J Biochem 205: 995–1001

Maccarrone M, Bladergroen MR, Rosato N and Finazzi-Agrò A (1995a) Role of lipid peroxidation in electroporation-induced cell permeability. Biochem Biophys Res Commun 209: 417–425

Maccarrone M, Veldink GA, Vliegenthart JFG and Finazzi-Agrò A (1995b) Inhibition of soybean lipoxygenase-1 by chain-breaking antioxidants. Lipids 30: 51–54

Maccarrone M, Nieuwenhuizen WF, Dullens HFJ, Catani MV, Melino G, Veldink GA, Vliegenthart JFG and Finazzi-Agrò A (1996) Membrane modifications in human erythroleukemia K562 cells during induction of programmed cell death by transforming growth factor β1 or cisplatin. Eur J Biochem 241: 297–302

Maccarrone M, Catani MV, Finazzi-Agrò A and Melino G (1997) Involvement of 5-lipoxygenase in programmed cell death of cancer cells. Cell Death Differ 4: 396–402

Maccarrone M, Fantini C, Ranalli M, Melino G and Finazzi-Agrò A (1998a) Activation of nitric oxide synthase is involved in tamoxifen-induced apoptosis of human erythroleukemia K562 cells. FEBS Lett 434: 421–424

Maccarrone M, Baroni A and Finazzi-Agrò A (1998b) Natural polyamines inhibit soybean (*Glycine max*) lipoxygenase-1, but not the lipoxygenase-2 isozyme. Arch Biochem Biophys 356: 35–40

Maccarrone M, Taccone-Gallucci M, Meloni C, Cococcetta N, Manca di Villahermosa S, Casciani U and Finazzi-Agrò A (1999a) Activation of 5-lipoxygenase and related cell membrane lipoperoxidation in hemodialysis patients. J Am Soc Nephrol 10: 1991–1996

Maccarrone M, Salucci ML, Melino G, Rosato N and Finazzi-Agrò A (1999b) The early phase of apoptosis in human neuroblastoma CHP100 cells is characterized by lipoxygenase-dependent ultraweak light emission. Biochem Biophys Res Commun 265: 758–762

Maccarrone M, Lorenzon T, Guerrieri P and Finazzi-Agrò A (1999c) Resveratrol prevents apoptosis in K562 cells by inhibiting lipoxygenase and cyclooxygenase activity. Eur J Biochem 265: 27–34

Maccarrone M, Bari M, Corasaniti MT, Nistico R, Bagetta G and Finazzi-Agrò A (2000a) HIV-1 coat glycoprotein gp120 induces apoptosis in rat brain neocortex by deranging the arachidonate cascade in favor of prostanoids. J Neurochem 75: 196–203

Maccarrone M, Van Zadelhoff G, Veldink GA, Vliegenthart JFG and Finazzi-Agrò A (2000b) Early activation of lipoxygenase in lentil (*Lens culinaris*) root protoplasts by oxidative stress induces programmed cell death. Eur J Biochem 267: 5078–5084

Maccarrone M, Ranalli M, Bellincampi L, Salucci ML, Sabatini S, Melino G and Finazzi-Agrò A (2000c) Activation of different lipoxygenase isozymes induces apoptosis in human erythroleukemia and neuroblastoma cells. Biochem Biophys Res Commun 272: 345–350

Maccarrone M, Salucci ML, Van Zadelhoff G, Malatesta F, Veldink G, Vliegenthart JFG and Finazzi-Agrò A (2001a) Tryptic digestion of soybean lipoxygenase-1 generates a 60 kDa fragment with improved activity and membrane binding ability. Biochemistry 40: 6819–6827

Maccarrone M, Melino G and Finazzi-Agrò A. (2001b) Lipoxygenases and their involvement in programmed cell death. Cell Death Differ 8: 776–784

Manev H, Uz T, Sugaya K and Qu T (2000) Putative role of neuronal 5-lipoxygenase in an aging brain. FASEB J 14: 1464–1469

Martín-Malo A, Carracedo J, Ramirez R, Rodriguez-Benot A, Soriano S, Rodriguez M and Aljama P (2000) Effect of uremia and dialysis modality on mononuclear cell apoptosis. J Am Soc Nephrol 11: 936–942

Matyshevskaia OP, Pastukh VN and Solodushko VA (1999) Inhibition of lipoxygenase activity reduces radiation-induced DNA fragmentation in lymphocytes. Radiat Biol Radioecol 39: 282–286

May C, Höhne M, Gnau P, Schwennesen K and Kindl H (2000) The N-terminal β-barrel structure of lipid body lipoxygenase mediates its binding to liposomes and lipid bodies. Eur J Biochem 267: 1100–1109

McGahon AJ, Nishioka WK, Martin SJ, Mahboubi A, Cotter TG and Green DR (1995) Regulation of the Fas apoptotic cell death pathway by Abl. J Biol Chem 270: 22625–22631

Metzler B, Hu Y, Sturm G, Wick G and Xu Q (1998) Induction of mitogen-activated protein kinase phosphatase-1 by arachidonic acid in vascular smooth muscle cells. J Biol Chem 273: 33320–33326

Mic FA, Molnar P, Kresztes T, Szegezdi E and Fesus L (1999) Clearance of apoptotic thymocytes is decreased by inhibitors of eicosanoid synthesis. Cell Death Differ 6: 593–595

Miller FD, Pozniak CD and Walsh GS (2000) Neuronal life and death: an essential role for the p53 family. Cell Death Differ 7: 880–888

Minor W, Steczko J, Stec B, Otwinowski Z, Bolin JT, Walter R and Axelrod B (1996) Crystal structure of soybean lipoxygenase L-1 at 1.4 Å resolution. Biochemistry 35: 10687–10701

Montero A and Badr KF (2000) 15-Lipoxygenase in glomerular inflammation. Exp Nephrol 8: 14–19

Nakano M, Ito T, Arimoto T, Ushijima Y and Kamiya K (1994) A simple luminescence method for detecting lipid peroxidation and antioxidant activity in vitro. Biochem Biophys Res Commun 202: 940–946

Nolte RT, Wisley GB, Westin S, Cobb JE, Lambert MH, Kurokawa R, Rosenfeld MG, Willson TM, Glass CK and Milburn MV (1998) Ligand binding and co-activator assembly of the peroxisome proliferator-activated receptor-γ. Nature 395: 137–143

O'Donnell VB, Spycher S and Azzi A (1995) Involvement of oxidants and oxidant-generating enzyme(s) in tumour-necrosis-factor-α-mediated apoptosis: role for lipoxygenase pathway but not mitochondrial respiratory chain. Biochem J 310: 133–141

Pastore D, Trono D, Padalino L, Simone S, Valenti D, Di Fonzo N and Passerella S (2000) Inhibition by α-tocopherol and L-ascorbate of linoleate hydroperoxidation and β-carotene bleaching activities in durum wheat semolina. J Cereal Sci 31: 41–54

Pérez-Gilabert M, Veldink GA and Vliegenthart JFG (1998) Oxidation of dilinoleoyl phosphatidylcholine by lipoxygenase-1 from soybeans. Arch Biochem Biophys 354: 18–23

Pinto MC, Garcia-Barrado JA and Macias P (1999) Resveratrol is a potent inhibitor of the dioxygenase activity of lipoxygenase. J Agric Food Chem 47: 4842–4846

Provost P, Samuelsson B and Radmark O (1999) Interaction of 5-lipoxygenase with cellular proteins. Proc Natl Acad Sci USA 96: 1881–1885

Quillet-Mary A, Jaffrezou JP, Mansat V, Bordier C, Naval J and Laurent G (1997) Implication of mitochondrial hydrogen peroxide generation in ceramide-induced apoptosis. J Biol Chem 272: 21388–21395

Ramachandran S, Carroll RT, Dunham WR and Funk MO (1992) Limited proteolysis and active-site labeling studies of soybean lipoxygenase 1. Biochemistry 31: 7700–7706

Refsgaard HH, Tsai L and Stadtman ER (2000) Modifications of proteins by polyunsaturated fatty acid peroxidation products. Proc Natl Acad Sci USA 97: 611–616

Ren Y and Savill J (1998) Apoptosis: The importance of being eaten. Cell Death Differ 5: 563–568

Rice RL, Tang DG, Haddad M, Honn KV and Taylor JD (1998) 12(S)-Hydroxyeicosa-tetraenoic acid increases the actin microfilament content in B16a melanoma cells: a protein kinase-dependent process. Int J Cancer 77: 271–278

Rotondo S, Rajtar G, Manarini S, Celardo A, Rotillo D, de Gaetano G, Evangelista V and Cerletti C (1998) Effect of trans-resveratrol, a natural polyphenolic compound, on human polymorphonuclear leukocyte function. Br J Pharmacol 123: 1691–1699

Rusterucci C, Montillet JL, Agnel JP, Battesti C, Alonso B, Knoll A, Bessoule JJ, Etienne P, Suty L, Blein JP and Triantaphylides C (1999) Involvement of lipoxygenase-dependent production of fatty acid hydroperoxides in the development of the hypersensitive cell death induced by cryptogein on tobacco leaves. J Biol Chem 274: 36446–36455

Sandstrom PA, Tebbey PW, Van Cleave S and Buttke TM (1994) Lipid hydroperoxides induce apoptosis in T cells displaying a HIV-associated glutathione peroxidase deficiency. J Biol Chem 269: 798–801

Sandstrom PA, Pardi D, Tebbey PW, Dudek RW, Terrian DM, Folks TM and Buttke TM (1995) Lipid hydroperoxide-induced apoptosis: lack of inhibition by Bcl-2 over-expression. FEBS Lett 365: 66–70

Schaul Y (2000) c-Abl: activation and nuclear targets. Cell Death Differ 7: 10–16

Schnurr K, Hellwing M, Seidemann B, Jungblut P, Kühn H, Rapoport SM and Schewe T (1996) Oxygenation of biomembranes by mammalian lipoxygenases: the role of ubiquinone. Free Rad Biol Med 20: 11–21

Sekharam M, Cunnick JM and Wu J (2000) Involvement of lipoxygenase in lysophosphatidic acid-stimulated hydrogen peroxide release in human HaCaT keratinocytes. Biochem J 346: 751–758

Shibata D, Steczko J, Dixon JE, Hermodson M, Yazdanparast R and Axelrod B (1987) Primary structure of soybean lipoxygenase-1. J Biol Chem 262: 10080–10085

Shureiqi I, Chen D, Lee JJ, Yang P, Newman RA, Brenner DE, Lotan R, Fischer SM and Lippman SM (2000) 15-LOX-1: a novel molecular target of nonsteroidal anti-inflammatory drug-induced apoptosis in colorectal cancer cells. J Natl Cancer Inst 92: 1136–1142

Soleas GJ, Diamandis EP and Goldberg DM (1997) Wine as a biological fluid: history, production, and role in disease prevention. J Clin Lab Anal 11: 287–313

Subbaramaiah K, Chung WJ, Michaluart P, Telang N, Tanabe T, inoue H, Jang M, Pezzuto JM and Dannenberg AJ (1998) Resveratrol inhibits cyclooxygenase-2 transcription and activity in phorbol ester-treated human mammary epithelial cells. J Biol Chem 273: 21875–21882

Sudharshan E and Appu Rao AG (1999) Involvement of cysteine residues and domain interactions in the reversible unfolding of lipoxygenase-1. J Biol Chem 274: 35351–35358

Sudharshan E, Srinivasulu S and Appu Rao AG (2000) pH-Induced domain interaction and conformational transitions of lipoxygenase-1. Biochim Biophys Acta 1480: 13–22

Sud'ina GF, Mirzoeva OK, Pushkareva MA, Korshunova GA, Sumbatyan NV and Varfolomeev SD (1993) Caffeic acid phenethyl ester as a lipoxygenase inhibitor with antioxidant properties. FEBS Lett 329: 21–24

Szekeres CK, Tang K, Trikha M and Honn KV (2000) Eicosanoid activation of extracellular signal-regulated kinase1/2 in human epidermoid carcinoma cells. J Biol Chem 275: 38831–38841

Tang, DG and Honn KV (1997) Apoptosis of W256 carcinosarcoma cells of the monocytoid origin induced by NDGA involves lipid peroxidation and depletion of GSH: role of 12-lipoxygenase in regulating tumor cell survival. J Cell Physiol 172: 155–170

Tang DG, Chen YQ and Honn KV (1996) Arachidonate lipoxygenases as essential regulators of cell survival and apoptosis. Proc Natl Acad Sci USA 93: 5241–5246

Tang DG, Guan K-L, Li L., Honn KV, Chen YQ, Rice RL, Taylor JD and Porter AT (1997) Suppression of W256 carcinosarcoma cell apoptosis by arachidonic acid and other polyunsaturated fatty acids. Int J Cancer 72: 1078–1087

Tang K, Nie D and Honn KV (1999) Role of autocrine motility factor in a 12-lipoxygenase dependent anti-apoptotic pathway. Adv Exp Med Biol 469: 583–590

Torres-Roca JF, Lecoeur H, Amatore C and Gougeon ML (1995) The early intracellular production of reactive oxygen intermediate mediates apoptosis in dexamethasone-treated thymocytes. Cell Death Differ 2: 309–319

Trono D, Pastore D and Di Fonzo N (1999) Carotenoid dependent inhibition of durum wheat lipoxygenase. J Cereal Sci 29: 99–102

Van Leyen K, Duvoisin RM, Engelhardt H and Wiedmann M (1998) A function for lipoxygenase in programmed organelle degradation. Nature 395: 392–395

Vanags DM, Larsson P, Feltenmark S, Jakobsson P-J, Orrenius S, Claesson H-E and Aguilar-Santelises M (1997) Inhibitors of arachidonic acid metabolism reduce DNA and nuclear fragmentation induced by TNF plus cycloheximide in U937 cells. Cell Death Differ 4: 479–486

Vernole P, Tedeschi B, Caporossi D, Maccarrone M, Melino G and Annicchiarico-Petruzzelli M (1998) Induction of apoptosis by bleomycin in resting and cycling human lymphocytes. Mutagenesis 13: 209–215

Wagenknecht B, Schulz JB, Gulbins E and Weller M (1998) Crm-A, bcl-2 and NDGA inhibit CD95L-induced apoptosis of malignant glioma cells at the level of caspase 8 processing. Cell Death Differ 5: 894–900

Wang H, Li J, Bostock RM and Gilchrist DG (1996) Apoptosis: a funtional paradigm for programmed plant cell death induced by host-selective phytotoxin and invoked during development. Plant Cell 8: 375–391

Wang M, Oppedijk BJ, Lu X, Van Duijn B and Schilperoort RA (1996) Apoptosis in barley aleurone during germination and its inhibition by abscisic acid. Plant Mol Biol 32: 1125–1134

Weichert H, Kolbe A, Kraus A, Wasternack C and Feussner I (2002) Metabolic profiling of oxylipins in germinating cucumber seedlings – lipoxygenase-dependent degradation of triacylglycerols and biosynthesis of volatile aldehydes. Planta 215: 612–619

Wratten ML, Van Ginkel G, Van't Veld AA, Bekker A, Van Faassen EE and Sevanian A (1992) Structural and dynamic effects of oxidatively modified phospholipids in unsaturated lipid membranes. Biochemistry 31: 10901–10907

Zhou YP, Teng D, Dralyuk F, Ostrega D, Roe MW, Philipson L and Polonsky KS (1998) Apoptosis in insulin-secreting cells. Evidence for the role of intracellular Ca^{2+} stores and arachidonic acid metabolism. J Clin Invest 101: 1623–1632

Chapter 21

Regulation of Photosynthetic Gene Expression by the Environment: From Seedling De-etiolation to Leaf Senescence

Christiane Reinbothe*
Lehrstuhl für Pflanzenphysiologie, Universität Bayreuth, Universitätsstraße 30, D-95447 Bayreuth, Germany

Steffen Reinbothe
Université Joseph Fourier et Centre National de la Recherche Scientifique (CNRS), UMR 5575, BP53, CERMO, F-38041 Grenoble cedex 9, France

Summary	334
I. Introduction	334
II. Control of Photosynthetic Gene Expression during Seedling Etiolation and De-etiolation	335
A. Skotomorphogenesis and Photomorphogenesis as Alternative Developmental Strategies to Adapt to the Light Environment	335
B. Phytochrome Action on Photosynthetic Genes	336
C. Chlorophyll Biosynthesis	339
1. Steps, Intermediates, Enzymes	339
2. Regulation of the C5-Pathway	339
a. Feedback Inhibition of 5-Aminolevulinic Acid Synthesis in Angiosperms	339
b. Light-Dependent Protochlorophyllide Reduction Catalyzed by NADPH: Protochlorophyllide Oxidoreductase	340
3. LHPP Provides the Link between Skotomorphogenesis and Photosynthesis of Higher Plants	341
a. Structural Components of LHPP	341
b. Developmental Expression of LHPP	342
4. Evolution of Protochorophyllide-Based Light-Harvesting	343
D. Plastid-Derived Factors Controlling Photosynthetic Gene Expression in Response to Light	344
1. Intermediates of the C5-Pathway	344
2. Redox Components Associated with Photosynthetic Electron Transport	345
E. Brassinosteroids as Factors Controling Morphogenesis	345
III. Photosynthetic Gene Expression during Leaf Senescence	346
A. Leaf Senescence as a Developmental Program	346
B. Chloroplast Destruction Occurring during Leaf Senescence	347
1. Breakdown of Plastid Proteins	347
2. Degradation of Chlorophyll	348
a. Induction of Chlorophyllase by Jasmonic Acid and Coronatine	349
b. Induction of Phaeophorbide a Oxygenase and Red Chlorophyll Catabolite Reductase during Senescence and Stress	349

*Author for correspondence: email:christiane.reinbothe@uni-bayreuth.de

    C. Leaf Senescence as a Cell Death Program 350
        1. Senescence-Associated Genes with Potential Functions in Signaling 351
        2. Hydrolytic Activities Induced during Senescence 352
            a. Proteolytic System 352
            b. Lipolytic Activities 352
        3. The Search for the Cell Death Factor 353
IV. Future Perspective 354
Acknowledgements 355
References 355

Summary

Both endogenous and exogenous factors are involved in modulating and coordinating gene expression during plant development. Among them are light and plant hormones such as ethylene, cytokinins, abscisic acid, gibberellins, and brassinosteroids. Light and brassinosteroids have received particular attention because of their obvious and pronounced effects on early plant development following seed germination. By contrast, much less is known about the terminal stage of plant development referred to as senescence and the factors controlling this cell death program. Plant hormones such as ethylene and jasmonic acid have, however, been implicated in the initiation and progression of leaf senescence.

At all stages of their life cycle, plants are prone to various forms of oxidative stress. Upon illumination, excited tetrapyrroles such as chlorophyll, heme, and their precursors as well as degradation products can transfer their excitation energy onto oxygen, leading to the formation of highly reactive singlet oxygen. Angiosperms being the most highly evolved group of plants must have evolved efficient strategies to prevent the accumulation of such potentially harmful compounds. It is the aim of this chapter to summarize current concepts on the regulation of plant gene expression by light and the plant hormone jasmonic acid, with particular emphasis on the mechanisms by which higher plants prevent photooxidative self-poisoning. Special reference is made to the plastid compartment, which is the major site of tetrapyrrole metabolism and a source of signals that coordinate nuclear gene expression in response to light.

I. Introduction

As an environmental cue, light influences plant growth and development. Direction, periodicity, quality, and/or quantity of light regulate physiological responses such as phototropism, circadian rhythms, morphogenesis, and leaf senescence (Kendrick and Kronenberg, 1994). To account for some or all of these different effects, light perception is mediated through the action of various photoreceptors, including phototropins, cryptochromes, and the phytochromes (Cashmore, 1999; Casal, 2000; Christie and Briggs, 2001; Nagy and Schaefer, 2002; Quail, 2002a). Expression of several thousand genes is regulated by light and multiple, highly coordinated signal transduction pathways exist which, in many cases, involve sets of different photoreceptors. In addition, plant hormones such as brassinosteroids regulate gene expression during

---

Abbreviations: ACD – accelerated cell death; 5-ALA – 5-aminolevulinic acid; bHLH – basic helix-loop-helix transcription factor; BIN – brassinosteroid-insensitive; BL – brassinolide; BR – brassinosteroid; BRI – brassinosteroid-insensitive; Chlase – chlorophyllase; Chl(ide) – chlorophyll(ide); COI – coronatine-insensitive; COP – constitutive photomorphogenic; CRY – cryptochrome; DET – de-etiolated; ELIP – early light-inducible protein; FAD – fatty acid desaturase; FLU – fluorescent; GUN – genome-uncoupled; HIR – high-irradiance response; HR – hypersensitive response; JA(-Me) – jasmonic acid (methyl ester); LAF – long after far red light; LHCII – light-harvesting chlorophyll *a*/*b* binding protein of photosystem II; LHPP – light-harvesting POR (NADPH:protochlorophyllide oxidoreductase) protochlorophyllide complex; LLS – lethal leaf spot; LSD – lesion-simulating disease; NCC – non-fluorescent chlorophyll catabolite; Pao – phaeophorbide *a* oxygenase; PCD – programmed cell death; PΦB – phytochromobilin; Pheide – phaeophorbide; PHY or phy – phytochrome; PIF3 – phytochrome-interacting factor 3; PKS1 – phytochrome kinase substrate 1; PLB – prolamellar body; POR – NADPH:protochlorophyllide oxidoreductase; PQ – plastoquinone; Pr/Pfr – red light/far red light-absorbing form of phytochrome; Proto (gen)IX – protoporphyrin(ogen) IX; PS – photosystem; RCC – red chlorophyll catabolite; RCD – radical-induced cell death; ROS – reactive oxygen species; SA – salicylic acid; SAG(s) – senescence-associated gene(s); SARK – senescence-associated receptor-like kinase; SIRK – senescence-induced receptor-like kinase; THI – thionin.

post-germination seedling development (see Altmann, 1999; Friedrichsen and Chory, 2001; Bishop and Koncz, 2002). Major signaling events involved in photomorphogenesis include light-dependent conformational changes in the phytochrome holoprotein, translocation of phytochrome to the nucleus and its binding to transcription factors, and transcription of photosynthetic genes. Coupled with these events, the degradation of positively acting intermediates is an important process whereby photomorphogenesis is repressed in the dark (Schaefer and Bowler, 2002; Fankhauser and Staiger, 2002).

Phytochrome is a chromoprotein that contains a linear tetrapyrrole chromophore called phytochromobilin (PΦB). Phytochromobilin is synthesized via the C5-pathway (Beale and Weinstein, 1990; von Wettstein et al., 1995; S. Reinbothe and C. Reinbothe, 1996b; Suzuki et al., 1997) from glutamate and leads to the formation of protoporphyrin IX. At this intermediate, the C5-pathway bifurcates into a Mg branch giving rise to chlorophyll Chl a and Chl b, and a Fe branch leading to heme and PΦB.

In angiosperms, Chl biosynthesis is strictly dependent on light (Granick, 1950). The only known step in Chl biosynthesis is catalyzed by the NADPH:protochlorophyllide (Pchlide) oxidoreductase (POR) (W.T. Griffiths, 1975, 1978; Apel et al., 1980). In etiolated plants, two isoforms of POR, termed PORA and PORB (Holtorf et al., 1995), establish a light-harvesting POR:Pchlide protein complex (LHPP) that is embedded into the prolamellar body of etioplasts (C. Reinbothe et al., 1999, 2003b,c; S. Reinbothe et al., 2003a). Etioplasts represent progenitors of the photosynthetically active chloroplast (Kirk and Tilney-Basset, 1978). Upon light perception, chlorophyllide (Chlide) a is made, and Chl a is assembled into the photosynthetic membrane complexes. Seedling de-etiolation can occur over a large range of different light intensities, including those available in the soil or under fallen leaves, while LHPP confers photoprotection to the plant (C. Reinbothe et al., 1999; Willows, 1999). Excess light energy is dissipated in the bulk of non-photoreducible Pchlide. The biosynthesis of Pchlide and its subsequent binding to LHPP is tightly controlled in time and space, and is subject to feedback inhibition in etiolated plants (Beale and Weinstein, 1990; Beale 1999). The inhibition of Pchlide production is released only when etiolated plants are exposed to light. Simultaneously, LHPP disintegrates and its protein constituents are degraded by plastid proteases (Kay and W.T. Griffiths, 1983; Häuser et al., 1984; S. Reinbothe at al., 1995a-c; C. Reinbothe et al., 1995d). Chlorophyllide a and Chlide b are in turn released and transiently bound to water-soluble chlorophyll proteins, thus ensuring that potentially phototoxic pigments are not present in free, excitable forms. In the final step, Chlides are esterified to Chl a and Chl b and assembled with the plastid- and nucleus-encoded Chl binding proteins into the photosynthetic membrane complexes.

During leaf senescence, plants encounter the challenge of having to sequester their potentially phototoxic Chls (Takamiya et al., 2000). Chlorophyll turnover is, in fact, a major aspect of chloroplast destruction during leaf senescence (S. Yoshida et al., 2003; H. Thomas et al., 2003). Interestingly, plant hormones such as jasmonic acid (JA) and ethylene have been reported to induce chlorophyllase (Chlase), one of the first enzymes involved in Chl breakdown (Benedetti et al., 1998; Jacobs-Wilk et al., 1999; Tsuchiya et al., 1999), and several key genes normally activated during leaf senescence (Semdner and Parthier, 1993; C. Reinbothe and S. Reinbothe, 1996a). The resulting Chlide is converted into linear products, one of which has been identified as the so-called red chlorophyll catabolite (RCC; Hörtensteiner et al., 1995). In a lesion mimic mutant of *Arabidopsis thaliana*, the gene encoding RCC reductase, which converts RCC to smaller, fluorescent products, was found to be defective (Mach et al., 2001). As a result of accumulation of free RCC or earlier break-down products, cell death occurred (Mach et al., 2001). Studies on targeted gene disruption and regulated expression of uroporphyrinogen III decarboxylase (G. Hu et al., 1998), coproporphyrinogen III oxidase (Ishikawa et al., 2001), and protoporphyrinogen oxidase (Molina et al., 1999) emphasize that tetrapyrrole synthesis and degradation need to be tighty controlled during the entire plant life cycle. It is the aim of the following sections to highlight the various principles that coordinate photosynthetic gene expression in relation to tetrapyrrole pigment metabolism during plant development.

II. Control of Photosynthetic Gene Expression during Seedling Etiolation and De-etiolation

A. Skotomorphogenesis and Photomorphogenesis as Alternative Developmental Strategies to Adapt to the Light Environment

Plant development can proceed in two fundamentally different ways. In the presence of light, plants undergo a series of temporally and spatially coordinated changes known as photomorphogenesis (Kendrick and Kronenberg, 1994). During this process leaves expand, stems undergo radial enlargement, pigments accumulate within developing chloroplasts, and numerous metabolic processes associated with photosynthetically competent cells become operational (Ma et al., 2001; Tepperman et al., 2001; Schroeder et al., 2002). In the absence of light, a radically different growth pattern is triggered, called skotomorphogenesis. Dark-grown seedlings display an apical hook, closed and unexpanded cotyledons, and elongated hypocotyls. Dark-grown angiosperm seedlings are devoid of Chl and are incapable of photosynthetic function (Kendrick and Kronenberg, 1994). If left to grow in the dark, they will eventually die. However, upon emergence from the soil or from underneath fallen leaves, the seedlings undergo de-etiolation, i.e. cotyledons open, expand, and begin to photosynthesize. At the same time, hypocotyl elongation is inhibited and cell differentiation is initiated in vegetative meristems. Proplastids present in leaf meristems differentiate into other forms such as etioplasts produced in dark-grown plants or chloroplasts in plants undergoing photomorphogenesis. Etioplasts transform into chloroplasts once the seedling de-etiolates (Virgin et al., 1963; Kahn, 1968; Kahn et al., 1970; Kirk and Tiley-Basset, 1978).

B. Phytochrome Action on Photosynthetic Genes

In plants, light-dependent responses are controlled by the concerted action of at least three types of photoreceptors—the red- and far red-light-absorbing phytochromes, the blue light-absorbing cryptochromes, and the blue light-absorbing phototropins (Quail, 2002a,b).

Phytochrome is presumably the best-characterized photoreceptor. It represents a chromoprotein that consists of the plastid-derived chromophore PΦB and a nucleus-encoded apoprotein (Fig. 1). Phytochrome apoproteins are typically encoded by a small family of genes, e.g. designated as *PHYA-PHYE* in *Arabidopsis* (Møller et al., 2002; Nagy and Schäfer, 2002; Quail, 2002a,b). They are assembled into their pigment-complexed homodimeric forms in the cytosol where the chromophore is covalently bound to the N-terminal half of each monomer (Montgomery and Lagarias, 2002). Phytochromobilin, synthesized within the chloroplast, must be exported to permit phytochrome assembly. Dimerization domains are located within the C-terminal half of the phytochrome apoprotein, as are other domains involved in the activation of signal transduction (Quail et al., 1995; Quail, 2002a).

Each phytochrome can exist in two photointerconvertible conformations: a red light-absorbing form denoted Pr and a far red light-absorbing form denoted Pfr (Fig. 1A and B). In the dark, de novo synthesized phytochrome is present in the Pr form. It is converted to Pfr in sunlight that contains a relatively larger proportion of red versus far red light (H. Smith, 1982). A fraction of *Arabidopsis* Pfr, in turn, translocates to the nucleus (summarized in Nagy and Schaefer, 2002), implicating signal transduction pathways coupling phytochrome action directly to nuclear gene activation. However, biochemical and pharmacological studies have revealed the operation of cytosolic signal transduction pathways as well, involving G-proteins, cGMP, calcium, and calmodulin (e.g. Shacklock et al., 1992; Bowler et al., 1994). Reverse genetic approaches have confirmed that G-proteins (Okamoto et al., 2001) and a cytoplasmically-localized calcium-binding protein act as negative regulators of phyA and cryptochrome responses (Guo et al., 2001).

→

Fig. 1. (Figure is on the following page) Structural domains and spectral properties of phytochromes. A, Biosynthetic pathway of the phytochrome chromophore. Intermediates are given in upper case letters. Enzymes are indicated in italics, and their *Arabidopsis* genes highlighted. Note that chromophore binding is autocatalytic, it requires phytochromes bilin lyase domain; phytochromes also possess PΦB isomerase activity. The arrow indicates the site of chromophore attachment to a cysteine in the bilin lyase domain. B, Absorption of red light triggers a Z to E isomerization in the C15 double bond between the C and D rings as indicated. The absorption spectra of Pr and Pfr are indicated. C, Structural domains of prokaryotic and plant phytochromes. Phytochrome consists of a photosensory domain and a regulatory out-put domain. The photosensory domain comprises four sub-domains called P1-4, whereas the regulatory out-put domain comprises four sub-domains designated R1-4. PAS is the acronym from the founding members of this protein domain (PER, period clock protein/ARNT, aromatic hydrocarbon receptor nuclear translocator/SIM, single minded protein). Cph1, Cyanobacterial phytochrome 1 from *Synechocystis sp. PCC6803*; RcaE, Regulator of chromatic adaptation E from *Fremyella diplosiphon*; Bph1, Bacteriophytochrome 1 from *Deinococcus radiodurans*; Ppr, PYP-phytochrome-related from *Rhodospirillum centenum*. Modified after Fankhauser (2001); with kind permission of Christian Fankhauser, Université de Genève, Geneva, Switzerland.

Chapter 21 Photosynthetic Gene Expression during Greening and Senescence

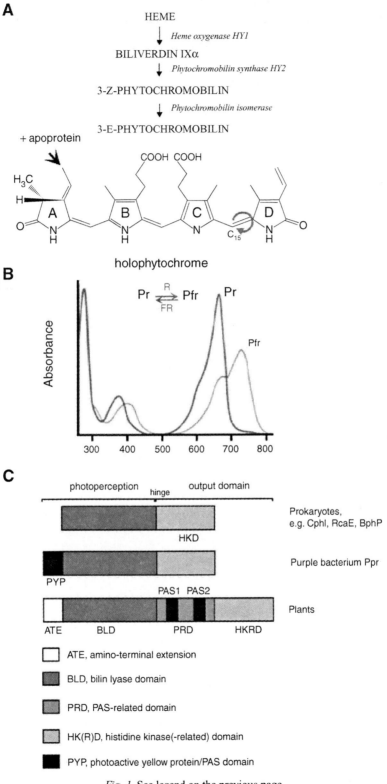

Fig. 1. See legend on the previous page.

Genetic screens have led to the identification of at least two major mutant classes that are impaired in phy signaling:

(i) mutants that are insensitive to light and
(ii) mutants that display constitutive photomorphogenesis in darkness (see Møller et al., 2002, for a review).

The most severe white light-insensitive mutants are the long hypocotyl, *hy*, mutants. They include true photoreceptor mutants, in which PΦB synthesis is perturbed (Davis et al., 1999; Muramato et al., 1999; Kohchi et al., 2001), and phytochrome apoprotein mutants that are no longer capable of expressing one or the other form of the phytochrome apoprotein. For example, *hy1* was found to lack heme oxygenase that converts heme to biliverdin IXα (Davis et al., 1999; Muramato et al., 1999), whereas *hy2* lacked 3Z-phytochromobilin synthase (alternatively named biliverdin reductase; Kohchi et al., 2001) (Fig. 1A). By contrast, *hy3* and *hy8* are deficient in the PHYB and PHYA apoproteins, respectively (Reed et al., 1993; Whitelam et al., 1993).

Hy5 is a signal transduction mutant. The *hy5* gene encodes a bZIP transcription factor (Oyama et al., 1997) that affects both phyA and phyB signaling. The severity of the *hy5* mutant demonstrates that the HY5 protein plays a key role in the control of photomorphogenesis, whereas other mutants may be specifically impaired in either phyA or phyB signaling (Møller et al., 2002; Nagy and Schäfer, 2002; Quail, 2002a).

Most other phyA signalling mutants are defective in genes encoding nuclear-localized proteins, such as FHY1, FHY3, SPA1, FAR1, LAF1, and EID1 (see Møller et al., 2002; Nagy and Schäfer, 2002). An exception to the rule appears to be PKS1, phytochrome kinase substrate 1, that is a constitutive cytoplasmic protein of unknown function (Fankhauser et al., 1999).

PKS1 as well as cryptochromes (CRY) 1 and 2 are substrates for in vitro phosphorylation by oat phyA. Within the C-terminal half of phytochrome, two regions have been identified that share similarity with bacterial histidine kinases operating as part of two component signaling systems (Quail, 2000; 2002a,b) (Fig. 1C). In bacterial phytochromes, the sensory histidine kinase domain autophosphorylates and relays a phosphor group to an aspartate residue of a response regulator that may be a transcriptional activator (Montgomery and Lagarias, 2002).

Phytochrome-interacting factor 3 (PIF3) is another interesting example of a putative signaling component (Quail, 2002a,b). It belongs to the family of basic helix–loop–helix (bHLH) transcription factors capable of binding to G-box promoter elements of light-regulated genes (Martinez-Garcia et al., 2000; Quail, 2000). As a major implication of this important finding, direct signaling pathways can be envisioned that could involve nuclear-localized Pfr and PIF3. Furthermore, phytochrome/PIF3 interactions were found to be red/far red light-reversible and only occurred when phytochrome was present in its Pfr form (Ni et al., 1999). Given that PIF3 can also bind to the G-boxes within the promoters of the *LHY* and *CCA1* genes (Martinez-Garcia et al., 2000), which encode Myb transcription factors, a role of PIF3 in controlling circadian rhythms has been proposed as well (Quail, 2000).

The second major mutant class comprises so-called de-etiolated (*det*), constitutively photomorphogenetic (*cop*), and *fusca* mutants. Eleven loci have been assigned to this large mutant class (*det1, cop1, cop8, cop9, cop10, cop11, cop16, fus5, fus8, fus11,* and *fus12*) (Hardtke and Deng, 2000). The phenotypes of these mutants indicate that these genes encode negative regulators of light signaling.

Considerable progress has been made towards deciphering the actual functions of many *det, cop,* and *fus* genes. A particularly interesting case is DET1. This protein was proposed to operate as a chromatin remodelling factor (Benvenuto et al., 2002). Like other *cop/fus* mutants, plants defective in DET1 display constitutive de-etiolation in darkness, implying that DET1 may play a role in the repression of light-inducible genes (Quail, 2002a). DET1 binds to nucleosome core particles via an interaction with the N-terminal tail of histone H2B (Benvenuto et al., 2002). Furthermore, DET1 is part of a larger complex containing UV–DDB1, which - in animal cells - is part of histone acetyltransferase complex (Schroeder et al., 2002). DET1 binds preferentially to non-acetylated H2B tails of light-inducible genes and is subsequently displaced by a light-dependent acetylation of the histone tails, thus permitting gene expression (Benvenuto et al., 2002).

Eight of the 11 *cop/det/fus* mutants are defective in components of the COP9 signalosome (Hardtke and Deng, 2000). Sequence analyses of its core subunits and associated proteins have unraveled their evolutionary relationship to the lid subcomplex of the 19S regulatory particle of the 26S proteasome, which degrades polyubiquitinated proteins (Hardtke and Deng, 2000). A role of the COP9 signalosome in the degradation of positive regulators of photomorphogenesis in the dark and negative regulators in the light is thus conceivable (Hardtke and Deng, 2000). Additional roles of the COP9 signalosome in auxin responses and pathogen defense have been proposed, explaining the pleiotropic character of most COP9-defective plants (Schwechheimer et al., 2001; Azevedo et al., 2002; Hellmann and Estelle, 2002).

The COP1 and COP10 proteins are not intrinsic components of the COP9 signalosome, but may play supplementary roles in regulating protein turnover (Hardtke and Deng, 2000; Hellmann and Estelle, 2002; Suzuki et al., 2002). Interestingly, COP10 resembles ubiquitin-conjugating E2 enzymes (Suzuki et al., 2002), whereas COP1 is similar to E3 ubiquitin ligases (Hardtke and Deng, 2000). Potential targets of COP1 have been identified (Hardtke and Deng, 2000) and include putative transcription factors (e.g. Holm et al., 2002) as well as CRY1 and CRY2 (Ahmad et al., 1998; H. Wang et al., 2001; H.Q. Yang et al., 2001).

C. Chlorophyll Biosynthesis

1. Steps, Intermediates, Enzymes

Chlorophyll and heme needed for the assembly and function of the photosynthetic apparatus are synthesized via the C5-pathway that is well established in most organisms (Beale and Weinstein, 1990; von Wettstein et al., 1995; S. Reinbothe and C. Reinbothe, 1996b; Suzuki et al., 1997) (Fig. 2). In higher plants and algae, the C5-pathway takes place in the chloroplast and involves both soluble and membrane-associated enzymes (Beale and Weinstein, 1990).

Tetrapyrrole synthesis is regulated by light, primarily at two points in the C5-pathway: the synthesis of 5-aminolevulinic acid (5-ALA) and the reduction of Pchlide to Chlide (S. Reinbothe and C. Reinbothe, 1996b; S. Reinbothe et al. 1996a). 5-Aminolevulinic acid is the first committed precursor of all tetrapyrroles in plants and its synthesis is generally accepted to be rate-limiting (Beale and Weinstein, 1990). In plants, 5-ALA formation requires the coordinated actions of glutamyl-tRNA synthetase, glutamyl–tRNA reductase, and glutamate-semialdehyde amino transferase, and appears to be regulated by a combination of factors including de novo enzyme synthesis and turnover (Beale, 1990), feedback inhibition by both heme and Pchlide (Beale, 1990; Beale and Weinstein, 1990; see below), and phytochrome (Huang et al., 1989). Porphobilinogen deaminase, 5-ALA dehydratase, coproporphyrinogen oxidase, and other enzymes catalyzing common steps of Chl and heme biosynthesis are presumed to be present at levels that are not rate limiting (Granick, 1950). However, their abundance and activity may be under the regulation of light quality and quantity, cell type, and age of the plant (A.G. Smith, 1986; Boese et al., 1991; Spano and Timko, 1991; Witty et al., 1993).

2. Regulation of the C5-Pathway

a. Feedback Inhibition of 5-Aminolevulinic Acid Synthesis in Angiosperms

In dark-grown angiosperms, Chl synthesis leads only to the formation of Pchlide, the immediate precursor of Chlide. Once a critical level of Pchlide has been reached, 5-ALA synthesis is rapidly switched off (Granick, 1950). Only after illumination, when Pchlide has been photoreduced to Chlide by virtue of the NADPH:Pchlide oxidoreductase (POR), does 5-ALA synthesis resume (S. Reinbothe et al., 1996a). Feedback control of Chl biosynthesis has been attributed to inhibition of 5-ALA synthesis. In analogy to its regulatory role in animals and yeast (Andrew et al., 1990; Labbe-Bois et al., 1990), heme has been proposed to act as a feedback inhibitor in plants as well (Pontoppidan and Kannangara, 1994).

Using a genetic approach, Meskauskine et al. (2001) identified an additional factor involved in the control of tetrapyrrole biosynthesis in *Arabidopsis*. They demonstrated that a negative regulator of tetrapyrrole biosynthesis exists, termed FLU (FLUORESCENT), which operates independently of heme and selectively affects only the Mg^{2+} branch of the C5-pathway. FLU is a nuclear-encoded plastid protein that, after import and processing, becomes tightly associated with plastid membranes. It is unrelated to any of the enzymes known to be involved in tetrapyrrole biosynthesis. Its predicted features suggest that FLU may mediate its regulatory effect through direct interaction with glutamyl-tRNA reductase (Meskauskine et al., 2001). A tetratrico-peptide repeat motif was identified that mediates protein-protein interactions in yeast two-hybrid screens (Meskauskine et al., 2002).

Recent work has shown that mutations in orthologous *flu* loci in barley give rise to *tigrina* phenotypes, comprising alternate necrotic and green leaf sections (Lee et al., 2003). In monocotyledonous plant species, leaf growth occurs via cell division at the leaf base followed by differentiation of tissues in the older parts. Excess Pchlide that accumulates in newly-formed tissues in the dark triggers photooxidative pigment bleaching during subsequent light periods if plants are grown in dark/light cycles. In tissues differentiated in the presence of light, no photooxidative damage occurs because Pchlide is continuously photoreduced to Chlide by POR

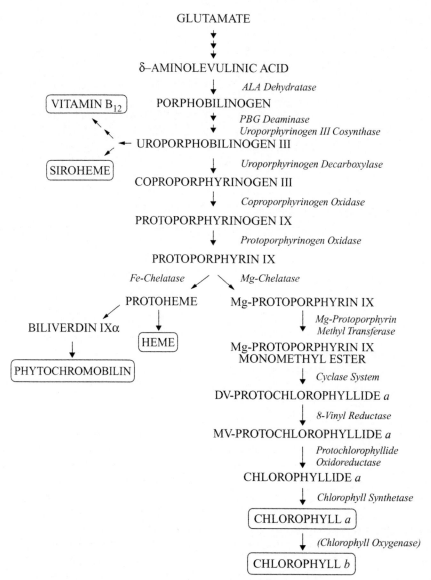

Fig. 2. The C5-pathway of tetrapyrrole synthesis. Intermediates and enzymes are highlighted in upper case letters and italics, respectively.

and converted to Chl (Gough and Kannangara, 1979; Casadoro et al., 1983).

In a recent report, another factor was described that negatively regulates the chlorophyll biosynthetic pathway in *Arabidopsis thaliana* (Huq et al., 2004). This factor is a bHLH transcriptional regulator related to the previously discussed phytochrome-interacting factor PIF3. It was proposed that PIF1 represses the expression of key enzymes operating in the C5-pathway in the dark and that this activity would be negatively regulated by light (Huq et al., 2004). Because PIF1 does not affect the expression of the FLU protein, additional signaling intermediates must exist that allow the plant to modulate the rate of chlorophyll biosynthesis in response to environmental light conditions.

b. Light-Dependent Protochlorophyllide Reduction Catalyzed by NADPH: Protochlorophyllide Oxidoreductase

The second regulatory point in the Mg branch of the C5-pathway occurs at the step of Pchlide reduction. In angiosperms, Pchlide reduction is a light-dependent reaction catalyzed by POR (W.T. Griffiths, 1975, 1978; Apel et al., 1980) (Fig. 3). Both POR and its pigment substrate Pchlide accumulate to high levels in the leaves

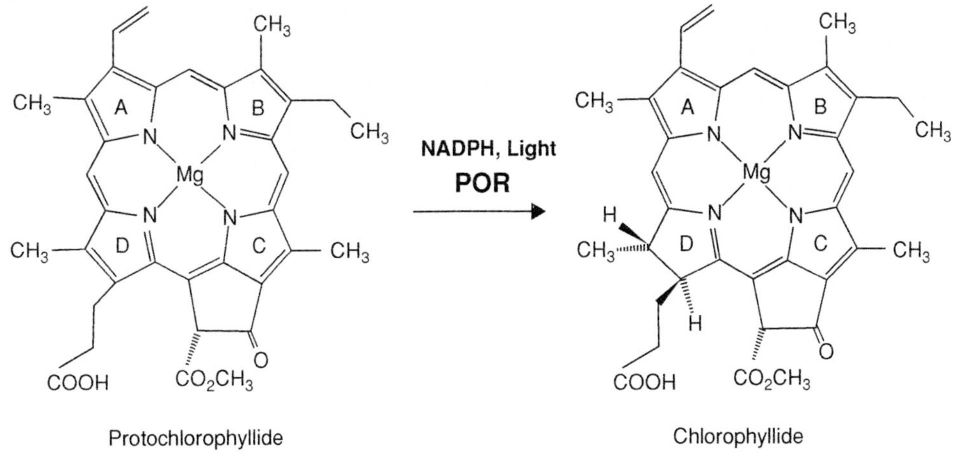

Chapter 21 Photosynthetic Gene Expression during Greening and Senescence

Fig. 3. Reaction catalyzed by POR. POR catalyzes the only known light-dependent step of chlorophyll biosynthesis in angiosperms, the region- and stereo-specific reduction of Pchlide to Chlide. NADPH is consumed and a hydrid transferred onto ring D of the macrocycle. A conserved Tyr, Tyr275 in case of the pea enzyme, then donates a proton to give rise to the fully saturated carbon-carbon bond.

of dark-grown plants (Boardman, 1962). Dark-stable Pchlide:NADPH:POR ternary complexes are poised such that absorption of a photon by the pigment itself leads to its immediate reduction resulting in the formation of Chlide *a* (see Lebedev and Timko, 1998, for a review). Chlide *a* is subsequently esterified and further modified to produce Chls *a* and *b* in conjunction with the formation of functional photosynthetic membrane complexes (reviewed by Sundqvist and Dahlin, 1997).

Eukaryotic POR is nuclear-encoded. It is synthesized as a larger precursor polypeptide in the cytosol (Apel, 1981). In etiolated angiosperms, POR is localized primarily in the prolamellar body (PLB) of etioplasts (Dehesh and Ryberg, 1985; Ryberg and Dehesh, 1986; Shaw et al., 1985). In light-grown seedlings, much smaller amounts of POR are found (Forreiter et al., 1990; Barthélemy et al., 2000). Despite its light requirement for catalysis, POR activity and level rapidly decrease as a result of proteolysis (Forreiter et al., 1990). At a stage when Chl accumulation reaches its maximum rate, apparently no POR protein is detectable.

In angiosperms such as barley (Holtorf et al., 1995), *Arabidopsis* (Armstrong et al., 1995; Oosawa et al., 2000), tobacco (Masuda et al., 2002), and *Amaranthus tricolor* (Iwamoto et al., 2001), as well as gymnosperms such as pine species (Forreiter and Apel, 1993; Skinner and Timko, 1998; Spano et al., 1992), *POR* gene families were identified that encode highly conserved POR polypeptides. PORA is a negatively light-regulated POR enzyme whose level drops as a result of the effect of light at the levels of transcription, mRNA stability, and plastid protein import (S. Reinbothe et al., 1995a-c, 1996a). PORB, the second POR protein identified in barley and *Arabidopsis* (Armstrong et al., 1995; Holtorf et al., 1995), was found to be consitutively expressed in dark-grown, illuminated, and light-adapted plants (S. Reinbothe et al., 1996a). Recent studies have led to the discovery of a third, positively light-regulated POR enzyme, termed PORC (Oosawa et al., 2000; Su et al., 2001; Pattanayak and Tripathy, 2002). Current work is focused on the question of whether plants with multiple *POR* genes have evolved specialized PORs with distinct biochemical and biophysical properties to perform unique functions during development or whether all of the different PORs accomplish redundant roles (Frick et al., 1995, 1999; Lebedev et al., 1995; Runge et al., 1996; Sperling et al., 1997, 1998, 1999; Lebedev and Timko, 1999; C. Reinbothe et al., 1999; Franck et al., 2000).

3. LHPP (Light-harvesting POR: protochlorophyllide complex) Provides the Link between Skotomorphogenesis and Photosynthesis of Higher Plants

a. Structural Components of LHPP

At least for dark-grown barley plants, PORA and PORB have been demonstrated to have unique functions. They structurally and functionally cooperate in terms of larger light-harvesting POR:Pchlide complexes (LHPP) in the prolamellar body of etioplasts (C. Reinbothe et al., 1999; S. Reinbothe et al., 2003a; C.

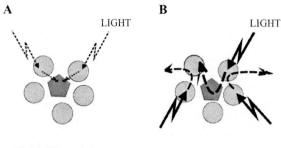

Fig. 4. A model of LHPP. LHPP is presumed to consist of two isoforms of the photoenzyme NADPH:Pchlide oxidoreductase, termed PORA (light grey circles) and PORB (dark-grey pentagons), Pchlide *b* and Pchlide *a*, respectively, NADPH, as well as galacto- and sulfolipids (not shown). All these constituents are held together in ring-like structures comprising 5 PORA-Pchlide *b*-NADPH ternary complexes and 1 PORB-Pchlide *a*-NADPH complex in the prolamellar body of etioplasts. By virtue of this organization, traces of light can be harnessed and used for driving Chlide *a* synthesis (A), whereas excess light energy is dissipated in the bulk of non-photoreducable Pchlide *b* (B).

Reinbothe et al., 2003b,c) (Fig. 4). In several aspects, LHPP resembles the major light-harvesting complex of photosystem II, LHCII (Kühlbrandt et al., 1994). By analogy to LHCII, that is composed of Chl *a* and Chl *b* (Kühlbrandt et al., 1994), LHPP contains two different types of pigments: Pchlide *a* and Pchlide *b* (C. Reinbothe et al., 1999, 2003b; S. Reinbothe et al., 2003a). However, in contrast to LHCII, these two pigments are bound to two different polypeptides: Pchlide *a* to PORB and Pchlide *b* to PORA (C. Reinbothe et al., 1999, 2003b; S. Reinbothe et al., 2003a). In the case of LHCII, a single polypeptide chain is sufficient to attain a three-dimensional structure capable of accommodating both Chl *a* and Chl *b* (Kühlbrandt et al., 1994). Monomeric LHCII further assembles into trimers (Dreyfuss and Thornbeer, 1994). Energy transfer between Chl *b* and Chl *a* can take place because of the different energy contents and life times of their excited states (Ide et al., 1987; Palsson et al., 1994). Presumably for the same reasons, energy transfer is also possible between Pchlide *b* and Pchlide *a* in LHPP and the prolamellar body (J.H.C. Smith and Benitez, 1954; Kahn et al., 1970; Mathis and Sauer, 1972; Vaughan and Sauer, 1974; Ignatov and Litvin, 1981; Fradkin et al., 1993). The native LHPP complex additionally contains NADPH as well as galacto- and sulfolipids that alter the spectral properties of bound pigments. The absorption maximum is shifted from ca. 630 nm to 650 nm (C. Reinbothe et al., 1999; S. Reinbothe et al., 2003a). In this wavelength region of the spectrum, Pr has its main absorption maximum as well (see Fig. 1),

and Pchlide *a* reduction and phytochrome action are thus tied during seedling development.

Recent biophysical studies have highlighted additional intermediates in the assembly pathway of the photosynthetically active thylakoid membrane complexes. Kóta et al. (2002) reported gross alterations in the secondary structure of membrane proteins affecting conformation, composition, and dynamics of lipid acyl chains as well as protein patterns. The authors suggested that the greening process may be accompanied by major reorganization events affecting both membrane protein assembly and the protein-lipid interphase. These conclusions may not be unanticipated because it is known that PORA and PORB, that are major constituents of the prolamellar body of the etioplast, undergo changes in their conformation (S. Reinbothe et al., 1995a-c; C. Reinbothe et al., 1995d), which eventually leads to their degradation by plastid proteases (see S. Reinbothe et al., 1996b, for a review). It was proposed that Chlide *a* and Chlide *b* may bind to other pigment carriers, such as the early light-inducible proteins – ELIPs (Kloppstech et al., 1984; Grimm and Kloppstech, 1987; Grimm et al., 1989; Adamska, 1997; Adamska et al., 1999) or PsbS (Funk et al., 1995a,b; Adamska et al., 1996), that belong to the same superfamily of light-harvesting proteins as LHCII and display conserved pigment binding domains/sites (B.R. Green et al., 1991; Jansson, 1994; B.R. Green and Durnford, 1996). ELIP expression has been demonstrated, for example, in light-exposed barley and pea plants; the time point of accumulation is consistent with a possible function in Chl(ide) storage. PsbS may also be involved in the greening process because the protein is expressed in etiolated plants and remains active in illuminated and green tissues (Funk et al., 1995a,b). An ortholog of water-soluble chlorophyll proteins of Brassicaceae (Satoh et al., 1998), that binds and sequesters Chlide in a nonhazardous form, has been identified (C. Reinbothe et al., 2004a).

b. Developmental Expression of LHPP

PORA and PORB expression is subject to developmental control (Dehesh et al., 1983; Z.H. He et al., 1994; Armstrong et al., 1995; Runge et al., 1996; Frick et al., 1999; Ougham et al., 2001). In barley and *Arabidopsis*, both *POR* genes are coordinately expressed during skotomorphogenesis (Armstrong et al., 1995; Frick et al., 1999). In monocotyledonous plant species, where leaf development is determined by basipetal cell growth in the leaf meristem and maturation of tissues in the older parts (see above), PORA and PORB expression are adjusted to plastid development. Proplastids present in the leaf base differentiate into etioplasts which, in turn, are transformed into chloroplasts in the presence of light. Consistent with the idea that skotomorphogenesis may be a developmental strategy to prepare the plant for light-harvesting, LHPP expression was found to be maximal in the leaf tip of etiolated barley plants whereas LHPP disintegrated upon light exposure (C. Reinbothe et al., 2004b). In dicotyledonous species such as *Arabidopsis*, a similar PORA and PORB expression pattern was observed (Armstrong et al., 1995) and a LHPP-like complex has recently been resolved (S. Reinbothe and C. Reinbothe, 2004).

Genetic and biochemical approaches have been used to address the actual functions of PORA and PORB in *Arabidopsis*. As a first approach, plants were grown under continuous far red light. Classically, phytochrome responses can be discerned by their wavelength and fluence or fluence-rate (i.e., intensity) requirements into three groups—very low fluence responses, low fluence responses, and high irradiance responses (HIR) (Shinomura et al., 2000; H. Smith, 2000; Nagy and Schaefer, 2002). The HIR of etiolated plants is mediated by phyA. It leads to a complete suppression of *PORA* gene expression (Runge et al., 1996). As a result, *Arabidopsis* plants were produced which lacked *PORA* transcript and PORA protein. Runge et al. (1996) observed that such plants displayed a dramatically increased light sensitivity during greening, as compared to plants grown in darkness, and concluded that PORA may be specifically needed for seedling de-etiolation. Later studies showed that overexpression of the cDNA encoding PORA leads to greater seedling survival rate after growth in far red light as compared to the non-transformed wild-type (Sperling, 1998; Sperling et al., 1997). Because overexpression of PORB cDNA permitted a better greening, it was proposed that PORA and PORB may accomplish redundant roles during seedling de-etiolation (Sperling et al., 1997, 1998, 1999). Apparently in line with this proposal are other studies showing that antisense inhibition of PORA and PORB expression results in abnormal etioplast development (Franck et al., 2000).

A breakthrough towards an understanding of PORA and PORB function during greening has recently been achieved. Frick et al. (2003) isolated a double knock-out mutant of *Arabidopsis* that lacks both PORB and PORC because of interruptions in their corresponding genes by a derivative of the maize Dissociation (Ds) transposable element. PORC has been identified by sequence homology as a consequence of the *Arabidopsis* genome project (Oosawa et al., 2000; Su et al., 2001; Pattanayak and Tripathy, 2002). Unlike PORA and PORB, PORC is expressed at barely detectable levels in dark-grown plants and is positively light-regulated during greening (Oosawa et al., 2000; Su et al., 2002; Frick et al., 2003). The spatial expression pattern of PORC appears similar to that of PORB in older seedlings and adult plants (Armstrong et al., 1995; Frick et al., 2003).

The phenotype of the *porB-porC* double mutant provides the first, though indirect, evidence for a photoprotective role of PORA in light-dependent Chl biosynthesis. Germinating *porB-porC* double mutants initially resemble wild-type seedlings at a time point when the wild type is known to transiently express PORA (Armstrong et al., 1995), but thereafter developed first *chlorina* and then *xantha* phenotypes as photomorphogenesis progressed and PORA expression was turned off (Frick et al., 2003). *PorB-porC* double mutants also suffered from a severe Chl *a* and almost complete Chl *b* deficiency and established only low amounts of thylakoid membrane stacking. Thus, PORA alone appears to be insufficient to sustain normal greening, but may play a major role in conferring photoprotection on PORB, as depicted by the LHPP model (see C. Reinbothe et al., 1999; S. Reinbothe et al., 2003a).

The functions of PORB and PORC in green plants have not been resolved. Unlike PORA, PORB and PORC may play redundant roles in maintaining light-dependent Chl biosynthesis in green plants (Frick et al., 2003). Alternatively, they could cooperate during growth and development. One possible scenario could be that PORB and PORC together maintain Pchlide reduction capacity, perhaps by forming larger complexes analogous to that of LHPP operating in etiolated plants (C. Reinbothe et al., 1999, 2003b).

4. Evolution of Protochlorophyllide-Based Light-Harvesting

Important clues concerning the evolution of Pchlide-based light-harvesting have come from studies of Chl biosynthesis mutants in cyanobacteria. Cyanobacteria, as the presumed endosymbiotic progenitors of chloroplasts, contain POR (Rowe and W.T. Griffiths, 1995; Suzuki and Bauer, 1995; Fujita et al., 1998) and additionally express a light-independent system for making Chl, which is composed of three subunits called ChlL, ChlN, and ChlB (Armstrong, 1998; Fujita et al., 1998). Mutants lacking POR exhibit a greater light sensitivity than the wild-type (Q. He et al., 1998), implying that POR may confer photoprotection onto the ChlL,B,N system. It has been proposed that ChlL, ChlB, and ChlN arose from gene duplications of nitrogenase subunits that, in their extant forms, are known to be oxygen-sensitive and to be rapidly inactivated if poised with oxygen (Fay, 1992; Burke et al., 1993). It is therefore conceivable that POR evolved to shield the Pchlide binding subunit of the ChlL,B,N system against oxygen and thereby help prevent the formation of singlet oxygen (S. Reinbothe et al., 1996b).

POR and ChlB,N,L coexist in gymnosperms (Forreiter and Apel, 1990; Spano et al., 1992). Frick et al. (2003) put forth the idea that the central role of photosynthesis in plant metabolism may have provided the necessary evolutionary pressure to retain more than one POR enzyme. While both Pchlide-reducing systems were retained in gymnosperms, successive loss of *chlL*, *B*, and *N* genes from the plastid DNA of the engulfed endosymbionts could have accompanied the evolution of angiosperms. Less evolved oxygenic photosynthetic organisms, such as the liverwort (*Marchantia paleacea*; Takio et al., 1998) and the green alga *Chlamydomonas reinhardtii* (Li and Timko, 1996), possess only a single POR protein. These differences amongst oxygenic photosynthesizers in the POR gene complement may be explained by species-specific environmental adaptations to the selective pressures encountered.

D. Plastid-Derived Factors Controlling Photosynthetic Gene Expression in Response to Light

Nuclear gene expression is controlled not only by light, sensed via phytochrome and other photoreceptors, but also by factors originating in the plastid (Oelmüller, 1989; Somanchi and Maxfield, 1999; Surpin et al., 2002). The existence of such factors, that are often collectively referred to as the plastid signal, was implied from studies with inhibitors and mutants of carotenoid biosynthesis. In either case, plants bleach upon high irradiance and at the same time show decreased expression of nuclear photosynthetic genes (Surpin et al., 2002).

1. Intermediates of the C5-Pathway

Recent evidence suggests the existence of at least two independent retrograde signaling pathways originating in the plastid that regulate the expression of photosynthetic genes. In a genetic screen, *genome uncoupled (gun)* mutants defective in nuclear-plastid crosstalk and nuclear gene expression were identified (Susek et al., 1993; Mochizuki et al., 2001). These mutants express nuclear-encoded photosynthetic genes in the absence of proper chloroplasts. Out of the 5 non-allelic *gun* loci known to date, 3 encode enzymes involved in tetrapyrrole synthesis: *GUN2* encodes heme oxygenase (allelic to *HY1*), *GUN3* encodes phytochromobilin synthase (allelic to *HY2*), and *GUN5* encodes the H-subunit of Mg-chelatase (Strand et al., 2003).

Mg-chelatase catalyzes the conversion of protoporphyrin (Proto) IX to Mg-ProtoIX. The enzyme is conserved in plants and *Rhodobacter capsulatus* (Gibson et al., 1996; Walker and Willows, 1997). Pioneering work with *Rhodobacter* has allowed assigning functions to its three subunits designated (b)ChlH, (b)ChlI, and (b)ChlD, respectively. Accordingly, subunit H is thought to bind Proto IX, whereas subunit I contains an ATP-binding site (Gibson et al., 1996; Karger et al., 2001).

Mutants deficient in the plant CHLH and CHLI subunits of Mg-chelatase have been identified in *Antirrhinum majus* (named olive; Hudson et al., 1993) and *Arabidopsis* (named ch42; Koncz et al., 1990), respectively. Unexpectedly, neither mutant type displayed *gun* phenotypes, although they expressed reduced levels of CHLI (Koncz et al., 1990; Hudson et al., 1993). These findings invoke a role of the H-subunit of Mg-chelatase in retrograde signaling. A possible mechanism could involve establishment of a porphyrin relay system that adapts nuclear photosynthetic gene expression to the need of Chl in the chloroplast (Strand et al., 2003). A *cis*-regulatory element responsive to the Mg-Proto IX-mediated signal has been identified in the promoter of the light-reponsive *LHCB1* gene (Strand et al., 2003). This G-box (CUF1) element is distinct from the GT-1 (G3M) motif operating in phy-mediated light perception, thus highlighting the existence of multiple, light-coordinated signaling pathways. Related CUF1 (CACGTA) elements are also present in the promoter regions of 18 genes known to respond to the plastid signal as well as in an additional 24 light-regulated genes (Strand et al., 2003).

In yet another exciting study, a genetic screen was performed to identify *phyA* signaling mutants. Møller et al. (2001) exploited the far-red light HIR of etiolated plants mediated by phyA to screen *Arabidopsis* for *long after far red light treatment* (*laf*) phenotypes. One of the identified mutant loci, *laf6*, encodes a protein related to ABC transporters. ABC transporters are implicated in metabolite transport across membranes (Holland and Blight, 1999; van den Brûle and Smart, 2001). Møller et al. (2001) observed that *laf6* plants accumulate ca. 2-fold higher levels of Proto IX in their plastid envelope. Protoporphyrin IX is synthesized from protoporphyrinogen (Protogen) IX by virtue of Protogen IX oxidase that is plastid envelope-bound (Matringe et al., 1992). These results may collectively suggest an interaction between the LAF6 protein, Protogen IX oxidase, and the H-subunit of Mg-chelatase in the crosstalk between the plastid and the nucleus. Protogen IX (Lehnen et al., 1990), Proto IX (Matringe et al., 1992), and Mg-Proto IX, as well as other intermediates and products of the C5-pathway such as heme (J. Thomas and Weinstein, 1990) and PΦB (Terry and Lagarias, 1991), are known to be exported from the plastid and could serve as plastid signals.

2. Redox Components Associated with Photosynthetic Electron Transport

A completely different retrograde signaling pathway has been proposed in which the redox state of the plastoquinone pool of the chloroplast may provide the ultimate signal (Escoubas et al., 1995; Durnford and Falkowski, 1997; Karpinski et al., 1997; Pfannschmidt et al., 2001a,b). Pfannschmidt et al. (2001b) performed experiments in which the authors modulated the redox poise of the plastoquinone pool and measured photosynthetic gene expression. Promoter-ß-glucuronidase constructs were prepared for four nucleus-encoded PS I genes: *PETH*, *PSAF*, *PSAD*, and *PETE*. These constructs were expressed in transgenic tobacco lines under different excitation pressures, i.e. illumination favoring either excitation of PS I or excitation of PSII, thereby affecting the PQ pool. Pfannschmidt et al. (2001b) observed that the *PSAF* and *PSAD* promoters exhibited higher activity in PS II than PS I light, but there was no reduction after shifting the transgenic lines from PS II to PS I light. The *PETE* promoter, like the *PSAF* and *PSAD* promoters, had a higher activity in PS II light compared with PS I light, but unlike the other two promoters, it was inactivated by a PS II to PS I shift. By contrast, the *PETH* promoter had similar activities under all applied light regimens. Additional experiments using electron transport inhibitors such as DCMU inhibiting electron flow from PS II to PQ, and DBMIB (2,5-dibromo-3-methyl-6-isopropyl-p-benzoquinone) blocking electron flow from plastoquinol to PS I, revealed that the *PSAD* and *PSAF* promoters respond to redox signals that originate between the PQ pool and PS I, or to electron transport capacity in general. Because DBMIB prevented the PS I-induced repression of *PETE*, it was suggested that the redox status of the PQ pool regulated this promoter. By contrast, Pursiheimo et al. (2001) provided evidence that LHCII kinase, and not the PQ pool, may be the primary redox sensor marking the beginning of a signaling cascade to the nucleus. The authors hypothesized that activation/inactivation of the LHCII kinase may occur via a "second loop" of redox regulation involving the cytochrome $b_6 f$ complex and the ferredoxin/thioredoxin system (Rintamäki et al., 2000). Thus, regulation of *LHCB* transcription may not be under direct control of the redox state of the PQ pool but may be mediated by LHCII kinase.

E. Brassinosteroids as Factors Controlling Morphogenesis

Brassinosteroids (BR) are key components in the control of seedling development after germination. As polyhydroxylated sterols, they are derived from the membrane sterol campesterol through a series of reductions, hydroxylations, epimerizations, and oxidations (see Clouse and Sasse, 1998; Altmann, 1999; Bishop and Yokota, 2001; Bishop and Koncz, 2002). There are two major branch points of the pathway termed early and late C-oxidation pathways that later converge to produce the end product brassinolide (BL).

Mutants that are either deficient in BR synthesis or BR signal transduction have been isolated. All BR mutants look de-etiolated in the dark and have short hypocotyls and open cotyledons (Kauschmann et al., 1996; Li et al., 1996; Szekeres et al., 1996; Li and Chory, 1997). In white light, these mutants are dark-green dwarfs with epinastic (curled down) leaves, reduced apical dominance, reduced fertility, and delayed senescence.

Det2 was one of the first *det* mutants characterized (Chory et al., 1991). Later studies showed that it is blocked in an early step of BL synthesis, leading from campesterol to campestanol (Li et al., 1996, 1997; Noguchi et al., 1999). As a result, cell elongation and growth of *det2* seedlings was drastically reduced, giving rise to the observed dwarf phenotype. As for most BR-deficient mutants, supplementing BRs to the growth medium rescued the growth inhibition of *det2* to wild type (Li et al., 1996).

BR-insensitive (*bri*) mutants are defective in components of BR signaling. Some of them have been isolated and characterized in *Arabidopsis* (e.g., *bri1*, *bin2*) and other plant species (summarized in Friedrichsen and Chory, 2001; Bishop and Koncz, 2002). It was demonstrated for example that *BRI1* encodes for all or part of the BR receptor. BRI1 consists of an extracellular ligand-binding domain and an intracellular receptor kinase domain (Li and Chory, 1997; Z.H. He et al., 2000; M. Oh et al., 2000; Z.Y. Wang et al., 2001; Li and Nam, 2002; Yin et al., 2002). Upon BR binding, autophosphorylation and activation of the intracellular kinase domain take place, which may be followed by phosphorylation of downstream components of the signal transduction pathways (Kang et al., 2001).

What is the link between growth regulation by BRs and light signaling? According to recent work,

BRs appear to be connected to light signaling via the *BAS1* gene product. BAS1 regulates levels of active BRs via C26-hydroxylation (Neff et al., 1999). In addition, BL biosynthesis is probably also controlled by direct interaction of cytochrome P450s with signaling proteins. Kang et al. (2001) found that the CYP92A6, a C-2 BR hydroxylase encoded by the *DDWF* (DARK-INDUCED *dwf*-LIKE-1) gene, interacts in the yeast two-hybrid system with the PRA2 light-repressible/dark-inducible small G protein. PRA2 is proposed to act as a connection between the light-signaling and BR-biosynthesis pathways by stimulating DDWF activity in the endoplasmic reticulum, leading to greater BL production and etiolation in the dark.

The dwarf phenotypes of BR-deficient and insensitive mutants reveal that BRs are essential for cell elongation (Y.X. Hu et al., 2000; Friedrichsen and Chory, 2001; Clouse, 2002). In plants, cell expansion requires coordination between changes in cell wall properties, synthesis and transport of new membrane and wall materials, as well as maintenance of osmotic potential (Taiz and Zeiger, 1991). The influx of water is the driving force for cell expansion, reducing the osmotic potential, which, in turn, is re-established by solute uptake into the cytoplasm and into the central vacuole. Based on these aspects it is not surprising that 7 of the BR-induced genes identified thus far encode cell wall-modifying enzymes (Yin et al., 2002). Their lack of expression in BR-deficient plants explains why cell elongation proceeds improperly. In addition, *DET3* has been shown to encode subunit C of the vacuolar H^+–ATPase (V-ATPase) (Schumacher et al., 1999). In eukaryotic cells, V-ATPases play central roles because of their primary function in the acidification of endomembrane compartments. Given that the V-ATPase (together with the H^+–pyrophosphatase) drives solute uptake into the vacuole, its de-regulation – as found in *det3* plants – further adds to the understanding of the observed cell expansion/elongation defects in *det* mutant plants (Schumacher et al., 1999).

III. Photosynthetic Gene Expression during Leaf Senescence

A. Leaf Senescence as a Developmental Program

The final stage of leaf development is senescence. It is characterized by the mobilization of nutrients from the leaf to sink tissues (Bleecker and Patterson, 1997). Leaf senescence is an active process that involves the coordinate expression of mRNAs, proteins, and other metabolic activities. It ultimately leads to the controlled destruction of chloroplasts and mitochondria, nuclear fragmentation, and chromatin decondensation, and terminates in cell death (for recent reviews, see Bleecker and Patterson, 1997; Buchanan-Wollaston, 1997; Nam, 1997; Weaver et al., 1997; H. Thomas et al., 2003; S. Yoshida, 2003; Lin and Wu, 2004). In the early stage of senescence, most of the changes occurring at the gene expression level are reversible. However, once a critical stage has been reached, senescence becomes irreversible. This "point of no return" defines the terminal stage of leaf senescence.

Experimentally, leaf senescence can be induced by different mechanisms referred to as age-associated senescence, dark-induced senescence, and hormone-induced senescence (Y. He et al., 2002; H. Thomas et al., 2003; S. Yoshida, 2003). Age-associated senescence is known to begin when leaf expansion is complete (Mae et al., 1987; Hensel et al., 1993; Crafts-Brandner et al., 1996). Its ultimate trigger has not yet been identified. Changes in photosynthesis and imbalanced source-to-sink relationships have been implicated in senescence induction (Gepstein, 1988; Jiang et al., 1993; Jiang and Rodermel, 1995). In line with this view is the rapid senescence-inducing effect of long-term darkness in many plant species. On the other hand, indirect, phytohormone-dependent changes in the expression of some key signal molecules may be light-controlled and lead to the induction of the senescence program.

Previous studies indicate that many of the genes that are induced during age-associated senescence (i.e., senescence occurring in leaves of non-stressed plants progressing through normal stages of development) are also induced by stresses and/or plant hormones (Azumi and Watanabe, 1991; Hsieh et al., 1995; Stirpe et al., 1996; Buchanan-Wollaston and Ainsworth, 1997; Park et al., 1998; Weaver et al., 1998). It thus seems likely that some of these genes function in both senescence and stress responses.

Jasmonic acid and ethylene are senescence-promoting as well as stress-related compounds (Sembdner and Parthier, 1993; Creelman and Mullet, 1995, 1997; Hadfield and Bennett, 1997; Bleecker and Kende, 2000; Ciardi and Klee, 2001; K.L.C. Wang et al., 2002). Jasmonic acid is widespread in the plant kingdom and accumulates transiently during both biotic and abiotic stresses. It appears that JA operates at many different levels, including early changes in photosynthetic gene expression in the nucleo-cytoplasmic compartment and delayed changes in the chloroplast itself (S. Reinbothe et al., 1994a,b; C. Reinbothe and S. Reinbothe, 1996a). Examples are the rapid polysomal discrimination of photosynthetic mRNAs versus stress-induced mRNAs occurring in methyl jasmonate (JA-Me)-treated barley leaves (S. Reinbothe et al., 1993a,b) and the delayed arrest of nuclear transcription and translation of PS genes (S. Reinbothe et al., 1994a,b; C. Reinbothe et al., 1997). JA-Me, the volatile methyl ester of JA, is an important signaling compound in local and systemic responses to stress and also in inter-plant communication (Farmer and Ryan, 1990; Creelman and Mullet, 1995, 1997). In the plastid compartment, early, JA-Me-induced changes in the processing pattern of rbcL transcripts, encoding the large subunit of ribulose-1,5-bisphosphate carboxylase/oxygenase, lead to a rapid cessation of plastid protein synthesis and cause a shut-down of photosynthesis (S. Reinbothe et al., 1993c). In addition, a role of JA in cell death containment during stress, pathogen defense, and senescence has been proposed (see section III.C.3.).

Ethylene has vital roles in many aspects of plant growth and development and is also implicated in plant responses to biotic and abiotic stresses, such as wounding, heat, pathogen attack, and (O_3). Ethylene biosynthesis genes are present in large multigene families, individual members of which respond differentially to developmental and environmental cues (K.L.C. Wang et al., 2002). For example, it has been reported that ozone (O_3)-induced leaf damage is preceded by a rapid increase in 1-aminocyclopropane-1-carboxylic acid (ACC) synthase activity, ACC content, and ethylene emission (Hadfield and Bennett, 1997; Lund et al., 1998; Drew et al., 2000; Overmyer et al., 2000; Moeder et al., 2002; Rao et al., 2002). Ethylene operates as a positive regulator of ROS production and cell death in compatible plant-bacteria interactions (Bent et al., 1992; Alvarez et al., 1998). Evidence is accumulating that ethylene also plays a role in the regulation of cell death in incompatible plant-pathogen interactions (Ciardi et al., 2001; Ordog et al., 2002), and a role of ethylene in the propagation of cell death during senescence has been proposed (Overmyer et al., 2003). In tomato, ethylene-receptor genes are induced in response to O_3 (Moeder et al., 2002), pathogen infection (Ciardi et al., 2000), and JA (Schenk et al., 2000), suggesting complex regulatory interactions between ethylene and JA perception and signaling (Bleecker and Kende, 2000; Ciardi and Klee, 2001). A possible scenario is that JA restricts ethylene-dependent cell death propagation by reducing ethylene-dependent ROS accumulation (Overmyer et al., 2003; see also section III.C.3. for details).

B. Chloroplast Destruction Occurring during Leaf Senescence

1. Breakdown of Plastid Proteins

The chloroplast is a major target of metabolic adjustment because all of the chlorophyll and more than 70% of leaf protein is found in this organelle. It is not exactly known which proteases are involved in mobilizing the nitrogen contained within chloroplast proteins. It has been discussed that proteins are either digested in the plastid compartment or in the vacuole. Experimental evidence supporting one or the other hypothesis has been obtained. Proteolytic activities degrading Rubisco as the most abundant plastid protein have been localized in the vacuole (Miller and Huffaker, 1981; Wittenbach et al., 1982; Thayer and Huffaker, 1984; T. Yoshida and Minamikawa, 1996). Since the decrease of Rubisco content is much faster than the decrease in the number of chloroplasts in leaves during senescence, it is unlikely that whole chloroplasts are taken up into the vacuole, as originally proposed by Wittenbach et al. (1982). Instead, the transport of Rubisco and/or its degradation products in terms of small spherical bodies through the cytoplasm to the vacuole were demonstrated recently. These small spherical bodies were named RBCs (Rubisco-containing bodies) and shown to contain another stromal protein, glutamine synthetase, but not thylakoid proteins (Chiba et al., 2003). This shows that RBCs differ functionally from the plastoglobuli that increase in number and size during senescence and that were demonstrated to contain thylakoid proteins (Guiamet et al., 1999). The concurrent export of proteins and chlorophyll derivates inside plastoglobuli and their transport to vacuoles was proposed by the same authors, but this hypothesis needs further experimental support.

Chloroplasts contain several proteases such as members of the stroma-localized Clp protease family and the thylakoid-associated FtsH and DegP protease families. Since most of the genes encoding Clp protease components are expressed constitutively, these proteases appear to serve housekeeping functions rather than specific roles during senescence (Nakabayashi et al., 1999; Zheng et al., 2002; Lin and Wu, 2004). However, at least ClpC1 as well as ClpD/ERD1 are up regulated during dark-induced leaf senescence. One might speculate that these proteins could play regulatory roles in the Clp protease complex during leaf senescence (Lin and Wu, 2004).

The expression pattern of the thylakoid-associated proteases FtsH and DegP is less well characterized.

DegP1, 5, and 8 are localized on the luminal side of the thylakoid membrane, making their function in the degradation of luminal proteins likely (Adam et al., 2001; Adam and Clarke, 2002). DegP2 was shown to be responsible for the initial degradation of the D1 protein and is localized on the stromal side of the thylakoid membrane (Haussuhl et al., 2001). FtsH is involved in the degradation of photosynthetic proteins (Chen et al., 2000; Takechi et al., 2000; Haussuhl et al., 2001). Interestingly, two of the eight chloroplast FtsH proteases and all four members of the DegP1 proteases were down regulated after prolonged dark incubation, a treatment known to induce senescence (Lin and Wu, 2004). By contrast, one protease belonging to the family of Lon proteases was up-regulated in response to dark treatment (Lin and Wu, 2004). Other, recently discovered plastid proteases, such as cysteine proteinase RD21A, SppA protease IV, and protease HhoA (Kleffman et al., 2004), await further investigation.

2. Degradation of Chlorophyll

Marked changes in leaf color occur every autumn. About 1.2 billion tons of Chl are estimated to be degraded globally each year, and the remaining carotenoids and/or accumulated anthocyanins turn leaves from green to yellow, orange, or red (Hörtensteiner et al., 1999; Matile et al., 1999; Takayami et al., 2000). The mechanism responsible for the degreening of plants and the degradation of Chl was unclear until recently. The identification of the intermediates and enzymes leading from Chl to the colorless final end products of the C5-pathway has allowed the construction of a basic pathway of Chl breakdown (Fig. 5). It is established that Chl *b* first needs to be converted to Chl *a* and that Chlase catalyzes the first committed step of Chl breakdown. Subsequently, the central Mg^{2+} atom is removed from the tetrapyrrole ring system by Mg dechelatase. The resulting phaeophorbide *a* is then cleaved and reduced to colorless, open

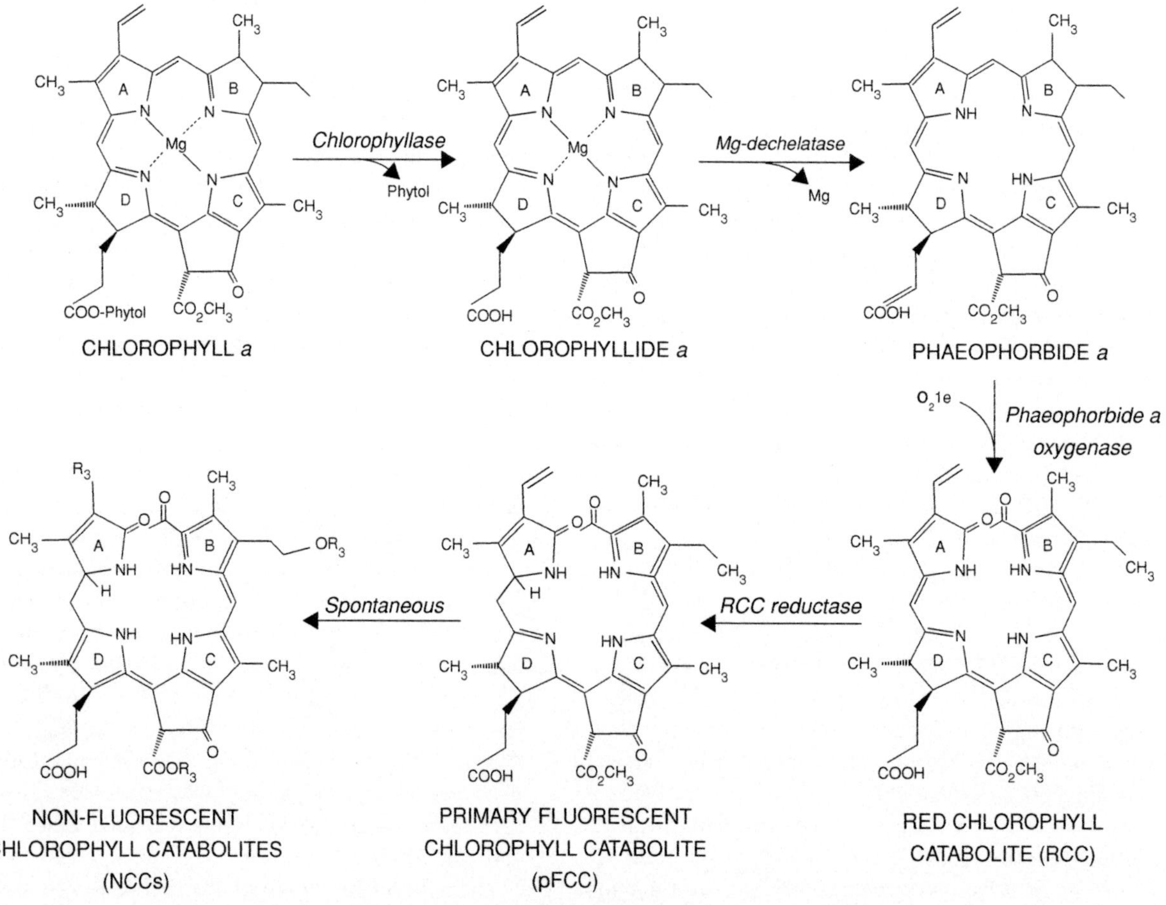

Fig. 5. The degradation pathway of Chl. Steps and enzymes are indicated.

tetrapyrrole compounds called nonfluorescent catabolites (NCCs) (Fig. 5). After further modifications, these compounds are transported to the vacuole. This degradation pathway of Chl was elucidated via cloning of the genes for different Chlase isoforms (Jacob-Wilk et al., 1999; Tsuchiya et al., 1999), of phaeophorbide *a* oxygenase (Pao) (Pruzinska et al., 2003), and the surprising result that conversion of the fluorescent Chl catabolites FCC to the first non-fluorescent Chl catabolite NCC is a spontaneous, non-enzymatic reaction (Oberhuber et al., 2003). Studies using mutants that are defective in one or the other catabolic enzymes operating in Chl breakdown (e.g. Mach et al., 2001; Pruzinska et al., 2003) have shown that many of the degradation products of the tetrapyrrole ring system are harmful phototoxins, necessitating Chl catabolism to be tightly controlled in time and space.

a. Induction of Chlorophyllase by Jasmonic Acid and Coronatine

The enzyme catalyzing the dephytylation of Chl, Chlase, was one of the first plant enzymes studied (Willstätter and Stoll, 1913). There are numerous reports describing Chlase and the regulation of its activity from plants and algae (Trebitsh et al., 1993; Khalyfa et al., 1995; Tsuchiya et al., 1997). A real breakthrough towards an understanding of Chlase structure and function was not made until 1999, when two independent groups succeeded in cloning *CHLASE* genes from *Chenopodium* (*caCHL*), *Arabidopsis* (*atCHL1*, *atCHL2*), and *Citrus* (*CHLASE1*) (Jacob-Wilk et al., 1999; Tsuchiya et al., 1999). Interestingly, two differentially expressed *CHLASE* genes were identified in *Arabidopsis* and termed *atCHL1* and *atCHL2* (Tsuchiya et al., 1999). While *atCHL1* was induced by JA-Me and the phytotoxin coronatine, a structural analog of JA-Me from *Pseudomonas* sp. (Weiler et al., 1994), *atCHL2* did not respond to JA-Me (Benedetti et al., 1998; Tsuchiya et al., 1999).

AtCHL1 and atCHL2 proteins, as well as their counterparts in other plant species, share conserved amino acid sequences and are approximately 318–347 amino acid residues in length (summarized in Takamiya et al., 2000). Surprisingly, no homologues were found in the sequenced genomes of cyanobacteria, the presumed endosymbiotic progenitors of chloroplasts. AtCHL1 and atCHL2 likewise possess a conserved lipase motif containing a putative active site serine residue (Ser162). In addition, aspartic acid and histidine residues have been identified to be essential in the active cleft of the enzyme, indicating that these amino acids, in conjunction with the conserved serine residue, could form a catalytic triad similar to that in serine esterases (including lipases) (Brady et al., 1990; Tsuchiya et al., 2003).

The predicted amino acid sequences of atCHL1 and atCHL2 contain transit peptides at their N-termini that do, however, differ considerably. While atCHL2 has a typical chloroplast transit peptide, the transit peptide of atCHL1 is reminiscent of that of proteins targeted to the endoplasmic reticulum. Although Chlase activity has been localized to the chloroplast envelope (Brandis et al., 1996; Matile et al., 1997), there is, to the best of our knowledge, no proof that atCHL1 and atCHL2 are plastid-bound. Further work will therefore be needed to localize atCHL1 and atCHL2 in chloroplasts. Apart from this, the plastid envelope localization of Chlase activity implies that pigment carrier proteins present in chloroplasts may transfer Chl *a* (and Chl *b*?) from the thylakoids to the envelope where their dephytylation and subsequent cleavage could occur. Potential Chl carrier proteins could belong to the same super-family of light-harvesting Chl binding proteins as discussed previously (see section II.C.3.a.).

b. Induction of Phaeophorbide a Oxygenase and Red Chlorophyll Catabolite Reductase during Senescence and Stress

One of the most remarkable steps in Chl degradation is the oxygenolytic cleavage of the macrocyclic ring by Pao (phaeophorbide *a* oxygenase), giving rise to the red chlorophyll catabolite RCC (Fig. 5). Production of RCC requires ferredoxin as an electron source, phaeophorbide (Pheide) *a*, and molecular oxygen; Pheide *b* is not accepted as a substrate (Hörtensteiner et al., 1995; Rodoni et al., 1997; Wüthrich et al., 2000). Pao activity is found only during senescence, suggesting that it may be one of the key regulators of Chl catabolism. Recent work has led to the identification of the *PAO* gene of *Arabidopsis thaliana* (*atPao*) (Pruzinska et al., 2003). AtPAO and related enzymes are Rieske:iron sulfur-proteins encoded by the *Arabidopsis accelerated cell death* (*ACD*) 1 gene and its orthologs, for example, the *lethal leaf spot* 1 (*LLS1*) gene of maize (Gray et al., 1997). Mutants lacking PAO exhibit spontaneously spreading cell death lesions (Fig. 6) and display constitutive defense gene activation, effects that are usually only observed when plants are challenged by pathogens (Gray et al., 1997; Pruzinska et al., 2003). In the maize *lls1* mutant lacking PAO, degradation of Pheide *a* is delayed, and *lls1* leaves stay green in darkness (Pruzinska et al., 2003). In white light, photoexcited excess Pheide *a* causes the spread

Fig. 6. Cell death phenotype in *lethal leaf spot 1* (*lls1*) lesion mimic maize plants. A, Typical *lls1* lesions develop in an apparently random location along a developmental gradient from the tip of leaves towards the base. In this example the light-dependent cell death phenotype is evident in the concentric ring appearance of the lesions in plants grown in the field with a long day photoperiod. Cell death is reduced during the dark and resumes in the light. B, The dependence of the *lls1* cell death phenotype on chlorophyll is evident by the suppression of cell death in albino sectors of an *albescent1* (*al1*)/*lls1* double mutant plant. Courtesy of John Gray, University of Toledo, Toledo, USA.

of cell death lesions (Gray et al., 1997; Pruzinska et al., 2003). By contrast, transgenic plants overexpressing ACD1/PAO do not develop lesions and are more resistant to pathogen infection than control plants (M. Yang et al., 2004).

RCC reductase catalyzes the reduction step, converting RCC to the primary fluorescent chlorophyll catabolite pFCC (Hörtensteiner et al., 1998) (Fig. 5). The responsible enzyme RCC reductase has been cloned recently; it is identical with the *ACD2* gene product of *Arabidopsis* (Mach et al., 2001). Lesion formation in *acd2* plants can be triggered by coronatine in a light-dependent manner (Mach et al., 2001). Coronatine-triggered and spontaneous lesion spreading in *acd2* plants requires protein translation, indicating that cell death occurs by an active process. RCC reductase activity has been purified from chloroplasts (Matile and Schellenberg, 1996; Rodoni et al., 1997), and the *ACD2* gene product contains a predicted chloroplast transit peptide, is processed in vivo, and co-purifies primarily with isolated chloroplasts (Mach et al., 2001).

C. Leaf Senescence as a Cell Death Program

Leaf senescence is a regulated process that in many aspects resembles programmed cell death (PCD) and apoptosis in animal/metazoan systems. Both cell death programs share a requirement of hormones or other bioactive compounds, are mediated by signal transduction machineries, and are accompanied by the activation of specific hydrolytic enzymes (Kuriyama and Fukuda et al., 2002). In either case, these events eventually lead to similar cytological phenotypes of dying cells, including cell condensation, cleavage of nuclear DNA, chromatin separation, and release of cytochrome *c* from the mitochondria into the cytosol (Balk and Leaver, 2001). It is appealing to hypothesize that the mechanisms underlying PCD and apoptosis in metazoans may also apply in plants. Recent work has shown that key components in either system include cysteine proteinases (called caspases and metacaspases, respectively), BCL-2-like (BLP) proteins, defender against death (DAD)-like proteins, ROS, and Ca^{2+} (see Hoeberichts and Woltering, 2002; H. Thomas et al., 2003; Watanabe and Lam, 2004).

Until recently, there was no evidence for the existence of caspases in plants. Caspases form core components of the apoptotic machinery in animals. They are involved in proteolytic cascades activated by extracellular and intracellular factors, including cytochrome *c* released from mitochondria as a result of the mitochondrial membrane permeability transition. There are no caspase orthologues in the sequenced *Arabidopsis* genome. However, pharmacological and biochemical studies (Bozhkov et al., 2004; Chichkova et al., 2004; Danon et al., 2004) as well as biocomputational work (Uren et al., 2000) have allowed the identification of a group of caspase-like proteins, designated metacaspases, that share the conserved catalytic cysteine and histidine diad with animal caspases. These caspases accumulate in senescing plant tissues

(Buchanan-Wollaston, 1997; C.M. Griffiths et al., 1997; Wagstaff et al., 2002).

Evidence from studies of animal apoptosis suggests that entry into PCD is dependent upon de-repression of pro-apoptotic genes that are located in the mitochondrial membrane (D.R. Green and Reed, 1998). Caspases are usually activated or inhibited by BCL-2-like proteins (BLPs) released from mitochondria. BLPs either inhibit (Bcl- and Bcl-X) or promote (Bax and Bak) apoptosis. The evidence for a common role of BLPs in plant and animal apoptosis is as follows (Kawai-Yamada et al., 2004; Coupe et al., 2004; see Watanabe and Lam, 2004, for a review): (i) proteins immunologically related to animal BCL2 have been detected in plants, (ii) transgenic plants expressing animal or viral regulators showed suppressed (Bcl- and Bcl-X) or enhanced (Bax and Bak) cell death execution, (iii) orthologs of the BI-1 gene have been identified in *Arabidopsis* and rice, and either protein substantially inhibited Bax-induced apoptosis in animal cells and yeast. Overexpression of BI-1 rescues plants expressing mammalian Bax from cell death. Finally, BI-1 expression was found to be up-regulated by pathogen inoculation in *Arabidopsis* (see Watanabe and Lam, 2004, for references).

Defender against apoptotic cell death (DAD1) is another highly conserved component of the cell death machinery in animal and plant kingdoms (Sugimoto et al., 2000). It was reported that DAD1 interacts with certain members of the BLP superfamily (Makishima et al., 2000), although the precise role of DAD in plant PCD is not yet known. In addition, other positive and negative regulators modulate cell death and defense responses, such as *R* genes and the *MLO* gene conferring resistance to the biotroph powdery mildew fungus *Blumeria graminis* f.sp. *hordei*. Barley lines carrying recessive mutations in the *MLO* locus show spontaneous leaf cell death and broad-spectrum resistance. It has been proposed that the MLO protein, similar to BI-1, may play a role in a survival pathway that interferes with penetration resistance during plant-biotrophic fungi interactions (summarized in Schulze-Lefert and Panstruga, 2003).

Mitochondrial membrane permeability transition and subsequent release of cytochrome *c* are hallmarks of apoptosis in plants and animals (see Danon et al., 2000; Hoeberichts and Woltering, 2002, for reviews). Mitochondrial permeability is stimulated by various signals, including (stress-induced) Ca^{2+} fluxes and increased levels of ROS. Conversely, loss of mitochondrial transmembrane potential leads to mass generation of ROS and thereby provides a powerful feedback amplification loop. SA-dependent formation of ROS triggers an increase in cytosolic Ca^{2+} (Kawano et al., 1998) and inhibits mitochondrial functions (Xie et al., 1999). There are also reports implying a role of ethylene as a stimulant of senescence-associated PCD in plants (Orzáez and Granell, 1997; Navarre and Wolpert, 1999). Evidence is accumulating for a role of calmodulin-like domain protein kinases in plant defense and stress responses (Blumwald et al., 1998). Some of these proteins activate NADPH oxidase that is an important source of ROS. ROS, particularly H_2O_2, have been implicated in activation of the NF-kB signaling pathway that plays an essential role in mediating both immune and inflammatory resposes, and tumor necrosic factor-induced apoptosis in animal cells (Perkins, 2000). In *Arabidopsis*, H_2O_2 is a potent activator of a MAPK cascade that induces specific stress responses in leaf cells (Kovtun et al., 2000). Taken together, these examples highlight an unanticipated complexity of cell death pathways in which common and unique players regulate PCD in plants and animals.

Cloning of senescence-associated genes (SAGs) and identification of *Arabidopsis* mutants and transformants in which senescence is delayed have shed additional light on the role of phytohormones such as ethylene and JA as signaling compounds (Noodén and Penny, 2001; Kuriyama and Fukuda, 2002). In the subsequent sections, potential components of signaling pathways as well as hydrolytic enzymes implicated in regulated organelle destruction during cell death will be discussed.

Nevertheless, it is also apparent that normal, developmentally-regulated leaf senescence is not quite the same as PCD. Only when leaf senescence is no longer reversible does PCD set in as a terminal stage. On the other hand, extreme environmental conditions, such as pest invasion, desiccation, drought, ozone, UV exposure, and the use of chemicals inducing PCD, are not causally related to leaf senescence but induce similar phenotypes of dying cells. Care must therefore be taken when discussing potential overlaps between PCD in reponse to harsh environmental conditions and PCD as the terminal stage of plant development.

1. Senescence-Associated Genes with Potential Functions in Signaling

A large number of SAGs have been cloned (see Buchanan-Wollaston, 1997; Biswal and Biswal, 1999; Chandlee, 2001). Among them are transcription factors belonging to the WRKY family (Eulgem et al., 2001) and senescence-induced or senescence-associated receptor-like kinases (named SIRK and SARK, respectively) (Hajouj et al., 2000; Robatzek and Somssich, 2001). Members of the WRKY family are characterized by the presence of a 60 amino acid motif, the WRKY domain (Eulgem et al., 2001). In *Arabidopsis*, two different WRKY proteins accumulate during leaf senescence (Hinderhofer and Zentgraf, 2001; Robatzek and Somssich, 2001). WRKY53 is expressed at the onset of senescence, whereas WRKY6 is up-regulated during both senescence and pathogen defense. Potential genes regulated by WRKY6 include calmodulin-regulated genes and different types of kinases including SIRK (Robatzek and Somssich, 2002). Although SIRK and SARK share similar structures, consisting of an extracellular leucine-rich domain, a transmembrane domain, and a Ser/Thr kinase domain (S. Yoshida, 2003), they are differentially expressed during senescence. While SARK in bean is induced during the early stage of age-dependent senescence, SIRK accumulates only in senescent leaves (Hinderhofer and Zentgraf, 2001; Robatzek and Somssich, 2002). SIRK and SARK action may involve the perception of extracellular messengers, conformational changes of the receptor across their respective target membranes, and the subsequent relay of phosphorylation cascades leading to senescence gene expression.

2. Hydrolytic Activities Induced during Senescence

a. Proteolytic System

Various salvage pathways operate to recycle nutrients from leaves undergoing senescence. In addition to protein degradation taking place in chloroplasts (see section III.B.1.), increased expression of polyubiquitin genes and genes encoding ubiquitin-conjugating enzymes have been reported (Belknap and Garbarino, 1996; Noodén et al., 1997). Interestingly, several *Arabidopsis* mutants showing delayed senescence are affected in proteolytic activities. For example, the *ore9* mutant is defective in an E3 ubiquitin ligase (S.A. Oh et al., 1997; Woo et al., 2001). E3 enzymes help choose which proteins should be ubiquitinated (for reviews see Glickman and Ciechanover, 2002; Vierstra, 2003). *Ore9* encodes an F-box protein that is part of the Skp1-cullin/CDC53-F-box (SCF) protein complex involved in various signal transduction pathways, including the light and phytochrome signals (Hellmann and Estelle, 2002). It has been proposed that the ORE9 protein may regulate the turnover of factors repressing leaf senescence (S. Yoshida, 2003). In another mutant, the *delayed leaf senescence* mutant *dls1*, an arginyl-tRNA:protein transferase was found to be impaired (S. Yoshida et al., 2002). Enzymes of this type are implicated in the N-end rule proteolytic pathway (Varshzavsky, 1997). To the best of our knowledge, no evidence exists to date for the operation of ubiquitin-dependent proteolysis in plastids, although recent proteomic approaches have identified an ubiquitin-specific protease 2 and 20S proteasome beta F1 subunits in *Arabidopsis* chloroplasts (Kleffmann et al., 2004).

b. Lipolytic Activities

Lipids can also be degraded and their constituents salvaged and converted to phloem-mobile sugars (sucrose) during leaf senescence (Thompson et al., 1998; Kaup et al., 2002). Phospholipase Dα and acyl hydrolases have been implicated in lipid degradation (Fan et al., 1997; Y. He et al., 2001; Y. He and Gan, 2002). Acyl hydrolases are involved in releasing oleic acid from triolein, and one of the identified *SAG*s, *SAG101*, has been shown to encode a senescence-induced acyl hydrolase (Y. He et al., 2001; Y. He and Gan, 2002). Antisense expression of *SAG101* slowed the progression of leaf senescence, whereas its overexpression enhanced leaf senescence (Y. He and Gan, 2002). It is tempting to speculate that SAG101 may also be involved in liberating α-linolenic acid from membrane lipids and that this compound, in turn, would be used for JA synthesis. If so, lipolytic membrane destruction would trigger the production of senescence-promoting JA. However, additional factors must be involved in regulating age-dependent senescence because senescence progression was not affected in JA-insensitive mutants (Y. He et al., 2001).

The plant vacuole is the counterpart of the lysosome that plays an especially important role for programmed cell death in animal cells. For example, the lysosome participates in autolytic cell destruction while caspases may play a key role in apoptosis (Ferri and Kroemer, 2001). Caspases are synthesized as pro-proteins that are cleaved and activated by removal of its pro-domains by virtue of cathepsin B and D in animal cells (Ferri and Kroemer, 2001; Kagedal et al., 2001). Interestingly, lysosomal cathepsin B and D as well as other proteases are stimulated by sphingosine, a well-known pro-apoptotic molecule (Ferri and Kroemer, 2001).

In an *accelerated cell death* mutant of *Arabidopsis*, *acd11*, a sphingosine-transfer protein was identified that may be involved in the activation of programmed cell death (Brodersen et al., 2002). Brodersen et al. (2002) proposed that loss of *ACD11* function would cause cells to collapse and that the ACD11 protein may negatively regulate PCD and defense in vivo. The PCD and defense pathways activated in *acd11* plants are salicylic acid (SA) dependent, but do not require intact JA or ethylene signaling pathways (Brodersen et al., 2002). SA is involved in PCD during the hypersensitive response (HR) by which plants respond to pathogens (Morel and Dangl, 1997). Light is required for PCD execution in *acd11* plants (Brodersen et al., 2002) and may lead to the production of reactive oxygen species. Application of an SA-analog to SA-deficient *acd11* plants induced cell death in the light, but not in the dark (Brodersen et al., 2002). In animal cells, oxidative stress and the tumor-suppressing protein p53 cause lysosomal breakage, which leads to autolytic degradation of cells (Zhao et al., 2001; Yuan et al., 2002). ACD11 protein may thus be operating in transferring sphingosine between membranes in vivo. Interestingly, JA signaling is required for fumonisin-B1-induced cell death (Asai et al., 2000). Fumonisin-B1 is a mycotoxin and sphingosine analog that induces PCD in both animal and plant systems (Gilchrist, 1997).

3. The Search for the Cell Death Factor

Cell death associated with leaf senescence, the HR occurring in plant-pathogen interactions, and other forms of cell death including sex determination in unisexual flowers and the formation of the vascular system (Pennell and Lamb, 1997) are regulated genetically. This is illustrated by studies on maize, tomato, and *Arabidopsis* mutants in which cell death is triggered spontaneously. These so-called lesion mimic mutants develop HR-like lesions and activate defense responses constitutively. They are represented by the *acd*- and *lsd*- mutants (Dietrich et al., 1994; Greenberg et al., 1994; Johal et al., 1995; Weymann et al. 1995) and can be separated into two classes (Dangl et al., 2000): (i) initiation mutants develop spontaneous lesions of determinate size similar in appearance to normal cell death triggered by pathogens during the HR and (ii) propagation mutants exhibit spreading cell death. In propagation mutants, extracellular superoxide radical (O_2^-) is necessary and sufficient to trigger spreading cell death. The containment of cell death to a limited number of surrounding cells indicates that a dialog of signals between dying cells and healthy neighbouring cells must exist that determines lesion size (Dangl et al., 2000). A question thus arises as to the nature of the factors and processes that propagate or halt the spread of cell death and hence regulate the extent of lesion propagation.

Of the many *Arabidopsis* cell death mutants described, seven corresponding genes have been cloned and characterized. Remarkably, four fall into the class of mutants that are impaired in tetrapyrrole metabolism, namely *acd1* encoding Pao, *acd2* encoding RCC reductase, *les22* encoding uroporphyrinogen decarboxylase, and *lin2* encoding coproporphyrinogen decarboxylase (Molina et al., 1999; Ishikawa et al., 2001; Mach et al. 2001; Pruzinska et al., 2003). Genes in a second class are involved in fatty acid metabolism as exemplified before for *acd11* (Mou et al., 2000; Kachroo et al., 2001). Finally, *lsd1* has been identified to encode a zinc-finger protein proposed to negatively regulate a cell death pathway by monitoring a superoxide-dependent signal (Jabs et al., 1996; Dietrich et al., 1997).

Reactive oxygen species (ROS) supposedly play important roles in lesion propagation and cell death (Vranová et al., 2002; Overmyer et al., 2003). Phytohormones such as ethylene, SA, and JA may additionally be involved in lesion formation, propagation, and containment. In a recent paper, op den Camp et al. (2003) reported that in the *flu* mutant of *Arabidopsis* (see also chapter II.C.2.a.), reactive oxygen species (in particular singlet oxygen) are produced because of the lack of sequestration of free Pchlide. The authors also measured increased levels of oxygenated membrane lipids, including 13S-hydroxyoctadecatrieonic acid in light-exposed *flu* plants. Pigment-sensitized lipid modifications thus may serve as a source of JA precursors, and the entire spectrum of cyclopentanone compounds emitted from affected cells could trigger cell death in neighbouring cells via an autocatalytic mechanism. In apparent contrast to this view are studies suggesting that JA is a factor involved in the containment of ROS-dependent lesion propagation (Overmyer et al., 2000; Rao et al., 2000). The JA-insensitive *jar1* and *coi1* mutants and JA-deficient *fad3-fad7-fad8* triple mutant all show an increased magnitude of ozone-induced oxidative burst, SA accumulation, and HR-like cell death. Pretreatment of the ozone-sensitive *Arabidopsis* accession Cvi-O with JA-Me completely abolished ozone-induced H_2O_2 accumulation, SA production, and defense gene activation, as evidenced by the lack of pathogenesis-related protein-1 mRNA accumulation (Overmyer et al., 2000; Rao et al., 2000). Furthermore, *jar1* exhibits a transient spreading cell death phenotype and a pattern of O_2^- accumulation similar to that observed in *rcd1* plants (Overmyer et al., 2000). *RCD1* defines a *radical-induced cell death* locus that mediates ozone and O_2^- sensitivity (Overmyer et al., 2000). Treatment of O_3-exposed *rcd1* mutant plants with JA halted spreading cell death, providing direct evidence for the role of JA in lesion containment (Overmyer et al., 2000). Similarly, pretreatment of tobacco with JA diminished O_3-dependent cellular damage (Overmyer et al., 2000; Rao et al., 2000). An explanation for these findings could be that JA-stimulated glutathione synthesis helps to limit ROS accumulation and cell death (Xiang and Oliver, 1998). It is also possible that lesion containment by JA could be achieved through regulation of ethylene receptors because it has been shown that JA induces ethylene receptor gene expression (Schenk et al., 2000). Increased receptor protein synthesis would decrease ethylene sensitivity and desensitize plants to ethylene. In this way, JA could affect ethylene-dependent lesion propagation by reducing ethylene-dependent ROS accumulation. As a result, halting of the lesion spread would occur (Overmyer et al., 2003). However, this hypothesis appears to be in clear contrast to the observation that treatment of tobacco leaves with JA-Me promotes H_2O_2 production (Dat et al., 2003). Therefore, the relationship between ROS and JA and their impact on the spread of cell death remain unresolved.

Other studies suggest indirect signaling pathways not involving JA. Mutants that constitutively overexpress the thionin (*THI 2.1*) gene, called *cet* mutants, have been isolated (Nibbe et al., 2002). These *cet* mutants spontaneously form microlesions. Lesion formation in *cet* mutants occurs independent of COI1-mediated JA signaling and, in case of *cet2* and *cet4.1*, also does not involve SA signaling. By contrast, COI1-mediated JA signaling in *cet3* requires SA. In SA-depleted transgenic *cet3* plants expressing the bacterial SA hydroxylase NahG, *THI2.1* expression was independent of lesion formation. In wild-type plants, NahG-dependent depletion of SA levels, however, abolished HR-like cell death (Nibbe et al., 2002). It thus appears that signals other than SA and JA are involved in the regulation of lesion initiation. Cloning of *cet* genes should unravel the nature of the cell death factor and the possible role of JA.

IV. Future Perspective

The results summarized in this review have unveiled a plethora of mechanisms governing plant gene expression in response to both exogenous and endogenous factors. Light and plant hormones, such as BRs, ethylene, SA, and JA, exert profound effects on early plant development following seed germination and the terminal stage of plant development, senescence. Both stages of development are intimately associated with the operation of photoreceptors and photosensory tetrapyrrole compounds of which some were discussed in this chapter.

Tetrapyrroles, such as Chl and heme, and most of their precursors and breakdown products are extremely phototoxic. Their synthesis and degradation are highly compartmentalized and regulated, and their levels are tightly adjusted to the respective needs during plant development. Tetrapyrroles are able to absorb light and donate active electrons that are used for different processes. In the absence of productive outlets for these active electrons, they can be donated to other compounds, including molecular oxygen, forming free radicals and singlet oxygen that cause pigment bleaching, protein denaturation, and membrane lipid peroxidation. If produced in excess, porphyrin precursors such as uroporphyrinogen III cause lesion mimic phenotypes, including spreading cell death lesions and induction of defense gene expression. Defects in Chl catabolism can also give rise to lesion mimic phenotypes. Evidence is thus accumulating for a general role of tetrapyrroles in controlling cell death in stress responses, senescence, and disease, and that there are similarities between PCD and apoptosis in plants and animals. Interestingly, tetrapyrroles also play important roles as intracellular signals coordinating post-germination greening. Mg-protoporphyrin IX is a plastid factor that controls nuclear gene expression. In mammalian systems, protoporphyrin IX, but not other porphyrins such as porphobilinogen, triggers apoptosis by induction of the mitochondrial permeability transition (Marchetti et al., 1996; Kroemer, 1999). Uroporphyrinogen III and RCC, intermediates in the biosynthesis and degradation pathways of Chl, are also able to specifically trigger cell death in plants. Interestingly, Mach et al. (2001) observed the ACD2 protein localizes to both mitochondria and chloroplasts. In mammalian systems, mitochondria integrate cell death signals and trigger cell death (Kroemer, 1999). It is appealing to hypothesize that both mitochondria and chloroplasts could be involved in the life/death decision in plants at different stages of development and in response to environmental cues. It should be the aim of future work to identify mitochondria- and chloroplast-derived cell death factors as well as signaling compounds in plants and to study their interaction. Approaches to be used to this end may involve genetic screens for hormone (ethylene, SA, JA)-insensitive plants, microarray analyses to comprehensively monitor alterations in gene expression during photosensitized, tetrapyrrole-mediated stress and defense responses, and two-hybrid screens for new partner proteins interacting with identified signaling compounds. In the post-genomic era of plant research, these genetic approaches will need to be complemented by biochemical techniques, permitting definite functions to be assigned to each component of the discovered signaling pathway. Manipulation of the tetrapyrrole pathway may allow the creation of agronomically important crop plants with increased stress and resistance properties.

Acknowledgments

We acknowledge the continuous interest and help of all of our co-workers. We are grateful to Dr. John Gray, The University of Toledo, Toledo, USA, for his inspiring comments on this chapter and especially for his generosity in providing the artwork in figure 6. We are indebted to L. Reinbothe for her encouragment and effort during the preparation of this chapter.

References

Adam Z and Clarke AK (2002) Cutting edge of chloroplast proteolysis. Trends Plant Sci 7: 452–456

Adam Z, Adamska I, Nakabayashi K, Ostersetzer O, Haussuhl K, Manuall A, Zheng B, Vallon O, Rodermel SR, Shinozaki K and Clarke AK (2001) Chloroplast and mitochondrial proteases in Arabidopsis. A proposed nomenclature. Plant Physiol 125: 1912–1918

Adamska I (1997) ELIPs – Light-induced stress proteins. Physiol Plant 100: 75–85

Adamska I, Funk C, Renger G and Andersson B (1996) Developmental regulation of the PsbS gene expression in spinach seedlings: the role of phytochrome. Plant Mol Biol 31: 793–802

Adamska I, Roobol-Boza M, Lindahl M and Andersson B (1999) Isolation of pigment-binding early light-inducible proteins from pea. Eur J Biochem 260: 453–460

Ahmad M, Jarillo JA, Smirnova O and Cashmore AR (1998) The CRY1 blue light photoreceptor of Arabidopsis interacts with phytochrome A. Mol Cell 1: 939–948

Altmann T (1999) Molecular physiology of brassinosteroids revealed by the analysis of mutants. Planta 208: 1–11

Alvarez ME, Pennell RI, Meijer PJ, Ishikawa A, Dixon RA and Lamb C (1998) Reactive oxygen intermediates mediate a systemic signal network in the establishment of plant immunity. Cell 92: 773–784

Andrew T, Riley PG and Dailey HA (1990) Heme response pathways in yeast. In: Dailey HA (ed) Biosynthesis of Heme and Chlorophyll, pp 163–200. McGraw–Hill, New York

Apel K (1981) The protochlorophyllide holochrome of barley (Hordeum vulgare L.): phytochrome-induced decrease of translatable mRNA coding for the NADPH:protochlorophyllide oxidoreductase. Eur J Biochem 120: 89–93

Apel K, Santel HJ, Redlinger TE and Falk H (1980) The protochlorophyllide holochrome of barley (Hordeum vulgare L.): isolation and characterization of the NADPH-protochlorophyllide oxidoreductase. Eur J Biochem 111: 251–258

Armstrong GA (1998) Greening in the dark: light-independent chlorophyll biosynthesis from anoxygenic photosynthetic bacteria to gymnosperms. J Photochem Photobiol B Biol 43: 87–100

Armstrong GA, Runge S, Frick G, Sperling U and Apel K (1995) Identification of NADPH: protochlorophyllide oxidoreductases A and B: a branched pathway for light-dependent chlorophyll biosynthesis in Arabidopsis thaliana. Plant Physiol 108: 1505–1517

Asai T, Stone JM, Heard JE, Kovtun Y, Yorgey P, Sheen J and Ausubel FM (2000) Fumonisin B1-induced cell death in Arabidopsis protoplasts requires jasmonate-, ethylene-, and salicylate-dependent signaling pathways. Plant Cell 12: 1823–1835

Azevedo C, Sadanandom A, Kitagawa K, Freialdenhoven A, Shirasu K and Schulze-Lefert P (2002) The RAR1 interactor SGT1, an essential component of R gene-triggered disease resistance. Science 295: 2073–2076

Azumi Y, Watanabe A (1991) Evidence for a senescence-associated gene induced by darkness. Plant Physiol 95: 577–583

Balk J and Leaver CJ (2001) The PET1-CMS mitochondrial mutation in sunflower is associated with premature programmed cell death and cytochrome c release. Plant Cell 13: 1803–1818

Barthélemy X, Bouvier G, Radunz A, Docquier S, Schmid GH and Franck F (2000) Localization of NADPH-protochlorophyllide reductase in plastids of barley at different greening stages. Photosynth Res 64: 63–76

Beale SI (1990) Biosynthesis of the tetrapyrrole pigment precursor, 6-aminolevulinic acid, from glutamate. Plant Physiol: 93: 1273–1279

Beale SI (1999) Enzymes of chlorophyll biosynthesis. Photosynth Res 60: 43–73

Beale SI and Weinstein JD (1990) Tetrapyrrole metabolism in photosynthetic organisms. In: Dailey HA (ed) Biogenesis of Heme and Chlorophylls, pp 287–391. McGraw-Hill, New York

Belknap WR and Garbarino JE (1996) The role of ubiquitin in plant senescence and stress responses. Trends Plant Sci 1: 331–335

Benedetti CE, Costa CL, Turcinelli SR and Arruda P (1998) Differential expression of a novel gene in response to coronatine, methyl jasmonate, and wounding in the coi1 mutant of Arabidopsis. Plant Physiol 116: 1037–1042

Bent AF, Innes RW, Ecker JR and Staskawicz BJ (1992) Disease development in ethylene-insensitive Arabidopsis thaliana infected with virulent and avirulent Pseudomonas and Xanthomonas pathogens. Mol Plant–Microbe Interact 5: 372–378

Benvenuto G, Formiggini F, Laflamme P, Malakhov M and Bowler C (2002) The photomorphogenesis regulator DET1 binds the aminoterminal tail of histone H2B in a nucleosome context. Curr Biol 12: 1529–1534

Bishop GJ and Koncz C (2002) Brassinosteroids and plant steroid hormone signaling. Plant Cell Suppl 14: 97–110

Bishop GJ and Yokota T (2001) Plants steroid hormones, brassinosteroids: Current highlights of molecular aspects on their synthesis/metabolism, transport, perception and response. Plant Cell Physiol 42: 114–120

Biswal UC and Biswal B (1999) Leaf senescence: physiology and molecular biology. Curr Sci 77: 775–782

Bleecker AB and Kende H (2000) Ethylene: a gaseous signal molecule in plants. Annu Rev Cell Dev Biol 16: 1–18

Bleecker A and Patterson S (1997) Last exit: senescence, abscission, and meristem arrest in Arabidopsis. Plant Cell 9: 1169–1179

Blumwald E, Aharon GS and Lam BCH (1998) Early signal transduction pathways in plant-pathogen interactions. Trends Plant Sci 3: 342–346

Boardman NK (1962) Studies on a protochlorophyll-protein complex. I. Purification and molecular weight determination. Biochem Biophys Acta 62: 63–79

Boese QF, Spano AJ, Li J and Timko MP (1991) Aminolevulinic acid dehydratase in pea (*Pisum sativum* L.). Identification of an unusual metal-binding domain in the plant enzyme. J Biol Chem 266: 17060–17066

Bowler C, Neuhaus G, Yamagata H and Chua NH (1994) Cyclic GMP and calcium mediate phytochrome phototransduction. Cell 77: 73–81

Bozhkov PV, Filonova LH, Suarez MF, Helmersson A, Smertenko AP, Zhivotovsky B and von Arnold S (2004) VEIDase is a principal caspase-like activity involved in plant programmed cell death and essential for embryonic pattern formation. Cell Death Diff 11: 175–182

Brady L, Brzozowski AM, Derewenda ZS, Dodson E, Dodson GO, Tolley S, Turkenburg JP, Christiansen L, Huge-Jensen B, Norskov L, Thim L and Menge U (1990) A serine protease triad forms the catalytic centre of a triacylglycerol lipase. Nature 343: 767–770

Brandis A, Vainstein A and Goldschmidt EE (1996) Distribution of chlorophyllase among components of chloroplast membranes in *Citrus sinensis* organs. Plant Physiol Biochem 34: 49–54

Brodersen P, Petersen M, Pike HM, Olszak B, Skov S, Odum N, Jorgensen LB, Brown RE and Mundy J (2002) Knock-out of Arabidopsis accelerated cell death11 encoding a sphingosine transfer protein causes activation of programmed cell death and defense. Genes Dev 16: 490–502

Buchanan-Wollaston V (1997) The molecular biology of leaf senescence. J Exp Bot 48: 181–199

Buchanan-Wollaston V, Ainsworth C (1997) Leaf senescence in *Brassica napus*: cloning of senescence related genes by subtractive hybridisation. Plant Mol Biol 33: 821–834

Burke DH, Hearst JE and Sidow A (1993) Early evolution of photosynthesis: clues from nitrogenase and chlorophyll iron proteins. Proc Natl Acad Sci USA 90: 7134–7138

Casadoro G, Byer-Hansen G, Kannangara G and Gough S (1983) An analysis of temperature and light sensitivity in tigrina mutants of barley. Carlsberg Res Commun 48: 95–129

Casal JJ (2000) Phytochromes, cryptochromes, phototropin: photoreceptor interactions in plants. Photochem Photobiol 71: 1–11

Cashmore AR, Jarillo JA, Wu YJ and Liu D (1999) Cryptochromes: blue light receptors for plants and animals. Science 284: 760–765

Chandlee JM (2001) Current molecular understanding of the genetically programmed process of leaf senescence. Physiol Plant 113: 1–8

Chen M, Choi Y, Voytas DF and Rodermel S (2000) Mutations in the Arabidopsis VAR2 locus cause leaf variegation due to the loss of a chloroplast FtsH protease. Plant J 22: 303–313

Chiba A, Ishida H, Nishizawa NK, Makino A and Mae T (2003) Exclusion of ribulose-1.5-bisphosphate carboxylase/oxygenase from chloroplasts by specific bodies in naturally senescing leaves of wheat. Plant Cell Physiol 44: 914–921

Chichova NV, Kim Sh, Titova ES, Kalkum M, Morozov VS, Rubtsov YP, Kalinina NO, Taliansky ME and Vartapetian AB (2004) A plant caspase-like protease activated during the hypersensitive response. Plant Cell 16: 157–171

Chory JP, Nagpal and Peto CA (1991) Phenotypic and genetic analysis of det2, a new mutant that affects light-regulated seedling development in Arabidopsis. Plant Cell 3: 445–459

Christie JM and Briggs WR (2001) Blue light sensing in higher plants. J Biol Chem 276: 11457–11460

Ciardi JA and Klee H (2001) Regulation of ethylene-mediated responses at the level of receptor. Ann Bot 88: 813–822

Ciardi JA, Tieman DM, Lund ST, Jones JB, Stall RE and Klee HJ (2000) Response to *Xanthomonas campestris* pv. *vesicatoria* in tomato involves regulation of ethylene receptor gene expression. Plant Physiol 123: 81–92

Ciardi JA, Tieman DM, Jones JB and Klee HJ (2001) Reduced expression of the tomato ethylene receptor gene LeETR4 enhances the hypersensitive response to *Xanthomonas campestris* pv. *vesicatoria*. Mol Plant–Microbe Interact 14: 487–495

Clouse SD (2002) Arabidopsis mutants reveal multiple roles for sterols in plant development. Plant Cell 14: 1995–2000

Clouse SD and Sasse JM (1998) Brassinosteroids: Essential regulators of plant growth and development. Annu Rev Plant Physiol Plant Mol Biol 49: 427–451

Coupe SA, Watson LM, Ryan DJ, Pinkney TT and Eason JR (2004) Molecular analysis of programmed cell death during senescence in *Arabidopsis thaliana* and *Brassica oleracea*: cloning broccoli LSD1, Bax inhibitor and serine palmitoyltransferase homologues. J Exp Bot 55: 59–68

Crafts-Brandner SJ, Klein RR, Klein P, Holzer R and Feller U (1996) Coordination of protein and mRNA abundances of stromal enzymes and mRNA abundances of the Clp protease subunits during senescence of *Phaseolus vulgaris* (L.) leaves. Planta 200: 312–318

Creelman RA and Mullet JE (1995) Jasmonic acid distribution and action in plants: Regulation during development and response to biotic and abiotic stress. Proc Natl Acad Sci USA 92: 4114–4119

Creelman RA and Mullet JE (1997) Biosynthesis and action of jasmonates in plants. Annu Rev Plant Physiol Plant Mol Biol 48: 355–381

Dangl JL, Dietrich RA and Thomas H (2000) Senescence and programmed cell death. In: Buchanan B, Gruissem W and Jones R (eds) Biochemistry and Molecular Biology of Plants, pp 1044–1100. American Society of Plant Physiologists, Rockville

Danon A, Delorme V, Mailhac N and Gallois P (2000) Plant programmed cell death: a common way to die. Plant Physiol Biochem 38: 647–655

Danon A, Rotari VI, Gordon A, Mailhac N and Gallois P (2004) Ultraviolet-C overexposure induces programmed cell death in Arabidopsis, which is mediated by caspase-like activities and which can be suppressed by caspase inhibitors, p35 and Defender against Apoptotic Death. J Biol Chem 279: 779–787

Dat JF, Pellinen R, Beeckman T, Van De Cotte B, Langebartels C, Kangasjarvi J, Inze D and van Breusegem F (2003) Changes in hydrogen peroxide homeostasis trigger an active cell death process in tobacco. Plant J 33: 621–632

Davis SJ, Kurepa J and Vierstra RD (1999) The *Arabidopsis thaliana* HY1 locus, required for phytochrome-chromophore biosynthesis, encodes a protein related to heme oxygenases. Proc Natl Acad Sci USA 96:6541–6546

Dehesh K and Ryberg M (1985) The NADPH-protochlorophyllide oxidoreductase is the major protein constituent of prolamellar bodies in wheat (*Triticum aestivum* L.). Planta 164: 396–399

Dehesh K, Häuser I, Apel K and Kloppstech K (1983) The distribution of NADPH-protochlorophyllide oxido-reductase in relation to chlorophyll accumulation along the barley leaf gradient. Planta 158: 134–139

Dietrich RA, Delaney TP, Uknes SJ, Ward ER, Ryals J and Dangl JL (1994) Arabidopsis mutants simulating disease resistance response. Cell 77: 565–577

Dietrich RA, Richberg MH, Schmidt R, Dean C and Dangl JL (1997) A novel zinc finger protein is encoded by the Arabidopsis LSD1 gene and functions as a negative regulator of plant cell death. Cell 88: 685–694

Drew MC, He CJ and Morgan PW (2000) Programmed cell death and aerenchyma formation in roots. Trends Plant Sci 5: 123–127

Dreyfuss BW and Thornber JP (1994) Assembly of the light-harvesting complexes (LHCs) of photosystem II. Monomeric LHCIIb complexes are intermediates in the formation of oligomeric LHC IIb complexes. Plant Physiol 106: 829–839

Durnford DG and Falkowski PG (1997) Chloroplast redox regulation of nuclear gene transcription during photoacclimation. Photosynth Res 53: 229–241

Escoubas JM, Lomas M, LaRoche J and Falkowski PG (1995) Light intensity regulation of cab gene transcription is signaled by the redox state of the plastoquinone pool. Proc Natl Acad Sci USA 92: 10237–10241

Eulgem T, Rushton PJ, Robatzek S and Somssich IE (2001) The WRKY superfamily of plant transcription factors. Trends Plant Sci 5: 199–206

Fan L, Zheng S and Wang X (1997) Antisense suppression of phospholipase D retards abscisic acid- and ethylene-promoted senescence of post-harvest Arabidopsis leaves. Plant Cell 9: 2183–2196

Farmer EE and Ryan CA (1990) Interplant communication: airborne methyl jasmonate induces synthesis of proteinase inhibitors in plant leaves. Proc Natl Acad Sci USA 87: 7713–7716

Fay P (1992) Oxygen relations of nitrogen fixation in cyanobacteria. Microbiol Rev 56: 340–369

Fankhauser C (2001) The phytochromes, a family of red/far-red absorbing photoreceptors. J Biol Chem 276: 11453–11456

Fankhauser C and Staiger D (2002) Photoreceptors in *Arabidopsis thaliana*: light perception, signal transduction and the entrainment of the endogenous clock. Planta 219: 1–16

Fankhauser C, Yeh KC, Lagarias JC, Zhang H, Elich TD and Chory J (1999) PKS1, a substrate phosphorylated by phytochrome that modulates light signaling in Arabidopsis. Science 284: 1539–1541

Ferri KF and Kroemer G (2001) Organelle-specific initiation of cell death pathways. Nat Cell Biol 3: E255–E263

Forreiter C and Apel K (1993) Light-independent and light-dependent protochlorophyllide-reducing activities and two distinct NADPH-protochlorophyllide oxidoreductase polypeptides in mountain pine (*Pinus mugo*). Planta 190: 536–545

Forreiter C, van Cleve B, Schmidt A and Apel K (1990) Evidence for a general light-dependent negative control of NADPH-protochlorophyllide oxidoreductase in angiosperms. Planta 183: 126–132

Fradkin LI, Domanskaya IN, Radyuk MS, Domanskii VP and Kolyago VM (1993) Effect of benzyladenine and irradiation on energy transfer from precursor to chlorophyll in greening barley leaves. Photosynthetica 29: 227–234

Franck F, Sperling U, Frick G, Pochert B, van Cleve B, Apel K and Armstrong GA (2000) Regulation of etioplast pigment-protein complexes, inner membrane architecture, and protochlorophyllide a chemical heterogeneity by light-dependent NADPH:protochlorophyllide oxidoreductases A and B. Plant Physiol 124: 1678–1696

Frick G, Apel K and Armstrong GA (1995) Light-dependent protochlorophyllide oxidoreductase, phytochrome and greening in *Arabidopsis thaliana*. In: Mathis P (ed) Photosynthesis: From Light to Biosphere, Vol. III, pp 893–898. Kluwer Academic Publishers, Dordrecht

Frick G, van Cleve B, Apel K and Armstrong GA (1999) Developmental regulation of the PORA and PORB genes in wild-type Arabidopsis seedlings grown in far-red light and in the dark-grown cop1 mutant. In: Garab G (ed) Photosynthesis: Mechanisms and Effects, Vol. IV, pp 3253–3256. Kluwer Academic Publishers, Dordrecht

Frick G, Su Q, Apel K and Armstrong GA (2003) An Arabidopsis porB porC double mutant lacking light-dependent NADPH:protochlorophyllide oxidoreductases B and C is highly chlorophyll-deficient and developmentally arrested. Plant J 25: 141–153

Friedrichsen D and Chory J (2001) Steroid signaling in plants: from the cell surface to the nucleus. Bioessays 23: 1028–1036

Fujita Y, Takagi H and Hase T (1998) Cloning of a gene encoding a protochlorophyllide reductase: the physiological significance of the co-existence of light-dependent and -independent protochlorophyllide reduction systems in the cyanobacterium, *Plectonema boryanum*. Plant Cell Physiol 39: 177–185

Funk C, Schröder WP, Napiwotzki A, Tjus SE, Renger G and Andersson B (1995a) The PSII-S protein of higher plants: A new type of pigment-binding protein. Biochemistry 34: 1135–1141

Funk C, Adamska I, Green BR, Andersson B and Renger G (1995b) The nuclear-encoded chlorophyll-binding PSII-S protein is stable in the absence of pigments. J Biol Chem 270: 30141–30147

Gepstein S (1988) Photosynthesis. In: Noodén L and Leopold A (eds) Senescence and Aging in Plants, pp 85–109. Academic Press, San Diego, CA

Gibson LC, Marrison JL, Leech RM, Jensen PE, Bassham DC, Gibson M and Hunter CN (1996) A putative Mg chelatase subunit from *Arabidopsis thaliana* cv C24: Sequence and transcript analysis of the gene, import of the protein into chloroplasts, and in situ localization of the transcript and protein. Plant Physiol 111: 61–71

Gilchrist DG (1997) Mycotoxins reveal connections between plants and animals in apoptosis and ceramide signaling. Cell Death Dif 4: 689–698

Glickman MH and Ciechanover A (2002) The ubiquitin-proteasome proteolysis pathway: destruction for the sake of construction. Physiol Rev 82: 373–428

Gough SP and Kannangara CG (1979) Biosynthesis of 5-aminolevulinate in tigrina mutants of barley. Carlsberg Res Commun 44: 403–416

Granick S (1950) The structural and functional relationship between heme and chlorophyll. Harvey Lect 44: 220–245

Gray J, Close PS, Briggs SP and Johal GS (1997) A novel suppressor of cell death in plants encoded by the Lls1 gene of maize. Cell 89: 25–31

Gray J, Janick-Buckner D, Buckner B, Close PS and Johal GS (2002) Light-dependent death of maize lls1 cells is mediated by mature chloroplasts. Plant Physiol 130: 1894–1907

Green BR and Durnford DG (1996) The chlorophyll-carotenoid proteins of oxygenic photosynthesis. Annu Rev Plant Physiol Plant Mol Biol 47: 685–714

Green BR, Pichersky E and Kloppstech K (1991) The chlorophyll a/b-binding light-harvesting antennas of green plants: the story of an extended gene family. Trends Biochem Sci 16: 181–186

Green DR and Reed JC (1998) Mitochondria and apoptosis. Science 281: 1309–1312

Greenberg JT, Guo A, Klessig DF and Ausubel FM (1994) Programmed cell death in plants: A pathogen-triggered response activated coordinately with multiple defense functions. Cell 77: 551–563

Griffiths CM, Hosken SE, Oliver D, Chojecki J and Thomas H (1997) Sequencing, expression pattern and RFLP mapping of a senescence-enhanced cDNA from *Zea mays* with high homology to oryzain gamma and aleurain. Plant Mol Biol 34: 815–821

Griffiths WT (1975) Characterization of the terminal stage of chlorophyll(ide) synthesis in etioplast membrane preparations. Biochem J 152: 623–635

Griffiths WT (1978) Reconstitution of chlorophyllide formation by isolated etioplast membranes. Biochem J 174: 681–692

Grimm B and Kloppstech K (1987) The early light-inducible proteins of barley. Characterization of two families of 2-h-specific nuclear coded chloroplast proteins. Eur J Biochem 167: 493–499

Grimm B, Kruse E and Kloppstech K (1989) Transiently expressed early light-inducible thylakoid proteins share transmembrane domains with light-harvesting chlorophyll binding proteins. Plant Mol Biol 13: 583–593

Guiamét JJ, Pichersky E and Noodén, LD (1999) Mass exodus from senescing soybean chloroplasts. Plant Cell Physiol 40: 986–992

Guo H, Mockler T, Duong H and Lin C (2001) SUB1, an Arabidopsis Ca^{2+}-binding protein involved in cryptochrome and phytochrome coaction. Science 291: 487–490

Hadfield KA and Bennett AB (1997) Programmed senescence of plant organs. Cell Death Diff 4: 662–670

Hajouj T, Michelis R and Gepstein S (2000) Cloning and characterization of a receptor-like protein kinase gene associated with senescence. Plant Physiol 124: 1305–1314

Hardtke CS and Deng XW (2000) The cell biology of the COP/DET/FUS proteins. Regulating proteolysis in photomorphogenesis and beyond? Plant Physiol 124: 1548–1557

Häuser I, Dehesh K and Apel K (1984) The proteolytic degradation in vitro of the NADPH-protochlorophyllide oxidoreductase of barley (*Hordeum vulgare* L.). Arch Biochem Biophys 228: 577–586

Haussuhl K, Andersson B and Adamska I (2001) A chloroplast DegP2 protease performs the primary cleavage of the photodamaged D1 protein in plant photosystem II. EMBO J 20: 713–722

He Q, Brune D, Nieman R and Vermaas W (1998) Chlorophyll a synthesis upon interruption and deletion of por coding for the light-dependent NADPH:protochlorophyllide oxidoreductase in a photosystem-I-less/chlL—strain of Synechocystis sp. PCC 6803. Eur J Biochem 253: 161–172

He Y and Gan S (2002) A gene encoding an acyl hydrolase is involved in leaf senescence in Arabidopsis. Plant Cell 14: 805–815

He Y, Tang W, Swain JD, Green AL, Jack TP and Gan S (2001) Networking senescence-regulating pathways by using Arabidopsis enhancer trap lines. Plant Physiol 126: 707–716

He Y, Fukushige H, Hildebrand DF and Gan S (2002) Evidence supporting a role of jasmonic acid in Arabidopsis leaf senescence. Plant Physiol 128: 876–884

He ZH, Li J, Sundqvist C and Timko MP (1994) Leaf developmental age control expression of genes encoding enzymes of chlorophyll and heme biosynthesis in pea (*Pisum sativum* L.). Plant Physiol 106: 537–546

He ZH, Wang ZY, Li JM, Zhu Q, Lamb C, Ronald P and Chory J (2000) Perception of brassinosteroids by the extracellular domain of the receptor kinase BRI1. Science 288: 2360–2363

Hellmann H and Estelle M (2002) Plant development: regulation by protein degradation. Science 297: 793–797

Hensel LL, Grbic V, Baumgarten DA and Bleecker AB (1993) Developmental and age-related processes that influence the longevity and senescence of photosynthetic tissues in Arabidopsis. Plant Cell 5: 553–564

Hinderhofer K and Zentgraf U (2001) Identification of a transcriptional factor specifically expressed at the onset of leaf senescence. Planta 213: 469–473

Hoeberichts FA and Woltering EJ (2002) Multiple mediators of plant programmed cell death: interplay of conserved cell death mechanisms and plant-specific regulators. BioEssays 25: 47–57

Holland IB and Blight MA (1999) ABC-ATPases, adaptable energy generators fuelling transmembrane movement of a variety of molecules in organisms from bacteria to humans. J. Mol. Biol. 293: 381–399

Holm M, Ma LG, Qu LJ and Deng XW (2002) Two interacting bZIP proteins are direct targets of COP1-mediated control of light-dependent gene expression in Arabidopsis. Genes Dev 16: 1247–1259

Holtorf H, Reinbothe S, Reinbothe C, Bereza B and Apel K (1995) Two routes of chlorophyllide synthesis that are differentially regulated by light in barley. Proc Natl Acad Sci USA 92: 3254–3258

Hörtensteiner S (1999) Chlorophyll breakdown in higher plants and algae. Cell. Mol Life Sci 56: 330–347

Hörtensteiner S, Vicentini F and Matile P (1995) Chlorophyll breakdown in senescent cotyledons of rape, *Brassica napus* L.: enzymatic cleavage of phaeophorbide a in vitro. New Phytol 129: 237–246

Hörtensteiner S, Wüthrich KL, Matile P, Ongania KH and Kräutler B (1998) The key step in chlorophyll breakdown in higher plants: cleavage of pheophorbide a macrocycle by a monooxygenase. J Biol Chem 273: 15335–15339

Hsieh HM, Liu WK and Huang PC (1995) A novel stress-inducible metallothionin-like gene from rice. Plant Mol Biol 28: 381–389

Hu G, Yalpani N, Briggs SP and Johal G (1998) A porphyrin pathway impairment is responsible for the phenotype of a dominant disease lesion mimic mutant of maize. Plant Cell 10: 1095–1105

Hu YX, Bao F and Li JY (2000) Promotive effect of brassinosteroids on cell division involves a distinct CycD3-induction pathway in Arabidopsis. Plant J 24: 693–701

Huang L, Bonner BA and Castelfranco PA (1989) Regulation of 5-aminolevulinic acid (ALA) synthesis in developing chloroplasts. 11. Regulation of ALA-synthesizing capacity by phytochrome. Plant Physiol 90: 1003–1008

Hudson A, Carpenter R, Doyle S and Coen ES (1993) Olive: a key gene required for chlorophyll biosynthesis in *Antirrhinum majus*. EMBO J 12: 3711–3719

Huq E, Al-Sady B, Hudson M, Kim Ch, Apel K and Quail P (2004) PHYTOCHROME-INTERACTING FACTOR 1 is a critical bHLH regulator of chlorophyll biosynthesis. Science 305: 1937–1941

Ide JP, Klug DR, Kühlbrandt W, Georgi L and Porter G (1987) The state of detergent-solubilized light harvesting chlorophyll-a/b protein complex as monitored by picosecond time-resolved fluorescence and circular dichroism. Biochim Biophys Acta 893: 349–364

Ishikawa A, Okamoto H, Iwasaki Y and Asahi T (2001) A deficiency of coproporphyrinogen III oxidase causes lesion formation in Arabidopsis. Plant J 27: 89–99

Jiang CZ and Rodermel SR (1995) Regulation of photosynthesis during leaf development in RbcS antisense DNA mutants of tobacco. Plant Physiol 107: 215–224

Jiang CZ, Rodermel SR and Shibles RM (1993) Photosynthesis, Rubisco activity and amount, and their regulation by transcription in senescing soybean leaves. Plant Physiol 101: 105–112

Ignatov NV and Litvin FF (1981) Energy migration in a pigmented protochlorophyllide complex. Biofizika 26: 664–668

Ishikawa A, Okamoto H, Iwasaki Y and Asahi T (2001) A deficiency of coproporphyrinogen III oxidase causes lesion formation in Arabidopsis. Plant J 27: 89–99

Iwamoto K, Fukuda H and Sugiyama M (2001) Elimination of POR expression correlates with red leaf formation in *Amaranthus tricolor*. Plant J 27: 275–284

Jabs T, Dietrich RA and Dangl JL (1996) Initiation of runaway cell death in an Arabidopsis mutant by extracellular superoxide. Science 273: 1853–1856

Jacob-Wilk D, Holland D, Goldschmidt EE, Riov J and Eyal Y (1999) Chlorophyll breakdown by chlorophyllase: isolation and functional expression of the Chlase1 gene from ethylene-treated Citrus fruit and its regulation during development. Plant J 20: 653–661

Jansson S (1994) The light-harvesting chlorophyll a/b binding proteins. Biochim Biophys Acta 1184: 1–19

Johal GS, Hulbert S and Briggs SP (1995) Disease lesion mimic mutations of maize: A model for cell death in plants. Bioessays 17: 685–692

Kachroo P, Shranklin J, Shah J, Whittle E and Klessig DF (2001) A fatty acid desaturase modulates the activation of defense signaling pathways in plants. Proc Natl Acad Sci 98: 9448–9453

Kågedal K, Zhao M, Svensson I and Brunk UT (2001) Sphingosine-induced apoptosis is dependent on lysosomal proteases. Biochem J 359: 335–343

Kahn A (1968) Developmental physiology of bean leaf plastids. Tube transformation and protochlorphyll(ide) photoconversion by a flash irradiation. Plant Physiol 43: 1781–1785

Kahn AA, Boardman NK and Thorne SW (1970) Energy transfer between protochlorophyllide molecules: Evidence for multiple chromophores in the photoactive protochlorophyllide-protein complex in vivo and in vitro. J Mol Biol 48: 85–101

Kang JG, Yun J, Kim DH, Chung KS, Fujioka S, Kim JI, Dae HW, Yoshida S, Takatsuto S, Song PS and Park CM (2001) Light and brassinosteroid signals are integrated via a dark-induced small G protein in etiolated seedling growth. Cell 105: 625–636

Karger GA, Reid JD and Hunter CN (2001) Characterization of the binding of deuteron-porphyrin IX to the magnesium chelatase H subunit and spectroscopic properties of the complex. Biochemistry 40: 9291–9299

Karpinski S, Escobar C, Karpinska B, Creissen G and Mullineaux PM (1997) Photosynthetic electron transport regulates the expression of cytosolic ascorbate peroxidase genes in Arabidopsis during excess light stress. Plant Cell 9: 627–640

Kaup MT, Froese CD and Thompson JE (2002) A role of diacylglycerol acyltransferase during leaf senescence. Plant Physiol 129: 1616–1626

Kauschmann A, Jessop A, Koncz C, Szekeres M, Willmitzer L and Altmann T (1996) Genetic evidence for an essential role of brassinosteroids in plant development. Plant J 9: 701–713

Kawai-Yamada M, Ohori Y and Ucimiya H (2004) Dissection of Arabidopsis Bax inhibitor -1 suppressing Bax-, hydrogen peroxide-, and salicylic acid- induced cell death. Plant Cell 16: 21–32

Kawano T, Sahashi N, Takahashi K, Uozumi N and Muto S (1998) Salicylic acid induces extracellular generation of superoxide followed by an increase in cytosolic calcium ion in tobacco suspension culture: the earliest events in salicylic acid signal transduction. Plant Cell Physiol 39: 721–730

Kay SA and Griffiths WT (1983) Light-induced breakdown of NADPH:protochlorophyllide oxidoreductase in vitro. Plant Physiol 72: 229–236

Kendrick RE and Kronenberg GH (1994) Photomorphogenesis in Plants. Martinus Nijhoff, Dordrecht

Khalyfa A, Kermasha S, Marsot P and Goetgebheur M (1995) Purification and characterization of chlorophyllase from alga *Phaeodactylum tricornutum* by preparative native electrophoresis. Appl Biochem Biotechnol 53: 11–27

Kirk JTO and Tilney-Basset RAE (1978) The Plastids: Their Chemistry, Structure, Growth and Inheritance. Elsevier North-Holland Biomedical Press, Amsterdam/New York

Kleffmann T, Russenberger D, von Zychlinske A, Christopher W, Sjölander K, Gruissem W and Baginsky S (2004) The *Arabidopsis thaliana* chloroplast proteome reveals pathway abundance and novel protein functions. Curr Biol 14: 354–362

Kloppstech K, Meyer G, Bartsch K, Hundrieser J and Link G (1984) Control of gene expression during the early stages of chloroplast development. In: Wiessner W, Robinson D and Starr R (eds) Compartments of Algal Cells and Their Interaction, pp 36–46. Springer Verlag, Berlin

Kohchi T, Mukougawa K, Frankenberg N, Masuda M, Yokota A and Lagarias JC (2001) The Arabidopsis hy2 gene encodes phytochromobilin synthase, a ferredoxin-dependent biliverdin reductase. Plant Cell 13: 425–436

Koncz C, Mayerhofer R, Koncz-Kalman Z, Nawrath C, Reiss B, Redei GP and Schell J (1990) Isolation of a gene encoding a novel chloroplast protein by T-DNA tagging in *Arabidopsis thaliana*. EMBO J 9: 1337–1346

Kóta Z, Horvath LJ, Droppa M, Horvath G, Farkas T and Pali T (2002) Protein assembly and heat stability in developing thylakoid membranes during greening. Proc Natl Acad Sci USA 99: 12149–12154

Kovtun Y, Chui WL, Tena G and Sheen J (2000) Functional analysis of oxidative stress-activated mitogen-activated protein kinase cascade in plants. Proc Natl Acad Sci USA 97: 2940–2945

Kroemer G (1999) Mitochondrial control of apoptosis: an overview. Biochem Soc Symp 66: 1–15

Kühlbrandt W, Wang DN and Fujiyoshi Y (1994) Atomic model of plant light-harvesting complex by electron crystallography. Nature 367: 614–621

Kuriyama H and Fukuda H (2002) Developmental programmed cell death in plants. Curr Opin Plant Biol 5: 568–573

Labbe-Bois R and Labbe P (1990) Regulation of gene expression by iron and heme. In: Dailey HA (ed) Biosynthesis of Heme and Chlorophyll, pp 235–285. McGraw–Hill, New York

Lebedev N and Timko MP (1998) Protochlorophyllide photoreduction. Photosynth Res 58: 5–23

Lebedev N and Timko MP (1999) Protochlorophyllide oxidoreductase B-catalyzed protochlorophyllide photoreduction in vitro: insight into the mechanism of chlorophyll formation in light-adapted plants. Proc Natl Acad Sci USA 96: 9954–9959

Lebedev N, van Cleve B, Armstrong G and Apel K (1995) Chlorophyll synthesis in a de-etiolated (det340) mutant of Arabidopsis without NADPH-protochlorophyllide (PChlide) oxidoreductase (POR) A and photoactive PChlide-F655. Plant Cell 7: 2081–2090

Lee KP, Kim C, Lee DW and Apel K (2003) TIGRINA d, required for regulating the biosynthesis of tetrapyrroles in barley, is an ortholog of the FLU gene of *Arabidopsis thaliana*. FEBS Lett 553: 119–24

Lehnen LP, Sherman TD, Becerril JM and Duke SO (1990) Tissue and cellular localization of acifluorfen-induced porphyrins in cucumber cotyledons. Pestic Biochem Physiol 37: 239–248

Li JM and Chory J (1997) A putative leucine-rich repeat receptor kinase involved in brassinosteroid signal transduction. Cell 90: 929–938

Li JM and Nam KH (2002) Regulation of brassinosteroid signaling by a GSK3/SHAGGY-like kinase. Science 295: 1299–1301

Li JM and Timko MP (1996) The pc-1 phenotype of *Chlamydomonas reinhardtii* results from a deletion mutation in the nuclear gene for NADPH:protochlorophyllide oxidoreductase. Plant Mol Biol 30: 15–37

Li JM, Nagpal P, Vitart V, McMorris TC and Chory J (1996) A role for brassinosteroids in light-dependent development of Arabidopsis. Science 272: 398–401

Lin J-F and Wu S-H (2004) Molecular events in senescing Arabidopsis leaves. Plant J 39: 612–628

Lund S, Stall RE and Klee HJ (1998) Ethylene regulates the susceptible response to pathogen infection in tomato. Plant Cell 10: 372–382

Ma LG, Li JM, Qu LJ, Hager J, Chen Z, Zhao HY and Deng XW (2001) Light control of Arabidopsis development entails coordinated regulation of genome expression and cellular pathways. Plant Cell 13: 2589–2607

Mach JM, Castillo AR, Hoogstraten R and Greenberg JT (2001) The Arabidopsis accelerated cell death gene ACD2 encodes red chlorophyll catabolite reductase and suppresses the spread of disease symptoms. Proc Natl Acad Sci USA 98: 771–776

Mae T, Makino A and Ohira K (1987) Carbon fixation and changes with senescence in rice leaves. In: Thomson W, Nothnegel E and Huffaker R (eds) Plant Senescence: Its Biochemistry and Physiology, pp 123–131. American Society of Plant Physiologists, Rockville, MD

Marchetti P, Hirsch T, Zamzami N, Castedo M, Decaudin D, Susin SA, Masse B and Kroemer G (1996) Mitochondrial permeability transition triggers lymphocyte apoptosis. J Immunol 157: 4830–4836

Martinez-Garcia JF, Huq E and Quail PH (2000) Direct targeting of light signals to a promoter element-bound transcription factor. Science 288: 859–863

Masuda T, Fusada N, Shiraishi T, Kuroda H, Awai K, Shimada H, Ohta H and Takamiya K (2002) Identification of two differentially regulated isoforms of protochlorophyllide oxidoreductase (POR) from tobacco revealed a wide variety of light- and development-dependent regulations of POR gene expression among angiosperms. Photosynth Res 74: 165–172

Mathis P and Sauer K (1972) Circular dichroism studies on the structure and the photochemistry of protochlorophyllide and chlorophyllide holochrome. Biochim Biophys Acta 267: 498–511

Matile P and Schellenberg M (1996) The cleavage of phaeophorbide a is located in the envelope of barley gerontoplasts. Plant Physiol Biochem 34: 55–59

Matile P, Schellenberger M and Vicentini F (1997) Localization of chlorophyllase in the chloroplast envelope. Planta 201: 96–99

Matile P, Thomas H and Hörtensteiner S (1999) Chlorophyll degradation. Annu Rev Plant Physiol Plant Mol Biol 50: 67–95

Matringe M, Camadro JM, Block MA, Joyard J, Scalla R, Labbe P and Douce R (1992) Localisation within chloroplasts of protoporphyrinogen oxidase, the target enzyme for diphenyletherlike herbicides. J Biol Chem 267: 4646–4651

Meskauskiene R, Nater M, Goslings D, Kessler F, op den Camp R and Apel K (2001) FLU: a negative regulator of chlorophyll biosynthesis in *Arabidopsis thaliana*. Proc Natl Acad Sci USA 98: 12826–12831

Meskauskiene R and Apel K (2002) Interaction of FLU, a negative regulator of tetrapyrrole biosynthesis, with the glutamyl-tRNA reductase requires the tetratricopeptide repeat domain of FLU. FEBS Lett 532: 27–30

Miller BL and Huffaker RC (1981) Partial purification and characterization of endoproteinases from senescing barley leaves. Plant Physiol 68: 930–936

Mochizuki N, Brusslan JA, Larkin R, Nagatani A and Chory J (2001) Arabidopsis genomes uncoupled 5 (GUN5) mutant reveals the involvement of Mg-chelatase H subunit in plastid-to-nucleus signal transduction. Proc Natl Acad Sci USA 98: 2053–2058

Moeder W, Barry CS, Tauriainen AA, Betz C, Tuomainen J, Utriainen M, Grierson D, Sandermann H, Langebartels C and Kangasjarvi J (2002) Ethylene synthesis regulated by biphasic induction of ACC synthase and ACC oxidase genes is required for H_2O_2 accumulation and cell death in ozone-exposed tomato. Plant Physiol 130: 1918–1926

Molina A, Volrath S, Guyer D, Maleck K, Ryals J and Ward E (1999) Inhibition of protoporphyrinogen oxidase expression in Arabidopsis causes a lesion-mimic phenotype that induces systemic acquired resistance. Plant J 17: 667–678

Møller SG, Kunkel T and Chua NH (2001) A plastidic ABC protein involved in intercompartmental communication of light signaling. Genes Dev 15: 90–103

Møller SG, Ingles PJ and Whitelam GC (2002) The cell biology of phytochrome signalling. New Phytol 154: 553–590

Montgomery BL and Lagarias JC (2002) Phytochrome ancestry: sensors of bilins and light. Trends Plant Sci 7: 357–366

Morel JB and Dangl JL (1997) The hypersensitive response and the induction of cell death in plants. Cell Death Diff 4: 671–683

Mou Z, He Y, Dai Y, Liu X and Li J (2000) Deficiency in fatty acid synthase leads to premature cell death and dramatic alterations in plant morphology. Plant Cell 12: 405–417

Muramoto T, Kohchi T, Yokota A, Hwang I and Goodman HM (1999) The Arabidopsis photomorphogenic mutant hy1 is deficient in phytochrome chromophore biosynthesis as a result of a mutation in a plastid heme oxygenase. Plant Cell 11: 335–348

Nagy F and Schäfer E (2002) Phytochromes control photomorphogenesis by differentially regulated, interacting signaling pathways in higher plants. Annu Rev Plant Biol 53: 329–355

Nakabayashi K, Ito M, Kiyosue T, Shinozaki K and Watanabe A (1999) Identification of clp genes expressed in senescing *Arabidopsis thaliana* leaves. Plant Cell Physiol 40: 504–514

Nam HG (1997) The molecular genetic analysis of leaf senescence. Curr Opin Biotechnol 8: 200–207

Navarre DA and Wolpert TJ (1999) Victorin induction of an apoptotic/senescence-like response in oats. Plant Cell 11: 137–144

Neff MM, Ngyuen SM, Malancharuvil EJ, Fujiko S, Noichi T, Seto H, Tsubuki M, Honda T, Takatsuto S, Yoshida S and Chory J (1999) BAS1: A gene regulating brassinosteroid levels and light responsiveness in Arabidopsis. Proc Natl Acad Sci USA 96: 15316–15323

Ni M, Tepperman JM and Quail PH (1999) Binding of phytochrome B to its nuclear signalling partner PIF3 is reversibly induced by light. Nature 400: 781–784

Nibbe M, Hilpert B, Wasternack C, Miersch O and Apel K (2002) Cell death and salicylate- and jasmonate-dependent stress responses in Arabidopsis are controlled by single cet genes. Planta 216: 120–128

Noguchi T, Fujioka S, Choe S, Takatsuto S, Yoshida S, Yuan H, Feldmann KA and Tax FE (1999) Brassinosteroid-insensitive dwarf mutants of Arabidopsis accumulate brassinosteroids. Plant Physiol 121: 743–752

Noodén LD and Penny JP (2001) Correlative controls of senescence and plant death in *Arabidopsis thaliana* (Brassicaceae). J Exp Bot 52: 2151–2159

Noodén LD, Guiamét JJ and John I (1997) Senescence mechanisms. Physiol Plant 101: 746–753

Oberhuber M, Berghold J, Breuker K, Hörtensteiner S and Kräutler B (2003) Breakdown of chlorophyll: a nonenzymatic reaction accounts for the formation of the colorless "nonfluorescent" chlorophyll catabolites. Proc Natl Acad Sci USA 100: 6910–6915

Oelmüller R (1989) Photooxidative destruction of chloroplasts and its effect on nuclear gene expression and extraplastidic enzyme levels. Photochem Photobiol 49: 229–239

Oh MH, Ray WK, Huber SC, Asara JM, Gage DA and Clouse SD (2000) Recombinant brassinosteroid insensitive 1 receptor-like kinase autophosphorylates on serine and threonine residues and phosphorylates a conserved peptide motif in vitro. Plant Physiol 124: 751–765

Oh, SA, Park J-H, Lee GI, Paek KH, Park SK and Nam HG (1997) Identification of three genetic loci controlling leaf senescence in *Arabidopsis thaliana*. Plant J 12: 527–535

Okamoto H, Matsui M and Deng XW (2001) Overexpression of the heterotrimeric G-protein α-subunit enhances phytochrome-mediated inhibition of hypocotyl elongation in Arabidopsis. Plant Cell 13: 1639–1651

Oosawa N, Masuda T, Awai K, Fusada N, Shimada H, Ohta H and Takamiya KI (2000) Identification and light-induced expression of a novel gene of NADPH-protochlorophyllide oxidoreductase isoform in *Arabidopsis thaliana*. FEBS Lett 474: 133–136

op den Camp R, Przybyla D, Ochsenbein C, Laloi C, Kim C, Danon A, Wagner D, Hidég E, Göbel C, Feussner I, Nater M and Apel K (2003) Rapid induction of distinct stess responses after the release of singlet oxygen in Arabidopsis. Plant Cell 15: 2320–2332

Ordog SH, Higgins VJ and Vanlerberghe GC (2002) Mitochondrial alternative oxidase is not a critical component of plant viral resistance but may play a role in the hypersensitive response. Plant Physiol 129: 1858–1865

Orzáez D and Granell A (1997) The plant homologue of the defender against apoptotic death gene is down-regulated during senescence of flower petals. FEBS Lett 404: 275–278

Ougham HJ, Thomas AM, Thomas BJ, Frick GA and Armstrong GA (2001) Both light-dependent protochlorophyllide oxidoreductase A and protochlorophyllide oxidoreductase B are down-regulated in the slender mutant of barley. J Exp Bot 52: 1447–1454

Overmyer K, Tuominen H, Kettunen R, Betz C, Langebartels C, Sandermann H and Kangasjärvi J (2000) The ozone-sensitive Arabidopsis rcd1 mutant reveals opposite roles for ethylene and jasmonate signaling pathways in regulating superoxide-dependent cell death. Plant Cell 12: 1849–1862

Overmyer K, Borsché M and Kangasjärvi J (2003) Reactive oxygen species and hormonal control of cell death. Trends Plant Sci 8: 335–342

Oyama T, Shimura Y and Okada K (1997) The Arabidopsis HY5 gene encodes a bZIP protein that regulates stimulus-induced development of root and hypocotyl. Genes Dev 11: 2983–2995

Palsson LO, Spangfort MD, Gulbinas V and Gillbro T (1994) Ultrafast chlorophyll b to chlorophyll excitation energy transfer in the isolated light harvesting complex, LHCII, of green plants: implications for the organisation of chlorophylls. FEBS Lett 339: 134–138

Park J-H, Oh SA, Kim YH, Woo HR and Nam HG (1998) Differential expression of senescence-associated mRNAs during leaf senescence induced by different senescence-inducing factors in Arabidopsis. Plant Mol Biol 37: 445–454

Pattanayak GK and Tripathy BC (2002) Catalytic function of a novel protein protochlorophyllide oxidoreductase C of *Arabidopsis thaliana*. Biochem Biophys Res Commun 291: 921–924

Pennell RI and Lamb C (1997) Programmed cell death in plants. Plant Cell 9: 1157–1168

Perkins ND (2000) The Rel/NF-kappaB family: friend and foe. Trends Biochem Sci 25: 434–440

Pfannschmidt T, Allen JF and Oelmüller R (2001a) Principles of redox control in photosynthesis gene expression. Physiol Plant 112: 1–9

Pfannschmidt T, Schütze K, Brost M and Oelmüller R (2001b) A novel mechanism of nuclear photosynthesis gene regulation by redox signals from the chloroplast during photosystem stoichiometry adjustment. J Biol Chem 276: 36125–36130

Pontoppidan B and Kannangara CG (1994) Purification and characterization of barley glutamyl-tRNA reductase, the enzyme that directs glutamate to chlorophyll biosynthesis. Eur J Biochem 225: 529–537

Pruzinska A, Tanner G, Anders I, Roca M and Hörtensteiner S (2003) Chlorophyll breakdown: phaeophorbide a oxygenase is a Rieske-type iron-sulfur protein, encoded by the accelerated cell death 1 gene. Proc Natl Acad Sci USA 100: 15259–15264

Pursiheimo S, Mulo, P, Rintamäki E and Aro E-M (2001) Coregulation of light-harvesting complex II phosphorylation and Lhcb accumulation in winter rye. Plant J 26: 317–327

Quail PH (2000) Phytochrome-interacting factors. Semin Cell Dev Biol 11: 457–466

Quail PH (2002a) Photosensory perception and signalling in plant cells: new paradigms? Curr Opin Cell Biol 14: 180–188

Quail PH (2002b) Phytochrome photosensory signalling networks. Nat Rev Mol Cell Biol 3: 85–93

Quail PH, Boylan, MT, Parks BM, Short TW, Xu Y and Wagner D (1995) Phytochromes: photosensory perception and signal transduction. Science 268: 675–680

Rao MV, Lee H, Creelman RA, Mullet JE, Davis KR (2000) Jasmonic acid signalling modulates ozone-induced hypersensitive cell death. Plant Cell 12: 1633–1646

Rao MV, Lee HI and Davis KR (2002) Ozone-induced ethylene production is dependent on salicylic acid, and both salicylic acid and ethylene act in concert to regulate ozone-induced cell death. Plant J 32: 447–456

Reed JW, Nagpal P, Poole DS, Furuya M and Chory J (1993) Mutations in the gene for the red/far-red light receptor phytochrome B alter cell elongation and physiological responses throughout Arabidopsis development. Plant Cell 5: 147–157

Reinbothe C and Reinbothe S (1996a) Jasmonates – secondary messengers in plant defense and stress reactions. In: Grillo S and Leone A (eds) Physical Stresses in Plants – Genes and their Products for Tolerance, pp 161–168. Springer, Berlin

Reinbothe S and Reinbothe C (1996b) The regulation of enzymes involved in chlorophyll biosynthesis. Eur J Biochem 237: 323–343

Reinbothe S and Reinbothe C (2004) Light-harvesting protochlorophyllide binding protein complex (LHPP) is developmentally expressed in angiosperms. In: Schnarrenberger K and Wittmann-Liebold B (eds) Proceedings of 12th International Congress on Genes, Gene Families and Isozymes. Monduzzi Editore, Bologna, Italy, in press

Reinbothe S, Reinbothe C and Parthier B (1993a) Methyl jasmonate represses translation initiation of a specific set of mRNAs in barley. Plant J 4: 459–467

Reinbothe S, Reinbothe C and Parthier B (1993b) Methyl jasmonate-regulated translation of nuclear-encoded chloroplast proteins in barley (*Hordeum vulgare* L. cv. Salome). J Biol Chem 268: 10606–10611

Reinbothe S, Reinbothe C, Heintzen C, Seidenbecher C and Parthier B (1993c) A methyl jasmonate-induced shift in the length of the 5′ untranslated region impairs translation of the plastid rbcL transcript in barley. EMBO J 12: 1505–1512

Reinbothe S, Reinbothe C, Lehmann J, Becker W, Apel K and Parthier B (1994a) JIP60, a methyl jasmonate-induced ribosome-inactivating protein involved in plant stress reactions. Proc Natl Acad Sci USA 91: 7012–7016

Reinbothe S, Mollenhauer B and Reinbothe C (1994b) JIPs and RIPs: The regulation of plant gene expression by jasmonates in response to environmental cues and pathogens. Plant Cell 6: 1197–1209

Reinbothe S, Runge S, Reinbothe C, van Cleve B and Apel K (1995a) Substrate-dependent transport of the NADPH:protochlorophyllide oxidoreductase into isolated plastids. Plant Cell 7: 161–172

Reinbothe S, Reinbothe C, Runge S and Apel K (1995b) Enzymatic product formation impairs both the chloroplast receptor binding function as well as translocation competence of the NADPH:protochlorophyllide oxidoreductase, a nuclear-encoded plastid protein. J Cell Biol 129: 299–308

Reinbothe S, Reinbothe C, Holtorf H and Apel K (1995c) Two NADPH:protochlorophyllide oxidoreductases in barley: Evidence for the selective disappearance of PORA during the light-induced greening of etiolated seedlings. Plant Cell 7: 1933–1940

Reinbothe C, Apel K and Reinbothe S (1995d) A light-induced protease from barley plastids degrades NADPH:protochlorophyllide oxidoreductase complexed with chlorophyllide. Mol Cell Biol 15: 6206–6212

Reinbothe S, Reinbothe C, Lebedev N and Apel K (1996a) PORA and PORB, two light-dependent protochlorophyllide-reducing enzymes of angiosperm chlorophyll biosynthesis. Plant Cell 8: 763–769

Reinbothe S, Reinbothe C, Apel K and Lebedev N (1996b) Evolution of chlorophyll biosynthesis: The challenge to survive photooxidation. Cell 86: 703–705

Reinbothe C, Parthier B and Reinbothe S (1997) Temporal pattern of jasmonate-induced alterations in gene expression of barley leaves. Planta 201: 281–287

Reinbothe C, Lebedev N and Reinbothe S (1999) A protochlorophyllide light-harvesting complex involved in de-etiolation of higher plants. Nature 397: 80–84

Reinbothe S, Pollmann S and Reinbothe C (2003a) In situ conversion of protochlorophyllide b to protochlorophyllide a in barley: Evidence for a novel role of 7-formyl reductase in the prolamellar body of etioplasts. J Biol Chem 278: 800–806

Reinbothe C, Buhr F, Pollmann S and Reinbothe S (2003b) In vitro reconstitution of LHPP with protochlorophyllides a and b. J Biol Chem 278: 807–815

Reinbothe C, Lepinat A, Deckers M, Beck E and Reinbothe S (2003c) The extra loop distinguishing POR from the structurally related short-chain alcohol dehydrogenases is dispensable for pigment binding, but needed for the assembly of LHPP. J Biol Chem 278: 816–822

Reinbothe C, Satoh H, Alcaraz JP and Reinbothe S (2004a) A novel role of water-soluble chlorophyll proteins in transitory storage of chorophyllide. Plant Physiol 134: 1355–1365

Reinbothe C, Pollmann S, Desvignes C, Weigele M, Beck E and Reinbothe S (2004b) LHPP, the light-harvesting NADPH:protochlorophyllide (Pchlide) oxidoreductase:Pchlide complex of etiolated plants, is developmentally expressed across the barley leaf gradient. Plant Sci, in press

Rintamäki E, Martinsuo P, Pursiheimo S and Aro E-M (2000) Cooperative regulation of light-harvesting complex II phosphorylation via the plastoquinol and ferredoxin-thioredoxin system in chloroplasts. Proc Natl Acad Sci USA 97: 11644–11649

Robatzek S and Somssich IE (2001) A new member of the Arabidopsis WRKY transcription factor family, AtWRKY6, is associated with both senescence- and defense-related processes. Plant J 28: 123–133

Robatzek S and Somssich IE (2002) Targets of AtWRKY6 regulation during plant senescence and pathogen defense. Gene Dev 16: 1139–1149

Rodoni S, Muhlecker W, Anderl M, Krautler B, Moser D, Thomas H, Matile P and Hörtensteiner S (1997) Chlorophyll breakdown in senescent chloroplasts: cleavage of pheophorbide a in two enzymic steps. Plant Physiol 115: 669–676

Rowe JD and Griffiths WT (1995) Protochlorophyllide reductase in photosynthetic prokaryotes and its role in chlorophyll synthesis. Biochem J 311: 417–424

Runge S, Sperling U, Frick G, Apel K and Armstrong GA (1996) Distinct roles for light-dependent NADPH:protochlorophyllide oxidoreductases (POR) A and B during greening in higher plants. Plant J 9: 513–523

Ryberg M and Dehesh K (1986) Localization of NADPH-protochlorophyllide oxidoreductase in dark-grown wheat (*Triticum aestivum*) by immuno-electron microscopy before and after transformation of the prolamellar bodies. Physiol Plant 66: 616–624

Satoh H, Nakayama K and Okada M (1998) Molecular cloning and functional expression of a water-soluble chlorophyll protein, a putative carrier of chlorophyll molecules in cauliflower. J Biol Chem 273: 30568–30575

Schäfer E and Bowler C (2002) Phytochrome-mediated photoperception and signal transduction. EMBO Rep 3: 1042–1048

Schenk PM, Kazan K, Wilson I, Anderson JP, Richmond T, Somerville SC and Manners JM (2000) Coordinated plant defense responses in Arabidopsis revealed by microarray analysis. Proc Natl Acad Sci USA 97: 11655–11660

Schroeder DF, Gahrtz M, Maxwell BB, Cook RK, Kan JM, Alonso JM, Ecker JR and Chory J (2002) De-etiolated1 (DET1) and damaged DNA binding protein1 (DDB1) interact to regulate Arabidopsis photomorphogenesis. Curr Biol 12: 1462–1472

Schulze-Lefert P and Panstruga R (2003) Establishment of biotrophy by parasitic fungi and reprogramming of host cells for disease resistance. Annu Rev Phytopathol 41: 641–667

Schumacher K, Vafeados D, McCarthy M, Sze H, Wilkins T and Chory J (1999) The Arabidopsis det3 mutant reveals a central role for the vacuolar H$^+$-ATPase in plant growth and development. Genes Dev 13: 3259–3270

Schwechheimer C, Serino G, Callis J, Crosby WL, Lyapina S, Deshaies RJ, Gray WM, Estelle M and Deng XW (2001) Interactions of the COP9 signalosome with the E3 ubiquitin ligase SCFTIR1 in mediating auxin response. Science 292: 1379–1382

Semdner G and Parthier B (1993) The biochemistry and physiological and molecular aspects of jasmonates. Annu Rev Plant Physiol Plant Mol Biol 44: 569–589

Shacklock PS, Read ND and Trewavas AJ (1992) Cytosolic free calcium mediates red light-induced photomorphogenesis. Nature 358: 753–755

Shaw P, Henwood J, Oliver R and Griffiths WT (1985) Immunogold localisation of protochlorophyllide oxidoreductase in barley etioplasts. Eur J Cell Biol 39: 50–55

Shinomura T, Uchida K and Furuya M (2000) Elementary processes of photoperception by phytochrome A for high irradiance response of hypocotyl elongation in *Arabidopsis thaliana*. Plant Physiol 122: 147–156

Skinner JS and Timko MP (1998) Loblolly pine (*Pinus taeda* L.) contains multiple-expressed genes encoding light-dependent NADPH:protochlorophyllide oxidoreductase (POR). Plant Cell Physiol 39: 795–806

Smith AG (1986) Enzymes for chlorophyll synthesis in developing pea. In: Akoyunoglou GA and Senger H (eds) Regulation of Chloroplast Differentiation, pp 49–54. Alan R Liss, New York

Smith H (1982) Light quality photoperception and plant strategy. Annu Rev Plant Physiol 33: 481–518

Smith H (2000) Phytochromes and light signal perception by plants – an emerging synthesis. Nature 407: 585–591

Smith JHC and Benitez A (1954) The effect of temperature on the conversion of protochlorophyll to chlorophyll a in etiolated barley leaves. Plant Physiol 29: 135–143

Somanchi A and Mayfield SP (1999) Nuclear-chloroplast signalling. Curr Opin Plant Biol 2: 404–409

Spano AJ and Timko MP (1991) Isolation, characterization and partial amino acid sequence of a chloroplast-localized porphobilinogen deaminase from pea (*Pisum sativum* L.). Biochim Biophys Acta 1076: 29–36

Spano AJ, He Z and Timko MP (1992) NADPH: protochlorophyllide oxidoreductases in white pine (*Pinus strobus*) and loblolly pine (*P. taeda*). Mol Gen Genet 236: 86–95

Sperling U (1998) The In vivo Functions of NADPH: Protochlorophyllide Oxidoreductases A and B as Studied by Genetic Manipulation of Their Expression in *Arabidopsis thaliana*. PhD Thesis. ETH Zürich, Switzerland

Sperling U, van Cleve B, Frick G, Apel K and Armstrong GA (1997) Overexpression of light-dependent PORA or PORB in plants depleted of endogenous POR by far-red light enhances seedling survival in white light and protects against photooxidative damage. Plant J 12: 649–658

Sperling U, Franck F, van Cleve B, Frick G, Apel K and Armstrong GA (1998) Etioplast differentiation in Arabidopsis: both PORA and PORB restore the prolamellar body membrane and photoactive protochlorophyllide-F655 to the cop1 photomorphogenic mutant. Plant Cell 10: 283–296

Sperling U, Frick G, van Cleve B, Apel K and Armstrong GA (1999) Pigment-protein complexes, plastid development and photo-oxidative protection: the effects of PORA and PORB overxpression on Arabidopsis seedlings shifted from far-red to white light. In: Argyroudi-Akoyunoglou JH and Senger H (eds) The Chloroplast: From Molecular Biology to Biotechnology, pp 97–102. Kluwer Academic Publishers, Dordrecht

Stirpe F, Barbieri L, Gorini P, Valbonesi P, Bolognesi A and Polito L (1996) Activities associated with the presence of ribosome-inactivating proteins increase in senescent and stressed leaves. FEBS Lett 382: 309–312

Strand A, Asami T, Alonso A, Ecker JR and Chory J (2003) Chloroplast to nucleus communication triggered by accumulation of Mg-protoporphyrinIX. Nature 423: 79–83

Su Q, Frick G, Armstrong G and Apel K (2001) PORC of *Arabidopsis thaliana*: a third light- and NADPH-dependent protochlorophyllide oxidoreductase that is differentially regulated by light. Plant Mol Biol 47: 805–813

Sugimoto A, Hozak RR, Nakashima T, Nishimoto T and Rothman JH (1995) Das.1, an endogenous programmed cell death suppressor in *Caenorhabditis elegans* and vertebrates. EMBO J 14: 4434–4441

Sundqvist C and Dahlin C (1997) With chlorophyll from prolamellar bodies to light-harvesting complexes. Physiol Plant 100: 748–759

Surpin M, Larkin RM and Chory J (2002) Signal transduction between the chloroplast and the nucleus. Plant Cell Vol 14: 327–338

Susek RE, Ausubel FM and Chory J (1993) Signal transduction mutants of Arabidopsis uncouple nuclear CAB and RBCS gene expression from chloroplast development. Cell 74: 787–799

Suzuki JY and Bauer CE (1995) A prokaryotic origin for light-dependent chlorophyll biosynthesis of plants. Proc Natl Acad Sci USA 92: 3749–3753

Suzuki JY, Bollovar DW and Bauer CE (1997) Genetic analysis of chlorophyll biosynthesis. Annu Rev Gen 31: 61–89

Suzuki G, Yanagawa Y, Kwok SF, Matsui M and Deng XW (2002) Arabidopsis COP10 is a ubiquitin-conjugating enzyme variant that acts together with COP1 and the COP9 signalosome in repressing photomorphogenesis. Genes Dev 16: 554–559

Szekeres M, Németh K, Koncz-Kálmán Z, Mathur J, Kauschmann A, Altmann T, Rédei GP, Nagy F, Schell J and Koncz C (1996) Brassinosteroids rescue the deficiency of CYP90, a cytochrome P450, controlling cell elongation and de-etiolation in Arabidopsis. Cell 85: 171–182

Taiz L and Zeiger E (2002) Plant Physiology, 3rd Edition. Sinauer Associates, Sunderland, MA

Takamiya K-i, Tsuchiya T and Ohta H (2000) Degradation pathway(s) of chlorophyll: what has gene cloning revealed? Trends Plant Sci 5: 426–431

Takechi K, Sodmergen, Murata M, Motoyoshi F and Sakamoto W (2000) The YELLOW VARIEGATED (VAR2) locus encodes a homologue of FtsH, an ATP-dependent protease in Arabidopsis. Plant Cell Physiol 41: 1334–1346

Takio S, Nakao N, Suzuki T, Tanaka K, Yamamoto I and Satoh T (1998) Light-dependent expression of protochlorophyllide oxidoreductase gene in the liverwort, *Marchantia paleacea* var. *diptera*. Plant Cell Physiol 39: 665–669

Tepperman JM, Zhu T, Chang HS, Wang X and Quail PH (2001) Multiple transcription-factor genes are early targets of phytochrome A signaling. Proc Natl Acad Sci USA 98: 9437–9442

Terry MJ and Lagarias JC (1991) Holophytochrome assembly. Coupled assay for phytochromobilin synthase in organello. J Biol Chem 266: 22215–22221

Thayer SS and Huffaker RC (1984) Vacuolar localization of endoproteinases EP1 and EP2 in barley mesophyll cells. Plant Physiol 75: 70–73

Thomas H, Ougham H, Wagstaff C and Stead A (2003) Defining senescence and death. J Exp Bot 54: 1127–1132

Thomas J and Weinstein JD (1990) Measurement of heme efflux and heme content in isolated developing chloroplasts. Plant Physiol 94: 1414–1423

Thompson JE, Froese CD, Madey E, Smith MD and Hong Y (1998) Lipid metabolism during plant senescence. Prog Lipid Res 37: 119–141

Trebitsh T, Goldschmidt EE and Riov J (1993) Ethylene induces de novo synthesis of chlorophyllase, a chlorophyll degrading enzyme in Citrus fruit peel. Proc Natl Acad Sci USA 90: 9441–9445

Tsuchiya T, Ohta H, Masuda T, Mikami B, Kita N, Shioi T and Takamiya K-i (1997) Purification and characterization of two isozymes of chlorophyllase from mature leaves of *Chenopodium album*. Plant Cell Physiol 38: 1026–1031

Tsuchiya T, Ohta H, Okawa K, Iwamatsu A, Shimada H, Masuda T and Takamiya K-i (1999) Cloning of chlorophyllase, the key enzyme in chlorophyll degradation: finding of a lipase motif and the induction by methyl jasmonate. Proc Natl Acad Sci USA 96: 15362–15367

Tsuchiya T, Suzuki T, Yamada T, Shimada H, Masuda T, Ohta H and Takamiya K-i (2003) Chlorophyllase as a serine hydrolase: identification of a putative catalytic triad. Plant Cell Physiol 44: 96–101

Uren AG, O'Rourke K, Aravind L, Pisabarro MT, Seshagiri S, Koonin EV and Dixit VM (2000) Identification of paracaspases and metacaspases. Two ancient families of caspase-like proteins, one of which plays a key role in MALT lymphoma. Mol Cell 6: 961–967

van den Brûle S and Smart CC (2002) The plant PDR family of ABC transporters. Planta 216: 95–106

Varshavsky A (1997) The N-end rule pathway of protein degradation. Gene Cell 2: 13–28

Vaughan GD and Sauer K (1974) Energy transfer from protochlorophyllide to chlorophyllide during photoconversion of etiolated bean holochrome. Biochim Biophys Acta 347: 383–394

Vierstra RD (2003) The ubiquitin/26S proteasome pathway, the complex last chapter in the life of many plant proteins. Trends Plant Sci 8: 135–142

Virgin HI, Kahn A and von Wettstein D (1963) The physiology of chlorophyll formation in relation to structural changes in chloroplasts. Photochem Photobiol 2: 83–91

von Wettstein D, Gough S and Kannangara CG (1995) Chlorophyll biosynthesis. Plant Cell 7: 1039–1057

Vranová E, Inzé D and van Breusegem F (2002) Signal transduction during oxidative stress. J Exp Bot 53: 1227–1236

Wagstaff C, Leverentz MK, Griffiths G, Thomas B, Chanasut U, Stead AD and Rogers HJ (2002) Cysteine protease gene expression and proteolytic activity during senescence of Alstroemeria petals. J Exp Bot 53: 233–240

Walker CJ and Willows RD (1997) Mechanism and regulation of Mg-chelatase. Biochem J 327: 321–333

Wang H, Ma LG, Li JM, Zhao HY and Deng XW (2001) Direct interaction of Arabidopsis cryptochromes with COP1 in mediation of photomorphogenic development. Science 294: 154–158

Wang KLC, Li H and Ecker JR (2002) Ethylene biosynthesis and signalling networks mediating responses to stress. Plant Cell 14: 131–151

Wang ZY, Seto H, Fujioka S, Yoshida S and Chory J (2001) BRI1 is a critical component of a plasma-membrane receptor for plant steroids. Nature 410: 380–383

Watanabe N and Lam E (2004) Recent advance in the study of caspase-like proteases and Bax inhibitor-1 in plants: their possible roles as regulator of programmed cell death. Mol Plant Path 5: 65–70

Weaver LM, Himelblau E and Amasino RM (1997) Leaf senescence: gene expression and regulation. In: Setlow JK (ed) Genetic Engineering, Vol 19: Principles and Methods, pp 215–234. Plenum Press, New York

Weaver LM, Gan S, Quirino B and Amasino RM (1998) A comparison of the expression patterns of several senescence-associated genes in response to stress and hormone treatments. Plant Mol Biol 37: 455–469

Weiler EW, Kutchan TM, Gorba T, Brodschelm W, Niesel U and Bublitz F (1994) The Pseudomonas phytotoxin coronatine mimics octadecanoid signalling molecules of higher plants. FEBS Lett 345: 9–13

Weymann K, Hunt M, Uknes S, Neuenschwander U, Lawton K, Steiner HY and Ryals J (1995) Suppression and restoration of lesion formation in Arabidopsis lsd mutants. Plant Cell 7: 2013–2022

Whitelam GC, Johnson E, Peng J, Carol P, Anderson ML, Cowl JS and Harberd NP (1993) Phytochrome A null mutants of *Arabidopsis* display a wild-type phenotype in white light. Plant Cell 5: 757–768

Willows R (1999) Making light of a dark situation. Nature 397: 27–28

Willstätter R and Stoll A (1913) Die Wirkungen der Chlorophyllase. In Untersuchungen über Chlorophyll, pp 172–187. Springer, Berlin

Wittenbach VA, Lin W and Herbert RR (1982) Vacuolar localization of proteases and degradation of chloroplasts in mesophyll protoplasts from senescing primary wheat leaves. Plant Physiol 69: 98–102

Witty M, Wallace-Cook ADM, Albrecht H, Spano AJ, Michel H, Shabanowitz J, Hunt DF, Timko MP and Smith AG (1993) Structure and expression of chloroplast-localized porphobilinogen deaminase from pea (*Pisum sativum* L.) isolated by redudant polymerase chain reaction. Plant Physiol 103: 139–147

Woo HR, Chung KM, Park J-H, Oh SA, Ahn T, Hong SH, Jang SK and Nam HG (2001) ORE9, an F-box protein that regulates leaf senescence in Arabidopsis. Plant Cell 13: 1779–1790

Wüthrich KL, Bovet L, Hunziker PE, Donnison IS and Hörtensteiner S (2000) Molecular cloning, functional expression and characterisation of RCC reductase involved in chlorophyll catabolism. Plant J: 189–198

Xiang C and Oliver DJ (1998) Glutathione metabolic genes coordinately respond to heavy metals and jasmonic acid in Arabidopsis. Plant Cell 10: 1539–1550

Xie Z and Chen Z (1999) Salicylic acid induces rapid inhibition of mitochondrial electron transport and oxidative phosphorylation in tobacco cells. Plant Physiol 120: 217–226

Yang HQ, Tang RH and Cashmore AR (2001) The signalling mechanism of Arabidopsis CRY1 involves direct interaction with COP1. Plant Cell 13: 2573–2587

Yang M, Wardzala E, Johal G and Gray J (2004) The wound-inducible Lls1 gene from maize is an ortholog of the Arabidopsis acd1 gene, and the LLS1 protein is present in non-photosynthetic tissues. Plant Mol Biol 54: 1765–1791

Yin Y, Wang ZY, Mora-Garcia S, Li J, Yoshida S, Asami T and Chory J (2002) BES1 accumulates in the nucleus in response to brassinosteroids to regulate gene expression and promote stem elongation. Cell 109: 181–191

Yoshida S (2003) Molecular regulation of leaf senescence. Curr Opin Plant Biol 6: 79–84

Yoshida T and Minamikawa T (1996) Successive amino-terminal proteolysis of large subunit of ribulose-1.5-bisphosphate carboxylase/oxygenase by vacuolar enzymes from French bean leaves. Eur J Biochem 238: 317–324

Yoshida S, Ito M, Callis J, Nishida I and Watanabe A (2002) A delayed leaf senescence mutant is defective in arginyl-tRNA:protein arginyltransferase, a component of the N-end rule pathway in Arabidopsis. Plant J 32: 129–137

Yuan X-M, Li W, Dalen H, Lotem J, Kama R, Sachs S and Brunk UT (2002) Lysosomal destabilization in p53-induced apoptosis. Proc Natl Acad Sci USA 99: 6286–6291

Zhao M, Eaton JW and Brunk UT (2001) Bcl-2 phosphorylation is required for inhibition of oxidative stress-induced lysosomal leak and ensuing apoptosis. FEBS Lett 509: 405–412

Zheng B, Halperin T, Hruskova-Heidingsfeldova O, Adam Z and Clarke AK (2002) Characterization of chloroplast Clp proteins in Arabidopsis: localization, tissue specificity and stress responses. Physiol Plant 114: 92–101

Author Index

A

Abadia, Anunciacion, 65–85
Abadia, Javier, 65–85
Adams III, William W., 39–48, 49–64
Allakhverdiev, Suleyman I., 193–203
Amiard, Veronique, 49–64
Asada, Kozi, 205–221

B

Baginsky, Sacha, 269–287
Baier, Margarete, 303–319
Balseris, Andrius, 155–173

C

Creissen, Gary P., 223–239

D

Demmig–Adams, Barbara, 39–48, 49–64
Dietz, Karl–Josef, 303–319

E

Ebbert, Volker, 39–48
Edelman, Marvin, 23–38
El-Tayeb, Mohammad A., 303–319
Endo, Tsuyoshi, 205–221

F

Finkemeier, Iris, 303–319
Förster, Britta, 11–22
Foyer, Christine H., 241–268

H

Häder, Donat-P., 87–105
Huner, Norman P.A., 155–173
Hurry, Vaughn M., 155–173

I

Ivanov, Alexander G., 155–173

J

Jansson, Stefan, 145–153
Jung, Hou-Sung, 127–143

K

Karpinski, Stanislaw, 223–239
Krol, Marianna, 155–173

L

Lamkemeyer, Petra, 303–319
Larkin, Robert M., 289–301
Li, Wen-Xue, 303–319
Link, Gerhard, 269–287

M

Maccarrone, Mauro, 321–332
Mattoo, Autar K., 23–38
Melis, Anastasios, 175–191
Michel, Klaus-Peter, 303–319
Morales, Fermin, 65–85
Mueh, Kristine E., 49–64
Mullineaux, Philip M., 223–239
Murata, Norio, 193–203

N

Nishiyama, Yoshitaka, 193–203
Niyogi, Krishna K., 127–143
Noctor, Graham, 241–268

O

Öquist, Gunnar, 155–173
Osmond, Barry, 11–22

P

Pistorius, Elfriede, 303–319
Pocock, Tessa, 155–173

R

Reinbothe, Christiane, 333–365
Reinbothe, Steffen, 333–365
Rosso, Dominic, 155–173

S

Sane, Prafullachandra V., 155–173
Savitch, Leonid V., 155–173
Stork, Tina, 303–319

T

Trebst, Achim, 241–268

V

Vener, Alexander V., 107–126

Y

Yamamoto, Harry Y., 1–10
Yokthongwattana, Kittisak, 175–191

Z

Zarter, C. Ryan, 39–48, 49–64

Subject Index

β-Car, 166, 167
β-carotene, 137, 139, 244, 246, 250, 328
β-carotene hydroxylase, 137, 139
γ-EC, 257
γ-glutamylcysteine synthetase, 237, 257, 259

2-cys peroxiredoxins, 251, 308, 313
2-Cys Prx protects, 251, 312
3-(3, 4-dichlorphenyl)-1, 1-dimethylurea (DCMU), 247
32 kDa protein, 24, 25
32*, 32, 33
43 kDa core antenna proteins, 246
47 kDa core antenna proteins, 246
5-LOX activating protein (FLAP), 326
505 *minus* 540 nm, 7
505-nm change, 8
6-palmitoylascorbic acid, 326, 328
77K fluorescence, 16
9-hydroperoxylinoleic acid, 326
9-hydroxylinoleic acid, 326
9-*cis*-epoxycarotenoid dioxygenase (NCED), 252

A

A-band, 160
A. thaliana, 112, 117, 120, 121
ABA, 252, 256
ABA biosynthesis, 252
ABC transporters, 344
ABI1, 233
ABI2, 233, 234
abiotic, 257
abiotic stresses, 242, 252, 256
abscisic acid, 224, 311
absorption, 31
"acceptor-side" mechanism, 195, 196
acetylated, 120
Achim Hager, 4
activation energy barrier, 159
activation of caspases, 324
acyl hydrolases, 352
acylation, 29
Adam Gilmore, 8
adaptability to light environments, 8
adaptation of the photosynthetic apparatus, 110
adaptive responses, 107, 116, 117
adenosine phosphosulfate reductase, 259
Advance Study Institute on the Biochemistry of Chloroplasts, 6
ageing, degenerative diseases, 254
aldolase, 258
algae, 115, 147, 243
alkyl hydroperoxides, 306, 312
Amaranthus, 25, 27
American Society of Plant Biologists, 9
Amyema, 18
analysis of plant pigments, 8
antenna, 243, 244, 305

antenna quenching, 156, 157
antennae-chlorophyll-deficient plants, 30
antheraxanthin, 4, 8, 116, 157
anti-phosphothreonine antibody, 112
anti-phosphotyrosine antibody, 111
antimycin, 255
antimycin A, 210
antioxidant defenses, 225, 226, 228–232, 234, 315
antioxidant network, 225–227, 231
antioxidants, 59, 243, 304, 308, 321
apoptosis, 321, 350
APX, 226, 231, 234
APX1, 230, 231
APX2, 229–231
Aquatic Environment, 87
Arabidopsis, 7, 112, 119, 121, 128, 129, 134, 135, 137–139, 146, 148, 150, 152, 157, 234, 243, 247, 250–252, 255
Arabidopsis thaliana, 15, 110, 156
arachidonate cascade, 323
AsA, 212, 214
ascorbate, 4, 127, 133, 137, 138, 212, 224–227, 233, 234, 245, 259, 260, 312
ascorbate oxidases, 253
ascorbate peroxidase (APX), 212, 225, 226, 244, 251
Ascorbate, glutathione, 242, 251
Ascorbate-dependent dioxygenases, 252
ascorbate-reducible transmembrane *b*-type cytochrome c, 254
ascorbate/DHA, 258
ascorbic acid, 326
ascorbic acid-deficient mutant, 231
assembly, 116, 119
assimilatory inhibition, 12
ATP, 25, 109, 112
atrazine, 26, 27, 36
atrazine resistance, 27
AtrbohD, 253
AtrbohF, 253
auxins, 234
availability, 7
azidoatrazine, 28

B

B_1-band, 160
B_2-band, 160
B-glucuronidase (*GUS*), 247
bacterium, 27
Barbara Demmig, 8
basic protein, 121
biogenesis, 107
biological clock, 34
biotic, 242, 256, 257
birch, 230
Blumeria gramis, 260
bound QB, 30
BR signaling, 345
Brassinosteroids (BR), 345

brown alga *Dictyota dichotoma*, 91
brown algae, 89
bZIP transcription factors, 233, 297, 338

C

C_3 and C_4 species, 13
C-band, 160
C-terminal sequence, 7
C. Freeman Allen, 7
C. reinhardtii, 111
C. reinhardtii, 112, 115, 120
C. Stacy French, 6
C5-pathway, 339
cab genes, 121
cadaverine, 328
caffeic acid, 325, 326
calcium, 230, 231, 252
calcium channels, 243
calcium release, 323
carbon, 156
carboxy terminus, 28
carboxy end, 28
carotene, 244, 246, 249
carotene biosynthesis, 250
carotenoid, 52–54, 87, 97, 127, 133, 136, 151, 230, 232, 244
carotenoid deficiency, 247
caspase, 350
caspase-like proteins, 350
catalase, 225, 228, 230, 232, 243, 257, 260
catalase deficiency, 257
catalase-peroxidases, 306
cauliflower, 255
ccr, 211
cell death, 321
cell growth, 243
cell signaling, 107, 121, 122
cell wall cleavage, 243
cell wall growth, 243
changing environment, 120
charge recombination, 159
Charles Reid Barnes Life Membership Award, 9
chilling, 116, 229, 258
chilling treatment, 116
Chl_Z^+, 167
Chlamydomas, 31, 32, 128, 132, 134, 135, 137, 148, 149, 150, 152
Chlamydomonas reinhardtii, 15, 17, 27, 110, 147, 157, 249
chloramphenicol, 97, 99
Chlorella vulgaris, 164
chlorophyll, 27
chlorophyll a, 146
chlorophyll a/b binding protein, 33
chlorophyll b, 146, 151
chlorophyll biosynthesis, 147
chlorophyll fluorescence, 50, 51, 89, 207
chlorophyll fluorescence quenching, 8
chlorophyll precursor signaling, 290
chlorophyll triplets, 28, 243, 244
Chlorophyllase, 349
Chlorophyte Prasiola, 97
Chlorophyte *Codium bursa*, 91
Chlorophyte *Enteromorpha*, 93

chlorophytes, 97
chloroplast, 24, 31, 36, 108, 109, 111, 113, 114, 121, 122, 251, 258
chloroplast development, 24
chloroplast envelope, 248
chloroplast lipid, 7
chloroplast membrane protein, 25
chloroplast movements, 12, 100, 131
chloroplast RNA, 24
chloroplast stroma, 242
chloroplast targeting, 121
chloroplast transcription, 256
chloroplast-encoded, 121
chlororespiration, 211
chlorosis, 116
chromatic adaptation, 33
chromatin condensation, 323
chronic photoinhibition, 88
circadian clock, 34
circadian controls, 247
circadian oscillation, 121
circadian rhythm, 121
Cladophora, 90
cleavage map, 27
Clinton "Chi Chichester, 3
cloned VDE, 7
Clp protease, 347
CO_2 compensation point, 14
CO_2 concentrating cells, 18
CO_2 concentrations, 309
CO_2 supply, 13
coastal areas, 88
cold acclimation, 156, 163
cold stress, 116, 163
cold-hard-band, 160
cold-sustained, 157
coleoptiles, 243
complex, 255
complex I (the major NADH-ubiquinone oxidoreductase) and complex III (ubiquinol-cytochrome c oxidoreductase), 254
complex I-deficient mutant, 255
confocal microscope, 13
conifers, 148
COP9 signalosome, 338
Corallina, 90, 95, 96
Corallina elongata, 92
Corallina officinalis, 91, 94–96
core proteins, 113
CP12 transcripts, 252
CP26, 134
CP29, 107, 108, 111, 112, 116, 120, 134, 135, 146
CP43, 107, 109, 111–113, 115–117, 119, 122
cross-talk, 34
C. Stacy French, 6
Cuscuta, 18
cyanobacteria, 89, 97, 98, 100, 115, 121, 147, 149, 151, 156, 247, 303, 304
cyclic Electron Flow, 207
cyclic electron transport, 133, 136
cyclic PS I electron transport, 33
cyclooxygenase, 326
cyclopentenone isoprostanes, 229

Subject Index

cyclophilin, 117, 118
cyclophilin, TLP40, 115
cyclosporin A, 118
cysteine rich domain, 7
Cyt b_{559}, 167
cytochrome c, 255, 351
cytochrome bf complex, 122
cytochromes, 247
cytoplasm, 253
cytoplasmic membrane, 304
cytoskeleton, 324
cytoskeleton organization, 323
cytosol, 122, 257, 258

D

D-*threo*-chloramphenicol, 27
D1, 26–28, 39, 44–46, 53, 55, 57–59, 107, 109, 111, 112, 114–117, 119, 121, 122, 244, 247, 249, 250
D1 degradation, 36
D1 mutants, 31
D1 polypeptide, 14
D1 protein, 15, 23, 35, 87, 88, 97, 99, 100, 198, 246
D1 protein phosphorylation, 113
D1 repair, 156
D1 repair cycle, 17
D1 turnover, 27, 115
D1/D2 heterodimer, 29, 31
D1/D2 heterodimer degradation, 31
(D1:1), 161
(D1:2), 161
D2, 27, 107, 109, 111–113, 115–117, 119, 122, 244
D2 protein, 31, 246
damage, 66–68, 70, 72, 75, 77, 78, 113
Dan Arnon, 6
darkness, 121
David Fork, 8
David I. Sapozhnikov, 3
Davis, 3
DBMIB, 33
dChl$_z$, 167
DCMU, 160
de-epoxidation, 251
de-epoxidation state (DES), 8
decylplastoquinone, 249
defense, 243
deficiencies, 66, 69, 72
deficiencies of Fe, N, 66
DegP2, 115
degradation, 26, 29, 114, 115, 119
degradation product, 28
degradation of Chl, 348
degradation of D1, 113, 119
dehydroascorbate, 212, 242, 245
dehydroascorbate peroxidase, 305
dehydroascorbic acid, 326
deletion mutation, 7
Department of Food Technology at the University of Illinois, 2
dephosphorylation, 33, 113–117, 122
depletion, 93
DHA, 212, 245, 253, 254, 259
DHA reductase, 212

DHA reduction, 244
DHAR, 251, 253, 254
DHAR activity, 259
diadinoxanthin cycle, 4
diatoxanthin, 60
Dictyota dichotoma, 92, 93
dicysteinyl motif, 259
diel solar cycle, 87
difference spectrum, 6
dimers of PS II, 115
direct mechanism, 157
disassembly, 116
disassembly of PS II, 113
dissipated thermally, 68, 74, 75
dissipation, 39, 43, 45
dissipation (NPQ), 44
dissolved and particulate organic carbon (DOC and POC), 88
disulfide, 272, 273, 275, 313
disulfide bridge, 275
dithiol, 273, 275
dithiol/disulfide, 273, 275, 277–279, 283
dithiol/disulfide exchange, 272, 275
dithiothreitol (DTT), 7, 100
diurnal changes, 100
diurnal cycles, 121
diurnal rhythm, 96
diuron, 24–26, 33
DNA, 115
DNA fragmentation, 323
domains, 8
"donor-side" mechanism, 196
Dorothea Siefermann, 7
down regulation of photosynthesis, 70, 73
drought, 65–67, 73–75, 229, 231, 233, 314
DTT, 99
DTT inhibition, 7
Dunaliella, 131
Dunaliella salina, 164
dynamic photoinhibition, 52, 53, 88

E

E.coli, 257
Early Light Inducible Proteins (ELIPs), 43, 55, 138, 147–150, 158
ecotypes, 15
ECS, 232
EDS1, 230
effective photosynthetic quantum yield, 94
effective photosynthetic yield, 88
electron spin resonance, 246
electron transport, 24, 56, 115
elevated temperature, 118
elevation of temperature, 117, 122
elicitor cryptogein, 229
elicitors, 230
elongation factors, 198
elongation step of translation, 198
embryonic axes, 243
emission peak temperatures (T_M), 159
endonuclease, 277, 278
energized photoreceptor, 35

energy dissipation, 39, 49–54, 56, 57, 59–61, 68, 72–75, 77, 78, 116, 157
energy distribution, 34
Enteromorpha, 94
environmental conditions, 117, 122
environmental stress, 194
enzymes, 117, 122, 304
EPR spectroscopy, 326
ESTs, 121
ethylene, 228, 230, 234, 243, 347
ETR1, 234
Euglena, 23
eukaryotes, 110, 115
eulittoral (intertidal zone), 88
excess light, 49, 50, 52, 53, 226, 229, 231
excitation energy distribution, 119
excitation pressure, 13
EXECUTER 1 (EX1), 247
exonucleases, 277, 278
expression, 122
expression dynamics, 272
external photoprotection, 12

F

*fad*B, 169
*fad*C, 169
far-red, 30, 33
fast induction kinetics, 95
Fast Repetition Rate Fluorometer, 19
fat (egger) plates, 9
fatty acid, 167
fatty acid β-oxidation, 242
fatty acid desaturases, 169
FCCP, 26
Fd, FTR, 244
Fe, 66, 69, 74, 77
Fe deficiencies, 66–68, 77
Fe chlorosis, 66
feedback, 70
feedback down-regulation of photosynthesis, 70, 72, 75
feedback de-excitation, 131, 132, 150
feedback down-regulation, 70, 75
Fenton reaction, 213
ferns, 115
ferredoxin, 108, 121, 210, 244
ferredoxin-NADP$^+$ reductase, 135
ferredoxin-plastoquinone oxidoreductase, 136
ferredoxin-quinone reductase, 210
flavonoids, 31
FLU (FLUORESCENT), 339
flu mutants, 247
fluidity and permeability of biomembranes, 323
fluorescence, 89
fluorescence lifetime approach, 18
fluorescence parameters, 87, 94, 95
folding, 119
folding catalysis, 117
FPLC, 117
FQR, 210
freshwater systems, 88

FTR, 244
FtsH, 115, 138, 139
funnel, 158

G

G-box, 297
G-protein, 168
"gain of function mutants, 17
gene expression, 122, 242, 260
gene regulatory systems, 274
genecological differentiation, 13
genecotypic differentiation, 15
genes, 146
genomes uncoupled, 293
genomic sequence, 110
geranyl-geraniol, 250
germinating seeds, 243
gibberellic acid, 243
GLDH, 255
glucose, 121, 254
glucose 6-phosphate dehydrogenase (G6PDH), phosphoribulokinase (PRK), fructose-1,6-*bis*phosphatase, 252
GLUT-type glucose transporters, 254
glutamine synthetase, 14
glutaredoxins (GRX), 225, 233, 244, 256
glutathione (GSH), 224–228, 231–234, 256, 257, 260, 272, 273, 276, 278, 281–283, 312
glutathione peroxidases (GPXs), 225, 244, 251, 256, 306
glutathione redox couple, 258
glutathione reductase, 234, 244
glutathione *S*-transferases, 251
glutathione synthesis, 258
glutathionylation, 233, 258, 260
GPX, 257
GR, 251, 259
grana, 115
grana regions, 32
granal membranes, 28
granal stacks, 13
gravitropic responses, 243
"green eukaryotes, 270
GRXs, 258, 259
GSH biosynthetic enzymes, 257
GSH synthesis, 257
GSH/GSSG, 259, 276, 278
GSH/GSSG couple, 256
GSH/GSSG ratio, 258
GSSG, 257, 258, 260, 272, 276, 278
GST1, 228, 232
GST2, 228
GSTs, 229, 234, 256
guard cell, 252
gun (*genomes uncoupled*), 247
GUN4, 294
GUN4 complex, 296
GUN5, 294

H

H_2O_2, 247, 251, 253, 257, 259, 305
H_2O_2 detoxification, 251

Subject Index

Halimeda, 91
HCO_3^-, 168
heat shock, 116–118
heat stress, 229, 233
heme c_i, 210
heme x, 210
hemi-parasite dodder, 18
herbicide receptor protein, 25
herbicide binding sites, 29
herbicide sensitivity, 24
herbicides, 26, 35, 247
high light, 52, 59, 60, 116, 120, 121, 228, 233, 308
high light intensities, 113
high light stress, 116, 242
high temperatures, 119
High-Light-Inducible Proteins (HLIPs), 43, 55, 147, 148, 151
homogentisate phytytransferase, 250
homogentisic acid, 248
homologous recombination, 243
Hordeum vulgare, 159
hormone, 224, 228, 230, 232, 234, 243, 250
hormone-induced senescence, 346
HPLC, 8
HPP-dioxygenase, 250
human health, 248
humic substances, 88
HY1, 293
HY2, 293
hydrogen peroxidases, 305
hydrogen peroxide, 197, 212, 224, 226, 227, 229, 242, 245, 254, 256, 260, 324
hydroperoxychromanone, 248
hydroperoxy-tocopherone, 249
hydroxyl, 260
hydroxyl radical, 197, 213, 224, 256, 305
hydroxyoctadecatrienoic acid, 229
hydroxyphenylpyruvate (HPP) dioxygenase, 248
hypersensitive response, 227, 324

I

IdiA protein, 305
illumination, 122
immunoblotting, 117
IMMUTANS, 212
in innate immune responses, 242
in situ measurement of xanthophyll cycle activity, 7
in vivo substrate of VDE, 7
indirect mechanism, 157
Inga sapindoides, 18
inhibition, 94, 321
inorganic particulate substances, 88
insertional mutagenesis, 135
Institute of Food Technologists, 9
intertidal, 15
intertidal algae, 15
intertidal zone, 89
intramembranal cycling, 29
iron, 66
iron chlorosis, 66
iron deficiency, 66
iron homeostasis, 305

iron sulfur clusters, 272
irradiance, 156
isoxaflutole, 249

J

jasmonic acid, 224, 228, 234, 346

K

K deficiency, 72
KCN, 255
kinases, 122, 276, 278, 281, 282
kinetic gas exchange and optical methods, 17
knockout plants, 122
knockouts, 108

L

L-galactono-1, 4-lactone, 253, 255
L-galactono-1, 4-lactone dehydrogenase, 255
Laminaria digitata, 93
Laser Induced Fluorescence Transient (LIFT) apparatus, 19
lateral migration, 113, 115
LCMS, 109–112
leaf disc O_2 electrode system, 16
leaves, 243
lentil root protoplasts, 325
Leo Vernon, 6
lesion mimic mutants, 227, 228
leucine zipper, 117
LH2 phosphorylation, 33
LHC, 148–152
(LHC) proteins, 146
Lhca, 150
Lhca1, 146
Lhca2, 146
Lhca5, 148, 150
Lhca6, 150
Lhcb1, 112, 146, 148
Lhcb2, 112, 146, 148
Lhcb3, 148, 150
Lhcb4, 146, 147, 149
Lhcb4.2, 120
Lhcb5, 147, 149, 150
Lhcb6, 148, 150
LHCI, 148–150
LHCII, 107–113, 116, 117, 119–122, 134, 135, 146, 148–150, 152
LHCII degradation, 120
LHCII kinase, 345
LHCII trimer, 158
LI818, 150
life cycle of the D1, 33
life history of D1, 29
light, 108, 119, 194, 247, 271, 273–281
light adaptations, 35
light energy, 116
light harvesting chlorophyll a/b protein, 36
light intensity, 26, 35
light meter, 36
light regulation, 279
light stress, 127, 130, 139

light-activated kinase, 114
light-dependent dynamics, 120
light-harvesting chlorophyll a/b-binding, 146
light-harvesting chlorophyll a/b binding protein (*Lhcb*), 247
light-harvesting polypeptide, 116
light-harvesting POR:Pchlide complexes (LHPP), 341
light-induced phosphorylation, 114, 121
light/dark regulation, 273
light/dark signaling, 257
light/dark transitions, 116
linolenic acid, 229
lipid, 167
lipid composition, 36
lipid environment, 36
lipid for de-epoxidation, 7
lipid hydroperoxides, 256
lipid peroxidation, 243, 304
lipid phase, 8
lipocalin, 7, 9
lipoic acid, 259
lipoperoxidation, 321
lipoxygenases, 321
liquid chromatography mass spectrometry, 109
liverworts, 115
localized, 8
long after far-red 6, 294
low temperatures, 116, 121, 246
"low-light" mechanisms, 196
low-temperature stress, 199
lumen, 4, 118
lutein, 3
lutein epoxide cycle, 4
lutein-epoxide, 18

M

MAAs, 98
macroalgae, 87–89, 94, 95, 98, 100
maize, 24, 29, 116, 120, 258
malate valve, 309
MALDI, 110
Malva, 157
mammalian phospholipid hydroperoxide glutathione peroxidases (PHGPX), 256
Mantoniella, 8
map-based cloning, 134
Marchantia polymorpha, 148
mass spectrometry, 109, 110
maximal fluorescence, 94
MDA, 212
MDHA, 245, 253, 254
MDHAR, 245, 251
Mehler reaction, 17, 229, 230
Mehler-peroxidase (water-water) cycle, 251
membrane attachment, 29
membrane protein, 24
mesophytic, 157
messenger, 8
metabolic regulation, 34
metabolism, 243
methionine, 120
Mg, 72

Mg-protoporphyrin IX, 131, 132, 292
Mg-protoporphyrin IX monomethyl ester, 290
MGDG, 7
micelle, 7
microarray, 139
microcystin, 117
mistletoes, 18
mitochondria, 254, 255, 323
mitochondrial membrane permeability transition, 351
mitosis, 243
mixing layer, 88
MK886, 326
mobile pool of LHCII, 119
molecular gear shift, 157
monodehydroascorbate peroxidase, 305
monodehydroascorbate radical, 212
monogalactosyldiacylglycerol (MGDG), 7
mosses, 115
multifunctional capabilities, 8
multiple functions, 8
mutants, 247, 282
mycosporine-like amino acids, 87, 97

N

N, 69–72, 77
N deficiency, 70–72, 74, 77
N-terminal acetylation, 111, 117, 120
N-terminal sequence, 7
N-terminally blocked, 120
N-terminus, 121
NAD(P)H Dehydrogenase, 207
NADH homeostasis, 255
NADH-MDHAR, 254
NADPH oxidases, 229, 234, 253
NADPH/NADP, 258
NADPH/NADP$^+$, 309
NADPH:Pchlide oxidoreductase (POR), 339
NDH, 207
necrotic cells, 324
Nernst equation, 258
nitric oxide, 243
nitrogen, 69, 70, 156
nitrogen deficiency, 70, 71
nitrogen nutrition in shade-sun acclimation, 15
nocturnal retention, 116
non expressor of PR1-1, 232
non-heme iron, 27
non-photochemical quenching (NPQ), 8, 89, 94, 156
nonphotochemical energy dissipation, 116
nonphotochemical quenching, 50, 51, 131
nordihydroguaiaretic acid (NDGA), 325
norflurazon, 247
Norris, 248, 250
NPQ, 40–45, 46, 50–53, 158
npq1 lor1, 165
npq4-1, 158
npq5, 157
NPR1, 232–234, 257, 260
npr1, nonexpressor of PR genes, 256
nuclear D1-kinase, 34
nuclear-encoded, 121

Subject Index

nucleobase sodium-dependent L-ascorbic acid transporters (NATs), 254
nucleus, 122, 257
nutrient, 65, 72, 77, 156
nutrient deficiencies, 77

O

O_2 photoreduction, 17
O_2/H_2O redox couple, 242
oceanic waters, 88
ODS-1, 8
okadaic acid, 117
Olga Koroleva, 4
Olle Björkman, 8
one-helix proteins, 138
optimal quantum yield, 88
organello run-on transcription, 280
organic peroxides, 244
oscillatory behaviour, 16
over-reduction, 208
oxidative burst, 253
oxidative damage, 69, 76, 77
oxidative load, 242
oxidative stress, 57, 74–77, 197, 224–226, 230, 234, 314
oxidative stress, serum starvation, 324
oxygen evolution, 26, 32, 113, 115, 118
oxygen evolving complex (OEC), 159
oxygen exchange, 89
oxygenic bacterial phototrophs, 34
oxygenic photosynthesis, 34
oxylipins, 312
oxyR, 257, 272
ozone, 93, 227–230, 233, 234, 243, 247, 252, 253

P

P, 71, 72
P deficiency, 77
P54, 278
P680, 244, 246
P680 triplets, 250
P680$^+$, 167
P700$^+$, 207
PAD4, 230
palmitic acid, 29
palmitoylation, 29, 168
PAM, 8
PAM fluorescence, 89, 91
PAM fluorescence parameters, 89
PAR, photosynthetic active radiation, 93
parsley, 232
pathogenesis related (PR) proteins, 252, 256
pea, 121
penetration into the water column, 88
peptidyl-prolyl isomerase, 117
peroxidases, 224, 227, 229
peroxinitrite, 306, 312
peroxiredoxins, 244, 251, 303, 306
PEST regions, 28
PEST-like region, 28

PGR5, 210
*pgr*5, 210
pgsA, 167
pH-dependent binding of VDE, 7
phaeophorbide *a* oxygenase, 349
phaeophytes, 89, 95, 97
phenylpropanoid pathway, 232
phosphatase, 119
phosphatase inhibitors, 116
phosphatidylglycerol (PG), 167
phospho-LHCII, 108
phosphoamino acid, 110
phosphoamino acid antibodies, 109
phospholipase Dα, 352
phosphopeptides, 117, 121
phosphoprotein, 109, 110, 112, 116, 117, 120, 121
phosphorus deficiencies, 72
phosphorylation, 26, 27, 32, 33, 119, 123, 276, 278, 281, 282
phosphorylation control, 281
phosphorylation of PS II, 113
phosphorylation site, 109–112, 116, 120
phosphorylation state, 120
Phosphothreonine antibodies, 109, 117
photo-oxidative damage, 242
photo-protective, 68
photoaffinity, 28
photoaffinity labeling, 25
photobiology, 30
photochemical efficiency, 89, 90, 91, 93, 116
photochemical quenching, 94, 156
photochemistry, 25
photodamage, 55, 57, 65, 70, 73, 75, 77, 78, 88, 156, 194
photodamaged D1, 114
photoinactivated D1, 114
photoinactivated PS II reaction, 18, 158
photoinactivation, 14, 113
photoinhibition, 23, 32, 34, 39, 40, 42, 45, 49–53, 55, 57, 58, 60, 65–67, 70–73, 75, 78, 88–90, 113, 156, 194, 246
photoinhibitory influences, 32
photoinhibitory print, 12
photomorphogenesis, 335, 336
photon yield, 51, 53
photon yield of photosynthesis, 50
photooxidative damage, 89
photooxidative stress, 308
photophosphorylation, 27
photoprotection, 14, 40, 49–53, 55–57, 59–61, 65, 66–69, 71–73, 75, 77, 116
photoprotective energy dissipation, 50, 51, 116
photoprotective roles of lutein, 18
photoreceptor, 30, 334
photoregulation, 25
photorespiration, 13, 17, 72, 74, 76, 77, 135, 156, 229, 230
photorespiratory carbon metabolism, 13
photorespiratory H_2O_2, 243
photorespiratory pathway, 242
photosensitizers, 30
photostasis, 13
photosynthesis, 8, 30, 50, 51, 53, 54, 57–61, 66, 70–78, 112, 116, 229, 242, 243, 251, 252, 254, 269–272, 282
photosynthesis herbicides, 25
photosynthetic acclimation, 130

photosynthetic apparatus, 116, 122
photosynthetic capacity, 41, 42, 52, 54, 55, 58, 156
photosynthetic cells, 243
photosynthetic CO_2 assimilation, 245
photosynthetic downregulation, 53
photosynthetic efficiency, 58
photosynthetic electron flow, 273, 281, 282
photosynthetic electron transport, 49, 54, 59, 61, 230, 269–271, 275–279, 282, 304
photosynthetic electron transport chain, 273
photosynthetic life forms, 8
photosynthetic machinery, 119
photosynthetic membrane, 120
photosynthetic proteins, 122
photosynthetic quantum yield, 92, 93, 99
photosystem, 229
photosystem I, 107, 229
photosystem II, 24, 49, 50, 88, 107, 242
photosystem II efficiency, 55, 59
photosystem stoichiometry, 121
photosystems, 34, 243
phototaxis, 12
phy signaling, 338
phycobilisomes, 147, 149, 151
physiological conditions, 122
physiology of the cycle, 8
phytochrome, 336
phytoene desaturase, 250
phytoplankton, 88, 97, 98, 100
Pigment bleaching, 87, 95
Pinus sylvestris, 161
plant, 116, 117, 119, 121
plant fitness, 8
plant hexose transporters, 254
plant productivity, 55, 58
plant's light meter, 36
plant-pathogen interactions, 243, 253
plasma membrane, 254
plasma membrane NADPH oxidase, 243
plastid gene expression, 270
plastid sigma factors, 283
plastid signal, 344
plastid signaling factor, 247
plastid transcription kinase, 281
plastid-to-nucleus, 290
plastoglobuli, 248
plastoquinol, 122
plastoquinone, 121, 122, 131, 132, 248–250, 345
plastoquinone antagonist, 33
plastosemiquinone, 30
polychromatic action spectrum, 97
polyubiquitin, 352
pool size of violaxanthin, 8
population of functional PS II centers, 14
pore-like structures, 323
porphyra, 100
porphyrin synthesis, 247
post-transcriptional processes, 273
post-translational pamitoylation, 29
posttranslational modification, 122

powdery mildew, 260
PP2A, 111, 115, 117, 118
PPIase, 117–119
PPIase protein-folding activity, 118
PPO herbicides, 247
PR proteins, 256
PR-1, 224, 232
precursor–product relationship, 28
primary producers, 87
primary productivity, 94
pro-oxidant, 308
processed D1, 29
prochlorophytes, 151
programmed cell death, 224, 228, 254, 328
programmed cell death (PCD), 321, 322, 350
propylgallate, 33
proteases, 114
protein degradation, 30
protein dephosphorylation, 117, 118
protein disulfide isomerases, 273–275
protein folding, 119
protein kinases, 34, 108, 110, 121, 229, 231, 233
protein modifications, 108
protein phosphatase, 114–119, 231, 233
protein phosphorylation, 39, 44, 45, 107–117, 120–122
protein synthesis, 122
protein turnover, 119
proteins, 122
proteolysis, 110, 113, 115
proteolytic degradation, 120
protochlorophyllide, 247
protoporphyrin IX, 247
protoporphyrinogen IX oxidase, 247
PrxQ, 308
PS I, 25, 33, 107, 108, 119, 120, 149–151
PS I excitation, 33
PS II, 27, 40, 107–109, 111–120, 122, 149–151, 156, 245–247, 249–251
PS II assembly, 114
PS II biogenesis, 118, 119
PS II core, 115–117
PS II core proteins, 33, 112, 113
PS II dephosphorylation, 119
PS II dimers, 116
PS II efficiency, 41, 42, 45, 50, 51, 53, 54, 57, 58, 60, 96
PS II monomers, 114, 119
PS II photochemical efficiency, 89, 90, 94, 99
PS II photoinhibition, 115
PS II protein turnover, 25
PS II proteins, 32, 35
PS II quenching centers, 158
PS II reaction center, 23, 27, 31, 155, 156, 244
PS II repair cycle, 115, 118
PS II stability, 116
PS II structure, 116
PS II supercomplexes, 116
PS II turnover, 107, 114–116
PS IIα, 159
PS IIβ centers, 159
PsaD, 107, 108, 111, 112, 120
*psb*A gene, 24, 27, 35, 161

Subject Index

*Psb*H, 107, 109, 111–113, 115–117
*Psb*S, 39–44, 50, 51, 54–56, 59, 61, 127, 132–135, 147, 148, 151, 152, 158
*psb*S mutants in *Arabidopsis*, 19
*Psb*S protein, 158
*Psb*T, 111
*Psb*Z, 158
PSI-H, 119
PSI-L, 119
pulse amplitude modulated (PAM) fluorescence, 88
pulse-amplitude modulated fluorometer, 8
purple bacteria, 244, 246
putrescine, 328
pyrazolate, 249
pyridoxamine, 247

Q

Q cycle, 207, 217, 254
Q_A, 156, 160, 161, 166, 244, 246
Q_B, 113, 161
Q-band, 160
Q_B binding niche, 36
Q_B niche, 30
Q_B pocket, 30
qE, 39, 40, 44, 46, 61
qI, 39, 40, 46
Qo, 122
Qo site, 108
quantitative relationship, 31
quantum yield, 30
quenching analysis, 95
Quercus rubra, 18
quiescence, 243
quinone, 26, 28
quinone anions, 26

R

R2K1, 162
R2S2C3, 162
radioactive labeling, 109
radioactive phosphate, 110
rapid proteolysis, 28
reaction center, 97
reaction center chlorophyll, P680, 248
reaction center quenching, 157
reactive nitrogen species, 231
reactive oxygen species (ROS), 18, 40, 46, 49, 54, 55, 58–60, 69, 128, 132, 197, 224, 269, 271, 273, 304, 353
reactive reduced species of oxygen, 213
red alga, 90
red algae, 89
red chlorophyll catabolite, 349
redox, 122
redox buffers, 243
redox chemistry, 242
redox potential, 155, 159, 254, 258
redox regulation, 23, 34
redox sensing, 303, 316

redox signaling, 314
redox signals, 121, 271
redox state, 111, 121, 244
redox status, 121
redox status of PS II, 31
redox-buffering capacity, 242
redox-controlled, 110
redox-dependent, 121
redox-dependent thylakoid kinases, 112
redox-induced, 108
reducing side, 25
regulatory function, 36
Reid Milner, 2
relaxation kinetics, 94, 96
release of cytochrome c, 324, 351
release of violaxanthin, 7
remote sensing, 19
repair, 193, 194
repair cycle, 114, 115, 156
repair mechanisms, 97
reprogramming of metabolism, 156
respiration, 151, 242, 254
respiratory, 304
respiratory chains, 304
resveratrol, 326
reversible binding, 118
Rhodobacter sphaeroides, 160
rhodophytes, 89, 94, 95–97, 100, 147, 149, 151
Rhodopseudomonas viridis, 27, 28
rhythmic behavior, 35
ribosomes, 115
rice, 116
Rieske iron-sulfur subunit, 135
Rieske:iron sulfur-proteins, 349
RNA, 272–274, 277–282
RNA binding, 275
RNA polymerases, 279
RNA-binding, 275, 277–279
RNA-degradation machinery, 277
Robin Hill, 6
root, 243
root hair growth, 243
ROS, 55, 57, 128, 132, 213, 223–231, 233, 234, 242, 243, 253, 259, 260, 271, 272, 276, 304, 324
Rubisco, 15, 17

S

S, 72
S deficiency, 72
salicylic acid, 224, 232, 233, 242, 243, 353
salinity, 65, 66, 75–77, 233
salt stress, 198
Samuel Cate Prescott Award, 9
SAR, 257, 260
Scenedesmus, 250
schlerophytic evergreens, 157
Scots pine (*Pinus sylvestris*), 148
SCPs, 151
SDS-PAGE, 24, 27, 33, 112

Secale cereale L, 156
seed plants, 115
seed production, 8
semiquinone anion radical, 30
senescence, 243, 333, 346
senescence-induced or senescence-associated receptor-like kinases, 351
sensor, 34
Ser/Thr phosphatase, 117
serine/threonine kinases, 111
shade-sun acclimation, 15, 17
shade-type, 36
shallow trap, 158
signal transducer, 246
signal transduction, 253, 259, 260, 321
signal-transducing molecules, 242
signal-transduction network, 8
signaling, 119
signaling molecules, 243
sinapate esters, 31
single turnover flash-yield of PS II, 16
singlet excited oxygen, 49, 56
singlet excited chlorophyll, 128
singlet excited oxygen, 56
singlet oxygen, 54, 128, 131, 136–138, 196, 197, 214, 242, 243, 246, 248, 250, 257, 260
sink limitation, 59, 70, 78
site-directed mutagenesis, 134, 168, 297
skotomorphogenesis, 335, 336
Small Cab-like Proteins (SCPs), 147
SOD, 212
sodium fluoride, 114
Solanum dulcamera, 15
solar radiation, 87, 88, 91–94
soluble phosphoprotein, 121
soluble sugars, 57
source sink relationships, 59
soybeans, 31
spectrophotometirc assay, 7
spermine, 327–328
spinach, 24, 116, 117, 158
Spirodela, 23–25, 27, 28, 31, 32, 34, 35
Spirodela oligorrhiza, 121
spruce needles, 243
Sputnik, 3
stabilization energy, 159
state transitions, 108, 119
stereospecific product feedback inhibition, 7
stomata, 229–231, 233
stomatal closure, 243
stomatal opening, 252, 253
streptomycin, 97, 99
stress, 116, 122
stress conditions, 113
stress response, 313
stress-enhanced proteins, 138
stroma, 4, 28, 115, 119, 121
strong illumination, 115
Stt7, 110, 122
Stt7 kinase, 108
subfreezing temperatures, 116
sublethal dose, 36

substrate stereospecificity of VDE, 7
subtidal, 88
subunits, L and M, 27
sucrose, 230
suicide polypeptide, 14
suicide program, 247
sulfate assimilation, 259
sulfenic acid, 313
sulfhydryl, 313
sulfinic acid, 313
sulfonic acid, 313
sulphur, 156
sun-type plastids, 36
sunlight, 30, 31
superoxide, 49, 54, 56, 59, 72, 212, 213, 242, 251, 254, 256, 260
superoxide anion, 213, 224
superoxide anion radicals, 212
superoxide dismutase, 212, 213, 225, 226, 232, 251, 305
superoxide dismutation, 253
superoxide radical, 197, 271
supralittoral, 88
sustained, 157
sustained phosphorylation, 116
sweet potato, 255
sxd, 250
synchronous BY-2 tobacco, 255
Synechococcus sp. PCC 7942, 161, 162
Synechocystis, 128, 137, 139, 147
Synechocystis 6803, 27
Synechocystis sp. PCC 6803, 167, 194
synergistic effects, 30, 31, 193, 199
synthesis, 25, 26, 29, 119, 124
systemic acquired resistance (SAR), 224, 227, 256

T

TAK, 110, 122
TAK kinases, 108
tautomycin, 117
temperature, 155, 156
terrestrialization, 8
Teruo Ogawa, 6
tetrapyrroles, 290
the apoplast, 253
thermal dissipation, 19, 39, 66, 68, 71, 74, 77, 89, 127, 128, 132, 252
thermal energy dissipation, 46, 71, 78
thermoluminescence (TL), 155, 159
thiol, 272, 273, 276
thiol redox state, 122
thiol-disulphide exchange, 257
thiol/disulfide, 257
thiol/disulfide buffer, 256
thiol/disulfide exchange, 257
thiol/disulphide status, 260
thioredoxin (TRX), 108, 121, 122, 244, 251, 272, 273, 275, 278, 279, 283, 314–316
thioredoxin peroxidase, 212
threonine, 111, 116, 120–122
threonine phosphorylation, 111
threonine residues, 112
thylakoid, 29, 36, 109, 111, 115, 117–123

Subject Index

thylakoid ascorbate transport, 251
thylakoid lumen, 117, 119
thylakoid membrane, 7, 24, 27, 110, 112, 248, 304, 314, 342
thylakoid phosphatase, 118
thylakoid phosphoproteins, 107
thylakoid protein phosphorylation, 108, 112
thylakoid-associated proteases, 347
tidal rhythm, 87–89
TLP20, 118
TLP40, 107, 114, 117–119, 122
TMP14, 107, 111, 112, 120, 121
tobacco, 24, 228, 230–232, 234, 250, 255
tobacco BY-2 cell cultures, 259
tobacco D1, 27
tocopherol, 127, 133, 137, 138, 225, 242, 246, 248
tocopherol cyclase, 250
tocopherol, ascorbate, and glutathione, 259
tocopherylquinone, 248, 249
tocoquinone, 245
Tommy Nakayama, 3
Tony San Pietro, 6
trans-membrane electron transport, 230
trans-membrane regions, 121
trans-membrane signaling, 119
transcription, 271, 273, 274, 276, 277, 279–283
transcription factors, 271
transcription-translation, 282
transit peptide, 120
translation, 273–277, 279, 283
translation initiation, 274, 275
translation of *psbA* mRNA, 198
transmembrane organization, 4
transparency, 88, 94
transthylakoid ΔpH, 157
trap limited, 158
Trevor W. Goodwin, 6
triazine-resistant, 25
triazine-susceptible, 25
triose phosphate isomerase, 258
triplet oxygen, 214, 243
triplet state, 248
Triticum aestivum, 156
trolox, 326, 327
TRX, 252, 257, 258
TRX f–activated enzymes, 259
TRX peroxidase, 256
TRX-modulated proteins, 259
TRX/TRX reductase, 259
TRX: thioredoxin, 273
trypsin, 24, 27, 112, 121
trypsin inhibitor, 259
TSP9, 107, 108, 110–112, 121, 122
turnover, 114
turnover of D1, 116
two component signaling systems, 338
type II Prx, 308, 311, 313
tyrosine phosphorylation, 111

U

ubiquinone-ubiquinol interconversions, 254
ubiquitin-conjugating enzymes, 352
ubisemiquinone, 254
Ulrich Schreiber, 9
ultrastructure, 36
ultraviolet, 87, 88, 97
Ulva, 99
Ulva rotundata, 15
University of California, 3
University of Hawaii, 4
unsaturated fatty acids, 214, 321
unsaturation, 167
unsaturation of fatty acids, 199
UV irradiation, 30
UV light, 324
UV radiation, 93–95, 100
UV-A, 31, 88, 93, 94, 97
UV-absorbing pigments, 31
UV-absorbing substances, 87, 97
UV-B, 23, 30, 89, 93, 94, 97
UV-B radiation, 31
UV-B irradiation, 31, 252
UV-B screening capacity, 31
UV-B sensitive, 31
UV-B tolerant, 31

V

VAZ pathway, 1, 3
VAZ scheme, 4
VAZ transmembrane pathway, 5
vertical distribution, 87, 88
vertical migrations, 88
very high light resistant (*VHLR*) mutants, 17
VHLR genes, 18
VIIIth International Congress on Photosynthesis, 4
violaxanthin, 151, 157, 251
violaxanthin cycle, 4
violaxanthin de-epoxidase, 4, 50, 132, 134, 135, 139, 151
violaxanthin-antheraxanthin-zeaxanthin scheme, 1
virus infection, 15
visible, 30
vitamin E, 248
vtc1, 231, 233, 234, 252
vtc1 Arabidopsis mutant, 253
vte1 mutant, 250

W

water availability, 156
water deficient conditions, 116
water stress, 14, 58
water-water cycle, 135, 212
wavelength, 33
winter, 43, 52–54, 57–60
winter rye, 116, 121, 163
wounding, 229, 243

X

xanthophyll, 151, 244
xanthophyll conversions, 145, 151, 152

xanthophyll cycle, 1, 14, 39, 40, 46, 51, 52–54, 57, 60, 61, 66, 68, 71, 72, 74, 76–78, 88, 99, 100, 116, 127, 133, 134, 146, 157, 232
xanthoxin, 252

Y

YAP-1, 257
ycf9 gene, 158
Yin and Yang, 19
Yin-Yang interpretation, 15

Z

Z + A, 50–52
Z_v-band, 160
Zea mays, 116
zeaxanthin, 4, 8, 39–42, 44, 45, 49–54, 56, 57, 59–61, 67, 69, 116, 132–134, 137, 151, 157, 251
zeaxanthin epoxidase, 137, 152